To my mother and father, Mary and Thomas Calinger,
for their affection, support and dedication to higher learning.

The pen-and-ink drawings above and on the cover are by Andrew N. Wyeth.

CLASSICS OF MATHEMATICS

Edited by RONALD CALINGER

MP MOORE PUBLISHING COMPANY, INC.
OAK PARK, ILLINOIS

Library of Congress Cataloging in Publication Data

Main entry under title:

Classics of mathematics.

1. Mathematics—History—Sources. I. Calinger, Ronald.

QA21.C55 510 80-15567

ISBN 0-935610-13-8

Classics of Mathematics, First Edition

Moore Publishing Company, Inc.
701 South Gunderson Avenue, Oak Park, Illinois 60304

Contents

III. THE HELLENISTIC AND ROMAN PERIODS

A. Euclid

B. Archimedes

VI. THE SCIENTIFIC REVOLUTION AT ITS ZENITH

C. The Arithmetization of Mathematical Analysis

D. Number Theory, Set Theory, and Symbolic Logic

Guide to Biographies

Preface

This anthology contains selections from writings of leading mathematicians from classical antiquity through the early 20th century. Since one of its goals is to present a broad coverage of mathematics in early civilizations, it also includes works by noted scholars such as Otto Neugebauer on ancient Mesopotamia and Sylvanus Morley on the Maya. While the selections are chiefly taken from authors of the ancient Mediterranean and modern Europe, some attention is given to medieval Islam, India, and China, as well as to the United States. Accessibility of sources was a major factor governing choices from medieval Islam, India, and China, while the volume's chronological limits restricted the number of works by Americans. Each of the selections presents a significant section of a book or article. Greek, Latin, German, French, and other foreign texts have been translated into English. An asterisk after a title refers the reader to the source. I especially drew upon the mathematical collections of H. Midonick, D. E. Smith, D. Struik, I. Thomas, J. van Heijenoort, and P. Wolff.

By its nature an anthology is both a sampler and an argument. As a sampler it attempts to be informative and representative. Consequently, it sacrifices completeness and, to a degree, reflects personal and arbitrary choices of the compiler. The choices here go beyond a merely perfunctory or idiosyncratic collection. Among the surfeit of possible source readings, many were obvious, such as Euclid's *Elements,* Archimedes' *On the Measurement of a Circle,* Descartes' *La Géométrie,* Newton's *Principia,* Euler's *Introductio in analysin infinitorum,* Gauss' *Disquisitiones arithmeticae,* Cauchy's *Calcul infinitésimal,* Cantor's *Grundlagen einer Allgemeinen Mannigfaltigkeitslehre,* and Hilbert's "Mathematical Problems: . . . Delivered at Paris in 1900." Others were based on considered selection criteria that strongly took into account recent research on the development of mathematics.

The texts that appear, the particular portions selected, the omission of other portions or other texts completely, and the placement of materials in this anthology may sometimes occasion surprise. My guiding principle throughout has been to illustrate major developments in mathematics and to do so insofar as possible at the level of comprehension of an intelligent non-specialist. A decade's experience in teaching the history of mathematics strongly influenced the choices. Nearly all selections come from pure mathematics. Accomplishments in geometry, algebra, the calculus, probability, the theory of numbers, and foundations are stressed. In order to present the evolution of mathematical ideas and methods in terms of their times and to avoid the misleading notion that they anticipate or lead inexorably to the present, I have divided the chapters chronologically rather than topically. As an aid to the reader seeking the latter ordering, there are topical subdivisions in the post-Renaissance chapters, when the materials become abundant. However, the reader should note that specialization did not begin to become the norm in mathematics until the 19th century.

The present anthology seeks to balance wide-range with a small number of significant topics treated in considerable depth. The decision was made to give the flavor of mathematics while recognizing that space was limited. I particularly regret that more material on applications of mathematics could not be included. The bibliographies will guide the interested reader into those areas.

As Sir James Lighthill and Morris Kline have persuasively argued, the close connections between mathematics and the physical world studied through their respective uses of analogy, experience, intuition, and reasoning are important for another reason besides the mathematization of the natural science that they produced after 1600. This interaction has also stimulated the most profound and fruitful developments within mathematics. These connections receive brief treatment in the chapter introductions, and I urge the reader to discover more about them as well as the applications of mathematics. Suggestions for further reading at the end of each chapter introduction can facilitate such inquiry.

This anthology is in the second instance an argument for a certain point of view or interpretation. The basic premise of the argument is that examining works of past masters in mathematics is productive of sound mathematics and, in a few instances, of mathematical innovation. Evidence for this argument appears in the following selections from Abel, Weierstrass, Lebesgue, and Hardy. To these able mathematicians, just as to Johann Bernoulli when he taught Leonhard Euler, an appeal to the classics of mathematical masters was essential for carrying forward the teaching and tasks of mathematics. Their writings reveal something of the imagination of authors working in different historical periods and circumstances, while allowing us to see the generally austere nature of mathematical thought from early times through the recent past. They allow the readers to examine live issues and methods; they are not collections of dead facts. The classics deal with fundamental problems (such as the rate of occurrence of prime numbers) that still fascinate mathematicians today, with tensions between theory and practice on the one hand and new theories and traditions in mathematics or society on the other, and with reasonings and techniques that continue to have wide application. They also reveal the exceptional writing skills of many master mathematicians.

Because a basic knowledge of the history of mathematics can add to the understanding of the classics contained herein, I have provided information on the nature of mathematics in different times and cultures, its changing social, intellectual, and institutional contexts, the place of beauty and economy in mathematical theorizing, and the human element in the discipline. Chapter introductions and biographies address these topics. To be sure, they contain a personal point of view. But because they contain comments to help the reader understand the continuing polemics in the history of mathematics, they also point to other views as well. The inclusion of biographies is not simply a flourish to offset the austerity of the discipline by briefly depicting the human qualities of its makers. Illustrating the process behind discovery makes it plain that great mathematicians do not make discoveries in an almost automatic fashion. Nor do these generally occur quickly. Henri Poincaré maintained that even apparent sudden bursts of mathematical illumination have been preceded by extended periods of gestation in the subconscious. Neither mathematical abstractions, their maturation, nor their formalization emerge without time. In that respect they are unlike Athena who stepped fully formed from the head of Zeus—an impression still given in many textbooks. There is abundant evidence in the biographies of mistakes, retracing of steps, difficulties faced in having new ideas accepted, and the refinement and consolidation of theories.

Those seeking more information on the history of mathematics might consult the bibliography and research manual written for the discipline by Kenneth O. May, the journal *Historia Mathematica,* and the textbooks of Carl Boyer, Howard Eves, Morris Kline, Edna Krammer, and J. F. Scott. Those wishing more biographical information might refer to Charles C. Gillispie's *Dictionary of Scientific Biography* (16 volumes).

I received much generous aid in the compilation of this anthology. Specific acknowledgments to publishers for permission to reprint are given in the pages of this volume. In addition, I wish to thank my good friend Jay Shafritz for suggesting this enterprise and his steady proddings for it to reach completion without undue delay. David Saunders and Saunders Maclane made helpful suggestions on a preliminary list of selections for the modern period. Several people reviewed groups of biographical sketches. Helena Pycior commented on figures living after 1800, Judy Grabiner on Cauchy, Leonard Borucki on mathematical statements in many of the biographies, and Frank Limouze on general historical data. My colleague Roderick Brumbaugh, whose name appears on the title page, discussed with me some of the problems of history generally and the limits of historical arguments. He also insisted upon precision of expression. Tom Carroll gave trenchant comments on two chapter introductions. Errors of fact or interpretation that remain are mine alone.

There are more people who merit recognition. Kurt Bing provided invaluable assistance in translations, and Marion T. Quiroga deserves thanks for the tedious work of typing and retyping the manuscript. She, with the help of Mary Jo Donnelly, Anna O'Neil, and Karen Wolf, proofread the galleys. Above all, I want to thank my wife Betty for her patience, stylistic comments, and her timely and perceptive criticisms of my studies of the human element and social context of mathematics. The presence of our son John Michael in this his first year not only brought affection and pleasant relaxation but also heightened my interest in the childhood of great mathematicians.

RONALD CALINGER

Introduction

It appears to me that if one wants to make progress in mathematics, one should study the masters and not the pupils.

—Niels Henrik Abel

From *A Mathematician's Apology**

G. H. HARDY

If intellectual curiosity, professional pride, and ambition are the dominant incentives to research, then assuredly no one has a fairer chance of gratifying them than a mathematician. His subject is the most curious of all—there is none in which truth plays such odd pranks. It has the most elaborate and the most fascinating technique, and gives unrivalled openings for the display of sheer professional skill. Finally, as history proves abundantly, mathematical achievement, whatever its intrinsic worth, is the most enduring of all.

We can see this even in semi-historic civilizations. The Babylonian and Assyrian civilizations have perished; Hammurabi, Sargon, and Nebuchadnezzar are empty names; yet Babylonian mathematics is still interesting, and the Babylonian scale of 60 is still used in astronomy. But of course the crucial case is that of the Greeks.

The Greeks were the first mathematicians who are still 'real' to us to-day. Oriental mathematics may be an interesting curiosity, but Greek mathematics is the real thing. The Greeks first spoke a language which modern mathematicians can understand; as Littlewood said to me once, they are not clever schoolboys or 'scholarship candidates', but 'Fellows of another college'. So Greek mathematics is 'permanent', more permanent even than Greek literature. Archimedes will be remembered when Aeschylus is forgotten, because languages die and mathematical ideas do not. 'Immortality' may be a silly word, but probably a mathematician has the best chance of whatever it may mean.

**Source:* From G. H. Hardy, *A Mathematician's Apology* (1967 edition), 80-81. Reprinted by permission of the Cambridge University Press.

From "Mathematics as an Element in the History of Thought"*

ALFRED NORTH WHITEHEAD

The science of pure mathematics, in its modern developments, may claim to be the most original creation of the human spirit. Another claimant for this position is music. But we will put aside all rivals, and consider the ground on which such a claim can be made for mathematics. The originality of mathematics consists in the fact that in mathematical science connections between things are exhibited which, apart from the agency of human reason, are extremely unobvious. Thus the ideas, now in the minds of contemporary mathematicians, lie very remote from any notions which can be immediately derived by perception through the senses; unless indeed it be perception stimulated and guided by antecedent mathematical knowledge. This is the thesis which I proceed to exemplify.

Suppose we project our imagination backwards through many thousands of years, and endeavour to realise the simple-mindedness of even the greatest intellects in those early societies. Abstract ideas which to us are immediately obvious must have been, for them, matters only of the most dim apprehension. For example take the question of number. We think of the number 'five' as applying to appropriate groups of any entities whatsoever—to five fishes, five children, five apples, five days. Thus in considering the relations of the number 'five' to the number 'three,' we are thinking of two groups of things, one with five members and the other with three members. But we are entirely abstracting from any consideration of any particular entities, or even of any particular sorts of entities, which go to make up the membership of either of the two groups. We are merely thinking of those relationships between those two groups which are entirely independent of the individual essences of any of the members of either group. This is a very remarkable feat of abstraction; and it must have taken ages for the human race to rise to it. During a long period, groups of fishes will have been compared to each other in respect to their multiplicity, and groups of days to each other. But the first man who noticed the analogy between a group of seven fishes and a group of seven days made a notable advance in the history of thought. He was the first man who entertained a concept belonging to the science of pure mathematics. At that moment it must have been impossible for him to divine the complexity and subtlety of these abstract mathematical ideas which were waiting for discovery. Nor could he have guessed that these notions would exert a widespread fascination in each succeeding generation. There is an erroneous literary tradition which represents the love of mathematics as a monomania confined to a few eccentrics in each generation. But be this as it may, it would have been impossible to anticipate the pleasure derivable from a type of abstract thinking which had no counterpart in the then-

**Source:* From Alfred North Whitehead, *Science and the Modern World,* Chapter II, 20-28. Reprinted by permission of the Macmillan Co. and Cambridge University Press.

existing society. Thirdly, the tremendous future effect of mathematical knowledge on the lives of men, on their daily avocations, on their habitual thoughts, on the organization of society, must have been even more completely shrouded from the foresight of those early thinkers. Even now there is a very wavering grasp of the true position of mathematics as an element in the history of thought. I will not go so far as to say that to construct a history of thought without profound study of the mathematical ideas of successive epochs is like omitting Hamlet from the play which is named after him. That would be claiming too much. But it is certainly analogous to cutting out the part of Ophelia. This simile is singularly exact. For Ophelia is quite essential to the play, she is very charming—and a little mad. Let us grant that the pursuit of mathematics is a divine madness of the human spirit, a refuge from the goading urgency of contingent happenings.

When we think of mathematics, we have in our mind a science devoted to the exploration of number, quantity, geometry, and in modern times also including investigation into yet more abstract concepts of order, and into analogous types of purely logical relations. The point of mathematics is that in it we have always got rid of the particular instance, and even of any particular sorts of entities. So that for example, no mathematical truths apply merely to fish, or merely to stones, or merely to colours. So long as you are dealing with pure mathematics, you are in the realm of complete and absolute abstraction. All you assert is, that reason insists on the admission that, if any entities whatever have any relations which satisfy such-and-such purely abstract conditions, then they must have other relations which satisfy other purely abstract conditions.

Mathematics is thought moving in the sphere of complete abstraction from any particular instance of what it is talking about. So far is this view of mathematics from being obvious, that we can easily assure ourselves that it is not, even now, generally understood. For example, it is habitually thought that the certainty of mathematics is a reason for the certainty of our geometrical knowledge of the space of the physical universe. This is a delusion which has vitiated much philosophy in the past, and some philosophy in the present. The question of geometry is a test case of some urgency. There are certain alternative sets of purely abstract conditions possible for the relationship of groups of unspecified entities, which I will call *geometrical conditions.* I give them this name because of their general analogy to those conditions, which we believe to hold respecting the particular geometrical relations of things observed by us in our direct perception of nature. So far as our observations are concerned, we are not quite accurate enough to be certain of the exact conditions regulating the things we come across in nature. But we can by a slight stretch of hypothesis identify these observed conditions with some one set of the purely abstract geometrical conditions. In doing so, we make a particular determination of the group of unspecified entities which are the *relata* in the abstract science. In the pure mathematics of geometrical relationships, we say that, if *any* group entities enjoy *any* relationships among its members satisfying *this* set of abstract geometrical conditions, then such-and-such additional abstract conditions must also hold for such relationships. But when we come to physical space, we say that some definitely observed group of physical entities enjoys some definitely observed relationships among its members which do satisfy this above-mentioned set of abstract geometrical conditions. We thence conclude that the additional relationships which we concluded to hold in *any* such case, must therefore hold in *this particular* case.

The certainty of mathematics depends upon its complete abstract generality.

But we can have no *a priori* certainty that we are right in believing that the observed entities in the concrete universe form a particular instance of what falls under our general reasoning. To take another example from arithmetic. It is a general abstract truth of pure mathematics that any group of forty entities can be subdivided into two groups of twenty entities. We are therefore justified in concluding that a particular group of apples which we believe to contain forty members can be subdivided into two groups of apples of which each contain twenty members. But there always remains the possibility that we have miscounted the big group; so that, when we come in practice to subdivide it, we shall find that one of the two heaps has an apple too few or an apple too many.

Accordingly, in criticising an argument based upon the application of mathematics to particular matters of fact there are always three processes to be kept perfectly distinct in our minds. We must first scan the purely mathematical reasoning to make sure that there are no mere slips in it—no casual illogicalities due to mental failure. Any mathematician knows from bitter experience that, in first elaborating a train of reasoning, it is very easy to commit a slight error which yet makes all the difference. But when a piece of mathematics has been revised, and has been before the expert world for some time, the chance of a casual error is almost negligible. The next process is to make quite certain of all the abstract conditions which have been presupposed to hold. This is the determination of the abstract premises from which the mathematical reasoning proceeds. This is a matter of considerable difficulty. In the past quite remarkable oversights have been made, and have been accepted by generations of the greatest mathematicians. The chief danger is that of oversight, namely, tacitly to introduce some condition, which it is natural for us to presuppose, but which in fact need not always be holding. There is another opposite oversight in this connection which does not lead to error, but only to lack of simplification. It is very easy to think that more postulated conditions are required than is in fact the case. In other words, we may think that some abstract postulate is necessary which is in fact capable of being proved from the other postulates that we have already on hand. The only effects of this excess of abstract postulates are to diminish our aesthetic pleasure in the mathematical reasoning, and to give us more trouble when we come to the third process of criticism.

This third process of criticism is that of verifying that our abstract postulates hold for the particular case in question. It is in respect to this process of verification for the particular case that all the trouble arises. In some simple instances, such as the counting of forty apples, we can with a little care arrive at practical certainty. But in general, with more complex instances, complete certainty is unattainable. Volumes, libraries of volumes, have been written on the subject. It is the battle ground of rival philosophers. There are two distinct questions involved. There are particular definite things observed, and we have to make sure that the relations between these things really do obey certain definite exact abstract conditions. There is great room for error here. The exact observational methods of science are all contrivances for limiting these erroneous conclusions as to direct matters of fact. But another question arises. The things directly observed are, almost always, only samples. We want to conclude that the abstract conditions, which hold for the samples, also hold for all other entities which, for some reason or other, appear to us to be of the same sort. This process of reasoning from the sample to the whole species is Induction. The theory of Induction is the despair of philosophy—and yet all our activities are based upon it. Anyhow, in criticising a mathematical conclusion as to a particular matter of fact, the real difficulties consist in finding out the abstract assumptions involved, and in estimating the evidence for their applicability to the particular case in hand.

It often happens, therefore, that in criticising a learned book of applied mathematics, or a memoir, one's whole trouble is with the first chapter, or even with the first page. For it is there, at the very outset, where the author will probably be found to slip in his assumptions. Farther, the trouble is not with what the author does say, but with what he does not say. Also it is not with what he knows he has assumed, but with what he has unconsciously assumed. We do not doubt the author's honesty. It is his perspicacity which we are criticising. Each generation criticises the unconscious assumptions made by its parents. It may assent to them, but it brings them out in the open.

The history of the development of language illustrates this point. It is a history of the progressive analysis of ideas. Latin and Greek were inflected languages. This means that they express an unanalysed complex of ideas by the mere modification of a word; whereas in English, for example, we use prepositions and auxiliary verbs to drag into the open the whole bundle of ideas involved. For certain forms of literary art—though not always—the compact absorption of auxiliary ideas into the main word may be an advantage. But in a language such as English there is the overwhelming gain in explicitness. This increased explicitness is a more complete exhibition of the various abstractions involved in the complex idea which is the meaning of the sentence.

By comparison with language, we can now see what is the function in thought which is performed by pure mathematics. It is a resolute attempt to go the whole way in the direction of complete analysis, so as to separate the elements of mere matter of fact from the purely abstract conditions which they exemplify.

The habit of such analysis enlightens every act of the functioning of the human mind. It first (by isolating it) emphasizes the direct aesthetic appreciation of the content of experience. This direct appreciation means an apprehension of what this experience is in itself in its own particular essence, including its immediate concrete values. This is a question of direct experience, dependent upon sensitive subtlety. There is then the abstraction of the particular entities involved, viewed in themselves, and as apart from that particular occasion of experience in which we are then apprehending them. Lastly there is the further apprehension of the absolutely general conditions satisfied by the particular relations of those entities as in that experience. These conditions gain their generality from the fact that they are expressible without reference to those particular relations or to those particular relata which occur in that particular occasion of experience. They are conditions which might hold for an indefinite variety of other occasions, involving other entities and other relations between them. Thus these conditions are perfectly general because they refer to no particular occasion, and to no particular entities (such as green, or blue, or trees) which enter into a variety of occasions, and to no particular relationships between such entities.

There is, however, a limitation to be made to the generality of mathematics; it is a qualification which applies equally to all general statements. No statement, except one, can be made respecting any remote occasion which enters into no relationship with the immediate occasion so as to form a constitutive element of the essence of that immediate occasion. By the 'immediate occasion' I mean that occasion which involves as an ingredient the individual act of judgment in question. The one excepted statement is:—If anything out of relationship, then complete ignorance as to it. Here by 'ignorance,' I mean *ignorance;* accordingly no advice can be given as to how to expect it, or to treat it, in 'practice' or in any other way. Either we know something of the remote occasion by the cognition which is itself an element of the immediate occasion, or we know nothing. Accordingly the full universe, disclosed for every variety of experience, is a universe in which every detail enters into its proper relationship with the immediate occasion. The generality of mathematics is the most complete generality consistent with the community of occasions which constitutes our metaphysical situation.

It is further to be noticed that the particular entities require these general conditions for their ingression into any occasions; but the same general conditions may be required by many types of particular entities. This fact, that the general conditions transcend any one set of particular entities, is the ground for the entry into mathematics, and into mathematical logic, of the notion of the 'variable.' It is by the employment of this notion that general conditions are investigated without any specification of particular entities. This irrelevance of the particular entities has not been generally understood: for example, the shape-iness of shapes, e.g., circularity and sphericity and cubicality as in actual experience, do not enter into the geometrical reasoning.

The exercise of logical reason is always concerned with these absolutely general conditions. In its broadest sense, the discovery of mathematics is the discovery that the totality of these general abstract conditions, which are concurrently applicable to the relationships among the entities of any one concrete occasion, are themselves inter-connected in the manner of a pattern with a key to it. This pattern of relationships among general abstract conditions is imposed alike on external reality, and on our abstract representations of it, by the general necessity that every thing must be just its own individual self, with its own individual way of differing from everything else. This is nothing else than the necessity of abstract logic, which is the presupposition involved in the very fact of inter-related existence as disclosed in each immediate occasion of experience.

The key to the patterns means this fact:—that from a select set of those general conditions, exemplified in any one and the same occasion, a pattern involving an infinite variety of other such conditions, also exemplified in the same occasion, can be developed by the pure exercise of abstract logic. Any such select set is called the set of postulates, or premises, from which the reasoning proceeds. The reasoning is nothing else than the exhibition of the whole pattern of general conditions involved in the pattern derived from the selected postulates.

The harmony of the logical reason, which divines the complete pattern as involved in the postulates, is the most general aesthetic property arising from the mere fact of concurrent existence in the unity of one occasion. Wherever there is a unity of occasion there is thereby established an aesthetic relationship between the general conditions involved in that occasion. This aesthetic relationship is that which is divined in the exercise of rationality. Whatever falls within that relationship is thereby exemplified in that occasion, whatever falls without that relationship is thereby excluded from exemplification in that occasion. The complete pattern of general conditions, thus exemplified, is determined by any one of many select sets of these conditions. These key sets are sets of equivalent postulates. This reasonable harmony of being, which is required for the unity of a complex occasion, together with the completeness of the realisation (in that occasion) of all that is involved in its logical harmony, is the primary article of metaphysical doctrine. It means that for things to be together involves that they are reasonably together. This means that thought can penetrate into every occasion of fact, so that by comprehending its key conditions, the whole complex of its pattern of conditions lies open before it. It comes to this:—provided we know something which is perfectly general about the elements in any occasion, we can then know an indefinite number of other equally general concepts which must also be exemplified in that same occasion. The logical harmony involved in the unity of an occasion is both exclusive and inclusive. The occasion must exclude the inharmonious, and it must include the harmonious.

Chapter I

Protomathematics in the Late Age of Stone and in Ancient Mesopotamia and Egypt

Introduction. Mathematical activity long antedates the emergence of mathematics as an independent, theoretical science in classical Greece. In the last 50 years anthropologists, ethnographers, historians, lexicographers, philologists, and psychologists have shown that rudimentary arithmetic and geometry are as old as civilization, going back to late Neolithic times in the fourth millennium B.C. in the Fertile Crescent and beyond.[1] The Neolithic ("New Stone," in Greek) period, which occurred among scattered peoples in Africa, Asia, and Europe, began about the mid-eighth millennium B.C. and was preceded by the Palaeolithic ("Old Stone," in Greek) period beginning about 1.75 million years ago. Numerals, the written signs for numbers, are one of 72 items occurring in every human culture known to ethnography, according to the anthropologist George P. Murdock. Psychological evidence suggests that human beings have always been aware of quantitative relations. Some animals and birds possess a numerical or quantitative sense, for example crows can distinguish between "one," "two," "three," and "four" hunters.[2] Like animals and birds early humans must have had some crude sense of quantity—the ability to distinguish between some and more, few and many, small and great. Similarly, it may plausibly be argued that organizing the binocular visual data of the space of everyday experience required a primitive intuitive geometry. Successful foraging or hunting also required a crude notion of far and near, that is of spatial displacement, while cave drawings show careful attention to form. A historical account of the origins of mathematics, therefore, should begin with a sketch of elementary mathematics in late Palaeolithic, Neolithic, and ancient Mesopotamian and Egyptian times.

During the Age of Stone, protomathematics, the purely perceptual stage of mathematics, was tied to a primitive, empirical physics. In its late-Palaeolithic period the notion of perceptible (concrete) number emerged without a conceptual basis or abstraction. It is difficult to overestimate how crucial this step was because number is the basis of mathematics and science. Certain tallying techniques of 100,000 years ago already exhibit a numerical sense. At that time Palaeolithic humans indicated the plurality of things by tallying sticks, which were notched strips of bone, ivory, or wood, or by crude graphical marks on the walls of caves. These representations suggest a concern with a totality rather than component parts. Such practices as letting a stroke mark correspond to an animal imply an intuition of equivalence or one-to-one correspondence, the basis of the notion of *cardinal number*.

Homo sapiens, appearing about thirty thousand years ago, used tallying techniques widely and developed the embryonic numerical sense still further. Two-counting was probably the most primitive number system. This and subsequent number systems were ordered and thus implied *ordinal number*. Later fingers and toes were used to obtain higher number bases, such as five, ten, and twenty. Most notable was the crude base-10 or decimal system. "Counting" words slowly progressed from adjectives to nouns. Adjectival number words did no more than designate the things that were enumerable. In the ancient Orient, adjectival forms were descriptive suffixes added to words for objects, such as trees, poles, or bowls, to make these words singular, dual, or larger plural forms (cat, catwo, cathree, etc.). Numbers as nouns (vocal numbers) were an enormous advance since such designation allowed numbers themselves to become independent of particular, concrete objects—an intermediate stage in the development of the completely abstract notion of number.

Practical needs—the counting of implements, weapons, animal skins, or animals, and basic trade and barter—primarily stimulated the growth of protomathematics in the Age of Stone but only slowly. Other stimuli were magic, ritual, aesthetics, and play. For millennia the practical societal needs of the nomadic Palaeolithic hunters, fisherman, and food-gatherers changed little. With its supple fingers and flexible thumb the human hand, which is well suited to hold and make tools as well as to count, enabled early humans to create material culture.

After the transition to *homo sapiens* changes in material culture accelerated, bringing to an end the Palaeolithic period. During the ensuing Neolithic period two fundamental developments in human culture further quickened material and intellectual change. The early Neolithic division of labor had sent men to the hunt and women to gather grains, nuts, and berries. From this women probably learned how to plant, care for, and grow seeds. This discovery was complemented by a spontaneous natural genetic mutation of hybrid cereals, especially of wheat. The two combined brought about what archaeologists term the Agricultural Revolution. The rise in agricultural productivity led to increased population and pressure on hunting and food-gathering space. Neolithic peoples began to become sedentary, at first in temporary settlements which they occupied during certain seasons and then in permanent villages or fixed abodes. Fixed abodes provided more efficiently for storage and allocation of the agricultural surplus. A few villagers were freed from work in the fields to plan and record sowings and harvestings. One consequence was calendar-making. Calendars emerged from a slow, cumulative process of seasonal experience. Humans now produced food, domesticated animals, and invented pottery.

The second Neolithic development was the advent of urban society bringing with it civilization. Urban communities first appeared in southern Mesopotamia (part of modern Iraq) and Egypt about 3000 B.C. A drying trend in climate forced increasing numbers of Neolithic nomads to settle near the water of the Tigris and Euphrates Rivers in Mesopotamia ("the land between the rivers," in Greek) and the Nile River in Egypt. The conditions were harsh and they had to overcome adversity, but through time these peoples brought out the potential of the rich alluvial plains. Some hunting was still possible but agriculture dominated. In both areas civilization was characterized by the invention of writing, the emergence of complex religious and socio-political institutions, the specialization of labor, and the discovery of new technologies in the smelting of copper and bronze—all of which transpired in urban settings. The emergence of urban settings larger than the previous Neolithic villages and of civilization lends credence to Aristotle's later belief that, "Man is by nature a political animal."

Ancient Mesopotamia and Egypt were theocratic states in which priest-scribes responded to increasingly complex practical needs arising from intensive agriculture, astral-religion, monumental building programs, and extensive trade and commerce. To the extent that they dealt successfully with these matters, the priest-scribes flourished and enjoyed prestige. In agriculture they worked to harness river waters for irrigation and to control flooding, an especially difficult task on the erratic Tigris and Euphrates. The priest-scribes also had to survey land, to assign fields to farmers, and to make observations and to keep careful records of the movements of heavenly bodies for calendric purposes. The astronomical records were important to their astral-religions in giving evidence of divine actions and in setting religious holidays. Among the buildings the priest-scribes planned were huge adobe granaries, magnificent palaces, ziggurats in Mesopotamia, and stone pyramids in Egypt. They also kept careful records for economies that were becoming more complex. To do their tasks well the priest-scribes developed an increasingly sophisticated protomathematics. As written documents show, their endeavors were influenced also by the play element or "mathematics for its own sake."

In ancient Mesopotamia protomathematics benefitted from cross-cultural influences and an extensive long-distance trade. During the early phase of an existence that lasted three millennia, the ancient Mesopotamian civilization was ruled in turn by the Sumerians in the south (to about 2340 B.C.), the Akkadians to the north (to about 2125 B.C.), and the city of Ur in Sumer (to about 2000 B.C.).[3] After 2000 B.C. the Amorites, a Semitic people, began a conquest that led to the establishment of their capital at Babylon less than a century later. The period from about 1900 to about 1600 B.C. is thus known as the Old Babylonian or Hammurabic Dynasty, the latter after the king famed for his legal code. In the traditional Mesopotamian society these successive people absorbed much of the intellectual and cultural heritage of their predecessors. Despite successive conquests Mesopotamia continued to be a crossroads for long-distance trade, especially by importing timber for building, metals for military and craft purposes, and stones for utilitarian and decorative uses. This trade added to the store of information and to pressures for improved computation and meticulous record-keeping.

Our knowledge of protomathematics in ancient Mesopotamia rests chiefly on over 300 mathematical tablets of baked clay. These are a small part of the total mathematical record and have been likened to a few torn pages from a book that was part of a large library. The tablets are divided into problem and table texts. Most date from the Old Babylonian Dynasty, when protomathematics stood at its peak in ancient Mesopotamia. The rest come from the Seleucid period beginning in 311 B.C. They are written in cuneiform, a script invented by the Sumerians. The Sumerians and their Mesopotamian successors wrote with a stylus that left wedge-shaped impressions on wet clay (hence the name cuneiform from the Latin *cuneus* for wedge and *forma* for shape). In addition to the mathematical tablets, tens of thousands of commercial tablets survive.

These sources show that the priest-scribes of ancient Mesopotamia contributed to the beginnings of arithmetic, algebra, and geometry. They made notable progress with number systems and computation. By 2350 B.C. the early Sumerians appear to have consistently gone beyond simple counting to the grouping of numbers in recording in such activities as commercial transactions, astronomical observations, and the measurement of time. They had two number systems—an indigenous sexagesimal (base-60) system built on a progression of tens and sixes (1, 10, 60, 600, 3600, . . .) and a decimal system, probably from another people. They emphasized the sexagesimal system perhaps because it permitted a greater facility with fractions—the bane of ancients and moderns alike—or because it corresponded to their division of the year into 360 days. As their empire grew, the Akkadians undertook extensive arithmetical computations: addition, subtraction, multiplication, and some division. They employed the early Sumerian script for numerals. By the Old Babylonian Dynasty basic numerals had emerged for one and powers of 60 and for 10. As the old script became fossilized and lost its concrete meaning, numerals became abstract symbols or ideograms. In the Old Babylonian Dynasty there was also an implicit positional or place-value notation, albeit without a zero symbol (so that the order of magnitude was judged by the context). In place-value notation the position in an expression governs value. Thus, in the modern decimal notation, $267 = 2 \times 10^2 + 6 + 10^1 + 7 \times 10^0$. No other ancient people, except perhaps a precursor of the Maya in Mesoamerica, made this breakthrough. The historian Otto Neugebauer has called positional notation "undoubtedly one of the most fertile inventions of humanity" comparable to "the invention of the alphabet."[4]

The problem and table texts, which became pervasive during the Old Babylonian Dynasty, reveal ingenuity and computational skill. The compilers of the multiplication tables to 60 showed an appreciation of structure and economy by listing the products from 1 to 20 followed by products for 30, 40, and 50. The remainder of the 59 products could be obtained by adding two of the tables' members. Division was ingeniously reduced to multiplication by tables of reciprocals. The priest-scribes gave reciprocals for all "regular" integers, that is those integers (a) whose reciprocals (1/a) were a finite sexagesimal fraction. They avoided the reciprocals of irregulars such as 7, 11, and 13 but had approximations. The problem and table texts demonstrate that the Babylonian priest-scribes were not only adept in fundamental addition, subtraction, multiplication, and division but also in summing basic arithmetical and geometrical progressions as well as in computing squares, cubes, and square roots. Using an iterative process probably beginning with a scribal rule attributed to Hero,[5] they found the equivalent of what in our notation is $\sqrt{2} \approx 1; 24, 51, 10$ (sexagesimal) ≈ 1.414 (decimal). The text Plimpton 322 also contains Pythagorean triples—that is triples of integers x, y, z satisfying the equations $x^2 + y^2 = z^2$.

The Old Babylonian cuneiform texts also cover elementary algebra and geometry. The algebra focussed on the solution of equations. It included methods to solve linear equations in two or three unknowns (using substitution), quadratic equations in two unknowns (in modern symbols $x^2 + y^2 = b$), and selected cubic equations (such as $x^3 + x = a$ in modern notation). These equations were expressed in words, and only positive rational roots were sought. Babylonian geometry was less successful, perhaps in part because clay tablets were a restricted medium for drawing accurate figures. It included correct formulas for calculating the areas of triangular and trapezoidal fields together with approximations for the volumes of cones and pyramids. Texts from Susa show an interest in polygons, such as the hexagon, inscribed in and circumscribed about a circle. The area of the circle was generally found by the rule $A = c^2/12$, where c is the circumference. This implies that *pi* equals 3. They also had a more accurate approximation of *pi* as 3; 7, 30 or $3\frac{1}{8}$.

Like its counterpart in Mesopotamia, the civilization of ancient Egypt lasted three millennia, beginning about 3100 B.C. Located in the narrow upper and lower Nile Valley, ancient Egypt was protected on both flanks by vast deserts and to the north by the Mediterranean Sea. It differed from Mesopotamia in its relative isolation and its less troubled dynamism. Its history included three kingdoms: the Old (*c.* 2700–*c.* 2200 B.C.), Middle (2052–1786 B.C.), and New (1575–1087 B.C.). The great pyramids were built during the Old Kingdom. The first writing of ancient Egypt was the cartoon-like hieroglyphics ("sacred carvings" in Greek) used in wall paintings and chiseled temple inscriptions. After 2500 B.C. Egyptian scholar-scribes added to hieroglyphics the simpler and faster cursive hieratic (sacred) and demotic (popular) scripts that were written with pen and ink on papyrus sheets.[6] Among the surviving papyri and ostraca only about a dozen deal with ancient Egyptian or pharaonic protomathematics. Of these the most important sources are two in hieratic script: the Moscow or Golenischev Papyrus dating from about 1850 B.C. and the more significant Rhind Papyrus prepared by the scribe A'hmosé about 1650 B.C.[7]

Based on the information from deciphered hieroglyphics inscriptions and mathematical papyri, the protomathematics of pharaonic Egypt did not match the Mesopotamian achievement except in geometry. (Among early ancient peoples the high level of achievement by the Mesopotamians was the exception; the Egyptians were more typical.) The rudimentary arithmetic of ancient Egypt had a decimal system of numbers, with values beyond 10 expressed by successive additions of numerals rather than in compact positional notation. The numerals were written from right to left. In the hieratic papyri there were tables for reducing almost all fractions (the sole exception was ⅔) to a sum of unit fractions, those with one in the numerator. As the A'hmosé papyrus demonstrates, scribes could add, subtract, multiply and divide positive integers and reducible fractions. Multiplication was done by continual doublings or duplation. Thus $23 \times 7 = 23 \times 1 + 23 \times 2 + 23 \times 4$. A'hmosé and the other scholar-scribes used a literal algebra that lacked symbols except for representing plus and minus by the legs of a person approaching and leaving. With this algebra they could solve linear equations in more than one unknown but only the simplest quadratic equations. To reach some solutions they utilized a trial-and-error procedure known as the "rule of false position." With this rule if $x + \frac{1}{2}x = 6$, one assumes that $x = 2$ giving $x + \frac{1}{2}x = 3$. This shows that the assumed value of x is only one-half the correct answer and then by adjustment $x = 4$.

The ancient Egyptian scribes also contributed to practical, mensurational geometry. Their rope-stretchers *(harpenodaptai)* made accurate surveys with knotted cords. For the mathematical system derived from and regulating these surveys the ancient Greeks coined the word geometry (*geo* = land and *metria* = measure). The Egyptian surveyors understood the Pythagorean relationship in a right triangle. In hieratic and demotic papyri scribes expressed correct rules for calculating the areas of triangles, rectangles, and isosceles trapezoids. In terms of modern notation, problem 50 of A'hmosé found the area of a circle as $A = (8/9\ d)^2 = (16/9\ r)^2 = 256/81\ r^2$. This means that *pi* equals 256/81 or 3.1605, a close approximation. The two outstanding achievements of the ancient Egyptian scribal-geometers appear in problems 10 and 14 of the Moscow Papyrus. Problem 10 may be translated as giving the area of a hemispheric surface as follows in modern terms: $A = 2\ (8/9\ d^2) \approx 2\pi r^2$. Problem 14 calculates the volume of a truncated pyramid (or frustum) by the equivalent of the modern formula $V = \frac{1}{3}h\ (a^2 + ab + b^2)$.

Although they had elementary algebra, arithmetic, and geometry, neither the priest-scribes of ancient Mesopotamia or Egypt passed the threshold from descriptive into theoretical mathematics. Having developed scientifc intuition they found an inexhaustible source of unsolved problems in experience, itself, and they drew upon the same experience to reach results that were original and new. From trial and error, acute observation, and occasional insights into specific relationships, they arrived at scribal rules. With these rules they, in effect, solved equations and found areas and volumes. However, they rarely generalized the results of these operations. Without such generalization they could not be expected to extract results from a general theory. Their invention of writing did initiate a literate tradition apparently indispensable for the emergence of mathematical theory. The ancient priest-scribes though had to respond almost entirely to the immediate problems posed by social needs which left little time for possible independent theorizing. Moreover their astral-religions probably impeded the beginnings of theoretical science. Even though the heliacal rising of the dog star was a predictable occurrence, nature for them was subject to the antic whims of gods. They could not divine its processes.

NOTES

1. The suggested readings following this introduction and subsequent footnotes cite selections from these studies.

2. O. Koehler, "The Ability of Birds to 'Count'" in James R. Newman (ed.), *The World of Mathematics* (New York: Simon and Shuster, 1956), vol. 1, pp. 488–496.

3. For general histories the reader might consult Samuel Noah Kramer, *The Sumerians: Their History, Culture and Character* (Chicago: The University of Chicago Press, 1963) and A. Leo Oppenheim, *Ancient Mesopotamia, Portrait of a Dead Civilization* (Chicago: The University of Chicago Press, 1964).

4. O. Neugebauer, *The Exact Sciences in Antiquity* (New York: Harper Torchbooks, 1962), p. 5.

5. Given $N = ab$, then

$\sqrt{N} \approx a_1 = \frac{1}{2}(a + b) = \frac{1}{2}(a + N/a)$: first approximation

$a_2 = \frac{1}{2}(a_1 + N/a_1)$: second, closer approximation

6. The writings of ancient Egypt were first deciphered in modern times by the French Egyptologist Jean François Champollion, who discovered the Rosetta Stone in 1799 with its trilingual key, a parallel inscription in hieroglyphics, demotic, and Greek.

7. Both of these papyri are discussed in Chapter I of this anthology.

SUGGESTIONS FOR FURTHER READING

Asger Aaboe, *Episodes from the Early History of Mathematics*. New York: Random House and the L. W. Singer Company, 1964.

A. B. Chace, L. S. Bull, H. P. Manning, and R. C. Archibald (eds.), *The Rhind Mathematical Papyrus*. 2 volumes. Oberlin, O.: Mathematical Association of America, 1927–1929.

S. Gandz, "Studies in Babylonian Mathematics" *Osiris*, 8: 12–40.

R. J. Gillings, *Mathematics in the Time of the Pharaohs*. Cambridge, Mass: The M.I.T. Press, 1972.

H. Goetsch, "Die Algebra der Babylonier," *Archive for History of Exact Sciences*, 5 (1968): 79–153.

P. Huber, "Bemerkungen über mathematische Keilschrifttexte" *Enseignement mathématique* 2nd ser., 3 (1957): 19-27.

Louis Charles Karpinski, *The History of Arithmetic*. New York: Russell and Russell, Inc., 1965.

K. Menninger, *Number Words and Number symbols: A Cultural History of Numbers*. Cambridge, Mass.: The M.I.T. Press, 1969.

Otto Neugebauer, *The Exact Sciences in Antiquity*. Princeton: Princeton University Press, 1952.

———, *A History of Ancient Mathematical Astronomy*. 3 volumes. New York: Springer-Verlag, 1975.

———, *Mathematische Keilschrifttexte*. 3 volumes in *Quellen und Studien zur Geschichte der Mathematik, Astronomie, und Physik*, Berlin: J. Springer, 1935–1937.

Otto Neugebauer and A. Sachs (eds.), *Mathematical Cuneiform Texts*. New Haven: American Oriental Society, 1945.

Richard A. Parker, *Demotic Mathematical Papyri*. Providence, R. I.: Brown University Press, 1972.

E. T. Peet, "Mathematics in Ancient Egypt," *Bulletin of the John Rylands Library* 15, no. 2 (1931); 409–441.

George Sarton, *The Study of the History of Mathematics and the History of Science*. (1st edition, 1936) New York: Dover Publicatons, 1954.

A. Seidenberg, "The Ritual Origins of Geometry," *Archives for History of Exact Sciences*, I (1960–1962): 488-527.

———, "The Ritual Origins of Counting," *Archives for History of Exact Sciences*, 2 (1962–1966): 1–40.

François Thureau-Dangin, "Sketch of the History of the Sexagesimal System," *Osiris* 7: 95–141.

B. L. van der Waerden, *Science Awakening*, I. trans. by Arnold Dresden. Groningen: P. Noordhoff, 1954.

Raymond Wilder, *Evolution of Mathematical Concepts*. New York: John Wiley & Sons, Inc., 1968.

1. From *The Exact Sciences in Antiquity**

O. NEUGEBAUER

BABYLONIAN MATHEMATICS

15. The following chapter does not attempt to give a history of Babylonian mathematics or even a complete summary of its contents. All that it is possible to do here is to mention certain features which might be considered characteristic of our present knowledge.

I have remarked previously that the texts on which our study is based belong to two sharply limited and widely separated periods. The great majority of mathematical texts are "Old-Babylonian"; that is to say, they are contemporary with the Hammurapi dynasty, thus roughly belonging to the period from 1800 to 1600 B.C. The second, and much smaller, group is "Seleucid", *i.e.,* datable to the last three centuries B.C. These dates are arrived at on quite reliable palaeographic and linguistic grounds. The more than one thousand intervening years influenced the forms of signs and the language to such a degree that one is safe in assigning a text to either one of the two periods.

So far as the contents are concerned, little change can be observed from one group to the other. The only essential progress which was made consists in the use of the "zero" sign in the Seleucid texts. It is further noticeable that numerical tables, especially tables of reciprocals, were computed to a much larger extent than known from the earlier period, though no new principle is involved which would not have been fully available to the Old-Babylonian scribes. It seems plausible that the expansion of numerical procedures is related to the development of a mathematical astronomy in this latest phase of Mesopotamian science.

For the Old-Babylonian texts no prehistory can be given. We know absolutely nothing about an earlier, presumably Sumerian, development. All that will be described in the subsequent sections is fully developed in the earliest texts known. It is customary to postulate a long development which is supposedly necessary to reach a high level of mathematical insight. I do not know on what experience this judgment is based. All historically well known periods of great mathematical discoveries have reached their climax after one or two centuries of rapid progress following upon, and followed by, many centuries of relative stagnation. It seems to me equally possible that Babylonian mathematics was brought to its high level in similarly rapid growth, based, of course, on the preceding development of the sexagesimal place value system whose rudimentary forms are already attested in countless economic texts from the earliest phases of written documents.

16. The mathematical texts can be classified into two major groups: "table texts" and "problem texts". A typical representative of the first class is the multiplication table discussed [in an earlier chapter]. The second class comprises a great variety of texts which are all more or less directly concerned with the formulation or solution of algebraic or geometrical problems. At present the number of problem texts known to us amounts to about one hundred tablets,

Source: From O. Neugebauer, *The Exact Sciences in Antiquity* (1957), 29-48. Reprinted by permission of the Brown University Press.

as compared with more than twice as many table texts. The total amount of Babylonian tablets which have reached museums might be estimated to be at least 500,000 tablets and this is certainly only a small fraction of the texts which are still buried in the ruins of Mesopotamian cities. Our task can therefore properly be compared with restoring the history of mathematics from a few torn pages which have accidentally survived the destruction of a great library.

17. The table texts allow us to reconstruct a small, however insignificant, bit of historical information. The archives from the City of Nippur, now dispersed over at least three museums, Philadelphia, Jena, and Istanbul, have given us a large percentage of table texts, many of which are clearly "school texts", *i.e.*, exercises written by apprentice scribes. This is evident, *e.g.*, from the repetition in a different hand of the same multiplication table on obverse and reverse of the same tablet. Often we also find vocabularies written on one side of a tablet which shows mathematical tables on the other side. These vocabularies are the backbone of the scribal instruction, necessary for the mastery of the intricacies of cuneiform writing in Akkadian as well as in Sumerian. Finally, many of our mathematical tables are combined with tables of weights and measures which were needed in daily economic life. There can be little doubt that the tables for multiplication and division were developed simultaneously with the economic texts. Thus we find explicitly confirmed what could have been concluded indirectly from our general knowledge of early Mesopotamian civilization.

18. Though a single multiplication table is rather trivial in content, the study of a larger number of these texts soon revealed unexpected facts. Obviously a complete system of sexagesimal multiplication tables would consist of 58 tables, each containing all products from 1 to 59 with each of the numbers from 2 to 59. Thanks to the place value notation such a system of tables would suffice to carry out all possible multiplications exactly as it suffices to know our multiplication table for all decimal products. At first this expectation seemed nicely confirmed except for the unimportant modification that each single tablet gave all products from 1 to 20 and then only the products for 30, 40, and 50. This is obviously nothing more than a space saving device because all 59 products can be obtained from such a tablet by at most one addition of two of its numbers. But a more disturbing fact soon became evident. On the one hand the list of preserved tables showed not only grave gaps but, more disconcertingly, there turned up tables which seemed to extend the expected scheme to an unreasonable size. Multiplication tables for 1,20 1,30 1,40 3,20 3,45 etc. seemed to compel us to assume the existence not of 59 single tables but of 3600 tables. The absurdity of this hypothesis became evident when tables for the multiples of 44,26,40 repeatedly appeared; obviously nobody would operate a library of $60^3 = 216000$ tablets as an aid for multiplication. And it was against all laws of probability that we should have several copies of multiplication tables for 44,26,40 but none for 11, 13, 14, 17, 19 etc.

The solution of this puzzle came precisely from the number 44,26,40 which also appears in another type of tables, namely, tables of reciprocals. Ignoring variations in small details, these tables of reciprocals are lists of numbers as follows

2	30	27	2,13,20
3	20	30	2
4	15	32	1,52,30
5	12	36	1,40
6	10	40	1,30
8	7,30	45	1,20
9	6,40	48	1,15
10	6	50	1,12
12	5	54	1,6,40
15	4	1	1
16	3,45	1,4	56,15
18	3,20	1,12	50
20	3	1,15	48
24	2,30	1,20	45
25	2,24	1,21	44,26,40

The last pair contains the number 44,26,40 and also all the other two-place numbers mentioned above occur as numbers of the second column. On the other hand, with one single exception to be mentioned presently, the gaps in our expected list of multiplication tables correspond exactly to the missing numbers in our above table of reciprocals. Thus our stock of multiplication tables is not a collection of tables for all products $a \cdot b$, for a and b from 1 to 59, but tables for the products $a \cdot \bar{b}$ where $\bar{b}$ is a number from the right-hand side of our last list. The character of these numbers $\bar{b}$ is conspicuous enough; they are the reciprocals of the numbers b of the left column, written as sexagesimal fractions:

$$\tfrac{1}{2} = 0;30$$
$$\tfrac{1}{3} = 0;20$$
$$\tfrac{1}{4} = 0;15$$
etc.
$$\frac{1}{1,21} = 0;0,44,26,40.$$

We can express the same fact more simply and historically more correctly in the following form. The above "table of reciprocals" is a list of numbers, b and $\bar{b}$, such that the products $b \cdot \bar{b}$ are 1 or any other power of 60. It is indeed irrelevant whether we write

$$2 \cdot 30 = 1,0$$
or
$$2 \cdot 0;30 = 1$$
or
$$0;2 \cdot 30 = 1$$
or
$$0;2 \cdot 0;30 = 0;1 \text{ etc.}$$

Experience with the mathematical problem texts demonstrates in innumerable examples that the Babylonian mathematicians made full use of this flexibility of their system.

Thus we have seen that the tables of multiplication combined with the tables of reciprocals form a complete system, designed to compute all products $a \cdot \bar{b}$ or, as we now can write, all sexagesimal divisions $\frac{a}{b}$ within the range of the above-given table of reciprocals. This table is not only limited but it shows gaps. There is no reciprocal for 7, for 11, for 13 or 14, etc. The reason is obvious. If we divide 7 into 1 we obtain the recurrent sexagesimal fraction 8,34,17,8,34,17, . . . ; similarly for $1/11$ the group 5,27,16,21,49 appears in infinite repetition. We have tables which laconically remark "7 does not divide", "11 does not divide", etc. This holds true for all numbers which contain prime numbers not contained in 60, *i.e.*, prime numbers different from 2, 3, and 5. We shall call these numbers "irregular" numbers in contrast to the remaining "regular" numbers whose reciprocals can be expressed by a sexagesimal fraction of a finite number of places.

We have mentioned one exception to our rule that all multiplication tables must concern numbers $\bar{b}$ or, as we shall call them now, regular numbers. This is the case of the first irregular number, namely 7, for which several multiplication tables are preserved. The purpose of this addition is clearly the completion of all tables $a \cdot b$ at least for the first decade, in which 7 would be the only gap because all the remaining numbers from 1 to 10 are regular. Thus we see that our original assumption was correct for the modest range from 1 to 10. Instead, however, of expanding this table up to 60, one chooses a much more useful sequence of numbers, namely, those which are needed not only for multiplication but also for division. The mere multiplications could always be completed by one simple addition from two different tables. This system of tables alone, as it existed in 1800 B.C., would put the Babylonians ahead of all numerical computers in antiquity. Between 350 and 400 A.D., Theon Alexandrinus wrote pages of explanations in his commentaries to Ptolemy's sexagesimal computations in the Almagest. A scribe of the administration of an estate of a Babylonian temple 2000 years before Theon would have rightly wondered about so many words for such a simple technique.

The limitations of the "standard" table of reciprocals which we reproduced above [p. 17] did not mean that one could not transgress them at will. We have texts from the same period teaching how to proceed in cases not contained in the standard table. We also have tables of reciprocals for a complete sequence of consecutive numbers, regular and irregular alike. The reciprocals of the irregular numbers appear abbreviated to three or four places only. But the real expansion came in the Seleucid period with tables of reciprocals of regular numbers up to 7 places for b and resulting reciprocals up to 17 places for $\bar{b}$. A table of this extent, containing the regular numbers up to about $17 \cdot 10^{12}$, can be readily used also for determining approximately the reciprocals of irregular numbers by interpolation. Indeed, in working with astronomical texts I have often used this table exactly for this purpose and I do not doubt that I was only repeating a process familiar to the Seleucid astronomers.

19. Returning to the Old-Babylonian period we find many more witnesses of the numerical skill of the scribes of this period. We find tables of squares and square roots, of cubes and cube roots, of the sums of squares and cubes needed for the numerical solution of special types of cubic equations, of exponential functions, which were used for the computation of compound interest, etc.

Very recently A. Sachs found a tablet which he recognized as having to do with the problem of evaluating the approximation of reciprocals of irregular numbers by a finite expression in sexagesimal fractions. The text deals with the reciprocals of 7, 11, 13, 14, and 17, in the last two cases in the form that $b \cdot \bar{b} = 10$ instead of $b \cdot \bar{b} = 1$ as usual. We here mention only the two first lines, which seem to state that

$$8,34,16,59 < \bar{7}$$

but

$$8,34,18 > \bar{7}.$$

Indeed, the correct expansion of $\bar{7}$ would be 8,34,17 periodically repeated. It is needless to underline the importance of a problem which is the first step toward a mathematical analysis of infinite arithmetical processes and of the concept of "number" in general. And it is equally needless to say that the new fragment raises many more questions than it solves. But it leaves no doubt that we must recognize an interest in problems of approximations for as early a period as Old-Babylonian times.

This is confirmed by a small tablet, now in the Yale Babylonian Collection. On it is drawn a square with its two diagonals. The side shows the number 30, the diagonal the numbers 1,24,51,10 and 42,25,35. The meaning of these numbers becomes clear if we multiply 1,24,51,10 by 30, an operation which can be easily performed by dividing 1,24,51,10 by 2 because 2 and 30 are reciprocals of one another. The result is 42,25,35. Thus we have obtained from $a = 30$ the diagonal $d = 42;25,35$ by using

$$\sqrt{2} = 1;24,51,10.$$

The accuracy of this approximation can be checked by squaring 1;24,51,10. One finds

$$1;59,59,59,38,1,40$$

corresponding to an error of less than $22/60^4$. Expressed as a decimal fraction we have here the approximation 1.414213 . . instead of 1.414214 . . . This is indeed a remarkably good approximation. It was still used by Ptolemy in computing his table of chords almost two thousand years later.

Another Old-Babylonian approximation of $\sqrt{2}$ is known to be 1;25. It is also contained in the approximation of $\sqrt{2}$ which we find in the Hindu Śulva-Sūtras whose present form might be dated to the 3rd or 4th century B.C. There we find

$$\sqrt{2} = 1 + \frac{1}{3} + \frac{1}{3 \cdot 4} - \frac{1}{3 \cdot 4 \cdot 34}$$

whose sexagesimal equivalent is

1;25 − 0;0,8,49,22, . . . = 1;24,51,10,37, ...

The possibility seems to me not excluded that both the main term and the subtractive correction are ultimately based on the two Babylonian approximations.

20. The above example of the determination of the diagonal of the square from its side is sufficient proof that the "Pythagorean" theorem was known more than a thousand years before Pythagoras. This is confirmed by many other examples of the use of this theorem in problem texts of the same age, as well as from the Seleucid period. In other words it was known during the whole duration of Babylonian mathematics that the sum of the squares of the lengths of the sides of a right triangle equals the square of the length of the hypotenuse. This geometrical fact having once been discovered, it is quite natural to assume that all triples of numbers l, b, and d which satisfy the relation $l^2 + b^2 = d^2$ can be used as sides of a right triangle. It is furthermore a normal step to ask the question: When do numbers l, b, d satisfy the above relation? Consequently it is not too surprising that we find the Babylonian mathematicians investigating the number-theoretical problem of producing "Pythagorean numbers". It has often been suggested that the Pythagorean theorem originated from the discovery that 3, 4 and 5 satisfy the Pythagorean relation. I see no motive which would lead to the idea of forming triangles with these sides and to investigate whether they are right triangles or not. It is only on the basis of our education in the Greek approach to mathematics that we immediately think of the possibility of a geometric representation of arithmetical or algebraic relations.

To say that the discovery of the geometrical theorem led naturally to the corresponding arithmetical problem is very different from expecting that the latter problem was actually solved. It is therefore of great historical interest that we actually have a text which clearly shows that a far reaching insight into this problem was obtained in Old-Babylonian times. The text in question belongs to the Plimpton Collection of Columbia University in New York.

As is evident from the break at the left-hand side, this tablet was originally larger; and the existence of modern glue on the break shows that the other part was lost after the tablet was excavated. Four columns are preserved, to be counted as usual from left to right. Each column has a heading. The last heading is "its name" which means only "current number", as is evident from the fact that the column of numbers beneath it counts simply the number of lines from "1st" to "15th". This last column is therefore of no mathematical interest. Columns II and III are headed by words which might be translated as "solving number of the width" and "solving number of the diagonal" respectively. "Solving number" is a rather unsatisfactory rendering for a term which is used in connection with square roots and similar operations and has no exact equivalent in our modern terminology. We shall replace these two headings simply by "b" and "d" respectively. The word "diagonal" occurs also in the heading of the first column but the exact meaning of the remaining words escapes us.

The numbers in columns I, II and III are transcribed in the following list. The numbers in [ ] are restored. The initial numbers "[1]" in lines 4 ff. are half preserved. . . . A "1" is completely preserved in line 14. In the transcription I have inserted zeros where they are required; they are not indicated in the text itself.

I	II (= b)	III (= d)	IV
[1,59,0,]15	1,59	2,49	1
[1,56,56,]58,14,50,6,15	56,7	3,12,1	2

[1,55,7,]41,15,33,45	1,16,41	1,50,49	3
[1,]5[3,1]0,29,32,52,16	3,31,49	5,9,1	4
[1,]48,54,1,40	1,5	1,37	5
[1,]47,6,41,40	5,19	8,1	6
[1,]43,11,56,28,26,40	38,11	59,1	7
[1,]41,33,59,3,45	13,19	20,49	8
[1,]38,33,36,36	9,1	12,49	9
1,35,10,2,28,27,24,26,40	1,22,41	2,16,1	10
1,33,45	45	1,15	11
1,29,21,54,2,15	27,59	48,49	12
[1,]27,0,3,45	7,21,1	4,49	13
1,25,48,51,35,6,40	29,31	53,49	14
[1,]23,13,46,40	56	53	15

This text contains a few errors. In II,9 we find 9,1 instead of 8,1 which is a mere scribal error. In II,13 the text has 7,12,1 instead of 2,41. Here the scribe wrote the square of 2,41, which is 7,12,1 instead of 2,41 itself. In III,15 we find 53 instead of 1,46 which is twice 53. Finally there remains an unexplained error in III,2 where 3,12,1 should be replaced by 1,20,25.

The relations which hold between these numbers are the following ones. The numbers b and d in the second and third columns are Pythagorean numbers; this means that they are integer solutions of

$$d^2 = b^2 + l^2$$

As b and d are known from our list, we can compute l and find

Line	l	Line	l
1	2,0	9	10,0
2	57,36	10	1,48,0
3	1,20,0	11	1,0
4	3,45,0	12	40,0
5	1,12	13	4,0
6	6,0	14	45,0
7	45,0	15	1,30
8	16,0		

If we then form the values of $\frac{d^2}{l^2}$ we obtain the numbers of column I. Thus our text is a list of the values of $\frac{d^2}{l^2}$, b, and d, for Pythagorean numbers. It is plausible to assume that the values of l were contained in the missing part. That they have been explicitly computed is obvious.

If we take the ratio $\frac{b}{l}$ for the first line we find $\frac{1,59}{2,0} = 0;59,30$ that is, almost 1. Hence the first right triangle is very close to half a square. Similarly one finds that the last right triangle has angles close to 30° and 60°. The monotonic decrease of the numbers in column I suggests furthermore that the shape of the triangles varies rather regularly between these two limits. If one investigates this general fact more closely, one finds that the values of $\frac{d^2}{l^2}$ in column I decrease almost linearly and that this holds still more accurately for the ratios $\frac{d}{l}$ themselves (Fig. 3).

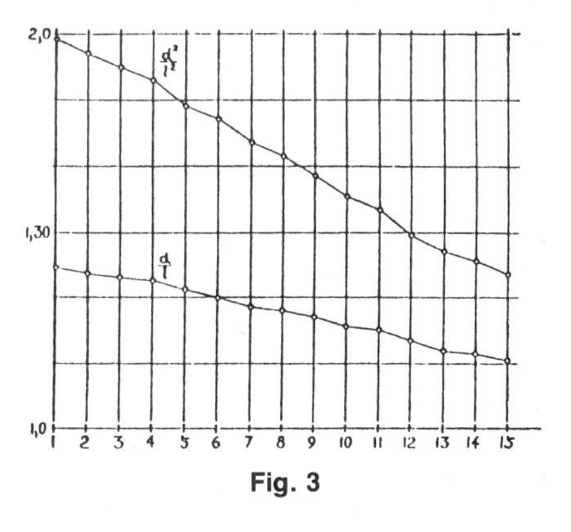

Fig. 3

This observation suggests that the ancient mathematician who composed this text was interested not only in determining triples of Pythagorean numbers but also in their ratios $\frac{d}{l}$. Let us investigate the mathematical character of this problem. We know that all Pythagorean triples are obtainable in the form

$$l = 2pq \quad b = p^2 - q^2 \quad d = p^2 + q^2$$

where p and q are arbitrary integers subject only to the condition that they are relatively prime and not simultaneously odd and $p > q$. Consequently we obtain for the ratio $\frac{d}{l}$ the expression

$$\frac{d}{l} = ½ (p \cdot \bar{q} + \bar{p} \cdot q)$$

where $\bar{p}$ and $\bar{q}$ are the reciprocals of p and q. This shows that $\frac{d}{l}$ are expressible as finite sexagesimal fractions, as is the case in our text, if and only if both p and q are regular numbers.

This fact can be easily checked in our list of numbers by computing the values of p and q which correspond to the l, b, and d of our text. Then one finds a very remarkable fact. The numbers p and q are not only regular numbers, as expected, but they are regular numbers contained in the "standard table" of reciprocals (p. 17) so well known to us from many tables of the same period. The only apparent exception is $p = 2,5$ but this number is again well known as the canonical example for the computation of reciprocals beyond the standard table. This seems to me a strong indication that the fundamental formula for the construction of triples of Pythagorean numbers was known. Whatever the case may be, the text in question remains one of the most remarkable documents of Old-Babylonian mathematics. We shall presently return to the question how a formula for Pythagorean numbers could have been found.

21. Pythagorean numbers were certainly not the only case of problems concerning relations between numbers. The tables for squares and cubes point clearly in the same direction. We also have examples which deal with the sum of consecutive squares or with arithmetic progressions. It would be rather surprising if the accidentally preserved texts should also show us the exact limits of knowledge which were reached in Babylonian mathematics. There is no indication, however, that the important concept of prime number was recognized.

All these problems were probably never sharply separated from methods which we today call "algebraic". In the center of this group lies the solution of quadratic equations for two unknowns. As a typical example might be quoted a problem from a Seleucid text. This problem requires the finding of a number such that a given number is obtained if its reciprocal is added to it.

Using modern notation we call the unknown number x, its reciprocal $\bar{x}$, and the given number b. Thus we have to determine x from

$$x\,\bar{x} = 1 \qquad x + \bar{x} = b.$$

In the text b has the value 2;0,0,33,20. The details of the solution are described step by step in the text as follows. From

$$\left(\frac{b}{2}\right)^2 = 1;0,0,33,20,4,37,46,40.$$

Subtract 1 and find the square root

$$\sqrt{\left(\frac{b}{2}\right)^2 - 1} = \sqrt{0;0,0,33,20,4,37,46,40} = 0;0,44,43,20.$$

The correctness of this result is checked by squaring. Then add to and subtract from $\frac{b}{2}$ the result. This answers the problem:

$$x = \frac{b}{2} + \sqrt{\ } = 1;0,0,16,40 + 0;0,44,43,20 = 1;0,45$$

$$\bar{x} = \frac{b}{2} - \sqrt{\ } = 1;0,0,16,40 - 0;0,44,43,20 = 0;59,15,33,20.$$

Indeed, x and $\bar{x}$ are reciprocal numbers and their sum equals the given number b.

This problem is typical in many respects. It shows, first of all, the correct application of the "quadratic formula" for the solution of quadratic equations. It demonstrates again the unrestricted use of large sexagesimal numbers. Finally, it concerns the main type of quadratic problems of which we have hundreds of examples preserved, a type which I call "normal form": two numbers should be found if (a) their product and (b) their sum or difference is given. It is obviously the purpose of countless examples to teach the transformation of more complicated quadratic problems to this "normal form"

$$x \cdot y = a$$
$$x \pm y = b$$

from which the solution then follows as

$$x = \frac{b}{2} + \sqrt{\left(\frac{b}{2}\right)^2 \mp a}$$
$$y = \pm\frac{b}{2} \mp \sqrt{\left(\frac{b}{2}\right)^2 \mp a}$$

simply by transforming the two original equations into two linear equations

$$x \pm y = b$$
$$x \mp y = \sqrt{b^2 \mp 4a}.$$

In other words, reducing a quadratic equation to its "normal form" means finally reducing it to the simplest system of linear equations.

The same idea can be used for finding three numbers, a, b, c, which satisfy the Pythagorean relation. Assume that one again started from a pair of linear equations

$$a = x + y$$
$$b = x - y$$

realizing that

$$a^2 = b^2 + c^2 \quad \text{if} \quad c^2 = 4xy.$$

Assuming that x and y are integers, then a and b will be integers; but $c = 2\sqrt{xy}$ will be an integer only if $\sqrt{xy}$ is an integer. This condition is satisfied if we assume that x and y are squares of integers

$$x = p^2 \qquad y = q^2$$

and thus we obtain the final result that a, b, and c form a Pythagorean triple if p and q are arbitrary integers $(p > q)$ and if we make

$$a = p^2 + q^2 \quad b = p^2 - q^2 \quad c = 2pq.$$

This is indeed the formula which we needed for our explanation of the text dealing with Pythagorean numbers.

22. It is impossible to describe in the framework of these lectures the details of the Babylonian theory of quadratic equations. It is not really necessary anyhow, since the whole material is easily available in the editions quoted in the bibliography to this chapter. A few features of this Babylonian algebra, however, deserve special emphasis because they are essential for the evaluation of this whole system of early mathematics.

First of all, it is easy to show that geometrical concepts play a very secondary part in Babylonian algebra, however extensively a geometrical terminology may be used. It suffices to quote the existence of examples in which areas and lengths are added, or areas multiplied, thus excluding any geometrical interpretation in the Euclidean fashion which seems so natural to us. Indeed, still more drastic examples can be quoted for the disregard of reality. We have many examples concerning wages to be paid for labor according to a given quota per man and day. Again, problems are set up involving sums, differences, products of these numbers and one does not hesitate to combine in this way the number of men and the number of days. It is a lucky accident if the unknown number of workmen, found by solving a quadratic equation, is an integer. Obviously the algebraic relation is the only point of interest, exactly as it is irrelevant for our algebra what the letters may signify.

Another important observation concerns the form in which all these algebraic problems are presented. The texts fall into two major classes. One class formulates the problem and then proceeds to the solution, step by step, using the special numbers given at the beginning. The text often terminates with the words "such is the procedure". The second class contains collections of problems only, sometimes more than 200 on a single tablet of the size of a small printed page. These collections of problems are usually carefully arranged, beginning with very simple cases e.g., quadratic equations in the normal form, and expanding step by step to more complicated relations, but all eventually reducible to the normal form. One standard form of such collections consists in keeping the condition $xy = 10{,}0$ fixed but varying the second equation to more and more elaborate polynomials, ending up, e.g., with expressions like

$$(3x + 2y)^2 + \\ {}^{2}/_{13} \left\{4\left[{}^{1}/_{7}\,((x + y) - ({}^{1}/_{2} + 1)\,(x - y))\right]^2 + (x + y)^2\right\} \\ = 4,45,0.$$

Investigating such series, one finds that they all have the same pair $x = 30$ $y = 20$ as solutions. This indicates that it was of no concern to the teacher that the result must have been known to the pupil. What he obviously had to learn was the method of transforming such horrible expressions into simpler ones and to arrive finally at the correct solutions. We have several tablets of the first class which solve one such example after another from corresponding collections of the second class.

From actually computed examples it becomes obvious that it was the general procedure, not the numerical result, which was considered important. If accidentally a factor has the value 1 the multiplication by 1 will be explicitly performed, obviously because this step is necessary in the general case. Similarly we find regularly a general explanation of the procedure. Where we would write $x + y$ the text would say "5 and 3, the sum of length and width". Indeed it is often possible to transform these examples directly into our symbolism simply by replacing the ideograms which were used for "length", "width", "add", "multiply" by our letters and symbols. The accompanying numbers are hardly more than a convenient guide to illustrate the underlying general process. Thus it is substantially incorrect if one denies the use of a "general formula" to Babylonian algebra. The sequences of closely related problems and the general rules running parallel with the numerical solution form de facto an instrument closely approaching a purely algebraic operation. Of course, the fact remains that the step to a consciously algebraic notation was never made.

23. The extension of this "Babylonian algebra" is truly remarkable. Though the quadratic equations form obviously the most significant nucleus a great number of related problems were also considered. Linear problems for several unknowns are common in many forms, e.g., for "inheritance" problems where the shares of several sons should be determined from linear conditions which hold between these shares. Similar problems arise from divisions of fields or from general conditions in the framework of the above mentioned collections of algebraic examples.

On the other hand we know from these same collections series of examples which are equivalent to special types of equations of fourth and sixth order. Usually these problems are easily reducible to quadratic equations for x^2 or x^3 but we have also examples which lead to more general relations of 5th and 3rd order. In the latter case the tables for $n^2 + n^3$ seem to be useful for the actual numerical solution of such problems, but our source material is too fragmentary to give a consistent description of the procedure followed in cases which are no longer reducible to quadratic equations.

There is finally no doubt that problems were also investigated which transcend, in the modern sense, the algebraic character. This is not only clear from problems which have to do with compound interest but also from numerical tables for the consecutive powers of given numbers. On the other hand we have texts which concern the determination of the exponents of given numbers. In other words one had actually experimented with special cases of logarithms without, however, reaching any general use of this function. In the case of numerical tables the lack of a general notation appears to be much more detrimental than in the handling of purely algebraic problems.

24. Compared with the algebraic and numerical component in Babylonian mathematics the role of "geometry" is rather insignificant. This is, in itself, not at all surprising. The central problem in the early development of mathematics lies in the numerical determination of the solution which satisfies certain conditions. At this level there is no essential difference between the division of a sum of money according to certain rules and the division of a field of given size into, say, parts of equal area. In all cases exterior conditions have to be observed, in one case the conditions of the inheritance, in another case the rules for the determination of an area, or the relations between measures or the customs concerning wages. The mathematical importance of a problem lies in its arithmetical solution; "geometry" is only one among many subjects of practical life to which the arithmetical procedures may be applied.

This general attitude could be easily exemplified by long lists of examples treated in the preserved texts. Most drastically, however, speak special texts which were composed for the use of the scribes who were dealing with mathematical problems and had to know all the numerical parameters which were needed in their computations. Such lists of "coefficients" were first identified by Professor Goetze of Yale University in two texts of the Yale Babylonian Collection. These lists contain in apparently chaotic order numbers and explanatory remarks for their use. One of these lists begins with coefficients needed for "bricks" of which there existed many types of specific dimensions, then coefficients for "walls", for "asphalt", for a "triangle", for a "segment of a circle", for "copper", "silver", "gold", and other metals, for a "cargo boat", for "barley", etc. Then we find coefficients for "bricks", for the "diagonal", for "inheritance", for "cut reed" etc. Many details of these lists are still obscure to us and demonstrate how fragmentary our knowledge of Babylonian mathematics remains in spite of the many hundreds of examples in our texts. But the point which interests us here at the moment becomes very clear, namely, that "geometry" is no special mathematical discipline but is treated on an equal level with any other form of numerical relation between practical objects.

These facts must be clearly kept in mind if we nevertheless speak about geometrical knowledge in Babylonian mathematics, simply because these special facts were eventually destined to play a decisive role in mathematical development. It must also be underlined that we have not the faintest idea about anything amounting to a "proof" concerning relations between geometrical magnitudes. Several tablets dealing with the division of areas show figures of trapezoids or triangles but without any attempt at being metrically correct. The description of geometry as the science of proving correct theorems from incorrect figures certainly fits Babylonian geometry so far as the figures are concerned and also with regard to the algebraic relations. But the real "geometric" part often escapes us. It is, for instance, not at all certain whether the triangles and trapezoids are right-angle figures or not. If the texts mention the "length" and "width" of such a figure it is only from the context that we can determine the exact meaning of these two terms. If the area of a triangle is found by computing $\frac{1}{2} a \cdot b$ it is plausible to assume that a and b are perpendicular dimensions, but there exist similar cases where only approximate formulae seem equally plausible.

There are nevertheless cases where no reasonable doubt can arise as to the correct interpretation of geometrical relations. The concept of similarity is utilized in numerous examples. The Pythagorean theorem is equally well attested; the same holds for its application to the determination of the height of a circular segment. On the other hand only a very crude approximation for the area of a circle is known so far, corresponding to the use of 3 for π. Several problems concerning circular segments and similar figures are not yet fully understood and it seems to me quite possible that better approximations of π were known and used in cases where the rough approximation would lead to obviously wrong results.

As in the case of elementary areas similar relations were known for volumes. Whole sections of problem texts are concerned with the digging of canals, with dams and similar works, revealing to us exact or approximate formulae for the corresponding volumes. But we have no examples which deal with these objects from a purely geometrical point of view.

24a. After completion of the manuscript, new discoveries were made which must be mentioned here because they contribute very essentially to our knowledge of the mathematics of the Old-Babylonian period. In 1936 a group of mathematical tablets were excavated by French archaeologists at Susa, the capital of ancient Elam, more than 200 miles east of Babylon. A preliminary report was published in the Proceedings of the Amsterdam Academy by E. M. Bruins in 1950 and the following remarks are based on this preliminary publication, though I restrict myself to the most significant results only. The texts themselves still remain unpublished, more than 20 years after their discovery.

The main contribution lies in the direction of geometry. One tablet computes the radius r of a circle which circumscribes an isosceles triangle of sides 50, 50, and 60 (result r = 31;15). Another tablet gives the regular hexagon, and from this the approximation $\sqrt{3} \approx$ 1;45 can be deduced. The main interest, however, lies in a tablet which gives a new list of coefficients similar to those mentioned above, p. 25. The new list contains, among others, coefficients concerning the equilateral triangle (confirming the above approximation $\sqrt{3} \approx$ 1;45), the square ($\sqrt{2} \approx$ 1;25), and the regular pentagon, hexagon, heptagon, and the circle. If A_n denotes the area, s_n the side of a regular n-gon, then one can explain the coefficients found in the list as follows:

$$A_5 = 1;40 \cdot s_5^2$$
$$A_6 = 2;37,30 \cdot s_6^2$$
$$A_7 = 3;41 \cdot s_7^2 .$$

If we, furthermore, call c_6 the circumference of the regular hexagon, c the periphery of the circle, then the text states

$$c_6 = 0;57,36 \cdot c.$$

Because $c_6 = \frac{3}{\pi} c$, the last coefficient implies the approximation

$$\pi \approx 3;7,30 = 3\tfrac{1}{8}$$

thus confirming finally my expectation that the comparison of the circumference of the regular hexagon with the circumscribed circle must have led to a better approximation of π than 3.

The relations for A_5, A_6, and A_7 correspond perfectly to the treatment of the regular polygon in Heron's Metrica XVIII to XX, a work whose close relationship to pre-Greek mathematics has become obvious ever since the decipherment of the Babylonian mathematical texts.

Also in many other respects do the tablets from Susa supplement and confirm what we knew from the contemporary Old-Babylonian sources in Mesopotamia proper. One example deals with the division of a triangle into a similar triangle and a trapezoid such that the product of the partial sides and of the partial areas are given values, the hypotenuse of the smaller triangle being known. This is a new variant of similar problems involving sums of areas and lengths or the product of areas. One of the tablets from Susa implies even a special problem of the 8th degree, whereas until now we had only the sixth degree represented in the Babylonian material. The new problem requires that one find the sides x and y of a rectangle whose diagonal is d, such that xy = 20,0 and $x^3 \cdot d$ =

14,48,53,20. This is equivalent to a quadratic equation for x^4

$$x^8 + a^2x^4 = b^2$$

$a = 20,0$ $b = 14,48,53,20$. The text proceeds to give the step-by-step solution of this equation, resulting in $x^4 = 11,51,6,40$ and finally leading to $x = 40$ [and] $y = 30$.

25. However incomplete our present knowledge of Babylonian mathematics may be, so much is established beyond any doubt: we are dealing with a level of mathematical development which can in many aspects be compared with the mathematics, say, of the early Renaissance. Yet one must not overestimate these achievements. In spite of the numerical and algebraic skill and in spite of the abstract interest which is conspicuous in so many examples, the contents of Babylonian mathematics remained profoundly elementary. In the utterly primitive framework of Egyptian mathematics the discovery of the irrationality of $\sqrt{2}$ would be a strange miracle. But all the foundations were laid which could have given this result to a Babylonian mathematician, exactly in the same arithmetical form in which it was obviously discovered so much later by the Greeks. And even if it were only due to our incomplete knowledge of the sources that we assume that the Babylonians did not know that $p^2 = 2q^2$ had no solution in integer numbers p and q, even then the fact remains that the consequences of this result were not realized. In other words Babylonian mathematics never transgressed the threshold of prescientific thought. It is only in the last three centuries of Babylonian history and in the field of mathematical astronomy that the Babylonian mathematicians or astronomers reached parity with their Greek contemporaries.

2. From the A'h-mosè or Rhind Papyrus*

JAMES R. NEWMAN

COMMENTARY

The oldest mathematical documents in existence are two Egyptian papyrus rolls dating from around the Twelfth Dynasty (2000-1788 B.C.). The earlier of the scrolls, the Golenischev—both are named after their former owners reposes in Moscow; the other, the Rhind papyrus, is in the British Museum. These remarkable texts make evident what has not always been acknowledged, namely, that the Egyptians possessed a good deal of arithmetic and geometric knowledge. Their methods were clumsy and they were incapable of grand generalizations—a preëminent ability of the Greeks. Yet it is nonsense to depreciate the real skill and imagination exhibited in these texts, to belittle the contribution made by the Egyptians to mathematical thought. Egyptian mathematics was precocious, as George Sarton has remarked;[1] its major achievements came early. It was also arrested in its development; after a short period of vigorous growth it made little further progress. The static character of

*Source: From James R. Newman (ed.), *The World of Mathematics* (1956), Volume 1, 169-178. Reprinted by permission of Simon and Schuster, Inc. and Scientific American, Inc.

Egyptian culture, the blight that fell upon Egyptian science around the middle of the second millennium, has often been emphasized but never adequately explained. Religious and political factors undoubtedly played a part in turning a dynamic society into one of stone.

The Rhind papyrus is described in the article which follows; the Golenischev deserves a note here. A scroll of the same length (544 cm.) as the Rhind, but only one quarter as wide (8 cm.), the Moscow papyrus is a collection of twenty-five problems rather than a treatise. The method of solving these problems agrees with rules given in the Rhind papyrus.[2] One of the problems indicates that the Egyptians may have known the formula for the volume of a truncated pyramid,

$$V = (h/3)\,(a^2 + ab + b^2),$$

where a and b are the lengths of the sides of the square (the base of the pyramid) and h is the height.[3] Sarton calls this solution the "masterpiece" of their geometry. It was indeed an impressive step forward, unsurpassed in three more millennia of Egyptian mathematics.[4]

THE RHIND PAPYRUS

In the winter of 1858 a young Scottish antiquary named A. Henry Rhind, sojourning in Egypt for his health, purchased at Luxor a rather large papyrus said to have been found in the ruins of a small ancient building at Thebes. Rhind died of tuberculosis five years later, and his papyrus was acquired by the British Museum. The document was not intact; evidently it had originally been a roll nearly 18 feet long and 13 inches high, but it was broken into two parts, with certain portions missing. By one of those curious chances that sometimes occur in archaeology, several fragments of the missing section turned up half a century later in the deposits of the New York Historical Society. They had been obtained, along with a noted medical papyrus, by the collector Edwin Smith. The fragments cleared up some points essential for understanding the whole work.

The scroll was a practical handbook of Egyptian mathematics, written about 1700 B.C. Soon after its discovery several scholars satisfied themselves that it was an antiquity of first importance, no less, as D'Arcy Thompson later said, than "one of the ancient monuments of learning." It remains to this day our principal source of knowledge as to how the Egyptians counted, reckoned and measured.

The Rhind was indited by a scribe named A'h-mosè (another, more sonorous form of his name is Aāh-mes) under a certain Hyksos king who reigned "somewhere between 1788 and 1580 B.C." A'h-mosè, a modest man, introduces his script with the notice that he copied the text "in likeness of writings of old made in the time of the King of Upper [and Lower] Egypt, [Ne-ma] 'et-Rê'." The older document to which he refers dates back to the 12th Dynasty, 1849-1801 B.C. But there the trail ends, for one cannot tell whether the writing from which A'h-mosè copied was itself a copy of an even earlier work. Nor is it clear for what sort of audience the papyrus was intended, which is to say we do not know whether "it was a great work or a minor one, a compendium for the scholar, a manual for the clerk, or even a lesson book for the schoolboy."

The Egyptians, it has been said, made no great contributions to mathematical knowledge. They were practical men, not much given to speculative or abstract inquiries. Dreamers, as Thompson suggests, were rare among them, and mathematics is nourished by dreamers—as it nourishes them. Egyptian mathematics nonetheless is not a subject whose importance the historian or student of cultural development can afford to disparage. And the Rhind Papyrus, though elementary, is a respectable mathematical accomplish-

The papyrus was originally a roll 13 inches high and almost 18 feet long. This photograph shows a small section of it about 4 inches high and 10 inches wide. Hieratic script reads from right to left and top to bottom.

ment, proffering problems some of which the average intelligent man of the modern world—38 centuries more intelligent, perhaps, than A'h-mosè—would have trouble solving.

Scholars disagree as to A'h-mosè's mathematical competence. There are mistakes in his manuscript, and it is hard to say whether he put them there or copied them from the older document. But he wrote a "fine bold hand" in hieratic, a cursive form of hieroglyphic; altogether it seems unlikely that he was merely an ignorant copyist.

It would be misleading to describe the Rhind as a treatise. It is a collection of mathematical exercises and practical examples, worked out in a syncopated, sometimes cryptic style. The first section presents a table of the division of 2 by odd numbers—from $^{2}/_{3}$ to $^{2}/_{101}$. This conversion was necessary because the Egyptians could operate only with unit fractions and had therefore to reduce all others to this form. With the exception of $^{2}/_{3}$, for which the Egyptians had a special symbol, every fraction had to be expressed as the sum of a series of fractions having 1 as the numerator. For example, the fraction $^{3}/_{4}$ was written as $^{1}/_{2}$, $^{1}/_{4}$ (note they did not use the plus sign), and $^{2}/_{61}$ was expressed as $^{1}/_{40}$, $^{1}/_{244}$, $^{1}/_{488}$, $^{1}/_{610}$.

It is remarkable that the Egyptians, who attained so much skill in their arithmetic manipulations, were unable to devise a fresh notation and less cumbersome methods. We are forced to realize how little we understand the circumstances of cultural advance: why societies move—or is it perhaps jump—from one orbit to another of intellectual energy, why the science of Egypt "ran its course on narrow lines" and adhered so rigidly to its clumsy rules. Unit fractions continued in use, side by side with improved methods, even among Greek mathematicians. Archimedes, for instance, wrote $^{1}/_{2}$, $^{1}/_{4}$ for $^{3}/_{4}$, and Hero, $^{1}/_{2}$, $^{1}/_{17}$, $^{1}/_{34}$, $^{1}/_{51}$ for $^{31}/_{51}$. Indeed, as late as the 17th century certain Russian documents are said to have expressed $^{1}/_{96}$ as a "half-half-half-half-half-third."

The Rhind Papyrus contains some 85 problems, exhibiting the use of fractions, the solution of simple equations and progressions, the mensuration of areas and volumes. The problems enable us to form a pretty clear notion of what the Egyptians were able to do with numbers. Their arithmetic was essentially additive, meaning that they reduced multiplication and division, as children and electronic computers do, to repeated additions and subtractions. The only multiplier they used, with rare exceptions, was 2. They did larger multiplications by successive duplications. Multiplying 19 by 6, for example, the Egyptians would double 19, double the result and add the two products, thus:

	1	19
	\ 2	38
	\ 4	76
Total	6	114

The symbol \ is used to designate the sub-multipliers that add up to the total multiplier, in this case 6. The problem 23 times 27 would, in the Rhind, look like this:

	\ 1	27
	\ 2	54
	\ 4	108
	8	216
	\ 16	432
Total	23	621

In division the doubling process had to be combined with the use of fractions. One of the problems in the papyrus is "the making of loaves 9 for man 10," meaning the division of 9 loaves among 10 men. This problem is not carried out without pain. Recall that except for $\frac{2}{3}$ the Egyptians had to reduce all fractions to sums of fractions with the numerator 1. The Rhind explains:

"The doing as it occurs: Make thou the multiplication $\frac{2}{3}$ $\frac{1}{5}$ $\frac{1}{30}$ times

1	$\frac{2}{3}$	$\frac{1}{5}$	$\frac{1}{30}$
\ 2	$1\frac{2}{3}$	$\frac{1}{10}$	$\frac{1}{30}$
4	$3\frac{1}{2}$	$\frac{1}{10}$	
\ 8	$7\frac{1}{5}$		

Total loaves 9; it, this is."

In other words, if one adds the fractions obtained by the indicated multiplications (2 + 8 = 10), he arrives at 9. The reader understandably, may find the demonstration baffling. For one thing, the actual working of the problem is not given. If 10 men are to share 9 loaves, each man, says A'h-mosè, is to get $\frac{2}{3}$, $\frac{1}{5}$, $\frac{1}{30}$ (*i.e.*, $\frac{27}{30}$) times 10 loaves; but we have no idea how the figure for each share was arrived at. The answer to the problem ($\frac{27}{30}$, or $\frac{9}{10}$) is given first and then verified, not explained. It may be, in truth, that the author had nothing to explain, that the problem was solved by trial and error—as, it has been suggested, the Egyptians solved all their mathematical problems.

An often discussed problem in the Rhind is: "Loaves 100 for man 5, $\frac{1}{7}$ of the 3 above to man 2 those below. What is the difference of share?" Freely translated this reads: "Divide 100 loaves among 5 men in such a way that the shares received shall be in arithmetical progression and that $\frac{1}{7}$ of the sum of the largest three shares shall be equal to the sum of the smallest two. What is the difference of the shares?" This is not as easy to answer as its predecessors, especially when no algebraic symbols or processes are used. The Egyptian method was that of "false position"—a mixture of trial and error and arithmetic proportion. Let us look at the solution in some detail:

"Do it thus: Make the difference of the shares $5\frac{1}{2}$. Then the amounts that the five men receive will be 23, $17\frac{1}{2}$, 12, $6\frac{1}{2}$, 1: total 60."

Now the assumed difference $5\frac{1}{2}$, as we shall see, turns out to be correct. It is the key to the solution. But how did the author come to this disingenuously "assumed" figure? Probably by trial and error. Arnold Buffum Chace, in his definitive study *The Rhind Papyrus*—from which I have borrowed shamelessly—proposes the following ingenious reconstruction of the operation:

Suppose, as a starter, that the difference between the shares were 1. Then the terms of the progression would be 1, 2, 3, 4, 5; the sum of the smallest two would be 3, and $\frac{1}{7}$ of the largest three shares would be $1\frac{5}{7}$ ($1\frac{1}{2}$, $\frac{1}{7}$, $\frac{1}{14}$ Egyptian style). The difference between the two groups (3 minus $1\frac{5}{7}$) would be $1\frac{2}{7}$, or $1\frac{1}{4}$, $\frac{1}{28}$. Next, trying 2 as the

Part of title page of the papyrus is reproduced in facsimile. Here the hieratic script reads from top to bottom and right to left. It has been translated: "Accurate reckoning of entering into things, knowledge of existing things all, mysteries . . . secrets all. Now was copied book this in year 33, month four of the inundation season [under the majesty of the] King of [Upper and] Lower Egypt, 'A-user-Rê', endowed with life, in likeness of writings of old made in the time of the King of Upper [and Lower] Egypt, [Ne-ma] 'et-Rê'. Lo the scribe A'h-mosè writes copy this."

difference between the successive shares, the progression would be 1, 3, 5, 7, 9. The sum of the two smallest terms would be 4; $^{1}/_{7}$ of the three largest terms would be 3, and the difference between the two sides, 1. The experimenter might then begin to notice that for each increase of 1 in the assumed common difference, the inequality between the two sides was reduced by $^{1}/_{4}$, $^{1}/_{28}$. Very well: to make the two sides equal, apparently he must multiply his increase 1 by as many times as $^{1}/_{4}$, $^{1}/_{28}$ is contained in $1^{1}/_{4}$, $^{1}/_{28}$. That figure is $4^{1}/_{2}$. Added to the first assumed difference, 1, it gives $5^{1}/_{2}$ as the true common difference. "This process of reasoning is exactly in accordance with Egyptian methods," remarks Chace.

Having found the common difference, one must now determine whether the progression fulfills the second requirement of the problem: namely, that the number of loaves shall total 100. In other words, multiply the progression whose sum is 60 (see above) by a factor to convert it into 100; the factor, of course, is $1^{2}/_{3}$. This the papyrus does: "As many times as is necessary to multiply 60 to make 100, so many times must these terms be multiplied to make the true series." (Here we see the essence of the method of false position.) When multiplied by $1^{2}/_{3}$, 23 becomes $38^{1}/_{3}$, and the other shares, similarly, become $29^{1}/_{6}$, 20, $10^{5}/_{6}$ and $1^{2}/_{3}$. Thus one arrives at the prescribed division of the 100 loaves among 5 men.

The author of the papyrus computes the areas of triangles, trapezoids and rectangles and the volumes of cylinders and prisms, and of course the area of a circle. His geometrical results are even more impressive than his arithmetic solutions, though his methods, as far as one can tell, are quite unrelated to the discipline today called geometry. "A cylindrical granary of 9 diameter and height 6. What is the amount of grain that goes into it?" In solving this problem a rule is used for determining the area of a circle which comes to Area = $(^{8}/_{9}d)^2$, where d denotes the diameter. Matching this against the modern formula, Area = πr^2, gives a value for π of 3.16—a very close approximation to the correct value. The Rhind Papyrus gives the area of a triangle as $^{1}/_{2}$ the base times the length of a line which may be

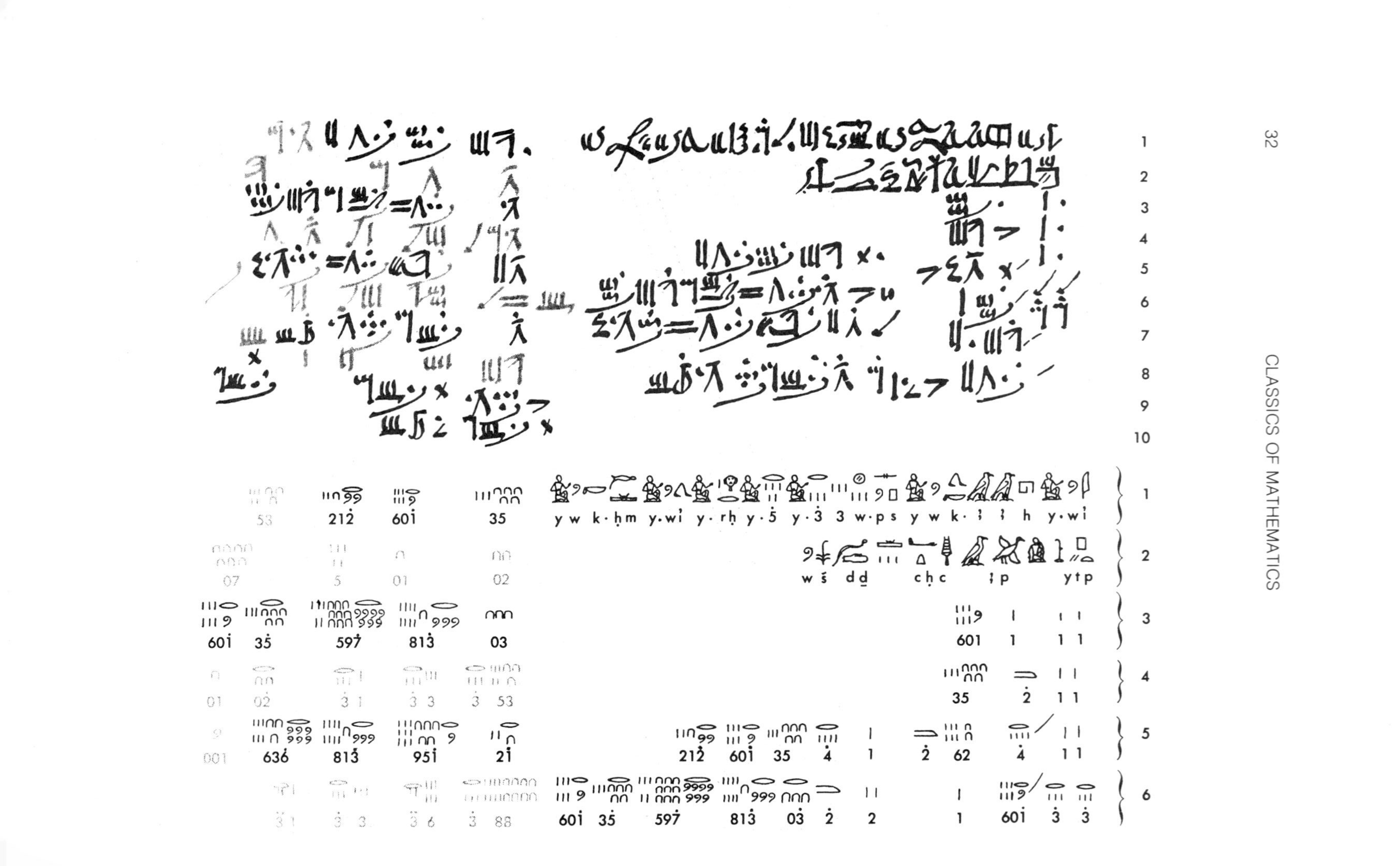

1 y w k · ḥm y · wi y · rḥ y · 5̇ y · 3̇ 3 w · ps y w k · ꜣ ꜣ h y · wi

53 212̇ 601̇ 35

2 wš ḏd cḥc ꜣp ytp

07 5 01 02

3 601̇ 35̇ 597̇ 813̇ 03 601 1 1 1

4 01 02̇ 3̇ 1 3̇ 3 3̇ 53 35 2̇ 1 1

5 001 636̇ 813̇ 951̇ 21̇ 212̇ 601̇ 35 4̇ 1 2̇ 62 4̇ 1 1

6 3̇ 1 3̇ 3 3̇ 6 3̇ 88 601̇ 35̇ 597̇ 813̇ 03̇ 2̇ 2 1 601̇ 3̇ 3̇

08 060.$\dot{1}$ 03$\dot{5}$ 56$\dot{2}$ 0$\dot{2}$ 63$\dot{6}$ 81$\dot{3}$ 95$\dot{1}$ 2$\dot{1}$ $\dot{3}$ 2 3$\dot{5}$ $\dot{5}$ $\dot{5}$ } 7

$\dot{4}$ 1 2 4 35 060$\dot{1}$ 03$\dot{5}$ 56$\dot{2}$ 0$\dot{2}$ $\dot{5}$ 1 dmd $\dot{2}$ 21$\dot{2}$ } 8

56$\dot{2}$ 562 $\dot{4}$ 035 $\dot{2}$ } 9

060.1 dmd 562 $\dot{4}$ } 10

iw·y h꜡·kwy sp·w 3 $\dot{3}$ · y $\dot{5}$ · y ḫr· y iw · y mḥ · kwy pty p꜡ ꜥḥꜥ ḏd św

Go down I times 3, ⅓ of me, ⅕ of me is added to me; return I, filled am I. What is the quantity saying it?

1 1 1 $\dot{4}$ 53 $\dot{1}06$ $\dot{2}12$

1 1 2 $\dot{2}$ $\dot{3}0$ $\dot{3}18$ $\dot{7}95$ 53 $\dot{1}06$

1 1 $\dot{3}$ $\dot{1}2$ $\dot{1}59$ $\dot{3}18$ $\dot{6}36$

$\dot{3}$ $\dot{3}$ $\dot{5}$ $\dot{2}0$ $\dot{2}65$ $\dot{5}30$ $\dot{1}060$

$\dot{5}$ $\dot{5}$

1 106 53 $\dot{1}06$ $\dot{2}12$

$\dot{2}$ 53 20 10 5 35

\ $\dot{4}$ 26 $\dot{2}$ 30 $\dot{3}18$ $\dot{7}95$ $\dot{5}3$ $\dot{1}06$

\ $\dot{1}06$ 1 35 $\dot{3}$ 3 $\dot{3}$ 1 $\dot{3}$ $\dot{2}0$ 10 70

\ $\dot{5}3$ 2 $\dot{1}2$ $\dot{1}59$ $\dot{3}18$ $\dot{6}36$ $\dot{2}$ 530

\ $\dot{2}12$ $\dot{2}$ 88 $\dot{3}$ 6 $\ddot{3}$ 3 $\dot{3}$ 1 $\ddot{3}$ 100 $\dot{4}$ 265

dmd 1 $\dot{2}0$ $\dot{2}65$ $\dot{5}30$ $\dot{1}060$ $\dot{4}$ 265

Total 53 4 2 1 80 dmd 1060

$\dot{4}$ $\dot{2}65$ Total

Problem 36 of the papyrus begins: "Go down I times 3, $^1/_3$ of me, $^1/_5$ of me is added to me; return I, filled am I. What is the quantity saying it?" The problem is then solved by the Egyptian method. On these pages is a facsimile of the problem as it appears in the papyrus. The hieratic script reads from right to left. The characters are reproduced in gray and black (the original papyrus was written in red and black). In the middle of the page is a rendering in hieroglyphic script, which also reads from right to left. Beneath each line of hieroglyphs is a phonetic translation. The numbers are given in Arabic with the Egyptian notation. Each line of hieroglyphs and its translation is numbered to correspond to a line of the hieratic. At the bottom of the page the phonetic and numerical translation has been reversed to read from left to right. Beneath each phonetic expression is its English translation. A dot above a number indicates that it is a fraction with a numerator of one. Two dots above a 3 represent $^2/_3$, the only Egyptian fraction with a numerator of more than one.

Readers who have the desire to trace the entire solution are cautioned that the scribe made several mistakes that are preserved in the various translations.

the altitude of the triangle, but, on the other hand—Egyptologists are not sure—may be its side. In an isosceles triangle, tall and with a narrow base, the error resulting from using the side instead of the altitude in computing area would make little difference. The three triangle problems in the Rhind Papyrus involve triangles of this type, but it is clear that the author had only the haziest notion of what triangles were like. What he was thinking of was (as one expert conjectures) "a piece of land, of a certain width at one end and coming to a point, or at least narrower at the other end."

Egyptian geometry makes a very respectable impression if one considers the information derived not only from the Rhind but also from another Egyptian document known as the Moscow Papyrus and from lesser sources. Its attainments, besides those already mentioned, include the correct determination of the area of a hemisphere (some scholars, however, dispute this) and the formula for the volume of a truncated pyramid, $V = (h/3)(a^2 + ab + b^2)$, where a and b are the lengths of the sides of the square and h is the height.

I should like to give one more example taken from the Rhind Papyrus, something by way of a historical oddity. Chace offers the following translation of the hard-to-translate Problem 79:

"Sum the geometrical progression of five terms, of which the first term is 7 and the multiplier 7.

"The sum according to the rule. Multiply 2801 by 7.

\ 1	2801
\ 2	5602
\ 4	11204
Total	19607

"The sum by addition

houses	7
cats	49
mice	343
spelt (wheat)	2401
hekat (half a peck)	16807
Total	19607"

This catalogue of miscellany provides a strange little prod to fancy. It has been interpreted thus: In each of 7 houses are 7 cats; each cat kills 7 mice; each mouse would have eaten 7 ears of spelt; each ear of spelt would have produced 7 hekat of grain. Query: How much grain is saved by the 7 houses' cats? (The author confounds us by not only giving the hekats of grain saved but by adding together the entire heterogeneous lot.) Observe the resemblance of this ancient puzzle to the 18th-century Mother Goose rhyme:

> As I was going to St. Ives
> I met a man with seven wives.
> Every wife had seven sacks,
> Every sack had seven cats,
> Every cat had seven kits.
> Kits, cats, sacks and wives,
> How many were there going to St. Ives?

(To this question, unlike the question in the papyrus, the correct answer is "one" or "none," depending on how it is interpreted.)

A considerable difference of opinion exists among students of ancient science as to the caliber of Egyptian mathematics. I am not impressed with the contention based partly on comparison with the achievements of other ancient peoples, partly on the wisdom of hindsight, that the Egyptian contribution was negligible, that Egyptian mathematics was consistently primitive and clumsy. The Rhind Papyrus, though it demonstrates the inability of the Egyptians to generalize and their penchant for clinging to cumbersome calculating processes, proves that they were remarkably pertinacious in solving everyday problems of arithmetic and mensuration, that they were not devoid of imagination in contriving algebraic puzzles, and that they were uncommonly skillful in making do with the awkward methods they employed.

It seems to me that a sound appraisal of Egyptian mathematics depends upon a much broader and deeper understanding of human culture than either Egyptologists or historians of science are wont to recognize. As to the question how Egyptian mathematics compares with Babylonian or Mesopotamian or Greek mathematics, the answer is comparatively easy and comparatively unimportant. What is more to the point is to understand why the Egyptians produced their particular kind of mathematics, to what extent it offers a culture clue, how it can be related to their social and political institutions, to their religious beliefs, their economic practices, their habits of daily living. It is only in these terms that their mathematics can be judged fairly.

NOTES

1. George Sarton, *A History of Science,* Cambridge (Mass.), 1952, p. 40.

2. O. Neugebauer, *The Exact Sciences in Antiquity,* Princeton, 1952, p. 78.

3. W. Struve, *Mathematischer Papyrus des Staatlichen Museums der Schönen Kunste in Moskau,* Berlin, 1930.

4. For an interesting survey of the beginnings of geometry, including the Babylonian, Egyptian, Indian, Chinese and Japanese contributions, see Julian Lowell Coolidge, *A History of Geometrical Methods,* Oxford, 1940, pp. 1–23. Another problem in the Moscow Papyrus which, as Coolidge mentions, has excited scholars, is that of finding the area of a basket, in connection with which exercise the Egyptians gave the excellent approximation, $\pi = (^{16}/_{9})^2$. See also B. L. Van der Waerden, *Science Awakening,* Groningen, 1954.

3. Problem No. 14 of the Moscow Papyrus*

BATTISCOMBE GUNN

T. ERIC PEET

Problem No. 14.

(Transcription, Pl. xxxvi.)

Example of calculating a truncated pyramid.

If you are told: A truncated pyramid of 6 *for the vertical height by* 4 *on the base by* 2 *on the top:*

You are to square this 4; *result* 16. *You are to double* 4; *result* 8. *You are to square this* 2; *result* 4. *You are to add the* 16 *and the* 8 *and the* 4; *result* 28. *You are to take* ⅓ *of* 6; *result* 2. *You are to take* 28 *twice; result* 56. *See, it is of* 56.

You will find (*it*) *right.*

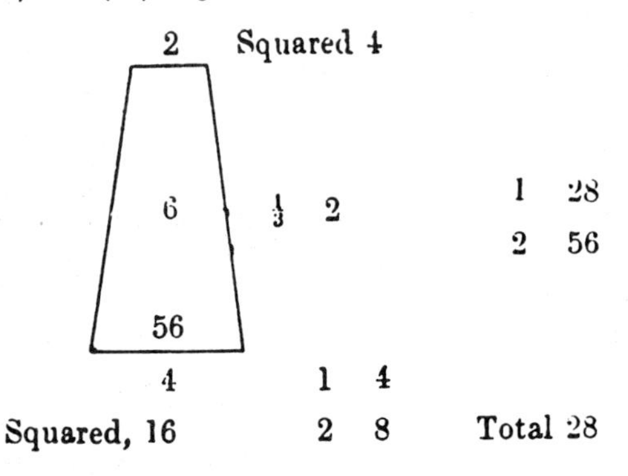

NOTES ON THE TEXT

The text is well preserved and presents no difficulties.

Figure. The solid is of course represented, as in the hieratic ideogram, as a simple trapezoid, and in the original is roughly drawn without regard to the proportions, as in the transcription.

COMMENTARY

The problem is to determine the volume of what we call a truncated pyramid, or frustum of a pyramid, the *data* being the vertical height *(stwti)* and the respective lengths of the sides of the two squares which bound the solid below and above.

**Source:* From Henrietta O. Midonick (ed.), *The Treasury of Mathematics* (1965), 520-521. Reprinted by permission of the Philosophical Library.

If we call the height h and the sides of the lower and upper squares a and b respectively, the working may be represented as follows:

Square a, result 16. Multiply a by b, result 8. Square b, result 4. Add these results, total 28.

Take one-third of h, result 2. Multiply 28 by this, result 56, which is the volume sought.

Expressing these operations by a general formula, we have.

$$V = (a^2 + ab + b^2)h/3,$$

which is exactly the formula used to-day to determine the volume of such solids.

The figure, and the numbers which accompany it, are quite straightforward. In the centre of the figure stands its height, 6. Below is the side a, namely 4, "squared, 16." Similarly above we have the side b, namely 2, "squared, 4." On the left (in the original), opposite the 6 inside the figure, we read $1/3$ 2, indicating that one-third of the height 6 is 2. Below on the left (in the original) is the multiplication of a and b, that is 4 multiplied by 2, and this is followed by the total, 28, of the 16, the 8 and the 4. Above this is the final step, the multiplication of 28 by 2, giving 56. This number, which is that of the required volume, is then inserted in the figure, near the base.

4. From the Old Testament of the *Bible*

• **Commentary on the Hebrews and Protomathematics**. Among the peoples of the ancient Fertile Crescent only the Babylonians and Egyptians made significant advances in protomathematics. Even so highly cultured a people as the Hebrews got no further than a crude level. If the statement in I Kings 7:23 was based upon or included some notion that a rough correspondence holds between the diameter of any circle and its circumference, of which the numbers in the text are an example, they would have set the value of the ratio we designate as pi at 3.0. That approximation was common among other ancient peoples including the Chinese. (Some occultists believe that the ancient Hebrews had the beginnings of a number mysticism known as the Cabbala that allowed a theosophical interpretation of Scripture at deeper levels. The Cabbala, however, incorporated later Gnostic elements and was probably not developed in written form until the third to the sixth century A.D.)

By the high Middle Ages the Judaeo-Christian tradition and Graeco-Roman learning occurring in a new, vigorous economic and technological environment were the two pillars of what we call "western civilization." To the Old Testament, as Hebrew-Scripture was designated, Latin Christians added the New Testament to form a single holy book. In these circumstances, where the *Bible* enjoyed unique authority, one verse from the Old Testament strongly supported the quantitative view of nature among men of learning in Europe. The verse, Solomon 11:20, is quoted below. Its influence upon European thinkers extended into the early modern period, when Gottfried Leibniz, one of the two discoverers of the calculus, cited it as an important verse.

• **An approximation of pi as 3.0**. Then he made the molten sea; it was round, ten cubits from brim to brim, and five cubits high, and a line of thirty cubits measured its circumference.—*I Kings 7:23* and repeated as *II Chronicles 4:2*.

• **On the quantitative approach to nature**. . . . thou hast arranged all things by measure and numbers and weight.—*Wisdom of Solomon 11:20*.

Chapter II

The Rise of Theoretical Mathematics in Ancient Greece

Introduction. During the three centuries after 600 B.C., the ancient Greeks (or Hellenes as they styled themselves) transformed mathematics from the empirical, trial-and-error protoscience they inherited into a rigorous, theoretical science. The Greeks achieved this monumental breakthrough by making original contributions to the nascent theory of numbers (arithmetic) and, above all, by inventing demonstrative geometry with its axiomatic, deductive methodology. In developing this methodology and in arranging mathematical theory systematically, they progressed from a crude, intuitive notion of proof to a formal one. A formal proof involves an explicitly logical step-by-step argument leading to a single, unambiguous conclusion. At the same time, Greek mathematicians ceased to be as anonymous as their predecessors in the ancient Near East. The Greeks favored crediting inventions to a single person, even when there was no sound historical basis for doing so. This was the convention among a people who believed strongly in heroes.

The men credited with creating, shaping, and practicing mathematics in ancient Greece are reported to have left voluminous writings. Unfortunately for our needs, only two fragments of their writings survive, namely those by Hippocrates of Chios on lunules and by Archytas of Tarentum on the duplication of the cube. Many manuscripts were lost when a number of great Greek libraries were destroyed in wars. Papyrus was also a fragile material that decomposed easily in the humid Aegean climate. Despite the paucity of primary sources, classical scholars and historians of mathematics have been able to reconstruct the outlines of the early development of theoretical mathematics in Greece. Among the evidence they have appealed to are comments on mathematical topics in the surviving writings of later major thinkers, such as Herodotus, Plato, and Aristotle. Their chief sources have been Euclid's *Elements*—a skillful compilation dating from about 295 B.C. which incorporated major writings from earlier Greek mathematicians—and the "Catalogue of Geometers" recorded by his commentator Proclus, who flourished in the 5th century A.D.

What were the antecedents of Greek theoretical mathematics? What were the chief influences upon its development in the late Archaic Period (the 6th century B.C.)? And what developments occurred during the Classical Period (the early-5th through the mid-4th century B.C.)? Through colonization, trade, and warfare the ancient Greeks had contacts with the ideas and learning of the peoples of Crete, Phoenicia, Mesopotamia, and Egypt. They compared, borrowed, and assimilated selectively earlier protomathematical lore. Herodotus (*c.* 480–*c.* 425 B.C.), an inquisitive Ionian traveler, recognized a debt to Egypt when he credited the scribes of Egypt with inventing mensurational geometry (*The Histories* II, 9). Later Aristotle even asserted (incorrectly) in *Metaphysics* (A 3) that the Egyptian scribal mathematicians had developed theoretical geometry. The Greeks were not merely cultural borrowers; there is truth as well as chauvinism in the proud boast of one of Plato's characters in the dialogue *Epinomis* (987 d): "Whatever the Greeks have acquired from foreigners they have in the end turned into something finer." What is more, ancient Mesopotamian and Egyptian protomathematics apparently did not have their greatest impact upon Greek mathematicians until after 300 B.C. in Ptolemaic Alexandria.

With its illustrative diagrams and systematic nature, demonstrative geometry was characteristic of a people who generally favored a natural visual approach to the physical world and stressed order. These traits were prominent among the ancient Greeks as their geometric pottery and their carefully sited and symmetrically designed cities show. Indeed, the two major formative influences on the development of theoretical mathematics were the fruitful, new conceptual and aesthetic foundations that the ingenious Greeks fashioned for their learning and culture. Greek learning blossomed in the 6th century B.C. following a fundamental shift of perspective in Ionia (the eastern Aegean area) from *mythos*, or an explanatory story that was usually spoken, to *logos*, or a reasoned account, with its relatively naturalistic and rational mode of inquiry.[1] This shift probably originated in the city of Miletus on the western coast of Asia Minor. Milesian sages speculated that the cosmos was in many respects knowable, rational, and simple. Afterward logical reasoning together with aesthetic beauty came to guide the imaginative Greek search for various truths and excellence. Theoretical mathematics was only one of the disciplines formed in that conceptual-aesthetic setting. Others were the initial stages of philosophy, critical history, rational diagnostic medicine, and political theory. Hellenic beauty was sensual and intellectual, appealing to the ear and mind in rhetoric, lyric poetry, drama, and abstract thought; to the eye in architecture and the idealized sculptures of the Classical Period—notably the Athena of Phidias, the Aphrodite of Praxiteles, the Doryphoros ("the spear-bearer") of Polycleitus, and the Diskobolos ("the discus-thrower") of Myron.

Emphasis upon theoretical cogency and aesthetic beauty reflected deeper currents in this remarkable civilization. The social and economic base was stable enough to provide pockets of leisure and security but strained enough to generate tensions for growth and change. Such a context encouraged confidence, curiosity, and innovations on intellectual as well as other fronts. During the late Archaic and Classical Periods symptoms of this vigor became manifest in many ways—territorially by colonization, politically by the evolution of democracy, socially by a prosperous economy and variegated religious cults, and intellectually by the creation of a rich literature.

By the 6th century B.C. the Greeks had become an agricultural and seafaring people who had molded a highly varied, littoral civilization extending beyond their Aegean cradle across the Mediterranean. The power of colonization among them is suggested by their word for colony—*apoika*, or away home. The *polis*, or roughly city-state, was the source of local patriotism and the political basis of their civilization. From the word *polis* comes our word "politics." Chief among the *poleis* were Sparta, Thebes and Athens. Most *poleis* differed from older Near Eastern cities in their tolerance of free and open debate. A limited form of democracy with adult, free, native male citizens possessing the franchise was another Greek invention, although not all of the aristocratically-dominated *poleis* favored it. Full democracy was most nearly approximated in Attica, the region around and including Athens.

The economic prosperity of the *poleis* derived principally from a combination of agriculture and commerce, not unlike the conditions that existed in Jeffersonian America. The Greek agricultural aristocrats combined the growing of wheat, barley, olives, and grapes with a highly successful trade in wine and olive oil. Greek industry also produced textiles, metals, tools, weapons, and pottery for trade. The underside of this economic prosperity was an extensive slavery that proved particularly inhumane in the silver mines of Laurium. While citizens of the Greek *poleis* did not identify themselves with a single country, nor even a confederation, they united around a common language and a religion of the Olympian gods. Each *polis* had a tutelary deity with Apollo being the most popular. Religious games were a show of their solidarity as well as impressive athletic and artistic events. By the end of the Archaic Period the staples of Greek education were the *Iliad* and *Odyssey* of Homer and the writings of Hesiod. Largely from these, humanism and secularism were interwoven into Greek society.

Theoretical learning was not the exclusive achievement of any one *polis* during the three centuries after 600 B.C., nor did it fare equally well in all parts of Hellas. Its first center in the early-6th century B.C. was in Ionia, especially Miletus, then a prosperous city that had long been free from external attack and whose confident citizens had contacts with earlier Near Eastern ideas and learning. The rising power of Persia ended its secure circumstances. Persian conquest of Miletus and nearby Ionian *poleis* by 540 forced the major schools to relocate in *Magna Graecia* ("Greater Greece"), essentially parts of what are now southern Italy and Sicily. The Persians attacked the Balkan mainland of Hellas itself between 491 and 480 B.C. Only after the Persians were repulsed did Athens move from a city-state into an empire on the Aegean Sea. As that empire reached its fullest extent and gained great wealth under Pericles about 450 B.C., Athens became the primary center for Hellenic intellectual and artistic creativity. To use the phrase attributed to Pericles by the historian Thucydides, it became the "School of Hellas." Athens retained that position even after its defeat by Sparta in the protracted, disastrous Peloponnesian War (*c.* 435–404 B.C.). The philosophical learning of democratic Athens did not penetrate authoritarian Sparta, Thessaly, and most of Boeotia, which speaks eloquently of the need for a proper socio-cultural climate for intellectual growth.

In ancient Greece the beginnings of the notion of demonstration and proof in mathematics—as opposed to mere opinion and probability—has been traced to the sixth century B.C., albeit faintly. The Neoplatonic commentator Proclus reported in the "Catalogue of Geometers" that Thales of Miletus (*c.* 625–*c.* 547 B.C.), one of the seven sages of antiquity, was said to have "proved" the proposition that the diameter bisects the circle (Euclid I. 17). However, Thales may have accomplished this by simply folding a papyrus drawing—rather like the technique that masons and carpenters used in their work. There is no known evidence to suggest that Thales had any notion of "equality" akin to Euclid's axiom of congruence, which states simply that "Things which coincide with one another are equal to one another." He may have indicated the need for demonstrative proofs, however, and, unlike earlier scribal mathematicians of the Near East, he formulated five general statements or propositions about geometric properties. Since little definite information about Thales survived even to Aristotle's time, the assertion about his initiating work on theoretical mathematics is of dubious authenticity.

With the dawn of the 5th century B.C., two schools in *Magna Graecia* founded by refugees from Ionia contributed to mathematical thought. Pythagoras of Samos (*c.* 560 B.C.–*c.* 480 B.C.) founded the first group at Croton in southern Italy. Pythagoras and his followers, who combined philosophy and religious mysticism in their cult, endeavored to establish mathematics as an abstract discipline. They made at least three basic contributions. First, they considered numbers to be essentially abstractions. Second, the limited "more or less" results of empirical science no longer sufficed for them. Instead they sought, but did not achieve, clear and unassailable proofs, because these were the best means in their scheme of things for the human intellect to fathom the ways of a geometer God. Their operational notion of proof, however, appears usually to have been quite informal. Consider their *arithmetica,* the beginnings of a theory of numbers, where they developed a theory of odd and even. To "show" that numbers were odd or even they may simply have pointed to dot or pebble representations of numbers. On the other hand, they did develop a rudimentary notion of proof by using chains of logical reasoning to investigate geometric theorems. Third, they cultivated mathematics as one contemplative exercise in a multifaceted search for eternal verities, rather than as a practical pursuit having applications in surveying, commerce, astronomy, architecture, and record-keeping, as had been the case with the Milesians, the ancient Babylonians, and the Egyptians.

The early Pythagoreans' pursuit of *arithmetica* and closely related studies produced significant results. In addition to having the theory of odd and even, they classified numbers as perfect, amicable, or figurate.[2] Figurate numbers refer to dot arrangements of numbers. The Pythagoreans found, for example, that adding successive integers (1, 2, 3, 4, . . .) gives triangles (3 ∴ , 6 ∴ , 10 ∴), or triangular numbers, while adding successive odd numbers (1, 3, 5, 7, . . .) gives squares 4 ::, 9 :::, 16 ::::), or square numbers. Having associated numbers with shapes, they investigated music and found relationships between numbers and harmonious sound. They discovered empirically that vibrating strings in proportions of 2:1, 3:2, and 4:3 give concord. They thereby reduced musical consonance to numerical ratios. These early discoveries of numerical relationships perhaps led to or reinforced their belief in the regularity of nature and the tantalizing corollary that this regularity was numerical. (Thus the belief arose among them that the ultimate stuff of the universe was atom-like numbers.[3]) The Pythagoreans expressed this belief in the generalization: "all is number." (By number they meant a positive "whole" number.) This generalization, which became the Pythagorean motto, included numerological assumptions and inspired a crude theory of integral proportion, that is, a theory of proportion based entirely on the ratios of whole numbers (commensurables).

The discovery—attributed to the Pythagorean leader Hippasus of Metapontum—of incommensurable ratios (that is, irrational numbers such as $\sqrt{2}$ and $\sqrt{5}$) proved an anomaly that must have disturbed at the least their leading intellects. While incommensurability does not appear to have provoked a foundational crisis at the time, it posed a fundamental problem for geometric algebra, because it suggested that geometric magnitudes (lengths, areas, and volumes) are *continuous* in character rather than *discrete.* In geometric algebra one seeks geometric solutions of algebraic equations. With Hippasus' discovery the Pythagoreans could no longer confidently manipulate geometric magnitudes in algebraic equations as if they were discrete numbers. How were they to deal with the multiplication and division of lengths or areas? To do this they devised an inventive mathematical procedure now known as the "application of areas." This procedure became the principal one in Greek geometrical algebra. An example, from a construction problem follows:

Given a line segment $\overline{BC}$ of length b, construct a rectangle with base $\overline{BC}$ which is equal to a square of edge c.

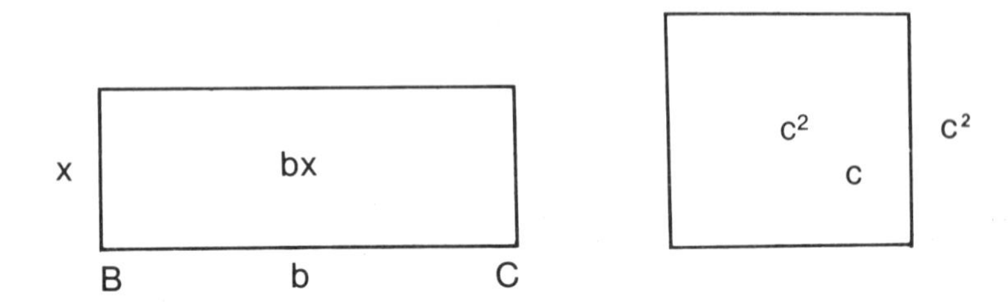

In modern notation this provides a solution for the equation $bx = c^2$ and corresponds to geometric division. In this division the given area c^2 has been applied to the given segment $\overline{BC}$. With the "application of areas" procedure geometers could compare any two geometric magnitudes because any rectilinear figure can be transformed into a rectangle of the same area and a predetermined height. Results from the Pythagorean geometric algebra later appeared in Books II and VI of Euclid's *Elements*.

Parmenides of Elea (c. 515–c. 450 B.C.) and his followers were the second group in *Magna Graecia* to influence the development of theoretical mathematics. They did so by making a strong, if not revolutionary, impact on the criteria for acceptable evidence and conclusive demonstration. For understanding the nature of the cosmos Parmenides contrasted the unreliability of the senses with the certainty of reasoned argument. Proceeding deductively from a limited notion of first principles of reasoning and often employing *reductio ad absurdum,* an indirect proof showing the impossibility of all alternatives, he espoused a species of monism which asserted that physical reality was necessarily a changeless, spherical Unity. In opposition to Heraclitus of Ephesus he argued that change was merely an illusion of the senses. Parmenides' pupil Zeno (c. 490–c. 425 B.C.) elaborated four ingenious *reductio ad absurdum* paradoxes that posed dilemmas for those who explained the universe in terms of plurality and motion. The paradoxes, known as (1) the Dichotomy or Stadium, (2) Achilles and his Opponent (the Tortoise), (3) the Flying Arrow, and (4) Moving Rows,[4] were pioneering analyses of the nature of the continuum and the nature of infinite processes. The first pair attacked the infinite divisibility of space; the second pair, unspecified notions of indivisible minima.

Theoretical mathematics experienced its greatest development in ancient Greece during the Classical Period. At this time Greek mathematicians investigated four major topics: (1) the origins of the theory of numbers, (2) the tradition in metrical geometry later attributed to Hero that sought to solve problems involving areas and volumes, (3) non-metrical geometry, especially three famous construction problems, and (4) the application of mathematics to music theory. The first and fourth topics were the especial, but not the exclusive, province of the Pythagoreans. The three famous construction problems of the third topic were (1) the quadrature of squaring of a circle, (2) the trisection of an arbitrary angle, and (3) the duplication of a cube—the so-called Delian problem.[5] Methods of solving these problems were restricted to constructions with straightedge and compass since, for aesthetic reasons, the straight line and circle were considered the only perfect curves. Fruitful study of these three problems outlived Classical Greece. The squaring of the circle long captured the attention of leading mathematicians; Gottfried Leibniz examined the arithmetical quadrature of the circle as part of his work leading to his invention of the first stage of the calculus in late 1675/76. The attendant problem, that is, the nature of *pi,* was not resolved until the 19th century, when mathematicians showed *pi* to be a transcendental number.

During the mid- and late-5th century B.C., Greek mathematicians imaginatively resolved challenging problems and questions of method. The greatest mathematician of the century was Hippocrates of Chios. As his biography in this anthology notes, Hippocrates successfully squared a lunule, or crescent shaped figure, and reduced the Delian problem to finding two mean proportionals between a given side and another twice its length. According to the 6th-century commentator Simplicius, Hippocrates deduced (but did not rigorously prove) that the ratio of the areas of two circles is the same as the ratio of the squares of their respective diameters (or radii). Presumably Hippocrates reached this conclusion after approximating the areas of two circles by inscribing regular polygons in them and then by increasing the number of sides of these polygons. Most historians trace the beginnings of a systematic ordering of geometric theorems, wherein a distinction is made between more and less fundamental theorems, to the lost *Elements* of Hippocrates. Although neither Hippocrates nor his immediate successors established what would today be considered adequate starting points for deductive proofs, they apparently carried out high order foundational studies that made possible the later findings of Plato's circle.

The atomist Democritus (*fl. c.* 430 B.C.), a wealthy citizen of Abdera in Thrace, made an important discovery in solid geometry that was later rigorously proved by Eudoxus, a member of Plato's circle. In the introduction to his treatise entitled *The Method,* Archimedes recorded that Democritus discovered the ratio between the volume of a pyramid (or cone) and a prism (or cylinder) of the same base and height is 1:3. Plutarch asserted that Democritus determined the volumes of pyramids and prisms by viewing them as solids composed of sections. The solids were theoretically sliced into thin sections that were squares or discs parallel to the base.

The early-4th century witnessed continuing progress and an incipient crisis in theoretical mathematics. The Pythagorean leader Archytas of Tarentum (*fl. c.* 375 B.C.), a teacher of Plato, mechanically solved the Delian problem of the duplication of the cube with a three dimensional figure and elaborated the Pythagorean *arithmetica.* Archytas also developed the theory of means and reportedly classified the four basic subjects in the Pythagorean program of studies; namely, geometry, arithmetic, music, and astronomy. These became the famous educational "quadrivium."

The foundational crisis appeared in Athens, which had become the center of mathematical studies in the Greek world by the early fourth century. In this city Theodorus of Cyrene (c. 465–c. 399 B.C.) and his talented pupil Theaetetus (c. 417–c. 369 B.C.), who was to be a colleague of Plato, generalized the theory of incommensurable line segments by finding that the ratios of the side of a three foot square ($\sqrt{3}$) through the side of a seventeen foot square ($\sqrt{17}$) are incommensurable with a unit. As we would say, they demonstrated that the square roots of the nonsquare integers from 3 through 17 are irrational. The square integers are, of course, 4, 9, and 16. No account of Theodorus' method of proof has survived and the topic at present is hotly disputed.[6] One interpretation is that a subtraction procedure which Aristotle called *antaneiresis* or *anthyphairesis* (reciprocal subtraction) was used. This procedure, which is similar to continued fractions, foreshadowed the "Euclidean" algorithm for finding the greatest common divisor for two numbers (Euclid VII. 1-3) or for two homogeneous geometric magnitudes (Euclid X. 2-4). If its division process ends at zero in a finite number of steps, the number is rational.[7] If the subtractions become periodic and go on indefinitely, the number is irrational. Whatever the method of demonstration, the generalization of Theodorus and Theaetetus invalidated those geometric proofs that utilized the Pythagorean theory of integral proportions. Thereby it presented a crisis in the existing foundations of geometry.

Theoretical mathematics in Classical Greece culminated in the mid-4th century B.C. in the provision of an axiomatic foundation for mathematical proofs and the resolution of two crucial problems, including the crisis posed by incommensurability. These developments occurred in Plato's circle at the Academy in Athens. Plato (c. 428–347 B.C.) and his student and rival Aristotle (384–322 B.C.) contributed principally to the starting points of proof theory. Plato wanted proof theory improved, so that it might contrast more sharply with probability and "opinion." The requisite starting points for formal proofs were still missing, however, as his treatment of the notion of hypothesis in *Meno* (86e ff.), *Phaedo* (92d), and the *Republic* (510c ff.) suggests. He stressed careful definitions in *Meno* and the mathematical excursus of the dialogue *Theaetetus*. Plato rated mathematical proof below the dialectic, because he believed that only the persistent, critical questioning of the dialectic could obtain genuine knowledge about the independent world of ideal Forms that he had posited. Mathematical training was to precede and prepare the student for dialectical studies. Plato thus advocated the study of mathematics and made it an integral part of higher learning. He condemned an ignorance of basic mathematics among the literate sector of society; in *Laws* VII (819-20) he called the ongoing fallacious notion that all magnitudes are commensurable a "disgrace."

For his part Aristotle rejected Plato's independent world of ideal Forms, rated the proofs of formal logic above the dialectic, and provided the first complete typology of the axiomatic foundations for proofs in formal logic. In *Posterior Analytics* Aristotle delineated those foundations in his theory of statements, wherein he posited three types of indemonstrable assumptions—careful definitions, common notions or axioms, and special notions or hypotheses (existence assumptions). He also developed syllogistic logic and used analogies in mathematical proofs. Otherwise in mathematics Aristotle accepted the concept of potential infinity but not of actual infinity and held that Eudoxus' theory of proportions had ended the crisis over incommensurables.

Eudoxus of Cnidus (c. 400–c. 347 B.C.), the mathematician of greatest note in Plato's circle, improved method but concentrated his efforts on problem-solving. Continuing the formalistic trend in Greek theoretical mathematics, he applied Aristotle's theory of statements to mathematical proof theory arriving at definitions, axioms, and postulates as its axiomatic starting points. Eudoxus solved the first of two major problems by devising a general theory of proportions that covered both commensurable and incommensurable quantities. His theory was more elegant than *anthyphairesis*. The key to it was his ponderous definition of proportionality for geometric ratios (Euclid V. 5). This definition, which is given in the section on Eudoxus in this anthology, gives the criteria for greater than ($>$) and less than ($<$) as well as equality ($=$). It treats geometric ratios in terms of continuously varying magnitudes. Thereby Eudoxus not only distinctly separated number and geometry but also gave a strong impetus to demonstrative geometry becoming the basis of rigorous mathematics.

Largely from the work of Hippocrates and Democritus, the computation of the areas of curvilinear figures and volumes of solids had become another major problem. In order to make these computations Eudoxus introduced a never-ending approximation process later known as the "method of exhaustion." This method involves inscribing and/or circumscribing a given figure or solid with rectilinear figures of increasing numbers of sides. It tacitly assumes the operation of the theory of limits. In addition, Eudoxus tacitly accepted in Euclid V.4 the essence of a basic assumption known as the Archimedean axiom: Given two points A and B on a straight line, no matter how close, and a point C to the right, no matter how far, then by measuring off the distance $\overline{AB}$ repeatedly one will eventually go beyond C. With these two tacit assumptions Eudoxus based his constructions and proofs on distances between points that could be made arbitrarily small, that is, infinitesimals. An explicit theory of limits and the Archimedean postulate were to be important to the infinitesimal calculus discovered two millennia later.

NOTES

1. Not all scholars agree with this interpretation. According to Heidegger, *mythos* and *logos* are the same.
2. For definitions, see the biography of Pythagoras in this anthology.
3. See Aristotle, *Metaphysics* I, v 986a and 986a 21.
4. For a description of each of these paradoxes see *Dictionary of Scientific Biography*, s.v., "Zeno of Elea."
5. According to Eratosthenes the people of the holy island of Delos were ordered by Apollo to double the size of a cube-shaped altar, while keeping its form, if they wished to escape a pestilence. As a result the duplication of the cube became known as the Delian problem.
6. For further information on this dispute see S. Heller, "Eine Bertrag zur Deutung der Theodoros-Stelle in Platons Dialog 'Theaetet'", *Centaurus* V (1956–58), pp. 13 ff.; W. R. Knorr, *The Evolution of the Euclidean Elements* (1975), chapters 3 and 4; A. Szabo, "Anfänge des euklidischen Axiomensystems," *Archive for History of Exact Sciences* I (1960–62), pp. 69 ff.; and B. L. van der Waerden, *Science Awakening* (1954), pp. 142 ff.
7. *Anthyphairesis* or the "Euclidean" algorithm operates as follows: Given two numbers a and b with $a > b$. Subtract b from a enough times until there is left a remainder $c < b$. Now repeat the procedure with c and b. If at some point the current smaller number goes exactly into the larger one or "measures it" to use Euclid's terms, the process terminates.

SUGGESTIONS FOR FURTHER READING

Asger Aaboe. *Episodes from the Early History of Mathematics.* New York: Random House and the L. W. Singer Company, 1964.

W. Burkert. *Lore and science in ancient Pythagoreanism.* Trans. by E. L. Milnar, Jr. Cambridge, Mass.: Harvard University Press, 1972.

Lucas N. H. Bunt, Phillip S. Jones, Jack D. Bedient. *The Historical Roots of Elementary Mathematics.* Englewood Cliffs, New Jersey: Prentice-Hall, Inc., 1976.

Ettorre Carruccio. *Mathematics and Logic in History and in Contemporary Thought.* Trans. by Isabel Quigly. London: Faber and Faber, 1964.

C. H. Edwards, Jr. *The Historical Development of the Calculus.* New York: Springer-Verlag, 1979.

D. H. Fowler. "Ratio in early Greek Mathematics." *Bulletin* (New Series) *of the American Mathematical Society* 1(1979), 807–846.

K. von Fritz. "The Discovery of Incommensurability by Hippasus of Metapontum." *Annals of Mathematics* 2(1945), 242–264. Reprinted in D. J. Furley and R. E. Allen, eds., *Studies in pre-Socratic Philosophy,* vol. I. London: Routledge and Kegan Paul, 1970, 382–412.

Sir Thomas L. Heath. *A History of Greek Mathematics.* 2 vols. Oxford: Oxford University Press, 1921.

———. *A Manual of Greek Mathematics.* Oxford: Oxford University Press, 1931; Dover paperback, 1963.

J. Hjelmslev. "Eudoxus' axiom and Archimedes' lemma." *Centaurus* 1(1950), 2-11.

Jean Itard. *Les Livres arithmetiques d'Euclide.* Paris: Hermann, 1961.

J. Klein. *Greek Mathematical Thought and the Origin of Algebra.* Trans. by E. Brann. Cambridge, Mass.: MIT Press, 1968.

W. R. Knorr. *The Evolution of the Euclidean Elements.* Dordrecht and Boston: Reidel, 1975.

Cornelius Lanczos. *Space Through the Ages.* New York: Academic Press, 1970.

G. E. R. Lloyd. *Magic, Reason and Experience.* Cambridge: Cambridge University Press, 1979, 66–79 and 102–125.

M. S. Mahoney. "Another Look at Greek Geometrical Analysis." *Archive for History of Exact Sciences* 5(1968), 318–348.

G. R. Morrow. *Proclus, A Commentary on the First Book of Euclid's Elements.* Princeton, N.J.: Princeton University Press, 1970.

Otto Neugebauer. *The Exact Sciences in Antiquity.* Princeton: Princeton University Press, 1952.

R. C. Redell. "Eudoxan Mathematics and the Eudoxan Spheres." *Archive for History of Exact Sciences* 20(1976), 1–19.

A. J. E. Smeur. "On the Value Equivalent to π in Ancient Mathematical Texts: A New Interpretation." *Archive for History of Exact Sciences* 6(1970), 249–270.

R. Smith. "The Mathematical Origins of Aristotle's Syllogistic." *Archive for History of Exact Sciences* 19(1978), 201–209.

A. Szabo. "Anfänge des euklidischen Axiomensystems." *Archive for History of Exact Sciences* l(1960), 37–106.

———. "Die fruhgriechische Proportionenlehre in Spiegel ihren Terminologie." 2(1965), 197–270.

A. E. Taylor. "Forms and Numbers: A Study in Platonic Metaphysics." *Mind* 35(1926), 419–440; *ibid.* 36(1927), 12–33.

Ivor Thomas, ed., *Greek Mathematical Works.* 2 vols. Cambridge, Mass.: Harvard University Press, 1951.

Sabetai Unguru. "History of Ancient Mathematics: Some Reflections on the State of the Art." *Isis* 70(1979), 555–565.

———. "On the Need to Rewrite the History of Greek Mathematics." *Archive for History of Exact Sciences* 15(1975), 67–114.

B. L. van der Waerden. *Science Awakening* I. Trans. by Arnold Dresden. Groningen: P. Noordhoff, 1954.

———. "Defense of a 'Shocking' Point of View." *Archive for History of Exact Sciences* 15(1976), 199–210.

W. C. Waterhouse. "The Discovery of Regular Solids." *Archive for History of Exact Sciences* 9(1972), 212–221.

Robert C. Yates. *The Trisection Problem.* Washington, D.C.: National Council of Teachers of Mathematics, 1971.

Chapter II
The Rise of Theoretical Mathematics in Ancient Greece

PROCLUS (410 or 412–485)

The Neoplatonic philosopher and commentator Proclus was born in Byzantium (now Istanbul, Turkey) where his father was an eminent lawyer. Sent as a youth to Alexandria to study rhetoric and Latin in preparation for the law, he received a "divine call" to philosophy while on a trip home and changed the focus of his studies to mathematics and the works of Aristotle. Becoming dissatisfied with his Alexandrian teachers, Proclus moved before he was twenty to Athens. Here the Platonic Academy had been revived by its director Plutarch of Athens. Until his death in 485, Proclus belonged to the Academy, initially as a student, then as a teacher, and finally as its director. His contemporaries knew him as a man of ample means who was abstemious in diet and, like the Pythagoreans, refrained from eating meat. He was a deeply religious man who scrupulously observed Egyptian and Greek holy days and who practiced necromancy and other forms of divination to heal disease.

Possessing an acute, logically clear, and orderly mind, Proclus turned his attention to the Neoplatonic philosophy which combined the teachings of Plato with a religious impulse for magic. He wrote an impressive series of commentaries on Plato's dialogues, including the *Timaeus* (his favorite work), the *Parmenides*, and the *Republic*. During the late Renaissance these commentaries influenced members of the Florentine Academy and Johann Kepler. Proclus also wrote broadly on all facets of Greek culture, contributing to science, religion, literature, and philosophy. His extant writings are significant sources of information on the last phase of ancient Greek culture.

Although not a creative mathematician, Proclus was a penetrating expositor and critic. In the history of mathematics he is best known for his *Commentary on the First Book of Euclid's Elements*, which derived partly from his lectures at the Academy. This text reveals his thorough grasp of existing mathematical method; clearly, he had studied the development of Greek mathematics in the thousand years from Thales to his time. In part, his text was probably based on another's writing that condensed the lost *History of Geometry* by Aristotle's pupil, Eudemus of Rhodes. One of its sections is referred to as the "Eudemian Summary" or "Catalogue of Geometers." Today, it is Proclus' *Commentary*, Pappus' *Mathematical Collection*, and the extant Greek classics themselves that are the main sources of the history of ancient Greek mathematics.

5. From "The Catalogue of Geometers"*

PROCLUS

Next we must speak of the development of geometry during the present era. The inspired Aristotle has said that the same beliefs have often recurred to men at certain regular periods in the world's history; the sciences did not arise for the first time among us or among the men of whom we know, but at countless other cycles in the past they have appeared and vanished and will do so in the future. But limiting our investigation to the origin of the arts and sciences in the present age, we say, as have most writers of history, that geometry was first discovered among the Egyptians and originated in the remeasuring of their lands. This was necessary for them because the Nile overflows and obliterates the boundary lines between their properties. It is not surprising that the discovery of this and the other sciences had its origin in necessity, since everything in the world of generation proceeds from imperfection to perfection. Thus they would naturally pass from sense-perception to calculation and from calculation to reason. Just as among the Phoenicians the necessities of trade and exchange gave the impetus to the accurate study of number, so also among the Egyptians the invention of geometry came about from the cause mentioned.

Thales, who had travelled to Egypt, was the first to introduce this science into Greece. He made many discoveries himself and taught the principles for many others to his successors, attacking some problems in a general way and others more empirically. Next after him Mamercus, brother of the poet Stesichorus, is remembered as having applied himself to the study of geometry; and Hippias of Elis records that he acquired a reputation as a geometer. Following upon these men, Pythagoras transformed mathematical philosophy into a form of liberal education, surveying its principles from the highest downwards and investigating its theorems in an immaterial and intellectual manner. He also discovered the theory of proportionals and the construction of the cosmic solids. After him Anaxagoras of Clazomenae applied himself to many questions in geometry, and so did Oenopides of Chios, who was a little younger than Anaxagoras. Both these men are mentioned by Plato in the *Erastae* as having got a reputation in mathematics. Following them Hippocrates of Chios, who invented the method of squaring lunules, and Theodorus of Cyrene became eminent in geometry. For Hippocrates wrote a book on Elements, the first of whom we have any record who did so.

Plato, who came after them, greatly advanced mathematics in general and geometry in particular because of his zeal for these studies. It is well-known that his writings are thickly sprinkled with mathematical terms and that he everywhere tries to arouse admiration for mathematics among students of philosophy. At this time also lived

Source: From Proclus, *A Commentary on the First Book of Euclid's Elements* trans. with Introduction and Notes, by Glenn R. Morrow (copyright © 1970 by Princeton University Press): pp. 51-57, partially changed. Footnotes omitted. Reprinted by permission of Princeton University Press.

Leodamas of Thasos, Archytas of Tarentum, and Theaetetus of Athens, who increased the number of theorems and arranged them in a more scientific system. Younger than Leodamas were Neoclides and his pupil, Leon, who added many discoveries to those of their predecessors, so that Leon was able to compile a book of Elements more carefully designed to take account of the number of propositions that had been proved and of their utility. He also discovered *diorismi*, whose purpose is to determine when a given problem is capable of solution and when it is not. Eudoxus of Cnidus, a little later than Leon and a member of Plato's circle, was the first to increase the number of so-called general theorems; he added to the three proportionals already known three more and multiplied the number of propositions concerning the "section," begun by Plato, employing the method of analysis for their solution. Amyclas of Heracleia, one of Plato's followers, Menaechmus, a student of Eudoxus and a member of Plato's circle, and his brother Dinostratus perfected geometry still further. Theudius of Magnesia had a reputation for excellence in mathematics as in the rest of philosophy, because he produced an admirable arrangement of the elements and generalized many special theorems. There was also Athenaeus of Cyzicus, who lived about this time and became famous in other branches of mathematics but most of all in geometry. These men lived together in the Academy, making their inquiries in common.

Hermotimus of Colophon continued the investigations begun by Eudoxus and Theaetetus, discovered many propositions of the *Elements*, and wrote some things about the theory of geometrical loci. Philippus of Mende, a pupil whom Plato had encouraged to study mathematics not only carried on his investigations according to Plato's instructions but also set himself to study all the problems that he thought would contribute to Plato's philosophy.

Those who have written histories bring to this point their account of the development of this science. Not long after these men came Euclid, who brought together the *Elements*, systematizing many of the theorems of Eudoxus, perfecting many of those of Theaetetus, and supplying irrefutable proofs of propositions that had been rather loosely proved by his predecessors. He lived in the time of Ptolemy the First, for Archimedes, who lived after the time of the first Ptolemy, mentions Euclid. It is also reported that Ptolemy once asked Euclid if there was not a shorter road to geometry than through the *Elements*, and Euclid replied that there was no royal road to geometry. He was, therefore, later than Plato's circle but earlier than Eratosthenes and Archimedes, for these two men were contemporaries, as Eratosthenes somewhere says. Euclid belonged to the persuasion of Plato and was at home in this philosophy; and this is why he thought the goal of the *Elements* as a whole to be the construction of the so-called Platonic figures.

Chapter II
The Rise of Theoretical Mathematics in Ancient Greece

PYTHAGORAS OF SAMOS (*c.* 560–*c.* 480 B.C.)

What is true for all other pre-Socratic Greeks is also true of Pythagoras; we know little about him. He was a mystic, geometer, philosopher, prophet, and sophist, that is, teacher of wisdom. Since his teachings were oral and the Pythagorean community required secrecy among its initiates, no firsthand written records of the master were prepared. Moreover, since the early Pythagoreans usually traced all their discoveries to their founder, it is difficult to sift out the basic achievements of Pythagoras himself in the second-hand sources and later commentaries.

The account of Pythagoras' life is hazy. He was born and grew up in Samos, an island near the coast of Asia Minor. As a young adult he probably traveled to Miletus and to Egypt, where the Persian King Cambyses captured him and took him to Babylon for seven years. From the Egyptian and Mesopotamian priest-scribes, he likely learned about mythical rites, numbers, music, and proto-science. Around 530 B.C. (or possibly 520 B.C.) he returned to Samos, but had to flee shortly thereafter, perhaps because he opposed the tyrant Polycrates.

Pythagoras next went to Croton, a Greek colony in southern Italy *(Magna Graecia)*, where he founded a religious and philosophical society. Like other contemporary mystery cults, its members followed an ascetic, monastic discipline, were vegetarians, and believed in the transmigration and reincarnation of the soul. Pleased with their hierarchical preferences, local aristocrats at first supported them amid a rising tide of democracy. About 500 B.C., the Crotoniates turned against Pythagoras and forced him to retire to Metapontum where he died. During the democratic revolution in *Magna Graecia* about 450 B.C. the "aristocratic" Pythagoreans were set upon and their meeting houses destroyed.

Following the death of Pythagoras, his followers split into two factions. The split showed the twofold aspect of Pythagorean doctrine, mystical–religious and scientific. The first faction, the *akousmatikoi*, accepted the words of the master as revelation. The second faction, the *mathematikoi*, founded by Hippasus, were probably Pythagoras' more advanced followers. They pursued new learning *(mathesis)*. As a result of their research, which helped to transform mathematics into a deductive science, they have been called the first group of theoretical mathematicians. From the term *mathematikoi*, we derive our word "mathematician".

Pythagoras and his followers were among the first, if not the first, to develop theoretical mathematics. They viewed numbers in some respects as abstractions and employed a rudimentary notion of proof to investigate geometrical theorems and principles by the use of chains of logical reasoning. Unlike the Egyptians and Mesopotamians, they cultivated mathematics not for its practical application but as a contemplative, religious exercise, which served as a means to penetrate the eternal verities of a geometer God.

Proceeding from the broad generalization that "all is number" (by number they meant integer), the Pythagoreans produced a theory of numbers *(arithmetica)* comprised of numerology and scientific speculation. In their numerology, even numbers were feminine and odd numbers masculine. The numbers also represented abstract concepts (*e.g.*, 1 stood for reason, 2 for opinion, 3 for harmony, 4 for justice, and so on). Their *arithmetica* had a theory of special classes of numbers. There were "perfect" numbers of two kinds. The first kind included only 10 *(tetractys)*, which was basic to the decimal system and the sum of the first four numbers $1 + 2 + 3 + 4 = 10$. The second kind of "perfect" numbers were those equal to the sum of their proper divisors, (*e.g.*, $6 = 1 + 2 + 3$ and $28 = 1 + 2 + 4 + 7 + 14$). By depicting numbers pictorially as dots drawn in the sand or groups of pebbles, they developed figurate numbers (triangular, square, rectangular, pentagonal, and higher numbers). From their study of an isosceles triangle with arms 1 and hypoteneuse $\sqrt{2}$ or the pentagon with value $\sqrt{5}$, they made their most important discovery, incommensurability or ratios that could not be expressed in terms of whole numbers.

The early Pythagorean achievements in plane geometry are still problematic. Pythagoras probably stated his theorem about the hypoteneuse of a right triangle but could not prove it. He and his followers investigated three of five regular polyhedra—the pyramid (4 faces), the hexahedron or cube (6 faces), and dodecahedron (12 faces). Possibly they studied the octahedron (8 faces) and icosahedron (20 faces) as well. They also developed a theory of means. For two numbers a and b they found the equivalent of the following three means: the arithmetic $\frac{(a + b)}{2}$, the geometric $(\sqrt{ab})$ and the harmonic or subcontrary (c, where $\frac{1}{c} = \frac{\frac{1}{a} + \frac{1}{b}}{2}$). Among the Pythagoreans the star pentagon was a secret identification symbol.

Although the subjects were not explicitly distinguished, the early Pythagorean program of study consisted of four subjects that later became the *quadrivium*—arithmetic, geometry, music, and astronomy. In each they emphasized the principle of beauty. They empirically discovered musical consonance by shortening the length of the string of a lyre. In astronomy, they believed that the planetary paths had to follow the simplest curve, that is, a circle. Some Pythagoreans, notably Philolaus, speculated that the earth revolves around a central fire—a view that presaged the origins of the heliocentric system.

6. From *On Marvels* 6*

APOLLONIUS PARADOXAGRAPHUS

Pythagoras, the son of Mnesarchus, first worked at mathematics and numbers, and later at one time did not hold himself aloof from the wonder-working of Pherecydes.

**Source:* From *Selections Illustrating the History of Greek Mathematics,* trans. by Ivor Thomas (1941), vol. I, 172. Reprinted by permission of Harvard University Press.

7. From Book VII of the *Elements:* Definitions*

EUCLID

1. An **unit** is that by virture of which each of the things that exist is called one.

2. A **number** is a multitude composed of units.

3. A number is a **part** of a number, the less of the greater, when it measures the greater;

4. but **parts** when it does not measure it.

5. The greater number is a **multiple** of the less when it is measured by the less.

6. An **even number** is that which is divisible into two equal parts.

7. An **odd number** is that which is not divisible into two equal parts, or that which differs by an unit from an even number.

8. An **even-times even number** is that which is measured by an even number according to an even number.

9. An **even-times odd number** is that which is measured by an even number according to an odd number.

10. An **odd-times odd number** is that which is measured by an odd number according to an odd number.

11. A **prime number** is that which is measured by an unit alone.

12. Numbers **prime to one another** are those which are measured by an unit alone as a common measure.

13. A **composite number** is that which is measured by some number.

14. Numbers **composite to one another** are those which are measured by some number as a common measure.

15. A number is said to **multiply** a

**Source:* Reprinted by permission from *The Thirteen Books of Euclid's Elements,* trans. by Sir Thomas L. Heath (1956 edition), vol. 2, 277-278. Copyright © 1956 Cambridge University Press.

number when that which is multiplied is added to itself as many times as there are units in the other, and thus some number is produced.

16. And, when two numbers having multiplied one another make some number, the number so produced is called **plane,** and its **sides** are the numbers which have multiplied one another.

17. And, when three numbers having multiplied one another make some number, the number so produced is **solid,** and its **sides** are the numbers which have multiplied one another.

18. A **square number** is equal multiplied by equal, or a number which is contained by two equal numbers.

19. And a **cube** is equal multiplied by equal and again by equal, or a number which is contained by three equal numbers.

20. Numbers are **proportional** when the first is the same multiple, or the same part, or the same parts, of the second that the third is of the fourth.

21. **Similar plane** and solid numbers are those which have their sides proportional.

22. A **perfect number** is that which is equal to its own parts.

8. From *Prior Analytics* i.23*

(Irrationality of the Square Root of 2)

ARISTOTLE

For all who argue *per impossibile* infer by syllogism a false conclusion, and prove the original conclusion hypothetically when something impossible follows from a contradictory assumption, as, for example, that the diagonal [of a square] is incommensurable [with the side] because odd numbers are equal to even if it is assumed to be commensurate. It is inferred by syllogism that odd numbers are equal to even, and proved hypothetically that the diagonal is incommensurate, since a false conclusion follows from the contradictory assumption.[1]

NOTE

1. It is generally believed that the Pythagoreans were aware of the irrationality of $\sqrt{2}$ (Theodorus, for example, when proving the irrationality of numbers began with $\sqrt{3}$), and that Aristotle has indicated the method by which they proved it. The proof, interpolated in the text of Euclid as x. 117 (Eucl., ed. Heiberg-Menge iii. 408-410), is roughly as follows. Suppose AC, the diagonal of a square, to be commensurable with its side AB, and let their ratio in its smallest terms be $a : b$.

Now $\quad AC^2 : AB^2 = a^2 : b^2$

and $\quad AC^2 = 2AB^2, a^2 = 2b^2$.

Hence a^2, and therefore a, is even.

Since $a : b$ is in its lowest terms it follows that b is odd.

Let $a = 2c$. Then $4c^2 = 2b^2$, or $b^2 = 2c^2$, so that b^2, and therefore b is even.

But b was shown to be odd, and is therefore odd and even, which is impossible. Therefore AC cannot be commensurable with AB.

*Source: Selections 8 (Aristotle), 9 (Iamblichus) and 10 (Porphyry) are from *Selections Illustrating the History of Greek Mathematics,* trans. by Ivor Thomas (1941), vol. I, 111 and 113. Reprinted by permission of Harvard University Press.

9. From *On Nicomachus' Introduction to Arithmetic*

(Arithmetic, Geometric and Harmonic Means)

IAMBLICHUS

In ancient days in the time of Pythagoras and the mathematicians of his school there were only three means, the arithmetic and the geometric and a third in order which was then called subcontrary, but which was renamed harmonic by the circle of Archytas and Hippasus, because it seemed to furnish harmonious and tuneful ratios.

10. From *Commentary on Ptolemy's Harmonics*

PORPHYRY

Archytas, in his discussion of means, writes thus:

"Now there are three means in music: first the arithmetic, secondly the geometric, and thirdly the subcontrary, the so-called harmonic. The arithmetic is that in which three terms are in proportion in virtue of some difference: the first exceeds the second by the same amount as the second exceeds the third.[1] And in this proportion it happens that the interval[2] between the greater terms is the lesser, while that between the lesser terms is the greater. The geometric mean is that in which the first term is to the second as the second is to the third. Here the greater terms make the same interval as the lesser.[3] The subcontrary mean, which we call harmonic, is such that by whatever part of itself the first term exceeds the second, the middle term exceeds the third by the same part of the third.[4] In this proportion the interval between the greater terms is the greater, that between the lesser terms is the lesser."

NOTES

1. *i.e.*, b is the arithmetic mean between a and c if

$$a - b = b - c.$$

2. The word *interval* is here used in the musical sense; mathematically it must be understood as the *ratio* between the two terms, not their arithmetical difference. Archytas asserts that

$$\frac{a}{b} < \frac{b}{c}.$$

and what Archytas says about the interval is contained in the definition.

3. *i.e.*, b is the geometric mean between a and c if

$$\frac{a}{b} = \frac{b}{c},$$

4. *i.e.*, b is the harmonic mean between a and c if

$$\frac{a-b}{a} = \frac{b-c}{c},$$

which can be written $\frac{1}{c} - \frac{1}{b} = \frac{1}{b} - \frac{1}{a}$,

so that $\frac{1}{c}, \frac{1}{b}, \frac{1}{a}$

form an arithmetical progression, and Archytas goes on to assert that

$$\frac{a}{b} > \frac{b}{c}.$$

11. From *Metaphysics* A5*

ARISTOTLE

In the time of these men [Leucippus and Democritus] and before them the so-called Pythagoreans applied themselves to mathematics and were the first to advance that science; and because they had been brought up in it they thought that its principles must be the principles of all existing things.

12. From *Commentary on Euclid i*

(Sum of the Angles of a Triangle)

PROCLUS

Eudemus the Peripatetic ascribes to the Pythagoreans the discovery of this theorem, that any triangle has its internal angles equal to two right after this fashion. Let ΑΒΓ be a triangle, and through A let ΔE be drawn parallel to ΒΓ. Now since ΒΓ, ΔE are parallel, and the alternate angles are equal, the angle ΔAB is equal to the angle ΑΒΓ, and ΕΑΓ is equal to ΑΓΒ. Let ΒΑΓ be added to both. Then the angles ΔAB, ΒΑΓ, ΓΑΕ, that is, the angles ΔAB, BAE, that is, two right angles, are equal to the three angles of the triangle. Therefore the three angles of the triangle are equal to two right angles.

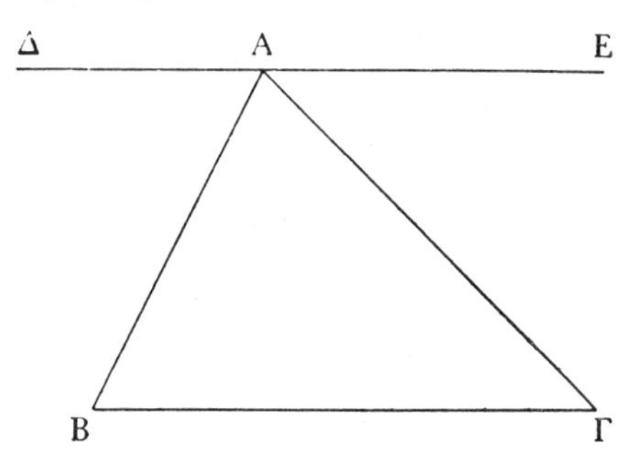

13. From *Convivial Questions* viii. 2.

(Pythagoras' theorem)

PLUTARCH

Among the most geometrical theorems, or rather problems, is this—given two figures, to apply a third equal to the one and similar to the other; it was in virtue of this discovery they say Pythagoras sacrificed. This is unquestionably more subtle and elegant than the theorem which he proved that the square on the hypotenuse is equal to the squares on the sides about the right angle.

*Source: Selections 11, 12, 13, 14, and 15 *Selections Illustrating the History of Greek Mathematics,* trans. by Ivor Thomas (1941), vol. I, 113, 177, 179, 215 and 217, 225. Reprinted by permission of Harvard University Press.

14. From *On Slips in Greetings* 5

LUCIAN

The triple interlaced triangle, the pentagram, which they (the Pythagoreans) used as a password among members of the same school, was called by them Health.[1]

NOTE

1. *Cf.* the scholium to Aristophanes, *Clouds* 609. The Pentagram is the star-pentagon, as in the [following] diagram. The fact that this was a familiar symbol among them lends some plausibility to the belief that they know how to construct the dodecahedron out of twelve pentagons.

15. From *Elements* X. Scholium

(The Irrational or Incommensurable)

EUCLID

The Pythagoreans were the first to make inquiry into commensurability, having first discovered it as a result of their observation of numbers; for though the unit is a common measure of all numbers they could not find a common measure of all magnitudes. The reason is that all numbers, of whatsoever kind, howsoever they be divided leave some least part which will not suffer further division; but all magnitudes are divisible *ad infinitum* and do not leave some part which, being the least possible, will not admit of further division, but that remainder can be divided *ad infinitum* so as to give an infinite number of parts, of which each can be divided *ad infinitum;* and, in sum, magnitude partakes in division of the principle of the infinite, but in its entirety of the principle of the finite, while number in division partakes of the finite, but in its entirety of the infinite. . . . There is a legend that the first of the Pythagoreans who made public the investigation of these matters perished in a shipwreck.

16. From *Elements* X. Definitions*

(Commensurable and Incommensurable)

EUCLID

1. Those magnitudes are said to be **commensurable** which are measured by the same measure, and those **incommensurable** which cannot have any common measure.

2. Straight lines are **commensurable in square** when the squares on them are measured by the same area, and **incommensurable in square** when the squares on them cannot possibly have any areas as a common measure.

3. With these hypotheses, it is proved that there exist straight lines infinite in multitude which are commensurable and incommensurable respectively, some in length only, and others in square also, with an assigned straight line. Let then the assigned straight line be called **rational,** and those straight lines which are the commensurable with it, whether in length and in square or in square only, **rational,** but those which are incommensurable with it **irrational.**

4. And let the square on the assigned straight line be called **rational** and those areas which are commensurable with it **rational,** but those which are incommensurable with it **irrational,** and the straight lines which produce them **irrational,** that is, in case the areas are squares, the sides themselves, but in case they are any other rectilineal figures, the straight lines on which are described squares equal to them.

Source: Reprinted from *The Thirteen Books of Euclid's Elements,* trans. by Sir Thomas L. Heath (1956 edition), vol 2, p. 10.

Chapter II
The Rise of Theoretical Mathematics in Ancient Greece

HIPPOCRATES OF CHIOS (c. 460–c. 380 B.C.)

Hippocrates of Chios,[1] the leading geometer of the fifth century B.C., was born on the Ionian Greek island of Chios, where he probably learned mathematics as a youth. As Chios is close to Samos, the birthplace of Pythagoras, Hippocrates may have early come under Pythagorean influence. He would be a lifelong fellow traveler of theirs. Hippocrates began his adult career in commerce. According to Aristotle, he was not a clever merchant and was defrauded of large sums of money and property by crooked Byzantine custom-house officials. It is more likely, however, that he lost his property when Athenian pirates captured his ship in the Samian War (440 B.C.). When he went to Athens to prosecute the pirates, he had to remain for a long time. We do not know the outcome of his complaint.

Hippocrates spent his most productive adult years in Athens in the second half of the fifth century B.C. There he became proficient in geometry, a subject that he taught to earn a living. Through this work he helped to make Athens the leading center of Greek mathematical research.

Upon arriving in Athens, Hippocrates encountered three special problems—the squaring or quadrature of the circle, the doubling or duplication of the cube, and the trisection of an angle—that had engaged the attention of Athenian Sophists and geometers. The aesthetic preferences of the ancient Greeks dictated that they solve these problems by use of straightedge and compass alone. Hippocrates addressed himself to the first two problems with good results. While he did not square the circle, he did succeed in solving a related problem when he squared a lunule. By proving that the area of a lunule, which resembles a crescent moon, equals the area of a triangle, he proved that the area of a curvilinear figure could be made equal to a rectilinear figure. He also showed that the problem of the duplication of a cube can be reduced to the simpler problem of finding two mean proportionals between a given straight line and another twice as long ($a : x = x : y = y : 2a$).[2] In studying the duplication of the cube, he probably introduced the method of reduction or geometrical analysis, wherein the things sought were taken up to an acknowledged first principle. He is credited with arranging theorems so that later ones could be proven on the basis of earlier ones and to have first composed a lost textbook, entitled *Elements of Geometry,* in the manner of Euclid.

NOTES

1. He should not be confused with his famous contemporary, the physician Hippocrates of Cos, who lived on another of the Ionian Dodecanese Islands. The name Hippocrates was not uncommon in ancient Greece.

2. The solution can be expressed in modern symbols as follows:

If $\frac{a}{x} = \frac{x}{y} = \frac{y}{2a}$,

then (1) $x^2 = ay$ and (2) $y^2 = 2ax$.

By substituting $y = \frac{x^2}{a}$ from equation (1) into equation (2), one obtains $\frac{x^4}{a^2} = 2ax$

or $x^4 = 2a^3x$

or $x^3 = 2a^3$, where x is the desired answer. It should be noted that this algebraic answer cannot be constructed by straightedge and compass, so the search for a purely geometrical solution continued.

17. From *Commentary on Aristotle's Physics* A 2*

PHILOPONUS

Hippocrates of Chios was a merchant who fell in with a pirate ship and lost all his possessions. He came to Athens to prosecute the pirates and, staying a long time in Athens by reason of the indictment, consorted with philosophers, and reached such proficiency in geometry that he tried to effect the quadrature of the circle. He did not discover this, but having squared the lune he falsely thought from this that he could square the circle also. For he thought that from the quadrature of the lune the quadrature of the circle also could be calculated.

18. From *Commentary on Aristotle's Physics* A 2*

(Quadrature of Lunules, the Crescent-Shaped Figures Between Two Intersecting Arcs of Circles)

SIMPLICIUS

Eudemus, however, in his *History of Geometry* says that Hippocrates did not demonstrate the quadrature of the lune on the side of a square[1] but generally, as one might say. For every lune has an outer circumference equal to a semicircle or greater or less, and if Hippocrates squared the lune having an outer circumference equal to a semicircle and greater and less, the quadrature would appear to be proved generally. I shall set out what Eudemus wrote word for word, adding only for the sake of clearness a few things taken from Euclid's *Elements* on account of the summary style of Eudemus, who set out his proofs

**Source:* Selections 17 and 18 are from *Selections Illustrating the History of Greek Mathematics,* trans. by Ivor Thomas (1941), vol. I, 235-245. Reprinted by permission of Harvard University Press.

in abridged form in conformity with the ancient practice. He writes thus in the second book of the *History of Geometry*.

"The quadratures of lunes, which seemed to belong to an uncommon class of propositions by reason of the close relationship to the circle, were first investigated by Hippocrates, and seemed to be set out in correct form; therefore we shall deal with them at length and go through them. He made his starting-point, and set out as the first of the theorems useful to his purpose, that similar segments of circles have the same ratios as the squares on their bases.[2] And this he proved by showing that the squares on the diameters have the same ratios as the circles.[3]

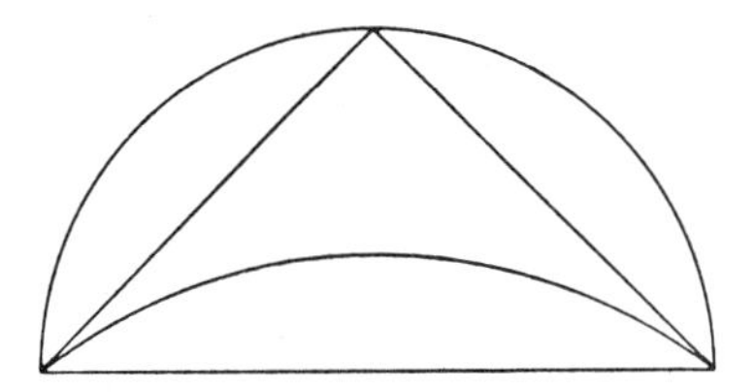

"Having first shown this he described in what way it was possible to square a lune whose outer circumference was a semicircle. He did this by circumscribing about a right-angled isosceles triangle a semicircle and about the base a segment of a circle similar to those cut off by the sides.[4] Since the segment about the base is equal to the sum of those about the sides, it follows that when the part of the triangle above the segment about the base is added to both the lune will be equal to the triangle. Therefore the lune, having been proved equal to the triangle, can be squared. In this way, taking a semicircle as the outer circumference of the lune, Hippocrates readily squared the lune.

"Next in order he assumes [an outer circumference] greater than a semicircle [obtained by] constructing a trapezium having three sides equal to one another while one, the greater of the parallel sides, is such that the square on it is three times the square on each of those sides, and then comprehending the trapezium in a circle and circumscribing about its greatest side a segment similar to those cut off from the circle by the three equal sides.[5] That the said segment[6] is greater than a semicircle is clear if a diagonal is drawn in the trapezium. For this diagonal, subtending two sides of the trapezium, must be such that the square on it is greater than double the square on one of the remaining sides. Therefore the square on BΓ is greater than double the square on either BA, AΓ, and therefore also on ΓΔ.[7]

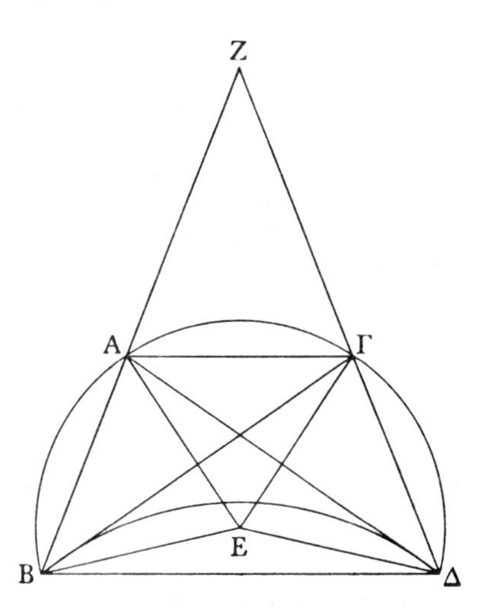

Therefore the square on BΔ, the greatest of the sides of the trapezium, must be less than the sum of the squares on the diagonal and that one of the other sides which is subtended by the said [greatest] side together with the diagonal.[8] For the squares on BΓ, ΓΔ are greater than three times, and the square on BΔ is equal to three times, the square on ΓΔ. Therefore the angle standing on the greatest side of the trapezium[9] is acute. Therefore the segment in which it is is greater than a semicircle. And this segment is the outer circumference of the lune.[10]

"If [the outer circumference] were less than a semicircle, Hippocrates solved this also, using the following preliminary construction. Let there be a cricle with diameter AB and centre K.

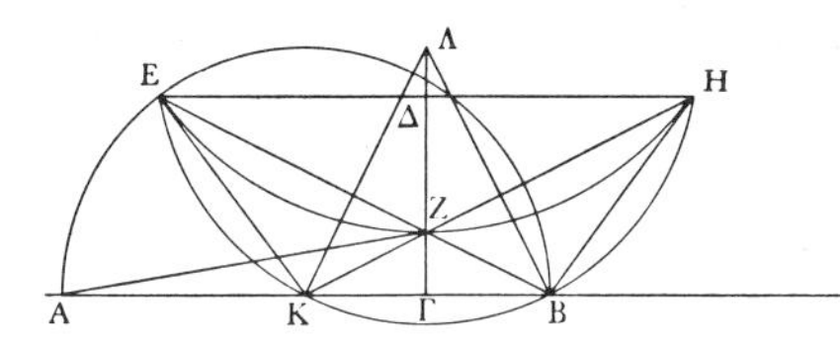

Let ΓΔ bisect BK at right angles; and let the straight line EZ be placed between this and the circumference verging towards B so that the square on it is one-and-a-half times the square on one of the radii.[11] Let EH be drawn parallel to AB, and from K let [straight lines] be drawn joining E and Z. Let the straight line [KZ] joined to Z and produced meet EH at H, and again let [straight lines] be drawn from B joining Z and H. It is then manifest that EZ produced will pass through B—for by hypothesis EZ verges towards B—and BH will be equal to EK.

"This being so, I say that the trapezium EKBH can be comprehended in a circle."

NOTES

1. As Alexander asserted. Alexander, as quoted by Simplicius *in Phys.* (ed. Diels 56. 1-57. 24), attributes two quadratures to Hippocrates.

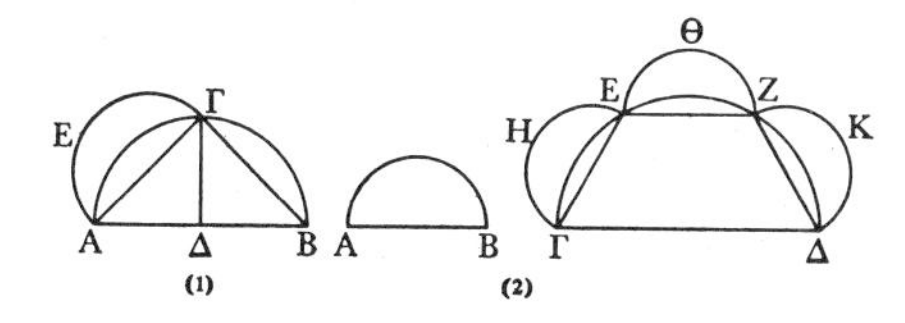

In the first, AB is the diameter of a circle, AΓ, ΓB are sides of a square inscribed in it, and AEΓ is a semicircle described on AΓ. Alexander shows that

lune AEΓ = triangle AΓΔ.

In the second, AB is the diameter of semicircle and on ΓΔ, equal to twice AB, a semicircle is described. ΓE, EZ, ZΔ are sides of a regular hexagon, and ΓHE, EΘZ, ZKΔ are semicircles described on ΓE, EZ, ZΔ. Alexander shows that

lune ΓHE + lune EΘZ + lune ZKΔ + semicircle AB + trapezium ΓEZΔ.

The proofs are easy. Alexander goes on to say that if the rectilineal figure equal to the three lunes ("for a rectilineal figure was proved equal to a lune") is subtracted, the circle will be squared. The fallacy is obvious and Hippocrates could hardly have committed it. This throws some doubt on the whole of Alexander's account, and Simplicius himself observes that Eudemus's account is to be preferred as he was "nearer to the times" of Hippocrates.

2. Lit. "as the bases in square."

3. This is Eucl. xii. 2. Euclid proves it by a method of exhaustion, based on a lemma or its equivalent which, on the evidence of Archimedes himself, can safely be attributed to Eudoxus. We are not told how Hippocrates effected the proof.

4. As Simplicius notes, this is the problem of Eucl. iii. 33 and involves the knowledge that similar segments contain equal angles.

5. Simplicius here inserts a proof that a circle can be described about the trapezium.

6. *i.e.*, the segment bounded by the outer circumference. Eudemus is going to show that the angle in it is acute and therefore the segment is greater than a semicircle.

7. A proof is supplied in the text, probably by Simplicius though Diels attributes it to Eudemus. The proof is that, since BΔ is parallel to AΓ but greater than it, ΔΓ and BA produced will meet in Z. Then ZAΓ is an isosceles triangle, so that the angle ZAΓ is acute, and therefore the angle BAΓ is obtuse.

8. *i.e.* $B\Delta^2 < B\Gamma^2 + \Gamma\Delta^2$.

9. *i.e.* the angle BΓΔ.

10. Simplicius notes that Eudemus has omitted the actual squaring of the lune, presumably as being obvious. Since

$$B\Delta^2 = 3BA^2$$

(segment on BΔ) = 3 (segment on BA)

= sum of segments on BA, AΓ, ΓΔ.

Adding to each side of the equation the portion of the trapezium included by the sides BA, AΓ and ΓΔ and the circumference of the segment on BΔ, we get trapezium ABΔΓ = lune bounded by the two circumferences and so the lune is "squared."

11. This is the first example we have had to record of the type of construction known

to the Greeks as *inclinations or vergings*. The general problem is to place a straight line so as to *verge towards* (pass through) a given point and so that a given length is intercepted on it by other lines. In this case the problem amounts to finding a length x such that, if Z be taken on ΓΔ so that $BZ = x$ and BZ be produced to meet the circumference in E, then $EZ^2 = \frac{3}{2}AK^2$, or $EZ = \sqrt{3/2}\, AK$. If this is done, $EB \cdot BZ = AB \cdot B\Gamma = AK^2$

or $(x + \sqrt{3/2}\, a) \cdot x = a^2$, where $AK = a$.

In other words, the problem amounts to solving the quadratic equation

$$x^2 + \sqrt{3/2}\, ax = a^2.$$

Chapter II
The Rise of Theoretical Mathematics in Ancient Greece

PLATO (*c.* 427–348/347 B.C.)

Plato, the central figure in Greek intellectual life in the early fourth century B.C., was the son of two aristocrats, Ariston and Perictione. On both sides the family was among the most distinguished in Athens. Ariston reportedly traced his ancestry to the god Poseidon, while the mother's family was related to Solon. The couple had four children. After Ariston died, Perictione married again. Her second spouse, who may have been her uncle Pyrilampes, was a major supporter of Pericles. Plato grew up in the house of his stepfather. A high-minded, aristocratic youth interested in poetry, art, and public affairs, he was no friend of egalitarianism or broad democracy.

The early life of Plato coincides with the tragic Peloponnesian War (435–404 B.C.) between Sparta and Athens. This protracted conflict shattered Athenian trade and empire, devasted Athens' population, and overturned its democracy. From age 18 to 23, Plato served in the Athenian calvary in Attica. By then he had grown into a handsome, physically dexterous man with a clear, thin voice. At about age 20 he met Socrates who became the greatest influence on this thought. For the next eight years, Plato was one of Socrates' few regular pupils. Probably he also attended the public lectures and disputes of the Sophists and the mathematical lessons of Theodorus. In 399 B.C. Socrates was tried, found guilty of impiety, and executed by a group of Athenian rulers known as "the Thirty." An illness prevented Plato from attending Socrates' last meeting with his friends. However Plato brought literary immortality to Socrates, who had never written a book, by making him the chief speaker in many of his philosophic dialogues.

After Socrates' execution, Plato and other disciples fled to Megara (a town situated between Athens and Corinth) to pursue truth without official interference and to master the art of argument. So began a decade of travels and studies, the *Wanderjahre*, for Plato. After briefly staying in Megara, he traveled about Greece. According to Cicero, Plato visited Egypt before journeying to southern Italy and Sicily. There the tyrant Dionysius the Elder may have briefly imprisoned him after inviting him to serve as a personal tutor. The falling out with Dionysius led to a cordial relationship with the Pythagorean leader, Archytas of Tarentum. Archytas improved his knowledge of the Pythagorean science-based education probably Plato's chief reason for the trip to the West. Afterward, his subsequent work showed a strong Pythagorean influence.

Plato's formative period ended with his return to Athens in 388 B.C. Committed to the educational program of Socrates, he made political philosophy his central concern. Political philosophy was a search for knowledge of moral absolutes and this knowledge had immediate ethical consequences. Like Socrates, Plato was convinced that the citizen, once in possession of absolute Good (essentially virtue or right conduct), would, on that account, make morally good choices. As a teacher, Plato endeavored to stage a confrontation between the citizen–learner and this knowledge. Knowledge of absolute Good, according to his doctrine, was acquired through Ideas or Forms. He divided the universe into two separate worlds: the ideal world of abstract, immutable Forms and the transitory material world of "sensible things." The Good was the supreme Form; beauty was another.

Shortly after returning to Athens in 388 B.C., Plato established a school called the Academy in the northeast of the city. He lived nearby, directing the affairs of the Academy and continuing to teach until his death. (He was only away from Athens in 361 B.C. on an ill-fated trip to Syracuse.) Inscribed above the gate to the Academy was the motto: "Let no man ignorant of geometry enter." In his most famous treatise, the *Republic*, Plato made the Pythagorean quadrivium—arithmetic, geometry, music, and astronomy—an integral part of the Greek higher curriculum and required students to complete their education with a thorough grounding in the dialectical method of Zeno and Socrates. During Plato's tenure as head of the Academy, Theaetetus, Aristotle, and Eudoxus taught at various times there.

Plato's contributions to theoretical mathematics are many. In addition to adding the subject to higher education, he identified for the educated public major problems that required further study. His *Parmenides* refers to Zeno's paradoxes; the *Theaetetus* to the theory of irrationals; and *Meno* to the necessity of framing sound definitions. He must have possessed some mathematical skill, because he is credited with treating numbers completely as abstractions in the ideal world and deriving the formula $(2n)^2 + (n^2 - 1)^2 = (n^2 + 1)^2$ for Pythagorean triples.

In the *Timaeus*, his chief scientific writing, Plato discussed the five regular polyhedra (that is, convex solids whose edges form congruent regular plane polygons). They are the cube or hexahedron (whose six faces are squares), the pyramid or tetrahedron (formed by four equilateral triangles), the octahedron (formed by eight equilateral triangles), dodecahedron (formed by twelve pentagons), and icosahedron (formed by twenty equilateral triangles). He also associated them with the four Empedoclean elements: earth (cube), air (octahedron), fire (pyramid), and water (icosahedron) plus the cosmos (dodecahedron). As a result, the regular polyhedra are today known as the Platonic or cosmic solids.

19. From the *Republic* VI.510*

(Approach to Mathematics)

PLATO

I think you know that those who deal with geometrics and calculations and such matters take for granted the odd and the even, figures, three kinds of angles and other things cognate to these in each field of inquiry; assuming these things to be known, they make them hypotheses, and henceforth regard it as unnecessary to give any explanation of them either to themselves or to others, treating them as if they were manifest to all; setting out from these hypotheses, they go at once through the remainder of the argument until they arrive with perfect consistency at the goal to which their inquiry was directed.

Yes, he said, I am aware of that.

Therefore I think you also know that although they use visible figures and argue about them, they are not thinking about these figures but of those things which the figures represent; thus it is the square in itself and the diameter in itself which are the matter of their arguments, not that which they draw; similarly, when they model or draw objects, which may themselves have images in shadows or in water, they use them in turn as images, endeavouring to see those absolute objects which cannot be seen otherwise than by thought.

20. From the *Republic* VII.522-528

(The Quadrivium)

PLATO

What sort of knowledge is there, Glaucon, which would draw the soul from becoming to being? And I have in mind another consideration: You will remember that our young men are to be warrior athletes?

Yes, that was said.

Then this new kind of knowledge must have an additional quality?

What quality?

It should not be useless to warriors.

Yes, if possible.

There were two parts in our former scheme of education, were there not?

Just so.

There was gymnastic which presided over the growth and decay of the body, and may therefore be regarded as having to do with generation and corruption?

True.

Then that is not the knowledge which we are seeking to discover?

No.

*Source: Selections 19 and 20 are from *The Dialogues of Plato*, trans. by Benjamin Jowett, 4th ed. (1953), vol. II. Reprinted by permission of the Clarendon Press, Oxford.

But what do you say of music, to the same extent as in our former scheme?

Music, he said, as you will remember, was the counterpart of gymnastic, and trained the guardians by the influences of habit, by harmony making them harmonious, by rhythm rhythmical, but not giving them science; and the words, whether fabulous or closer to the truth, were meant to impress upon them habits similar to these. But in music there was nothing which tended to that good which you are now seeking.

You are most accurate, I said, in your reminder; in music there certainly was nothing of the kind. But what branch of knowledge is there, my dear Glaucon, which is of the desired nature; since all the useful arts were reckoned mean by us?

Undoubtedly; and yet what study remains, distinct both from music and gymnastic and from the arts?

Well, I said, if nothing remains outside them, let us select something which is a common factor in all.

What may that be?

Something, for instance, which all arts and sciences and intelligences use in common, and which everyone has to learn among the first elements of education.

What is that?

The little matter of distinguishing one, two, and three—in a word, number and calculation:—do not all arts and sciences necessarily partake of them?

Yes.

Then the art of war partakes of them?

To be sure.

Then Palamedes, whenever he appears in tragedy, proves Agamemnon ridiculously unfit to be a general. Did you never remark how he declares that he had invented number, and had measured out the camping-ground at Troy, and numbered the ships and everything else; which implies that they had never been numbered before, and Agamemnon must be supposed literally to have been incapable of counting his own feet—how could he if he was ignorant of number? And if that is true, what sort of general must he have been?

I should say a very strange one, if this was as you say.

Can we deny that a warrior should have a knowledge of arithmetic?

Certainly he should, if he is to have the smallest understanding of military formations, or indeed, I should rather say, if he is to be a man at all.

I should like to know whether you have the same notion which I have of this study?

What is your notion?

It appears to me to be a study of the kind which we are seeking, and which leads naturally to reflection, but never to have been rightly used; for it has a strong tendency to draw the soul towards being.

How so? he said.

I will try to explain my meaning, I said; and I wish you would share the inquiry with me, and say 'yes' or 'no' when I attempt to distinguish in my own mind what branches of knowledge have this attracting power, in order that we may have clearer proof that arithmetic is, as I suspect, one of them.

Explain, he said.

Do you follow me when I say that objects of sense are of two kinds? some of them do not invite the intelligence to further inquiry because the sense is an adequate judge of them; while in the case of other objects sense is so untrustworthy that inquiry by the mind is imperatively demanded.

You are clearly referring, he said, to the appearance of objects at a distance, and to painting in light and shade.

No, I said, you have not quite caught my meaning.

Then what things do you mean?

When speaking of uninviting objects, I mean those which do not pass straight from one sensation to the opposite; inviting objects are those which do; in this latter case the sense coming upon the object, whether at a distance or near, does not give one particular impression more strongly than its opposite.

An illustration will make my meaning clearer:—here are three fingers—a little finger, a second finger, and a middle finger.

Very good.

You may suppose that they are seen quite close: And here comes the point.

What is it?

Each of them equally appears a finger, and in this respect it makes no difference whether it is seen in the middle or at the extremity, whether white or black, or thick or thin, or anything of that kind. In these cases a man is not compelled to ask of thought the question what is a finger? for the sight never intimates to the mind that a finger is the opposite of a finger.

True.

And therefore, I said, there is nothing here which is likely to invite or excite intelligence.

There is not, he said.

But is this equally true of the greatness and smallness of the fingers? Can sight adequately perceive them? and is no difference made by the circumstance that one of the fingers is in the middle and another at the extremity? And in like manner does the touch adequately perceive the qualities of thickness or thinness, or softness or hardness? And so of the other senses; do they give perfect intimations of such matters? Is not their mode of operation on this wise—the sense which is concerned with the quality of hardness is necessarily concerned also with the quality of softness, and only intimates to the soul that the same thing is felt to be both hard and soft?

It is, he said.

And must not the soul be perplexed at this intimation which this sense gives of a hard which is also soft? What, again, is the meaning of light and heavy, if the sense pronounces that which is light to be also heavy, and that which is heavy, light?

Yes, he said, these intimations which the soul receives are very curious and require to be explained.

Yes, I said, and in these perplexities the soul naturally summons to her aid calculation and intelligence, that she may see whether the several objects announced to her are one or two.

True.

And if they turn out to be two, is not each of them one and different?

Certainly.

And if each is one, and both are two, she will conceive the two as in a state of division, for if they were undivided they could only be conceived of as one?

True.

The eye, also, certainly did see both small and great, but only in a confused manner; they were not distinguished.

Yes.

Whereas on the contrary the thinking mind, intending to light up the chaos, was compelled to reconsider the small and great viewing them as separate and not in that confusion.

Very true.

Is it not in some such way that there arises in our minds the inquiry 'What is great?' and 'What is small?'

Exactly so.

And accordingly we made the distinction of the visible and the intelligible.

A very proper one.

This was what I meant just now when I spoke of impressions which invited the intellect, or the reverse—those which strike our sense simultaneously with opposite impressions, invite thought; those which are not simultaneous with them, do not awaken it.

I understand now, he said, and agree with you.

And to which class do unity and number belong?

I do not know, he replied.

Think a little and you will see that what has preceded will supply the answer; for if simple unity could be adequately perceived by the sight or by any other sense, then, as we were saying in the case of the finger, there would be nothing to attract towards being; but when something contrary to unity is always seen at the same time, so that there seems to be no more reason for calling it one than the opposite, some discriminating power becomes necessary, and in such a case the soul in perplexity, is obliged to rouse her power of thought and to ask: "What *is* absolute unity?" This is the way in which the study of the one has a power of drawing and converting the mind to the contemplation of true being.

And surely, he said, this occurs notably in the visual perception of unity; for we see the same thing at once as one and as infinite in multitude?

Yes, I said; and this being true of one must be equally true of all number?

Certainly.

And all arithmetic and calculation have to do with number?

Yes.

And they appear to lead the mind towards truth?

Yes, in a very remarkable manner.

Then this is a discipline of the kind for which we are seeking; for the man of war must learn the art of number or he will not know how to array his troops, and the philosopher also, because he has to rise out of the sea of change and lay hold of true being, or be for ever unable to calculate and reason.

That is true.

But our guardian is, in fact, both warrior and philosopher?

Certainly.

Then this is a kind of knowledge which legislation may fitly prescribe; and we must endeavour to persuade those who are to be the principal men of our State to go and learn arithmetic, and take up the study in no amateurish spirit but pursue it until they can view the nature of numbers with the unaided mind; nor again, like merchants or retail-traders, with a view to buying or selling, but for the sake of their military use, and of the soul herself, because this will be the easiest way for her to pass from becoming to truth and being.

That is excellent, he said.

Yes, I said, and now having spoken of it, I must add how charming the science is! and in how many ways it conduces to our desired end, if pursued in the spirit of a philosopher, and not of a shopkeeper!

How do you mean?

I mean that arithmetic has, in a marked degree, that elevating effect of which we were speaking, compelling the soul to reason about abstract number, and rebelling against the introduction of numbers which have visible or tangible bodies into the argument. You know how steadily the masters of the art repel and ridicule anyone who attempts to divide the perfect unit when he is calculating, and if you divide, they multiply,[1] taking care that the unit shall continue one and not appear to break up into fractions.

That is very true.

Now, suppose a person were to say to them: O my friends, what are these wonderful numbers about which you are reasoning, in which, as you say, there is a unity such as you demand, and each unit is equal, invariable, indivisible,—what would they answer?

They would answer, as I should conceive, that they were speaking of those numbers which can only be grasped by thought, and not handled in any other way.

Then you see that this study may be truly called necessary for our purpose, since it evidently compels the soul to use the pure intelligence in the attainment of pure truth?

Yes; that is a marked characteristic of it.

And have you further observed, that those who have a natural talent for calculation are generally quick at every other kind of study; and even the dull, if they have been trained and exercised in this, although they may derive no other advantage from it, always become much quicker than they would otherwise have been?

Very true, he said.

And indeed, you will not easily find a

study of which the learning and exercise require more pains, and not many which require as much.

You will not.

And, for all these reasons, arithmetic is a kind of knowledge in which the best natures should be trained, and which must not be given up.

I agree.

Let this then be adopted as one of our subjects of education. And next, shall we inquire whether the kindred science also concerns us?

You mean geometry?

Exactly so.

Clearly, he said, we are concerned with that part of geometry which relates to war; for in pitching a camp, or taking up a position, or closing or extending the lines of an army, or any other military manoeuvre, whether in actual battle or on a march, it will make all the difference whether a general is or is not a geometrician.

Yes, I said, but for that purpose a very little of either geometry or calculation will be enough; the question relates rather to the greater and more advanced part of geometry—whether that tends in any degree to make more easy the vision of the Idea of good; and thither, as I was saying, all things tend which compel the soul to turn her gaze towards that place where is the full perfection of being, which she ought, by all means, to behold.

True, he said.

Then if geometry compels us to view being, it concerns us; if becoming only, it does not concern us?

Yes, that is what we assert.

Yet anybody who has the least acquaintance with geometry will not deny that such a conception of the science is in flat contradiction to the ordinary language of geometricians.

How so?

They speak, as you doubtless know, in terms redolent of the workshop. As if they were engaged in action, and had no other aim in view in all their reasoning, they talk of squaring, applying, extending and the like, whereas, I presume, the real object of the whole science is knowledge.

Certainly, he said.

Then must not a further admission be made?

What admission?

That the knowledge at which geometry aims is knowledge of eternal being, and not of aught which at a particular time comes into being and perishes.

That, he replied, may be readily allowed, and is true.

Then, my noble friend, geometry will draw the soul towards truth, and create the spirit of philosophy, and raise up that which is now unhappily allowed to fall down.

Nothing will be more likely to have such an effect.

Then nothing should be more sternly laid down than that the inhabitants of your fair city should by no means remain unversed in geometry. Moreover the science has indirect effects, which are not small.

Of what kind? he said.

There are the military advantages of which you spoke, I said; and further, we know that for the better apprehension of any branch of knowledge, it makes all the difference whether a man has a grasp of geometry or not.

Yes indeed, he said, all the difference in the world.

Then shall we propose this as a second branch of knowledge which our youth will study?

Let us do so, he replied.

And suppose we make astronomy the third—what do you say?

I am strongly inclined to it, he said; the observation of the seasons and of months and years is as essential to the general as it is to the farmer or sailor.

I am amused, I said, at your fear of the world, lest you should appear as an ordainer of useless studies; and I quite admit that it is by no means easy to believe that in every man there is an eye of the soul which, when by other pursuits lost and dimmed, is purified and reillumined by these studies; and is more precious far than ten thousand bodily eyes, for by it alone is truth seen. Now there are two classes of persons: some who will agree with you and will take your words as a revelation; another class who have never perceived this truth will probably find them unmeaning, for they see no noticeable profit which is to be obtained from them. And therefore you had better decide at once with which of the two you are proposing to argue. You will very likely say with neither, and that your chief aim in carrying on the argument is your own improvement, while at the same time you would not grudge to others any benefit which they may receive.

I should prefer, he said, to speak and inquire and answer mainly on my own behalf.

Then take a step backward, for we have gone wrong in the order of the sciences.

What was the mistake? he said.

After plane geometry, I said, we proceeded at once to solids in revolution, instead of taking solids in themselves; whereas after the second dimension the third, which is concerned with cubes and dimensions of depth, ought to have followed.

That is true, Socrates; but so little seems to have been discovered as yet about these subjects.

Why, yes, I said, and for two reasons:—in the first place, no government patronizes them; this leads to a want of energy in the pursuit of them, and they are difficult; in the second place, students cannot learn them unless they have a director. But then a director can hardly be found, and even if he could, as matters now stand, the students, who are very conceited, would not attend to him. That, however, would be otherwise if the whole State were to assist the director of these studies by giving honour to them; then disciples would show obedience,[2] and there would be continuous and earnest search, and discoveries would be made; since even now, disregarded as they are by the world, and maimed of their fair proportions, because those engaged in the research have no conception of its use, still these studies force their way by their natural charm, and it would not be surprising if they should some day emerge into light.[3]

Yes, he said, there is a remarkable charm in them. But I do not clearly understand the change in the order. By geometry, I suppose that you meant the theory of plane surfaces?

Yes, I said.

And you placed astronomy next, and then you made a step backward?

Yes, and my haste to cover the whole field has made me less speedy; the ludicrous state of research in solid geometry, which, in natural order, should have followed, made me pass over this branch and go on to astronomy, or motion of solids.

True, he said.

Then assuming that the science now omitted would come into existence if encouraged by the State, let us take astronomy as our fourth study.

NOTES

1. Meaning either (1) that they integrate the number because they deny the possibility of fractions; or (2) that division is regarded by them as a process of multiplication, for the fractions of one continue to be units.

2. [Or, "be persuaded of the importance of the study."]

3. [Or, "if the problems should be solved."]

21. From the *Timaeus* 53-56*

PLATO

The four elements and the regular solids. Geometrically, solids are bounded by planes, and the most elementary plane figure is the triangle. Two types of triangle are chosen as the basic constituents of all solid bodies, and four basic solids are constructed from them. Transformation of the elements one into another is accounted for by three of them being built up from the same type of basic triangle: the fourth (earth) being built up from triangles of the other type cannot be transformed into the remaining three.

In the first place it is clear to everyone that fire, earth, water, and air are bodies, and all bodies are solids. All solids again are bounded by surfaces, and all rectilinear surfaces are composed of triangles. There are two basic types of triangle, each having one right angle and two acute angles: in one of them these two angles are both half right angles, being subtended by equal sides, in the other they are unequal, being subtended by unequal sides. This we postulate as the origin of fire and the other bodies, our argument combining likelihood and necessity; their more ultimate origins are known to god and to men whom god loves. We must proceed to enquire what are the four most perfect possible bodies which, though unlike one another, are some of them capable of transformation into each other on resolution. If we can find the answer to this question we have the truth about the origin of earth and fire and the two mean terms between them; for we will never admit that there are more perfect visible bodies than these, each in its type. So we must do our best to construct four types of perfect body and maintain that we have grasped their nature sufficiently for our purpose. Of the two basic triangles, then, the isosceles has only one variety, the scalene an infinite number. We must therefore choose, if we are to start according to our own principles, the most perfect of this infinite number. If anyone can tell us of a better choice of triangle for the construction of the four bodies, his criticism will be welcome; but for our part we propose to pass over all the rest and pick on a single type, that of which a pair compose an equilateral triangle. It would be too long a story to give the reason, but if anyone can produce a proof that it is not so we will welcome his achievement. So let us assume that these are the two triangles from which fire and the other bodies are constructed, one isosceles and the other having a greater side whose square is three times that of the lesser. We must now proceed to clarify something we left undetermined a moment ago. It appeared as if all four types of body could pass into each other in the process of change; but this appearance is misleading. For, of the four bodies that are produced by our chosen types of triangle, three are composed of the scalene, but the fourth alone from the isosceles. Hence all four cannot pass into each

*Source: From Plato, *Timaeus and Critias*, trans. by Desmond Lee (Penguin Classics, 1965), pp. 72-77.

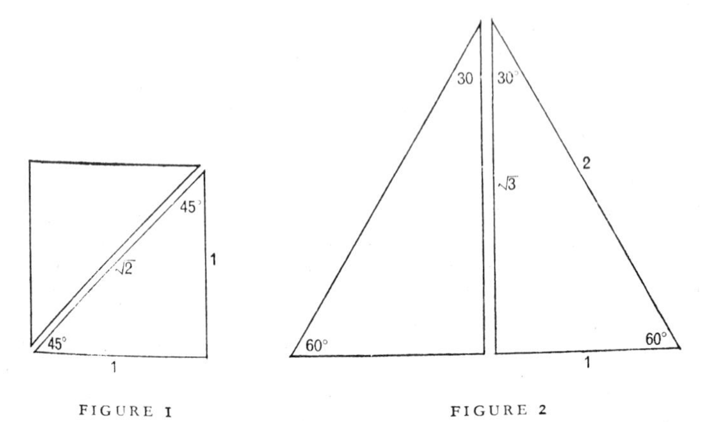

FIGURE 1

FIGURE 2

The two basic triangles. Cornford suggests that their selection is determined by 'the choice of the regular solids' for the four elements; but there is an interesting alternative suggestion in Toulmin and Goodfield, *The Architecture of Matter* (Pelican), p. 80.

other on resolution, with a large number of smaller constituents forming a lesser number of bigger bodies and vice versa; this can only happen with three of them. For these are all composed of one triangle, and when larger bodies are broken up a number of small bodies are formed of the same constituents, taking on their appropriate figures; and when small bodies are broken up into their component triangles a single new larger figure may be formed as they are unified into a single solid.[1]

So much for their transformation into each other. We must next describe what geometrical figure each body has and what is the number of its components. We will begin with the construction of the simplest and smallest figure. Its basic unit is the triangle whose hypotenuse is twice the length of its shorter side. If two of these are put together with the hypotenuse as diameter of the resulting figure, and if the process is repeated three times and the diameters and shorter sides of the three figures are made to coincide in the same vertex, the result is a single equilateral triangle composed of six basic units. And if four equilateral triangles are put

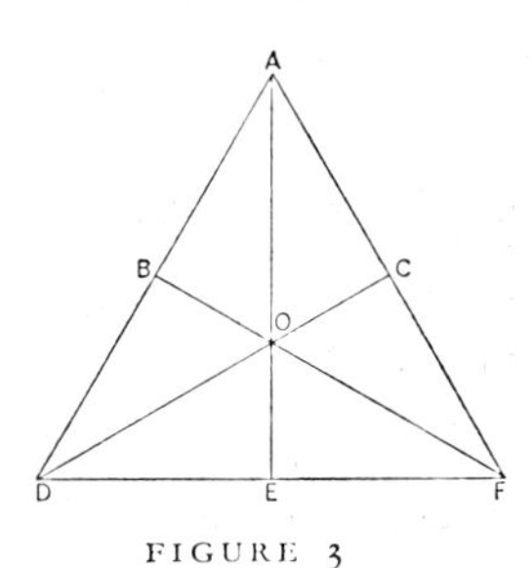

FIGURE 3

'Two of these': e.g. ABO, ACO. 'The resulting figure': e.g. ABOC. The three figures ABOC, DBOE, FEOC coincide at the same vertex O, and produce the equilateral triangle ADF.

together, three of their plane angles meet to form a single solid angle, the one which comes next after the most obtuse of plane angles:[2] and when four such angles have been formed the result is the simplest solid figure, which divides the surface of the sphere circumscribing it into equal and similar parts. [*See Figure 3 above.*]

The second figure is composed of the same basic triangles put together to form eight equilateral triangles, which yield a single solid angle from four planes. The formation of six such solid angles completes the second figure.

The third figure is put together from one hundred and twenty basic triangles, and has twelve solid angles, each

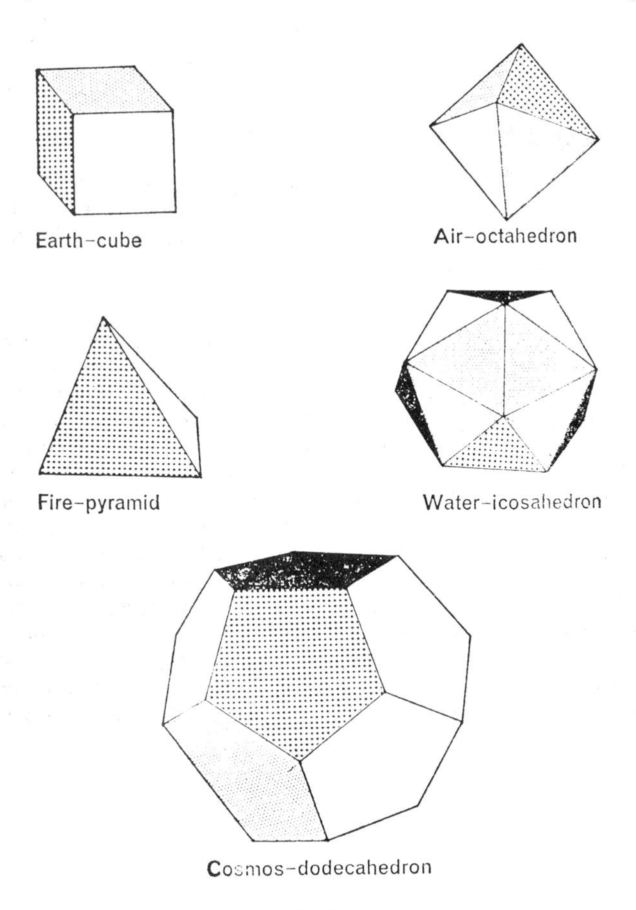

FIGURE 4

The four figures are the pyramid, the octahedron, the icosahedron, and the cube; the fifth the dodecahedron. The dodecahedron cannot be constructed out of the basic triangles, and because it approaches the sphere most nearly in volume is associated here with 'the whole (spherical) heaven', just as it is associated in the *Phaedo* 110B with the spherical earth. Exactly how Plato supposed god used it for 'arranging the constellations' (literally 'embroidering with figures') we are not told.

bounded by five equilateral plane triangles, and twenty faces, each of which is an equilateral triangle.

After the production of these three figures the first of our basic units is dispensed with, and the isosceles triangle is used to produce the fourth body. Four such triangles are put together with their right angles meeting at a common vertex to form a square. Six squares fitted together complete eight solid angles, each composed by three plane right angles. The figure of the resulting body is the cube, having six plane square faces.

There still remained a fifth construction, which the god used for arranging the constellations on the whole heaven.

With all this in mind, one might properly ask whether the number of worlds is finite or indefinite. The answer is that to call it indefinite is to express an indefinite opinion where one needs definite information, but that to pause at this point and ask whether one ought to say that there is really one world or five is reasonable enough. Our own view is that the most likely account reveals that there is a single, divine world; different considerations might lead to a different view, but they may be dismissed.[3]

NOTES

1. The three sentences are very compressed and to some extent anticipate what we shall shortly be told about the distribution of regular solids between the elements. The process of transformation is thought of as the breaking down of a regular solid into its constituent triangles, which can then rejoin to form a solid of different figure. From this process the cube (earth) must be excluded as its constituent triangle is of a different type to that of the other three. The description of the process of transformation is somewhat obscure but will be elaborated later.

The exclusion of earth from the cycle of transformation seems to be due solely to the assignation to it of the cube, and not to be based on any facts of observation.

2. The triangles are equilateral, so each solid angle contains $3 \times 60° = 180°$: the phrase 'the one which comes next after . . .' means the 'least angle which is not less than 180°, another way of saying it is itself 180°'.

3. The 'fifth construction' has been mentioned almost as an afterthought, and this paragraph seems to be a footnote suggested by it. Its point and precise meaning are obscure.

Chapter II
The Rise of Theoretical Mathematics in Ancient Greece

EUDOXUS (*c.* 400–*c.* 347 B.C.)

Eudoxus, a principal founder of classical Greek exact sciences, ranks second only to Archimedes as a geometer in antiquity. During his lifetime he also won acclaim as an astronomer, philosopher, geographer, physician, and orator. His friends called him Eudoxus, the renowned, and later Eratosthenes referred to him as "godlike." Born on the prosperous island–city of Cnidus on the Black Sea about 400 B.C., Eudoxus died in his native town 53 years later after a distinguished career as teacher and law giver.

Eudoxus began his higher education by studying the quadrivium under Archytas of Tarentum and medicine with Philiston in Sicily. At age 23, he arrived in Athens to attend Plato's Academy, where he hoped to learn philosophy and rhetoric. However, he was so poor he was forced to live in the nearby seaport of Piraeus, where food and lodging were cheap. To be present at Plato's discussions, which he found stimulating, he had to walk daily from Piraeus to Athens and back, a walk that took two hours each way. After two months of studies, he returned to Cnidus. About 370 B.C., his friends provided the means that enabled him to travel to Egypt, where he remained for 16 months. (According to the Roman philosopher Seneca writing in the 1st century A.D., he learned astronomy from the priests at Heliopolis, made astronomical observations, and composed his eight-year calendric cycle, *Oktaetris*. Seneca's assertions are dubious at best, however.) From Egypt, Eudoxus traveled to Cyzicus on the south shore of the Sea of Marmara, where he established a successful school.

About 365, Eudoxus paid a second visit to Athens, this time as a master teacher accompanied by several disciples. Plato held a banquet in his honor. Though they differed on many points, the two men respected each other. Mutual influences are difficult to determine, but surely Plato opposed the Cnidian's doctrine of Forms, which held that Forms or Ideas were not "ideal" but were actually "blended with observable things." In the period following their debates concerning the Good, Plato also wrote *Philebos*, criticizing Eudoxian hedonism, the doctrine that pleasure, correctly understood, was the highest good *(summum bonum)*. Plato's criticism was on intellectual not moral grounds. Eudoxus had not advocated a dissolute life. Pleasure was rational and included honor, justice, and moderation in all things. Aristotle reported in the *Ethics* (X.2) that Eudoxus was an upright man known for his moderation and strength of character.

After the second visit to Athens, Eudoxus returned to Cnidus, where he wrote textbooks and lectured on cosmology, meteorology, and theology. Although none of his writings have survived, it is believed that his mathematical works provide the basis for Books V, VI, and XII of Euclid's *Elements*. He also wrote at least four books on astronomy: *On Speeds*, *Enoptron* ("Mirrors"), *Phaenomena*, and *Disappearances of the Sun* (which perhaps dealt with eclipses), and a geographical treatise entitled *Ges periodos* ("Tour of the Earth"), which included historical, political, ethnographic, and religious detail about the known regions of the Earth.

In his most notable work in mathematics, Eudoxus resolved two major problems and improved method. The discovery by the Pythagoreans and Theaetetus of several incommensurable quantities (or irrationals) had led to a temporary paralysis in Greek number- and proof-theory, because the Pythagorean theory of proportions dealt only with commensurable quantities or rationals. Eudoxus surmounted the impass by devising a general theory of proportionality that treated commensurable and incommensurable quantities alike. In effect, he rigorously defined real numbers. His celebrated formulation of ratios in terms of continuously varying magnitudes appears in Euclid V, Definition 5. His magnitudes were not numbers but comprised entities such as line segments, areas, angles, volumes, and time. The effect of his introduction of magnitudes was to separate sharply number and geometry and to make geometry the basis for rigorous mathematics for the next two millennia.

Eudoxus also introduced the "method of exhaustion" to calculate the areas of plane curvilinear figures and the volumes of solids bounded by curved surfaces. Archimedes believed that Eudoxus used this new method to prove for the first time that the areas of two circles are to each other as the squares of their respective diameters (Euclid XII.2) and that the volume of a pyramid (or circular cone) is one-third the volume of a prism (or cylinder) that has the same base and equal height (Euclid XII.7 and 10). These proofs show that the Eudoxian "method of exhaustion" tacitly assumed the operation of the theory of limits. That is, it forfeited absolute accuracy in mathematical statement for a never-ending approximation process that approaches a goal or limit as close as the inquirer desires. In method, Eudoxus first formally systematized Aristotle's theory of statements with its axioms, postulates, and definitions into what came to be known as the "Euclidean" axiomatic method.

In classical Greece, Eudoxus first applied spherical geometry to astronomy. His work culminated in an ingenious geocentric system with 27 rotating homocentric spheres to describe the motions of celestial bodies. His astronomical model was abstract and elegant though not exact. It might have disappeared had not Aristotle adopted it and turned its geometrical spheres into a cumbrous physical mechanism in his cosmology.

22. From Book V of the *Elements:* Definitions*

(Theory of Proportions)

EUCLID

1. A magnitude is a part of a magnitude, the less of the greater, when it measures the greater.

2. The greater is a multiple of the less when it is measured by the less.

3. A ratio is a sort of relation in respect of size between two magnitudes of the same kind.

4. Magnitudes are said to have a ratio to one another which are capable, when multiplied, of exceeding one another.

5. Magnitudes are said to be in the same ratio, the first to the second and the third to the fourth, when, if any equimultiples whatever be taken of the first and third, and any equimultiples whatever of the second and fourth, the former equimultiples alike exceed, are alike equal to, or alike fall short of, the latter equimultiples respectively taken in corresponding order.

23. From Book XII.2 of the *Elements*

(Method of Approximation, the so-called Method of Exhaustion)

EUCLID

Circles are to one another as the squares on the diameters.

Let *ABCD, EFGH* be circles, and *BD, FH* their diameters; I say that, as the circle *ABCD* is to the circle *EFGH*, so is the square on *BD* to the square on *FH*.

For, if the square on *BD* is not to the square on *FH* as the circle *ABCD* is to the circle *EFGH*, then, as the square on *BD* is to the square on *FH*, so will the circle *ABCD* be either to some less area than the circle *EFGH*, or to a greater.

First, let it be in that ratio to a less area *S*.

Let the square *EFGH* be inscribed in the circle *EFGH;* then the inscribed square is greater than the half of the circle *EFGH*, inasmuch as, if through the points *E, F, G, H* we draw tangents to the circle, the square *EFGH* is half the square circumscribed about the circle, and the circle is less than the circumscribed square; hence the inscribed square *EFGH* is greater than the half of the circle *EFGH*.

Let the circumferences *EF, FG, GH, HE* be bisected at the points *K, L, M, N*, and let *EK, KF, FL, LG, GM, MH, HN, NE* be joined; therefore each of the triangles *EKF, FLG, GMH, HNE* is also greater than the half of the segment of the circle about it, inasmuch as, if

Source: Selections 22 and 23 are reprinted by permission from *The Thirteen Books of Euclid's Elements,* trans. by Sir Thomas L. Heath (1956 edition), vol. II, 113-114 and vol. III, 371-373 (respectively).

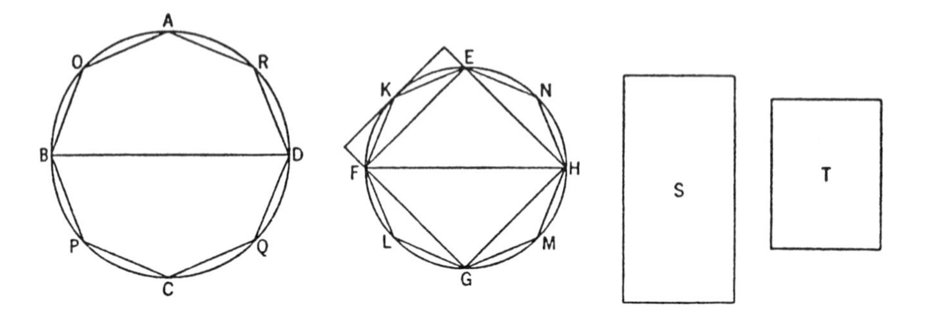

through the points *K, L, M, N* we draw tangents to the circle and complete the parallelograms on the straight lines *EF, FG, GH, HE,* each of the triangles *EKF, FLG, GMH, HNE* will be half of the parallelogram about it, while the segment about it is less than the parallelogram; hence each of the triangles *EKF, FLG, GMH, HNE* is greater than the half of the segment of the circle about it.

Thus, by bisecting the remaining circumferences and joining straight lines, and by doing this continually, we shall leave some segments of the circle which will be less than the excess by which the circle *EFGH* exceeds the area *S*.

For it was proved in the first theorem of the tenth book that, if two unequal magnitudes be set out, and if from the greater there be subtracted a magnitude greater than the half, and from that which is left a greater than the half, and if this be done continually, there will be left some magnitude which will be less than the lesser magnitude set out.

Let segments be left such as described, and let the segments of the circle *EFGH* on *EK, KF, FL, LG, GM, MH, HN, NE* be less than the excess by which the circle *EFGH* exceeds the area *S*.

Therefore the remainder, the polygon *EKFLGMHN*, is greater than the area *S*.

Let there be inscribed, also, in the circle *ABCD* the polygon *AOBPCQDR* similar to the polygon *EKFLGMHN;* therefore, as the square on *BD* is to the square on FH, so is the polygon *AOBPCQDR* to the polygon *EKFLGMHN*.

But, as the square on *BD* is to the square on *FH*, so also is the circle *ABCD* to the area *S;* therefore also, as the circle *ABCD* is to the area *S*, so is the polygon *AOBPCQDR* to the polygon *EKFLGMHN;* therefore, alternately, as the circle *ABCD* is to the polygon inscribed in it, so is the area *S* to the polygon *EKFLGMHN*.

But the circle *ABCD* is greater than the polygon inscribed in it; therefore the area *S* is also greater than the polygon *EKFLGMHN*.

But it is also less: which is impossible.

Therefore, as the square on *BD* is to the square on *FH*, so is not the circle *ABCD* to any area less than the circle *EFGH*.

Similarly we can prove that neither is the circle *EFGH* to any area less than the circle *ABCD* as the square on *FH* is to the square on *BD*.

I say next that neither is the circle *ABCD* to any area greater than the circle *EFGH* as the square on *BD* is to the square on *FH*.

For, if possible, let it be in that ratio to a greater area *S*.

Therefore, inversely, as the square on *FH* is to the square on *DB*, so is the area *S* to the circle *ABCD*.

But, as the area *S* is to the circle *ABCD*, so is the circle *EFGH* to some area less than the circle *ABCD;* therefore also, as the square on *FH* is to the square on *BD*, so is the circle *EFGH* to some area less than the circle *ABCD*: which was proved impossible.

Therefore, as the square on *BD* is to the square on *FH*, so is not the circle *ABCD* to any area greater than the circle *EFGH*.

And it was proved that neither is it in that ratio to any area less than the circle *EFGH*; therefore, as the square on *BD* is to the square on *FH*, so is the circle *ABCD* to the circle *EFGH*.

Therefore etc. Q. E. D.

Chapter II
The Rise of Theoretical Mathematics in Ancient Greece

ARISTOTLE (384–322 B.C.)

Aristotle, the greatest systematic philosopher of ancient times, was born in Stagira, an Ionian Greek colony bordering on Macedon. His father, Nicomachus, was personal physician to a Macedonian king; his mother, Phaestis, came from the city of Chalcis. His later interests in biology may have derived from his father's profession, and he may have received some medical training early. When both parents died in his boyhood, Aristotle, a brother, and a sister were raised by a relative.

At age 17, Aristotle came to Athens, where, like a number of other Ionians and Macedonians, he entered Plato's Academy. He encountered a stimulating, prosperous city. Athens had recovered from the havoc and defeat of the Peloponnesian War three decades earlier and had restored democracy. Athens continued to be the chief Greek intellectual and cultural center with a thriving commerce. Aristotle's career would be associated with the city, where he would always be an outsider, a non-citizen hampered by legal restrictions.

Aristotle's arrival in Athens in 367 B.C. marks the beginning of three periods of mature activity. First, he spent 20 years at Plato's Academy, where he was known as an eager young man with a keen mind and insatiable thirst for knowledge. Plato is said to have called him "The Mind;" others are reported to have dubbed him "The Reader." After completing the course of studies at the Academy, he lectured there on rhetoric and pursued research and writing, largely in physical science and cosmology. In mathematics, he kept abreast of recent developments but was not a major contributor to new knowledge.

About the time Plato died in 347 B.C., Aristotle left Athens and stayed away for 12 years for reasons that are not clear. Perhaps the choice of Speusippus to succeed Plato as head of the Academy caused it, or perhaps it was the result of growing anti-Macedonian feelings in Athens that the orator Demosthenes led. During this second period, Aristotle spent three happy years continuing his philosophical inquiries and teaching in Asia Minor, where he married Pythias. In 345-44 B.C., he went to the nearby island of Lesbos where he conducted extensive biological observations. In 343-42 B.C., he went to Macedon to tutor the 13-year-old prince, the future Alexander the Great. Even after the tutorial program ended in 340 B.C., Alexander retained Aristotle as a trusted friend and counsel.

When Alexander put Athens under Macedonian control in 335 B.C., Aristotle returned there and set up a school called the Lyceum. In the Lyceum, which resembled present-day graduate schools by developing specialists, Aristotle introduced written examinations into western education. Because the teacher and students often walked about in the open air when discussing issues, they were nicknamed the Peripatetics. When Alexander died in 323 B.C., threats were raised against Aristotle in Athens because of his pro-Macedonian views. Like Socrates earlier, he was charged with impiety. Stating that he did not want Athens to have the disgrace of murdering a second philosopher, he withdrew in voluntary exile to Chalcis, where he died the following year.

In mathematics, which he classified as an autonomous deductive science along with physics and metaphysics, Aristotle contributed chiefly to foundations. He ranked proof-theory above the Socratic dialectic as a means to obtain truths. He improved the means to develop these proofs more rigorously by introducing syllogistic logic and a controlled use of analogies. In the Greek movement to axiomatize mathematics, he constructed a theory of statements beginning with "common notions" or axioms, "special notions" or postulates, and careful definitions that underlay formal logical reasoning. Also in mathematics, he gave a *reductio ad absurdum* proof of the irrationality of $\sqrt{2}$, recognized that the new Eudoxian theory of proportions covered rational and irrational quantities, and accepted potential infinity (increasing or decreasing without end) while rejecting actual infinity (the infinite as completed totality). Finally, in his book *Physics* he defined space, time, and motion as continuous and dismissed Zeno's paradoxes on motion as fallacies.

With his stress on logic and classification, Aristotle sought to make all knowledge his province: astronomy, biology, and physics as well as ethics, politics, and poetry. The *Politics, Meteorologica, Nichomachean Ethics, Posterior Analytics,* and *Rhetoric* are among his major books.

24. From *Posterior Analytics* i.10*

(First Principles or Theory of Statements)

ARISTOTLE

I mean by the first principles in every genus those elements whose existence cannot be proved. The meaning both of these primary elements and of those deduced from them is assumed; in the case of first principles, their existence is also assumed, but in the case of the others deduced from them it has to be proved. Examples are given by the unit, the straight and triangular; for we must assume the existence of the unit and magnitude, but in the case of the others it has to be proved.

Of the first principles used in the demonstrative sciences some are peculiar to each science, and some are common, but common only by analogy, inasmuch as they are useful only in so far as they fall within the genus coming under the science in question.

Examples of peculiar first principles are given by the definitions of the line and the straight; common first principles are such as that, when equals are taken from equals, the remainders are equal. Only so much of these common first

*Source: From *Selections Illustrating the History of Greek Mathematics*, trans. by Ivor Thomas (1941), vol. I, 419-423. Reprinted by permission of Harvard University Press.

principles is needed as falls within the genus in question; for such a first principle will have the same force even though not applied generally but only to magnitudes, or by the arithmetician only to numbers.

Also peculiar to a science are the first principles whose existence it assumes and whose essential attributes it investigates, for example, in arithmetic units, in geometry points and lines. Both their existence and their meaning are assumed. But of their essential attributes, only the meaning is assumed. For example, arithmetic assumes the meaning of odd and even, square and cube, geometry that of irrational or inflection or verging,[1] but their existence is proved from the common first principles and propositions already demonstrated. Astronomy proceeds in the same way. For indeed every demonstrative science has three elements: (1) that which it posits (the genus whose essential attributes it examines); (2) the so-called common axioms, which are the primary premises in its demonstrations; (3) the essential attributes, whose meaning it assumes. There is nothing to prevent some sciences passing over some of these elements; for example, the genus may not be posited if it is obvious (the existence of number, for instance, and the existence of hot and cold are not similarly evident); or the meaning of the essential attributes might be omitted if that were clear. In the case of the common axioms, the meaning of taking equals from equals is not expressly assumed, being well known. Nevertheless in the nature of the case there are these three elements, that about which the demonstration takes place, that which is demonstrated and those premises by which the demonstration is made.

That which necessarily exists from its very nature and which we must necessarily believe is neither hypothesis nor postulate. For demonstration is a matter not of external discourse but of meditation within the soul, since syllogism is such a matter. And objection can always be raised to external discourse but not to inward meditation. That which is capable of proof but assumed by the teacher without proof is, if the pupil believes and accepts it, *hypothesis*, though it is not hypothesis absolutely but only in relation to the pupil; if the pupil has no opinion on it or holds a contrary opinion, the same assumption is a *postulate*. In this lies the distinction between hypothesis and postulate; for a postulate is contrary to the pupil's opinion, demonstrable, but assumed and used without demonstration.

The *definitions* are not hypotheses (for they do not assert either existence or non-existence), but it is in the premises of a science that hypotheses lie. Definitions need only to be understood; and this is not hypothesis, unless it be contended that the pupil's hearing is also a hypothesis. But hypotheses lay down facts on whose existence depends the existence of the fact inferred. Nor are the geometer's hypotheses false, as some have maintained, urging that falsehood must not be used, and that the geometer is speaking falsely in saying that the line which he draws is a foot long or straight when it is neither a foot long nor straight. The geometer draws no conclusion from the existence of the particular line of which he speaks, but from what his diagrams represent. Furthermore, all hypotheses and postulates are either universal or particular, but a definition is neither.

NOTE

1. Euclid does not define κεκλάσθαι "to be inflected," or νεύειν, "to verge."

25. From *Metaphysics* 1066-1067*

(The Infinite, the essence of the Archimedean Postulate)

ARISTOTLE

The infinite is either that which is incapable of being traversed because it is not its nature to be traversed (this corresponds to the sense in which the voice is 'invisible'), or that which admits only of incomplete traverse or scarcely admits of traverse, or that which, though it naturally admits of traverse, is not traversed or limited; further, a thing may be infinite in respect of addition or of subtraction or of both. The infinite cannot be a separate, independent thing. For if it is neither a spatial magnitude nor a plurality, but infinity itself is its substance and not an accident, it will be indivisible; for the divisible is either magnitude or plurality. But if indivisible, it is not infinite, except as the voice is invisible; but people do not mean this, nor are we examining this sort of infinite, but the infinite as untraversable. Further, how can an infinite exist by itself, unless number and magnitude also exist by themselves,—since infinity is an attribute of these? Further, if the infinite is an accident of something else, it cannot be *qua* infinite an element in things, as the invisible is not an element in speech, though the voice is invisible. And evidently the infinite cannot exist actually. For then any part of it that might be taken would be infinite; for 'to be infinite' and 'the infinite' are the same, if the infinite is substance and not predicated of a subject. Therefore it is either indivisible, or if it is secable, it is divisible into ever divisible parts; but the same thing cannot be many infinites, yet as a part of air is air, so a part of the infinite would be infinite, if the infinite is a substance and a principle. Therefore it must be insecable and indivisible. But the actually infinite cannot be indivisible; for it must be a quantity. Therefore infinity belongs to a subject incidentally. But if so, as we have said, it cannot be it that is a principle, but rather that of which it is an accident—the air or the even number.

This inquiry is universal; but that the infinite is not *among sensible things*, is evident from the following argument. If the definition of a body is 'that which is bounded by planes', there cannot be an infinite body either sensible or intelligible; nor a separate and infinite number, for number or that which has a number can be completely enumerated. The truth is evident from the following concrete argument. The infinite can neither be composite nor simple. For (1) it cannot be a composite body, since the elements are limited in multitude. For the contraries must be equal and no *one* of them must be infinite; for if one of the two bodies falls at all short of the other in potency, the finite will be destroyed by the infinite. And that *each* should be infinite is impossible. For body is that which has extension in all directions, and the infinite is the boundlessly extended, so that the infinite body will be infinite in every direction. Nor (2) can the infinite body be one and simple—neither, as some say, something which is apart from the elements,

**Source:* From W. D. Ross (ed.), *Aristotle: Selections* (1955), 83-86. Reprinted by permission of Charles Scribner's Sons.

from which they generate these (for there is no such body apart from the elements; for everything can be resolved into that of which it consists, but no such product of analysis is observed except the simple bodies), nor fire nor any other of the elements. For apart from the question how any of them could be infinite, the All, even if it is finite, cannot either be or become one of them, as Heraclitus says all things sometime become fire. The same argument applies to the One, which the natural philosophers posit besides the elements. For everything changes from the contrary, *e.g.*, from hot to cold.

Further, every sensible body is somewhere, and whole and part have the same proper place, e.g., the whole earth and part of the earth. Therefore if (1) the infinite body is homogeneous, it will be unmovable or it will be always moving. But the latter is impossible; for why should it rather move down than up or anywhere else? E.g. if there is a clod which is part of an infinite body, where will this move or rest? The proper place of the body which is homogeneous with it is infinite. Will the clod occupy the whole place, then? And how? *This is impossible*. What then is its rest or its movement? It will either rest everywhere, and then it cannot move; or it will move everywhere, and then it cannot be still. But (2) if the infinite body has unlike parts, the proper places of the parts are unlike also, and, firstly, the body of the All is not one except by contact, and, secondly, the parts will be either finite or infinite in variety of kind. *Finite* they cannot be; for then those of one kind will be infinite in quantity and those of another will not (if the All is infinite), *e.g.*, fire or water would be infinite, but such an infinite part would be destruction to its contrary. But if the parts are *infinite* and simple, their places also are infinite and the elements will be infinite; and if this is impossible, and the places are finite, the All also must be limited.

In general, there cannot be an infinite body and also a proper place for all bodies, if every sensible body has either weight or lightness. For it must move either towards the middle or upwards, and the infinite—either the whole or the half—cannot do either; for how will you divide it? Or how will part of the infinite be up and part down, or part extreme and part middle? Further, every sensible body is in a place, and there are six kinds of place, but these cannot exist in an infinite body. In general, if there cannot be an infinite place, there cannot be an infinite body; *and there cannot be an infinite place*, for that which is in a place is somewhere, and this means either up or down or in one of the other directions, and each of these is a limit.

The infinite is not the same in the sense that it is one thing whether exhibited in distance or in movement or in time, but the posterior among these is called infinite in virtue of its relation to the prior, *i.e.*, a movement is called infinite in virtue of the distance covered by the spatial movement or alteration or growth, and a time is called infinite because of the movement which occupies it.

26. From the *Metaphysics* (1068b–1069a) and *Physics* (230a–240a)*

(On the Continuous and Zeno's Paradoxes)

ARISTOTLE

Things which are in one place (in the strictest sense) are *together in place*, and things which are in different places are *apart*. Things whose extremes are together *touch*. That at which the changing thing, if it changes continuously according to its nature, naturally arrives before it arrives at the extreme into which it is changing, is *between*. That which is most distant in a straight line is *contrary in place*. That is *successive* which is after the beginning (the order being determined by position or form or in some other way) and has nothing of the same class between it and that which it succeeds, e.g., lines succeed a line, units a unit, or one house another house. (There is nothing to prevent a thing of some *other* class from being between.) For the successive succeeds something and is something later; "one" does not succeed "two," nor the first day of the month the second. That which, being successive, touches, is *contiguous*. Since all change is between opposites, and these are either contraries or contradictories, and there is no middle term for contradictories, clearly that which is *between* is between contraries. The *continuous* is a species of the contiguous or of that which touches; two things are called continuous when the limits of each, with which they touch and are kept together, become one and the same, so that plainly the continuous is found in the things out of which a unity naturally arises in virtue of their contact. And plainly the successive is the first of these concepts; for the successive does not necessarily touch, but that which touches is successive. And if a thing is continuous, it touches, but if it touches, it is not necessarily continuous; and in things in which there is no touching, there is no organic unity. Therefore a point is not the same as a unit; for contact belongs to points, but not to units, which have only succession; and there is something between two of the former, but not between two of the latter. . . .

If the terms "continuous," "in contact," and 'in succession' as defined above—things being 'continuous' if their extremities are one, 'in contact' if their extremities are together, and 'in succession' if there is nothing of their own kind intermediate between them—nothing that is continuous can be composed of indivisibles; e.g., a line cannot be composed of points, the line being continuous and the point indivisible. For the extremities of two points can neither be *one* (since of an indivisible there can be no extremity as distinct from some other part) nor *together* (since that which has no parts can have no extremity, the extremity and the thing of which it is the extremity being distinct).

Moreover if that which is continuous is composed of points, these points must

**Source:* From W. D. Ross (ed.), *Aristotle: Selections* (1955), 88-89. Reprinted by permission of Charles Scribner's Sons.

be either *continuous* or *in contact* with one another: and the same reasoning applies in the case of all indivisibles. Now for the reason given above they cannot be continuous; and one thing can be in contact with another only if whole is in contact with whole or part with part or part with whole. But since an indivisible has no parts, they must be in contact with one another as whole with whole. And if they are in contact with one another as whole with whole, they will not be continuous; for that which is continuous has distinct parts; and these parts into which it is divisible are different in this way, *i.e.*, spatially separate.

Nor again can a point be *in succession* to a point or a moment to a moment in such a way that length can be composed of points or time of moments; for things are in succession if there is nothing of their own kind intermediate between them, whereas that which is intermediate between points is always a line and that which is intermediate between moments is always a period of time.

Again, if length and time could thus be composed of indivisibles, they could be divided into indivisibles, since each is divisible into the parts of which it is composed. But, as we saw, no continuous thing is divisible into things without parts. Nor can there be any thing of another kind intermediate between the points or between the moments; for if there could be any such thing it is clear that it must be either indivisible or divisible, and if it is divisible it must be divisible either into indivisibles or into divisibles that are infinitely divisible, in which case it is continuous.

Moreover it is plain that everything continuous is divisible into divisibles that are infinitely divisible; for if it were divisible into indivisibles, we should have an indivisible in contact with an indivisible, since the extremities of things that are continuous with one another are one, and such things are therefore in contact.

The same reasoning applies equally to magnitude, to time, and to motion; either all of these are composed of indivisibles and are divisible into indivisibles, or none. This may be made clear as follows. If a magnitude is composed of indivisibles, the motion over that magnitude must be composed of corresponding indivisible motions; *e.g.*, if the magnitude ABC is composed of the indivisibles A, B, C, each corresponding part of the motion DEF of Z over ABC is indivisible. Therefore since where there is motion there must be something that is in motion, and where there is something in motion there must be motion, the actual state of motion will also be composed of indivisibles. So Z traverses A when its motion is D, B when its motion is E, and C similarly when its motion is F. Now a thing that is in motion from one place to another cannot at the moment when it was in motion both be in motion and at the same time have completed its motion at the place to which it was in motion; *e.g.*, if a man is walking to Thebes he cannot be walking to Thebes and at the same time have completed his walk to Thebes; and, as we saw, Z traverses the partless section A in virtue of the presence of the motion D. Consequently, if Z actually passes through A *after* being in process of passing through, the motion must be divisible; for at the time when it was passing through, it neither was at rest nor had completed its passage but was in an intermediate state; while if it is passing through and has completed its passage *at the same moment*, then that which is walking will at the moment when it is walking have completed its walk and will be in the place to which it is walking, that is to say it will have completed its motion at the place to which it is in motion.

And if a thing is in motion over the whole ABC and its motion is the three D, E, and F, and if it is not in motion at all over the partless section A but has completed its motion over it, then the motion will consist not of motions but of starts, and it will be possible for a thing to have completed a motion without being in motion; for on this assumption it has completed its passage through A without passing through it. So it will be possible for a thing to have completed a walk without ever walking; for on this assumption it has completed a walk over a particular distance without walking over that distance. Since, then, everything must be either at rest or in motion, and Z is therefore at rest in each of the sections A, B, and C, it follows that a thing can be continuously at rest and at the same time in motion; for, as we saw, Z is in motion over the whole ABC and at rest in any part (and consequently in the whole) of it. Moreover if the indivisibles composing DEF are motions, it would be possible for a thing in spite of the presence in it of motion to be not in motion but at rest, while if they are not motions, it would be possible for motion to be composed of something other than motions.

And if length and motion are thus indivisible, it is neither more nor less necessary that time also be similarly indivisible, that is to say composed of indivisible moments; for if the whole distance is divisible and an equal velocity will cause a thing to pass through less of it in less time, the time must also be divisible, and conversely, if the time in which a thing is carried over the section A is divisible, this section A must also be divisible.

And since every magnitude is divisible into magnitudes—for we have shown that it is impossible for anything continuous to be composed of indivisible parts, and every magnitude is continuous—it necessarily follows that the quicker of two things traverses a greater magnitude in an equal time, an equal magnitude in less time, and a greater magnitude in less time, in conformity with the definition sometimes given of the "quicker." Suppose that A is quicker than B. Now since of two things that which changes sooner is quicker, in the time FG, in which A has changed from C to D, B will not yet have arrived at D but will be short of it; so that in an equal time the quicker will pass over a greater magnitude. More than this, it will pass over a greater magnitude in *less* time; for in the time in which A has arrived at D, B being the slower has arrived, let us say, at E. Then since A has occupied the whole time FG in arriving at D, it will have arrived at H in less time than this, say FK. Now the magnitude CH that A has passed over is greater than the magnitude CE, and the time FK is less than the whole time FG; so that the quicker will pass over a greater magnitude in less time. And from this it is also clear that the quicker will pass over an equal magnitude in less time than the slower. For since it passes over the greater magnitude in less time than the slower and (regarded by itself) passes over LM the greater in more time than LN the less, the time PR in which it passes over LM will be more than the time PS in which it passes over LN; so that, the time PR being less than the time PT in which the slower passes over LN, the time PS will also be less than the time PT; for it is less than the time PR, and that which is less than something else that is less than a thing, is also itself less than that thing. Hence it follows that the quicker will traverse an equal magnitude in less time than the slower. Again, since the motion of anything must always occupy either an equal time or less or more time in comparison with that of another thing, and since, whereas a thing is slower if its motion occupies more time and of equal velocity if its motion occupies an equal time, the quicker is neither of equal velocity nor slower, it follows that the motion of the quicker can occupy neither an equal time nor more time. It can only be, then, that it occupies less time, and thus we get the necessary consequence that the quicker will pass over an equal magnitude (as well as a greater) in less time than the slower.

And since every motion is in time and

a motion may occupy any time, and the motion of everything that is in motion may be either quicker or slower, both quicker motion and slower motion may occupy any time; and this being so, it necessarily follows that time also is continuous. By continuous I mean that which is divisible into divisibles that are infinitely divisible; and if we take this as the definition of 'continuous', it follows necessarily that time is continuous. For since it has been shown that the quicker will pass over an equal magnitude in less time than the slower, suppose that A is quicker and B slower, and that the slower has traversed the magnitude CD in the time FG. Now it is clear that the quicker will traverse the same magnitude in less time than this; let us say in the time FH. Again since the quicker has passed over the whole CD in the time FH, the slower will in the same time pass over CK, say, less than CD. And since B, the slower, has passed over CK in the time FH, the quicker will pass over it in less time; so that the time FH will again be divided. And if this is divided the magnitude CK will also be divided just as CD was; and again if the magnitude is divided, the time will also be divided. And we can carry on this process for ever, taking the slower after the quicker and the quicker after the slower alternately, and using what has been demonstrated at each stage as a new point of departure; for the quicker will divide the time and the slower will divide the length. If, then, this alternation always holds good, and at every turn involves a division, it is evident that all time must be continuous. And at the same time it is clear that all magnitude is also continuous; for the divisions of which time and magnitude respectively are susceptible are the same and equal.

Moreover the current popular arguments make it plain that, if time is continuous, magnitude is continuous also, inasmuch as a thing passes over half a given magnitude in half the time taken to cover the whole; in fact without qualification it passes over a less magnitude in less time; for the divisions of time and of magnitude will be the same. And if either is infinite, so is the other, and the one is so in the same way as the other; *i.e.*, if time is infinite in respect of its extremities, length is also infinite in respect of its extremities; if time is infinite in respect of divisibility, length is also infinite in respect of divisibility; and if time is infinite in both respects, magnitude is also infinite in both respects.

Hence Zeno's argument makes a false assumption in asserting that it is impossible for a thing to pass over or severally to come in contact with infinite things in a finite time. For there are two senses in which length and time and generally anything continuous are called "infinite;" they are called so in respect either of divisibility or of their extremities. So while a thing in a finite time cannot come in contact with things quantitatively infinite, it can come in contact with things infinite in respect of divisibility; for in this sense the time itself is also infinite; and so we find that the time occupied by the passage over the infinite is not a finite but an infinite time, and the contact with the infinite is made by means of moments not finite but infinite in number.

The passage over the infinite, then, cannot occupy a finite time, and the passage over the finite cannot occupy an infinite time; if the time is infinite the magnitude must be infinite also, and if the magnitude is infinite, so also is the time. This may be shown as follows. Let AB be a finite magnitude, and let us suppose that it is traversed in infinite time C, and let a finite period CD of the time be taken. Now in this period the thing in motion will pass over a certain segment of the magnitude; let BE be the segment that it has thus passed over (this will be either an exact measure of AB or less or greater than an exact measure; it makes no difference which it is). Then, since a magnitude equal to BE will always be passed over in an equal time, and BE measures the whole magnitude, the whole time occupied in passing over AB will be finite; for it will be divisible into periods equal in number to the segments into which the magnitude is divisible. Moreover if it is the case that infinite time is not occupied in passing over every magnitude, but it is possible to pass over some magnitude, say BE, in a finite time, and if this BE measures the whole of which it is a part, and if an equal magnitude is passed over in an equal time, then it follows that the time like the magnitude is finite. That infinite time will not be occupied in passing over BE is evident if the time be taken as limited in one direction; for as the part will be passed over in less time than the whole, the time occupied in traversing this part must be finite, the limit in one direction being given. The same reasoning will also show the falsity of the assumption that infinite length can be traversed in a finite time. It is evident, then, from what has been said that neither a time nor a surface nor in fact anything continuous can be indivisible. . . .

Zeno's reasoning is fallacious, when he says that if everything when it occupies an equal space is at rest, and if that which is in locomotion is always occupying such a space at any moment, the flying arrow is therefore motionless. This is false, for time is not composed of indivisible moments any more than any magnitude is composed of indivisibles.

Zeno's arguments about motion, which cause so much disquietude to those who try to solve the problems that they present, are four in number. The first asserts the non-existence of motion on the ground that that which is in locomotion must arrive at the halfway stage before it arrives at the goal. This we have discussed above.

The second is the so-called "Achilles," and it amounts to this, that in a race the quickest runner can never overtake the slowest, since the pursuer must first reach the point whence the pursued started, so that the slower must always hold a lead. This argument is the same in principle as that which depends on bisection, though it differs from it in that the spaces with which we successively have to deal are not divided into halves. The result of the argument is that the slower is not overtaken; but it proceeds along the same lines as the bisection-argument (for in both a division of the space in a certain way leads to the result that the goal is not reached, though the "Achilles" goes further in that it affirms that even the quickest runner in legendary tradition must fail in his pursuit of the slowest), so that the solution must be the same. And the axiom that that which holds a lead is never overtaken is false; it is not overtaken, it is true, while it holds a lead; but it is overtaken nevertheless if it is granted that it traverses the finite distance prescribed. These, then, are two of his arguments.

The third is that already given above, to the effect that the flying arrow is at rest, which result follows from the assumption that time is composed of moments; if this assumption is not granted, the conclusion will not follow.

The fourth argument is that concerning the two rows of bodies, each row being composed of an equal number of bodies of equal size, passing each other on a racecourse as they proceed with equal velocity in opposite directions, the one row originally occupying the space between the goal and the middle point of the course and the other that between the middle point and the starting-post. This, he thinks, involves the conclusion that half a given time is equal to double that time. The fallacy of the reasoning lies in the assumption that a body occupies an equal time in passing with equal velocity a body that is in motion and a body of equal size that is not; which is false. For instance (so runs the argument) let A, A . . . be the stationary bodies of equal size, B, B . . . the bodies, equal in number and in size to A, A . . . originally occupying the half of the course from the starting-post to the middle of the A's, and C, C . . . those originally occupying the other half from the goal to the middle of the A's, equal in number, size, and velocity to B, B. . . . Then three consequences follow: First, as the B's and the C's pass one another, the first B reaches the last C at the same moment as the first C reaches the last B. Secondly at this moment the first C has passed all the A's, whereas the first B has passed only half the A's, and has consequently occupied only half the time occupied by the first C, since each of the two occupies an equal time in passing each A. Thirdly at the same moment all the B's have passed all the C's; for the first C and the first B will simultaneously reach the opposite ends of the course, since (so says Zeno) the time occupied by the first C in passing each of the B's is equal to that occupied by it in passing each of the A's because an equal time is occupied by both the first B and the first C in passing all the A's. This is the argument, but it presupposes the aforesaid fallacious assumption.

Chapter III

Mathematics in the West During Hellenistic and Roman Times

Introduction. Progress continued in theoretical mathematics in the Mediterranean region during Hellenistic (338–133 B.C.) and Roman times (133 B.C.–A.D. 476). Although progress was intermittent and quality uneven, theoretical mathematics in antiquity reached its zenith during the 3rd century B.C. Geometers, mostly of Greek descent, were major participants in a late flowering of Classical Greek learning. By the first years of the 3rd century Euclid had consolidated and given definitive form to the work of Plato's circle and their Hellenic sources; the *Elements* was a vital link in the transmission of mathematical ideas from the Classical Greeks to geometers of later times. Shortly after Euclid came two brilliant geometers: Archimedes, who refined the method of exhaustion applying it to a wide range of problems, and Apollonius, who thoroughly developed the theory of conic sections. The polymath Eratosthenes, who flourished at the same time, invented a "sieve" for finding prime numbers. This level of mathematical achievement was not equalled again until the Scientific Revolution of the 17th century.

After the 3rd-century efflorescence, the final seven centuries of antiquity registered several lesser but still notable accomplishments in mathematics. They were lesser in terms of originality, depth, and completeness of thought. Because few primary sources have survived from later antiquity, our knowledge is drawn mainly from scholarly commentaries on Greek mathematical classics. Written by such authors as Theon, Pappus, Proclus, and Eutocius during the 4th through 6th centuries A.D., these show continuing attention to geometry albeit with few new theoretical insights. Apparently two major related trends developed in geometry during later antiquity; both emphasized applied mathematics. The first, concentrating almost exclusively on practical application, failed to stimulate further developments within mathematics. The writings of Hero (or Heron) of Alexandria on metrical geometry dating from the late first century A.D. epitomize this trend. In the second trend, geometry was intimately connected with the disciplines of cartography and astronomy. The resulting symbiotic development of geometry and astronomy produced an important new branch of mathematics. Begun by Hipparchus of Nicaea in the 2nd century B.C., the symbiosis culminated in the applied trigonometry of Claudius Ptolemy in the 2nd century A.D. Finally, the 3rd century A.D. experienced a brief revival of theoretical mathematics. It was highlighted by the significant contributions of Diophantus to algebra and the origins of number theory together with the clear and concise work of the versatile Pappus in geometrical analysis.

Changing political and social circumstances during Hellenistic and Roman times influenced opportunities for and, to a degree, directions in new mathematical studies. The Hellenistic Age began with the rise to eminence of Macedon in the Balkans—a major change in political power. Continuing warfare among the Greek *poleis,* even after the Peloponnesian War, and a loss of markets in the West had made the Greek *poleis* unstable and ripe for external conquest by the mid-4th century B.C. Philip II (reigned 359–336 B.C.), an able and highly ambitious king of Macedon, successfully exploited this situation. After meddling in the affairs of central Greece for over a decade, he defeated a coalition of Theban and Athenian forces at Chaeronea in 336 B.C. This defeat marked the end of Classical Greek freedom and autonomy. Two years later, while preparing to invade the vast but weak Persian Empire, the hard-driving Philip fell victim to an assassin. His 20-year-old son Alexander (reigned 336–323 B.C.) succeeded him.

Possessed of vaulting ambition, military genius, mastery of detail, and courage, Alexander set out on an unprecedented program of conquest beginning with a campaign against Persia. First he had to establish control over Greece, where he was patron to his former teacher Aristotle in Athens. Perhaps at Aristotle's suggestion, Alexander included geometers and natural philosophers on his military expeditions. The geometers were to work on catapult engineering problems and the natural philosophers to gather data to send to Aristotle. After winning many battles in Asia Minor, Alexander was welcomed in late 332 B.C. as a liberator-god in Egypt. There the victor laid plans for a city to be called Alexandria at the westernmost mouth of the Nile. Proceeding eastward, he defeated the main Persian army under King Darius III at Gaugemala near Nineveh in 331 B.C. and captured the chief Persian capital, Persepolis, with its royal treasury a year later. Alexander's comprehension of supply problems and his skillful use of mule pack-trains partly made possible these long-distance campaigns. After crossing the Indus River his weary troops refused to go farther, and the disappointed monarch was forced to return to the west. While in Babylon in 323 B.C. Alexander died suddenly, perhaps from malaria, alcoholism, or poison, at age 33. Political unity was never achieved in his new empire.

Life in the Successor-Empires[1] differed from that in Classical Greece in at least three important ways. First, vast monarchies whose populations were concentrated in the large commercial cities replaced the *polis* as the basic political unit. Significantly, the ethnic variety in these cities, which were usually capitals, stimulated intellectual and cultural contacts and borrowing among Greeks and non-Greeks. (Only the urban poor had little or no opportunity to participate in these cultural exchanges.) The new monarchs recruited Greek and Macedonian foreigners into their bureaucracies and the professions, while the local urban upper class and many of the newly emergent middle class sought Greek learning and manners. Soon a form of Greek common speech—koinê—became the universal language of the cities facilitating a two-way transmission of culture that favored Greek ideas and institutions. These intellectual and cultural interactions outside Greece produced the second distinguishing feature of Hellenistic life: a cosmopolitan outlook that appreciated the unity of the Mediterranean world and thus departed from the Greek ethnocentrism with its qualitative distinction between Greek and "barbarian." A third distinguishing feature, ruler worship, was of Oriental origin and did not take root in Macedon and Greece.

After the Successor-Empires were established, the intellectual center of mathematics shifted from Athens to the new Greek city of Alexandria in Egypt. More precisely, the shift occurred after about 295 B.C., when Ptolemy I Soter (Savior) began building the famous library. Athens, located on the conquered and depopulated Balkan mainland of Greece, had been relegated to the periphery of the world of theoretical and natural knowledge. Three other great Mediterranean cities supported by thriving commerce and led by enlightened—or simply ambitious—rulers had supplanted it. Two of these—Syracuse in Sicily, the so-called Athens of the West, and Pergamum, the capital of the Attalid dynasty—only briefly nurtured theoretical mathematics. By comparison Alexandria, the capital and chief seaport of Ptolemaic Egypt, not only emerged as the chief mathematical center but also retained that status for the remainder of the Hellenistic and the entire Roman eras. Almost every noteworthy mathematician of both eras either studied, taught, or lived there. (The existence in some instances of gaps of over a century between leading names suggests that the tradition was not exactly continuous.)

Alexandria achieved and maintained its position as the mathematical center by depending upon kingly patronage and individual reputations. These in turn rested on the city's economic and intellectual strengths. While all Hellenistic empires produced surpluses of wealth, the Ptolemies ruled over the richest, containing the largest and most cosmopolitan city.[2] The perennial fertility of the Nile Valley, an industrious peasantry, and an effective royal monopoly over grain sales account for much of this wealth. As Aristotle had correctly noted in the *Politics* (1330a), the great cities of antiquity often lived at the expense of the peasants. Using the accumulated riches and the favorable location of Alexandria, the early Ptolemies oversaw the building of their capital city into the commercial and cultural hub of the ancient world. They fostered (and taxed) an increasing trade between Mediterranean Europe and the East. From the first a ruling elite of Macedonians and Greeks dominated commercial life with the support of thinly Hellenized Jewish merchants and native Egyptians. There was also a large number of slaves. Within the populous, polyglot city there was a transmission and occasional fusion of knowledge. Intellectual exchanges were enriched by trade with Phoenicians—and especially Babylonians, or Chaldaeans, who brought their religion and flourishing astronomy. At the least Alexandria required accountants and geometers to work with architect-engineers in the city's impressive building program.

In Greek Alexandria the schooling rested on the Athenian model provided by Plato in the *Republic* (especially Book VII), Aristotle in the *Politics* (1337b), and Isocrates in his essays on oratory. The first two authors had accorded a fundamental role to mathematics in higher learning and, indeed, by the third century B.C. Alexandria was truer to that tradition than Athens, whose Academy and Lyceum had turned more to philosophical speculation. The commercial prosperity of Alexandria, the intellectual cross-fertilization, and the Athenian model in schooling do not themselves explain its longstanding prominence in mathematics, however. To take but one example, Greek schooling spread throughout the Hellenistic and Roman worlds.

The early Ptolemies' keen patronage of literature, general scholarship, natural philosophy, and the visual arts was the most essential element in providing support for certain mathematicians who were mainly responsible for establishing Alexandrian mathematical primacy. By building and endowing the library and museum they outdistanced their rival patrons, the Attalids of Pergamum. Sometime after 295 B.C., Ptolemy I Soter (305–285 B.C.), like Alexander a former student of Aristotle, commissioned another Aristotelian Demetrius of Phalerum, to commence work on the library. The idea of libraries probably came from Assyria and Babylon because few Greek tyrants before Alexander's time collected books and from Aristotle, who amassed the first sizable private library. Whatever the motivations were, Demetrius planned the two institutions recruiting scholars in literary, medical, and scientific disciplines from across the Greek world.

Both the library and the museum were well-established in the Brucheion section in Alexandria during the reign of Ptolemy Euergetes (246–221 B.C.), who is said to have fancied himself a mathematician. By that time the library was part of the museum complex. The museum drew upon some Athenian traditions and added distinctive features. Like Plato's Academy and Aristotle's Lyceum, the museum housed a community of scholars who to an extent lived, worked, and ate together. It also included a group of aristocratic patrons and patronesses who joined in discussions and arguments. Unlike the Academy and Lyceum, the museum was primarily concerned with research rather than teaching, and it was entirely financed by the Ptolemies. The royal support was not merely financial; under the aegis of the early Ptolemies the members of the museum pursued their studies in a lively intellectual atmosphere largely free of political or religious direction. (Only satires against the king were not permitted.) The fortunes of museum and its members, therefore, fluctuated from one generation of the monarchy to another. The early Ptolemies provided a model for royal patronage of the sciences not unlike that adopted in the late-17th century by the French king Louis XIV, when he founded and fully financed the Paris Academy of Sciences.

The conjectured motives of the Ptolemies for being patrons of the sciences shed some light on the application and status of mathematics in late antiquity. The inventor Philo of Byzantium (*fl. c.* 250 B.C.) spoke to this issue in his textbook entitled *Mechanics.* In chapter three of the fourth book (On Catapults), Philo remarked that his application of mathematics to perfecting the weapons of war stemmed from the Ptolemies' interest in the technical arts. Such practical considerations seem the exception however. Philo and others refer more to the Ptolemies' love of fame, seeing their occasional subsidizing of prominent theoretical mathematicians as adding to the luster of their court. Royal patronage for whatever reason was especially welcome in a time when mathematics was not yet a profession offering regular opportunities for earning a livelihood. For the most part antiquity lacked a rationale that could change this situation, such as the belief that the systematic pursuit of mathematics and natural science is a key to material progress.

By the 3rd century B.C. mathematics had emerged as an independent discipline based on Aristotle's original systematic distinction between *mathematikē* and *physikē*. Plato's term *ta mathemata* ("learning") in the *Republic* had included both what we call mathematics and his studies of the Good. Still earlier the Pythagorean *mathematikoi* had studied an even broader range of subjects. Aristotle had organized mathematics around a primary abstract branch (arithmetic and plane and solid geometry) and a "lesser" physical branch (optics, music, mechanics, and astronomy). Afterwards mathematical practitioners increasingly pursued interests different from those of traditional philosophy, especially ethics but sometimes portions of natural philosophy as well. In essence they added the title of mathematician to that of philosopher.

In the brilliant late flowering of Classical Greek culture in the third century, there were correspondences between abstract mathematics, literature and the visual arts. In literature and mathematics there was a strong impulse to produce compilations of Classical sources. The resources of the Alexandrian library and museum made it a natural center for this activity. The Ptolemies, who favored literature in their patronage, had made their library the center of literary studies in the Mediterranean world. They outbid their rivals to buy the best of Classical Greek, ancient Egyptian, Hebrew, and Babylonian manuscripts; their holdings came to number some 700 thousand scrolls by the Roman period. The Ptolemies even instructed sea captains to seek out new books. At the library was a large staff of copyists as well as grammarians, lexicographers, and textual critics who took up the task of critically examining compilations of different handwritten versions of the same text. Their goal was to determine or restore originals. The many versions of Homer's *Iliad* and *Odyssey,* the continuing staples of Greek education, drew considerable attention. From previous comparative studies at the Lyceum and these at the library was born the science of philology. For their part, visual artists expanded the boundaries of their sense of beauty by gradually moving from the general or idealized representation of the 5th century B.C. to a more individual, emotional, and realistic representation. Similarly by addressing the aesthetic impulse, mathematicians pursued more harmonious patterns of mathematical ideas in their proofs and gained a deeper appreciation of the canons of beauty as expressed in geometrical shapes and musical laws.

The 3rd century burst of activity in abstract mathematics began with the work of Euclid (*c.* 330–*c.* 270 B.C.), the founder of the great mathematical tradition in Alexandria. His *Elements* with its thirteen "books"[3] treating plane geometry, number theory, and solid geometry was a definitive work of Alexandrian compilation assembling the mathematical findings of Plato's circle and their Hellenic predecessors. Euclid's compilation was much more than the sum of its parts and was a source of at least three major innovations in method and proof theory. Since the subsequent biography of Euclid (*see* pp. 105–107) adumbrates the subjects, sources, strengths, and weaknesses of the *Elements,* this account will concentrate upon three innovations.

Building upon Aristotle's theory of statements and probably Eudoxus' application of it to mathematics, Euclid brought the axiomatic-deductive method of demonstrative geometry to a new standard of mathematical rigor and elegance. Proceeding from starting points consisting of 23 (sometimes unsatisfactory) definitions, 5 postulates, and 5 axioms, he derived mostly by synthetic deductive reasoning an orderly chain of 467 propositions. In the *Elements* Euclid employed—and perhaps helped to create—high standards of consistency in the demonstration of propositions, displayed an austere economy in basing proofs on a near minimum of assumptions (as Aristotle had urged in *Posterior Analytics*), systematically arranged propositions in a natural progression, and treated each detail with great power. The axiomatic-deductive method had by then reached a stage that allowed geometers to secure geometric exactness and abstract mathematical certainty. Its proofs stood in contradistinction to the probabilistic nature of conclusions in pre-Eudoxian mathematics and the simply persuasive nature of the dialectic in rhetoric and philosophy. Thus, a goal which Plato had set earlier was achieved.

There were two other innovations in the *Elements.* Euclid was probably the author of the ingenious proof of Pythagoras' theorem given as I.47 and a more general, though not complete, proof in VI.31. Pythagoras' theorem set the metric for Euclidean geometry that may be expressed in modern notation as $ds^2 = dx^2 + dy^2$. The final innovation was Euclid's new theory of parallels. By a brilliant stroke, he based it on a postulate, the fifth, rather than a theorem and made the essential characteristic of parallelism *nonsecancy* or *nonmeeting* rather than equidistance.

Through the centuries geometers would patiently study the *Elements* and elaborate upon it. From the start they concentrated on a single problem—the question of whether the parallel postulate was independent. Not only did this pose an immediate foundational problem but eventually raised questions about the concepts, methods, and physical application of geometry. In regard to foundations, axiomatic systems must possess three properties: completeness (no fundamental assumptions are lacking), consistency (no internal contradictions are possible), and independence (no postulate can be proved from the others). Perhaps because of Euclid's proof of the converse of the parallel postulate in I.27, many ancients believed that the parallel postulate too might be dependent, that is to say, it was not a self standing assumption but a theorem requiring proof. Ptolemy and Proclus strongly supported this position but never satisfactorily established it. Later al-Khayyāmī and Nasir-al-Din al-Tusi imaginatively connnected the parallel postulate with its consequences regarding the sum of the angles in a quadrilateral and triangle respectively.

The challenged status of the parallel postulate remained a "blot on geometry" into modern times. In the late-17th century, John Wallis and Giuseppe Vitale attempted, as the ancients had, to deduce the fifth postulate from the rest of Euclid. In the early-18th century, Girolamo Saccheri argued *reductio ad absurdum* that a denial of the fifth postulate leads to a contradiction in Euclid. Johann Lambert, following Saccheri's lead, investigated the postulate's consequences for angle sums in triangles in his *Theory of Parallels* (publ. posth., 1786). The young Kant, in turn, followed Lambert. He speculated that there must be another consistent geometry besides Euclid's that applied to the physical universe but soon abandoned this position. The problem of parallels was finally resolved in the 19th century with the invention of non-Euclidean geometries by Carl Gauss, Nikolai Lobachevsky, Janos Bolyai, and Bernhard Riemann. Their work demonstrated that substituting a contrary for the fifth postulate (namely, that there are either numerous parallels to a given line or none) still produced a consistent geometry (Eugenio Beltrami).

Soon after the invention of rich alternatives to Euclid broke the rigidity in geometry, other fundamental developments occurred in the discipline. In the early 20th century, David Hilbert, the leading champion of axiomatic systems, showed that Euclid's system was not complete. In his proofs Euclid had appealed in a few instances to mathematical intuition as well as deductive reason. The use of pseudo-Riemannian geometry in Einstein's general relativity was important in the development of a powerful differential geometry with its manifolds, tensors, and geodesics. Moreover, Einstein's relativity affirms that differential geometry with the curvature tensor describes the large space of the universe more accurately than does Euclid's plane geometry.

There was more to expand upon in Euclid's *Elements* than the theory of parallels and axiomatics. Modern mathematicians, for example, have reserved their highest praise for the rigor and subtlety of the following two books: the fifth on Eudoxus' general theory of proportions which elegantly covers commensurables and incommensurables and the tenth on Theaetetus' theory of irrationals—and have appealed to them in refining number theory. In the 3rd century B.C., two other topics experienced notable developments. The method of exhaustion was advanced beyond its embryonic stage in Book XII, while the theory of divisibility, especially as it appears in Books VII and IX, led to advances in the formation of number theory. Euclid's theory of divisibility rests on his algorithm (VII. 1-3), which covers the essential conditions for the divisibility of integers, and primes, those numbers divisible only by 1 and themselves. The boldest proposition concerning primes, IX.20, holds in modern terms that the number of primes is infinite.

Shortly after Euclid came Archimedes (c. 287–212 B.C.), the most original and profound mathematician of antiquity. He was also a talented engineer and perhaps the greatest physicist of his age. Like Newton, he possessed genius in both theory and experiment. The interaction between these two aspects of his genius was perhaps the key to his mathematical discoveries. Like most mathematicians of his day Archimedes visited Alexandria, but only briefly. He stayed in Syracuse, the largest Greek city in Sicily, writing his precise, lucid treatises in the Sicilian-Doric dialect. These treatises may be loosely grouped as follows—I (Areas and volumes of figures bounded by curved lines and surfaces): *On the Quadrature of the Parabola, On the Sphere and Cylinder, On Spirals, On Conoids and Spheroids,* and *On the*

Measurement of the Circle—II (Geometry applied to statical and hydrostatical problems): *On the Equilibrium of Plane Figures* (Books I and II), *On the Method of Mechanical Theorems,* and *On Floating Bodies*—and III (Miscellaneous): *The Sandreckoner, The Cattle Problem,* and a fragment of *Stomachion.*

As the first two groups of writings demonstrate, Archimedes principally studied geometry, wherein he concentrated on making the method of exhaustion into a powerful mathematical technique that had wide application. Taking *Elements* X.1 (concerning the infinite subdivision of a line and the basis of Eudoxus–Euclid's embryonic method of exhaustion), he perceptively made it into a postulate which he stated as assumption 5 of *On the Sphere and Cylinder.* Building upon this postulate Archimedes produced two types of the method of exhaustion that have been labelled "compression" and "approximation" methods by the historian E. J. Dijksterhuis. Archimedes almost exclusively used the "compression" method which he divided into two forms—decreasing differences and decreasing ratios. Both forms of "compression" depend upon the successive inscribing and circumscribing of regular polygons. In each, the area or volume of a curvilinear figure was compressed between the inscribed and circumscribed polygons. Examples of the decreasing difference form occur in the short treatise entitled *On the Measurement of the Circle* (Proposition 1 and 3), while the ratio form $\frac{\text{circumscribed polygon}}{\text{inscribed polygon}}$ occurs in book 1 of *On the Sphere and Cylinder* (Propositions 13, 14, 33, 34, 42 and 44). Archimedes employs the separate "approximation" method on only one occasion—in the parabola treatise (Proposition 18–24). Proposition 23 shows that he derived the elementary identity $1 + \frac{1}{4} + (\frac{1}{4})^2 + \ldots + (\frac{1}{4})^n + \frac{1}{3} \cdot (\frac{1}{4})^n = \frac{4}{3}$ preparatory to showing that any segment of a parabola equals $\frac{4}{3}$ times a triangle of the same base and equal height. As with all other exhaustion demonstrations, he proves proposition 23 by a double *reductio ad absurdum* argument. His discovery of this theorem, however, was made in a strikingly original way. He imagined an abstract mental balance and an ideal center of gravity, mechanical constructs ingeniously used to solve geometrical problems—a procedure he also embraced in *On the Equilibrium of Plane Figures* and described in *On the Method. On the Method* details his heuristic infinitesimal technique for determining unknown volumes and areas based on placing abstract geometric magnitude, "indivisible" slices, at a known distance from a fulcrum on a given side of a weightless lever and then applying the law of the lever to attain a balance.

Although his method of exhaustion is impressive, it is a mistake to believe that Archimedes had discovered the infinitesimal calculus. While his method of compression gives the least upper bound *(supremum)* and greatest lower bound *(infimum),* nowhere does he explicitly introduce limit concepts; and he avoids the use of infinite series. Nor does he recognize the inverse relationship between area and tangent problems, both of which he examined. Finally, there is no general computational algorithm for the curvilinear volumes and areas, the center of gravity of the triangle, and the area enclosed by his spiral. He failed to establish a connection among these problems, namely in modern notation that all of them depend upon the integral $\int x^2 dx$. Dependence upon geometrical algebra, the lack of an adequate number theory, the power of the deductive method, and the absence of symbolic notation all prevented Archimedes from developing general procedures and exploiting analogies in the area of the calculus.

Among Archimedes' other mathematical accomplishments were his work on curves and the extension of numerical calculation. The ancient Greeks were only interested in a few curves and these were usually static depending upon uniform linear or uniform circular motion. Perhaps from his study of a parallelogram of velocities applied to tangents, Archimedes discovered his spiral, a transcendental curve composed of both types of uniform motion. In terms of modern polar coordinates, the equation for his spiral is $r = a\Theta$. The pursuit of precise numerical computation pervades Archimedes' mathematical studies. A case in point is the famous proposition 3 of *On the Measurement of the Circle,* where he first establishes explicitly an accurate inequality for the ratio of the circumference of a circle to its diameter, a ratio today called π. Working with regular 96-gons, Archimedes first found by manual calculation that $\frac{265}{153} < \sqrt{3} < \frac{1351}{780}$ and next that $\frac{6336}{2017\frac{1}{4}} < \pi < \frac{14688}{4673\frac{1}{2}}$ which he rounded to $3\frac{10}{71} < \pi < 3\frac{1}{7}$. He thus recognized the importance of small-scale precision. As the historian W. R. Knorr has argued, Archimedes probably obtained even more accurate approximations to π by following a compression procedure that began with the inscription and circumscription of decagons in and about a circle and ended with regular polygons of 640 sides.

One of Archimedes' correspondents was Eratosthenes of Cyrene (c. 276–c. 195 B.C.), the chief Librarian of Alexandria. A polymath whose breadth of learning in such fields as historical criticism, chronology, philosophy, and poetry almost rivalled Aristotle's, Eratosthenes gained the nickname *Beta* (Number Two), which perhaps meant that a poll of scholars would accord him the "vote of Themistocles" in every branch of knowledge.[4] He is best known for founding mathematical (as opposed to merely descriptive) geography and for his accurate calculation of the circumference of the earth (*see* pp. 142–147). In pure mathematics he developed a sieve to find prime numbers. According to Euclid IX.20, the number of primes is infinite and thus no complete catalogue of them is possible. This did not prevent mathematicians from seeking a catalogue beginning with Eratosthenes. His sieve involves a simple procedure: writing out the positive integers and then crossing out all multiples of 2, 3, 5, 7, etc. until only the primes remain. His sieve identified the primes but did not provide a simple rule to determine how often they occurred. That rule, the prime number theorem, was not discovered until the mid-18th century by Leonhard Euler. Euler's prime number theorem indicates that the probability of a large number n being prime is about $1/\log n$.

The last great geometer of antiquity was Apollonius of Perga (c. 246–c. 174 B.C.). Visiting Pergamum and living most of his adult life in Alexandria, Apollonius had access to the two chief libraries of Hellenistic times. King Attalus I (241–197 B.C.) of Pergamum had tried to turn his kingdom into a second Athens. He commissioned a large library and attracted Athenian scholars to his court. Close contact with a major library such as Pergamum's, was important for all leading Hellenistic mathematicians. Perhaps it was at the court of Attalus that Apollonius held discussions preparatory to his writing the *Conics,* a seminal work in mathematics.

The *Conics,* the first known exhaustive monograph on a specific mathematical subject, consists of eight books. In preparing the first four, Apollonius must have drawn upon the corpus of elementary theorems on conic sections, including that given in Euclid's lost *Conics.* The last four books, of which the eighth is lost, are more advanced. Of these Book V on normals to curves, when drawn as maxima or minima, has evoked the most admiration. In striving for generality, Apollonius proved innovative in method. Instead of producing three conic "sections" by cutting a cone orthogonally with a right, acute, or obtuse angle, he adopted the radically different and more general approach of generating all three curves from the double oblique circular cone. Using this approach he found an ellipse (hyperbola) to be the locus of a moving point P in such a way that the sum (difference) of its distance from two given points, the *foci,* remains constant. He thus determined the three conic curves by an appeal to the "method of application of areas." From the application of areas he coined the terms parabola, hyperbola, and ellipse for these three curves.

Apollonius contributed to every area of pure and applied mathematics, including astronomy. (For his improved inequality for π, work on large numbers, and new geometrical models in astronomy, *see* pp. 148–154.)

By the death of Apollonius an intense period of mathematical achievement appears to have ended. Later antiquity saw mathematics develop only sporadically. The connections between theoretical speculation and general applications that had proved so fruitful for both were reduced, if not lost. Mathematics was more and more focused on limited applications in surveying, geography, and astronomy. The next mathematician of note was Hero of Alexandria. At the Museum he taught computational geometry, arithmetic, geodesy, physics, mechanics, and pneumatics (the theory and use of air pressure) writing textbooks on each of these subjects. His books, which comprise two categories—technical and mathematical—differ in style from the *Pneumatica* with its concise, clear technical descriptions to the *Mechanica* and *Dioptra* with their discursive style and theoretical component. These books reveal that their author had a practical purpose in seeking to test new effects rather than to confirm or refute physical theories. The author emerges at the least as a clever inventor whose new apparatus possibly included an improved dioptra for surveyors, a screw cutter, a simple steam engine, war engines, and parlor magic toys, such as trick jars and puppets that moved when a fire was lit. Hero's major mathematical books were *Definitiones, Metrica,* and *Geometrica.* In these and *Dioptra* he drew upon the mathematical lore of ancient Egypt and Babylon to give prescriptions to find geometrical areas that were of use to surveyors, masons, and carpenters. The *Dioptra* contains Hero's prescription for finding the area of a triangle which in modern symbols is $A = \sqrt{s(s - a)(s - b)(s - c)}$, where the semiperimeter $s = \dfrac{a + b + c}{2}$. (According to the Arab al-Biruni, Archimedes knew this rule earlier.)

The emergence of trigonometry was the chief mathematical development of late antiquity. Evolving as a means by which Hellenistic astronomers might "save the appearances," that is, predict planetary positions irrespective of a physical mechanism, trigonometry at first belonged to astronomy rather than to mathematics. Its methods and proofs, therefore, while more stringent than those of contemporary philosophy did not have the rigor of geometry. The three astronomers primarily responsible for founding trigonometry are Hipparchus (*fl.* mid-2nd century B.C.), Menelaus (late-1st century Christian era), and Ptolemy (*c.* A.D. 100–178), who brought it to a mature stage in the *Syntaxis Mathematica* (the *Almagest*).

We are poorly informed about the exact steps in the development of trigonometry prior to Ptolemy because most of the writings of the other two men are lost. In his studies Hipparchus employed numerical methods from Babylonian astronomy and probably adopted Apollonius' geometrical model of epicycles and eccentric circles (to explain the motion of the sun, moon, and planets). Building upon this base, Hipparchus began to transform astronomy from a descriptive qualitative to a predictive quantitative science. As an aid to prediction, he prepared the first table of chords subtended by arcs of a circle. In effect, these chords were the modern sine function. Starting with a circle divided into 360° in the Babylonian manner, Hipparchus used linear interpolation to compute chords at 7½° intervals.

This was an embryonic stage of plane trigonometry. Menelaus of Alexandria subsequently prepared a table of chords, now lost, and wrote the three books of his *Sphaerica,* wherein he founded the more difficult spherical trigonometry to handle great circle arcs and launched the study of trigonometry as an independent discipline. His *Sphaerica* has survived only in an Arabic translation. Its third book opens with the transversal proposition known as "Menelaus theorem," which Ptolemy established as being fundamental for sphaerical astronomy. With Ptolemy's *Almagest* trigonometry attained the form that it kept for almost 1500 years. The *Almagest* presented his geocentric system of deferent circles and epicycles—a model not abandoned until the work of Copernicus and Kepler. Ptolemy mostly developed the plane trigonometry required to compute the positions of celestial bodies in the first two books of the *Almagest.* (For his table of chords and computation of irrational numbers, *see* pp. 155–159.) He also presented what is now known as "Ptolemy's theorem," which allows one to find chord $(\alpha - \beta)$, when given chords α and β. (In modern terms, chord α is roughly equivalent to sin α.)

By the 2nd century B.C., a major shift in the political, economic, and social life of the Mediterranean world had influenced the course of learning, including a diminution in the pursuit of theoretical mathematics. Governmental corruption and cycles of famine in the Hellenistic era helped ease the way for a new power, Rome, to gain prominence. Rome, founded (according to legend) in 753 B.C., had survived centuries of peninsular struggles and decades of civil strife before it conquered Carthage in three Punic Wars ending in 146 B.C. and then rapidly extended its power through the Eastern Mediterranean. Macedon and Pergamum were incorporated as provinces by 129 B.C., only Ptolemaic Egypt briefly remained outside its control. With a society dependent on war, Rome became in one sense a vast parasite whose inhabitants lived from plunder and taxes. Without a sound economic base Rome lagged while the Mediterranean fringes prospered. In the city a large impoverished mob grew in size while on the large agricultural estates *(latifundia)* slavery increased with the seizure of war captives. Operating within this changing urban setting, the Romans attained a relatively brief high point in their culture during the 1st century B.C. (the end of the Republic and the Principate of Augustus). They were, to be sure, influenced by the Classical Greek and the Hellenistic culture but offered a different spirit and emphasis to suit their ethos founded in duty, prudence, and tradition. Their cultural peak rests especially on two disciplines—the law, particularly as expounded in the orations and treatises of Cicero and a literature that flowered in the poetry of Horace and Vergil, two clients of Augustus, and Ovid, whom Augustus exiled because his love elegies suggested loose sexual codes. The chief expression of the cultural peak in the visual arts was the massive architecture–engineering projects that constructed the Pantheon, Colosseum, and Ara Pacis (Altar of Peace).

Absorbed in law, government, and military conquest, the Romans saw no value in mathematical theory and encouraged only a limited range of its practical applications. One result was that the Romans did not produce one eminent mathematician in their thousand year history. Even after they imported Greek teachers, many Romans remained ambivalent to Classical Greek and early Hellenistic learning directing their antipathy especially to theory. Roman society was based on the family and the education of its upper class on rhetoric, literature and *sapientia,* the utilitarian, more than *philosophia.* Among the Romans Greek *logos* became *ratio.* This applied to mathematics as well as philosophy. Cicero succinctly contrasted the Greek and Roman views of mathematics at the start of his *Tusculan Disputations* (I, ii):

> With the Greeks (demonstrative) geometry was regarded with the utmost respect, and consequently none were held in greater honour than mathematicians, but we Romans have restricted this art to the practical purposes of measuring and reckoning.

The measuring was primarily associated with land surveys and vast engineering projects, such as roads, aqueducts, and public buildings, while the reckoning entailed the Roman numerals that lacked place-value notation as well as the more efficient abacus in different forms. Presumably, the limited mathematical residue from Classical Greece and early Hellenistic times sufficed to solve the problems that the Romans faced.[5]

The 1st century B.C., which saw the Roman conquest of Egypt, brought changes that affected the course of mathematics. During a skirmish between the troops of Julius Caesar and local forces in Alexandria in 48 B.C., a fire destroyed much of the famous library but not the entire museum complex. While Cleopatra received the Pergamum library as a replacement, it was significantly housed in the daughter library at the temple of Serapis rather than at the museum. This relocation suggests that the power of religious officials had grown in Alexandria and indicates that a shift of interest toward religion and away from theoretical mathematics had occurred.

During the late-1st century B.C., the long embattled Roman Republic ended and the Empire emerged largely through the efforts of the shrewd Julius Caesar and his heir Octavian, called Augustus. Once in power, Augustus confiscated the Alexandrian treasury for his own, using it to support the visual arts and literature. While theoretical mathematics had been excluded from royal patronage probably since the later Ptolemies, the full loss became evident in Roman times. During Augustan times there was no small cadre of mathematical experts such as early Alexandria had. Without them the oral tradition, indispensable to train students in geometric methods and formulation of new problems, was lost. Moreover, the term mathematics came to mean something different in the Roman empire. As astrology became almost universal, the Latin term *mathematicii* came to mean astrologers, while the term geometer referred to a person whom we now call a mathematician. This new meaning was not without effect. No less a figure than Ptolemy was known equally for the *Tetrabiblos* on astrological influences as for the *Almagest.* The Roman distinction between mathematician and geometer lasted in a fashion until the 17th and 18th centuries.

The birth of Jesus of Nazareth (*c.* 6 B.C.) marks nearly the midpoint in the Religious Revolution stretching from Confucius and Buddha in the early-5th century B.C. to Mohammed in the 6th century A.D. The growth of Christianity, its superceding of pagan religions, and the response of early Church Fathers to Classical Greek and Alexandrian mathematics are treated in the next section of this anthology.

As indicated earlier, the development of mathematics during the late Roman imperial period centered in Alexandria, not in Rome. With a flourishing grain trade until the 4th century A.D. and a large library in the Serapeum, Alexandria still gave sporadic support to mathematicians. After the activities of Hero in the first century and Ptolemy in the second, the city witnessed a brief mathematical revival in the turbulent century beginning about 225 as a result of the accomplishments of Diophantus (*fl. c.* 250) and Pappus (*fl.* 300–350). Drawing upon materials from Classical Greece, early Alexandria, ancient Egypt, and Babylon Diophantus contributed to algebra and computational arithmetic in his *magnum opus, Arithmetica* in 13 books. In this unsystematic collection of 189 problems, he introduced abbreviations to assist in problem-solving and made original contributions to the indeterminate equations now bearing his name. (*See* pp. 160–165.) The chief writing of

Pappus, an accomplished geometer and astronomer, is the *Synagoge* or *Collection* in at least eight books. The *Collection,* essentially a handbook to be read with original sources, deals with attempts to square the circle and trisect an angle by Archimedean methods (Book I), isoperimetry (Book V), and geometric *loci* with respect to three or four lines—these can be depicted by conic sections—and to six or more lines—these depict the higher curve of a cubic equation (Book VII). Book VII is also important for its survey of many works by earlier geometers, such as Euclid and Apollonius, that are otherwise lost. With Diophantus and Pappus the evolution of original ideas in mathematics in antiquity essentially ended. They did not have worthy successors in pure mathematics until the 17th century, when Pierre Fermat substantially expanded upon propositions from the *Arithmetica* and René Descartes from the *Collection.*

The history of mathematics in antiquity is the history of the ideas, practical experiences, and social circumstances of a very few exceptional individuals. In 4th- and early 5th-century Alexandria, these individuals were teachers and commentators. Although 3rd-century Alexandria had original lines of mathematical inquiry—despite the war, malaria epidemics, and famine as well as corruption, intrigue, and religious turmoil that occurred within and about the city—the later period did not. The Alexandrian economy and society deteriorated as trade declined and religious hostilities increased. Particularly damaging was the sacking and burning of the over 300-thousand-volume library at the temple of Serapis in 392 when the Roman emperor Theodosius proscribed pagan religion. About this time the museum had two pagan members who were teacher–commentators in mathematics. They were Theon (*fl.* 350–400) and his learned daughter Hypatia (*c.* 370–415). Theon prepared an extensive commentary on Ptolemy's *Almagest* replete with sexagesimal calculations and a famous recension of Euclid's *Elements.* Besides writing a commentary on Diophantus' *Arithmetica* and another on Apollonius' *Conics* (both no longer extant), Hypatia was a Neoplatonic leader and a friend of the Roman prefect Orestes, an enemy of the local Christian bishop. Probably for the last two activities she was murdered by a mob of fanatic Christians.

NOTES

1. There were four of these ruled by the Antagonids in Macedon, the Ptolemies in Egypt, the Seleucids in Asia Minor and Persia, and the Attalids in Pergamum.

2. Within a century of its founding Alexandria had a population in the hundreds of thousands. By the time of Caesar Augustus it had over 600,000 and perhaps as many as 1,000,000 residents.

3. The Hellenistic meaning of the word book is often not understood. In those times scribes wrote on rolls of papyrus. A standard roll consisted of twenty sheets glued together to form a strip from 15 to 20 feet long. Works longer than a standard roll were divided into "books," with one book consisting of a cylinder about 6 inches in diameter and containing the equivalent of from ten to twenty thousand words of modern English text. These books were kept in pigeonholes at libraries with a wooden tag on the outer end.

4. The Greek defeat of the Persians in the sea battle at Salamis depended greatly upon the stratagems, fortitude, and valor of the Athenean leader Themistocles, who in turn received wise counsel from Aristides. After the victory the Greek generals returned to the Isthmus to vote on who was the bravest in battle. In the casting of ballots "each voted for himself as the most valorous and for Themistocles as the second." *See* Plutarch, *Parallel Lives:* "Themistocles" (Sections 12–17).

5. Aesthetics produced a rare instance of Roman originality subsequently of importance to theoretical mathematics. A Roman mosaic of the third century contains a coiled ribbon that may be described as a Möbius band. In its simplest form a Möbius band is a strip of paper with a half twist (180°) and joined at the ends. Possessed of only one side and one edge, this band was to be important in modern topology. *See* Lorraine L. Larison, "The Möbius Band in Roman Mosaics," in *American Scientist* (September–October 1973), 544-547.

SUGGESTIONS FOR FURTHER READING

Asger Aaboe, *Episodes from the Early History of Mathematics.* New York: Random House and the L. W. Singer Company, 1964.

Apollonius of Perga, *Conics,* 3 vols., translated by R. Catesby Taliaferro. Annapolis, Md.: Classics of the St. Johns Program, 1939.

Lucas N. H. Bunt, Philip S. Jones, Jack D. Bedient, *The Historical Roots of Elementary Mathematics.* Englewood Cliffs, N.J.: Prentice-Hall, 1976.

Marshall Clagett, *Archimedes in the Middle Ages.* Madison, Wisc.: University of Wisconsin Press, 1964. Vol. I.

_________, *Greek Science in Antiquity.* New York: Abelard Schuman, 1959; 3rd Edition, Collier, 1969.

J. L. Coolidge, *History of Conic Sections and Quadric Surfaces.* New York: Oxford University Press, 1945.

_________, *A History of Geometrical Methods.* New York: Dover reprint, 1963.

E. J. Dijksterhuis, *Archimedes.* New York: Humanities Press, 1957.

Howard Eves, *A Survey of Geometry.* Boston: Allyn and Bacon, 1963; rev. ed., 1972.

S. H. Gould, "The Method of Archimedes," *American Mathematical Monthly,* 62 (1955), 473–476.

Jeremy Gray, *Ideas of Space.* Oxford: Clarendon Press, 1979.

T. L. Heath, *Apollonius of Perga, Treatise on Conic Sections.* New York: Barnes and Noble, 1961.

_________, *Diophantus of Alexandria.* New York: Cambridge University Press, 1910.

_________, *A History of Greek Mathematics.* New York: Oxford University Press, 1921.

_________, *A Manual of Greek Mathematics.* New York: Oxford University Press, 1931.

_________, *The Thirteen Books of Euclid's Elements.* New York: Cambridge University Press, 1926.

_________, *The Works of Archimedes.* New York: Cambridge University Press, 1897.

J. Hintikka and U. Remes, *The Method of Analysis: Its Geometrical Origin and its General Significance.* Dordrecht: Reidel, 1974.

W. R. Knorr, "Archimedes and the Measurement of the Circle: A New Interpretation," *Archive for History of Exact Sciences,* 15 (1976), 115–140.

_________, "Archimedes and the Elements: Proposal for a Revised Chronological Ordering of the Archimedean Corpus," *Archive for History of Exact Sciences,* 19(1978), 211–290.

G. E. R. Lloyd, *Greek Science After Aristotle.* New York: W. W. Norton & Company Inc., 1973.

Michael Moffatt, *The Ages of Mathematics.* Garden City, N.Y.: Doubleday & Company, Inc. 1977. Vol. I.

Ian Mueller, *Philosophy of Mathematics and Deductive Structure in Euclid's Elements.* Cambridge, Mass.: M.I.T. Press, 1981.

Otto Neugebauer, *A History of Ancient Mathematical Astronomy.* New York: Springer-Verlag, 1975. 3 volumes.

Edward Alexander Parsons, *The Alexandrian Library.* New York: Elsevier, 1952.

Olaf Pederson, *A Survey of the Almagest.* Copenhagen: Odense University Press, 1974. *Acta Historica Scientiarum Naturalium et Medicinalium,* Volume 30.

Proclus, *A Commentary on the First Book of Euclid's Elements,* translated by Glenn R. Morrow. Princeton: Princeton University Press, 1970.

A. Wayne Roberts and Dale E. Varberg, *Faces of Mathematics.* New York: Harper & Row, 1978.

George Sarton, *A History of Science.* Cambridge, Mass.: Harvard University Press, 1952 and 1959. 2 volumes.

W. H. Stahl, *Roman Science.* Madison, Wisc.: University of Wisconsin Press, 1962.

Noel M. Swerdlow, "Ptolemy on Trial," *The American Scholar* (1979), 523–531.

Ivor Thomas (ed.), *Selections Illustrating the History of Greek Mathematics.* Cambridge, Mass.: Harvard University Press, 1939–41. 2 volumes.

G. J. Toomer (ed. and trans.), *Diocles on Burning Mirrors.* New York: Springer-Verlag, 1976.

B. L. van der Waerden, *Science Awakening.* Groningen: P. Noordhoff, 1954.

Paul Ver Eecke (ed.), *Pappus d'Alexandrie: La Collection Mathématique.* Paris: Albert Blanchard, 1933.

Chapter III
The Hellenistic and Roman Periods

EUCLID (*c.* 330–270 B.C.)

Only two things are known about the life of Euclid, the most celebrated geometer of all time. If he was younger than the first pupils of Plato but older than Archimedes, which there is good reason to believe, then he flourished about 295 B.C. Secondly, Euclid taught at the museum, or "temple of the Muses," in Alexandria.[1] The Aristotelian Demetrius of Phalerum had proposed the building of the museum and the great library in Alexandria to Ptolemy I (Soter), who reigned from 305 to 285 B.C. Once construction of these institutions was begun, Demetrius probably invited Euclid to join a group of eminent scholars working at the museum. The museum was a research institute stressing scientific and literary studies. Based on the facts that Demetrius was an exile from Athens and the materials in Euclid's *Elements* are heavily indebted to the work of Plato's circle, it seems likely that Euclid had received his mathematical education in Athens at the Platonic Academy before coming to Alexandria. In his commentary on the first book of the *Elements*, Proclus conjectured that Euclid "was a Platonist," because he encouraged the study of mathematics and ended the *Elements* with the five regular, or "Platonic," solids. The testimony of so ardent a Neoplatonist as Proclus is not conclusive on this point, however.

There are a number of anecdotes about Euclid. In one, Ptolemy Soter asked him if there was no shorter way to the study of geometry than the *Elements*. Euclid replied, "there is no royal road to geometry." A second story, reported by Stobaeus, concerns a pupil who asked Euclid at the end of his first lesson in geometry what gain he would get by learning such things, whereupon Euclid summoned a slave and exclaimed, "Give him three obols (coins) since he must needs gain out of what he learns." Euclid is thought to have been a well-disposed and exacting scholar who did not vaunt his knowledge. Pappus, the 4th-century-A.D. geometer, praised him for his "scrupulous fairness and his exemplary kindness toward all who advance mathematical science to however small an extent."

Euclid's fame rests almost exclusively upon his great work, the *Elements (Stoichia)*, which consists of 13 books. The *Elements* became a classic soon after its publication. Archimedes, for example, refers to it as the standard textbook of mathematics, which it was for over 2,000 years. As a result, Euclid's name became a synonym for geometry until the 20th century. The *Elements* has had an immense influence upon the western mind. No other single scientific, philosophical, or literary book—except for the Bible—has had so many translations, editions, and commentaries. In the sciences its geometrical conception of mathematics greatly influenced Medieval

THIRTEEN BOOKS OF THE *ELEMENTS**

	Book	Subject
Plane Geometry	I	The Geometry of Straight Lines and Plane Rectilinear Figures
	II	Transformation of Areas
	III-IV	Major Propositions about Circles
	V	The Theory of Proportion applied to Commensurable and Incommensurable Magnitudes.
	VI	The Application of this General Theory of Proportion to Similar Figures
Arithmetic	VII	Pythagorean Theory of Numbers
	VIII	Series of Numbers in Continued Proportion
	IX	Miscellany on the Theory of Numbers, including Products and Primes.
Plane Geometry	X	The Classification of Incommensurables (Irrational Magnitudes)
Solid Geometry	XI	The Geometry of Three Dimensions, particularly Parallelepipeds
	XII	The Method of Exhaustion
	XIII	The Inscription of the Five Regular Polyhedra in a Sphere

*Some old editions of the *Elements* contain two more books with additional results on regular solids. Both post-date Euclid, however, Book XIV was written by Hypsicles (*fl. c.* 150 B.C.), while parts of Book XV may date from the 6th century A.D.

natural philosophy as well as Newton's *Principia* (1687) and Kant's *Critique of Pure Reason* (1781).

The *Elements* was on the whole the culmination of the classical Greek tradition in theoretical mathematics. In this work Euclid compiled and codified materials from his Hellenic predecessors, particularly the Pythagoreans, Hippocrates of Chios, Theaetetus, Plato, Eudoxus, and Aristotle. His text shows him to be a talented editor but the quality of his exposition is uneven. In those books where he has excellent sources, he excels. This is true for Book V, which is based on Eudoxus, and Book X, which is based on Theaetetus. In Book VIII, however, whose source is the early Pythagoreans, there are cumbrous enunciations, some repetition, and even logical fallacies. For critics, this raises questions about Euclid's ability as a geometer. Another, and perhaps equally plausible, interpretation is that he wanted to include the traditional teaching of arithmetic in Books VII through IX with few revisions.

Euclid was not merely a compiler of the *Elements;* he also made original contributions to geometry. The most important was his refinement of the axiomatic method with its notion of proof and a strictly logical ordering of theorems. From explicitly stated (although sometimes vague) definitions as well as 5 axioms (common notions) and 5 postulates, he derived by deductive argument an orderly chain of 467 theorems. Displaying, the austerity that he was to make canonical in mathematics, he presented a minimum of assumptions to achieve these proofs, employing very little that was superfluous, and he proceeded from the simple to the more complex in a natural progression. Put another way, his disposition and arrangement of materials and the power with which he treats each detail are remarkable. Euclid is generally credited with two other innovations. He probably developed the ingenious proof of proposition I.47, the Pythagorean theorem, and a more general, although not complete, proof of this theorem in VI.31. He also probably developed the theory of parallels in the *Elements*, because Aristotle spoke of the failure to find such a theory in his day. Euclid brilliantly chose nonsecancy rather than equidistance between straight lines as the test of parallelism.

The *Elements* was immensely successful. Its widespread adoption coupled with the destruction of ancient libraries led to the disappearance of all previous compilations on theoretical mathematics. Not even its immediate predecessor by Theudius of Magnesia survived. With the loss of these compilations and original source works, the *Elements* became the first major source of mathematical knowledge in the western world. Moreover, until the 19th century mathematicians believed that there could be no other consistent geometry that applied to our physical world besides Euclid's.[2]

Euclid wrote on all branches of mathematics known in his time. His writings include *Data, On the Division of Figures, Conics* (which perhaps served as the basis for the first three books of Apollonius' *Conic Sections*), *Porisms, Fallacies* (Pseudaria), *Phaenomena* (a text on astronomy), *Catoptrics* (a text on mirrors), *Elements of Music*, a lost book *On the Balance*, and possibly *Book on the Heavy and the Light* (a text on dynamics).

The Arabs believed that the name Euclid came from *ucli* (key) and *dis* (measure) and thus disclosed the "key of geometry." They also asserted that the words "Let no one come to our school who has not learned the *Elements* of Euclid" were posted over the door of every Greek school. This notice modified the motto above the gate to Plato's Academy by replacing the word "geometry" with the words "the *Elements* of Euclid."

NOTE

1. Euclid is often regarded as the founder of a school of mathematics at Alexandria which was unrivaled in antiquity. This claim cannot at present be authenticated because no germane source materials have survived.
2. During that century Gauss, Lobachevsky, Bolyai, and Riemann invented non-Euclidean geometries by departing from the fifth, or parallel, postulate and the movement "away from Euclid" began.

27. From Book I of the *Elements:* Definitions, Postulates, Axioms, and Propositions 1-13*

EUCLID

DEFINITIONS

1. A **point** is that which has no part.
2. A **line** is breadthless length.
3. The extremities of a line are points.
4. A **straight line** is a line which lies evenly with the points on itself.
5. A **surface** is that which has length and breadth only.
6. The extremities of a surface are lines.
7. A **plane surface** is a surface which lies evenly with the straight lines on itself.
8. A **plane angle** is the inclination to one another of two lines in a plane which meet one another and do not lie in a straight line.
9. And when the lines containing the angle are straight, the angle is called **rectilineal**.
10. When a straight line set up on a straight line makes the adjacent angles equal to one another, each of the equal angles is **right**, and the straight line standing on the other is called a **perpendicular** to that on which it stands.
11. An **obtuse angle** is an angle greater than a right angle.
12. An **acute angle** is an angle less than a right angle.
13. A **boundary** is that which is an extremity of anything.
14. A **figure** is that which is contained by any boundary or boundaries.
15. A **circle** is a plane figure contained by one line such that all the straight lines falling upon it from one point among those lying within the figure are equal to one another;
16. And the point is called the **center** of the circle.
17. A **diameter** of the circle is any straight line drawn through the center and terminated in both directions by the circumference of the circle, and such a straight line also bisects the circle.
18. A **semicircle** is the figure contained by the diameter and the circumference cut off by it. And the center of the semicircle is the same as that of the circle.
19. **Rectilineal figures** are those which are contained by straight lines, **trilateral** figures being those contained by three, **quadrilateral** those contained by four, and **multilateral** those contained by more than four straight lines.
20. Of trilateral figures, an **equilateral triangle** is that which has its three sides equal, an **isosceles triangle** that which has two of its sides alone equal, and a **scalene triangle** that which has its three sides unequal.
21. Further, of trilateral figures, a **right-angled triangle** is that which has a right angle, an **obtuse-angled triangle** that which has an obtuse angle, and an **acute-angled triangle** that which has its three angles acute.
22. Of quadrilateral figures, a **square** is that which is both equilateral and right-angled; an **oblong** that which is

**Source:* Reprinted by permission from *The Thirteen Books of Euclid's Elements* trans. by Sir. Thomas L. Heath (1956 edition), 150-153 and 240-276 with notes deleted. Reprinted by permission of Cambridge University Press.

right-angled but not equilateral; a **rhombus** that which is equilateral but not right-angled; and a **rhomboid** that which has its opposite sides and angles equal to one another but is neither equilateral nor right-angled. And let quadrilaterals other than these be called **trapezia**.

23. **Parallel** straight lines are straight lines which, being in the same plane and being produced indefinitely in both directions, do not meet one another in either direction.

POSTULATES

Let the following be postulated:

1. To draw a straight line from any point to any point.
2. To produce a finite straight line continuously in a straight line.
3. To describe a circle with any center and distance.
4. That all right angles are equal to one another.
5. That, if a straight line falling on two straight lines make the interior angles on the same side less than two right angles, the two straight lines, if produced indefinitely, meet on that side on which are the angles less than the two right angles.

COMMON NOTIONS

1. Things which are equal to the same thing are also equal to one another.
2. If equals be added to equals, the wholes are equal.
3. If equals be subtracted from equals, the remainders are equal.

[7] 4. Things which coincide with one another are equal to one another.

[8] 5. The whole is greater than the part.

PROPOSITIONS

PROPOSITION 1

On a given finite straight line to construct an equilateral triangle.

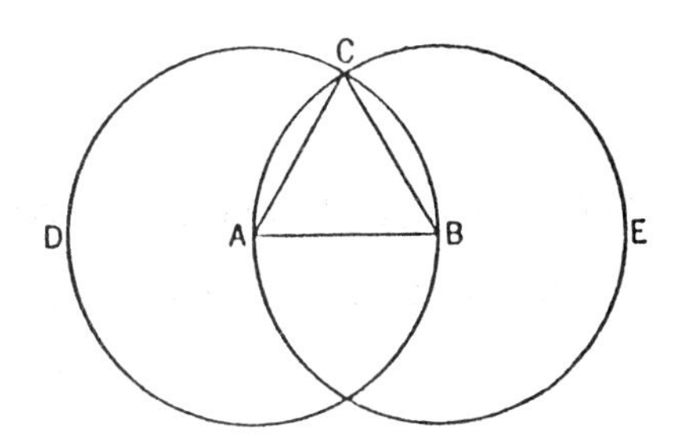

Let *AB* be the given finite straight line.

Thus it is required to construct an equilateral triangle on the straight line *AB*.

With center *A* and distance *AB* let the circle *BCD* be described [Post. 3]; again, with center *B* and distance *BA* let the circle *ACE* be described [Post. 3]; and from the point *C*, in which the circles cut one another, to the points *A, B* let the straight lines *CA, CB* be joined [Post. 1].

Now, since the point *A* is the center of the circle *CDB, AC* is equal to *AB* [Def. 15].

Again, since the point *B* is the center of the circle *CAE, BC* is equal to *BA* [Def. 15].

But *CA* was also proved equal to *AB;* therefore each of the straight lines *CA, CB* is equal to *AB*.

And things which are equal to the same thing are also equal to one another; therefore *CA* is also equal to *CB* [*C.N.* 1].

Therefore the three straight lines *CA, AB, BC* are equal to one another.

Therefore the triangle *ABC* is equilateral; and it has been constructed on the given finite straight line *AB*. (Being) what it was required to do.

PROPOSITION 2

To place at a given point (as an extremity) a straight line equal to a given straight line.

Let *A* be the given point, and *BC* the given straight line.

Thus it is required to place at the point *A* (as an extremity) a straight line equal to the given straight line *BC*.

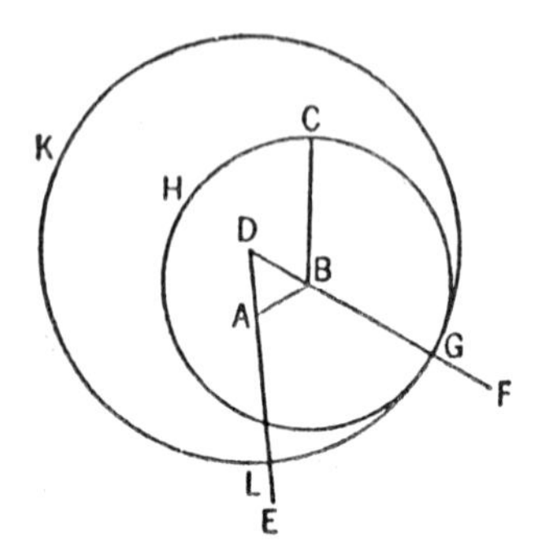

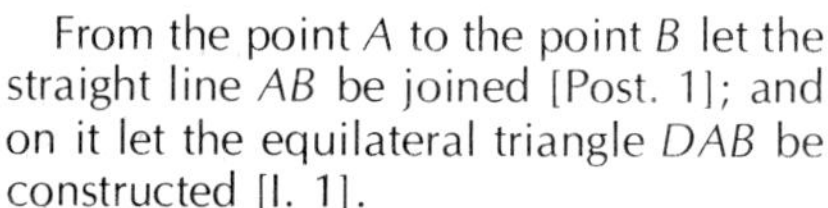

From the point *A* to the point *B* let the straight line *AB* be joined [Post. 1]; and on it let the equilateral triangle *DAB* be constructed [I. 1].

Let the straight lines *AE, BF* be produced in a straight line with *DA, DB* [Post. 2]; with center *B* and distance *BC* let the circle *CGH* be described [Post. 3]; and again, with center *D* and distance *DG* let the circle *GKL* be described [Post. 3].

Then, since the point *B* is the center of the circle *CGH, BC* is equal to *BG*.

Again, since the point *D* is the center of the circle *GKL, DL* is equal to *DG*.

And in these *DA* is equal to *DB*; therefore the remainder *AL* is equal to the remainder *BG* [*C.N.* 3].

But *BC* was also proved equal to *BG*; therefore each of the straight lines *AL, BC* is equal to *BG*.

And things which are equal to the same thing are also equal to one another [*C.N.* 1]; therefore *AL* is also equal to *BC*.

Therefore at the given point *A* the straight line *AL* is placed equal to the given straight line *BC*. (Being) what it was required to do.

PROPOSITION 3

Given two unequal straight lines, to cut off from the greater a straight line equal to the less.

Let *AB, C* be the two given unequal straight lines, and let *AB* be the greater of them.

Thus it is required to cut off from *AB* the greater a straight line equal to *C* the less.

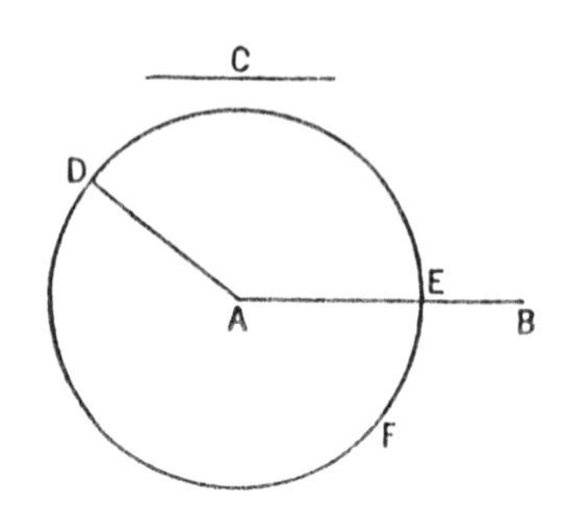

At the point *A* let *AD* be placed equal to the straight line *C* [I. 2]; and with center *A* and distance *AD* let the circle *DEF* be described [Post. 3].

Now, since the point *A* is the center of the circle *DEF, AE* is equal to *AD* [Def. 15].

But *C* is also equal to *AD*.

Therefore each of the straight lines *AE, C* is equal to *AD*; so that *AE* is also equal to *C* [*C.N.* 1].

Therefore, given the two straight lines *AB, C,* from *AB* the greater *AE* has been cut off equal to *C* the less. (Being) what it was required to do.

PROPOSITION 4

If two triangles have the two sides equal to two sides respectively, and have the angles contained by the equal straight lines equal, they will also have the base equal to the base, the triangle will be equal to the triangle, and the remaining angles will be equal to the remaining angles respectively, namely those which the equal sides subtend.

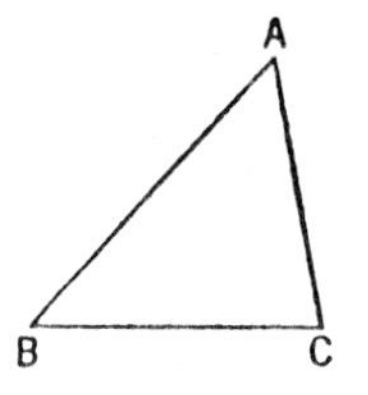

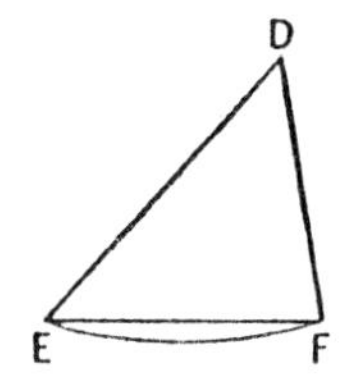

Let *ABC, DEF* be two triangles having the two sides *AB, AC* equal to the two sides *DE, DF* respectively, namely *AB* to *DE* and *AC* to *DF*, and the angle *BAC* equal to the angle *EDF*.

I say that the base *BC* is also equal to the base *EF*, the triangle *ABC* will be equal to the triangle *DEF*, and the remaining angles will be equal to the remaining angles respectively, namely those which the equal sides subtend, that is, the angle *ABC* to the angle *DEF*, and the angle *ACB* to the angle *DFE*.

For, if the triangle *ABC* be applied to the triangle *DEF*, and if the point *A* be placed on the point *D* and the straight line *AB* on *DE*, then the point *B* will also coincide with *E*, because *AB* is equal to *DE*.

Again, *AB* coinciding with *DE*, the straight line *AC* will also coincide with *DF*, because the angle *BAC* is equal to the angle *EDF;* hence the point *C* will also coincide with the point *F*, because *AC* is again equal to *DF*.

But *B* also coincided with *E;* hence the base *BC* will coincide with the base *EF*.

[For if, when *B* coincides with *E* and *C* with *F*, the base *BC* does not coincide with the base *EF*, two straight lines will enclose a space: which is impossible.

Therefore the base *BC* will coincide with *EF*] and will be equal to it [*C.N.* 4]. Thus the whole triangle *ABC* will coincide with the whole triangle *DEF*, and will be equal to it.

And the remaining angles will also coincide with the remaining angles and will be equal to them, the angle *ABC* to the angle *DEF*, and the angle *ACB* to the angle *DFE*.

Therefore etc. (Being) what it was required to prove.

PROPOSITION 5

In isosceles triangles the angles at the base are equal to one another, and, if the equal straight lines be produced further, the angles under the base will be equal to one another.

Let *ABC* be an isosceles triangle having the side *AB* equal to the side *AC;* and let the straight lines *BD, CE* be produced further in a straight line with *AB, AC* [Post. 2].

I say that the angle *ABC* is equal to the angle *ACB*, and the angle *CBD* to the angle *BCE*.

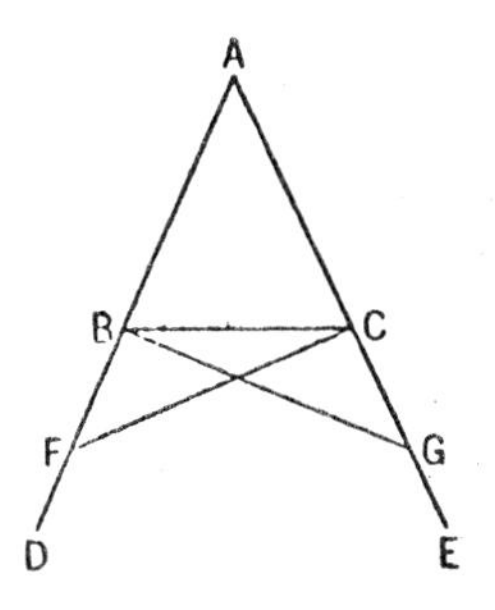

Let a point *F* be taken at random on *BD;* from *AE* the greater let *AG* be cut off equal to *AF* the less [I. 3]; and let the straight lines *FC, GB* be joined [Post. 1].

Then, since *AF* is equal to *AG* and *AB* to *AC*, the two sides *FA, AC* are equal to the two sides *GA, AB*, respectively; and they contain a common angle, the angle *FAG*.

Therefore the base *FC* is equal to the base *GB*, and the triangle *AFC* is equal to the triangle *AGB*, and the remaining angles will be equal to the remaining angles respectively, namely those which the equal sides subtend, that is, the angle *ACF* to the angle *ABG*, and the angle *AFC* to the angle *AGB* [I. 4].

And, since the whole *AF* is equal to the whole *AG*, and in these *AB* is equal to *AC*, the remainder *BF* is equal to the remainder *CG*.

But *FC* was also proved equal to *GB;* therefore the two sides *BF, FC* are equal to the two sides *CG, GB* respectively; and the angle *BFC* is equal to the angle *CGB*, while the base *BC* is common to them; therefore the triangle *BFC* is also equal to the triangle *CGB*, and the remaining angles will be equal to the remaining angles respectively, namely those which the equal sides subtend; therefore the angle *FBC* is equal to the angle *GCB*, and the angle *BCF* to the angle *CBG*.

Accordingly, since the whole angle *ABG* was proved equal to the angle *ACF*, and in these the angle *CBG* is equal to the angle *BCF*, the remaining angle ABC is equal to the remaining angle ACB; and they are at the base of the triangle ABC.

But the angle FBC was also proved equal to the angle GCB; and they are under the base.

Therefore etc. Q. E. D.

PROPOSITION 6

If in a triangle two angles be equal to one another, the sides which subtend the equal angles will also be equal to one another.

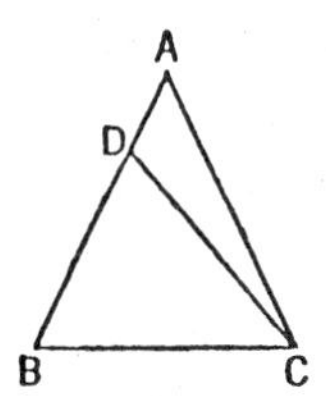

Let ABC be a triangle having the angle ABC equal to the angle ACB; I say that the side AB is also equal to the side AC.

For, if AB is unequal to AC, one of them is greater.

Let AB be greater; and from AB the greater let DB be cut off equal to AC the less; let DC be joined.

Then, since DB is equal to AC, and BC is common, the two sides DB, BC are equal to the two sides AC, CB respectively; and the angle DBC is equal to the angle ACB; therefore the base DC is equal to the base AB, and the triangle DBC will be equal to the triangle ACB, the less to the greater: which is absurd.

Therefore AB is not unequal to AC; it is therefore equal to it.

Therefore etc. Q. E. D.

PROPOSITION 7

Given two straight lines constructed on a straight line (from its extremities) and meeting in a point, there cannot be constructed on the same straight line (from its extremities), and on the same side of it, two other straight lines meeting in another point and equal to the former two respectively, namely each to that which has the same extremity with it.

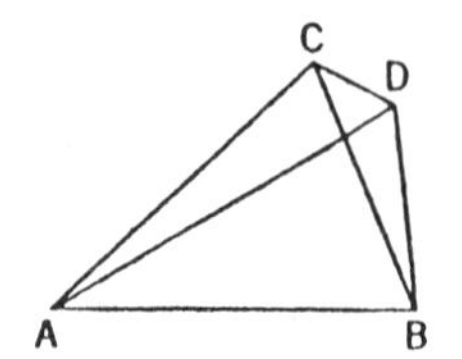

For, if possible, given two straight lines AC, CB constructed on the straight line AB and meeting at the point C, let two other straight lines AD, DB be constructed on the same straight line AB, on the same side of it, meeting in another point D and equal to the former two respectively, namely each to that which has the same extremity with it, so that CA is equal to DA which has the same extremity A with it, and CB to DB which has the same extremity B with it; and let CD be joined.

Then, since AC is equal to AD, the angle ACD is also equal to the angle ADC [I. 5]; therefore the angle ADC is greater than the angle DCB; therefore the angle CDB is much greater than the angle DCB.

Again, since CB is equal to DB, the angle CDB is also equal to the angle DCB.

But it was also proved much greater than it: which is impossible.

Therefore etc. Q. E. D.

PROPOSITION 8

If two triangles have the two sides equal to two sides respectively, and have also the base equal to the base, they will also have the angles equal which are contained by the equal straight lines.

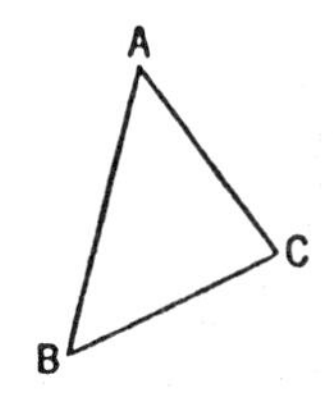

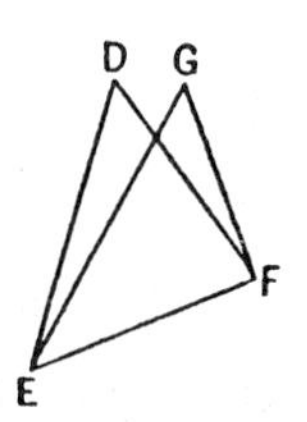

Let ABC, DEF be two triangles having the two sides AB, AC equal to the two sides DE, DF respectively, namely AB to

DE, and *AC* to *DF;* and let them have the base *BC* equal to the base *EF;* I say that the angle *BAC* is also equal to the angle *EDF*.

For, if the triangle *ABC* be applied to the triangle *DEF*, and if the point *B* be placed on the point *E* and the straight line *BC* on *EF*, the point *C* will also coincide with *F*, because *BC* is equal to *EF*.

Then, *BC* coinciding with *EF, BA, AC* will also coincide with *ED, DF;* for, if the base *BC* coincides with the base *EF*, and the sides *BA, AC* do not coincide with *ED, DF* but fall beside them as *EG, GF*, then, given two straight lines constructed on a straight line (from its extremities) and meeting in a point, there will have been constructed on the same straight line (from its extremities), and on the same side of it, two other straight lines meeting in another point and equal to the former two respectively, namely each to that which has the same extremity with it.

But they cannot be so constructed [I. 7].

Therefore it is not possible that, if the base *BC* be applied to the base *EF*, the sides *BA, AC* should not coincide with *ED, DF;* they will therefore coincide, so that the angle *BAC* will also coincide with the angle *EDF*, and will be equal to it.

If therefore etc. Q. E. D.

PROPOSITION 9

To bisect a given rectilineal angle.

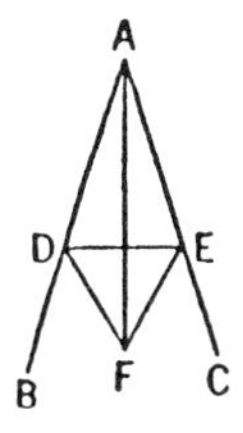

Let the angle *BAC* be the given rectilineal angle.

Thus it is required to bisect it.

Let a point *D* be taken at random on *AB;* let *AE* be cut off from *AC* equal to *AD* [I. 3]; let *DE* be joined, and on *DE* let the equilateral triangle *DEF* be constructed; let *AF* be joined.

I say that the angle *BAC* has been bisected by the straight line *AF*.

For, since *AD* is equal to *AE*, and *AF* is common, the two sides *DA, AF* are equal to the two sides *EA, AF* respectively.

And the base *DF* is equal to the base *EF;* therefore the angle *DAF* is equal to the angle *EAF* [I. 8].

Therefore the given rectilineal angle *BAC* has been bisected by the straight line *AF*. Q. E. F.

PROPOSITION 10

To bisect a given finite straight line.

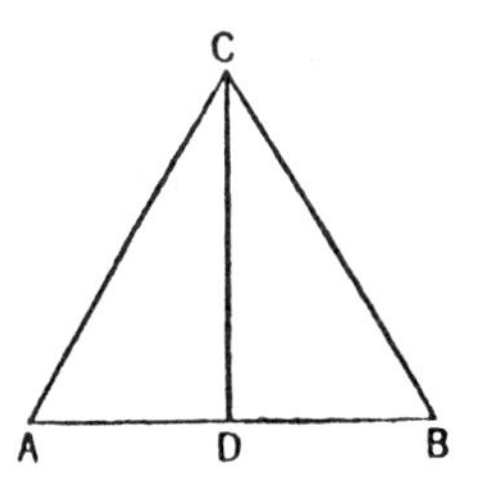

Let *AB* be given the finite straight line.

Thus it is required to bisect the finite straight line *AB*.

Let the equilateral triangle *ABC* be constructed on it [I. 1], and let the angle *ACB* be bisected by the straight line *CD* [I. 9]; I say that the straight line *AB* has been bisected at the point *D*.

For, since *AC* is equal to *CB*, and *CD* is common, the two sides *AC, CD* are equal to the two sides *BC, CD* respectively; and the angle *ACD* is equal to the angle *BCD;* therefore the base *AD* is equal to the base *BD* [I. 4].

Therefore the given finite straight line *AB* has been bisected at *D*. Q. E. F.

PROPOSITION 11

To draw a straight line at right angles to a given straight line from a given point on it.

Let *AB* be the given straight line, and *C* the given point on it.

Thus it is required to draw from the

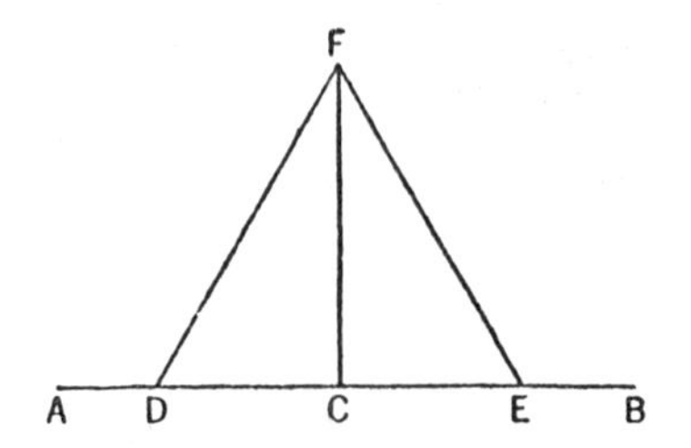

point C a straight line at right angles to the straight line AB.

Let a point D be taken at random on AC; let CE be made equal to CD [I. 3]; on DE let the equilateral triangle FDE be constructed [I. 1], and let FC be joined; I say that the straight line FC has been drawn at right angles to the given straight line AB from C the given point on it.

For, since DC is equal to CE, and CF is common, the two sides DC, CF are equal to the two sides EC, CF respectively; and the base DF is equal to the base FE; therefore the angle DCF is equal to the angle ECF [I. 8]; and they are adjacent angles.

But, when a straight line set up on a straight line makes the adjacent angles equal to one another, each of the equal angles is right [Def. 10]; therefore each of the angles DCF, FCE is right.

Therefore the straight line CF has been drawn at right angles to the given straight line AB from the given point C on it.

Q. E. F.

PROPOSITION 12

To a given infinite straight line, from a given point which is not on it, to draw a perpendicular straight line.

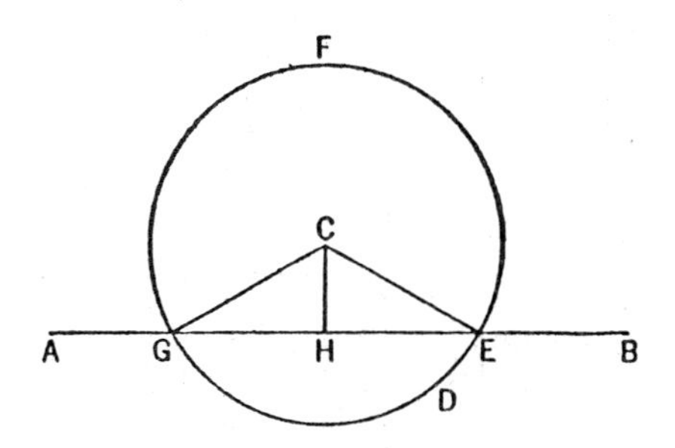

Let AB be the given infinite straight line, and C the given point which is not on it; thus it is required to draw to the given infinite straight line AB, from the given point C which is not on it, a perpendicular straight line.

For let a point D be taken at random on the other side of the straight line AB, and with center C and distance CD let the circle EFG be described [Post. 3]; let the straight line EG be bisected at H [I. 10], and let the straight lines CG, CH, CE be joined [Post. 1].

I say that CH has been drawn perpendicular to the given infinite straight line AB from the given point C which is not on it.

For, since GH is equal to HE, and HC is common, the two sides GH, HC are equal to the two sides EH, HC respectively; and the base CG is equal to the base CE; therefore the angle CHG is equal to the angle EHC [I. 8]. And they are adjacent angles.

But, when a straight line set up on a straight line makes the adjacent angles equal to one another, each of the equal angles is right, and the straight line standing on the other is called a perpendicular to that on which it stands [Def. 10].

Therefore CH has been drawn perpendicular to the given infinite straight line AB from the given point C which is not on it.

Q. E. F.

PROPOSITION 13

If a straight line set up on a straight line make angles, it will make either two right angles or angles equal to two right angles.

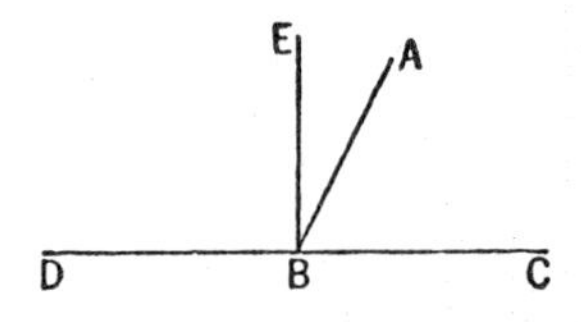

For let any straight line AB set up on the straight line CD make the angles CBA, ABD; I say that the angles CBA,

ABD are either two right angles or equal to two right angles.

Now, if the angle *CBA* is equal to the angle *ABD*, they are two right angles [Def. 10].

But, if not, let *BE* be drawn from the point *B* at right angles to *CD* [I. 11]; therefore the angles *CBE, EBD* are two right angles.

Then, since the angle *CBE* is equal to the two angles *CBA, ABE,* let the angle *EBD* be added to each; therefore the angles *CBE, EBD* are equal to the three angles *CBA, ABE, EBD* [C.N. 2].

Again, since the angle *DBA* is equal to the two angles *DBE, EBA,* let the angle *ABC* be added to each; therefore the angles *DBA, ABC* are equal to the three angles *DBE, EBA, ABC* [C.N. 2].

But the angles *CBE, EBD* were also proved equal to the same three angles; and things which are equal to the same thing are also equal to one another [C.N. 1]; therefore the angles *CBE, EBD* are also equal to the angles *DBA, ABC*.

But the angles *CBE, EBD* are two right angles; therefore the angles *DBA, ABC* are also equal to two right angles.

Therefore etc. Q. E. D.

28. From Book I of the *Elements:* Propositions 27-32*

(Theory of Parallels)

EUCLID

PROPOSITION 27

If a straight line falling on two straight lines make the alternate angles equal to one another, the straight lines will be parallel to one another.

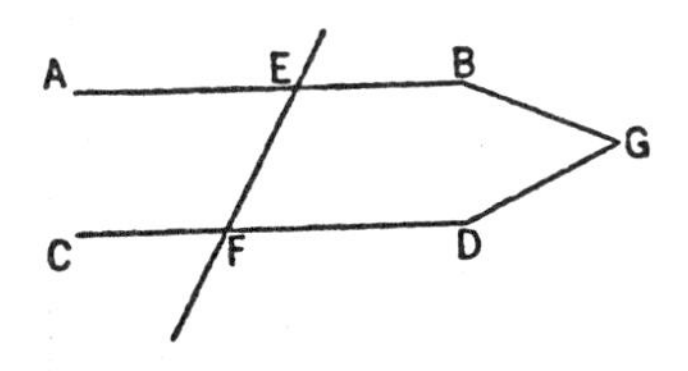

For let the straight line *EF* falling on the two straight lines *AB, CD* make the alternate angles *AEF, EFD* equal to one another; I say that *AB* is parallel to *CD*.

For, if not, *AB, CD* when produced will meet either in the direction of *B, D* or towards *A, C*.

Let them be produced and meet, in the direction of *B, D*, at *G*.

Then, in the triangle *GEF*, the exterior angle *AEF* is equal to the interior and opposite angle *EFG:* which is impossible [I. 16].

Therefore *AB, CD* when produced will not meet in the direction of *B, D*.

Similarly it can be proved that neither will they meet towards *A, C*.

But straight lines which do not meet in either direction are parallel [Def. 23] therefore *AB* is parallel to *CD*.

Therefore etc. Q. E. D.

PROPOSITION 28

If a straight line falling on two straight lines make the exterior angle equal to the interior and opposite angle on the same side, or the interior angles on the

**Source:* Reprinted by permission from *The Thirteen Books of Euclids Elements*, trans. by Sir Thomas L. Heath (1956 edition), vol. 1, 307-317 with notes deleted.

same side equal to two right angles, the straight lines will be parallel to one another.

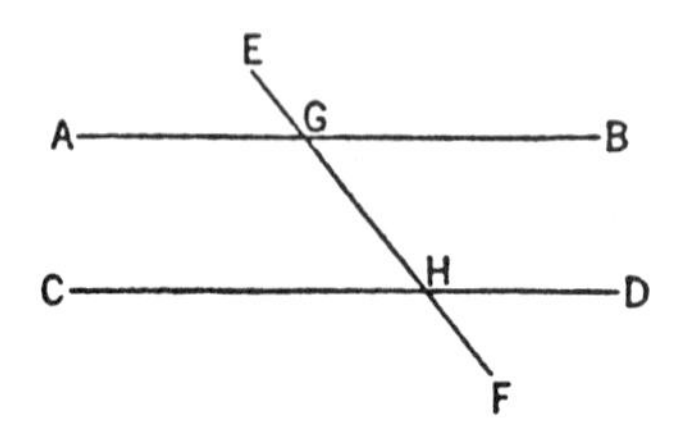

For let the straight line *EF* falling on the two straight lines *AB, CD* make the exterior angle *EGB* equal to the interior and opposite angle *GHD*, or the interior angles on the same side, namely *BGH, GHD*, equal to two right angles; I say that *AB* is parallel to *CD*.

For, since the angle *EGB* is equal to the angle *GHD*, while the angle *EGB* is equal to the angle *AGH* [I. 15], the angle *AGH* is also equal to the angle *GHD;* and they are alternate; therefore *AB* is parallel to *CD* [I. 27].

Again, since the angles *BGH, GHD* are equal to two right angles, and the angles *AGH, BGH* are also equal to two right angles [I. 13], the angles *AGH, BGH* are equal to the angles *BGH, GHD*.

Let the angle *BGH* be subtracted from each; therefore the remaining angle *AGH* is equal to the remaining angle *GHD;* and they are alternate; therefore *AB* is parallel to *CD* [I. 27].

Therefore etc. Q. E. D.

PROPOSITION 29

A straight line falling on parallel straight lines makes the alternate angles equal to one another, the exterior angle equal to the interior and opposite angle, and the interior angles on the same side equal to two right angles.

For let the straight line *EF* fall on the parallel straight lines *AB, CD;* I say that it makes the alternate angles *AGH, GHD* equal, the exterior angle *EGB* equal to the interior and opposite angle

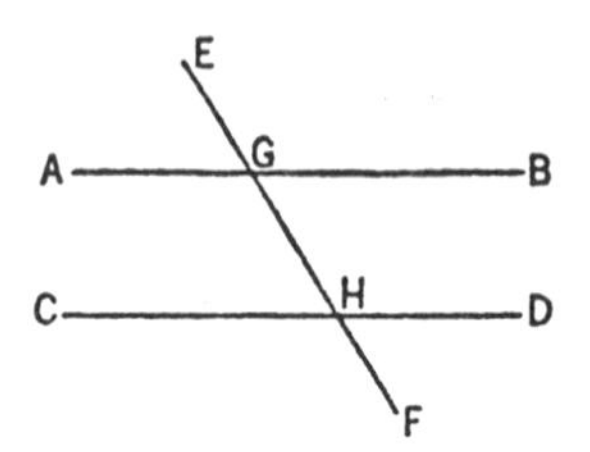

GHD, and the interior angles on the same side, namely *BGH, GHD*, equal to two right angles.

For, if the angle *AGH* is unequal to the angle *GHD*, one of them is greater.

Let the angle *AGH* be greater.

Let the angle *BGH* be added to each; therefore the angles *AGH, BGH* are greater than the angles *BGH, GHD*.

But the angles *AGH, BGH* are equal to two right angles [I. 13]; therefore the angles *BGH, GHD* are less than two right angles.

But straight lines produced indefinitely from angles less than two right angles meet [Post. 5]; therefore *AB, CD*, if produced indefinitely, will meet; but they do not meet, because they are by hypothesis parallel.

Therefore the angle *AGH* is not unequal to the angle *GHD*, and is therefore equal to it.

Again, the angle *AGH* is equal to the angle *EGB* [I. 15] therefore the angle *EGB* is also equal to the angle *GHD* [*C.N.* 1].

Let the angle *BGH* be added to each; therefore the angles *EGB, BGH* are equal to the angles *BGH, GHD* [*C.N.* 2].

But the angles *EGB, BGH* are equal to two right angles [I. 13]; therefore the angles *BGH, GHD* are also equal to two right angles.

Therefore etc. Q. E. D.

PROPOSITION 30

Straight lines parallel to the same straight line are also parallel to one another.

Let each of the straight lines *AB, CD* be parallel to *EF;* I say that *AB* is also parallel to *CD*.

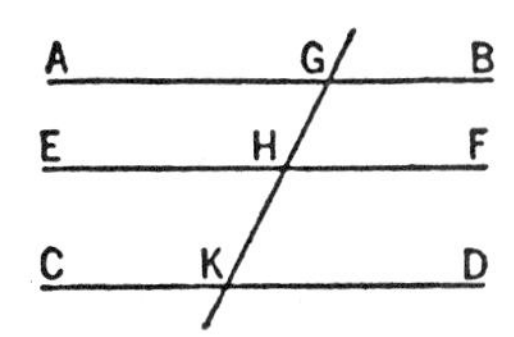

For let the straight line *GK* fall upon them.

Then, since the straight line *GK* has fallen on the parallel straight lines *AB, EF*, the angle *AGK* is equal to the angle *GHF* [I. 29].

Again, since the straight line *GK* has fallen on the parallel straight lines *EF, CD*, the angle *GHF* is equal to the angle *GKD* [I. 29].

But the angle *AGK* was also proved equal to the angle *GHF;* therefore the angle *AGK* is also equal to the angle *GKD* [*C.N.* 1]; and they are alternate.

Therefore *AB* is parallel to *CD*. Q. E. D.

PROPOSITION 31

Through a given point to draw a straight line parallel to a given straight line.

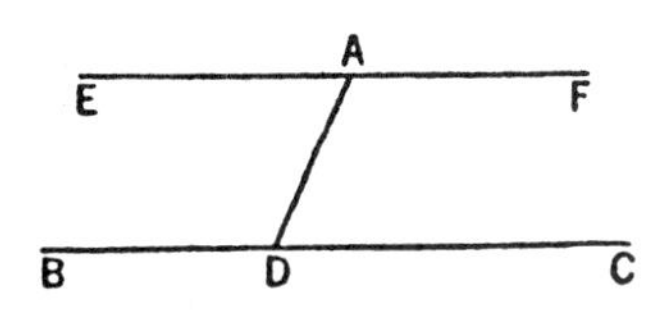

Let *A* be the given point, and *BC* the given straight line; thus it is required to draw through the point *A* a straight line parallel to the straight line *BC*.

Let a point *D* be taken at random on *BC*, and let *AD* be joined; on the straight line *DA*, and at the point *A* on it, let the angle *DAE* be constructed equal to the angle *ADC* [I. 23]; and let the straight line *AF* be produced in a straight line with *EA*.

Then, since the straight line *AD* falling on the two straight lines *BC, EF* has made the alternate angles *EAD, ADC* equal to one another, therefore *EAF* is parallel to *BC* [I. 27].

Therefore through the given point *A* the straight line *EAF* has been drawn parallel to the given straight line *BC*.
Q. E. F.

PROPOSITION 32

In any triangle, if one of the sides be produced, the exterior angle is equal to the two interior and opposite angles, and the three interior angles of the triangle are equal to two right angles.

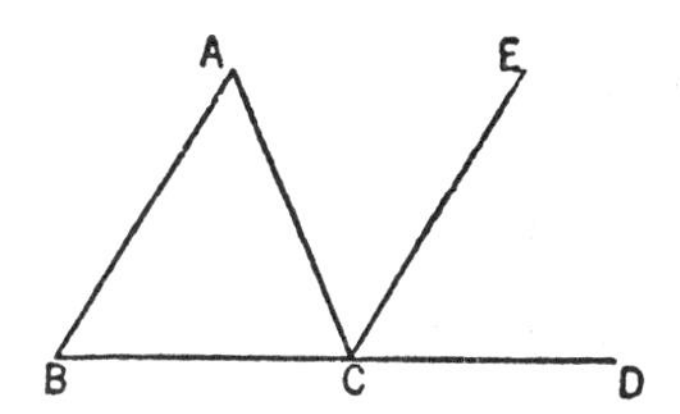

Let *ABC* be a triangle, and let one side of it *BC* be produced to *D;* I say that the exterior angle *ACD* is equal to the two interior and opposite angles *CAB, ABC*, and the three interior angles of the triangle *ABC, BCA, CAB* are equal to two right angles.

For let *CE* be drawn through the point *C* parallel to the straight line *AB* [I. 31].

Then, since *AB* is parallel to *CE*, and *AC* has fallen upon them, the alternate angles *BAC, ACE* are equal to one another [I. 29].

Again, since *AB* is parallel to *CE*, and the straight line *BD* has fallen upon them, the exterior angle *ECD* is equal to the interior and opposite angle *ABC* [I. 29].

But the angle *ACE* was also proved equal to the angle *BAC;* therefore the whole angle *ACD* is equal to the two interior and opposite angles *BAC, ABC*.

Let the angle *ACB* be added to each; therefore the angles *ACD, ACB* are equal to the three angles *ABC, BCA, CAB*.

But the angles *ACD, ACB* are equal to two right angles [I. 13]; therefore the angles *ABC, BCA, CAB* are also equal to two right angles.

Therefore etc. Q. E. D.

29. From Book I of the *Elements*: Proposition 47*

(Pythagorean Theorem)

EUCLID

PROPOSITION 47

In right-angled triangles the square on the side subtending the right angle is equal to the squares on the sides containing the right angle.

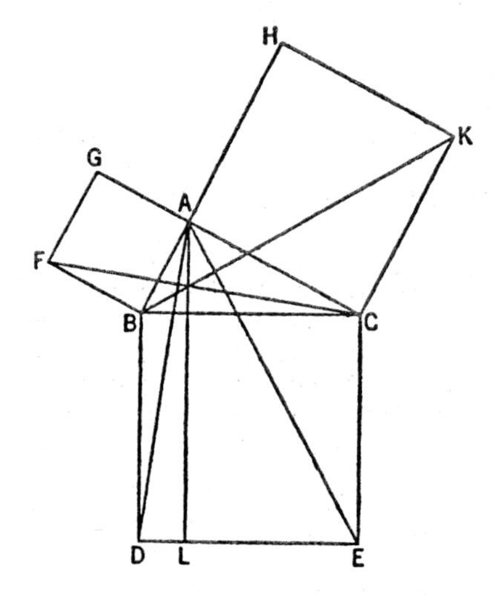

Let *ABC* be a right-angled triangle having the angle *BAC* right; I say that the square on *BC* is equal to the squares on *BA, AC*.

For let there be described on *BC* the square *BDEC*, and on *BA, AC* the squares *GB, HC* [I. 46]; through *A* let *AL* be drawn parallel to either *BD* or *CE*, and let *AD, FC* be joined.

Then, since each of the angles *BAC, BAG* is right, it follows that with a straight line *BA*, and at the point *A* on it, the two straight lines *AC, AG* not lying on the same side make the adjacent angles equal to two right angles; therefore *CA* is in a straight line with *AG* [I. 14].

For the same reason *BA* is also in a straight line with *AH*.

And, since the angle *DBC* is equal to the angle *FBA*: for each is right: let the angle *ABC* be added to each; therefore the whole angle *DBA* is equal to the whole angle *FBC* [*C.N.* 2].

And, since *DB* is equal to *BC*, and *FB* to *BA*, the two sides *AB, BD* are equal to the two sides *FB, BC* respectively; and the angle *ABD* is equal to the angle *FBC;* therefore the base *AD* is equal to the base *FC*, and the triangle *ABD* is equal to the triangle *FBC* [I. 4].

Now the parallelogram *BL* is double of the triangle *ABD*, for they have the same base *BD* and are in the same parallels *BD, AL* [I. 41].

And the square *GB* is double of the triangle *FBC*, for they again have the same base *FB* and are in the same parallels *FB, GC* [I. 41].

[But the doubles of equals are equal to one another.]

Therefore the parallelogram *BL* is also equal to the square *GB*.

**Source:* Reprinted by permission from *The Thirteen Books of Euclid's Elements*, trans. by Sir Thomas L. Heath (1956 edition), vol. 1, 349-350.

Similarly, if *AE, BK* be joined, the parallelogram *CL* can also be proved equal to the square *HC;* therefore the whole square *BDEC* is equal to the two squares *GB, HC* [C.N. 2].

And the square *BDEC* is described on *BC*, and the squares *GB, HC* on *BA, AC*.

Therefore the square on the side *BC* is equal to the squares on the sides *BA, AC*.

Therefore etc. Q. E. D.

30. From Book Seven of the *Elements* Propositions 1 and 2*

(Euclidean algorithm)

EUCLID

PROPOSITION 1

Two unequal numbers being set out, and the less being continually subtracted in turn from the greater, if the number which is left never measures the one before it until an unit is left, the original numbers will be prime to one another.

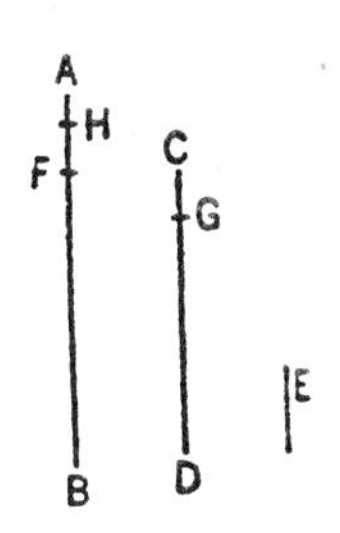

For, the less of two unequal numbers *AB, CD* being continually subtracted from the greater, let the number which is left measure the one before it until an unit is left; I say that *AB, CD* are prime to one another, that is, that an unit alone measures *AB, CD*.

For, if *AB, CD* are prime to one another, some number will measure them.

Let a number measure them, and let it be *E;* let *CD*, measuring *BF*, leave *FA* less than itself, let *AF*, measuring *DG*, leave *GC* less than itself, and let *GC*, measuring *FH*, leave an unit *HA*.

Since, then, *E* measures *CD*, and *CD* measures *BF*, therefore *E* also measures *BF*.

But it also measures the whole *BA;* therefore it will also measure the remainder *AF*.

But *AF* measures *DG;* therefore *E* also measures *DG*.

But it also measures the whole *DC;* therefore it will also measure the remainder *CG*.

But *CG* measures *FH;* therefore *E* also measures *FH*.

But it also measures the whole *FA;* therefore it will also measure the remainder, the unit *AH*, though it is a number: which is impossible.

Therefore no number will measure the numbers *AB, CD;* therefore *AB, CD* are prime to one another [VII. Def. 12].

Q. E. D.

PROPOSITION 2

Given two numbers not prime to one another, to find their greatest common measure.

**Source:* Reprinted by permission from *The Thirteen Books of Euclid's Elements*, trans. by Sir Thomas L. Heath (1956 edition), vol. 2, 295-299.

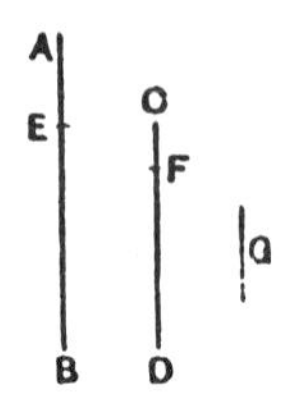

Let *AB, CD* be the two given numbers not prime to one another.

Thus it is required to find the greatest common measure of *AB, CD*.

If now *CD* measures *AB*—and it also measures itself—*CD* is a common measure of *CD, AB*.

And it is manifest that it is also the greatest; for no greater number than *CD* will measure *CD*.

But, if *CD* does not measure *AB*, then, the less of the numbers *AB, CD* being continually subtracted from the greater, some number will be left which will measure the one before it.

For an unit will not be left; otherwise *AB, CD* will be prime to one another [VII. I], which is contrary to the hypothesis.

Therefore some number will be left which will measure the one before it.

Now let *CD*, measuring *BE*, leave *EA* less than itself, let *EA*, measuring *DF*, leave *FC* less than itself, and let *CF* measure *AE*.

Since then, *CF* measures *AE*, and *AE* measures *DF*, therefore *CF* will also measure *DF*.

But it also measures itself; therefore it will also measure the whole *CD*.

But *CD* measures *BE;* therefore *CF* also measures *BE*.

But it also measures *EA;* therefore it will also measure the whole *BA*.

But it also measures *CD;* therefore *CF* measures *AB, CD*.

Therefore *CF* is a common measure of *AB, CD*.

I say next that it is also the greatest.

For, if *CF* is not the greatest common measure of *AB, CD*, some number which is greater than *CF* will measure the numbers *AB, CD*.

Let such a number measure them, and let it be *G*.

Now, since *G* measures *CD*, while *CD* measures *BE, G* also measures *BE*.

But it also measures the whole *BA;* therefore it will also measure the remainder *AE*.

But *AE* measures *DF;* therefore *G* will also measure *DF*.

But it also measures the whole *DC;* therefore it will also measure the remainder *CF*, that is, the greater will measure the less: which is impossible.

Therefore no number which is greater than *CF* will measure the numbers *AB, CD;* therefore *CF* is the greatest common measure of *AB, CD*.[1]

PORISM. From this it is manifest that, if a number measure two numbers, it will also measure their greatest common measure. Q. E. D.

NOTE

1. Here we have the exact method of finding the greatest common measure given in the text-books of algebra, including the *reductio ad absurdum* proof that the number arrived at is not only a common measure but the *greatest* common measure. The process of finding the greatest common measure is simply shown thus:

$$\begin{array}{l} b)\ a\ (p \\ \quad \underline{pb} \\ \quad\ c)\ b\ (q \\ \qquad \underline{qc} \\ \qquad\ d)\ c\ (r \\ \qquad\quad \underline{rd} \end{array}$$

We shall arrive, says Euclid, at some number, say d, which measures the one before it, i.e. such that $c = rd$. Otherwise the process would go on until we arrived at unity. This is impossible because in that case a, b would be prime to one another, which is contrary to the hypothesis.

Next, like the text-books of algebra, he goes on to show that d will be *some* common measure of a, b. For d measures c; therefore it measures $qc + d$, that is, b, and hence it measures $pb + c$, that is, a.

Lastly, he proves that d is the *greatest* common measure of a, b as follows.

Suppose that e is a common measure greater than d.

Then e, measuring a, b, must measure $a - pb$, or c.

31. From Book Nine of the *Elements:* Proposition 14*

(Fundamental Theorem in the Theory of Numbers)

EUCLID

PROPOSITION 14

If a number be the least that is measured by prime numbers, it will not be measured by any other prime number except those originally measuring it.

For let the number *A* be the least that is measured by the prime numbers *B, C, D;* I say that *A* will not be measured by any other prime number except *B, C, D.*

For, if possible, let it be measured by the prime number *E*, and let *E* not be the same with any one of the numbers *B, C, D.*

Now, since *E* measures *A*, let it measure it according to *F;* therefore *E* by multipling *F* has made *A*.

A ———————

E ———

F ———

B —

C ——

D ———

And *A* is measured by the prime numbers *B, C, D.*

But, if two numbers by multiplying one another make some number, and any prime number measure the product, it will also measure one of the original numbers [VII. 30]; therefore *B, C, D* will measure one of the numbers *E, F.*

Now they will not measure E; for *E* is prime and not the same with any one of the numbers *B, C, D.*

Therefore they will measure *F*, which is less than *A:* which is impossible, for *A* is by hypothesis the least number measured by *B, C, D.*

Therefore no prime number will measure *A* except *B, C, D.* Q. E. D.

Source: Reprinted by permission from *The Thirteen Books of Euclid's Elements,* trans. by Sir Thomas L. Heath (1956 edition), vol. 2, 402-403.

32. From Book Nine of the *Elements*: Proposition 20*

(Infinitude of Primes)

EUCLID

PROPOSITION 20

Prime numbers are more than any assigned multitude of prime numbers.

Let *A, B, C* be the assigned prime numbers; I say that there are more prime numbers than *A, B, C*.

For let the least number measured by *A, B, C* be taken, and let it be *DE;* let the unit *DF* be added to *DE*.

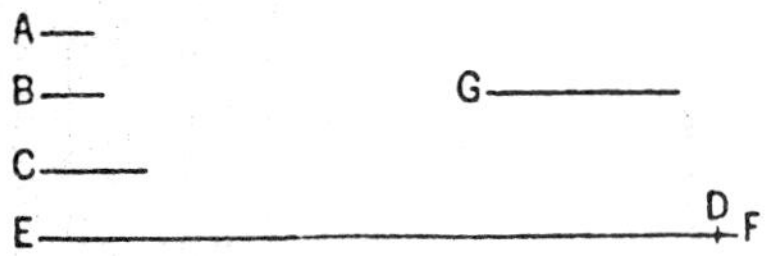

Then *EF* is either prime or not.

First, let it be prime; then the prime numbers *A, B, C, EF* have been found which are more than *A, B, C*.

Next, let *EF* not be prime; therefore it is measured by some prime number [VII. 31].

Let it be measured by the prime number *G*.

I say that *G* is not the same with any of the numbers *A, B, C*.

For, if possible, let it be so.

Now, *A, B, C* measure *DE;* therefore *G* also will measure *DE*.

But it also measures *EF*.

Therefore *G*, being a number, will measure the remainder, the unit *DF:* which is absurd.

Therefore *G* is not the same with any one of the numbers *A, B, C*.

And by hypothesis it is prime.

Therefore the prime numbers *A, B, C, G* have been found which are more than the assigned multitude of *A, B, C*.

Q. E. D.

**Source:* Reprinted by permission from *The Thirteen Books of Euclid's Elements,* trans. by Sir Thomas L. Heath (1956 edition), vol. 2, 412. Copyright © 1956 Cambridge University Press.

Chapter III
The Hellenistic and Roman Periods

Archimedes (c. 287–212 B.C.)

Archimedes was the greatest mathematician and physicist of antiquity. The son of the astronomer Pheidias, he grew up in Syracuse, a Greek settlement in Sicily. As a young man he seems to have visited Egypt and to have studied mathematics in Alexandria (the chief center of scientific learning in the West after the decline of Athens by 332 B.C.). In Alexandria he probably studied under the pupils of Euclid and became the friend of Conon of Samos and of Eratosthenes, to whom he later dedicated the *Method*. After completing his studies, Archimedes returned to Syracuse, where he occasionally worked for King Hieron II who may have been his kinsman. He composed his writings in Syracuse and died during the city's capture by the Romans in 212 B.C. amid the second Punic War. The Roman soldiers feared him for the devices he invented for defending the city. These probably included battering rams, catapults, cranes, a compound pulley to move ships on shore easily, and perhaps burning (paraboloidal) mirrors. According to both Livy and Plutarch, a Roman soldier killed him while he was engaged in mathematical research. As he had requested, a cylinder circumscribing a sphere and the ratio of the two volumes (his favorite theorem) were inscribed on his grave marker.

There are a number of anecdotes about Archimedes. The Roman architect Vitruvius reported that King Hieron asked Archimedes to check whether a goldsmith had fraudulently alloyed his gold crown or wreath with some baser metal, and to do so without destroying the workmanship. Soon after he made a discovery that so excited him that he ran naked through the streets shouting "Eureka!" While bathing he had noticed that his body was partly buoyed up by the water. Thus he was able to determine the relative contents of the metals in the crown by water displacement. A second anecdote from Pappus and Plutarch claims that after he discovered how to move a given weight with a given force using leverage, he boasted to King Hieron: "Give me a place to stand on, and I can move the earth."

A questionable legend has grown regarding Archimedes' attitude toward mechanical inventions, even though he was famous for these. Besides the military weapons noted above, he early invented a water screw as a means to draw water out of the Nile for irrigation. According to Cicero (in the *Republic*), Archimedes constructed a model planetarium to demonstrate the Eudoxian system of astronomy with the apparent motion of the sun, moon, and planets about the fixed earth. Nevertheless, in his eulogy to Archimedes, Plutarch claimed that Archimedes disdained the practical as "sordid and ignoble." This appears to be more a prejudice of Plutarch, though Archimedes wrote only the lost work *On Sphere-making* about his inventions.

All of the mathematical sciences were Archimedes' province—arithmetic, geometry, astronomy, hydrostatics, and mechanics. Unlike Euclid and Apollonius, he wrote no textbooks but prepared small tracts instead. In his writings he displayed rigor and a sound knowledge of Euclid's *Elements.* His best-known writing is the *Sandreckoner;* other tracts are *On the Sphere and Cylinder, On the Measurement of the Circle,* and *On the Quadrature of the Parabola.* In examining the areas and volumes of figures bounded by curved lines and surfaces, he refined Eudoxus' "method of exhaustion," essentially the theory of limits. In Proposition 3 of *On the Measurement of the Circle,* he computed the value of the ratio of the circumference of a circle to its diameter to be less than $3\ ^1/_7$ but greater than $3\ ^{10}/_{71}$ by inscribing and circumscribing a circle with a regular polygon of 96 sides. (This ratio was not denoted by π until early modern times.) In Assumption 5 of *On the Sphere and Cylinder,* he stated Archimedes' postulate: given two points A and B and a third C, no matter how distant, *i.e.,* $A \quad B \qquad\qquad C$, then by measuring off $\overline{AB}$ again and again, one eventually goes beyond C. This postulate is important to the modern treatment of infinite divisibility and the continuum. He also found the formulas for the surface area and volume of a sphere—in modern notations $S = 4\pi r^2$ [Proposition 33] and $V = {}^4/_3\ \pi r^3$ [Proposition 34].

Archimedes' other mathematical achievements are numerous. He first successfully applied geometry to statical and hydrostatical problems; he also gave proofs (albeit incomplete ones) of the law of the lever for commensurable and incommensurable magnitudes. He worked on the origins of trigonometry and probably derived Heron's formula for the area of a triangle, $\sqrt{s(s-a)(s-b)(s-c)}$, where s is the semiperimeter and a, b, and c are the sides. Using the process of *neusis* (verging), he mechanically trisected a given angle as follows:

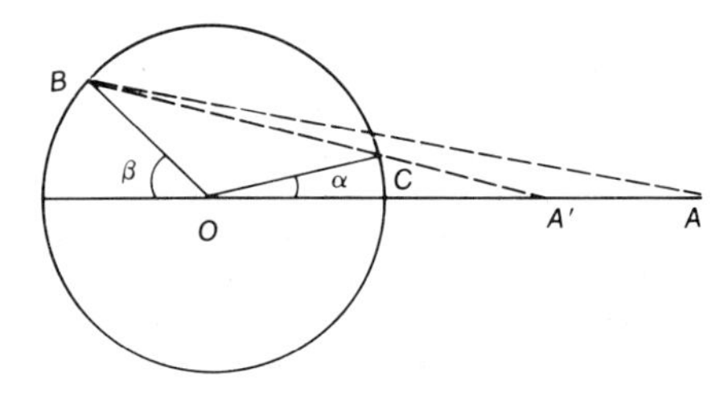

Given circle O with radius OB and an angle β. Extend the diameter to point A. Let line BA sweep down until $CA' = OC$ (the radius). Then angle $\alpha = {}^1/_3$ angle β.

In one section of the *Sandreckoner,* Archimedes turned from the then dominant geometrical branch of mathematics to arithmetic. He gave a system for representing large numbers, counting in modern notation in terms of $P = 10^8$ (octads). He projected numbers as large as P^P in order to express the number of grains of sand to fill "a mass equal in magnitude to the universe," that is the volume of a sphere bounded by the fixed stars. From his computations he found the approximate number of grains of sand needed to fill the universe was a mere 10^{63}.

33. From *On the Sphere and Cylinder* I: Assumptions*

ARCHIMEDES

[ASSUMPTIONS]

1. *Of all lines which have the same extremities the straight line is the least.*

2. Of other lines in a plane and having the same extremities, [any two] such are unequal whenever both are concave in the same direction and one of them is either wholly included between the other and the straight line which has the same extremities with it, or is partly included by, and is partly common with, the other; and that [line] which is included is the lesser [of the two].

3. Similarly, of surfaces which have the same extremities, if those extremities are in a plane, the plane is the least [in area].

4. Of other surfaces with the same extremities, the extremities being in a plane, [any two] such are unequal whenever both are concave in the same direction and one surface is either wholly included between the other and the plane which has the same extremities with it, or is partly included by, and partly common with, the other; and that [surface] which is included is the lesser [of the two in area].

5. [**Postulate of Archimedes**] Further, of unequal lines, unequal surfaces, and unequal solids, the greater exceeds the less by such a magnitude as, when added to itself, can be made to exceed any assigned magnitude among those which are comparable with [it and with] one another.

**Source:* Reprinted by permission from Thomas L. Heath (ed.), *The Works of Archimedes* (Dover Edition, 1953), 3-4.

34. From *On the Spheres and Cylinder* I: Propositions 33 and 34*

(Surface and Volume of a Sphere)

ARCHIMEDES

PROPOSITION 33

The surface of any sphere is equal to four times the greatest circle in it.

Let C be a circle equal to four times the great circle.

Then, if C is not equal to the surface of the sphere, it must either be less or greater.

I. Suppose C less than the surface of the sphere.

**Source:* Reprinted by permission from Thomas L. Heath (ed.), *The Works of Archimedes* (Dover Edition, 1953), 39-43.

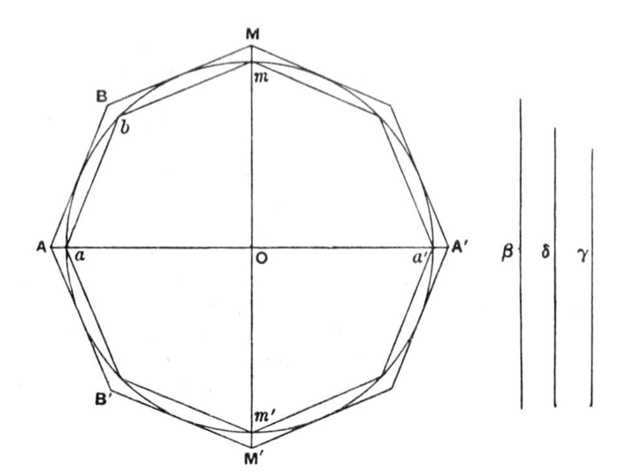

It is then possible to find two lines β, γ, of which β is the greater, such that $\beta : \gamma <$ (surface of sphere) : C [Prop. 2]. Take such lines, and let δ be a mean proportional between them.

Suppose similar regular polygons with $4n$ sides circumscribed about and inscribed in a great circle such that the ratio of their sides is less than the ratio $\beta : \delta$ [Prop. 3].

Let the polygons with the circle revolve together about a diameter common to all, describing solids of revolutions as before.

Then (surface of outer solid) : (surface of inner solid) = (side of outer)^2 : (side of inner)^2 [Prop. 32] $< \beta^2 : \delta^2$, or $\beta : \gamma <$ (surface of sphere) : C, *a fortiori*.

But this is impossible, since the surface of the circumscribed solid is greater than that of the sphere [Prop. 28], while the surface of the inscribed solid is less than C [Prop. 25].

Therefore C is not less than the surface of the sphere.

II. Suppose C greater than the surface of the sphere.

Take lines β, γ, of which β is the greater, such that $\beta : \gamma < C$: (surface of sphere).

Circumscribe and inscribe to the great circle similar regular polygons, as before, such that their sides are in a ratio less than that of β to δ, and suppose solids of revolution generated in the usual manner. Then, in this case, (surface of circumscribed solid) : (surface of inscribed solid) $< C$: (surface of sphere).

But this is impossible, because the surface of the circumscribed solid is greater than C [Prop. 30], while the surface of the inscribed solid is less than that of the sphere [Prop. 23].

Thus C is not greater than the surface of the sphere.

Therefore, since it is neither greater nor less, C is equal to the surface of the sphere.

PROPOSITION 34

Any sphere is equal to four times the cone which has its base equal to the greatest circle in the sphere and its height equal to the radius of the sphere.

Let the sphere be that of which $ama'm'$ is a great circle.

If now the sphere is not equal to four times the cone described, it is either greater or less.

I. If possible, let the sphere be greater than four times the cone.

Suppose V to be a cone whose base is equal to four times the great circle and whose height is equal to the radius of the sphere.

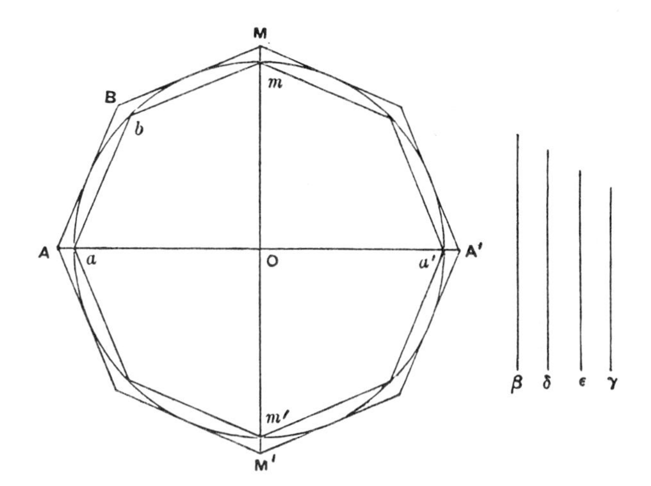

Then, by hypothesis, the sphere is greater than V; and two lines β, γ can be found (of which β is the greater) such that $\beta : \gamma <$ (volume of sphere) : V.

Between β and γ place two arithmetic means δ, ϵ.

As before, let similar regular polygons with sides $4n$ in number be circumscribed about and inscribed in the great circle, such that their sides are in a ratio less than $\beta : \delta$.

Imagine the diameter aa' of the circle to be in the same straight line with a diameter of both polygons, and imagine the latter to revolve with the circle about aa', describing the surfaces of two solids of revolution. The volumes of these solids are therefore in the triplicate ratio of their sides [Prop. 32].

Thus (vol. of outer solid) : (vol. of inscribed solid) $< \beta^3 : \delta^3$, by hypothesis, $< \beta : \gamma$, *a fortiori* (since $\beta : \gamma > \beta^3 : \delta^3$), $<$ (volume of sphere) : V, *a fortiori*.

But this is impossible, since the volume of the circumscribed solid is greater than that of the sphere [Prop. 28], while the volume of the inscribed solid is less than V [Prop. 27].

Hence the sphere is not greater than V, or four times the cone described in the enunciation.

II. If possible, let the sphere be less than V. In this case we take β, γ (β being the greater) such that $\beta : \gamma < V$: (volume of sphere).

The rest of the construction and proof proceeding as before, we have finally (volume of outer solid) : (volume of inscribed solid) $< V$: (volume of sphere).

But this is impossible, because the volume of the outer solid is greater than V [Prop. 31, Cor.], and the volume of the inscribed solid is less than the volume of the sphere.

Hence the sphere is not less than V.

Since then the sphere is neither less nor greater than V, it is equal to V, or to four times the cone described in the enunciation.

35. From *Measurement of a Circle:* Proposition 3*

(Approximation of π Using in Essence Upper and Lower Limits)

ARCHIMEDES

The ratio of the circumference of any circle to its diameter is less than $3\frac{1}{7}$ *but greater than* $3\frac{10}{71}$.

[In view of the interesting questions arising out of the arithmetical content of this proposition of Archimedes, it is necessary, in reproducing it, to distinguish carefully the actual steps set out in the text as we have it from the intermediate steps (mostly supplied by Eutocius) which it is convenient to put in for the purpose of making the proof easier to follow. Accordingly all the steps not actually appearing in the text have been enclosed in square brackets, in order that it may be clearly seen how far Archimedes omits actual calculations and only gives results. It will be observed that he gives two fractional approximations to $\sqrt{3}$ (one being less and the other greater than the real value) without any explanation as to how he arrived at them; and in like manner approximations to the square roots of several large numbers which are not complete squares are merely stated. . . .]

I. Let AB be the diameter of any circle, O its centre, AC the tangent at A; and let the angle AOC be one-third of a right angle.

Then $OA : AC\ [= \sqrt{3} : 1] > 265 : 153$ . . . . (1),

and $OC : CA\ [= 2 : 1] = 306 : 153$ . . . . (2).

First, draw OD bisecting the angle AOC and meeting AC in D.

Now $CO : OA = CD : DA$, [Eucl. VI. 3]

so that $[CO + OA : OA = CA : DA$, or]

$CO + OA : CA = OA : AD$.

Therefore [by (1) and (2)]

$OA : AD > 571 : 153$ . . . . (3).

Hence $OD^2 : AD^2\ [= (OA^2 + AD^2) : AD^2$

$> (571^2 + 153^2) : 153^2]$

$> 349450 : 23409$,

so that $OD : DA > 591\frac{1}{8} : 153$ . . . . (4).

Secondly, let OE bisect the angle AOD, meeting AD in E.

[Then $DO : OA = DE : EA$,

so that $DO + OA : DA = OA : AE$.]

Therefore $OA : AE\ [> (591\frac{1}{8} + 571) : 153$, by (3) and (4)]

$> 1162\frac{1}{8} : 153$ . . . . (5).

[It follows that

$$OE^2 : EA^2 > \{(1162\tfrac{1}{8})^2 + 153^2\} : 153^2$$
$$> (1350534\tfrac{33}{64} + 23409) : 23409$$
$$> 1373943\tfrac{33}{64} : 23409.]$$

*Source: Reprinted by permission from Thomas L. Heath (ed.), *The Works of Archimedes* (Dover Edition, 1953), 93-98.

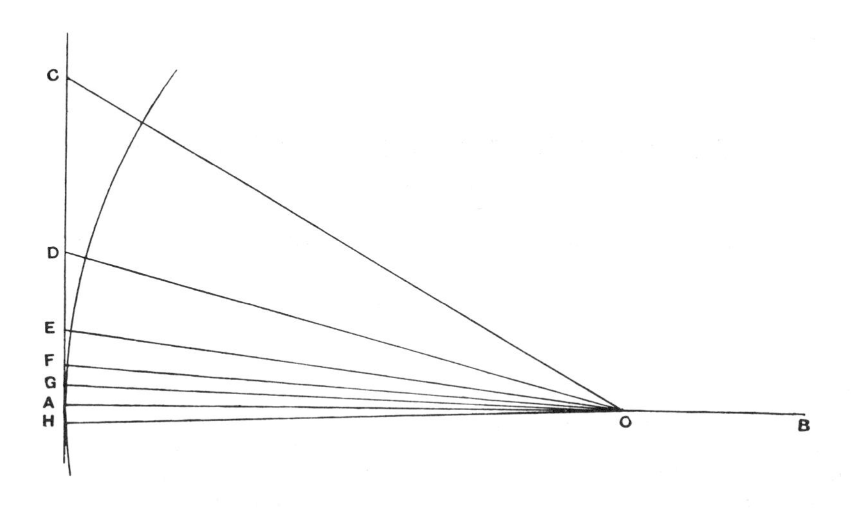

Thus $\qquad OE : EA > 1172\frac{1}{8} : 153$ .... (6).

Thirdly, let OF bisect the angle AOE and meet AE in F.

We thus obtain the result [corresponding to (3) and (5) above] that

$$OA : AF \; [> (1162\tfrac{1}{8} + 1172\tfrac{1}{8}) : 153]$$
$$> 2334\tfrac{1}{4} : 153 \qquad \text{(7).}$$

[Therefore $\qquad OF^2 : FA^2 > \{(2334\frac{1}{4})^2 + 153^2\} : 153^2$
$$> 5472132\tfrac{1}{16} : 23409.]$$

Thus $\qquad OF : FA > 2339\frac{1}{4} : 153$ .... (8).

Fourthly, let OG bisect the angle AOF, meeting AF in G.

We have then

$$OA : AG \; [> (2334\tfrac{1}{4} + 2339\tfrac{1}{4}) : 153, \text{ by means of (7) and (8)}]$$
$$> 4673\tfrac{1}{2} : 153.$$

Now the angle AOC, which is one-third of a right angle, has been bisected four times, and it follows that

$$\angle AOC = \tfrac{1}{48} \text{ (a right angle).}$$

Make the angle AOH on the other side of OA equal to the angle AOG, and let GA produced meet OH in H.

Then $\qquad \angle GOH = \frac{1}{24}$ (a right angle).

Thus GH is one side of a regular polygon of 96 sides circumscribed to the given circle.

And, since $\qquad OA : AG > 4673\frac{1}{2} : 153,$

while $\qquad AB = 2OA, \quad GH = 2AG,$

it follows that

$$AB : (\text{perimeter of polygon of 96 sides}) \; [> 4673\tfrac{1}{2} : 153 \times 96]$$
$$> 4673\tfrac{1}{2} : 14688.$$

But
$$\frac{14688}{4673\frac{1}{2}} = 3 + \frac{667\frac{1}{2}}{4673\frac{1}{2}}$$
$$\left[< 3 + \frac{667\frac{1}{2}}{4672\frac{1}{2}}\right]$$
$$< 3\tfrac{1}{7}.$$

Therefore the circumference of the circle (being less than the perimeter of the polygon) is *a fortiori* less than $3\frac{1}{7}$ times the diameter AB.

II. Next let AB be the diameter of a circle, and let AC, meeting the circle in C, make the angle CAB equal to one-third of a right angle. Join BC.

Then $$AC : CB \; [= \sqrt{3} : 1] < 1351 : 780.$$

First, let AD bisect the angle BAC and meet BC in d and the circle in D. Join BD.

Then $$\angle BAD = \angle dAC$$
$$= \angle dBD,$$

and the angles at D, C are both right angles.

It follows that the triangles ADB, $[ACd]$, BDd are similar.

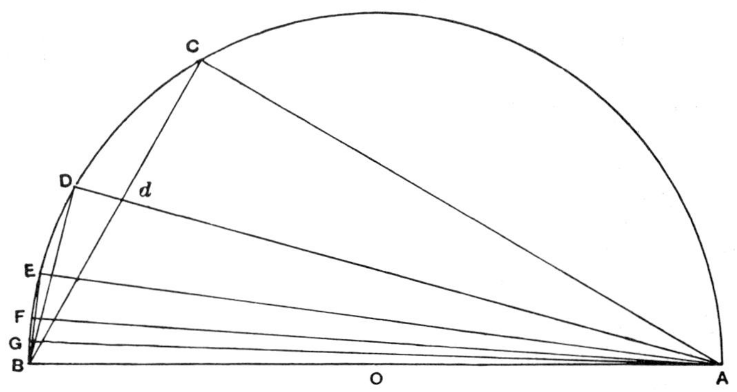

Therefore $$AD : DB = BD : Dd$$
$$[= AC : Cd]$$
$$= AB : Bd \qquad \text{[Eucl. VI. 3]}$$
$$= AB + AC : Bd + Cd$$
$$= AB + AC : BC$$

or $$BA + AC : BC = AD : DB.$$

[But $$AC : CB < 1351 : 780, \text{ from above,}$$

while $$BA : BC = 2 : 1$$
$$= 1560 : 780.]$$

Therefore $$AD : DB < 2911 : 780 \qquad \ldots\ldots (1).$$

[Hence $$AB^2 : BD^2 < (2911^2 + 780^2) : 780^2$$
$$< 9082321 : 608400.]$$

Thus $$AB : BD < 3013\tfrac{3}{4} : 780 \qquad \ldots\ldots (2).$$

Secondly, let AE bisect the angle BAD, meeting the circle in E; and let BE be joined.

Then we prove, in the same way as before, that
$$AE : EB [= BA + AD : BD$$
$$< (3013\tfrac{3}{4} + 2911) : 780, \text{ by (1) and (2)]}$$
$$< 5924\tfrac{3}{4} : 780$$
$$< 5924\tfrac{3}{4} \times {}^4/_{13} : 780 \times {}^4/_{13}$$
$$< 1823 : 240 \qquad \ldots\ldots (3).$$

[Hence $$AB^2 : BE^2 < (1823^2 + 240^2) : 240^2$$
$$< 3380929 : 57600.]$$

Therefore $$AB : BE < 1838\tfrac{9}{11} : 240 \qquad \ldots\ldots (4).$$

Thirdly, let AF bisect the angle BAE, meeting the circle in F.

Thus $$AF : FB [= BA + AE : BE$$
$$< 3661\tfrac{9}{11} : 240, \text{ by (3) and (4)]}$$
$$< 3661\tfrac{9}{11} \times {}^{11}/_{40} : 240 \times {}^{11}/_{40}$$
$$< 1007 : 66 \qquad \ldots\ldots (5).$$

[It follows that

$$AB^2 : BF^2 < (1007^2 + 66^2) : 66^2$$
$$< 1018405 : 4356.]$$

Therefore $AB : BF < 1009^1/_6 : 66$ .... (6).

Fourthly, let the angle BAF be bisected by AG meeting the circle in G.

Then $AG : GB$ [$= BA + AF : BF$]

$$< 2016^1/_6 : 66, \text{ by (5) and (6).}$$

[And $AB^2 : BG^2 < \{(2016^1/_6)^2 + 66^2\} : 66^2$

$$< 4069284^1/_{36} : 4356.]$$

Therefore $AB : BG < 2017\frac{1}{4} : 66$,

whence $BG : AB > 66 : 2017\frac{1}{4}$ .... (7).

[Now the angle BAG which is the result of the fourth bisection of the angle BAC, or of one-third of a right angle, is equal to one-fortyeighth of a right angle.

Thus the angle subtended by BG at the centre is

$$^1/_{24} \text{ (a right angle).]}$$

Therefore BG is a side of a regular inscribed polygon of 96 sides.

It follows from (7) that

(perimeter of polygon) : AB [$> 96 \times 66 : 2017\frac{1}{4}$]

$$> 6336 : 2017\frac{1}{4}.$$

And $$\frac{6336}{2017\frac{1}{4}} > 3^{10}/_{71}.$$

Much more then is the circumference to the diameter

$$< 3^1/_7 \text{ but } > 3^{10}/_{71}.$$

36. From *Quadrature of the Parabola:* Introduction, Propositions 20, 23, and 24*

ARCHIMEDES

INTRODUCTION

Archimedes to Dositheus greeting.

When I heard that Conon, who was my friend in his lifetime, was dead, but that you were acquainted with Conon and withal versed in geometry, while I grieved for the loss not only of a friend but of an admirable mathematician, I set myself the task of communicating to you, as I had intended to send to Conon, a certain geometrical theorem which had not been investigated before but has now been investigated by me, and which I first discovered by means of mechanics and then exhibited by means of geometry. Now some of the earlier geometers tried to prove it possible to find a rectilineal area equal to a given circle and a given segment of a circle; and after that they endeavoured to square the area bounded by the section of the whole cone and a straight line, assuming lemmas not easily conceded, so that it was recognised by most people that the problem was not solved. But I am not aware that any one of my predecessors has attempted to square

*Source: Reprinted by permission from Thomas L. Heath (ed.), *The Works of Archimedes* (Dover Edition, 1953), 233-234 and 248-252.

the segment bounded by a straight line and a section of a right-angled cone [a parabola], of which problem I have now discovered the solution. For it is here shown that every segment bounded by a straight line and a section of a right-angled cone [a parabola] is four-thirds of the triangle which has the same base and equal height with the segment, and for the demonstration of this property the following lemma is assumed: that the excess by which the greater of (two) unequal areas exceeds the less can, by being added to itself, be made to exceed any given finite area. The earlier geometers have also used this lemma; for it is by the use of this same lemma that they have shown that circles are to one another in the duplicate ratio of their diameters, and that spheres are to one another in the triplicate ratio of their diameters, and further that every pyramid is one third part of the prism which has the same base with the pyramid and equal height; also, that every cone is one third part of the cylinder having the same base as the cone and equal height they proved by assuming a certain lemma similar to that aforesaid. And, in the result, each of the aforesaid theorems has been accepted no less than those proved without the lemma. As therefore my work now published has satisfied the same test as the propositions referred to, I have written out the proof and send it to you, first as investigated by means of mechanics, and afterwards too as demonstrated by geometry. Prefixed are, also, the elementary propositions in conics which are of service in the proof. *Farewell.*

PROPOSITION 20

If Qq be the base, and P the vertex, of a parabolic segment, then the triangle PQq is greater than half the segment PQq.

For the chord Qq is parallel to the tangent at P, and the triangle PQq is half the parallelogram formed by Qq, the

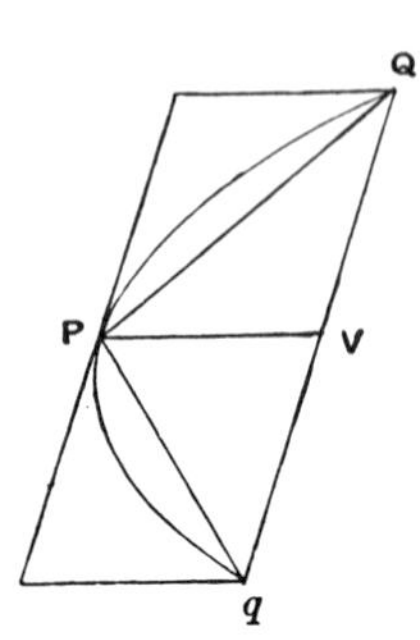

tangent at P, and the diameters through Q,q.

Therefore the triangle PQq is greater than half the segment.

Cor. It follows that *it is possible to inscribe in the segment a polygon such that the segments left over are together less than any assigned area.*

PROPOSITION 23

Given a series of areas A, B, C, D, . . . Z, of which A is the greatest, and each is equal to four times the next in order, then

$$A + B + C + \ldots + Z + \tfrac{1}{3}Z = \tfrac{4}{3}A.$$

Take areas $b, c, d, \ldots$ such that $b = \frac{1}{3}B$, $c = \frac{1}{3}C$, $d = \frac{1}{3}D$, and so on.

Then, since $b = \frac{1}{3}B$, and $B = \frac{1}{4}A$, $B + b = \frac{1}{3}A$. Similarly $C + c = \frac{1}{3}B$.

Therefore

$$B + C + D + \ldots + Z + b + c + d + \ldots + z = \tfrac{1}{3}(A + B + C + \ldots + Y).$$

But
$$b + c + d + \ldots + y = \tfrac{1}{3}(B + C + D + \ldots + Y).$$

Therefore, by subtraction,

$$B + C + D + \ldots + Z + z = \tfrac{1}{3}A$$

or

$$A + B + C + \ldots + Z + \tfrac{1}{3}Z = \tfrac{4}{3}A.$$

[The algebraical equivalent of this result is of course

$$1 + \tfrac{1}{4} + (\tfrac{1}{4})^2 + \ldots + (\tfrac{1}{4})^{n-1}$$
$$= \tfrac{4}{3} - \tfrac{1}{3}(\tfrac{1}{4})^{n-1}$$
$$= \frac{1 - (\frac{1}{4})^n}{1 - \frac{1}{4}} \;]$$

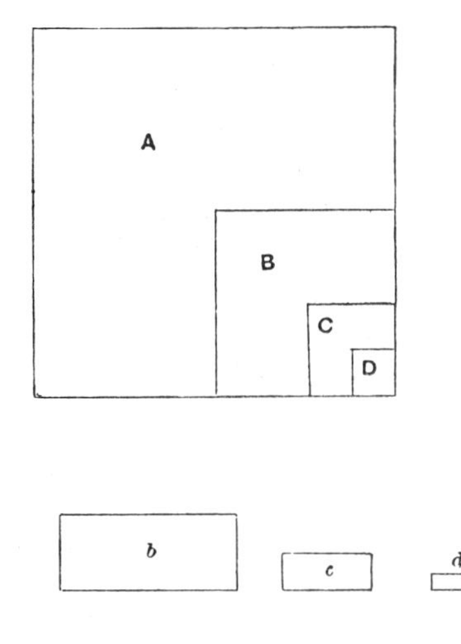

PROPOSITION 24

Every segment bounded by a parabola and a chord Qq is equal to four-thirds of the triangle which has the same base as the segment and equal height.

Suppose $K = \frac{4}{3}\,\Delta PQq$, where P is the vertex of the segment; and we have then to prove that the area of the segment is equal to K.

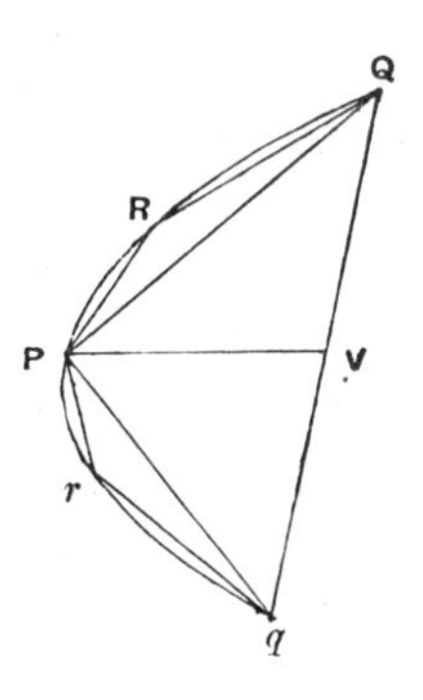

For, if the segment be not equal to K, it must either be greater or less.

I. Suppose the area of the segment greater than K.

If then we inscribe in the segments cut off by PQ, Pq triangles which have the same base and equal height, i.e. triangles with the same vertices R, r as those of the segments, and if in the remaining segments we inscribe triangles in the same manner, and so on, we shall finally have segments remaining whose sum is less than the area by which the segment PQq exceeds K.

Therefore the polygon so formed must be greater than the area K; which is impossible, since [Prop. 23]

$$A + B + C + \ldots + Z < \tfrac{4}{3} A,$$

where $A = \Delta\ PQq$.

Thus the area of the segment cannot be greater than K.

II. Suppose, if possible, that the area of the segment is less than K.

If then $\Delta PQq = A$, $B = \frac{1}{4}A$, $C = \frac{1}{4}B$, and so on, until we arrive at an area X such that X is less than the difference between K and the segment, we have

$$\begin{aligned} A + B + C + \ldots + X + \tfrac{1}{3}X &= \tfrac{4}{3} A \qquad \text{[Prop. 23]} \\ &= K. \end{aligned}$$

Now, since K exceeds $A + B + C + \ldots + X$ by an area less than X, and the area of the segment by an area greater than X, it follows that

$$A + B + C + \ldots + X > \text{(the segment)};$$

which is impossible, by Prop. 22 above.

Hence the segment is not less than K.

Thus, since the segment is neither greater nor less than K,

$$\text{(area of segment } PQq) = K = \tfrac{4}{3}\,\Delta PQq.$$

37. From *On the Equilibrium of Planes I*: Propositions 6 and 7*

(Principle of the Lever)

ARCHIMEDES

PROPOSITION 6

Commensurable *magnitudes balance at distances reciprocally proportional to their weights.*

Let A, B be commensurable magnitudes with centres [of gravity] A, B, and let EΔ be any distance, and let

$$A : B = \Delta\Gamma : \Gamma E;$$

it is required to prove that the centre of gravity of the magnitude composed of both A, B is Γ.

Since

$$A : B = \Delta\Gamma : \Gamma E,$$

and A is commensurate with B, therefore ΓΔ is commensurate with ΓE, that is, a straight line with a straight line [Eucl. x. 11]; so that EΓ, ΓΔ have a common measure. Let it be N, and let ΔH, ΔK be each equal to EΓ, and let EΛ be equal to ΔΓ. Then since $\Delta H = \Gamma E$, it follows that $\Delta\Gamma = EH$; so that $\Lambda EE = H$. Therefore $\Lambda H = 2\Delta\Gamma$ and $HK = 2\Gamma E$; so that N measures both ΛH and HK, since it measures their halves [Eucl. x. 12]. And since

$$A : B = \Delta\Gamma : \Gamma E,$$

while

$$\Delta\Gamma : \Gamma E = \Lambda H : HK—$$

for each is double of the other—therefore

$$A : B = \Lambda H : HK.$$

Now let Z be the same part of A as N is of ΔH;

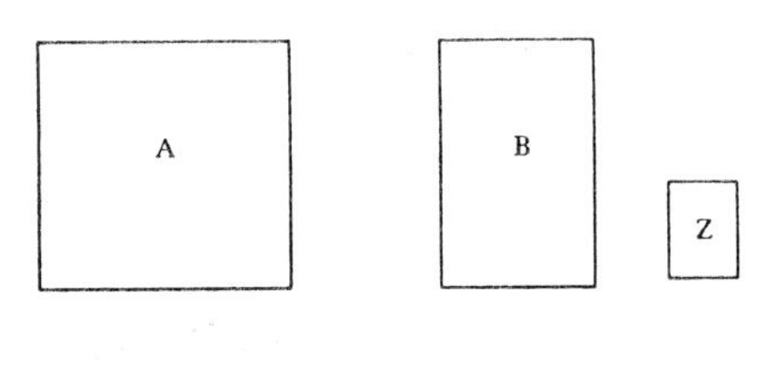

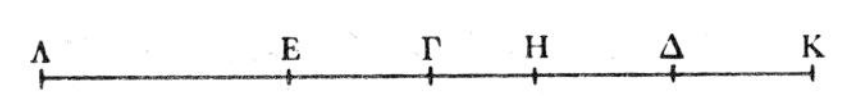

then

$$\Delta H : N = A : Z. \qquad \text{[Eucl. v., Def. 5]}$$

And

$$KH : \Lambda H = B : A; \text{ [Eucl. v. 7, coroll.]}$$

therefore, *ex aequo*,

$$KH : N = B : Z; \quad \text{[Eucl. v. 22]}$$

therefore Z is the same part of B as N is of KH. Now A was proved to be a multiple of Z; therefore Z is a common measure of A, B. Therefore, if ΛH is divided into segments equal to N and A into segments equal to Z, the segments in ΛH equal in magnitude to N will be equal in number to the segments of A equal to Z. It follows that, if there be placed on each of the segments in ΛH a

**Source:* From *Selections Illustrating the History of Greek Mathematics,* trans. by Ivor Thomas (1941), vol. II, 209-217. Reprinted by permission of Harvard University Press.

magnitude equal to Z, having its centre of gravity at the middle of the segment, the sum of the magnitudes will be equal to A, and the centre of gravity of the figure compounded of them all will be E; for they are even in number, and the numbers on either side of E will be equal because $\Lambda E = HE$.

Similarly it may be proved that, if a magnitude equal to Z be placed on each of the segments [equal to N] in KH, having its centre of gravity at the middle of the segment, the sum of the magnitudes will be equal to B, and the centre of gravity of the figure compounded of them all will be Δ. Therefore A may be regarded as placed at E, and B at Δ. But they will be a set of magnitudes lying on a straight line, equal one to another, with their centres of gravity at equal intervals, and even in number; it is therefore clear that the centre of gravity of the magnitude compounded of them all is the point of bisection of the line containing the centres [of gravity] of the middle magnitudes. And since $\Lambda E = \Gamma\Delta$ and $E\Gamma = \Delta K$, therefore $\Lambda\Gamma = \Gamma K$; so that the centre of gravity of the magnitude compounded of them all is the point Γ. Therefore if A is placed at E and B at Δ, they will balance about Γ.

PROPOSITION 7

And now, if the magnitudes be incommensurable, they will likewise balance at distances reciprocally proportional to the magnitudes.

Let (A + B), Γ be incommensurable magnitudes,[1] and let ΔE, EZ be distances, and let

$$(A + B) : \Gamma = E\Delta : EZ;$$

I say that the centre of gravity of the magnitude composed of both (A + B), Γ is E.

For if (A + B) placed at Z do not balance Γ placed at Δ, either (A + B) is too much greater than Γ to balance or less. Let it [first] be too much greater, and let there be subtracted from (A + B) a magnitude less than the excess by which (A + B) is too much greater than Γ to balance, so that the remainder A is commensurate with Γ. Then, since A, Γ are commensurable magnitudes, and

$$A : \Gamma < \Delta E : EZ,$$

A, Γ will not balance at the distances ΔE, EZ, A being placed at Z and Γ at Δ. By the same reasoning, they will not do so if Γ is greater than the magnitude necessary to balance (A + B).[2]

NOTES

1. As becomes clear later in the proof, the first magnitude is regarded as made up of two parts—A, which is commensurate with Γ and B, which is not commensurate; if (A + B) is too big for equilibrium with Γ, then B is so chosen that, when it is taken away, the remainder A is still too big for equilibrium with Γ. Similarly if (A + B) is too small for equilibrium.

2. The proof is incomplete and obscure; it may be thus completed.
Since

$$A : \Gamma < \Delta E : EZ,$$

Δ will be depressed, which is impossible, since there has been taken away from (A + B) a magnitude less than the deduction necessary to produce equilibrium, so that Z remains depressed. Therefore (A + B) is not greater than the magnitude necessary to produce equilibrium; in the same way it can be proved not to be less; therefore it is equal.

38. From *The Sand-Reckoner:* Introduction and Section on Large Numbers*

ARCHIMEDES

THERE are some, king Gelon, who think that the number of the sand is infinite in multitude; and I mean by the sand not only that which exists about Syracuse and the rest of Sicily but also that which is found in every region whether inhabited or uninhabited. Again there are some who, without regarding it as infinite, yet think that no number has been named which is great enough to exceed its multitude. And it is clear that they who hold this view, if they imagined a mass made up of sand in other respects as large as the mass of the earth, including in it all the seas and the hollows of the earth filled up to a height equal to that of the highest of the mountains, would be many times further still from recognising that any number could be expressed which exceeded the multitude of the sand so taken. But I will try to show you by means of geometrical proofs, which you will be able to follow, that, of the numbers named by me and given in the work which I sent to Zeuxippus, some exceed not only the number of the mass of sand equal in magnitude to the earth filled up in the way described, but also that of a mass equal in magnitude to the universe. Now you are aware that "universe" is the name given by most astronomers to the sphere whose centre is the centre of the earth and whose radius is equal to the straight line between the centre of the sun and the centre of the earth. This is the common account as you have heard from astronomers. But Aristarchus of Samos brought out a book consisting of some hypotheses, in which the premisses lead to the result that the universe is many times greater than that now so called. His hypotheses are that the fixed stars and the sun remain unmoved, that the earth revolves about the sun in the circumference of a circle, the sun lying in the middle of the orbit, and that the sphere of the fixed stars, situated about the same centre as the sun, is so great that the circle in which he supposes the earth to revolve bears such a proportion to the distance of the fixed stars as the centre of the sphere bears to its surface. Now it is easy to see that this is impossible; for, since the centre of the sphere has no magnitude, we cannot conceive it to bear any ratio whatever to the surface of the sphere. We must however take Aristarchus to mean this: since we conceive the earth to be, as it were, the centre of the universe, the ratio which the earth bears to what we describe as the "universe" is the same as the ratio which the sphere containing the circle in which he supposes the earth to revolve bears to the sphere of the fixed stars. For he adapts the proofs of his results to a hypothesis of this kind, and in particular he appears to suppose the magnitude of the sphere in which he represents the earth as moving to be equal to what we call the "universe."

I say then that, even if a sphere were

Source: Reprinted by permission from Thomas L. Heath (ed.), *The Works of Archimedes* (Dover Edition, 1953), 221-222 and 227-229.

made up of the sand, as great as Aristarchus supposes the sphere of the fixed stars to be, I shall still prove that, of the numbers named in the *Principles,*[1] some exceed in multitude the number of the sand which is equal in magnitude to the sphere referred to . . . .

ORDERS AND PERIODS OF NUMBERS

I. We have traditional names for numbers up to a myriad (10,000); we can therefore express numbers up to a myriad myriads (100,000,000). Let these numbers be called numbers of the *first order.*

Suppose the 100,000,000 to be the unit of the *second order,* and let the *second order* consist of the numbers from that unit up to $(100{,}000{,}000)^2$.

Let this again be the unit of the *third order* of numbers ending with $(100{,}000{,}000)^2$; and so on, until we reach the 100,000,000*th order* of numbers ending with $(100{,}000{,}000)^{100{,}000{,}000}$, which we will call P.

II. Suppose the numbers from 1 to P just described to form the *first period.*

Let P be the unit of the *first order of the second period,* and let this consist of the numbers from P up to 100,000,000 P.

Let the last number be the unit of the *second order of the second period,* and let this end with $(100{,}000{,}000)^2\,P$.

We can go on in this way till we reach the 100,000,000*th order of the second period* ending with $(100{,}000{,}000)^{100{,}000{,}000}\,P$, or P^2.

III. Taking P^2 as the unit of the *first order of the third period,* we proceed in the same way till we reach the 100,000,000*th order of the third period* ending with P^3.

IV. Taking P^3 as the unit of the *first order of the fourth period,* we continue the same process until we arrive at the 100,000,000*th order of the* 100,000,000*th period* ending with $P^{100{,}000{,}000}$. This last number is expressed by Archimedes as "a myriad-myriad units of the myriad-myriad-th order of the myriad-myriad-th period (α'ι μυριακισμυριοστᾶς περιόδου μυριακισμυριοστῶν 'αριθμῶν μυρίαι μυριάδες)," which is easily seen to be 100,000,000 times the product of $(100{,}000{,}000)^{99{,}999{,}999}$ and $P^{99{,}999{,}999}$, *i.e.* $P^{100{,}000{,}000}$.

[The scheme of numbers thus described can be exhibited more clearly by means of *indices* as follows.

FIRST PERIOD

First order.	Numbers from	1 to 10^8.
Second order.	" "	10^8 to 10^{16}.
Third order.	" "	10^{16} to 10^{24}.
⋮		
(10^8)*th order.*	" "	$10^{8\cdot(10^8-1)}$ to $10^{8\cdot 10^8}$ (P, say).

SECOND PERIOD

First order.	Numbers from	$P.1$ to $P.10^8$.
Second order.	" "	$P.10^8$ to $P.10^{16}$.
⋮		
(10^8)*th order.*	" "	$P.10^{8\cdot(10^8-1)}$ to $P.10^{8\cdot 10^8}$ (or P^2).

⋮

(10^8)TH PERIOD

First order.	" "	$P^{10^8-1}.1.$ to $P^{10^8-1}.10^8$.
Second order.	" "	$P^{10^8-1}.10^8$ to $P^{10^8-1}.10^{16}$.
⋮		
(10^8)*th order.*	" "	$P^{10^8-1}\,.10^{8\cdot(10^8-1)}$ to $P^{10^8-1}.10^{8\cdot 10^8}$ (*i.e.* P^{10^8}).

The prodigious extent of this scheme will be appreciated when it is considered that the last number in the *first period* would be represented now by 1 followed by 800,000,000 ciphers, while the last number of the 10^8)*th period* would require 100,000,000 times as many ciphers, *i.e.,* 80,000 million millions of ciphers.]

OCTADS

Consider the series of terms in continued proportion of which the first is 1 and the second 10 [*i.e.* the geometrical progression 1, 10^1, 10^2, 10^3, . . .]. The *first octad* of these terms [*i.e.* 1, 10^1, 10^2, . . . 10^7] fall accordingly under the *first order of the first period* above described, the *second octad* [*i.e.* 10^8, 10^9, . . . 10^{15}] under the *second order of the first period*, the first term of the octad being the unit of the corresponding order in each case. Similarly for the *third octad*, and so on. We can, in the same way, place any number of octads.

THEOREM

If there be any number of terms of a series in continued proportion, say A_1, A_2, A_3, . . . A_m, . . . A_n, . . . A_{m+n-1}, . . . of which $A_1 = 1$, $A_2 = 10$ [so that the series forms the geometrical progression 1, 10^1, 10^2, . . .10^{m-1}, . . . 10^{n-1}, . . . 10^{m+n-2}, . . .], *and if any two terms as A_m, A_n be taken and multiplied, the product $A_m \cdot A_n$ will be a term in the same series and will be as many terms distant from A_n as A_m is distant from A_1; also it will be distant from A_1, by a number of terms less by one than the sum of the numbers of terms by which A_m and A_n respectively are distant from A_1.*

Take the term which is distant from A_n by the same numbers of terms as A_m is distant from A_1. This number of terms is m (the first and last being both counted). Thus the term to be taken is m terms distant from A_n, and is therefore the term A_{m+n-1}.

We have therefore to prove that

$$A_m \cdot A_n = A_{m+n-1}.$$

Now terms equally distant from other terms in the continued proportion are proportional.

Thus

$$\frac{A_m}{A_1} = \frac{A_{m+n-1}}{A_n}.$$

But

$$A_m = A_m \cdot A_1, \text{ since } A_1 = 1.$$

Therefore

$$A_{m+n-1} = A_m \cdot A_n \quad \ldots\ldots\ldots\ldots\ldots\ldots (1).$$

The second result is now obvious, since A_m *is* m terms distant from A_1, A_n is n terms distant from A_1, and A_{m+n-1} is $(m + n - 1)$ terms distant from A_1.

APPLICATION TO THE NUMBER OF THE SAND

By Assumption 5 [of this treatise].

(diam. of poppy-seed) $\not<$ 1/40 (finger-breadth); and, since spheres are to one another in the triplicate ratio of their diameters, it follows that (sphere of diam. 1 finger-breadth)

$\not>$ 64,000 poppy-seeds
$\not>$ 64,000 × 10,000
$\not>$ 640,000,000
$\not>$ 6 units of *second order* + 40,000,000 units of *first order*
(*a fortiori*) $<$ 10 units of *second order* of numbers.
} grains of sand.

We now gradually increase the diameter of the supposed sphere, multiplying it by 100 each time. Thus, remembering that the sphere is thereby multiplied by 100^3 or 1,000,000, the number of grains of sand which would be contained in a sphere with each successive diameter may be arrived at as follows.

Diameter of sphere.	**Corresponding number of grains of sand.**	
(1) 100 finger-breadths	<1,000,000 × 10 units of *second order*	
	<(7th term of series) × (10th term of series)	
	<16th term of series	[*i.e.*, 10^{15}]
	<[10^7 or] 10,000,000 units of the *second order*.	
(2) 10,000 finger-breadths	<1,000,000 × (last number)	
	<(7th term of series) × (16th term)	
	<22nd term of series	[*i.e.*, 10^{21}]
	<[10^5 or] 100,000 units of *third order*.	

(3) 1 stadium (<10,000 finger-breadths)	<100,000 units of *third order*.	
(4) 100 stadia	<1,000,000 × (last number)	
	<(7th term of series) × (22nd term)	
	<28th term of series	[10^{27}]
	<[10^3 or] 1,000 units of *fourth order*.	
(5) 10,000 stadia	<1,000,000 × (last number)	
	<(7th term of series) × (28th term)	
	<34th term of series	[10^{33}]
	<10 units of *fifth order*.	
(6) 1,000,000 stadia	<(7th term of series) × (34th term)	
	<40th term	[10^{39}]
	<[10^7 or] 10,000,000 units of *fifth order*.	
(7) 100,000,000 stadia	<(7th term of series) × (40th term)	
	<46th term	[10^{45}]
	<[10^5 or] 100,000 units of *sixth order*.	
(8) 10,000,000,000 stadia	<(7th term of series) × (46th term)	
	<52nd term of series	[10^{51}]
	<[10^3 or] 1,000 units of *seventh order*.	

But, by the propostion above, . . .

(diameter of "universe") < 10,000,000,000 stadia.

Hence *the number of grains of sand which could be contained in a sphere of the size of our "universe" is less than* 1,000 *units of the seventh order of numbers* [or 10^{51}].

From this we can prove further that *a sphere of the size attributed by Aristarchus to the sphere of the fixed stars would contain a number of grains of sand less than* 10,000,000 *units of the eighth order of numbers* [or $10^{56+7} = 10^{63}$].

For, by hypothesis,

(earth) : ("universe") = ("universe"): (sphere of fixed stars).

And . . .

(diameter of "universe") < 10,000 (diam. of earth);

whence

(diam. of sphere of fixed stars) < 10,000 (diam. of "universe").

Therefore

(sphere of fixed stars) < $(10,000)^3$. ("universe").

It follows that the number of grains of sand which would be contained in a sphere equal to the sphere of the fixed stars

<$(10,000)^3$ × 1,000 units of *seventh order*

<(13th term of series) × (52nd term of series)

<64th term of series [*i.e.*, 10^{63}]

<[10^7 or] 10,000,000 units of *eighth order* of numbers.

CONCLUSION

I conceive that these things, king Gelon, will appear incredible to the great majority of people who have not studied mathematics, but that to those who are conversant therewith and have given thought to the question of the distances and sizes of the earth the sun and moon and the whole universe the proof will carry conviction. And it was for this reason that I thought the subject would be not inappropriate for your consideration.

NOTE

1. This was apparently the title of the work sent to Zeuxippus.

39. From *The Cattle Problem**

(Indeterminate Analysis)[1]

ARCHIMEDES

A problem which Archimedes solved in epigrams, and which he communicated to students of such matters at Alexandria in a letter to Eratosthenes of Cyrene.

If thou art diligent and wise, O stranger, compute the number of cattle of the Sun, who once upon a time grazed on the fields of the Thrinacian isle of Sicily, divided into four herds of different colours, one milk white, another a glossy black, the third yellow and the last dappled. In each herd were bulls, mighty in number according to these proportions: Understand, stranger, that the white bulls were equal to a half and a third of the black together with the whole of the yellow, while the black were equal to the fourth part of the dappled and a fifth, together with, once more, the whole of the yellow. Observe further that the remaining bulls, the dappled, were equal to a sixth part of the white and a seventh, together with all the yellow. These were the proportions of the cows: The white were precisely equal to the third part and a fourth of the whole herd of the black; while the black were equal to the fourth part once more of the dappled and with it a fifth part, when all, including the bulls, went to pasture together. Now the dappled in four parts[2] were equal in number to a fifth part and a sixth of the yellow herd. Finally the yellow were in number equal to a sixth part and a seventh of the white herd. If thou canst accurately tell, O stranger, the number of cattle of the Sun, giving separately the number of well-fed bulls and again the number of females according to each colour, thou wouldst not be called unskilled or ignorant of numbers, but not yet shalt thou be numbered among the wise. But come, understand also all these conditions regarding the cows of the Sun. When the white bulls mingled their number with the black, they stood firm, equal in depth and breadth,[3] and the plains of Thrinacia, stretching far in all ways, were filled with their multitude. Again, when the yellow and the dappled bulls were gathered into one herd they stood in such a manner that their number, beginning from one, grew slowly greater till it completed a triangular figure, there being no bulls of other colours in their midst nor none of them lacking. If thou art able, O stranger, to find out all these things and gather them together in your mind, giving all the relations, thou shalt depart crowned with glory and knowing that thou hast been adjudged perfect in this species of wisdom.[4]

NOTES

1. It is unlikely that the epigram itself, first edited by G. E. Lessing in 1773, is the work of Archimedes, but there is ample evidence from antiquity that he studied the actual problem. See T. L. Heath, *The Works of Archimedes,* pp. 319-326.

2. *i.e.,* a fifth and a sixth both of the males and of the females.

**Source:* Reprinted by permission from Thomas L. Heath (ed.), *The Works of Archimedes* (Dover Edition, 1953), 319-326.

3. At a first glance this would appear to mean that the sum of the number of white and black bulls is a square, but this makes the solution of the problem intolerably difficult. There is, however, an easier interpretation. If the bulls are packed together so as to form a square figure, their number need not be a square, since each bull is longer than it is broad. The simplified condition is that the sum of the number of white and black bulls shall be a rectangle.

4. If

X, x are the numbers of white bulls and cows respectively,						
Y, y	"	"	" black	"	"	"
Z, z	"	"	" yellow	"	"	"
W, w	"	"	" dappled	"	"	"

the first part of the epigram states that

(a)
$$X = (1/2 + 1/3)\,Y + Z \quad \ldots\ldots \quad (1)$$
$$Y = (1/4 + 1/5)\,W + Z \quad \ldots\ldots \quad (2)$$
$$W = (1/6 + 1/7)\,X + Z \quad \ldots\ldots \quad (3)$$

(b)
$$x = (1/3 + 1/4)\,(Y + y) \quad \ldots\ldots \quad (4)$$
$$y = (1/4 + 1/5)\,(W + w) \quad \ldots\ldots \quad (5)$$
$$w = (1/5 + 1/6)\,(Z + z) \quad \ldots\ldots \quad (6)$$
$$z = (1/6 + 1/7)\,(X + x) \quad \ldots\ldots \quad (7)$$

The second part of the epigram states that

$$X + Y = \text{a rectangular number} \quad \ldots\ldots \quad (8)$$
$$Z + W = \text{a triangular number} \quad \ldots\ldots \quad (9)$$

This was solved by J. F. Wurm, and the solution is given by A. Amthor, *Zeitschrift für Math. u. Physik. (Hist.-litt. Abtheilung)*, xxv. (1880), pp. 153-171, and by Heath, *The Works of Archimedes*, pp. 319-326. For reasons of space, only the results can be noted here.

Equations (1) to (7) give the following as the values of the unknowns in terms of an unknown integer n:

$$X = 10366482n \qquad x = 7206360n$$
$$Y = 7460514n \qquad y = 4893246n$$
$$Z = 4149387n \qquad z = 5439213n$$
$$W = 7358060n \qquad w = 3515820n.$$

We have now to find a value of n such that equation (9) is also satisfied—equation (8) will then be simultaneously satisfied. Equation (9) means that

$$Z + W = \frac{p(p+1)}{2}$$

Chapter III
The Hellenistic and Roman Periods

ERATOSTHENES (*c.* 276–*c.* 195 B.C.)

A young contemporary of Archimedes, Eratosthenes was born in Cyrene (now Shabhat, Libya) on the south coast of the Mediterranean Sea. Going to Athens to study philosophy (*c.* 260 B.C.), he found the Platonic, Aristotelian, Epicurean, and Stoic Schools in constant dispute with one another. He especially admired two teachers—the Peripatetic Ariston of Chios, who criticized the moral casuistry of the Stoics and held to absolutist ethics, and Arcesilaus, who reorganized the Academy by requiring each student first to learn mathematics thoroughly and then to study the dialectical method before addressing more abstract problems. Arcesilaus was known for not indoctrinating students in his own views. Eratosthenes received his higher education at the Academy, where Autolycus taught him mathematics.

Around 246 B.C., Ptolemy III invited Eratosthenes to Alexandria. He was to spend the rest of his life there. For a while, he tutored the crown prince. When the chief librarian died (*c.* 235 B.C.), Ptolemy appointed him to head the famous library.

In a city where polymaths were greatly admired, Eratosthenes wrote substantial works on geography, mathematics, astronomy, philosophy, chronology, and literary criticism, composed poetry, and was a distinguished athlete. He was known as a learned, industrious, and conscientious man with artistic tastes. Students at the Alexandrian Museum called him "Pentathlus," the champion of the five sports. According to *Suda Lexikon,* he was described as "All-Rounder," "Another Plato," and "Beta." The nickname "Beta" may suggest that he was second only to Archimedes or that he attained superior rank in many fields instead of the highest rank in one. Strabo, who was critical of his geographical work, called him a mathematician among geographers and a geographer among mathematicians. But this appears an erroneous assessment, since Archimedes saw fit to dedicate his treatise *On Method* to Eratosthenes.

In old age Eratosthenes became almost blind from opthalmia and is said to have committed suicide by voluntary starvation.

The *Geography* (3 books) of Eratosthenes contained his procedure and calculations for measuring the circumference of the Earth. He recorded the angle of the shadow made by a gnomon at Alexandria at noon on the summer solstice, a time when the sun was known to cast no shadow at Syene (modern Aswan, Egypt). The inclination of the sun's rays with the vertical expressed in the Alexandrian system of angle measurement was $1/50$ of four right angles, *i.e.,* $7° 12'$. (The familiar 360° was unknown to him.)

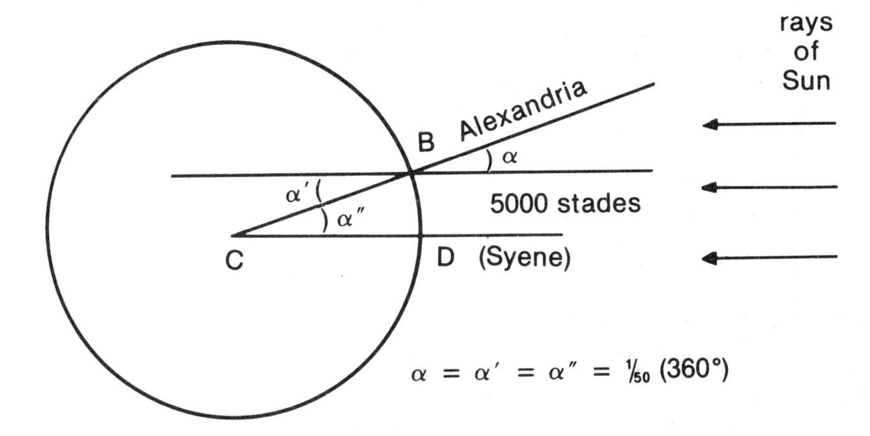

The distance from Syene to Alexandria was known. To make his calculation Eratosthenes also employed Euclid I.15 (equality of vertical angles, *i.e.*, $<\alpha = <\alpha'$) and I.29 (equality of alternate interior angles, *i.e.*, $<\alpha' = <\alpha''$). Expressed in modern notation, the calculation to determine the circumference can be stated as follows:

$$\frac{C}{360^\circ} = \frac{5{,}000 \text{ stades}}{1/50\,(360^\circ)} = \frac{5{,}000 \text{ stades}}{7^\circ 12'}$$

$$C = 250{,}000 \text{ stades}$$

Since he knew that 5,000 stades was only an estimate of the distance between Alexandria and Syene, he later took the liberty to increase his result by 2,000 stades. The accuracy of his result depends upon which of several possible stades he employed as a unit of measurement, because the stade varies from $7^1/_2$ to 10 stades to the Roman mile. His 252,000 stades were probably equivalent to 29,000 English miles which is a close approximation of the earth's actual circumference of 24,888 miles.

Eratosthenes' chief work in mathematics was probably the book *Platonicus,* in which he described a mechanical solution to the Delian problem of doubling the cube. Using an apparatus with three movable rulers, he found mean proportionals in continued proportion in order to solve the Delian problem. He also discussed the generation of mean proportionals. His best known achievement in mathematics was his discovery in arithmetic of a method for finding all the prime numbers less than a given number N. This method is now known as the sieve of Eratosthenes.

40. From *Introduction to Arithmetic I:* Chapters XII-XIII*

(Sieve for Finding Primes)

NICOMACHUS OF GERASA

CHAPTER XII

[1] The secondary, composite number[1] is an odd number, indeed, because it is distinguished as a member of this same class, but it has no elementary quality, for it gets its origin by the combination of something else. For this reason it is characteristic of the secondary number to have, in addition to the fractional part with the number itself as denominator, yet another part or parts with different denominators, the former always, as in all cases, unity, the latter never unity, but always either that number or those numbers by the combination of which it was produced. For example, 9, 15, 21, 25, 27, 33, 35, 39; each one of these is measured by unity, as other numbers are, and like them has a fractional part with the same denominator as the number itself, by the nature of the class common to them all; but by exception and more peculiarly they also employ a part, or parts, with a different denominator; 9, in addition to the ninth part, has a third part besides; 15 a third and a fifth besides a fifteenth; 21 a seventh and a third besides a twenty-first, and 25, in addition to the twenty-fifth, which has as a denominator 25 itself, also a fifth, with a different denominator.

[2] It is called secondary, then, because it can employ yet another measure along with unity, and because it is not elementary, but is produced by some other number combined with itself or with something else; in the case of 9, 3; in the case of 15, 5 or, by Zeus, 3; and those following in the same fashion. And it is called composite for this, or some such, reason: that it may be resolved into those numbers out of which it was made, since it can also be measured by them. For nothing that can be broken down is incomposite, but by all means composite.

CHAPTER XIII

[1] Now while these two species of the odd are opposed to each other a third one[2] is conceived of between them, deriving, as it were, its specific form from them both, namely the number which is in itself secondary and composite, but relatively to another number is prime and incomposite. This exists when a number, in addition to the common measure, unity, is measured by some other number and is therefore able to admit of a fractional part, or parts, with denominator other than the number itself, as well as the one with itself as denominator. When this is compared with another number of similar properties, it is found that it cannot be measured by a measure common to the other, nor does it have a fractional

Source: From Nicomachus of Gerasa, *Introduction to Arithmetic* (*Great Books* edition) trans. by Luther d'Ooge, Frank Egleston Robbins, and Louis Charles Karpinski, vol. II, 818-819. Reprinted by arrangement with the University of Michigan Press.

part with the same denominator as those in the other. As an illustration, let 9 be compared with 25. Each in itself is secondary and composite, but relatively to each other they have only unity as a common measure, and no factors in them have the same denominator, for the third part in the former does not exist in the latter nor is the fifth part in the latter found in the former.

[2] The production of these numbers is called by Eratosthenes the "sieve," because we take the odd numbers mingled together and indiscriminate and out of them by this method of production separate, as by a kind of instrument or sieve, the prime and incomposite by themselves, and the secondary and composite by themselves, and find the mixed class by themselves.

[3] The method of the "sieve" is as follows. I set forth all the odd numbers in order, beginning with 3, in as long a series as possible, and then starting with the first I observe what ones it can measure, and I find that it can measure the terms two places apart, as far as we care to proceed. And I find that it measures not as it chances and at random, but that it will measure the first one, that is, the one two places removed, by the quantity of the one that stands first in the series, that is, by its own quantity, for it measures it 3 times; and the one two places from this by the quantity of the second in order, for this it will measure 5 times; and again the one two places farther on by the quantity of the third in order, or 7 times, and the one two places still farther on by the quantity of the fourth in order, or 9 times, and so *ad infinitum* in the same way.

[4] Then taking a fresh start I come to the second number and observe what it can measure, and find that it measures all the terms four places apart, the first by the quantity of the first in order, or 3 times; the second by that of the second, or 5 times; the third by that of the third, or 7 times; and in this order *ad infinitum.*

[5] Again, as before, the third term 7, taking over the measuring function, will measure terms six places apart, and the first by the quantity of 3, the first of the series, the second by that of 5, for this is the second number, and the third by that of 7, for this has the third position in the series.

[6] And analogously throughout, this process will go on without interruption, so that the numbers will succeed to the measuring function in accordance with their fixed position in the series; the interval separating terms measured is determined by the orderly progress of the even numbers from 2 to infinity, or by the doubling of the position in the series occupied by the measuring term, and the number of times a term is measured is fixed by the orderly advance of the odd numbers in series from 3.

NOTES

1. Cf. Euclid, *Elements.* VII, Def. 14.
2. Cf. Euclid, *Elements,* VII, Def. 13.

41. From *On the Circular Motion of the Heavenly Bodies* i.10.52*

(Estimate of the Circumference of the Earth)

CLEOMEDES[1]

Such then is Posidonius's method of investigating the size of the earth, but Eratosthenes' method depends on a geometrical argument, and gives the impression of being more obscure. What he says will, however, become clear if the following assumptions are made. Let us suppose, in this case also, first that Syene and Alexandria lie under the same meridian circle; secondly, that the distance between the two cities is 5,000 stades; and thirdly, that the rays sent down from different parts of the sun upon different parts of the earth are parallel; for the geometers proceed on this assumption. Fourthly, let us assume that, as is proved by the geometers, straight lines falling on parallel straight lines make the alternate angles equal, and fifthly, that the arcs subtended by equal angles are similar, that is, have the same proportion and the same ratio to their proper circles—this also being proved by the geometers. For whenever arcs of circles are subtended by equal angles, if any one of these is (say) one-tenth of its proper circle, all the remaining arcs will be tenth parts of their proper circles.

Anyone who has mastered these facts will have no difficulty in understanding the method of Eratosthenes, which is as follows. Syene and Alexandria, he asserts, are under the same meridian. Since meridian circles are great circles in the universe, the circles on the earth which lie under them are necessarily great circles also. Therefore, of whatever size this method shows the circle on the earth through Syene and Alexandria to be, this will be the size of the great circle on the earth. He then asserts, as is indeed the case, that Syene lies under the summer tropic. Therefore, whenever the sun, being in the Crab at the summer solstice, is exactly in the middle of the heavens, the pointers of the sundials necessarily throw no shadows, the sun being in the exact vertical line above them; and this is said to be true over a space 300 stades in diameter. But in Alexandria at the same hour the pointers of the sundials throw shadows, because this city lies farther to the north than Syene. As the two cities lie under the same meridian great circle, if we draw an arc from the extremity of the shadow of the pointer to the base of the pointer of the sundial in Alexandria, the arc will be a segment of a great circle in the bowl of the sundial, since the bowl lies under the great circle. If then we conceive straight lines produced in order from each of the pointers through the earth, they will meet at the centre of the earth. Now since the sundial at Syene is vertically under the sun, if we conceive a straight line drawn from the sun to the top of the pointer of the sundial, the line stretching from the sun to

*Source: From *Selections Illustrating the History of Greek Mathematics,* trans. by Ivor Thomas (1941), vol. II, 267, 269, 271, and 273. Reprinted by permission of Harvard University Press.

the centre of the earth will be one straight line. If now we conceive another straight line drawn upwards from the extremity of the shadow of the pointer of the sundial in Alexandria, through the top of the pointer to the sun, this straight line and the aforesaid straight line will be parallel, being straight lines drawn through from different parts of the sun to different parts of the earth. Now on these parallel straight lines there falls the straight line drawn from the centre of the earth to the pointer at Alexandria, so that it makes the alternate angles equal; one of these is formed at the centre of the earth by the intersection of the straight lines drawn from the sundials to the centre of the earth; the other is at the intersection of the top of the pointer in Alexandria and the straight line drawn from the extremity of its shadow to the sun through the point where it meets the pointer. Now this latter angle subtends the arc carried round from the extremity of the shadow of the pointer to its base, while the angle at the centre of the earth subtends the arc stretching from Syene to Alexandria. But the arcs are similar since they are subtended by equal angles. Whatever ratio, therefore, the arc in the bowl of the sundial has to its proper circle, the arc reaching from Syene to Alexandria has the same ratio. But the arc in the bowl is found to be the fiftieth part of its proper circle. Therefore the distance from Syene to Alexandria must necessarily be a fiftieth part of the great circle of the earth. And this distance is 5,000 stades. Therefore the whole great circle is 250,000 stades. Such is the method of Eratosthenes.[2]

NOTES

1. Cleomedes probably wrote about the middle of the first century B.C. His handbook *De motu circulari corporum caelestium* is largely based on Posidonius.

2. The attached figure will help to elucidate Cleomedes. S is Syene and A Alexandria; the centre of the earth is O. The sun's rays at the two places are represented by the broken straight lines. If α be the angle made by the sun's rays with the pointer of the sundial at Alexandria (OA produced), the angle SOA is also equal to α, or one-fiftieth of four right angles. The arc SA is known to be 5,000 stades and it follows that the whole circumference of the earth must be 250,000 stades.

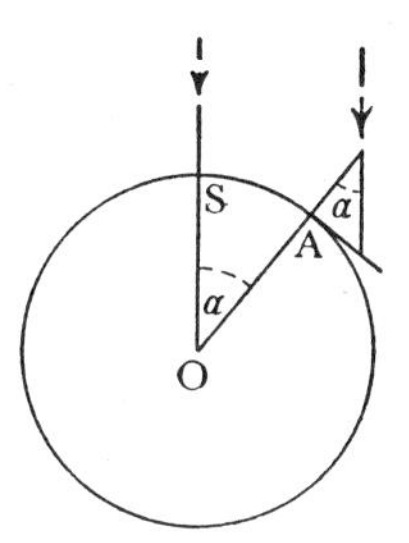

Chapter III
The Hellenistic and Roman Periods

Apollonius (c. 246–c. 174 B.C.)

Although Apollonius ranks with Archimedes, Eudoxus, and Euclid as one of the foremost mathematicians of antiquity, we know very little about his life. It is believed that he was born at Perga, a small Greek city in southern Asia Minor. The best evidence for his life comes from the introductions to the individual books of his *Conics*. These indicate that he resided for some time in Alexandria. He may have studied there as a young man under the successors of Euclid, becoming famous as an astronomer during the reign of Ptolemy Philopator (221–205 B.C.). He visited Pergamum (in western Asia Minor) and Ephesus. At Pergamum, which had a recently-founded center of learning and a library patterned after those in Alexandria, he met the geometer Eudemus to whom he later dedicated the first three books of his *Conics*. After this visit he returned to Alexandria, where he composed the *Conics* in his later years. His age at the time of its completion is suggested by the preface to Book II, which mentions his adult son Apollonius, and the preface to Book IV, which states that Eudemus was dead.

The prefaces to the *Conics* reveal that Apollonius was associated with many other leading mathematicians of his day. After Eudemus died, he dedicated Book IV and the remaining books to Attalus (probably a mathematician of Macedonian origin and not the king of the same name). He also mentions Philonides, the Epicurean geometer whom he met in Ephesus, and Conon, a correspondent of Archimedes. In the preface to Book I, Apollonius recounts that he had composed an early version of Conics for the geometer Naucrates, who was visiting him in Alexandria.

The modern fame of Apollonius rests chiefly on the *Conics*, which appeared in eight books containing 487 propositions. In the preface to Book I, Apollonius points out the limitations and inadequacies of his predecessors, a step which seems to justify the appearance of his new textbook, although some readers, including Pappus, have judged him boastful and envious. Books I–IV have survived in the Greek original, V–VII only in Arabic translation, and Book VIII is lost. To appreciate his achievement in this work, one needs to know the stage that the study of conic sections had reached before his work began. Eudoxus' student Menaechmus had begun the investigation of the mathematical properties of conic sections by the mid-4th century B.C. In Books I–IV, Apollonius drew upon the corpus of elementary theorems on conic sections, primarily from Euclid and Archimedes. The last four books, more specialized than the others, present extensions and his own original ideas.

The *Conics* quickly became an authoritative text. Its author not only demonstrated a mastery of earlier sources but also presented sets of theorems previously existing only in disconnected and haphazard ways in a comprehensive and rationally ordered exposition. He also coined the terms "ellipse" (meaning "falling short"), "parabola" (meaning "exact application"), and "hyperbola" (meaning "excess"). He borrowed these terms from early Pythagorean terminology from the method of "application of areas." Because his treatise became canonical and eliminated preceding writings on the subject, its superiority to them in mathematical consistency, generality, and rigor cannot be judged by direct comparison. According to Geminus, contemporary admirers of the *Conics* gave Apollonius the title "The Great Geometer" as a result of their admiration for the *Conics*. Later mathematicians and astronomers have concurred. Kepler drew upon the *Conics* to explain the orbits of the planets. Descartes, who admired its thorough treatment of second order curves, used it to develop analytical geometry. Halley showed his esteem by preparing a new edition.

Apollonius contributed to all known branches of geometry and astronomy. According to Pappus, he also wrote *Cutting off of a Ratio* (which has survived), as well as *Cutting off of an Area, Determinate Section, Inclinations, On the Burning Glass, Plane Loci,* and *Tangencies* (which are lost). In his work he extended the theory of irrationals presented in Book X of Euclid, calculated the value of π to more precise limits than the $3^{10}/_{71}$ to $3^{1}/_{7}$ proposed by Archimedes and (like Archimedes) developed a method for expressing large numbers. In effect, Apollonius expressed these numbers in terms of a place-value system with base 10,000 thereby overcoming the limitations of the Greek alphabetic numeral system. To account for celestial motions, he invented the models of eccentrics and epicycles later used in Ptolemy's *Almagest*. These innovations made possible a rational astronomy rather than the speculative cosmogony of his predecessors. Though Ptolemy does not refer to his lunar theory in the *Almagest*, Apollonius was fatuously nicknamed epsilon, "ϵ", because the shape of this letter reminded another ancient astronomer of the moon about which Apollonius knew so much.

42. From *Conics:* Introduction to Book One*

APOLLONIUS

Apollonius to Eudemus, greeting.

If you are in good health and matters are in other respects as you wish, it is well; I am pretty well too. During the time I spent with you at Pergamum, I noticed how eager you were to make acquaintance with my work in conics; I have therefore sent to you the first book, which I have revised, and I will send the remaining books when I am satisfied with them. I suppose you have not forgotten hearing me say that I took up this study at the request of Naucrates the geometer, at the time when he came to

**Source:* Selections 42 and 43 are from *Selections Illustrating the History of Greek Mathematics,* trans. by Ivor Thomas (1941), vol. II, 281-295 and 305-309 with the Greek originals deleted. Reprinted by permission of Harvard University Press.

Alexandria and stayed with me, and that, when I had completed the investigation in eight books, I gave them to him at once, a little too hastily, because he was on the point of sailing, and so I was not able to correct them, but put down everything as it occurred to me, intending to make a revision at the end. Accordingly, as opportunity permits, I now publish on each occasion as much of the work as I have been able to correct. As certain other persons whom I have met have happened to get hold of the first and second books before they were corrected, do not be surprised if you come across them in a different form.

Of the eight books the first four form an elementary introduction. The first includes the methods of producing the three sections and the opposite branches [of the hyperbola] and their fundamental properties, which are investigated more fully and more generally than in the works of others. The second book includes the properties of the diameters and the axes of the sections as well the asymptotes, with other things generally and necessarily used in determining limits of possibility; and what I call diameters and axes you will learn from this book. The third book includes many remarkable theorems useful for the syntheses of solid loci and for determining limits of possibility; most of these theorems, and the most elegant, are new, and it was their discovery which made me realize that Euclid had not worked out the synthesis of the locus with respect to three and four lines, but only a chance portion of it, and that not successfully; for the synthesis could not be completed without the theorems discovered by me. The fourth book investigates how many times the sections of cones can meet one another and the circumference of a circle; in addition it contains other things, none of which have been discussed by previous writers, namely, in how many points a section of a cone or a circumference of a circle can meet [the opposite branches of hyperbolas].

The remaining books are thrown in by way of addition: one of them discusses fully *minima* and *maxima*, another deals with equal and similar sections of cones, another with theorems about the determinations of limits, and the last with determinate conic problems. When they are all published it will be possible for anyone who reads them to form his own judgement. Farewell.

DEFINITIONS

If a straight line be drawn from a point to the circumference of a circle, which is not in the same plane with the point, and be produced in either direction, and if, while the point remains stationary, the straight line be made to move round the circumference of the circle until it returns to the point whence it set out, I call the surface described by the straight line *a conical surface;* it is composed of two surfaces lying vertically opposite to each other, of which each extends to infinity when the straight line which describes them is produced to infinity; I call the fixed point the vertex, and the straight line drawn through this point and the centre of the circle I call the *axis*.

The figure bounded by the circle and the conical surface between the vertex and the circumference of the circle I term a *cone*, and by the *vertex of the cone* I mean the point which is the vertex of the surface, and by the *axis* I mean the straight line drawn from the vertex to the centre of the circle, and by the *base* I mean the circle.

Of cones, I term those *right* which have their axes at right angles to their bases, and *scalene* those which have their axes not at right angles to their bases.

In any plane curve I mean by a *diameter* a straight line drawn from the curve which bisects all straight lines drawn in the curve parallel to a given straight line, and by the *vertex of the curve* I mean the extremity of the straight line on the curve, and I describe each of the parallels as being drawn *ordinatewise* to the diameter.

Similarly, in a pair of plane curves I mean by a *transverse diameter* a straight line which cuts the two curves and bisects all the straight lines drawn in either curve parallel to a given straight line, and by the *vertices of the curves* I mean the extremities of the diameter on the curves; and by an *erect diameter* I mean a straight line which lies between the two curves and bisects the portions cut off between the curves of all straight lines drawn parallel to a given straight line; and I describe each of the parallels as drawn *ordinate-wise to the diameter*.

By *conjugate diameters* in a curve or pair of curves I mean straight lines of which each, being a diameter, bisects parallels to the other.

By an *axis* of a curve or pair of curves I mean a straight line which, being a diameter of the curve or pair of curves, bisects the parallels at right angles.

By *conjugate axes* in a curve or pair of curves I mean straight lines which, being conjugate diameters, bisect at right angles the parallels to each other.

43. From *Conics:* Propositions 7 and 11*

(A Novel Method of Construction of Sections)

APOLLONIUS

PROPOSITION 7
[Construction of Sections]

IF a cone be cut by a plane through the axis, and if it be also cut by another plane cutting the plane containing the base of the cone in a straight line perpendicular to the base of the axial triangle,[1] *or to the base produced, a section will be made on the surface of the cone by the cutting plane, and straight lines drawn in it parallel to the straight line perpendicular to the base of the axial triangle will meet the common section of the cutting plane and the axial triangle and, if produced to the other part of the section, will be bisected by it; if the cone be right, the straight line in the base will be perpendicular to the common section of the cutting plane and the axial triangle; but if it be scalene, it will not in general be perpendicular, but only when the plane through the axis is perpendicular to the base of the cone.*

Let there be a cone whose vertex is the point A and whose base is the circle BΓ, and let it be cut by a plane through the axis, and let the section so made be the triangle ABΓ. Now let it be cut by another plane cutting the plane containing the circle BΓ in a straight line ΔE which is either perpendicular to BΓ or to BΓ produced, and let the section made on the surface of the cone be ΔZE[2]; then the common section of the cutting plane and of the triangle ABΓ is ZH. Let any point Θ be taken on ΔZE,

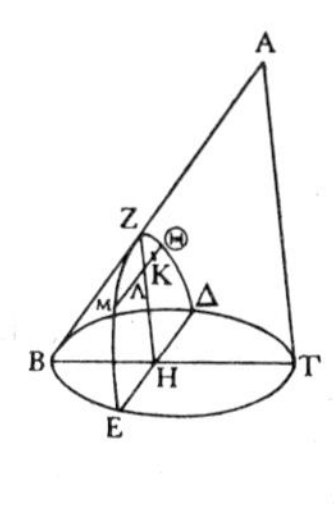

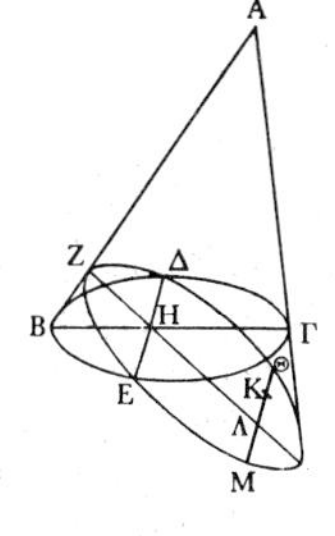

and through Θ let ΘK be drawn parallel to ΔE. I say that ΘK intersects ZH and, if produced to the other part of the section ΔZE, it will be bisected by the straight line ZH.

For since the cone, whose vertex is the point A and base the circle BΓ, is cut by a plane through the axis and the section so made is the triangle ABΓ, and there has been taken any point Θ on the surface, not being on a side of the triangle ABΓ, and ΔH is perpendicular to BΓ, therefore the straight line drawn through Θ parallel to ΔH, that is ΘK, will meet the triangle ABΓ and, if produced to the other part of the surface, will be bisected by the triangle. Therefore, since the straight line drawn through Θ parallel to ΔE meets the triangle ABΓ and is in the plane containing the section ΔZE, it will fall upon the common section of the cutting plane and the triangle ABΓ. But the common section of those planes is ZH; therefore the straight line drawn through Θ parallel to ΔE will meet ZH; and if it be produced to the other part of the section ΔZE it will be bisected by the straight line ZH.

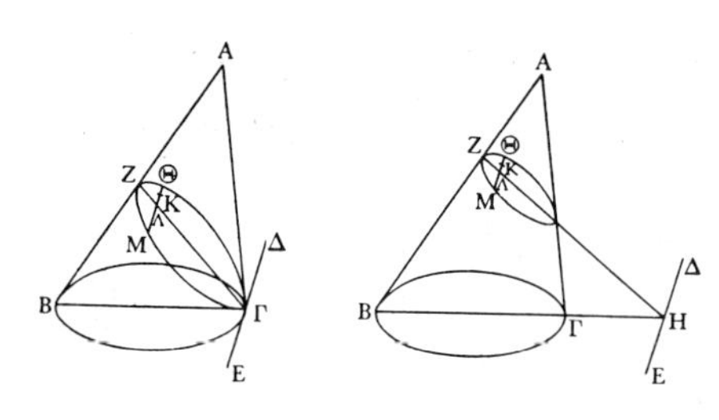

Now the cone is right, or the axial triangle ABΓ is perpendicular to the circle BΓ, or neither.

First, let the cone be right; then the triangle ABΓ will be perpendicular to the circle BΓ [Eucl. xi. 18]. Then since the plane ABΓ is perpendicular to the plane BΓ, and ΔE is drawn in one of the planes perpendicular to their common section BΓ, therefore ΔE is perpendicular to the triangle ABΓ [Eucl. xi. Def. 4]; and therefore it is perpendicular to all the straight lines in the triangle ABΓ which meet it [Eucl. xi. Def. 3]. Therefore it is perpendicular to ZH.

Now let the cone be not right. Then, if the axial triangle is perpendicular to the circle BΓ, we may similarly show that ΔE is perpendicular to ZH. Now let the axial triangle ABΓ be not perpendicular to the circle BΓ. I say that neither is ΔE perpendicular to ZH. For if it is possible, let it be; now it is also perpendicular to BΓ; therefore ΔE is perpendicular to both BΓ, ZH. And therefore it is perpendicular to the plane through BΓ, ZH [Eucl. xi. 4]. But the plane through BΓ, HZ is ABΓ; and therefore ΔE is perpendicular to the triangle ABΓ. Therefore all the planes through it are perpendicular to the triangle ABΓ [Eucl. xi. 18]. But one of the planes through ΔE is the circle BΓ; therefore the circle BΓ is perpendicular to the triangle ABΓ. Therefore the triangle ABΓ is perpendicular to the circle BΓ; which is contrary to hypothesis. Therefore ΔE is not perpendicular to ZH.

Corollary. From this it is clear that ZH is a diameter of the section ΔZE, inasmuch as it bisects the straight lines drawn parallel to the given straight line ΔE, and also that parallels can be bisected by the diameter ZH without being perpendicular to it.

PROPOSITION 11
[Application of Areas]

Let a cone be cut by a plane through the axis, and let it be also cut by another plane cutting the base of the cone in a straight line perpendicular to the base of the axial triangle, and further let the diameter of the section be parallel to one side of the axial triangle; then if any straight line be drawn from the section of the cone parallel to the common section of the cutting plane and the base of the cone as far as the diameter of the section, its square will be equal to the rectangle bounded by the intercept made by it on the diameter in the direc-

tion of the vertex of the section and a certain other straight line; this straight line will bear the same ratio to the intercept between the angle of the cone and the vertex of the segment as the square on the base of the axial triangle bears to the rectangle bounded by the remaining two sides of the triangle; and let such a section be called a parabola.

For let there be a cone whose vertex is the point A and whose base is the circle $B\Gamma$, and let it be cut by a plane through the axis, and let the section so made be the triangle $AB\Gamma$, and let it be cut by another plane cutting the base of the cone in the straight line ΔE perpendicular to $B\Gamma$, and let the section so made on the surface of the cone be ΔZE, and let ZH, the diameter of the section, be parallel to $A\Gamma$, one side of the axial triangle and from the point Z let $Z\Theta$ be drawn perpendicular to ZH, and let $B\Gamma^2 : BA \cdot A\Gamma = Z\Theta : ZA$, and let any point K be taken at random on the section, and through K let $K\Lambda$ be drawn parallel to ΔE. I say that $K\Lambda^2 = \Theta Z \cdot Z\Lambda$.

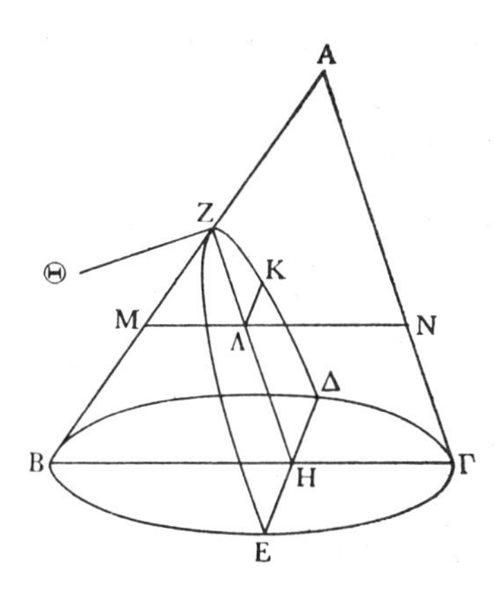

For let MN be drawn through Λ parallel to $B\Gamma$; but $K\Lambda$ is parallel to ΔE; therefore the plane through $K\Lambda$, MN is parallel to the plane through $B\Gamma$, ΔE [Eucl. xi. 15], that is to the base of the cone. Therefore the plane through $K\Lambda$, MN is a circle, whose diameter is MN [Prop. 4]. And $K\Lambda$ is perpendicular to MN, since ΔE is perpendicular to $B\Gamma$ [Eucl. xi. 10];

therefore $M\Lambda \cdot \Lambda N = K\Lambda^2$.

And since $B\Gamma^2 : BA \cdot A\Gamma = \Theta Z : ZA$,

while $B\Gamma^2 : BA \cdot A\Gamma = (B\Gamma : \Gamma A)(B\Gamma : BA)$,

therefore $\Theta Z : ZA = (B\Gamma : \Gamma A)(\Gamma B : BA)$.

But $B\Gamma : \Gamma A = MN : NA$
$= M\Lambda : \Lambda Z$, [Eucl. vi. 4]

and $B\Gamma : BA = MN : MA$
$= \Lambda M : MZ$ [*ibid.*]
$= N\Lambda : ZA$. [Eucl. vi. 2]

Therefore $\Theta Z : ZA = (M\Lambda : \Lambda Z)(N\Lambda : ZA)$.

But $(M\Lambda : \Lambda Z)(\Lambda N : ZA) = M\Lambda \cdot \Lambda N : \Lambda Z \cdot ZA$.

Therefore $\Theta Z : ZA = M\Lambda \cdot \Lambda N : \Lambda Z \cdot ZA$.

But $\Theta Z : ZA = \Theta Z \cdot Z\Lambda : \Lambda Z \cdot ZA$,

by taking a common height $Z\Lambda$;

therefore $M\Lambda \cdot \Lambda N : \Lambda Z \cdot ZA = \Theta Z \cdot Z\Lambda : \Lambda Z \cdot ZA$.

Therefore $M\Lambda \cdot \Lambda N = \Theta Z \cdot Z\Lambda$. [Eucl. v. 9]

But $M\Lambda \cdot \Lambda N = K\Lambda^2$;

and therefore $K\Lambda^2 = \Theta Z \cdot Z\Lambda$.

Let such a section be called a *parabola*, and let ΘZ be called the *parameter of the ordinates* to the diameter ZH, and let it also be called the *erect side (latus rectum)*.[3]

NOTES

1. Literally, "the triangle through the axis."

2. This applies only to figures 1 and 2.

3. A *parabola* because the square on the ordinate KΛ is *applied* to the parameter ΘZ in the form of the rectangle ΘZ . ZΛ, and is exactly equal to this rectangle. It was Apollonius's most distinctive achievement to have based his treatment of the conic sections on the Pythagorean theory of the *application of areas*.

Chapter III
The Hellenistic and Roman Periods

Claudius Ptolemy (c. A.D. 100–178)

We know little of the life of Claudius Ptolemy, the last major Hellenistic astronomer. One source states that he was born in Ptolemais Hermiou (in middle Egypt near Akhmīn), which seems plausible. His Latin name, "Ptolemaus," suggests that he lived in Egypt and was a descendant of Greek or hellenized forebears. His first name indicates that he was a Roman citizen, a status probably conferred on one of his ancestors by either Claudius or Nero. Although nothing definite is known about how he acquired his education or who his colleagues were, he could have studied under Theon in Alexandria during the decade of the 120s. The internal evidence of his writings suggests that Ptolemy spent his entire adult career in Alexandria, where he continued the highest tradition of classic Greek scholarship at the museum (an active center of learning). He served, but was not a member of, the Hellenistic royal court at Alexandria.

The development of classical Greek and Hellenistic astronomy and trigonometry culminated in the writings of Ptolemy. His later reputation rests mainly on his chief work, *Mathematikos Syntaxeos Biblion Proton (The Thirteen Books of Mathematical Compilation)*, which the Arabs called *Almagest* meaning the very great or greatest. Covering the whole of mathematical astronomy as conceived by the ancients, the *Almagest* presents a geocentric model of the universe, with celestial motions explained with epicycles (that is, revolving circles whose centers also revolve on the circumference of another circle) and eccentric circles. Its author assumed from Aristotelian physics that observable motions in the heavens were reducible to uniform circular motion. The Ptolemaic system of the universe was the standard theory of astronomy in Europe and the Near East until the time of Copernicus in the 16th century.

The *Almagest* continued and completed the pioneering work of Hipparchus, Menelaus, and others in plane and spherical trigonometry and numerical calculations. Ptolemy developed trigonometry by the use of chords of arcs that were predecessors of the sine, cosine, and tangent functions. In calculating trigonometric tables he worked with a circle whose diameter was divided into 120 parts (120^p), while he also adopted the Babylonian sexagesimal place values system both for expressing angles and for calculating. He found the values of some chords (crds) by using elementary geometry, especially the Pythagorean theorem. By combining these values with an approximation procedure, Ptolemy computed a table of chords at ½° intervals and to three sexagesimal places. His chord functions may be related to the modern sine function by $\sin \alpha = \frac{1}{2 \cdot 60} \text{Crd } 2\alpha$.

In his chord computations, Ptolemy found accurate values for several irrational numbers. For example:

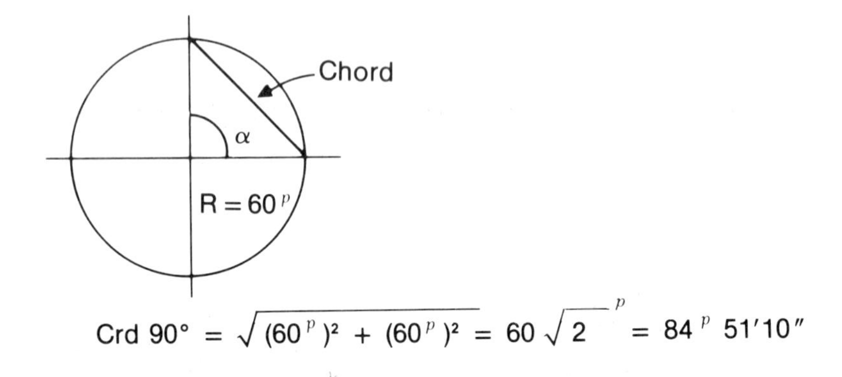

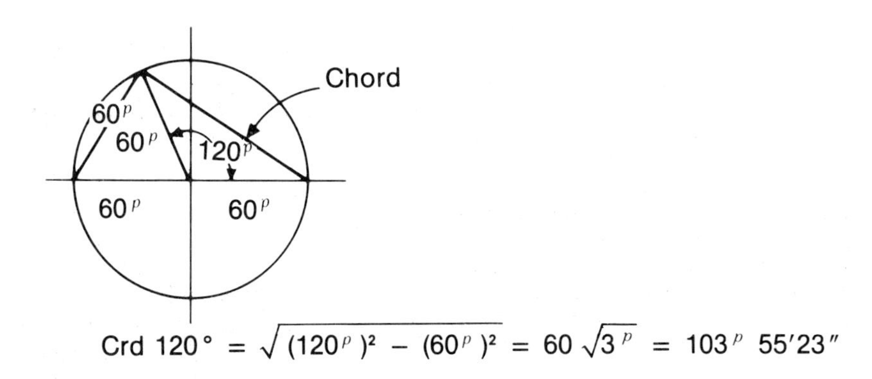

He also found Crd 36°, the side of a regular inscribed decagon, equals $30^p(\sqrt{5} - 1) = 37^p\,4'\,55''$. In Book IX of the *Almagest*, Ptolemy proved that for any arc S less than 180° $(\text{Crd S})^2 + \text{Crd}(180 - \text{S})^2 = (120^p)^2$. This is equivalent to the modern relationship in a unit circle $\sin^2\alpha + \cos^2\alpha = 1$, where α is any acute angle.

Ptolemy wrote widely on other sciences. In addition to his first work, *Almagest*, he authored in astronomy *Canobic Inscription* and *Handy Tables*, two systematic continuations of the *Almagest*, as well as *Tetrabiblos*, *Analemma*, and *Planetary Hypotheses*. In related fields he wrote *Optics*, which examines astronomical refraction, and *Harmonics*. Late in his career he wrote *Geography*, which places the prime meridian in the Canary Islands. In the 16th century, the map projections of the *Geography* stimulated the development of cartography.

44. From the *Syntaxis* or *Almagest* i*

(Trigonometry: Table of Sines)

CLAUDIUS PTOLEMY

10. ON THE LENGTHS OF THE CHORDS IN A CIRCLE

With a view to obtaining a table ready for immediate use, we shall next set out the lengths of these [chords in a circle], dividing the perimeter into 360 segments and by the side of the arcs placing the chords subtending them for every increase of half a degree, that is, stating how many parts they are of the diameter, which it is convenient for the numerical calculations to divide into 120 segments. But first we shall show how to establish a systematic and rapid method of calculating the lengths of the chords by means of the uniform use of the smallest possible number of propositions, so that we may not only have the sizes of the chords set out correctly, but may obtain a convenient proof of the method of calculating them based on geometrical considerations.[1] In general we shall use the sexagesimal system for the numerical calculations owing to the inconvenience of having fractional parts, especially in multiplications and divisions, and we shall aim at a continually closer approximation, in such a manner that the difference from the correct figure shall be inappreciable and imperceptible.

(ii.) *sin* 18° *and sin* 36°

32. First, let ΑΒΓ be a semicircle on the diameter ΑΔΓ and with centre Δ, and from Δ let ΔΒ be drawn perpendicular to ΑΓ, and let ΔΓ be bisected at

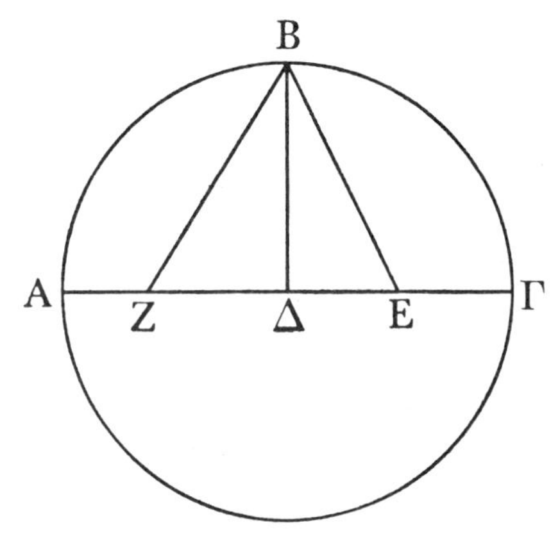

E, and let EB be joined, and let EZ be placed equal to it, and let ZB be joined. I say that ZΔ is the side of a decagon, and BZ of a pentagon.[2]

For since the straight line ΔΓ is bisected at E, and the straight line ΔZ is added to it,

$$\Gamma Z \,.\, Z\Delta + E\Delta^2 = EZ^2 \quad \text{[Eucl. ii. 6]}$$
$$= BE^2,$$

since EB = ZE.

But $E\Delta^2 + \Delta B^2 = EB^2$; [Eucl. i. 47]

therefore $\Gamma Z \,.\, Z\Delta + E\Delta^2 = E\Delta^2 + \Delta B^2$.

When the common term $E\Delta^2$ is taken away, the remainder $\Gamma Z \,.\, Z\Delta = \Delta B^2$

i.e., $= \Delta\Gamma^2$;

therefore ZΓ is divided in extreme and mean ratio at Δ [Eucl. vi., Def. 3]. Therefore, since the side of the hexagon and the side of the decagon inscribed in the same circle when placed in one straight line are cut in extreme and

*Source: From *Selections Illustrating the History of Greek Mathematics,* trans. by Ivor Thomas (1941), vol. II, 413-425 with Greek originals deleted. Reprinted by permission of Harvard University Press.

mean ratio [Eucl. xiii. 9], and ΓΔ, being a radius, is equal to the side of the hexagon [Eucl. iv. 15, coroll.], therefore ΔZ is equal to the side of the decagon. Similarly, since the square on the side of the pentagon is equal to the rectangle contained by the side of the hexagon and the side of the decagon inscribed in the same circle [Eucl. xiii. 10], and in the right-angled triangle BΔZ the square on BZ is equal [Eucl. i. 47] to the sum of the squares on BΔ, which is a side of the hexagon, and ΔZ, which is a side of the decagon, therefore BZ is equal to the side of the pentagon.

Then since, as I said, we made the diameter[3] consist of 120^p, by what has been stated ΔE, being half of the radius, consists of 30^p and its square of 900^p, and BΔ, being the radius, consists of 60^p and its square of 3600^p, while EB^2, that is EZ^2, consists of 4500^p; therefore EZ is approximately $67^p\ 4'\ 55''$,[4] and the remainder ΔZ is $37^p\ 4'\ 55''$. Therefore the side of the decagon, subtending an arc of 36° (the whole circle consisting of 360°), is $37^p\ 4'\ 55''$ (the diameter being 120^p). Again, since ΔZ is $37^p\ 4'\ 55''$, its square is $1375^p\ 4'\ 15''$, and the square on ΔB is 3600^p, which added together make the square on BZ $4975^p\ 4'\ 15''$, so that BZ is approximately $70^p\ 32'\ 3''$. And therefore the side of the pentagon, subtending 72° (the circle consisting of 360°), is $70^p\ 32'\ 3''$ (the diameter being 120^p).

Hence it is clear that the side of the hexagon, subtending 60° and being equal to the radius, is 60^p. Similarly, since the square on the side of the square,[5] subtending 90°, is double of the square on the radius, and the square on the side of the triangle, subtending 120°, is three times the square on the radius, while the square on the radius is 3600^p, the square on the side of the square is 7200^p and the square on the side of the triangle is 10800^p. Therefore the chord subtending 90° is approximately $84^p\ 51'\ 10''$ (the diameter consisting of 120^p), and the chord subtending 120° is $103^p\ 55'\ 23''$.[6]

(iii.) $\sin^2\theta + \cos^2\theta = 1$

35. The lengths of these chords have thus been obtained immediately and by themselves,[7] and it will be thence clear that, among the given straight lines, the lengths are immediately given of the chords subtending the remaining arcs in the semicircle, by reason of the fact that the sum of the squares on these chords is equal to the square on the diameter; for example, since the chord subtending 36° was shown to be $37^p\ 4'\ 55''$ and its square $1375^p\ 4'\ 15''$, while the square on the diameter is 14400^p, therefore the square on the chord subtending the remaining 144° in the semicircle is $13024^p\ 55'\ 45''$ and the chord itself is approximately $114^p\ 7'\ 37''$, and similarly for the other chords.[8]

We shall explain in due course the manner in which the remaining chords obtained by subdivision can be calculated from these, setting out by way of preface this little lemma which is exceedingly useful for the business in hand.

(iv.) *"Ptolemy's Theorem"*

36. Let ABΓΔ be any quadrilateral inscribed in a circle, and let AΓ and BΔ be joined. It is required to prove that the rectangle contained by AΓ and BΔ is equal to the sum of the rectangles contained by AB, ΔΓ and AΔ, BΓ.

For let the angle ABE be placed equal to the angle ΔBΓ. Then if we add the angle EBΔ to both, the angle ABΔ = the

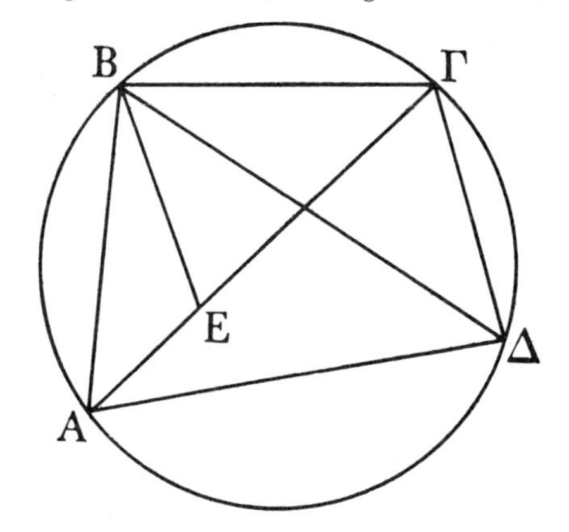

angle EBΓ. But the angle BΔA = the angle BΓE [Eucl. iii. 21], for they subtend the same segment; therefore the triangle ABΔ is equiangular with the triangle BΓE.

∴ BΓ : ΓE = BΔ : ΔA; [Euclid. vi. 4]
∴ BΓ . AΔ = BΔ . ΓE. [Eucl. vi. 6]

Again, since the angle ABE is equal to the angle ΔBΓ, while the angle BAE is equal to the angle BΔΓ [Eucl. iii. 21], therefore the triangle ABE is equiangular with the triangle BΓΔ;

∴ BA : AE = BΔ : ΔΓ; [Eucl. vi. 4]
∴ BA . ΔΓ = BΔ . AE. [Eucl. vi. 6]

But it was shown that

BΓ . AΔ = BΔ . ΓE;
and ∴ AΓ . BΔ = AB . ΔΓ + AΔ . BΓ;
[Eucl. ii. 1]

which was to be proved.

NOTES

1. . . . Ptolemy meant more than a graphical method; the phrase indicates a *rigorous proof* by means of geometrical considerations, as will be seen when the argument proceeds: It may be inferred, therefore, that when Hipparchus proved "by means of lines" (*On the Phaenomena of Eudoxus and Aratus,* ed. Manitius 148-150) certain facts about the risings of stars, he used rigorous, and not merely graphical calculations; in other words, he was familiar with the main formulae of spherical trigonometry.

2. *i.e.,* ZΔ is equal to the side of a regular decagon, and BZ to the side of a regular pentagon, inscribed in the circle ABΓ.

2. Following the usual practice, I shall denote *segments* of the diameter by ^{p}, sixtieth parts of a τμῆμα by the numeral with a single accent, and second-sixtieths by the numeral with two accents. As the circular associations of the system tend to be forgotten, and it is used as a general system of enumeration, the same notation will be used for the *squares* of parts.

4. Theon's proof that $\sqrt{4500}$ is approximately $67^p\ 4'\ 55''$ has already been given (vol. i. pp. 56-61).

5. This is, of course, the square itself; the Greek phrase is not so difficult. We could translate, "the second power of the side of the square," but the notion of powers was outside the ken of the Greek mathematician.

6. Let AB be a chord of a circle subtending an angle α at the centre O, and let AKA′ be drawn perpendicular to OB so as to meet OB in K and the circle again in A′. Then

$$\sin\alpha\,(=\sin AB) = \frac{AK}{AO} = \frac{\tfrac{1}{2}AA'}{AO}.$$

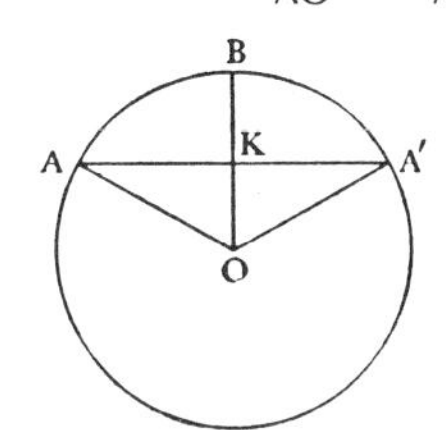

And AA′ is the chord subtended by double of the arc AB, while Ptolemy expresses the lengths of chords as so many 120th parts of the diameter; therefore sin α is half the chord subtended by an angle 2α at the centre, which is conveniently abbreviated by Heath to ½(crd. 2α), or, as we may alternatively express the relationship, sin AB is "half the chord subtended by double of the arc AB," which is the Ptolemaic form; as Ptolemy means by this expression precisely what we mean by sin AB, I shall interpolate the trigonometrical notation in the translation wherever it occurs. It follows that cos α [$=\sin(90-\alpha)$] = ½ crd. $(180° - 2\alpha)$, or, as Ptolemy says, "half the chord subtended by the remaining angle in the semicircle." Tan α and the other trigonometrical ratios were not used by the Greeks.

In the passage to which this note is appended Ptolemy proves that

side of decagon (=crd. 36°=2 sin 18°)=$37^p\ 4'\ 55''$,
side of pentagon (=crd. 72°=2 sin 36°)=$70^p\ 32'\ 3''$,
side of hexagon (=crd. 60°=2 sin 30°)=60^p,
side of square (=crd. 90°=2 sin 45°)=$84^p\ 51'\ 10''$,
side of equilateral triangle (=crd. 120°=2 sin 60°)=$103^p\ 55'\ 23''$.

7. *i.e.,* not deduced from other known chords.

8. *i.e.,* crd. 144°(=2 sin 72°)=$114^p\ 7'\ 37''$. If the given chord subtends an angle 2θ at the centre, the chord subtended by the remaining arc in the semicircle subtends an angle $(180 - 2\theta)$, and the theorem asserts that

$$(\text{crd. } 2\theta)^2 + (\text{crd. } \overline{180 - 2\theta})^2 = (\text{diameter})^2,$$

$$\text{or}\quad \sin^2\theta + \cos^2\theta = 1.$$

Chapter III
The Hellenistic and Roman Periods

Diophantus (fl. A.D. 250)

Diophantus ranks with Pappus as one of the two major mathematicians of the late Alexandrian period. He was probably Greek, although we know virtually nothing about the origins of his life. Even the dating of Diophantus' mature activity in the mid-third century A.D. is based not on personal or contemporary records but chiefly on a letter by Michael Psellus in the eleventh century. Psellus wrote that Anatolius, who became bishop of Laodicea in A.D. 270, dedicated a tract on Egyptian computation to his friend Diophantus. This dating agrees with a supposition that the Dionysius to whom Diophantus dedicated *Arithmetica* is St. Dionysius, who led the Christian school in Alexandria after 231 and became bishop there in 247. The only other information on Diophantus' life comes from an arithmetical epigram in the *Greek Anthology*. The epigram states that his boyhood lasted $1/6$ of his life, that his beard grew after $1/12$ more, that he married after $1/7$ more and had a son five years later, that the son lived only half as long as the father, and that the father died four years later. This gives the following equation: $1/6x + 1/12x + 1/7x + 5 + 1/2x + 4 = x$. The solution is $x = 84$, the age of Diophantus at his death. If the epigrammatic data and the solution are correct, then his boyhood lasted 14 years; he grew a beard at 21; he married at 33 and had a son who died at 42 when the father was 80.

Only fragments of the writings of Diophantus still exist. His *Moriastica* and *Porismata* are completely lost. The first seven of the thirteen books of his *Arithmetica* and a fragment of his tract entitled, *On Polygonal Numbers,* are all that have survived.

Diophantus' most important work was the *Arithmetica.* Despite its title, this text is not about the Pythagorean *arithmetica* (elementary number theory) but logistic or computational arithmetic, an area that Plato had attempted to ban from mathematics. In the *Arithmetica,* Diophantus was clever and imaginative in solving each of 189 heterogeneous problems by separate methods; but he rarely approached the deeper level of generalization. His text is thus more like the procedural texts of ancient Egypt and Babylon than the orderly, deductive geometry of Euclid, Archimedes, and Apollonius.

The *Arithmetica* is one of the earliest texts on algebra extant. It mainly states problems that can be solved as equations. While the equations are mostly linear and quadratic, there is one special cubic equation and some of higher degree that were reduced to a lower degree by a skillful choice of numerical values. To help compute solutions, Diophantus devised and introduced a more comprehensible notation. Capitalized abbreviations of Greek words represented the exponents from 2 through 6 as follows:

Δ^{υ} (for *dynamis*, "power") = n^2
κ^{υ} (for *cubos* ($\kappa\upsilon'\beta o\varsigma$) = n^3
$\Delta^{\upsilon}\Delta$ (for *dynamodynamis* or square-square) = n^4
$\Delta\kappa^{\upsilon}$ (for square-cube) = n^5
$\kappa^{\upsilon}\kappa$ (for cubocube) = n^6

The symbol "ς," the final sigma in the word *arithmos* (number), stood for the unknown "x" of our notation. Diophantus indicated addition by grouping together all of the terms to be added at the beginning of a problem while the terms to be subtracted followed the symbol ⋀, a stenographic abbreviation (in his times) for the Greek verb "to want." The symbol $\overset{\circ}{M}$ indicates units and that a pure number followed. Using the Greek alphabetic numerals and his notation, Diophantus would represent $x^6 - 2x^5 + x^3 - 2x - 3$ as follows: $\kappa^{\upsilon}\kappa\alpha\kappa^{\upsilon}\alpha$ ⋀ $\Delta\kappa^{\upsilon}\beta\varsigma\beta\overset{\circ}{M}\bar{\gamma}$. His notation marked a fundamental step from verbal toward symbolic algebra.

Diophantus accepted only positive rational roots for his equations. He ignored irrationals. Negative quantities were not yet known in the West.

The *Arithmetica* has substantially influenced mathematics, although that influence was not felt until much later because the Hellenistic culture was waning in Diophantus' time. In books 2 through 7, he pursued extensively indeterminate equations for the first time. As a result, he is considered the founder of indeterminate analysis, a branch of algebra now called Diophantine analysis. His propositions on prime numbers and representation of numbers as the sum of two (V, 9) three (V, 11) and four (IV, 29 and 30 and V, 14) squares were later examined, proved, and generalized by mathematicians from Fermat through Euler and Gauss in founding modern number theory.

45. From the Dedication to *Arithmetica**

(Algebraic Notation)

DIOPHANTUS

DEDICATION

Knowing that you are anxious, my most esteemed Dionysius, to learn how to solve problems in numbers, I have tried, beginning from the foundations on which the subject is built, to set forth the nature and power in numbers.

Perhaps the subject will appear to you rather difficult, as it is not yet common knowledge, and the minds of beginners are apt to be discouraged by mistakes; but it will be easy for you to grasp, with your enthusiasm and my teaching; for keenness backed by teaching is a swift road to knowledge.

As you know, in addition to these things, that all numbers are made up of some multitude of units, it is clear that their formation has no limit. Among them are—

squares, which are formed when any number is multiplied by itself; the number itself is called the *side of the square*[1];

cubes, which are formed when squares are multiplied by their sides,

square-squares, which are formed

*Source: From *Selections Illustrating the History of Greek Mathematics,* trans. by Ivor Thomas (1941), vol. II, 519, 521, and 523. Reprinted by permission of Harvard University Press.

when squares are multiplied by themselves;

square-cubes, which are formed when squares are multiplied by the cubes formed from the same side;

cube-cubes, which are formed when cubes are multiplied by themselves;

and it is from the addition, subtraction, or multiplication of these numbers or from the ratio which they bear one to another or to their own sides that most arithmetical problems are formed; you will be able to solve them if you follow the method shown below.

Now each of these numbers, which have been given abbreviated names, is recognized as an element in arithmetical science; the *square* [of the unknown quantity][2] is called *dynamis* and its sign is Δ with the index υ, that is Δ^{υ};

the cube is called *cubus* and has for its sign κ with the index υ, that is κ^{υ};

the square multiplied by itself is called *dynamodynamis* and its sign is two deltas with the index υ, that is $\Delta^{\upsilon}\Delta$;

the square multiplied by the cube formed from the same root is called *dynamocubus* and its sign is $\Delta\kappa$ with the index υ, that is $\Delta\kappa^{\upsilon}$;

the cube multiplied by itself is called *cubocubus* and its sign is two kappas with the index υ, $\kappa^{\upsilon}\kappa$.

The number which has none of these characteristics, but merely has in it an undetermined multitude of units, is called *arithmos,* and its sign is S [x].[3]

There is also another sign denoting the invariable element in determinate numbers, the unit, and its sign is M with the index O, that is $\overset{\circ}{\mathrm{M}}$.

As in the case of numbers the corresponding fractions are called after the numbers, a *third* being called after 3 and a *fourth* after 4, so the functions named above will have reciprocals called after them:

arithmos [x]	*arithmoston* $\left[\frac{1}{x}\right]$,
dynamis [x^2]	*dynamoston* $\left[\frac{1}{x^2}\right]$,
cubus [x^3]	*cuboston* $\left[\frac{1}{x^3}\right]$,
dynamodynamis [x^4]	*dynamodynamoston* $\left[\frac{1}{x^4}\right]$,
dynamocubus [x^5]	*dynamocuboston* $\left[\frac{1}{x^5}\right]$,
cubocubus [x^6]	*cubocuboston* $\left[\frac{1}{x^6}\right]$.

And each of these will have the same sign as the corresponding process, but with the mark x to distinguish its nature.

THE MINUS SIGN AND MULTIPLICATION INVOLVING IT

A *minus* multiplied by a *minus* makes a *plus*,[4] a *minus* multiplied by a *plus* makes a *minus*, and the sign of a *minus* is a truncated Ψ turned upside down, that is ⋔.[5]

NOTES

1. Or "square root."

2. It is not here stated in so many words, but becomes obvious as the argument proceeds that *δύναμις* and its abbreviation are restricted to the square of the *unknown* quantity; the square of a determinate number is *τετράγωνος*. There is only one term, *κύβος*, for the cube both of a determinate and of the unknown quantity. The higher terms, when written in full as *δυναμοδύναμις*, *δυναμόκυβος* and *κυβόκυβος*, are used respectively for the fourth, fifth and sixth powers both of determinate quantities and of the unknown, but their abbreviations, and that for *κύβος*, are used to denote powers of the unknown only.

3. I am entirely convinced by Heath's argument, based on the Bodleian MS. of Diophantus and general considerations, that this symbol is really the first two letters of *ἀριθμός*; this suggestion brings the symbol into line with Diophantus's abbreviations for *δύναμις*, *κύβος*, and so on. It may be declined throughout its cases, *e.g.*, $S^{\tilde{\omega}\nu}$ for the genitive plural.

Diophantus has only one symbol for an unknown quantity, but his problems often lead to subsidiary equations involving other unknowns. He shows great ingenuity in isolating these subsidiary unknowns. In the translation I shall use different letters for the different unknowns as they occur, for example, x, z, m.

Diophantus does not admit negative or zero values of the unknown, but positive fractional values are admitted.

4. Lit. "a deficiency multiplied by a deficiency makes a forthcoming."

5. The sign has nothing to do with Ψ, but I see no reason why Diophantus should not have described it by means of Ψ, and cannot agree with Heath (*H.G.M.* ii. 459) that "the description is evidently interpolated." But Heath seems right in his conjecture, first made in 1885, that sign ⋔ is a compendium for the root of the verb *λείπειν*, and is, in fact, a Λ with an I placed in the middle. When the sign is resolved in the manuscripts into a word, the dative *λείΨει* is generally used, but there is no conclusive proof that Diophantus himself used this non-classical form.

46. From the *Arithmetica**

(Origins of the Theory of Numbers)

DIOPHANTUS

II. 8

8. To divide a given square number into two squares.[1]

Given square number 16. x^2 one of the required squares. Therefore $16 - x^2$ must be equal to a square.

Take a square of the form $(mx - 4)^2$, m being any integer and 4 the number

*Source: From Thomas L. Heath, *Diophantus of Alexandria: A Study in the History of Greek Algebra* (1964), 144-145, 188-189, and 208-209. Reprinted by permission of Cambridge University Press.

which is the square root of 16, e.g. take $(2x - 4)^2$, and equate it to $16 - x^2$.

Therefore $4x^2 - 16x + 16 = 16 - x^2$, or $5x^2 = 16x$, and $x = 16/5$.

The required squares are therefore $256/25$, $144/25$.

IV. 29 and 30

29. To find four square numbers such that their sum added to the sum of their sides makes a given number.[2]

Given number 12.

Now $x^2 + x + 1/4 =$ a square.

Therefore the sum of four squares + the sum of their sides + 1 = the sum of four other squares = 13, by hypothesis.

Therefore we have to divide 13 into four squares; then, if we subtract $1/2$ from each of their sides, we shall have the sides of the required squares.

Now $13 = 4 + 9 = (64/25 + 36/25) + (144/25 + 81/25)$, and the sides of the required squares are $11/10$, $7/10$, $19/10$, $13/10$, the squares themselves being $121/100$, $49/100$, $361/100$, $169/100$.

30. To find four squares such that their sum *minus* the sum of their sides is a given number.

Given number 4.

Now $x^2 - x + 1/4 =$ a square.

Therefore (the sum of four squares) − (sum of their sides) + 1 = the sum of four other squares = 5, by hypothesis.

Divide 5 into four squares, as $9/25$, $16/25$, $64/25$, $36/25$.

The sides of these squares *plus* $1/2$ in each case are the sides of the required squares.

Therefore sides of required squares are $11/10$, $13/10$, $21/10$, $17/10$, and the squares themselves $121/100$, $169/100$, $441/100$, $289/100$.

V. 2

2. To divide unity into three parts such that, if we add the same number to each of the parts, the results are all squares.

Necessary condition. The given number must not be 2 or any multiple of 8 increased by 2.

Given number 3. Thus 10 is to be divided into three squares such that each > 3.

Take $1/3$ of 10, or $3\frac{1}{3}$, and find x so that $\frac{1}{9x^2} + 3\frac{1}{3}$ may be a square, or $30x^2 + 1 =$ a square $= (5x + 1)^2$, say.

Therefore $x = 2$, $x^2 = 4$, $1/x^2 = 1/4$, and $1/36 + 3\frac{1}{3} = 121/36 =$ a square.

Therefore we have to divide 10 into three squares each of which is as near as possible to $121/36$.

Now $10 = 3^2 + 1^2 =$ the sum of the three squares 9, $16/25$, $9/25$.

Comparing the sides 3, $4/5$, $3/5$ with $11/6$, or (multiplying by 30) 90, 24, 18 with 55, we must make each side approach 55.

[Since then $55/30 = 3 - 35/30 = 4/5 + 31/30 = 3/5 + 37/30$], we put for the sides of the required numbers

$$3 - 35x,\ 31x + 4/5,\ 37x + 3/5.$$

The sum of the squares $= 3555x^2 - 116x + 10 = 10$.

Therefore $x = 116/3555$, and this solves the problem.

NOTES

1. It is to this proposition that Fermat appended his famous note in which he enunciates what is known as the "great theorem" of Fermat. The text of the note is as follows:

"On the other hand it is impossible to separate a cube into two cubes, or a biquadrate into two biquadrates, or generally *any power except a square into two powers with the same exponent.* I have discovered a truly marvellous proof of this, which however the margin is not large enough to contain."

2. On this problem Bachet observes that Diophantus appears to assume, here and in some problems of Book v., that any number not itself a square is the sum of two or three or four squares. He adds that he has verified this statement for all numbers up to 325, but would like to see a scientific proof of the theorem. These remarks of Bachet's are the occasion for another of Fermat's famous notes: "I have been the first to discover a most beautiful theorem of the greatest gener-

ality, namely this: Every number is either a triangular number or the sum of two or three triangular numbers; every number is a square or the sum of two, three, or four squares; every number is a pentagonal number or the sum of two, three, four or five pentagonal numbers; and so on *ad infinitum,* for hexagons, heptagons and any polygons whatever, the enunciation of this general and wonderful theorem being varied according to the number of the angles. The proof of it, which depends on many various and abstruse mysteries of numbers, I cannot give here; for I have decided to devote a separate and complete work to this matter and thereby to advance arithmetic in this region of inquiry to an extraordinary extent beyond its ancient and known limits."

Chapter IV

Arabic Primacy with Hindu, Chinese, and Maya Contributions

Introduction. Classical Greek and Hellenistic mathematics was transmitted to four successors: Rome, Byzantium, medieval Christian Europe, and medieval Islam. But in Rome, Byzantium, and the Latin West mathematics languished. From the 9th to the 15th centuries extensive mathematical activity revived only in the large cosmopolitan cities of Islam. Though Baghdad was chief among these urban centers, important mathematician-astronomers also emerged at the capital cities of royal courts located on the eastern and western periphery of Islam beginning in the 10th century. These capitals ranged from central Asia (areas now part of the Soviet Union) to Cordova in Moorish Spain. Arabic thinkers cultivated mathematics in at least two ways: (1) by the preservation and transmission of older knowledge and (2) by original contributions to arithmetic, algebra, and trigonometry. This introduction will survey the period of Arabic primacy in mathematics and will cover a second major tradition upon which Arabic authors drew, namely the Indic.[1]

Between the 7th and 9th centuries, Islamic civilization spread throughout the Mediterranean. Islam, "submission to the will of Allah," began with the Prophet Muḥammad (*c.* 570–632), who preached an uncompromising monotheism. His brotherhood of followers named Muslims (in Arabic "true believers") were to unite around his holy book, the *Quran,* a series of visionary revelations. After his message initially failed to attract followers in pagan Mecca, Muhammad fled to Medina in 622. That flight—the *Hegira*—eventually marked the beginning date of the Muslim calendar. Once Medina was won over and Mecca subdued, Islam spread rapidly. The previously rival tribes of the Arabian peninsula were soon unified into a potent Muslim force that embarked upon expansion. Their military campaigns were swift and effective not only because of the evangelistic militancy of the new religion. Other factors that proved to their advantage were a war between the Byzantine and Persian empires that left both exhausted, an enmity between Greek and Semitic (Egyptian and Syrian) Christians that caused the latter to view the Arabs as deliverers from the Byzantine army as well as polytheistic tendencies, and the firm Muslim political organization of the *caliphs* (or rulers) who claimed to be the true successors of Muḥammad. Moreover, the *Quran* had forged the Arabic language into a powerful instrument of unity. Barely over a century after the prophet's death, Muslim conquests stretched from Mesopotamia across the former Persian empire to the Indus Valley in the east, through Syria, Alexandria, and Egypt (wrested from the Byzantine empire in the center), through North Africa, the Strait of Gibraltar,[2] and Spain (taken from the Visigoths in the west).

By the early-9th century, a vital Islamic civilization had taken shape. Several conditions, mostly results of the conquests, were responsible. The Umayyad caliphs, who ruled from 662 to 749 from their capital in Damascus, oversaw the gradual construction of a new urban constellation dedicated to increasing trade and commerce which, not only revitalized Aleppo, Antioch, and Damascus but also turned military camps and towns into new cities of over one hundred thousand inhabitants. The Arabs now assumed new roles in the Mediterranean world. Within the cities they became a governing elite; within the surrounding countryside they became large landowners. These developments, in turn, supported the formation of a vigorous high culture based upon the Arabic language. The term Arab came to designate any person who spoke and wrote in the language, not simply a single ethnic group. Drawing upon knowledge from the past (Hellenistic, Aramaic, and Indo-Persian, including neo-Babylonian) together with the exchange of ideas among different peoples (Zoroastrians, Hindus, Jews, and Christians from Syria, Egypt, and Byzantium), the new culture was rich and eclectic. At first, the Muslims had not encouraged cultural interchanges. They had kept their distance from conquered peoples, tolerating religions based on a sacred book, including Christianity, if their members paid taxes and did not proselytize. Widespread intermarriage with conquered peoples (a practice initially forbidden), numerous converts, and urban growth lent a cosmopolitan quality to Islamic society. Under the Umayyads the Arabs, who had no prior high culture, proved acquisitive pupils as the processes of borrowing and superceding began. Schools were built and literacy increased. The study of language was dominant in the nascent culture. The sister sciences of lexicography and philology were emphasized. These were needed to make the classical language of the *Quran* available to converts without vernacular corruption.

The early members of a new line of caliphs, the Abbāsids, who reigned from 758 to 1258, completed the societal and intellectual framework within which mathematics flourished among the Arabs. The founder of the new dynasty, al-Mansūr (754–775), who put to death Umayyad claimants to the caliphate, built a new capital, Baghdad (762), near the ruins of the Sāssānid capital of Ctesiphon and Babylon on the Tigris. The new capital was to give legitimacy and glory to the new line. The site was chosen because of its excellence as a military camp, its central location in a fertile region growing rice, barley, wheat, millet, and corn, its accessibility to imported food from as far away as Armenia, and its potential as a trade center allowing exchanges with regions as distant as China with its porcelains and silks. The Abbāsid empire was neo-Muslim and more international than the Umayyad. Beginning with al-Mansūr, its bureaucracy followed Persian procedures. Under Hārūn al-Rashīd (786–809), who conducted diplomacy with his contemporary Charlemagne, Baghdad grew in splendor. Shaharazād's *The Thousand and One Nights* is set in the city and has al-Rashīd as a major character. From his early reign al-Rashīd encouraged learning—a movement also promoted by libraries at mosques and in the homes of rich dignitaries who wished to follow the caliph's example. Paper, a product recently introduced from China, replaced papyrus and parchment.

Following a victory over his powerful Byzantine enemy, al-Rashīd, who is often portrayed as the ideal of Islamic kingship, extended royal munificence to attract poets, musicians, and a number of scholars to his court to add to its high culture. At this time Nestorian (east Syrian) physicians from the clinical school at Jundishāpūr, Persia—who the invalid al-Mansūr had originally invited—became influential and began to introduce ancient Greek ideas and writings into Baghdad. After the demise of Alexandria, Antioch and Nestorian Christian monasteries in Syria and the cities of Jundishāpūr and Harrān in Persia had preserved ancient learning. An orthodox

Muslim with a court dominated by orthodox theologians, al-Rashīd and his court found unacceptable the segments of Greek literature which referred to many gods.

Under al-Ma'mūn (813–833), the elder son of al-Rashīd, Islamic society was more firmly established and its energies turned toward educational developments in which mathematics was important. The Baghdad of al-Ma'mūn became a prosperous commercial and dynamic intellectual center whose population reached almost one-half million. Attaining a splendor seldom seen before, it stood alone in wealth and culture as a rival to Byzantium in the Mediterranean and Middle East. Within it al-Ma'mūn institutionalized royal patronage of theoretical and natural knowledge—a vital step for the fruitful study of mathematics. Drawn to his father's library and probably inspired by Syrian and Persian models, this keen-minded caliph had constructed around 815 a "House of Wisdom" (Bayt al-Ḥikmah), a combined academy, library, and translation bureau.[3] Next to it was built an astronomical observatory. These were the first major centers in Islam concerned with theoretical and natural knowledge. Within them astronomy, astrology, cosmology, geography, natural philosophy, optics, mathematics, metaphysics, alchemy, and medicine were investigated with royal financing. Mathematics was closely linked with the first six subjects. Among the Arabs astronomy was especially important for determining the precise hours for daily prayers and the direction of Mecca. Astronomers also faced the challenge of reckoning by and improving upon the Muslim lunar calendar and often had to assist in drawing up horoscopes which nearly all medieval Arabic rulers consulted. Under the Abbāsids most mathematicians were also astronomers.

Significantly, al-Ma'mūn wanted his academy to translate almost the whole of available Classical Greek and Hellenistic manuscripts into Arabic. Consequently, its first director, the Nestorian physician Hunayn ibn Ishāq (809 or 810–877), headed missions to Constantinople and Sicily to acquire these writings. There was at least one pressing reason for the sudden shift in interest to Greek thought in Islam. Earlier contacts with the Indo-Persian culture had presented no challenge to Muslim intellectuals who were at that time mainly theologians. In debates with Christians and Jews, two minorities in Damascus and Baghdad, Muslim theologians were often bested, however, because they could not defend their principles with logical arguments and "proofs" in the Greek manner. This failure had the potential to undermine religious law in Islamic society, the very basis of authority for the caliphate. Thus the turning toward Greek theoretical and natural knowledge may have begun partly as an attempt by Muslim intellectuals to safeguard caliphal interests. In addition, al-Ma'mūn, an unorthodox Muslim, was the son of a Persian slave. Surrounded by Persian advisers, he may have seen Greek logic with its mathematical proof theory as another path to truth complementing the path of faith, prophetic traditions *(hadīth)* and the *Quran*.

The translation of Sanskrit and Greek texts into Arabic and Syriac was a key feature of the cultivation of mathematics under the early Abbāsids. These translations which facilitated the absorption of Indic and Greek scientific ideas into Islam were made through the 10th century. Al-Mansūr had inaugurated this tradition. When in either A.H. 154 or 156 (771 or 773) an embassy from Sind presented him with a copy of the cryptic Sanskrit astronomical text *Mahāsiddhānta,*[4] al-Mansūr had al-Fazārī and a Hindu pandit in the embassy render it into Arabic. The result was *Zīj al-Sindhind.* (Zīj means a set of astronomical tables, while sindhind is a corrupted form of the Sanskrit *siddhānta.*) Al-Fazārī also translated sections of another Indic mathematical treatise, entitled *Āryabhatīya.* His work established a model for Arabic translation and introduced Arabic natural philosophers to Indic astronomical parameters and computational procedures. It was, however, derivative and not very systematic. Alone it did not stimulate original mathematical contributions. That occurred only after major Classical Greek and Hellenistic mathematical manuscripts were translated, selectively assimilated, reworked and revised. Among the first to be translated were Euclid's *Elements* in 800 and Ptolemy's influential *Mathematical Syntaxis* in 827.

A school of translators who not only rendered Greek writings into Arabic but also demonstrated a critical understanding of (and occasional deviation from) Greek methods first appeared in the House of Wisdom. Three of its most active members, the Banū Mūsā (sons of Mūsā—Muḥammad, Ahmad, and al-Hasan) organized this school. The Banū Mūsā influenced Ḥunayn ibn Ishāq and Thābit ibn Qurra (836–901). Hunayn, who concentrated on medical translations, rendered Plato's *Timaeus* and Aristotle's *Analytica posteriora* into Arabic. Thābit translated several treatises of Archimedes, the *Conics* of Apollonius and the *Introduction to Arithmetic* of Nichomachus and wrote scholarly commentaries on Euclid's *Elements* and Ptolemy's *Almagest.* Thābit's contemporary, Qustā ibn Lūqā, made the works of Hero and Diophantus available in Arabic. Most of the Hellenistic Greek mathematical works are now known only from these translations.

The Banū Mūsā stressed plane and solid geometry at the House of Wisdom and led astronomical observations. Their chief geometrical treatise, *Book on the Measurement of Plane and Spherical Figures,* demonstrated that they were not only disciples of the Greeks but also original mathematicians. In large measure this treatise examines the contributions of Euclid and Archimedes to the method of exhaustion. The Banū Mūsā employed a method similar to Archimedes' heuristic procedure of infinitesimals to find the area of curvilinear figures and offered abbreviated, less rigorous proofs that depend on what they called the "rule of contraries" and Euclid XII.16. In defining the area and volumes of curvilinear figures they made an important departure from the Greek legacy in geometry. Rather than employing the Greek technique of comparisons to other figures, they refer directly to arithmetical operations and numerical values. They thereby took a step toward extending the number system to include irrationals as well as integers and rational fractions. This represents an initial stage in the generalization of the concept of number to include (positive) real numbers, a stage not achieved by the Greeks. It is difficult to believe that only a century after Arab conquerors thought there was no number larger than ten thousand and at a time when Charlemagne's nobles were learning to write their names, mathematics of this caliber was achieved in Baghdad.[5]

In addition to scholarly translations, a fusion of elements of Greek learning with Islamic culture was taking place. Arab mathematicians also introduced the footnote into scholarship. Al-Kindī (c. 801–c. 873), the Latin Alkindus, who is known as "the first Philosopher of the Arabs," was crucial to both. In al-Ma'mūn's Baghdad the confrontation between Islamic theologians and Greek philosophy had produced an impasse, the theologians wishing philosophy to be subservient to religion. A generalist of broad intellectual interests, al-Kindī attempted to harmonize strands of Greek thought with that of Islam. Influenced by Aristotle's *Metaphysics,* he wrote in *First Philosophy* that knowledge results from cumulative efforts over the centuries and thus recognizes truths regardless of its origins. Although he cited few Greek sources besides Plato and Aristotle, al-Kindī also wrote in *First Philosophy:* "We endeavor in this book, as is our habit in all subjects, to recall that concerning which the Ancients have said everything in the past." One result of this approach was the explicit use of footnotes, a convention not employed by the ancient Greeks.

Al-Kindī's interest in Greek mathematics was second only to his attention to medicine. His *Letter on . . . the Book of Aristotle* held the study of the mathematical sciences to be preparatory to gaining a knowledge of nearly all other subjects—a view not unlike that of Plato's *Republic*. Neoplatonic ideas drawn from Plotinus and his disciple Porphyry permeate his writings, including fourteen treatises on mathematics. Plotinus and Porphyry situated an ineffable One above all else and proposed mathematics and metaphysics as the means to understand the world of Plato's ideal Forms. Later Arabs carried Neoplatonism beyond al-Kindī's initial stage. That tradition was more fully developed and reconciled with Islam by al-Fārābi, a Turk, and ibn-Sīnā (the Latin Avicenna), a Persian.

In discussions and debates al-Kindī could be harsh, referring to his theological and intellectual opponents as ignorant, bigoted, and narrow-minded. These charges provoked ill-feeling and following a plot against him by two of the Banu-Mūsā he lost caliphal patronage and underwent a public humiliation. He spent his final years in seclusion.

Accounts of the history of mathematics in medieval Islam often begin with Muḥammad ibn Mūsā al-Khwārizmī (*fl.* 800–847), whom the historian al-Tabari believed had Zoroastrian forebears. Al-Khwārizmī's prominence in the history of mathematics stems from two principal contributions: he effectively transmitted Hindi (Gwalior) numerals and he initiated the study of algebra among the Arabs. In both numeration and algebra al-Khwārizmī appears to have skillfully synthesized Greek and Indic traditions. He and other mathematicians in early Baghdad were largely drawn to number from the Greek legacy. The Neoplatonic and Neopythagorean notion of "unity in multiplicity" made a particular contribution. The doctrine that numbers and figures are generated from and ontologically linked to the One, was congenial to Muslims for whom all multiplicity came from the Creator Who is One. To represent numbers al-Khwārizmī adopted the decimal Hindi (now called Hindu–Arabic) numerals for 1 to 9, a zero sign or "cipher" as the Arabs called it, and place-value arrangement. These numerals had originally been introduced into Baghdad by al-Fazārī's translations.[6] In an elementary treatise al-Khwārizmī first presented the decimal place-value system to the Arabic learned world in a systematic fashion. He used the new numerals in the four basic arithmetical operations, employed them to extract roots, and explained a variety of applications for them.

Although this treatise was more responsible than any other for spreading Hindu–Arabic numerals through the Muslim world (and to Christian Europe in the High Middle Ages), Arab thinkers accepted them only very slowly. Among the first supporters, al-Kindī wrote four treatises supporting Hindu numerals, entitled *Hisabu'l hindi*. Then, in the 10th century, a group of scholars in Basra known as "The Brethren of Purity" expounded Neopythagorean–Hermetic ideas regarding number in clear and simple *Epistles*. They wrote: "The science of number is the 'root' of the sciences, the foundation of wisdom, the source of knowledge, and the pillar of meaning."[7] Still, the Arabs long retained a variety of expressions for numerals. These expressions included the Greek alphabetic numerals and a sexagesimal modified place-value system in astronomy. Furthermore, two sets of Hindu-Arabic numerals with slightly different shapes existed in the Muslim world by the second half of the 9th century. Distinct from the east Arabic characters, the *hurūf al-ghubār* (letters of dust) were developed in Spain for use with a sand abacus. Our modern numerals are quite similar to the *ghubār* numerals. Probably Gerbert, later Pope Sylvester II, first attempted to introduce these to the Latin West in the final years of the 10th century and the translators Adelard of Bath and Robert of Chester repeated the attempt in the 12th century with only a modest impact. Leonardo Fibonacci met with more success in the 13th century.

In the Muslim world, the high point in the treatment of decimal numeration did not occur for two centuries after Leonardo. Seeking more precise trigonometric tables for astronomy, the erudite Persian Jamshīd al-Kāshī (or al-Kashānī, d. 1429) introduced finite decimal fractions methodically and first explained in detail all arithmetical operations with them in his computational masterpiece *Risala al-muhītīyya* ("The Treatise on the Circumference," 1424)[8] and the encyclopedic *Miftah al-hisab* ("The Key to Arithmetic"). The Latin West did not achieve a comparable level of operations with decimal fractions until the work of Simon Stevin in the late-16th century.

Al-Khwārizmī's most important work is the *Kitab al-mukhtasar fī hīsab al-jabr wa'l-muqābalah* ("The Compendious Book on Calculation by Completion and Balancing," *c.* 830). This elementary, practical text on linear and quadratic equations is known in English simply as the *Algebra* from its contracted title. The two operations required to solve its equations, *al-jabr* (whence "algebra") and *al-muqābalah,* refer respectively to the restoration or completion of an equation by eliminating negative quantities and the balancing of the two sides of an equation by removing equal positive quantities of the same power. Perhaps al-Khwārizmī's most notable contribution to the theory of equations is his demonstration that quadratic equations can have two roots. His choice to recognize only positive real roots, including irrational ones, was usually followed by later Muslim algebraists who also ignored equations that did not have positive real roots. (*See* pp. 183–187 for more on the *Algebra.*)

Although some scholars remain skeptical, al-Khwārizmī must have drawn extensively upon Greek, Indic and neo-Babylonian sources in preparing the *Algebra.* In general, the Greek legacy offered a proof theory so rigorous that it could be carried to a stultifying extreme, while the Indic offered an antidote with its appeal to the "indefinite." In particular, the use of geometric figures to explain equations in the introduction of the *Algebra* argues for a familiarity with Euclid's *Elements,* while the book's complete statement in words with no symbols (even for numerals), its use of the "rule of three" to handle ratios and proportions, and its mensuration methods suggest that al-Khwārizmī drew more heavily upon Sanskrit sources, expecially the writings of Brahmagupta. The laborious exposition of the text entirely in words strongly suggests that Diophantus' *Arithmetica* with its abbreviations and stenographic notation was not available.

The place of al-Khwārizmī in the history of algebra is often misunderstood. He did not found the subject nor even carry it as far as the Greeks and Hindus had. He was, however, unusually influential in the development of algebra in medieval Islam and the later medieval Latin West. He began a tradition of distinguished Muslim algebraists. In the late-9th century, Thābit ibn Qurra geometrically solved quadratic equations and what amounted to selected cubic equations in his efforts to solve the classical problems of the trisection of an angle and the construction of two mean proportionals. His contemporary al-Mahani after studying Archimedes' work on spheres solved by the intersection of conic sections a specific cubic equation of the type $x^3 + r = px^2$. In the late-10th century Abū'l-Wafā even solved a quartic equation (in modern form $x^4 + px^3 = q$) geometrically by the intersection points of a parabola and hyperbola, while al-Kūhi thoroughly studied trinomial or three-termed equations. The 11th century began with al-Haytham (Latin Alhazen) solving specific cubic equations geometrically. It continued with al-Khāyyamī's pursuit of an analogous solution of all forms of cubics that have positive real roots—a topic that will be discussed below. In the West al-Khwārizmī's *Algebra* was first translated into Latin as the *Liber Alghoarismi* by John of Seville (*c.* 1135) and then the *Liber Algorismus* by John of Sacrobosco (*c.* 1250). These translations and the growing use of Hindu–Arabic numerals made his name synonymous with the "new arithmetic" in Christian Europe. Any treatise on that topic was given the Latin form of his name, *algorismus,* from which comes our word algorism and its corrupted form algorithm.

One of the most active members of the early House of Wisdom, al-Khwārizmī also contributed substantially to geography. Arab intellectuals, like their ancient Greek counterparts, believed that the earth was a sphere. With this in mind al-Ma'mūn commissioned one of the earliest geodesic surveys to measure the length of a degree of meridian near Palmyra—an enterprise possibly suggested by a reading of Eratosthenes. It is likely that al-Khwārizmī participated in this survey whose results when properly multiplied give a circumference of the earth of 20,400 miles. He also must have worked with other scholars on a world map for al-Ma'mūn and wrote a book known as the *Geography,* which gave latitudes and longitudes of cities and regions. The original title of his *Geography* was *Kitāb sūrat al-ard* ("Book of the Form of the Earth"). It carefully revised Ptolemy's *Geography* and gave more accurate coordinates for many Islamic cities.

In Islam the development of trigonometry was closely linked to the study of astronomy and benefited, as had algebra, from the integration of Indic ideas and computational methods with Greek models. Before the 9th century, Islamic astronomy had been based on Sāssānid Persian and Indic astronomical tables, particularly those translated by al-Fazārī. Even as al-Khwārizmī improved upon al-Fazārī's *Zij,* the Indic influence was being superceded. This occurred when not only Ptolemy's *Mathematical Syntaxis* was translated several times but also his *Tetrabiblos* and his astronomical tables, known as *Canones procheroi.* From the late-9th century the Muslims continued the tradition of Ptolemy in astronomy. His *Mathematical Syntaxis* became known as *al-Majisti* or in Latin *Almagest* ("the Greatest"). Muslim astronomers thereafter worked with epicycles in preparing tables and until the 11th century generally accepted the geocentric model of the universe.

The Muslims had two types of trigonometric reckoning from which to choose—Ptolemy's chords of twice the angle and the Hindu sine of an arc. The Hindu sine was not the modern ratio but a half chord. Astronomers needed one of these methods to calculate the different chords or arcs lying between 0 and 90° at intervals of one degree or less. The Muslims adopted and improved upon the Hindu method. Two men introduced the Hindu sine and trigonometry among the Arabs—the learned Thābit ibn Qurra (836–901) and al-Battāni (858–929; Latin Albategnius). In his study of sundials and shadows, Thābit (whom the Banū Mūsā had invited to Baghdad) formulated the equivalent of theorems for sines and cosines for spherical triangles as well. In a famous *Zij,* al-Battāni employed the sine, cosine, and versed sine (that is, 1 − cosine), as well as clumsy gnomonics to represent our tangents and cotangents. Al-Battāni also elegantly solved problems in spherical trigonometry by applying the principle of orthographic projection—a principle later developed by Regiomontanus in the West.

Abū'l Wafā (940–997 or 998) and al-Bīrūnī (973–1050) were the next to substantially improve trigonometry. Abū'l Wafā gave Muslim trigonometry a more systematic form making use of all six trigonometric quantities, including the secant and cosecant, which he introduced. In the Hindu manner he could calculate sine values from cosine values by using an identity such as $\cos^2 a + \sin^2 a = 1$. He further developed a method of repeated divisions that allowed him to compute a new sine table at intervals of one-half degree that was accurate to the fourth sexagesimal place. As his treatises known as *Shadows* and *Chords* show, al-Bīrūnī also computed the six trigonometric quantities, found the value for $\cos 3\theta$, and worked with Menelaus' theorem in spherical trigonometry. When his royal patron in Khwārazm south of the Aral Sea was deposed, al-Bīrūnī fled for a time to India. While there he gained a moderate command of Sanskrit that allowed him to correct corruptions in earlier translations of Sanskrit texts with the help of Hindu pandits. A critical thinker, he questioned Ptolemy's geocentric model of the universe suggesting instead a sun-centered model. Al-Bīrūnī's views in astronomy and even more the subsequent criticisms of Ptolemaic astronomy with important departures in planetary theory by al-Tūsī and his disciples most likely influenced Nicholas Copernicus.

After the mid-11th century the Abbāsid caliphate swiftly disintegrated. The Abbāsid lands, which lacked cohesion from the early-10th century, had long suffered from heavy taxes that discouraged farming and industry, strife that left farm lands desolate, and recurring epidemics of plague, smallpox, and malaria. To these internal problems were now added three major waves of invasion from east and west. The Seljuk Turks attacked from central Asia in 1071. The year 1096 brought the first of nine Crusades from western Europe. The last official Crusade was in 1291 but the hostilities long continued. The third group of invaders, the Mongols, wreaked the most havoc, destroying many of the cultural centers of eastern Islam. Ghengiz Khan (*c.* 1155–1227) entered Abbāsid lands about 1206. In 1258 his grandson, Hūlāgū, destroyed Baghdad and put to death the caliph, his family, and top officials. By that time the center of learning in Islam was located in Cordoba, Seville, and Toledo, the major cities of Muslim Spain.[9]

Although official financial support and encouragement for mathematics dwindled during this chaotic period, episodes of original and relatively extensive mathematical activity continued in eastern Islam through the 13th century. Official schools and patronage of large-scale building projects had been two important sources of support. These schools now displayed a marked lack of interest in mathematics.[10] Support from the caliphal retinue for the construction of mosques, schools, observatories, and bridges[11]—and with them ingenious mosaics—decreased sharply. These losses were partly offset by patronage of astronomy, astrology, calendrics, and theoretical mathematics by some Seljuk and Mongol sultans or their courts. These conquerors also attempted to provide islands of tranquillity for their savants to pursue research. That tranquillity, however, proved all too transitory amid increasing tumult. Probably the two foremost universal mathematicians in medieval Islam now appeared. They were the Persians 'Umar al-Khayyāmī (*c.* 1048–*c.* 1131), who was attached to the Seljuk courts, and Nasīr al-Dīn al-Tūsi (1201–1274), who constructed for Hūlāgū a major observatory at Marāgha.

Although the English-speaking world primarily knows al-Khayyāmī as the poet Omar Khayyam because of Edward Fitzgerald's translation of the *Rubaīyāt,* to his contemporaries al-Khayyāmī was primarily a master of the mathematical sciences. Probably the most outstanding of the medieval Muslim algebraists, he presented the first satisfactory geometrical theory of cubic equations in the *Risāla* (or "Algebra," *c.* 1079). To do this he appealed to two sections of ancient Greek mathematics, namely its geometrical algebra and the attempts to double the cube and trisect an angle by the intersection of conic sections. Among the Muslims Thābit and al-Haytham had earlier examined these subjects. Building upon this base, al-Khayyāmī constructed fourteen types of cubic equations and solved them by the intersection or contact point(s) of the following curves: cubics that are written in modern notation as $x^3 + b^2x = b^2c$ by a parabola and semicircle, those of form $x^3 + ax^2 = c^3$ by a hyperbola and parabola, and those of form $x^3 + ax^2 + b^2x = b^2c$ by an ellipse and hyperbola. He even solved one quartic that in modern notation is $(100 - x^2)(10 - x)^2 = 8100$ by the intersection points of a hyperbola and circle. Al-Khayyāmī dealt only with the sections of these curves falling within the first quadrant (*i.e.,* the solution or roots had to be positive real numbers). He showed that cubic equations may have no such roots, only one root, or two roots. His geometrical solution and classification of cubics subsequently influenced Sciopione del Ferro, René Descartes, and Isaac Newton. (*See* pp. 188–189 for al-Khayyāmī's inventive contributions to astronomy, computation, the binomial theorem, the Eudoxian theory of ratios, number theory, and the theory of parallels.)

Nasīr al-Dīn al-Tūsī rejuvenated the serious study of mathematics in eastern Islam during the 13th century. Some of his efforts were directed to finding institutional and financial support for the discipline. As a young man he had earned a reputation as a distinguished scholar by reviving the Peripatetic ideas of Avicenna and advancing astronomy. That reputation reached beyond Persia to as far as China. When Hūlāgū, who esteemed astrology and astronomy, conquered northern Persia in 1256, he took al-Tūsi into his service. Within three years al-Tūsi convinced the Mongol chieftain to construct and financially endow an observatory at Marāgha. Under al-Tūsī's direction that observatory became a *madrasa,* a center for higher instruction in natural and theoretical knowledge and for collaborative studies. To assist in mathematical instruction he prepared a series of recensions of writings by Euclid, Archimedes, Menelaus, and Ptolemy.

Al-Tūsī contributed to arithmetic, geometry, and trigonometry. In the first two subjects he continued the work of al-Khayyāmī as he demonstrated the commutative property for real numbers, including irrationals, and attempted to prove Euclid's parallel postulate. His most original mathematical contribution was in trigonometry. Expanding upon the ideas of Abū'l Wafā and al-Bīrūnī, he was the first Muslim to separate trigonometry from astronomy and to establish it as an independent branch of pure mathematics in his *Shal al-qita* ("Book of the Principal of Transversal"). In this book he introduced the six cases for a spherical right triangle. If a and b are the legs and c the hypotenuse of a spherical triangle, then

$$\begin{aligned} \sin b &= \sin c \sin B \\ \sin b &= \tan a \cot A \\ \cot A &= \tan B \cot c \end{aligned} \qquad \text{and} \qquad \begin{aligned} \cos A &= \cos a \sin B \\ \cos c &= \cot A \cot B \\ \cos c &= \cos a \cos b. \end{aligned}$$

Apart from the Greek legacy, early medieval Muslim mathematicians were most indebted to the mathematics and astronomy from India. Hindu sages developed these two disciplines, often in an interlocking fashion, from the mid-4th until the 12th centuries of the Christian era. Besides making progress in trigonometry and inventing a decimal positional numeral system, they made what were perhaps their principal mathematical contributions in arithmetic, where they proved extremely skillful in computation, and in algebra, where they solved indeterminate or Diophantine equations. In computing and solving equations the Hindu masters not only discovered facile procedures for operating with fractions, rationals, and irrationals, but also introduced negative numbers and incorporated operations with zero. Negative numbers, which were first utilized to represent debts, were only slowly accepted.

Although the surviving literary evidence concerning mathematics in classical and early medieval India is scant and fragmentary,[12] it sheds much light on the development of the discipline. In part it reveals the pressures for innovation and growth, the chief strength—and a serious shortcoming—in Hindu mathematics. Most extant writings treat mathematical astronomy and contain only a few sections on mathematics. Devising and simplifying computations and approximations for a more accurate astronomy and, to a lesser extent, for the needs of commerce are prominent. Within the framework of these utilitarian concerns, the authors seem motivated by simple delight in numerical manipulations. Thus, their innovation almost never departed from computations and various reductions of these to sets of rules. The Hindus, like the Babylonians and Chinese, made this work the heart of their mathematics. The word *ganaka,* a calculator, came to denote "astronomer" as well as "mathematician."

Except for its computational originality, Hindu astronomy was characteristically repetitive and failed to recognize the possible relationships of some problems in physics. Astronomers preserved older knowledge, including contradictory doctrines. The Hindu cultural preference was to subordinate internal consistency, because they were reluctant to part with any segment of their tradition. Tolerance of inconsistency together with the lack of a critical methodology to replace personal bias undoubtedly prevented the evolution of fundamental theory in astronomy and, it would seem, in mathematics. Indeed, Hindu astronomer–mathematicians normally ignored or eschewed theory inherited in the deductive, axiomatic geometry of the Greeks; no example of a demonstration appears in their work. Hindu geometry did not proceed beyond an elementary mensurational stage employed in surveying and architecture.

The texts in mathematical astronomy pose some problems for historians of mathematics. Consider their style of presentation. To facilitate memorization, the texts were composed in verses as were the pre-Socratic natural philosophical works of the Greeks. Versification with its substitution of terms led to imprecision in technical terminology, while attempts to follow the meter often resulted in the omission of important mathematical prescriptions. When the texts were copied without comprehension or care, unstated procedures for solving problems were lost. Today these procedures can only be guessed at, based on the solutions achieved. Another problem arises from the incomplete investigation of foreign influences and how the Hindus modified imported thought. Located at a crossroads between China and the lands of the Middle East and western Mediterranean, India suffered many military intrusions and had long-range trade and intellectual contacts. Since Alexander the Great's invasion of the Indus region in 326 B.C., for example, contacts with the West were frequent. During India's classical and medieval times, the dominant influences upon Hindu astronomy were the Greco–Babylonian (*c.* 200–400) and the Greek (*c.* 400–1200). While the Hindus built their trigonometry on Greek spherical astronomy, they apparently drew mostly upon Aristotelian sources with only a few additions from a corrupted form of Ptolemy. The Hindus probably remained ignorant of most of Ptolemaic astronomy.

India's Classical Age began under the imperial Gupta (320–550). This dynasty unified northern India through conquests, a relatively small and effective administration, and the fostering of unity through a cultural revival that produced at least five astronomical *Siddhāntas.* The Gupta Empire was the largest in India since Ashoka's six centuries earlier. With Rome waning and China experiencing internal divisions and barbarian invasions, it was the most stable empire in the world. Although the Gupta rulers practiced religious tolerance, they preferred Hinduism over Buddhism and gave patronage to the brahman (or top) caste. Derived from the Aryan priestly class, the brahmans had accepted the Buddhist principle of *ahimsa* (nonviolence and respect for all life) and were vegetarians.[13] Accompanying the resurgence of Hinduism was an increased use of Sanskrit, the sacred language of the brahman priests.

The Gupta rulers' patronage of literature, the fine arts, and scholarship led to a cultural peak. Sanskrit literature was highlighted by the poetry and drama of Kalidasa (*c.* 400–455), the "Indian Shakespeare," and a fascinating array of fairy tales and fables that later influenced Boccacio, Chaucer, and the brothers Grimm. Painting and sculpture proliferated as central elements of temple decorations, often including erotic scenes. Such scenes on temple and monastery frescoes were prepared to reveal the beauty of the human form. To support general scholarship the Guptas founded and endowed a major center of learning at Nalands.

Among the foremost intellectual achievements in Gupta India was the evolution of astronomy beyond the crude, often inaccurate calendric stage contained in the *Vedas,* the *Brahmanas,* and the canonical astronomical works of the Jaina priests. The oldest of these texts probably date from about 1000 B.C. They treated *yugas* (periods), variations in *rtus* (seasons) in order to determine seasonal sacrifices, equinoxes, solstices, *paksas* (half lunations), *adhimsas* (intercalary months) for lunar years of 351 days and the civil year of 360 days, *naksatras* (constellations), planetary conjunctions and the computation of solar and lunar eclipses. They also dealt with cosmology. By the end of the fifth century Gupta pandits prepared as many as eighteen reformed astronomical compendia. They were composed of cryptic rules written in Sanskrit verse. These compendia were called *siddhāntas,* a term meaning "final conclusion" or "solution." Portions of five have survived. Four *siddhāntas* were original with the Hindus, namely the *Surya* (System of the Sun), *Vasistha, Paitāmaha,* and *Romaka* (or Roman). As Al-Bīrūnī noted, the fifth and most accurate, the *Pauliśa,* was a translation from a Greek text. The use in each of the equivalent of linear zigzag functions, sexagesimal fractions, and epicycles shows that their authors were influenced by Babylonian materials from the Seleucid period and Hellenistic manuscripts. The Ujjain astronomer and astrologer Varāhamira (*fl.* 6th century) preserved the main features of each in the *Pañcasiddhāntikā,* which the Brown historians Otto Neugebauer and David Pingree have recently translated into English.

During late Gupta and subsequent medieval times, Hindu mathematics developed largely within the intellectual confines of paksas, or schools of astronomy. Two of the most famous schools were the Brāhmapaksa (founded *c.* 400), which centered in the beautiful city of Ujjain (or Ujjayinī), and the Āryapaksa (founded *c.* 500), which was long influential in southern India. Within these mathematics progressed episodically. Our concern is with three of their members: Āryabhata I (b. 476), who founded the Āryapaksa, as well as Brahmagupta (598–*c.* 665) and Bhāskara II (1115–*c.* 1185), both of whom belonged to the Brāhmapaksa. The last two are perhaps the most outstanding Hindu mathematicians of the period.

The mathematical ideas of Āryabhata I appeared in a slender volume on astronomy, entitled *Āryabhatīya* (499), written when he was 23. A resident of Pataliputra, the imperial capital of the Guptas, he attempted to recast and codify the results of the *Siddhāntas.* To do so he emphasized trigonometry and prepared a table of sines. The rough methods of the *Romaka* and *Pauliśa* for predicting eclipses had employed a predecessor of the sine function, while a table in the *Paitāmaha* essentially listed values for sines. As subsequent writings by Varāhamira confirm, the Hindus generally based sine tables on a radius equal to 120 parts, which was double the value used by Ptolemy to compute chords. The relation between the Hindu sine function and the Ptolemaic chord is as follows: $\mathrm{Sin}_{120}\theta = \mathrm{Crd}_{60}2\theta$. Āryabhata was the first to associate the half chord with half the arc of a full chord. He also invented the equivalent of the cosine and versed sine. Āryabhata's treatment of π was inconsistent, yielding four values. Three of them are $3\,3/14$, $3\,27/191$, and $\sqrt{10}$, the Jaina value. Appealing to the method of exhaustion, he achieved a greater accuracy with his fourth approximation. To do this he divided a circle with a diameter of 20,000 parts into 21,600 minutes of arc. He then found the perimeter of the resulting polygon as follows: "Add 4 to 100, multiple by 8, and add 62,000." This gave 62,832 divided by 20,000 or $\pi = 3.1416$. Unfortunately the Hindu pandits neglected this last value and chose to use $\sqrt{10}$ instead.

A "cookbook" form of algebra is also found in the *Āryabhatīya's* 66 elaborate verbal rules and 118 verses. The absence of symbols or a general method make the algebraic procedures difficult to follow. Nevertheless, Āryabhata provided a series of steps that led to the solution of specific linear and quadratic equations. Significantly, he also solved some linear Diophantine equations by means of continued fractions. Diophantine equations have two or more variables, for example $ax \pm by = c$, and their solutions must be positive fractions or integers. While Diophantus had sought only one rational solution, Āryabhata and his Hindu successors devised methods to obtain all integral solutions. Diophantine equations were quite important in Hindu astronomical calculations; their solutions indicated when constellations were to appear in the skies.

The eminence of Brahmagupta in Hindu mathematics rests on his work in the *Brāhmasphutsiddhānta.* This volume was composed when he was thirty and resided in a capital city called Bhillamāla (modern Bhinmal in southern Rājasthān). Among its twenty-four chapters are four and a half on geometry, trigonometry, algebra, and arithmetic. Brahmagupta's ability in addressing these subjects varies from pedestrian in geometry to brilliant in algebra. In the former his value for π was no better than $\sqrt{10}$: sometimes his work was formulaic. These included "Heron's formulas" for finding the area of triangles and quadrilaterals, $A = \sqrt{(s-a)(s-b)(s-c)(s-d)}$, where s is the semiperimeter. Brahmagupta exhibited higher ability in spherical trigonometry, where he expanded upon materials from the *Paitāmaha* and *Āryabhatīya.* He devised an extraordinary rule for determining solar altitude when the sun is in the northern hemisphere and the angle of the shadow and the east-west line is 45°. In modern symbols this rule is equivalent to

$$\sin \alpha = \sqrt{\frac{(R/2^2 - \mathrm{Sin}^2\eta)\cdot 12^2}{72 + S_0{}^2} + \left(\frac{12 \cdot S_0 \cdot \mathrm{Sin}\eta}{72 + S_0{}^2}^{2}\right)} + \frac{12 \cdot S_0 \cdot \mathrm{Sin}\eta}{72 + S_0{}^2}.$$

Brahmagupta again displayed great inventiveness in the theory of equations and numerical computation. Limited to an algebra consisting of words and abbreviations but no symbols, he nevertheless solved quadratic equations of the type $px^2 + qx + r = 0$ and proposed the first general solution of linear Diophantine equations. He also examined quadratic Diophantine equations, identifying those of type $y^2 = ax^2 + 1$ as essential to a general solution. Brahmagupta introduced negative numbers and the zero into the four arithmetical operations. He correctly noted that a positive divided (or multiplied) by a positive or a negative by a negative gives an affirmative answer, that a positive divided (or multiplied) by a negative or vice-versa gives a negative answer, and that a negative subtracted from zero results in a positive number. Uninhibited by Greek formalism, he was not afraid to introduce operations with zero, a step that involves dangerous subtleties. This freedom led him to assume that a cipher divided by a cipher is nought ($\frac{0}{0} = 0$) and that a positive or negative divided by a cipher is a fraction ($\frac{a}{0}$). At the same time he apparently recognized that zero as a divisor raised ambiguities. While multiplication by real numbers, except for zero, gives unique results, the results with zero are not unique. Today we say that division by zero leads to an undefined result.

By 600 the Hindu pandits essentially had a decimal positional system of numbers with a zero that was often represented by a dot. As the *Brāhmasphutsiddhānta,* epigraphs, and stone inscriptions reveal, word- and alphabetic-numerals continued in use. A widely-accepted symbolic notation was still in the making. The decimal positional system had evolved slowly. From Vedic times the Indians had a decimal, versified system, albeit without place-value. Fascinated by large numbers, the ancient Indians enthusiastically pursued the study of number. They were not unlike the ancient Egyptian scribal mathematicians in this regard. By Ashoka's rule in the 3rd century B.C. India had two sets of numerals—Kharosuthi, which resembled Roman numerals, and Brahmī, which had separate symbols for each digit from 1 through 9. The Netherlands historian Hans Freudenthal maintains that the Hindus knew place-value before they acquired neo-Babylonian and Hellenistic materials containing the sexagesimal positional system and an omicron for a medial zero. The British historian Joseph Needham conjectures another possible foreign source of the zero in the Indochinese studies of the Taoist "emptiness." Even without a foreign source, the concept of zero was congenial to the "void" of Hindu philosophy. By the late-9th century the Brahmī symbols evolved into the decimal positional Gawalior numerals which in turn were the basis for the west Arabic or ghubār numerals.

Bhaskara II was the last major Hindu astronomer–mathematician of the medieval period. Unlike prior Hindu masters, he may have been influenced by Islamic astronomy—a supposition based on his advanced knowledge of trigonometry in the *Siddhāntaśiromani* (1150). This supposition has yet to be substantiated, however. Among the trigonometric rules that he states is the equivalent of the following:

$$\text{Sin}\,(\alpha + \beta) = \frac{\text{Sin}\,\alpha \cdot \text{Cos}\,\beta}{R} + \frac{\text{Cos}\,\alpha \cdot \text{Sin}\,\beta}{R} .$$

Hindu traditions influenced him more than imported ideas. (*See* pp. 196–205 for his improvement upon Brahmagupta's work with zero, negative numbers, and indeterminate equations as well as his pioneering attempts at providing an algebraic symbolism.) Finally, his *Līlāvatī* contains an interesting method of multiplication known as the *gelosia,* or grating, method. It operates as follows: To multiply 149 by 27, first draw a two-by-three grid with diagonals beginning in the square on the bottom right. Next, multiply the numbers at the head of each column and record the results in the appropriate squares. Then sum the results beginning in the lower right corner and proceeding along the diagonal rows. The answer is 4023.

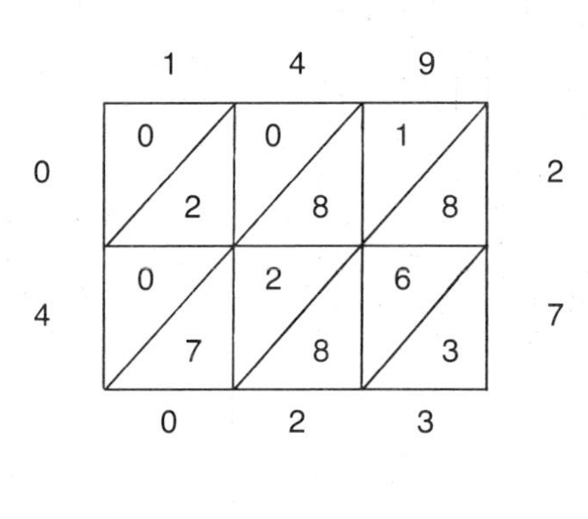

NOTES

1. The two concluding selections in this chapter (51 and 52) treat the Chinese *Nine Chapters on the Mathematical Art* and Maya calendrics and vigesimal, place-value numeration.

2. The name Gibraltar comes from "jebel" = mount and Tarik, the name of the Arab commander who crossed it in 711.

3. The "House of Wisdom" is often referred to as the "Dār al-Hikma."

4. This work was based largely on the *Brāhmasphutasiddhānta* of Brahmagupta. Al-Fazārī's *Zīj* strongly influenced astronomy in western Europe through its translation into Latin by Adelard of Bath in 1126.

5. The Banū Mūsā efforts in plane and solid geometry were not lost in Islam or the Latin West. Thābit ibn Qurra, another student of Archimedes, made novel contributions to the method of exhaustion. In computing the area of parabolas and other curvilinear figures by a process involving upper and lower integral sums, he introduced the division of the segment of integration into unequal parts. (See Khalil Jaouiche, *Le Livre du Qarastūn de Tābit ibn Qurra,* Leiden: Brill, 1976, pp. 101 ff.). Thābit, al-Khayyāmī, al-Karazi, and al-Tūsi pursued studies that linked geometry and irrational numbers. Their studies often criticized segments of Euclid's general theory of proportions in books five and six of the *Elements.* Finally, in the Latin West Leonardo Fibonacci drew upon ideas from the translated writings—the *Verba filiorum*—of the Banū Mūsā in preparing the *Practica geometriae* (1220/1221).

6. The Syrian bishop Severus Sebokht had spread a knowledge of Hindu numerals and methods of calculation to western Asia as early as the late-7th century.

7. Seyyed Hossein Nasr, *Science and Civilization in Islam* (New York: New American Library, 1968), p. 153.

8. Al-Kāshī, who had Ulugh Beg as a patron at Samarkand, found the ratio of the circumference of a circle to its radius to be 6.283185307 . . . . When divided by 2, this gives $\pi = 3.14149265 \ldots$, which agrees with the modern value.

9. Among their permanent residents in the late-12th century were two prominent Muslim Aristotelians, the astronomer al-Bitrūjī and the commentator ibn Rushd (Latin Averroes), as well as the Jewish scholar Moses Maimonides. Their temporary residents included many scholars from northern Europe through whom much of Arabic learning was transmitted to Christian Europe. These scholars were especially drawn to the famous library in Cordoba that held over 400,000 volumes.

10. Nasr, *Science and Civilization in Islam,* p. 150.

11. For an account of the role of bridges in Islamic culture see Ivo Andric, *The Bridge on the Drina* (London: George Allen and Unwin Ltd., 1959; Signet Edition, 1967).

12. India's classical period lasted from A.D. 320 to about 700. The following six centuries of fragmentation were a time of divisive Islamic influence, especially by the Turks. After this Medieval period came consolidation and expansion under the Mughal Empire.

13. According to the Chinese monk Fa-hsien, who provides much of our written knowledge of Gupta India, "Throughout the country no one kills any living thing, or drinks wine, or eats onions, or garlic . . . they do not keep pigs or fowls, there are no dealings in cattle, no butchers' shops or distilleries in their market places." See Vincent A. Smith, *The Oxford History of India* (Oxford: The Clarendon Press, 1958) 3rd edition, p. 171.

SUGGESTIONS FOR FURTHER READING

Tian-se Ang, "Chinese Interest in Right-Angle Triangles," *Historia Mathematica* 5(1978), 253–266.

D. M. Bose, S. N. Sen, B. V. Subbarayappa, eds., *A Concise History of Science in India.* New Delhi: Indian National Science Academy, 1971.

W. E. Clark (ed.), *The Aryabhatiya of Aryabhata*. Chicago: Open Court, 1930.

Ali Abdullah al-Daffa, *The Muslim Contribution to Mathematics*. London: Croom Helm Ltd., 1977.

B. Datta and A. N. Singh, *History of Hindu Mathematics*. Bombay: Asia Publishing House, 1962.

Philip Hitti, *History of the Arabs*. London: Macmillan, 1937. Tenth Edition, 1970.

________, *Makers of Arab History*. New York: St. Martin's Press, 1968.

L. C. Jain, "On Certain Mathematical Topics of the Dhavalā Texts," *Indian Journal of History of Science* 11(1976), 85–111.

E. S. Kennedy, trans. and commentator, Ahmad al-Bīrūnī's *The Exhaustive Treatise on Shadows*. Aleppo: University of Aleppo, 1976. 2 vols.

Harold Lamb, *Omar Khayyam, A Life*. New York: Doubleday, 1936.

Lam Lay-Yong, "The Chinese Connection between the Pascal Triangle and the Solution of Numerical Equations of Any Degree," *Historia Mathematica* 7 (1980), 407–424.

________, "Chu Shih-Chieh's Suan-hsueh ch'i-meng," *Archive for History of Exact Sciences* 21 (1979), 1–31.

________, *A Critical Study of Yang Hui Suan Fa: A Thirteenth-Century Chinese Mathematical Treatise*. Singapore: Singapore University Press, 1977.

Ulrich Libbrecht, *Chinese Mathematics in the Thirteenth Century*. Cambridge, Mass.: The MIT Press, 1973.

Wang Ling and Joseph Needham, "Horner's Method in Chinese Mathematics; Its Origins in the Root-Extraction Procedures of the Han Dynasty," *T'oung Pao* 43(1955), 345–401.

Floyd G. Lounsbury, "Maya Numeration, Computation, and Calendrical Astronomy," in C. C. Gillispie, ed., *Dictionary of Scientific Biography*. New York: Charles Scribner's Sons, 1978. Vol. 15, 759–818.

Yoshio Mikami, *The Development of Mathematics in China and Japan*. New York: Hafner, 1913. Chelsea Reprint, 1961.

Sylvanus Griswold Morley, *The Ancient Maya*. Palo Alto: Stanford University Press, 1947. 3rd edition, 1956.

Shigeru Nakayama, "Japanese Scientific Thought" in C. C. Gillespie (ed.), *Dictionary of Scientific Biography*. New York: Charles Scribner's Sons, 1978. Vol. XV, 728–758.

Seyyed Hossein Nasr, *Science and Civilization in Islam*. New York: New American Library, 1968.

Joseph Needham in collaboration with Wang Ling, *Science and Civilisation in China*. Cambridge: Cambridge University Press, 1959. Volume 3.

David Pingree, *Census of the Exact Sciences in Sanskrit*. Philadelphia: American Philosophical Society, vol. 1 (1970), vol. 2 (1971), and vol. 3 (1976).

________, "The Fragments of the Works of al-Fazārī," *Journal of Near Eastern Studies* 29 (1970), 103–123.

________, "The Greek Influence on Early Islamic Mathematical Astronomy," *Journal of Near Eastern Studies* 93 (1973), 32–43.

________, "History of Mathematical Astronomy in India," in C. C. Gillespie, ed., *Dictionary of Scientific Biography*. New York: Charles Scribner's Sons, 1978. Vol. XV, 533–633.

P. M. Pullan, *The History of the Abacus.* New York: Frederick A. Praeger, 1969.

A. I. Sabra (ed.), *Omar Khayyam, Explanation of the Difficulties in Euclid's Postulates.* Alexandria: publisher unknown, 1961.

S. Saidan, *The Arithmetic of al-Uglidsi.* Boston: Reidel, 1978.

Nathan Sivin, *Cosmos and Computation in Early Chinese Mathematical Astronomy.* Leiden: E. J. Brill, 1969.

David Eugene Smith and Yoshio and Mikami, *A History of Japanese Mathematics.* Chicago: Open Court, 1914.

C. N. Srinivasiengar, *The History of Ancient Indian Mathematics.* Calcutta: The World Press Private Ltd., 1967.

D. J. Struik, "Omar Khayyam, Mathematician," *The Mathematics Teacher* 51 (1958), 280–285.

Frank J. Swetz and T. I. Kao, *Was Pythagoras Chinese? An Examination of Right Triangle Theory in Ancient China.* University Park, Pa.: Pennsylvania State University Press, 1977.

B. L. van der Waerden, "Pell's Equation in Greek and Hindu Mathematics," *Russian Mathematical Surveys* 31 (1976), 210–225.

Donald Blackmore Wagner, "An Early Chinese Derivation of the Volume of a Pyramid: Liu Hui, Third Century A. D.," *Historia Mathematica* 6 (1979) 164–188.

Kiyosi Yabuuti, "Sciences in China from the Fourth to the End of the Twelfth Century, in Guy S. Metraux and Francois Crouzet (eds.), *The Evolution of Science.* New York: Mentor, 1963.

Mohammad Yadegari, "The Binomial Theorem: A Widespread Concept Medieval Islamic Mathematics," *Historia Mathematica* 7, 4, 1980, 401–406.

Adolf P. Youschkevitch, *Les mathematiques Arabes (VIIIe - XVe siecles),* trans. by M. Cazenave and K. Jaouiche. Paris: Vrin, 1976.

Chapter IV
Arabic Primacy with Hindu, Chinese, and Maya Contributions

(Muhammad ibn Mūsā) al-Khwārizmī (fl. 800-847)

Although his ancestors came from Khorezm (which corresponds to the modern city of Khiva and its surrounding district in the northwestern Uzbek Soviet Socialist Republic south of the Aral Sea in central Asia), al-Khwārizmī probably grew up near Baghdad and was an orthodox Muslim. Under Caliph al-Ma'mūn (reigned 813-833), a patron of humanistic and scientific learning, he became a member of the "House of Wisdom" (Bayt al-Hikmah), a kind of science academy in Baghdad. He thus lived and worked at the intellectual and cultural center of Islamic civilization, when it was beginning to assimilate Greek and Hindu science. For al-Ma'mūn he prepared an astronomical treatise, which he based on the Sanskrit astronomical research of Brahmagupta, and dedicated his book *Algebra* to that caliph. He was probably one of the astronomers called to cast a horoscope for Caliph al-Wāthiq (d. 847) when the latter was dying.

Al-Khwārizmī conducted research on topics other than astronomy and algebra. His text *Geography,* which he seems to have based partly on Ptolemy's *Geography,* contained more accurate maps of Islamic areas than Ptolemy's. (Islamic geography was important since faithful Muslims faced Mecca at prayer.) He also authored treatises on Hindu numerals and on the Jewish calendar and three lost books entitled *Book on the Construction of the Astrolabe, On the Sundial,* and *Chronicle* (a historical text on the events of the early Islamic period). His scientific and mathematical writings—all mediocre at best—were uncommonly influential because they transmitted new knowledge to his successors in the high period of Islamic culture.

The *Algebra (Hisab al-jabr wa'l-muqābalah)* was al-Khwārizmī's major work. This text on elementary practical mathematics contained sections on algebra, practical mensuration, and legacies. Our concern here is with the first section. His book was not only the first in Arabic to treat algebra but also gave the subject its name. The word *al-jabr* in the title meant "restoration" or "completion" and referred to the process of removing negative quantities, as when the expression $40 + x^2 = 18 + 10x$ is converted to $22 + x^2 = 10x$. Al-Khwārizmī solved problems that could be reduced to six standard forms of linear and quadratic equations. He divided these six forms into two groups of three types; in modern notation they are as follows:

Group 1	Group 2
$ax^2 = bx$	$ax^2 + bx = c$
$ax^2 = b$	$ax^2 + c = bx$
$ax = b$	$ax^2 = bx + c$, where a, b, and c are positive whole numbers.

These forms show that he recognized neither negative numbers nor zero as a coefficient. His *Algebra* suffered in that he used no symbols but expressed everything in words.

Al-Khwārizmī contributed to areas of mathematics besides algebra. His astronomical treatise contained an early sine table with base 150 (a common Hindu parameter). Though elementary, his treatise on Hindu numerals was important because it first systematically expounded the decimal system with digits 1 to 9, 0, and place value. His treatise introduced those numerals to Islamic works from which they were put into Latin translation in 12th-century Europe. Because European scholars linked al-Khwārizmī to the "new arithmetic" involving Hindu numerals, any treatise dealing with that topic was given the Latin form of his name, *algorismus*, a form subsequently corrupted to the modern algorithm.

47. From *The Book of Algebra and Almucabola**

(Quadratic Equations in Algebra: Verbal Form)

AL-KHWĀRIZMĪ

CONTAINING DEMONSTRATIONS OF THE RULES OF THE EQUATIONS OF ALGEBRA[1]

. . . Furthermore I discovered that the numbers of restoration and opposition are composed of these three kinds: namely, roots, squares, and numbers.[2] However, number alone is connected neither with roots nor with squares by any ratio. Of these, then, the root is anything composed of units which can be multiplied by itself, or any number greater than unity multiplied by itself: or that which is found to be diminished below unity when multiplied by itself. The square is that which results from the multiplication of a root by itself.

Of these three forms, then, two may be equal to each other, as for example:

Squares equal to roots,
Squares equal to numbers, and
Roots equal to numbers.[3]

CHAPTER I. CONCERNING SQUARES EQUAL TO ROOTS[4]

The following is an example of squares equal to roots: a square is equal to 5 roots. The root of the square then is 5, and 25 forms its square which, of course, equals five of its roots.

Another example: the third part of a square equals four roots. Then the root of the square is 12 and 144 designates its square. And similarly, five squares equal 10 roots. Therefore one square equals two roots and the root of the square is 2. Four represents the square.

In the same manner, then, that which

**Source:* The English translation is originally from L. C. Karpinski's *Robert of Chester's Latin translation of the Algebra of Khowarizmi* (1915). It appeared with added notes in D. J. Struik (ed.), *A Source Book in Mathematics, 1200-1800* (1969), 56-60, and is reprinted by permission of Harvard University Press.

involves more than one square, or is less than one, is reduced to one square. Likewise you perform the same operation upon the roots which accompany the squares.

CHAPTER II. CONCERNING SQUARES EQUAL TO NUMBERS

Squares equal to numbers are illustrated in the following manner: a square is equal to nine. Then nine measures the square of which three represents one root.

Whether there are many or few squares, they will have to be reduced in the same manner to the form of one square. That is to say, if there are two or three or four squares, or even more, the equation formed by them with their roots is to be reduced to the form of one square with its root. Further, if there be less than one square, that is, if a third or a fourth or a fifth part of a square or root is proposed, this is treated in the same manner.

For example, five squares equal 80. Therefore one square equals the fifth part of the number 80 which, of course, is 16. Or, to take another example, half of a square equals 18. This square therefore equals 36. In like manner all squares, however many, are reduced to one square, or what is less than one is reduced to one square. The same operation must be performed upon the numbers which accompany the squares.

CHAPTER III. CONCERNING ROOTS EQUAL TO NUMBERS

The following is an example of roots equal to numbers: a root is equal to 3. Therefore nine is the square of this root.

Another example: four roots equal 20. Therefore one root of this square is 5. Still another example: half a root is equal to ten. The whole root therefore equals 20, of which, of course, 400 represents the square.

Therefore roots and squares and pure numbers are, as we have shown, distinguished from one another. Whence also from these three kinds which we have just explained, three distinct types of equations are formed involving three elements, as

A square and roots equal to numbers,
A square and numbers equal to roots, and
Roots and numbers equal to a square.

CHAPTER IV. CONCERNING SQUARES AND ROOTS EQUAL TO NUMBERS

The following is an example of squares and roots equal to numbers: a square and 10 roots are equal to 39 units.[5] The question therefore in this type of equation is about as follows: what is the square which combined with ten of its roots will give a sum total of 39? The manner of solving this type of equation is to take one-half of the roots just mentioned. Now the roots in the problem before us are 10. Therefore take 5, which multiplied by itself gives 25, an amount which you add to 39, giving 64. Having taken then the square root of this which is 8, subtract from it the half of the roots, 5, leaving 3. The number three therefore represents one root of this square, which itself, of course, is 9. Nine therefore gives that square.

Similarly, however many squares are proposed all are to be reduced to one square. Similarly also you may reduce whatever numbers or roots accompany them in the same way in which you have reduced the squares.

The following is an example of this reduction: two squares and ten roots equal 48 units. The question therefore in this type of equation is something like this: what are the two squares which when combined are such that if ten roots of them are added, the sum total equals 48? First of all it is necessary that the two squares be reduced to one. But since one square is the half of two, it is at once evident that you should divide by two all the given terms in this problem. This gives a square and 5 roots equal to 24 units. The meaning of this is about as follows: what is the square which amounts to 24 when you add to it 5 of its roots? At the outset it is necessary, recalling the rule above given, that you take one-half of the roots. This gives two and one-half which multiplied by itself gives 6¼. Add this to 24, giving 30¼. Take then of this total the square root, which is, of course, 5½. From this subtract half of the roots, 2½, leaving 3, which expresses one root of the square, which itself is 9.

. . . .

CHAPTER VI. GEOMETRICAL DEMONSTRATIONS[6]

We have said enough, says al-Khwārizmī, so far as numbers are concerned, about the six types of equations. Now, however, it is necessary that we should demonstrate geometrically the truth of the same problems which we have explained in numbers. Therefore our first proposition is this, that a square and 10 roots equal 39 units.

The proof is that we construct [Fig. 1] a square of unknown sides, and let this square figure represent the square (second power of the unknown) which together with its root you wish to find. Let the square, then, be *ab*, of which any side represents one root. When we multiply any side of this by a number (or numbers) it is evident that that which results from the multiplication will be a number of roots equal to the root of the same number (of the square). Since then ten roots were proposed with the square, we take a fourth part of the number ten and apply to each side of the square an area of equidistant sides, of which the length should be the same as the length of the square first described and the breadth 2½, which is a fourth part of 10. Therefore four areas of equidistant sides are applied to the first square, *ab*. Of each of these the length is the length of one root of the square *ab* and also the breadth of each is 2½, as we have just said. These now are the areas *c, d, e, f*. Therefore it follows from what we have said that there will be four areas having sides of unequal length, which also are regarded as unknown. The size of the areas in each of the four corners, which is found by multiplying 2½ by 2½, completes that which is lacking in the larger or whole area. Whence it is we complete the drawing of the larger area by the addition of the four products, each 2½ by 2½; the whole of this multiplication gives 25.

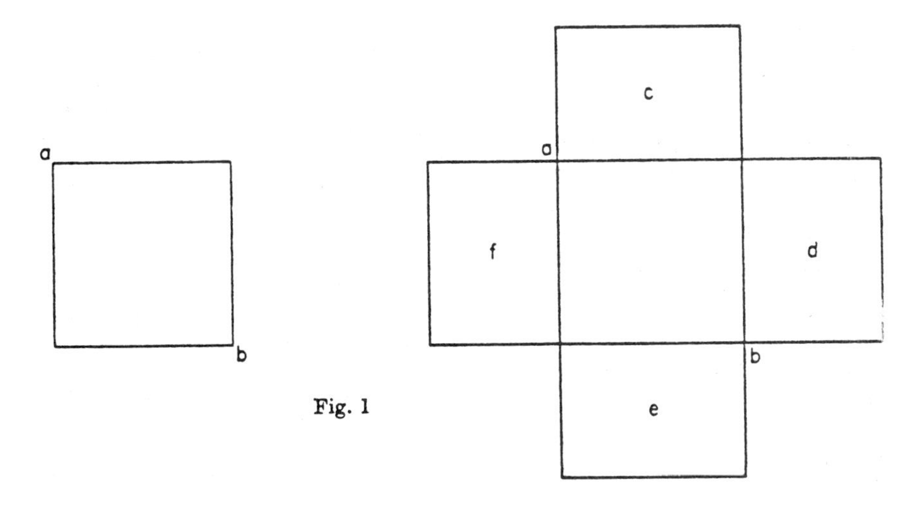

Fig. 1

And now it is evident that the first square figure, which represents the square of the unknown [x^2], and the four surrounding areas [10x] make 39. When we add 25 to this, that is, the four smaller squares which indeed are placed at the four angles of the square *ab*, the drawing of the larger square, called *GH*, is completed [Fig. 2]. Whence also the sum total of this is 64, of which 8 is the root, and by this is designated one side of the completed figure. Therefore when we subtract from eight twice the fourth part of 10, which is placed at the extremities of the larger square *GH*, there will remain but 3. Five being subtracted from 8, 3 necessarily remains, which is equal to one side of the first square *ab*.

This three then expresses one root of the square figure, that is, one root of the proposed square of the unknown, and 9 the square itself. Hence we take half of ten and multiply this by itself. We then add the whole product of the multiplication to 39, that the drawing of the larger square *GH* may be completed; for the lack of the four corners rendered incomplete the drawing of the whole of this square. Now it is evident that the fourth part of any number multiplied by itself and then multiplied by four gives the same number as half of the number multiplied by itself. Therefore if half of the root is multiplied by itself, the sum total of this multiplication will wipe out, equal, or cancel the multiplication of the fourth part by itself and then by four.

The remainder of the treatise deals with problems that can be reduced to one of the six types, for example, how to divide 10 into two parts in such a way that the sum of the products obtained by multiplying each part by itself is equal to 58: $x^2 + (10 - x)^2 = 58$, $x = 3$, $x = 7$. This is followed by a section on problems of inheritance.

NOTES

1. *Jabr* is the setting of a bone, hence reduction or restoration; *muqābala* is confrontation, opposition, face-to-face (explanation by Professor E. S. Kennedy).

2. The term "roots" *(radices)* stands for multiples of the unknown, our *x;* the term "squares" *(substantiae)* stands for multiples of our x^2; "numbers" *(numeri)* are constants.

3. In our notation, $x^2 = ax$, $x^2 = b$, $x = c$.

4. Latin: *de substantiis numeros coaequantibus*. The examples are $x^2 = 5x$, $\frac{1}{3}x^2 = 4x$, $5x^2 = 10x$.

5. This example, $x^2 + 10x = 39$, answer $x = 3$, "runs," as Karpinski notices in his introduction to this translation, "like a thread of gold through the algebras for several cen-

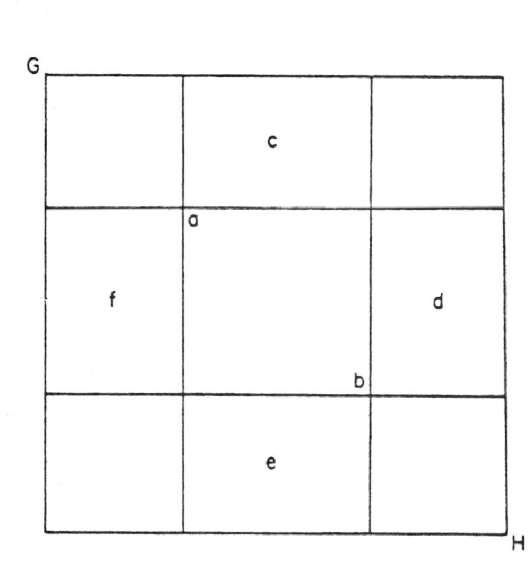

Fig. 2

turies, appearing in the algebras of Abu Kamil, Al-Karkhi and Omar al-Khayyami, and frequently in the works of Christian writers," and it still graces our present algebra texts. The solution of this type, $x^2 + ax = b$, is, as we can verify, based on the formula $x = \sqrt{(a/2)^2 + b} - a/2$.

6. For these geometric demonstrations we must go back, as said, to Euclid's *Elements* (Book VI, Prop. 28, 29; see also Book II, Prop. 5, 6). See also on this subject the introduction to the *Principal works of Simon Stevin,* vol. IIB (Swets-Zeitlinger, Amsterdam, 1958). . . .

Chapter IV
Arabic Primacy with Hindu, Chinese, and Maya Contributions

'Umar al-Khayyāmī (c. 1048–c. 1131)

The Persian poet, astronomer, and mathematician 'Umar al-Khayyāmī (also known as Omar Khayyam) was born and received his early education in Nīshāpūr (now Neyshābūr, Khorāsān province, Iran). The epithet al-Khayyāmī suggests that his father, Ibrāhīm, was probably a tentmaker. Shortly before 'Umar's birth, the Seljuk Turks had conquered Persia and established an unstable military empire there. Some sources state that he spent his later youth in Balkh (now in Afghanistan), where he continued his education. By age 17, he was well versed in philosophy and may have become a tutor.

In al-Khayyāmī's time, teaching did not provide enough funds or leisure for scholarly research. To have opportunities for extended learning, a scholar had to be attached to the court of a sovereign or magnate; even then study could be impeded by the indifference of the patron, court intrigues, or the disruptions of war. About 1070, in search of such opportunity, al-Khayyāmī went to Samarkand, where the chief justice, Abū Tāhir, became his patron. Shortly afterward, the Seljuk sultan, Malik Shāh, invited him to Isfahan to help plan and then to supervise an astronomical observatory. He remained at the observatory for nearly 18 years and was one of eight astronomers commissioned by Malik Shāh to reform the Muslim solar calendar. About 1079 al-Khayyāmī proposed a new calendar that was more accurate than the later Gregorian one. He also served as court astrologer in Isfahan, even though he rejected judicial astrology.

In 1092 al-Khayyāmī fell from favor after Malik Shāh died and the vizier was murdered. The observatory lost its support, and orthodox Muslims who were critical of the freethinking—as exemplified in al-Khayyāmī's poetry—became influential at court. Al-Khayyāmī had written a collection of quatrains in Persian—the Rubā'īyāt. Recent scholarship suggests that of over 1,000 quatrains published under his name only 102 were actually written by him. The authentic verses reveal a skeptical man examining questions about the nature of the physical universe, the passing of time, and man's relation to God. To clear himself of the charge of atheism—explicit or implicit—al-Khayyāmī reportedly embarked upon a pilgrimage to Mecca. When he returned, he attempted to persuade the new sultan to support the observatory; he wrote a propagandistic work that portrayed ancient Iranian rulers as magnanimous in support of education, building, and scholars. However, he only occasionally served and taught at the Isfahan court. After 1118 he lived for some time in the new Seljuk capital, Merv (near modern Mary, Turkmen S.S.R.).

A polymath, al-Khayyāmī contributed to not only mathematics, astronomy, and philosophy, but also to jurisprudence, history, and medicine. As a philosopher, he followed the Arabic Aristotelianism of ibn Sina (Avicenna), developing the doctrine of conceptualism as his contemporary Peter Abelard was doing in Latin Christendom. For his theistic freethought, love of justice, philosophical poetry, skepticism, and epicurean spirit that verges on hedonism, he has been called, anachronistically, the Persian Voltaire.

Working in the mainstream of Muslim mathematics, al-Khayyāmī advanced arithmetic, algebra, and geometry. In doing that he drew upon Hellenistic and ancient Oriental sources. The numerals he employed were those of the Hindu decimal positional system. To extract square and cube roots, he appealed to the ancient Chinese method given in the *Mathematics in Nine Sections* (second or first century B.C.). (Ruffini and Horner rediscovered this method in Europe in the early 19th century.) While extracting roots, he studied the binomial expansion $(a + b)^n$ and is said to have derived a table for binomial coefficients up to $n = 12$. He observed the binomial property now expressed as:

$$C_n^m = C_{n-1}^{m-1} + C_{n-1}^m$$

al-Khayyāmī's two chief writings on mathematics are the algebraical *Risāla* and the commentaries on Euclid. The *Risāla*, a treatise on cubic equations, presents one of the first definitions of algebra and contains one of the first attempts to classify the normal forms of 25 linear, quadratic, and cubic equations. (Normal forms of equations are those with positive coefficients.) Its author supposed that cubic equations that are not reducible to quadratic equations had to be solved by studying the intersections of corresponding conic sections. In essence, he developed a geometric theory of cubic equations. Although he failed to discover a general arithmetical solution for cubics in terms of quadratic radicals, al-Khayyāmī did not despair. He observed that "Perhaps someone else who comes after us may find it. . . ." (In the 16th century Tartaglia and Cardano arrived at the solution.)

In the commentaries on Euclid's *Elements*, al-Khayyāmī improved upon the Eudoxian theory of ratios (Book V) and the theory of parallel lines. By extending arithmetical language to ratios and by placing irrational magnitudes and numbers on the same operational scale, he developed a broader concept of number than the ancient Greeks had. To establish the fifth or parallel postulate of Euclid, he constructed a birectangle now known as "Saccheri's quadrilateral." Assuming that the two lower angles are right angles, he then investigated whether the upper angles were acute, right, or obtuse. His conjectures on acute and obtuse angles are similar to Lobachevsky's and Riemann's first theorems in their non-Euclidean geometries.

48. From the *Algebra**

'UMAR AL-KHAYYĀMĪ

CHAPTER I DEFINITIONS

Algebra. By the help of God and with His precious assistance, I say that Algebra is a scientific art. The objects with which it deals are absolute numbers and measurable quantities which, though themselves unknown, are related to "things" which are known, whereby the determination of the unknown quantities is possible. Such a thing is either a quantity or a unique relation, which is only determined by careful examination. What one searches for in the algebraic art are the relations which lead from the known to the unknown, to discover which is the object of Algebra[1] as stated above. The perfection of this art consists in knowledge of the scientific method by which one determines numerical and geometric unknowns.

Measurable Quantities. By measurable quantities I mean continuous quantities of which there are four kinds, viz., line, surface, solid, and time, according to the customary terminology of the Categories[2] and what is expounded in metaphysics.[3] Some consider space a subdivision of surface, subordinate to the division of continuous quantities, but investigation has disproved this claim. The truth is that space is a surface only under circumstances the determination of which is outside the scope of the present field of investigation. It is not customary to include "time" among the objects of our algebraic studies, but if it were mentioned it would be quite admissible.

The Unknown. It is a practice among algebraists in connection with their art to call the unknown which is to be determined a "thing,"[4] the product obtained by multiplying it by itself a "square,"[5] and the product of the square and the "thing" itself a "cube." The product of the square multiplied by itself is "the square of the square," the product of its cube multiplied by its square "the cube of the square," and the product of a cube into itself "a cube of the cube," and so on, as far as the succession is carried out.[6] It is known from Euclid's book, the *Elements,*[7] that all the steps are in continuous proportion; *i.e.,* that the ratio of one to the root is as the ratio of the root to the square and as the ratio of the square to the cube.[8] Therefore, the ratio of a number to a root is as the ratio of roots to squares, and squares to cubes, and cubes to the squares of the squares, and so on after this manner.[9]

Sources. It should be understood that this treatise cannot be comprehended except by those who know thoroughly Euclid's books, the *Elements* and the *Data,* as well as the first two books from Apollonius' work on *Conics.* Whoever lacks knowledge of any one of these books cannot possibly understand my work, as I have taken pains to limit myself to these three books only.

Algebraic Solutions. Algebraic solu-

**Source:* From Henrietta O. Midonick (ed.), *The Treasury of Mathematics* (1965), 584-594. Reprinted by permission of Philosophical Library, Inc. Translated from the Arabic, with explanatory notes by Daoud S. Kasir.

tions are accomplished by the aid of equations; that is to say, by the well-known method of equating these degrees one with the other. If the algebraist were to use the square of the square in measuring areas, his result would be figurative[10] and not real, because it is impossible to consider the square of the square as a magnitude of a measurable nature. What we get in measurable quantities is first one dimension, which is the "root"[11] or the "side"[12] in relation to its square; then two dimensions, which represent the surface and the (algebraic) square representing the square surface; and finally, three dimensions, which represent the solid.[13] The cube in quantities is the solid bounded by six squares, and since there is no other dimension, the square of the square does not fall under measurable quantities. This is even more true in the case of higher powers.[14] If it is said that the square of the square is among measurable quantities, this is said with reference to its reciprocal value in problems of measurement and not because it in itself is measurable. This is an important distinction to make.

The square of the square is, therefore, neither essentially nor accidentally a measurable quantity, and is not as even and odd numbers, which are accidentally included in measurable quantities, depending on the way in which they represent continuous measurable quantities as discontinuous.

What is found in the books of algebra relative to these four geometric quantities—namely, the absolute numbers, the "sides," the squares, and the cubes—are three equations containing numbers, sides, and squares. We, however, shall present methods by which one is able to determine the unknown quantities in equations including four degrees concerning which we have just said that they are the only ones that can be included in the category of measurable quantities, namely, the number, the thing, the square, and the cube.

The demonstration[15] (of solutions) depending on the properties of the circle—that is to say, as in the two works of Euclid, the *Elements* and the *Data*—is easily effected; but what we can demonstrate only by the properties of conic sections should be referred to the first two books on conics.[16] When, however, the object of the problem is an absolute number,[17] neither we, nor any of those who are concerned with algebra, have been able to prove this equation—perhaps others who follow us will be able to fill the gap—except when it contains only the three first degrees, namely, the number, the thing, and the square.[18] For the numerical demonstration given in cases that could also be proved by Euclid's book, one should know that the geometric proof of such procedure does not take the place of its demonstration by number, if the object of the problem is a number and not a measurable quantity. Do you not see how Euclid proved certain theorems relative to proportions of geometric quantities in his fifth book, and then in the seventh book gave a demonstration of the same theorems for the case when their object is a number?[19]

CHAPTER II
TABLE OF EQUATIONS

The equations among those four quantities are either simple or compound. The simple equations are of six species:

1. A number equals a root.
2. A number equals a square.
3. A number equals a cube.
4. Roots equal a square.
5. Squares equal a cube.
6. Roots equal a cube.

Three of these six species are mentioned in the books of the algebraists. The latter said a thing is to a square as a square is to a cube. Therefore, the equality between the square and the cube is equivalent to the equality of the thing to the square. Again they said that a number is to a square as a root is to a cube. And they did not prove by geometry. As for the number which is equal to a cube there is no way of determining its side when the problem is numerical except by previous knowledge of the order of cubic numbers. When the problem is geometrical it cannot be solved except by conic sections. . . .

CHAPTER V
PRELIMINARY THEOREMS FOR THE CONSTRUCTION OF CUBIC EQUATIONS

After presenting those species of equations which it has been possible to prove by means of the properties of the circle, *i.e.,* by means of Euclid's book, we take up now the discussion of the species which cannot be proved except by means of the properties of conics. These include fourteen species: one simple equation, namely, that in which *a number is equal to a cube;* six trinomial equations; and seven tetranomial equations.

Let us precede this discussion by some propositions based on the book of *Conics* so that they may serve as a sort of introduction to the student and so that our treatise will not require familiarity with more than the three books already mentioned, namely, the two books of Euclid on the *Elements* and the *Data,* and the first two parts of the book on *Conics.*

Between two given lines it is required to find two other lines such that all four will form a continuous proportion.[20]

Let there be two straight lines (given) *AB, BC* (Fig. 1), and let them enclose the right angle *B*. Construct a parabola the vertex of which is the point *B,* the axis *BC,* and the parameter *BC*. Then the position of the conic *BDE* is known because the positions of its vertex and axis are known, and its parameter is given. It is tangent to the line *BA,* because the angle *B* is a right angle and it is equal to the angle of the ordinate, as was shown in the figure of the thirty-third proposition in the first book on

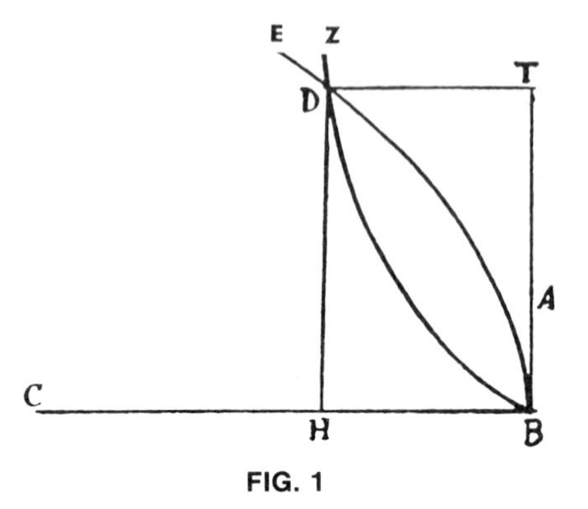

FIG. 1

Conics.[21] In the same manner construct another parabola, with vertex *B,* axis *AB,* and parameter *AB*. This will be the conic *BDZ,* as was shown also by Apollonius in the fifty-sixth proposition of the first book.[22] The conic *BDZ* is tangent to the line *BC*. Therefore, the two (parabolas) necessarily intersect. Let *D* be the point of intersection. Then the position of point *D* is known because the position of the two conics is known. Let fall from the point *D* two perpendiculars, *DH* and *DT,* on *AB* and *BC* respectively. These are known in magnitude, as was shown in the *Data.*[23] I say that the four lines *AB, BH, BT, BC* are in continuous proportion.

Demonstration. The square of *HD* is equal to the product of *BH* and *BC,* because the line *DH* is the ordinate of the parabola *BDE*. Consequently *BC* is to *HD,* which is equal to *BT,* as *BT* to *HB*. The line *DT* is the ordinate of the parabola *BDZ*. The square of *DT* (which is equal to *BH*) is equal to the product of *BA* and *BT*. Consequently *BT* is to *BH* as *BH* to *BA*. Then the four lines are in continuous proportion and the line *DH* is of known magnitude, as it is drawn from the point the position of which is known, to a line whose position is known, at an angle whose magnitude is known. Similarly, the length of *DT* is known. Therefore, the two lines, *BH* and *BT,* are known and are the means of the proportion between the two lines,

AB and *BC;* that is to say, *AB* to *BH* is as *BH* to *BT* and is as *BT* to *BC*. That is what we wanted to demonstrate.

Given (Fig. 2) *the rectangular parallelepiped ABCDE, whose base is the square AD, and the square MH, construct on MH a rectangular parallelepiped equal to ABCDE.*

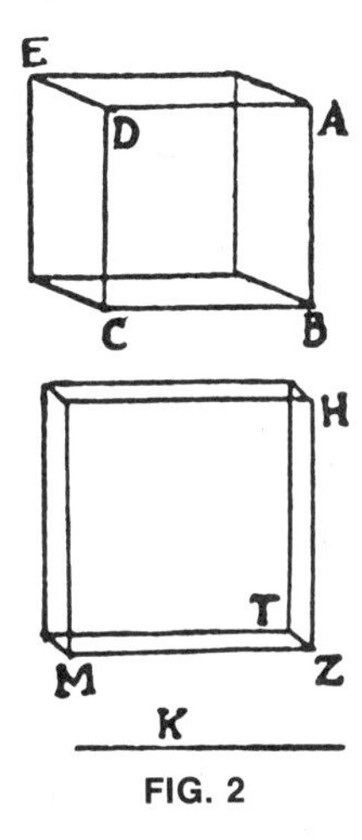

FIG. 2

Let *AB* to *MZ* be as *MZ* to *K* and let *AB* to *K* be as *ZT* to *ED*.[24] Then make *ZT* perpendicular to the plane *MH* at the point *Z* and complete the solid *MZTH*. Then I say that this solid is equal to the given solid.

Demonstration. The square *AC* to the square *MH* is as *AB* to *K*. Then the square *AC* to the square *MH* is as *ZT,* which is the height of the solid *MTH* to *ED,* which is the height of the solid *BE*. Therefore the two solids are equal, for their bases are reciprocally proportional to their heights, as it was demonstrated in the eleventh book of the *Elements*.

Whenever we speak of "a solid" we mean a rectangular parallelepiped and whenever we say "plane" we refer to the rectangle.

Given the solid ABCD (Fig. 3), *whose base AC is a square, it is required to construct a solid whose base is a square, whose height is equivalent to a given line ET, and which is equal to the solid ABCD.*

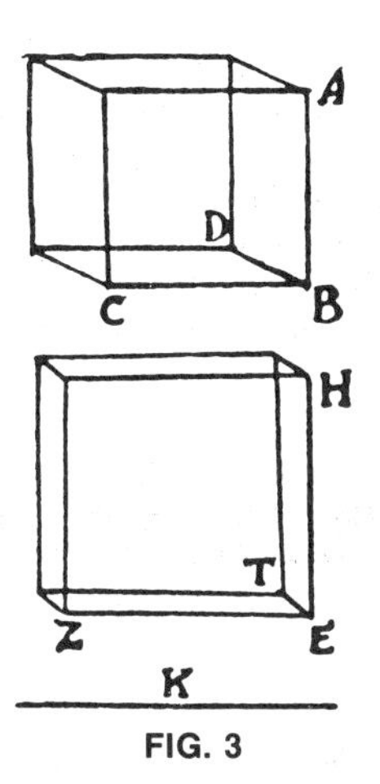

FIG. 3

Let *ET* be to *BD* as *AB* to *K* and take between *AB* and *K* a mean proportional line *EZ*.[25] Make *EZ* perpendicular to *ET* and complete *ZT*. Then make *EH* perpendicular to the plane *TZ* and equal to *EZ,* and complete the solid *HETZ*. Then I say the volume of solid *T,* whose base is the square *HZ* and height the given line *ET,* is equal to the volume of given solid *D*.

Demonstration. The square *AC* to the square *HZ* is as *AB* to *K*. Consequently, the square *AC* to the square *HZ* is as *ET* to *BD*. Therefore the bases of the two solids are also reciprocally proportional to their heights and the solids then are equal. That is what we wanted to demonstrate.

After these preliminary proofs we shall be able to give the solution of the third species of the simple equation, *a cube is equal to a number.*[26]

Let the number be equal to the solid *ABCD* (Fig. 4), and its base *AC* the square of one, as we have said previously. Its length is equal to the given number. It is desired to construct a cube equal to this solid. Take between the two lines *AB* and *BD* two mean proportionals.[27] These are known in magnitude, as has been demonstrated.[28] They are *E* and *Z*. Then draw *HT* equal to the line *E* and construct on it the cube *THKL*. This cube and its side are

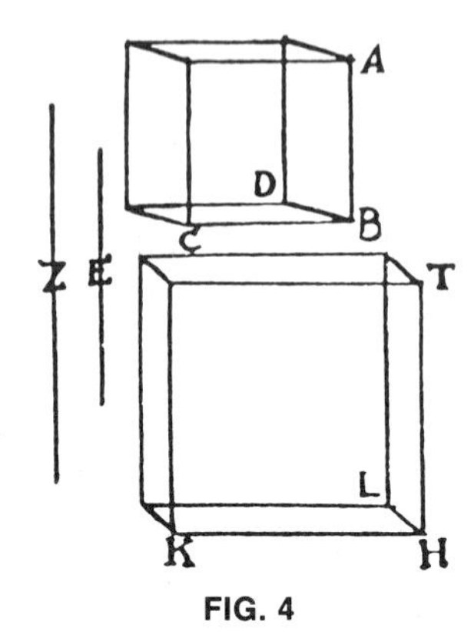

FIG. 4

known in magnitude. Then I say that this cube is equal to the solid *D*.

Demonstration. The square *AC* to the square *TK* is twice the ratio *AB* to *HK*,[29] and twice the ratio of *AB* to *HK* is equal to the ratio *AB* to *Z*. But the first is to the third of the four lines as *HK*, the second, is to *BD*, the fourth. The bases *(TK, AC)* of the cube *L* and the solid *D* are then reciprocally proportional to their height. Then the solids are equal, which is what we wanted to demonstrate.

NOTES

1. The author refers here to the algebraic relations existing between the known and the unknown quantities which the algebraist has to establish. For other Arabic definitions of algebra see *Haji Khalpha, Mohammed ibn Musa,* by Karpinski, p. 67; *Mukadamat ibn Khaldun,* p. 422 (Egypt).

2. *Category* of Aristotle, cap. 6; phys. Iv, cap. 4 ult. According to Aristotle's definitions, point, line, and surface are first principles and must be assumed. Heath, *Euclid,* vol. I, pp. 155-156, 158, 159, 165, 170.

3. Here the author is referring to his book on metaphysics. See A. Christensen: *Un Traite de Metaphysique d'Omar. Le Monde Oriental,* 1908, vol. I, pp. 1-16.

4. The Arabic word *shai* (literally a "thing") here means the "unknown." Latin translators used the word *res.*

5. *Mal* (literally, "substances") is the word used by the author to indicate the second power of the unknown. Gherardo of Cremona (*c.* 1150) used *census,* which has the same meaning.

6. $x \cdot x = x^2$, $x \cdot x \cdot x = x^3$, $x^2 \cdot x^2 = x^4$, $x^2 \cdot x^3 = x^5$, $x^3 \cdot x^3 = x^6$.

7. *Elements of Euclid,* Heath, vol. II, p. 390.

8. *E.g.,* $1 : x = x : x^2 = x^2 : x^3$.

9. *E.g.,* $a : ax = ax : ax^2 = ax^2 : ax^3 = ax^3 : ax^4$, etc.

10. The literal translation of the Arabic word *majaz* is "path, way." The author means that a quantity raised to the fourth power cannot be represented geometrically and therefore has no real geometric meaning, while algebraically it has. In other words, the author meant by *majaz,* "hypothetical." See *Akrabu'l-Mawarid* (Beirut, 1889).

11. The Arab writers used also the word "root" for the first power of the unknown quantity in an equation.

12. *Dulu* (literally, side), meaning an unknown quantity, is represented geometrically by a line, while *mal* (square) is represented by a surface, and *muk'ab* (cube) by a solid.

13. The author's idea of dimension conforms fundamentally with that of Aristotle, Proclus, and al-Nairizi. See Heath, *Euclid,* vol. I, pp. 157-159.

14. It was Descartes who first defeated this method of reasoning, which had been universally accepted before him.

15. The author refers here to the demonstration of processes which constitute the solution.

16. The author is referring here to the above mentioned works of Apollonius.

17. Or "if it is required to satisfy the proposed equation by a whole number." The author means here that cubic equations can be solved geometrically, but that neither he nor his predecessors could solve them algebraically. It was to Archimedes, al-Mahani, Thabit ibn Qorra, and al-Khazin that he referred here. It was not until the 16th century that Cardan and Tartaglia succeeded in solving cubic equations algebraically. Smith, *History of Mathematics,* vol. II. pp. 455 (footnote 2) and 459.

18. For example, as in $x^2 - 7x + 12 = 0$, where 12 is the number, x is the "thing," and x^2 is the square.

19. Compare, for example, Prop. 5 of Book V with Prop. 8 of Book VII.

20. $AB : x = x : y = y : \underline{BC}$.

In the parabola BDE $\overline{HD}^2 = BT^2 = BH \cdot BC$, hence $BC : BT = BT : HB$.

In the parabola BDZ $\overline{DT}^2 = HB^2 = BA \cdot BT$, hence $BT : HB = HB : BA$.

Consequently $BC : BT = BT : HB = HB : BA$ or $AB : BH = BH : BT = BT : BC$.

This is the second of the two constructions of the problem attributed to Menæchmus.

21. Apollonius, ed. Oxford, 1710, fol., p. 57. The author refers here to the 32nd proposition of this book.

22. Apollonius, ed. Oxford, Book I. proposition 52.

23. See propositions 30, 25, 26.

24. That is, construct K from the proportion $AB : MZ = MZ : K$, in which $\overline{AB}^2$, MZ are known and ZT as the third term of $AB : K = ZT : ED$. The volume of $ABCDE = \overline{AB}^2 \cdot DE$. But $\overline{MZ}^2 \cdot ZT =$ volume of the parallelepiped on $MH = \overline{AB}^2 \cdot DE$, since the first proportion gives $\overline{MZ}^2 = AB \cdot K$ and the second $ZT = \frac{AB \cdot ED}{K}$. But $\overline{AB}^2 \cdot DE =$ volume of parallelepiped $ABCDE$. Hence the two solids are equal.

25. Construct K from the proportion

$$ET : BD = AB : K,$$

and EZ from

$$AB : EZ = EZ : K.$$

It follows that $\overline{AB}^2 : \overline{EZ}^2 = ET : BD$.

Then $\overline{AB}^2 \cdot BD = \overline{EZ}^2 \cdot ET$.

Or solid D = solid T.

26. $a = x^3$.

27. $AB = E^2 : Z$; $\overline{AB}^2 = \frac{AB \cdot E^2}{Z}$ (E, Z mean proportional lines)

$$\overline{AB}^2 : E^2 = AB : Z$$

But $AB : E = Z : BD$ gives

$$AB : Z = E : BD$$

Hence $\overline{AB}^2 : E^2 = E : BD$

$\therefore$ $E^2 = \overline{AB}^2 : BD$, etc.

28. See . . . footnote 1 above.

29. That is $(AB : HK)$ $(AB : HK)$ or $\overline{AB}^2 : \overline{HK}^2$

Chapter IV
Arabic Primacy with Hindu, Chinese, and Maya Contributions

Bhāskara II (also known as Bhāskarācharya or Bhāskara the Learned; 1115–c. 1185)

Bhāskara II was the leading Indian astronomer and mathematician of the 12th century. A son of the Brahman Maheśvara, he was probably born in Vijayapura (now Bijapur in Mysore). He served as head of the astronomical observatory at Ujjain (the chief mathematical center of ancient and medieval India). He was also a noted astrologer. According to tradition, his astrological meddling and an unfortunate twist of fate deprived his daughter of her only chance for marriage and happiness. Perhaps to console her, he named his first book *Līlāvatī* after her.

Bhāskara wrote at least two books on mathematics. The first—*Līlāvatī* ("The Beautiful")—has thirteen chapters on arithmetic, geometry, and algebra. The second—*Bījagaṇita* ("Seed Counting")—contains twelve chapters on algebra. In these books, Bhāskara skillfully used the decimal system. In the decimal system he treated zero as a complete number and not simply as a positional notation. Like earlier Indian mathematicians he discussed arithmetic operations with zero. He went beyond them, however, in examining the universal problem of division by zero. Although he stated that a quantity divided by zero gives an infinite quantity (for example, in modern symbols $\frac{3}{0} = \infty$), and that $\frac{a}{0} \times 0 = a$, he knew that the reverse process of multiplication by zero did not produce a unique value. His inclusion of negative numbers in computations and recognition that the square roots of numbers have a positive and a negative value helped lead to the subsequent acceptance of negative numbers, even though he noted that "people do not approve of negative solutions." He grasped the convention of signs (minus times minus equals plus, and plus times minus equals minus).

In algebra, Bhāskara improved upon the work of the Indian mathematician Brahmagupta (598–c. 665). He introduced some symbolic notation into his own basically rhetorical problems by representing negative quantities by a superior dot over positive quantities and by representing unknown quantities by the initial syllables of the words for colors. He solved first and second degree indeterminate equations and reduced quadratic equations to a single type to solve them. Like many later mathematicians Bhāskara was intrigued by what is now known as Pell's equation (in modern symbols, $ax^2 + 1 = y^2$) and provided a solution. In the case when $a = 8$, $x = 6$ and $y = 17$ satisfy the equation. In geometry he studied regular polygons with as many as 384 sides and obtained an approximation of π as 3.14166.

Bhāskara wrote two books on mathematical astronomy—*Siddhāntaśi-*

romani ("Head Jewel of Accuracy," 1150) and *Karaṇakutūhala* ("Calculation of Astronomical Wonders," 1183). These texts cover astronomical observations, lunar and solar eclipses, planetary positions and conjunctions, heliacal risings and settings, cosmography, geography, spherical trigonometry, astronomical equipment, and mathematical techniques used in astronomy.

49. From *Līlāvatī**

(Arithmetic and Geometry)

BHĀSKARA II

CHAPTER III, Section VI RULE OF PROPORTION

74. *Rule of three inverse.*[1]

If the fruit diminish as the requisition increases, or augment as that decreases, they, who are skilled in accounts, consider the rule of three terms to be inverted.[2]

When there is diminution of fruit, if there be increase of requisition, and increase of fruit if there be diminution of requisition, then the inverse rule of three is [employed]. For instance,

75. When the value of living beings[3] is regulated by their age; and in the case of gold, where the weight and touch are compared; or when heaps[4] are subdivided; let the inverted rule of three terms be [used].

76. *Example.* If a female slave sixteen years of age, bring thirty-two [*nishcas*], what will one aged twenty cost? If an ox, which has been worked a second year, sell for four *nishcas* what will one, which has been worked six years, cost?

1st Qu. Statement: 16 32 20. Answer: 25-$^3/_5$ *nishcas.*

2d Qu. Statement: 2 4 6. Answer: 1-$^1/_3$ *nishca.*

77. *Example.* If a *gadyánaca* of gold of the touch of ten may be had for one *nishca* [of silver], what weight of gold of fifteen touch may be bought for the same price?

Statement: 10 1 15. Answer $^2/_3$.

78. *Example.* A heap of grain having been meted with a measure containing seven *ad'hacas,* if a hundred such measures were found, what would be the result with one containing five *ad'hacas?*

Statement: 7 100 5. Answer 140.

80. *Example.* If the interest of a hundred for a month be five, say what is the interest of sixteen for a year? Find likewise the time from the principal and interest; and knowing the time and produce, tell the principal sum.

```
            1   12
Statement: 100  16
            5
```

Answer: the interest is 9-$^3/_5$.

```
                                  1
To find the time; Statement: 100  16
                              5   48/5
```

Answer: months 12.

Source: The selections from *Līlāvatī* (49) and *Bījagṇita* (50) are taken from the English translation from the Sanskrit by Henry Thomas Colebrooke in 1817. They appeared in Henrietta Midonick, *The Treasury of Mathematics,* 117-140.

To find the principal; Statement: 1 12
100
5 48/5

Answer: principal 16.

81. *Example.* If the interest of a hundred for a month and one-third, be five and one fifth, say what is the interest of sixty-two and a half for three months and one fifth?

Statement: $\frac{4}{3}$ $\frac{16}{5}$ / $\frac{100}{1}$ $\frac{125}{2}$ / $\frac{26}{5}$

Answer: interest 7-4/5.

82. *Example of the rule of seven:* If eight, best, variegated, silk scarfs, measuring three cubits in breadth and eight in length, cost a hundred [*nishcas*]; say quickly, merchant, if thou understand trade, what a like scarf, three and a half cubits long and half a cubit wide, will cost.

Statement: 3 1/2
8 7/2
8 1
100

Answer: *Nishca* 0, *drammas* 14, *panas* 9, *cacini* 1, *cowryshells* 6-2/3.

83. *Example of the rule of nine:* If thirty benches, twelve fingers thick, square of four wide, and fourteen cubits long, cost a hundred [*nishcas*]; tell me, my friend, what price will fourteen benches fetch, which are four less in every dimension?

Statement: 12 8
16 12
14 10
30 14
100

Answer: *Nishcas* 16-2/3.

CHAPTER IV, SECTION VI PERMUTATION AND COMBINATION

114. *Example:* In a pleasant, spacious and elegant edifice, with eight doors,[5] constructed by a skilful architect, as a palace for the lord of the land, tell me the permutations of apertures taken one, two, three, &c.[6] Say, mathematician, how many are the combinations in one composition, with ingredients of six different tastes, sweet, pungent, astringent, sour, salt and bitter,[7] taking them by ones, twos, or threes, &c.

Statement [1st Example]: 8 7 6 5 4 3 2 1
1 2 3 4 5 6 7 8.

Answer: the number of ways in which the doors may be opened by ones, twos, or threes, &c. is 8, 28, 56, 70, 56, 28, 8, 1. (1 2 3 4 5 6 7 8) And the changes on the apertures of the octagon palace amount to 255.[8]

Statement 2d example: 6 5 4 3 2 1
1 2 3 4 5 6.

Answer: the number of various preparations with ingredients of divers tastes is 6, 15, 20, 15, 6, 1.[9] (1 2 3 4 5 6)

CHAPTER V, PROGRESSIONS SECTION I, ARITHMETICAL PROGRESSION

115. *Rule:* Half the period, multiplied by the period added to unity, is the sum of the arithmeticals one, &c. and is named their addition. This, being multiplied by the period added to two, and being divided by three, is the aggregate of the additions.

123. *Rule:*[10] half a stanza. The sum divided by the period, and the first term subtracted from the quotient, the remainder, divided by half of one less than the number of terms, will be the common difference.

124. *Example:* On an expedition to seize his enemy's elephants, a king marched two *yojánas* the first day. Say, intelligent calculator, with what increasing rate of daily march did he proceed, since he reached his foe's city, a distance of eighty *yojánas,* in a week?

Statement: First term 2; Com. diff.? Period 7; Sum 80.

Answer: Com. diff. 22/7.

125. *Rule:*[11] From the sum of the progression multiplied by twice the common increase, and added to the square of the difference between the first term and half that increase, the square root being extracted, this root less the first term and added to the [above-mentioned] portion of the increase, being divided by the increase, is pronounced to be the period.

126. *Example:* A person gave three *drammas* on the first day, and continued to distribute alms increasing by two [a day]; and he thus bestowed on the priests three hundred and sixty *drammas:* say quickly in how many days?

Statement: First term 3; Com. diff. 2; Period? Sum. 360.

Answer: Period 18.

SECTION II
GEOMETRICAL PROGRESSION

127. *Rule:* a couplet and a half. The period being an uneven number, subtract one, and note "multiplicator;" being an even one, halve it, and note "square:" until the period be exhausted. Then the produce arising from multiplication and squaring [of the common multiplier] in the inverse order from the last, being lessened by one, the remainder divided by the common multiplier less one, and multiplied by the initial quantity, will be the sum of a progression increasing by a common multiplier.

128. *Example:* A person gave a mendicant a couple of cowry shells first; and promised a two-fold increase of the alms daily. How many *nishcas* does he give in a month?

Statement: First term, 2; Two-fold increase, 2; Period, 30.

Answer, 2147483646 cowries; or 104857 *nishcas,* 9 *drammas,* 9 *panas,* 2 *cacinis,* and 6 shells.

CHAPTER VI
PLANE FIGURE

158. *Example.* Where the difference of the side and upright is seven and hypotenuse is thirteen, say quickly, eminent mathematician, what are the side and upright?[12]

Statement. Difference of side and upright 7. Hypotenuse 13. Proceeding as directed, the side and upright come out 5 and 12. See

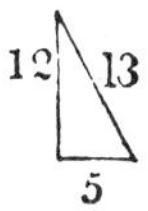

12 13
5

159. *Rule.*[13] The product of two erect bambus being divided by their sum, the quotient is the perpendicular from the junction [intersection] of threads passing reciprocally from the root [of one] to the tip [of the other.] The two bambus, multiplied by an assumed base, and divided by their sum, are the portions of the base on the respective sides of the perpendicular.

160. *Example.* Tell the perpendicular drawn from the intersection of strings stretched mutually from the roots to the summits of two bambus fifteen and ten cubits high standing upon ground of unknown extent.

Statement Bambus 15, 10. The perpendicular is found 6.

Next to find the segments of the base: let the ground be assumed 5; the segments come out 3 and 2. Or putting 10, they are 6 and 4. Or taking 15, they are 9 and 6. See the figures

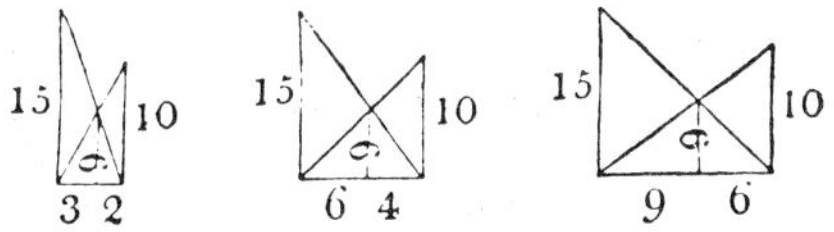

15 10 6 3 2 · 15 10 6 6 4 · 15 10 6 9 6

In every instance the perpendicular is the same: viz. 6.

The proof is in every case by the rule of three: if with a side equal to the base, the bambu be the upright, then with the segment of the base what will be the upright?[14]

161. *Aphorism.*[15] That figure, though rectilinear, of which sides are proposed by some presumptuous person, wherein one side[16] exceeds or equals the sum of

the other sides, may be known to be no figure.

162. *Example:* Where sides are proposed two, three, six, and twelve in a quadrilateral, or three, six and nine in a triangle, by some presumptuous dunce, know it to be no figure.

CHAPTER XII
PULVERIZER[17]

248–252. *Rule.* In the first place, as preparatory to the investigation of a pulverizer,[18] the dividend, divisor and additive quantity[19] are, if practicable, to be reduced by some number.[20] If the number, by which the dividend and divisor are both measured, do not also measure the additive quantity, the question is an ill put [or impossible] one.

249–251. The last remainder, when the dividend and divisor are mutually divided, is their common measure.[21] Being divided by that common measure, they are termed reduced quantities.[22] Divide mutually the reduced dividend and divisor, until unity be the remainder in the dividend. Place the quotients one under the other; and the additive quantity beneath them, and cipher at the bottom.[23] By the penult multiply the number next above it and add the lowest term. Then reject the last and repeat the operation until a pair of numbers be left. The uppermost of these being abraded[24] by the reduced dividend, the remainder is the quotient. The other [or lowermost] being in like manner abraded by the reduced divisor, the remainder is the multiplier.

252. Thus precisely is the operation when the quotients are an even number.[25] But, if they be odd, the numbers as found must be subtracted from their respective abraders, the residues will be the true quotient and multiplier.

253. *Example.* Say quickly, mathematician, what is that multiplier, by which two hundred and twenty-one being multiplied, and sixty-five added to the product, the sum divided by a hundred and ninety-five becomes exhausted?

Statement: Dividend 221 Additive 65.
Divisor 195

Here the dividend and divisor being mutually divided, the last of the remainders (or divisors) is 13. By this common measure, the dividend, divisor and additive, being reduced to their least terms, are Divd. 17 Addve. 5.
Divr. 15

The reduced dividend and divisor being divided reciprocally, and the quotients put one under the other, the additive under them, and cipher at the bottom, the series which results is 1 Then multi-
7
5
0
plying by the penult the number above it and proceeding as directed, the two quantities are obtained 40 These being
35
abraded by the reduced dividend and divisor 17 and 15, the quotient and multiplier are obtained 6 and 5. Or, by the subsequent rule (§262), adding them to their abraders multiplied by an assumed number, the quotient and multiplier [putting 1] are 23 and 20; or, putting 2, they are 40 and 35: and so forth.[26]

CHAPTER XIII

277. Joy and happiness is indeed ever increasing in this world for those who have *Līlāvatī* clasped to their throats, decorated as the members are with neat reduction of fractions, multiplication and involution, pure and perfect as are the solutions, and tasteful as is the speech which is exemplified.

NOTES

1. *Vyasta-trairasica* or *Viloma-trairasica*, rule of three terms inverse.

2. The method of performing the inverse

rule has been already taught (§70). "In the inverse method, the operation is reversed." That is, the fruit to be multiplied by the argument and divided by the demand.

When fruit increases or decreases, as the demand is augmented or diminished, the direct rule *(crama-trairasica)* is used. Else the inverse.

3. Slaves and cattle. The price of the older is less; of the younger, greater.

4. When heaps of grain, which had been meted with a small measure, are again meted with a larger one, the number decreases; and when those, which had been meted with a large measure, are again meted with a smaller one, there is increase of number.

5. *Muc'ha,* aperture for the admission of air: a door or window; . . . a portico or terrace.

6. The variations of one window or portico open (or terrace unroofed) and the rest closed; two open, and the rest shut; and so forth.

7. *Amera-cosha* 1.3.18.

8. An octagon building, with eight doors (or windows; porticos or terraces;) facing the eight cardinal points of the horizon, is meant.

9. Total number of possible combinations, 63.

10. The first term, period and sum being known, to find the common difference which is unknown.

11. The first term, common difference and sum being known, to find the period which is unknown.

12. This example of a case where the difference of the sides is given, is omitted by Suryadasa, but noticed by Ganesa. Copies of the text vary; some containing and others omitting, the instance.

13. Having taught fully the method of finding the sides in a right-angled triangle, the author next propounds a special problem. To find the perpendicular, the base being unknown.

14. On each side of the perpendicular, are segments of the base relative to the greater and smaller bambus, and larger or less analogously to them. Hence this proportion. "If with the sum of the bambus, this sum of the segments equal to the entire base be obtained, then, with the smaller bambu, what is had?" The answer gives the segment, which is relative to the least bambu. Again: "if with a side equal to the whole base, the higher bambu be the upright, then with a side equal to the segment found as above, what is had?" The answer gives the perpendicular let fall from the intersection of the threads. Here a multiplicator and a divisor equal to the entire base are both cancelled as equal and contrary: and there remain the product of the two bambus for numerator and their sum for denominator. Hence the rule.

15. The aphorism explains the nature of impossible figures. . . . In a triangle or other plane rectilinear figure, one side is always less than the sum of the rest. If equal, the perpendicular is nought, and there is no complete figure. If greater, the sides do not meet.

16. The principal or greatest side.

17. *Cuttaca-vyavahara* or *cuttacad'hyaya* determination of a grinding or pulverizing multiplier, or quantity such, that a given number being multiplied by it, and the product added to a given quantity, the sum (or, if the additive be negative, the difference) may be divisible by a given divisor without remainder.

In Brahmagupta's work the whole of algebra is comprised under this title of *Cuttacad'hyaya,* chapter on the pulverizer.

18. *Cuttaca* or *Cutta,* from *cutt,* to grind or pulverize; (to multiply: all verbs importing tendency to destruction also signifying multiplication.

The term is here employed in a sense independent of its etymology to signify a multiplier such, that a given dividend being multiplied by it, and a given quantity added to (or subtracted from) the product, the sum (or difference) may be measured by a given divisor.

The derivative import is, however, retained in the present version to distinguish this from multiplier in general; *cuttaca* being restricted to the particular multiplier of the problem in question.

19. *Cshepa,* or *cshepaca,* or *yuti,* additive. From *cship* to cast or throw in, and from *yu* to mix. A quantity superinduced, being either affirmative or negative, and consequently in some examples an additive, in others a subtractive, term.

20. *Visudd'hi,* subtractive quantity, contradistinguished from *cshepa* additive, when this is restricted to an affirmative one.

21. *Apavartana,* abridgment; abbreviation. Depression or reduction to least terms; division without remainder: also the number which serves to divide without residue; the common measure, or common divisor of equal division.

22. *Drid'ha,* firm: reduced by the common divisor to the least term.

23. *Tashta,* abraded; from *tacsh,* to pare or abrade: divided, but the residue taken, disregarding the quotient: reduced to a residue. As it were a residue after repeated subtractions. *Tacshana,* the abrader; the divisor employed in such operation.

24. *P'hala-valli,* the series of quotients; to be reduced by the operation forthwith directed to only two terms.

25. Even, as 2, 4, 6, &c.

26. Putting 3, they are 57 and 50.

50. From *Bījagṇita*

(Algebra)

BHĀSKARA II

CHAPTER I, SECTION II LOGISTICS OF NEGATIVE AND AFFIRMATIVE QUANTITIES

ADDITION

3. *Rule for addition of affirmative and negative quantities: half a stanza.* In the addition of two negative or two affirmative[1] quantities, the sum must be taken: but the difference of an affirmative and a negative quantity is their addition.

4. *Example.* Tell quickly the result of the numbers three and four, negative or affirmative, taken together: that is, affirmative and negative, or both negative or both affirmative, as separate instances: if thou know the addition of affirmative and negative quantities.

The characters, denoting the quantities known and unknown,[2] should be first written to indicate them generally; and those, which become negative, should be then marked with a dot over them.

Statement: 3.4. Adding them, the sum is found 7.

Statement: $\dot{3}.\dot{4}$. Adding them, the sum is $\dot{7}$.

Statement: $3.\dot{4}$. Taking the difference, the result of addition comes out $\dot{1}$.

Statement: $\dot{3}.4$. Taking the difference, the result of addition is 1.

So in other instances, and in fractions likewise.

SUBTRACTION

5. *Rule for subtraction of positive and negative quantities: half a stanza.* The quantity to be subtracted being affirmative, becomes negative; or, being negative, becomes affirmative: and the addition of the quantities is then made as above directed.

6. *Example: half a stanza.* Subtracting two from three, affirmative from affirmative, and negative from negative, or the contrary, tell me quickly the result.

Statement: 3.2. The subtrahend, being affirmative, becomes negative; and the result is 1.

Statement: $\dot{3}.\dot{2}$. The negative subtrahend becomes affirmative; and the result is $\dot{1}$.

Statement: $3.\dot{2}$. The negative subtrahend becomes affirmative; and the result is 5.

Statement: $\dot{3}.2$. The affirmative subtrahend becomes negative; and the result is $\dot{5}$.

MULTIPLICATION

7. *Rule for multiplication [and division] of positive and negative quantities: half a stanza.* The product of two quantities both affirmative, is positive.[3] When a positive quantity and a negative one are multiplied together, the product is negative. The same is the case in division.

SECTION III CIPHER

12. *Rule for addition and subtraction of cipher: part of a stanza.* In the addition of cipher, or subtraction of it, the quantity, positive or negative, remains

the same. But, subtracted from cipher, it is reversed.

13. *Example:* half a stanza. Say what is the number three, positive, or [the same number] negative, or cipher, added to cipher, or subtracted from it?

Statement: 3.$\dot{3}$.0. These, having cipher added to, or subtracted from, them, remain unchanged: 3.$\dot{3}$.0.

Statement: 3.$\dot{3}$.0. Subtracted from cipher, they become $\dot{3}$.3.0.

14. *Rule:* (completing the stanza, §12.) In the multiplication and the rest of the operations[4] of cipher, the product is cipher; and so it is in multiplication by cipher: but a quantity, divided by cipher, becomes a fraction the denominator of which is cipher.[5]

15. *Example: half a stanza.* Tell me the product of cipher multiplied by two;[6] and the quotient of it divided by three, and of three divided by cipher; and the square of nought; and its root.

Statement: Multiplicator 2. Multiplicand 0. Product 0.

[Statement: Multiplicator 0. Multiplicand 2. Product 0.]

Statement: Dividend 0. Divisor 3. Quotient 0.

Statement: Dividend 3. Divisor 0. Quotient the fraction $^3/_0$.

This fraction, of which the denominator is cipher, is termed an infinite quantity.[7]

16. In this quantity consisting of that which has cipher for its divisor, there is no alteration, though many be inserted or extracted; as no change takes place in the infinite and immutable GOD, at the period of the destruction or creation of worlds, though numerous orders of beings are absorbed or put forth.

Statement: 0. Its square 0. Its root 0.

SECTION IV
ARITHMETICAL OPERATIONS ON UNKNOWN QUANTITIES

17. "So much as" and the colours "black, blue, yellow and red,"[8] and others besides these, have been selected by venerable teachers for names of values[9] of unknown quantities, for the purpose of reckoning therewith.

18. *Rule for addition and subtraction:* Among quantities so designated, the sum or difference of two or more which are alike must be taken: but such as are unlike, are to be separately set forth.

19. *Example.* Say quickly, friend, what will affirmative one unknown with one absolute, and affirmative pair unknown less eight absolute, make, if addition of the two sets take place? and what will they make, if the sum be taken inverting the affirmative and negative signs?

Statement: *ya* 1 *ru* 1
ya 3 *ru* $\dot{7}$. Answer:
the sum is *ya* 2 *ru* $\dot{8}$ the known quantities. Hence the one side being divided by the residue of the first (letter or) colour, a value of the (letter or) colour which furnishes the divisor is obtained. If there be many such sides, by so treating those that constitute equations, by pairs, other values are found.

CHAPTER VII
VARIETIES OF QUADRATICS

178. *Example from ancient authors:* The square of the sum of two numbers, added to the cube of their sum, is equal to twice the sum of their cubes. Tell the numbers, mathematician!

The quantities are to be so put by the intelligent algebraist, as that the solution may not run into length. They are accordingly put *ya* 1 *ca* $\dot{1}$ and *ya* 1 *ca* 1. Their sum is *ya* 2. Its square *ya v* 4. Its cube *ya gh* 8. The square of the sum added to the cube is *ya gh* 8 *ya v* 4. The cubes of the two quantities respectively are *ya gh* 1 *ya v. ca bh* $\dot{3}$ *ca v. ya bh* 3 *ca gh* $\dot{1}$ cube of the first; and *ya gh* 1 *ya v. ca bh* 3 *ca v. ya bh* 3 *ca gh* 1 cube of the second; and the sum of these is *ya gh* 2 *ca v. ya bh* 6; and doubled, *ya gh* 4 *ca v. ya bh* 12. Statement for equal subtraction: *ya gh* 8 *ya v* 4 *ca v. ya bh* 0 After equal subtraction *ya gh* 4 *ya v* 0 *ca v. ya bh* 12
made, depressing both sides by the common divisor *ya,* and superadding

unity, the root of the first side of equation is *ya* 2 *ru* 1. Roots of the other side (*ca v* 12 *ru* 1) are investigated by the rule of the affected square, and are L2 G7 or *L*28 *G*97. "Least" root is a value of *ca*. Making an equation of a "greatest" root with *ya 2 ru* 1, the value of *ya* is obtained: viz. 3 or 48. Substitution being made with the respective values, the two quantities come out 1 and 5, or 20 and 76, and so forth.

CHAPTER IX
CONCLUSION

A particle of tuition conveys science to a comprehensive mind; and having reached it, expands of its own impulse. As oil poured upon water, as a secret entrusted to the vile, as alms bestowed upon the worthy, however little, so does science infused into a wise mind spread by intrinsic force.

It is apparent to men of clear understanding, that the rule of three terms constitutes arithmetic; and sagacity, algebra. Accordingly I have said in the chapter on Spherics:

224. "The rule of three terms is arithmetic; spotless understanding is algebra.[10] What is there unknown to the intelligent? Therefore, for the dull alone, it[11] is set forth."

225. To augment wisdom and strengthen confidence, read, do read, mathematician, this abridgment elegant in style, easily understood by youth, comprising the whole essence of computation, and containing the demonstration of its principles, replete with excellence and void of defect.

NOTES

1. *Rina* or *cshaya,* minus; literally debt or loss: negative quantity.

D'hana or *swa,* plus; literally wealth or property: affirmative or positive quantity.

2. *Rasi,* quantity, is either *vyacta,* absolute, specifically known, (which is termed *rupa,* form, species;) or it is *avyacta,* indistinct, unapparent, unknown *(ajnyata).* It may either be a multiple of the arithmetical unit, or a part of it, or the unit itself.

3. The sign only of the product is taught. All the operations upon the numbers are the same which were shown in simple arithmetic (*Līlavatī* § 14–16).

4. Square and square-root.

5. As much as the divisor is diminished, so much is the quotient raised. If the divisor be reduced to the utmost, the quotient is to the utmost increased. But, if it can be specified, that the amount of the quotient is so much, it has not been raised to the utmost: for a quantity greater than that can be assigned. The quotient therefore is indefinitely great, and is rightly termed infinite.

6. Or else multiplying two.

7. *Ananta-rasi,* infinite quantity. *C'ha-hara,* fraction having cipher for its denominator.

8. *Yavat-tavat,* correlatives, quantum, tantum; quot, tot: as many, or as much, of the unknown, as this coefficient number. *Yavat* is relative of the unknown; and *tavat* of its coefficient.

The initial syllables of the *Sanscrit* terms enumerated in the text are employed as marks of unknown quantities; viz. *ya, ca, ni, pi, lo,* (also *ha, swe, chi,* &c. for green, white, variegated and so forth). Absolute number is denoted by *ru,* initial of *rupa* form, species. The letters of the alphabet are also used (ch. 6), as likewise the initial syllables of the terms for the particular things (§111).

9. *Mana, miti, unmana* or *unmiti,* measure or value.

10. *Vija.*

11. The solution of certain problems set forth in the section. The preceding stanza . . . premises, "I deliver for the instruction of youth a few answers of problems found by arithmetic, algebra, the pulverizer, the affected square, the sphere, and [astronomical] instruments."

Chapter IV
Arabic Primacy with Hindu, Chinese, and Maya Contributions

Zhang Chang (Chang Ts'ang; fl. 165–142 B.C.)

The Chinese soldier, statesman, and scholar Zhang Chang began his career as a civil servant. After serving with distinction in the military during the wars of Liu Bang (Kao Tsu), the first Han emperor, he rose through the governmental ranks to become prime minister about 176 B.C. He remained prime minister until his death in 142 B.C. His contemporaries knew him as an able financial administrator and as a man of learning in mathematics, astronomy, and astrology. He may have lived more than one hundred years.

Zhang Chang is generally believed to have compiled and written a commentary on *Jiu zhang xuan shu (Nine Chapters on the Mathematical Art)*. Its sources were probably remnants of mathematical texts that survived the burning of books ordered by the Qin dynasty emperor Qin Shi Huang about 214 B.C. In his drive for the political unification of China, this emperor had proscribed all books having reference to past history and brutally ordered the killing (sometimes by live burial) of scholars and philosophers with unorthodox views as a means of breaking old traditions and the influence of scholars and pedants. Nevertheless, people preserved fragments of the old books, often at great risk to their lives. The source materials for the *Nine Chapters* were in existence before the end of the classical Zhou dynasty (1122-256 B.C.). During the early-3rd century B.C. royal tutors taught its methods.

The *Nine Chapters* strongly influenced Chinese mathematics for more than a thousand years. In A.D. 263, mathematician Liu Hui provided a new edition with extensive annotations. During the 7th century, Liu's edition became a standard textbook for Chinese mathematical students. In 1261, the outstanding algebraist, Yang Hui prepared a commentary and added three chapters to the book. The *Nine Chapters* was not intended as a theoretical work. Rather it was a practical handbook for architects, engineers, surveyors, and tradesmen. It has many problems on building canals and dikes, town walls, barter, taxation, and public services.

The *Nine Chapters* contains 246 problems grouped into nine sections. Each chapter begins with a general rule *(Shu)* that allows one to arrive at a solution *(Chao)* for all problems in the section. The book covers methods of multiplication and division, proportions, arithmetical and geometrical progressions, and the extraction of approximate values of square and cube roots. By the 1st century A.D., the *Nine Chapters* included the essentials of the so-called "Horner method" for finding square roots. Horner's method isolates the root between two positive numbers and successively diminishes the difference between these numbers. The first chapter of the book is devoted to plane mensuration with rules for finding the areas of rectangles, trapezoids, triangles, and circles for land surveying. Translated into modern notation, the areas of circles are taken to be $\frac{3}{4}d^2$ and $\frac{1}{12}c^2$, where d equals the diameter and c the circumference. The value of π was initially taken to be 3. (Later Liu Hui improved this approximation to 3.14.) The fifth chapter of the book deals with determining the volumes of solid figures; such as prisms, pyramids, the cylinder, and the circular cone for engineers. Its last chapter on right angles *(gou gu)* applies the Pythagorean theorem.

The *Nine Chapters* has at least two general innovations. First, it employs the "Rule of False Position" to solve equations. This rule is a Chinese algebraic invention, although it was also known to ancient Egyptian scribes. An example of this problem-solving technique follows: given $x + \frac{1}{3}x = 8$. Guess that $x = 3$. This gives $3 + 1 = 4$. Since $\frac{4}{8} = \frac{1}{2}$, the answer or actual x is $2 \times 3 = 6$. The Chinese mainly applied this rule for solving equations of the type $ax = b$. During the Middle Ages Arabic mathematicians transmitted the "Regula Falsae Positionis" to Europe. Second, the *Nine Chapters* recognizes both positive *(zheng)* and negative *(fu)* numbers as solutions to equations. This is the first known appearance of negative quantities in any civilization. (Liu Hiu represented positive and negative numbers by red and black calculating rods, respectively.)

51. From *The Development of Mathematics in China and Japan**

(Problems in the *Nine Chapters*)

YOSHIO MIKAMI

1. From the sixth chapter *jun shu* on pursuit and alligation

"While a good runner walks 100 paces (or a), a bad runner goes 60 paces (or b). Now the latter goes 100 paces (or c) in advance of the former, who then pursues the other. In how many paces will the two come together?"

The answer is: $a \times c \div (a - b) = 250$ paces.

"A hare runs 100 paces (a) ahead of a dog. The latter pursues the former for 250 paces (b), when the two are 30 paces (c) apart. In how many further paces will the dog overtake the hare?"

The answer is: $cb \div (a - c) = 107\frac{1}{7}$ paces.

2. From the seventh chapter *yin bu zu* on excess and deficiency

"When buying things in companionship, if each gives 8 pieces, the surplus is 3; if each gives 7, the deficiency is 4. It is required to know the number of persons and the price of the things bought."

**Source:* From Yoshio Mikami, *The Development of Mathematics in China and Japan* (1913), 16 and 18-20. For a fine brief introduction to the *Nine Chapters* see Joseph Needham, *Science and Civilisation in China* (Cambridge: Cambridge University Press, 1975), vol. 3, 24-27.

"When buying hens in companionship, if each gives 9, then 11 will be surplus; and if each gives 6, then 16 will be the deficiency. What will be the number of persons and the price of the hens?"

These problems are equivalent to the solution of the equations

$$y = ax - b \text{ and } y = a'x + b'.$$

The rule for solution is given as follows:

"Arrange the rates (a and a') forwarded by the partners in buying things. What surpasses (b), or is deficient (b'), be each arranged below these rates, and then cross multiply with them. Add the products together, and one gets the *shih*. The surplus and deficiency being added, make the *fa*. If fractional, first make both members equidenominated. Then the *shih* and the *fa* being divided by the difference of the rates, the quotients represent the price of the things bought and the number of persons, respectively."

According to this rule we first form the arrangement (1), from which by cross multiplication follows (2); and then by addition we have (3). Thus

$$(1)\ \begin{matrix} a\ a' \\ b\ b' \end{matrix} \quad (2)\ \begin{matrix} ab'\ a'b \\ b\ \ b' \end{matrix} \quad (3)\ \begin{matrix} ab + a'b \\ b + b' \end{matrix}$$

And now we have to take

$$\text{price} = \frac{ab' + ab'}{a - a'}, \quad \text{persons} = \frac{b + b'}{a - a'}.$$

"Of two water-weeds, the one grows 3 feet and the other one foot on the first day. The growth of the first becomes every day half of that of the preceding day, while the other grows twice as much as on the day before. In how many days will the two grow to equal heights?"

This question is solved by applying the method of the surplus and deficiency. For in two days there is a deficiency of 1·5 feet; and in 3 days a surplus of 1·75 feet.

The answer is thus given to be $2\ ^6/_{13}$ days, when both grow to the same height of 4 feet and $8\ ^6/_{13}$ decimal parts.

3. From the eighth chapter *fang cheng* on the way of calculating by tabulation (This chapter treats simultaneous linear equations using both positive and negative numbers.)

The first problem in this section reads:

"There are three classes of corn, of which three bundles of the first class, two of the second and one of the third make 39 measures. Two of the first, three of the second and one of the third make 34 measures. And one of the first, two of the second and three of the third make 26 measures. How many measures of grain are contained in one bundle of each class?"

"Rule. Arrange the 3, 2 and 1 bundles of the three classes and the 39 measures of their grains at the right. Arrange other conditions at the middle and at the left."

The arrangement then takes the form shown in the accompanying diagram.

1	2	3	1st class
2	3	2	2nd "
3	1	1	3rd "
26	34	39	measures

The text proceeds: "With the first class in the right column multiply currently the middle column, and directly leave out."

This means to subtract the terms in the right column as often as possible from the corresponding terms of the middle column thus multiplied. The arrangement after such an operation becomes as annexed.

1	0	3	1st class
2	5	2	2nd "
3	1	1	3rd "
26	24	39	measures

"Again multiply the next, and directly leave out."

This teaches to repeat the operation with the left column. The arrangement now becomes as represented in the diagram.

0	0	3	1st class
4	5	2	2nd "
8	1	1	3rd "
39	24	39	measures

"Then with what remains of the second class in the middle column, directly leave out."

That is, the 2nd class from the left column is to be eliminated by applying the process as above described. The result is as shown here.

		36	third class
		99	measures

"Of the quantities that do not vanish, make the upper the *fa*, the divisor, and the lower the *shih*, the dividend, i. e., the dividend for the third class.

"To find the second class, with the divisor multiply the measure in the middle column and leave out of it the dividend for the third class. The remainder, being divided by the number of bundles of the second class, gives the dividend for the 2nd class.

"To find the first class, also with the divisor multiply the measures in the right column and leave out from it the dividends for the third and second classes. The remainder being divided by the number of bundles of the first class, gives the dividend for the first class.

"Divide the dividends of the three classes by the divisor, and we get their respective measures."

The above process, as will be seen on a first glance, does not deviate seriously from our procedure in solving the simultaneous system

$$3x + 2y + 1z = 39,$$
$$2x + 3y + 1z = 34,$$
$$1x + 2y + 3z = 26.$$

The only difference is that the expressions are arranged in vertical columns instead of writing in horizontal lines as we do nowadays. The Chinese manipulation appears however without doubt to have been carried on with the calculating pieces, not in a written scheme.

The answer for the above problem is given as $9\ ^1/_4$, $4\ ^1/_4$, and $2\ ^3/_4$ measures of grain, respectively.

The above process is directly applicable only in the case when the terms in a column are all subtractible from other columns. But how had done the ancient Chinese to proceed in the case, for which such is not the case?

The following is one of such examples.

"There are three kinds of corn. The grains contained in two, three and four bundles, respectively, of these three classes of corn, are not sufficient to make a whole measure. If however we add to them one bundle of the 2nd, 3rd, and 1st classes, respectively, then the grains would become full one measure in each case. How many measures of grain does then each one bundle of the different classes contain?"

1		2	1st class
	3	1	2nd "
4	1		3rd "
1	1	1	measures

The arrangement that corresponds to this problem will be as annexed.

The empty blanks have no numbers to be arranged therein.

The answer is given as $^9/_{25}$, $^7/_{25}$, $^4/_{25}$ measures of the three classes.

Chapter IV

Arabic Primacy with Hindu, Chinese, and Maya Contributions

The Civilization of the Maya: Commentary

The Maya civilization was perhaps the earliest and most advanced culture in pre-Columbian America. According to archaeologists and historians, it extended from southern Mexico (chiefly Yucatan) to Belize, Guatemala, and Northern Honduras. It is believed to have flourished from the late-3rd century to the early-9th century of our Christian era (the Maya "Classic Age"). The societies of this Classic period were primitive kingdoms and incipient city–states ruled by "divine kings" who had a priestly class in their courts. Maya life was greatly influenced by religion. The chief function of the priests was to study the earth and the heavens in order to gain insights about supernatural sanctions. They were to know especially how to avoid the displeasure of gods who were believed to control health, disease, and nature. The religious ceremonies included the sacrificing of young men and women. Among the priests were sky-watchers, scribal experts in calendar-making and computation, and planners of huge palaces and temples. The palaces and temples were built of stucco and stone and were erected with forced labor. The temples were generally truncated pyramids. Presumably the Classic Maya had a warrior or noble class (evidence of wars of conquest exists), and the mass of the population must have had to work hard to support these kings, priests, and warriors. Although most were farmers who grew beans, squash, peppers, hemp, cotton, and cacao, there were many artisans—spinners, weavers, potters, sculptors, carpenters, masons—and merchants who traveled as far as Cuba. The Classic Maya also had subject populations and slaves.

The Classic kingdoms collapsed during the 9th century. After two centuries of decline, the Maya civilization experienced a lesser flowering in the "Mexican" or "early postclassic" period of Toltec domination. The "Early Postclassic" period lasted from the 10th to the end of the 12th century; a "Late Postclassic" period dates from the 13th to the 16th century. Concentrated political power and "high" culture had apparently disappeared, and monuments of the Classic Age were falling into ruins. By the time the Spanish arrived, the Maya no longer lived in impressive cities and towns; most of the people dwelt in straw-covered wooden huts clustered in villages. The reason for the decline is not entirely clear. Probable causes are exhaustion of the soil, changes in climate, feuding or warlike tendencies, and perhaps a plague.

The study of the achievements of the Maya numerators, sky-watchers, and calendar priests rests on relatively few written records. Neither the Classic nor Postclassic Maya developed an alphabet. Instead, they reduced picture writing to abbreviated forms. The primary sources that have survived are inscribed stone monuments and hieroglyphic codices. The most famous codex is the (presumably) 12th-century Dresden Codex (named for the European library in which it is housed). The Maya did not leave behind a single treatise on mathematical or astronomical theories and methods nor on mathematical proofs or algorithms. The cumbersome system of writing with glyphs must have hindered, if not prohibited, the preparation and transmission of such mathematical disquisitions. Not a single name of the contributors to Maya mathematics, astronomy, or calendrics has survived. Thus the men of Maya science remain totally anonymous.

The achievements of the Classic Maya in mathematics and calendrical astronomy were remarkable nevertheless. Like the ancient Babylonian scribes, their learned men had a number system with a place-value notation that was used in commercial records, taxes or levies of tribute, census, mensuration, eclipse-possible records in astronomy, and other governmental or religious functions. Their place-value system, which was perhaps seven centuries old even when the Classic period began in the 3rd century, had a symbol for zero (an ornate shell). The zero symbol was used as we use it today. The Maya numeration was vigesimal, having bar-and-dot numerals for the numbers one through twenty. A repetition of dots represented units up to four; bars were for multiples of five; and combinations of bars and dots represented intervening numbers. The choice of base-twenty perhaps evolved from counting on both fingers and toes. The word for twenty in the different Maya languages is often "man," referring to the totality of these digits. The Maya number system also had a decimal substratum. Besides having words for *twenty* or *score*, the Maya had words for multiples of twenty and powers of twenty. The names of Maya numerical units go up to 20^6, while their calendar cycles involve numerals as large as 360×20^{12} units. Historical and archaeological records suggest that the Classic Maya had the most highly developed number system in pre-Columbian America. They lacked a fractional arithmetic, however. In calendrical astronomy the Maya priests developed a rather accurate "calendar year" consisting of eighteen named months of twenty days each and a residue period of five days.

52. From *The Ancient Maya**

SYLVANUS GRISWOLD MORLEY

H. G. Wells in his *Outline of History* says that the invention of a graphic system is the true measure of civilization, and Edward Gibbon, in his *Decline and Fall of the Roman Empire,* claims that the use of letters is the principal characteristic which distinguishes civilized people from savages. By such standards, the Maya were the most civilized people of the New World in pre-Columbian times, since they alone originated a system of writing.

THE DEVELOPMENT OF WRITING

Writing seems to have passed

**Source:* This selection is taken from Sylvanus G. Morley, *The Ancient Maya* (3rd edition, 1956) and is reprinted by permission of Stanford University Press.

everywhere through three stages of development:

1. *Pictorial or representative writing,* wherein a picture of the idea is portrayed. Thus a deer hunt is represented by the picture of a deer and a man throwing a spear at it. This is called pictographic writing.

2. *Ideographic writing,* wherein characters stand for ideas rather than representing pictures of them. In ideographic writing the characters employed are usually little more than conventionalized symbols. In Chinese writing, the ideograph for "trouble" is the conventionalized symbol for a woman, repeated twice, standing under a gate.

3. *Phonetic writing,* wherein the characters have lost all resemblance to the objects they originally portrayed, and denote only sounds. Phonetic writing may be further divided into *(a)* syllabic writing, in which each character stands for a syllable, and *(b)* alphabetic writing, in which each of the characters stands for a single sound. Egyptian hieroglyphic writing is an example of the former; modern alphabets are examples of the latter.

Maya hieroglyphic writing is ideographic, since its characters represent ideas rather than pictures or sounds. It has been thought by some that there are phonetic elements included in Maya writing.

MAYA WRITING ONE OF THE EARLIEST EXAMPLES OF A GRAPHIC SYSTEM

One of the important facts about the Maya hieroglyphic writing is that, barring such purely pictorial efforts as the paleolithic cave paintings or the American Indian pictographs, it may represent the earliest stage of a formal graphic system that has come down to us. This does not mean that the Maya hieroglyphic writing is the oldest graphic system known. Although the earliest Egyptian and Sumerian inscriptions go back to the fourth millennium before Christ, the earliest known Maya writing was done after the beginning of the Christian Era. However, early Egyptian hieroglyphics had already advanced to a semiphonetic stage. In addition to the many ideographs present, perhaps half the characters are phonetic, mostly syllabic. A similar condition obtains in the earliest cuneiform writing.

The Maya "Rosetta Stone" is the *Relación de las cosas de Yucatán,* written about 1566 by Bishop Diego de Landa. . . .

STORY TOLD BY THE MAYA INSCRIPTIONS

The Maya inscriptions treat primarily of chronology, astronomy, and religious matters. They are not records of personal glorification, like the inscriptions of Egypt, Assyria, and Babylonia. They are so completely impersonal that it is unlikely that the name glyphs of specific men were ever recorded upon the monuments. . . .

MAYA CALENDAR

(Permutations of the Long Count in Maya Calendrics, that is tables of multiples including 52 civil or calendar years—52 × 365 = 18,980—which also equals 73 sacred years or almanacs—73 × 260)

The Maya had two calendars whose dates their priests sought to relate. The first calendar was the sacred year or *tzolkin* of 260 days. Comprised of 13 weeks of 20 days each, it determined the pattern of ceremonial life. The second was the civil calendar or *haab* of 365 days. It contained 19 months—18 months of 20 days each and a final month of 5 days. Relating the dates of one to the other over long periods of time involved determining the least common denominator of 260 and 365. Both numbers divisible by 5; 260 gives a quotient of 52, and 365 gives a quotient of 73, so the least common multiple of 260 and 365 is 5 × 52 × 73, or 18,980. Therefore, Wheel *A* will make 73 revolutions and Wheel *B* will make 52 revolutions before the two wheels have returned to their original positions, a total of 18,980 elapsed days, or about 52 years.

Once every 52 civil years, then, any given year-bearer coincided with the first day of the year. Thus any Maya who lived more than 52 years began to see New Year's Days of the same name repeat themselves. We do not know the ancient Maya name or hieroglyph for this 52-year period, but modern students of the Maya calendar have called it the Calendar Round.

None of the peoples of Mesoamerica who borrowed their calendars from the Maya made use of time periods longer than this 18,980-day period. The Aztecs, for example, conceived time as an endless succession of these 52-year periods and gave to them the name *xiuhmolpilli,* meaning "year bundle" or complete round of the years. . . .

MAYA ARITHMETIC

In order to escape rapidly mounting calendric chaos, the Maya priests devised a simple numerical system which even today stands as one of the brilliant achievements of the human mind.

Some time during the fourth or third centuries before Christ, the priests devised a system of numeration by position, involving the conception and use of the mathematical quantity of zero, a notable intellectual accomplishment.[1]

The unit of the Maya calendar was the day or *kin.* The second order of units, consisting of 20 kins, was called the *uinal.* In a perfect vigesimal system of numeration, the third term should be 400 (20 × 20 × 1) but at this point the Maya introduced a variation for calendric reckoning. The third order of the Maya system, the *tun,* was composed of 18 (instead of 20) uinals, or 360 (instead of 400) kins. This was a closer approximation to the length of the solar calendar.

Above the third order the unit of progression is uniformly 20, as will be seen from the numerical values of the nine known orders of time periods:

20 *kins*	= 1 *uinal* or 20 days
18 *uinals*	= 1 *tun* or 360 days
20 *tuns*	= 1 *katun* or 7,200 days
20 *katuns*	= 1 *baktun*[2] or 144,000 days
20 *baktuns*	= 1 *pictun* or 2,880,000 days
20 *pictuns*	= 1 *calabtun* or 57,600,000 days
20 *calabtuns*	= 1 *kinchiltun* or 1,152,000,000 days
20 *kinchiltuns*	= 1 *alautun* or 23,040,000,000 days

MAYA GLYPH FORMS

Every Maya hieroglyph occurs in two forms in the inscriptions: (1) the normal form and (2) a head variant, this being the head of a deity, man, animal, bird, serpent, or some mythological creature. Very rarely there is a third form where the glyph is a full figure.

The glyphs for the nine time periods are given in Figure 22, with normal forms at the left and head variants at the right. Head variants have not yet been identified for the last three periods.

MAYA NUMERICAL NOTATIONS

The ancient Maya used two types of notation in writing their numbers: (1) bar-and-dot numerals, and (2) head-variant numerals. In the first notation, the dot • has a numerical value of 1 and the bar ▬ a numerical value of 5, and by varying combinations of these two symbols, the numbers from 1 to 19 were written (Fig. 23). The numbers above 19 were indicated by their positions and will be described later.

Maya bar-and-dot notation was sim-

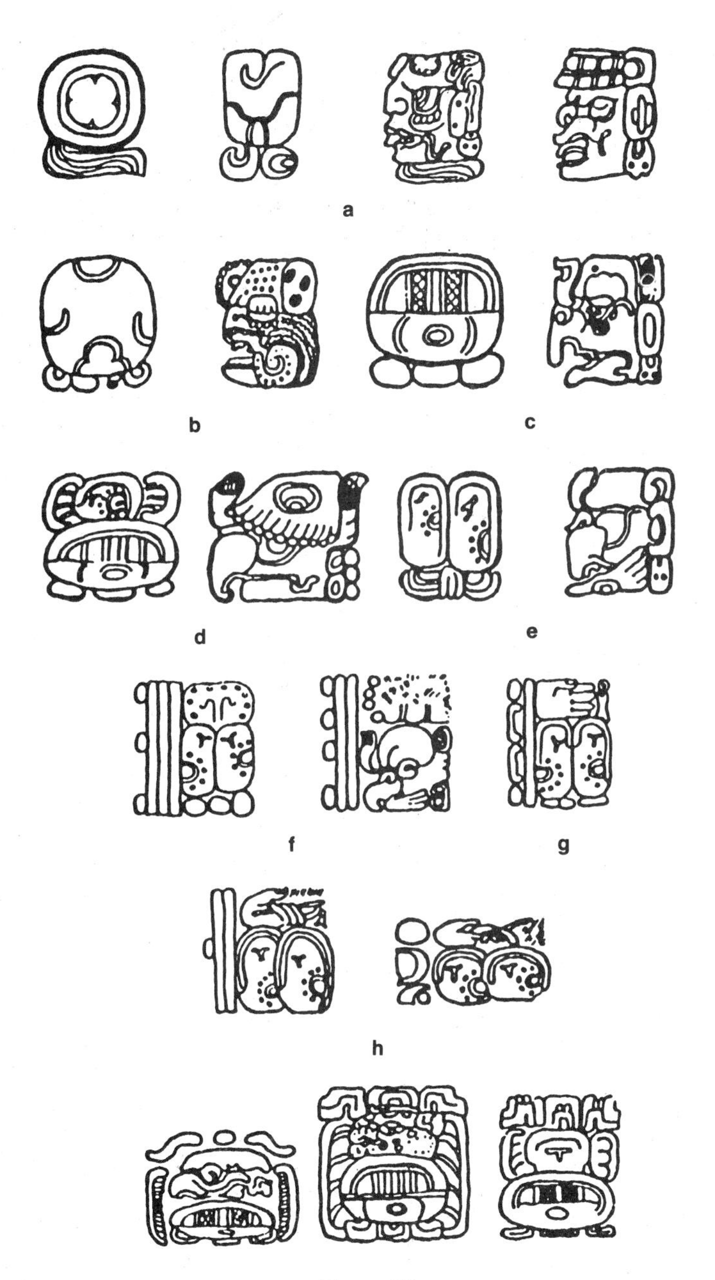

Figure 22

GLYPHS FOR THE NINE KNOWN MAYA TIME-PERIODS: *(A) KIN; (B) UINAL; (C) TUN; (D) KATUN; (E) BAKTUN; (F) PICTUN; (G) CALABTUN; (H) KINCHILTUN; (I) ALAUTUN* OR INITIAL SERIES INTRODUCING-GLYPH.

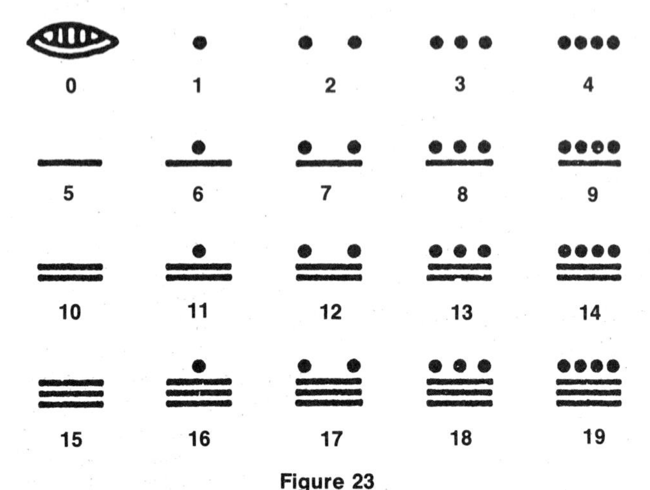

Figure 23

GLYPHS FOR 0 AND THE NUMBERS 1 TO 19 INCLUSIVE.

pler than Roman notation and superior in two respects. To write the numbers from 1 to 19 in Roman notation, it is necessary to employ the symbols I, V, and X, and the processes of addition and subtraction: VI is V plus I, but IV is V minus I. In order to write the same numbers in Maya bar-and-dot, it is necessary to employ only the dot and the bar, and one arithmetical process, that of addition.

The second notation employed in writing Maya numbers used different types of human heads to represent the numbers from 1 to 13, and zero. The Maya head notation is comparable to our Arabic notation, where ten symbols represent zero and the nine digits. These Maya head-variant numerals are heads of the patron deities of the first fourteen numbers (Plate 30*a*).[3] The head variant for 10 is the death's head, and in forming the head variants for the numbers from 14 to 19 the fleshless lower jaw (Plate 30*a*) is used to represent the value of 10. For example, if the jaw is applied to the lower part of the head for 6, which is characterized by a pair of crossed sticks in the large-eye socket, the resulting head will be that for 16. It is probable that the heads representing numbers 1 to 13 are those of the *Oxlahuntiku* or Thirteen Gods of the Upper World.

In writing bar-and-dot numbers above 19, the ancient Maya used a positional system of numeration. In our decimal system, the postions to the left of the decimal point increase by tens. In the Maya vigesimal system the values of the positions increase by twenties from bottom to top. An exception is made in counting time, when, as already noted, the third position is 18 instead of 20 times the second.

To illustrate this, let us see how the Maya would have written the number 20, which is 1 complete unit of the second order and no units of the first order. This necessitates a symbol for zero in the lowest position to show that no units of the first order are involved; for this we shall use the conventionalized shell, one of the commonest symbols for zero. Thus, by placing a shell in the lowest position to denote 0 units of the first order and a dot • in the second position to denote 1 unit of the second order, the number 20 can be written (Fig. 24). The manner of writing other numbers is also shown in Figure 24, including two numbers written in the

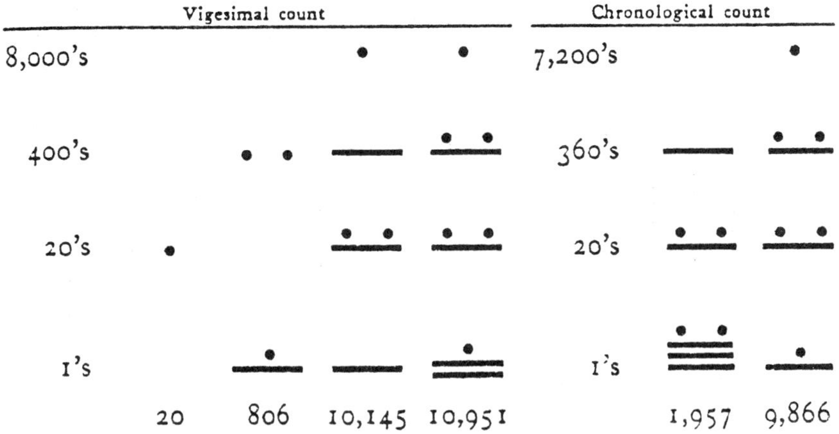

Vigesimal count					Chronological count		
8,000's					7,200's		
400's					360's		
20's					20's		
1's					1's		
	20	806	10,145	10,951		1,957	9,866

Figure 24

EXAMPLES OF MAYA POSITIONAL MATHEMATICS.

chronological count. The simplicity of Maya addition is also obvious in Figure 24: 10,951, the sum of the numbers in the two preceding columns, is obtained simply by combining the dots and bars of 806 and 10,145 into a new figure.

NOTES

1. **Editor's note.** The invention of place-value notation in Mesoamerica may have been the work of earlier peoples other than the Maya, such as those in Oaxaca, Tabasco, and Vera Cruz.

2. The period of the fifth order, the *baktun,* was originally called the "cycle" by modern investigators. The ancient name for this period, however, was probably *baktun* as given above.

3. The head-variant for the number 11 has not yet been surely identified.

Chapter V

The Medieval–Renaissance–Reformation Periods in Europe

Introduction
by
Joseph E. Brown
Rensselaer Polytechnic Institute

Strong contrasts characterized the European medieval world. Among them a poverty of mathematical thought contrasted with the richness of general culture. Chronologically, that culture synthesizing Roman, Germanic, and above all Christian elements began within the Roman Empire (*c.* 400) and merged with the exuberant artistic, literacy and religious movements of the Renaissance–Reformation (*c.* 1400-1600). The pagan Romans had cared little for mathematics apart from its applications to architecture and engineering. Seriously deficient in mathematical vocabulary (for example, there is no term for parabola), the Latin language itself was one reflection of their disinterest. Prospects for theoretical mathematics further deteriorated when Christians came to the fore. Among the devout, occupied with preparations for the next world, intellectual energies were expended mainly on sacred sciences. In the interest of religious integrity, the early Church Fathers had often criticized pagan Graeco–Roman learning. Nevertheless, a fateful accommodation between pagan learning and Christian belief was formed though not without tension and suspicion. Origen, an Alexandrian Greek who was perhaps the most brilliant of the Church Fathers, had insisted that his pupils receive a mathematical education. The Roman–African, St. Augustine of Hippo (d. 430), whose Latin works such as *The City of God* became magisterial, reinvoked Biblical authority to assert that God had formed the universe "in number, weight and measure." Such scriptural and patristic warrant assured some place for rudimentary mathematics in a Christian scheme of education. Unfortunately, by mid-5th century the conditions necessary for the education that either Origen or Augustine advocated were disappearing. As a cultural separation developed between Latin West and Greek Byzantium the major portion of Greek mathematics disappeared. Constantinople, founded in 330 to oversee an Eastern Christian empire, produced a distinctive Byzantine culture and survived until 1453.

Rome's urban culture had long been weakened by trading deficits, swollen military costs, out-migration of the skilled and the wealthy, and the influx of barbarian tribes who controlled the West by the year 500. Boethius, a scholarly patrician Roman in the service of the Ostrogoth chieftain at Ravenna, attempted to salvage portions of the Greek legacy before his political execution in 524 or 525. He prepared Latin paraphrases of Nicomachus of Gerasa's *Arithmetica* and Euclid's *Elements* that partly preserved the axioms, definitions, and theorem enunciations but discarded the proofs altogether.

Stripped of theory and geometric proofs, mathematics was esteemed merely as an illuminator of Biblical passages among the literate elite in monasteries derived from St. Benedict of Nursia (fl. 529). Monks charged with keeping the devil at bay also kept alive the mensurational skills of ancient surveyors *(agrimensores)* and the calculating techniques of the *computus* tradition. In the fragmented, warrior society, monasteries were enclaves of stability and organization. Monks Christianized the barbarian migrants who spread across Europe in two waves during the 5th–6th and 9th–10th centuries bringing new blood and opening new lands. Monks made possible the revival of learning under the Frankish overlord Charlemagne, who was crowned Roman Emperor by Pope Leo III (Christmas 800). Although the title was a misnomer on both counts, it forecast a reconstituted, unified commonwealth, Europe. Equally farsighted was his program that brought learned monks to court for the rejuvenation and promulgation of Latin literacy. This barely literate, saddle-calloused warrior who supervised the egg count at the imperial hen house and whipped a schoolboy for a grammatical solecism seemed an unlikely sponsor of reform in education. Yet his monks at the palace school who carefully collected and meticulously copied rare Latin works from antiquity took the most essential step for the recovery of Latin letters. The typography of this page, descended from their standardized calligraphy, memorializes the achievement of that humble community of scholars.

The reemergence of mathematics followed that of grammar. Another Benedictine, Gerbert of Aurilac, the first teacher of Arabic calculational methods learned from travels in Moorish Catalonia, became also the first mathematician pope as Sylvester II (d. 1003). For the cumbersome system of counting individual pebbles *(calculi)* one-at-a-time on the counting board *(abacus),* he substituted the single counter *(apex)* marked with numerical symbols equivalent to "1" through "9." This simplification achieved considerable economy, required that some arithmetic operations be done mentally preparing for place-value calculation, and, most importantly, began the flow of Arabic mathematics into Europe. Disciples in Lorraine (northeastern France) continued Gerbert's interest in computing.

About 1050 the High Middle Ages began. The start of a new millennium in Europe saw a milder climate, major economic and technological progress, reform in the church (such as at the monastery of Cluny) and the Crusades. The Crusades stimulated European trade and commerce in the Mediterranean and expanded contacts with Arabic culture. The High Middle Ages also were a time of expansion on the continent particularly evident in the growth of new towns. These towns became the centers for a new social class whose position was based on commercial wealth rather than land. The momentum for new learning now came from the towns, not the monasteries. Urban and ecclesiastical administration required a trained corps of clerical and curial experts, a need that prompted the rise of universities. The new universities (such as those at Bologna, Paris and Oxford) trained students in practical and professional skills as well as scholastic theology. They quickly became centers of refined and sophisticated scholarship.

By the 12th century, the story of mathematics, like that of European society, was one of recovery. The basic subject was arithmetic called *algorismus* after al-Khwārizmī. Employing Hindu–Arabic numerals, a place-value system, and well developed operational rules, it superceded the abacus and moved calculation to paper (also borrowed from the Arabs). Algebra, the second basic advance, took its name from the Arabic title of al-Khwārizmī's work, although the latter was not fully translated or exploited. It gave rules for solving six types of problems including those with an unknown of the second degree of the (modern) form $ax^2 + bx = c$. Demonstrative geometry was the third acquisition. Euclid became a household word among scholars and his *Elements* was early translated into four versions.

The first quarter of the 13th century produced two original mathematicians, Leonardo Fibonacci of Pisa and Jordanus de Nemore (probably Nemours, southeast of Paris). Extensively traveled, Leonardo had mastered Arabic algebra and considerable Archimedean geometry, a singular achievement for a Latin. Jordanus, somewhat less gifted, drew on Leonardo and Arabic algebra for which he employed letter symbols but more as verbal shorthand than for operational utility.

Early agrimensorial methods, enriched by Euclidean theory in the 13th century, incorporated some Archimedean materials in the following two centuries. Leonardo was the pioneer and the *Practica geometriae* (1346) of Dominicus de Clavisio, ingenious in its use of simple instruments and proportionality, became the standard work. Meantime proportionality theory had grown significantly. Jordanus, in a creative deviation from current understanding treated ratios as rational fractions for operational manipulations. Even more significant than the acceptance of ratios as a kind of number was the fourteenth century development of a "calculus of ratios." Its baroque but logically sound vocabulary effectively converted ratios into exponents before that operator was invented. The principal figures were the Oxonian, Thomas Bradwardine, and the Parisian, Nicole Oresme. Oresme also applied geometry to represent "intensity" of a quality or movement, adumbrating coordinate geometry and graphing.

Despite curricular study, creative breakthroughs and new lines of development the Latin Schoolmen did not develop an appreciation of mathematics *per se*. Neither the old nor the new works were fully exploited. Oresme valued mathematics as a refutation of the claims of court astrologers. The Franciscan encyclopedist, Roger Bacon (d. 1292), praised mathematics highly but understood it little and based his technological optimism on mastery of occult forces rather than the mathematization of engineering. Thirteenth century optimism, also evident in the ambitions of Crusades and the encyclopedic *Summa* of St. Thomas Aquinas, turned to skepticism and pessimism in the next century. It was a time of economic contraction, crop failures, chronic and brutal warfare, polarization of church hierarchy and members around rival popes and the calamitous Black Death.

By mid-15th century, Europe had rebounded economically and demographically and was building a more spectacular Renaissance around values more secular and less clerical. Its new ethos, "Humanism," remained strongly Christian but added a careful study of pagan literary classics while downgrading scholastic theology and philosophy, including its logic and science. Humanism affected students, courtiers, artists and cultivated merchants. Art was its major glory. Painting was naturalistic and individualistic and its perspective techniques showed the movement of medieval optics toward projective geometry, evident in the works of Leonardo da Vinci (d. 1519). By Gerard Mercator's map of 1569, geometer cartographers had learned how to project a spherical surface onto a plane. Overseas discoveries that paralleled the recoveries of ancient learning prompted such inventiveness. Old and new flourished in the symbiotic relationship of trade, art and science that existed in such Italian city states as Venice, Milan, Florence, and Rome, at the crossroads of Mediterranean–European trade. Florentines reconstituted a latter-day version of Plato's Academy in 1462. The Platonic revival stimulated interest in mathematics and a new round of translations—this time from Greek originals—by such competent, professional mathematicians as Francisco Maurolico and Federigo Commandino (both d. 1573).

These works were spread by the new printing press (from the 1450s). Some mathematicians, such as Peter Bienewicz (Apian) and Johann Müller (Regiomontanus), established their own presses. Regiomontanus (who in conjunction with his teacher Georg Puerbach had improved trigonometric theory and tables) died prematurely in 1476 working on a reform of the Julian calendar urgently needed by navigators. Before the advent of printing, seamen, accountants, and others had to learn computation largely by oral tradition. Schoolboys in the Baltic trading towns of the Hanseatic League went to a *Rechenschule.* Italians had a *scuole d'abaco* in which the algebraic methods of Leonardo Fibonacci (largely ignored in universities) were transmitted.

Exploitation of algebra and computation were prominent aspects of the revolutionary science of the 16th century. Architects of that new science were Paracelsus (d. 1541) in chemistry, Nicolause Copernicus (d. 1543) in astronomy, Andreas Vesalius (d. 1564) in anatomical medicine, and Girolamo Cardano (d. 1576) in algebra. Cardano's general solution of equations of the third degree and selected higher ones went beyond the work of Niccolò Tartaglia. In 1597, François Viete's analytical methods offered an almost fully developed operational symbolism employing vowels to represent unknown quantities and consonants for knowns and constants.

By 1600 European mathematics had easily outstripped its classical sources. John Napier published his work on logarithms on 1614, and Henry Brigg's recommendation for a decimal base and their extension beyond trigonometry to ordinary number calculation showed how powerful and practical European computation had become. Even more original was the work of Johann Kepler, who in 1615 used infinitesimals to make cubic measurements of wine vats.

If mathematics had undergone important changes since 1050 so had Europe. The commonwealth of Christendom had given way to rival national states and, from 1517, to Protestant and Catholic communions. The Florentine Niccolò Machiavelli's politics of naked realism had challenged the Platonic ideal visions of his contemporaries. Still others had visions of magical conquests abetted by mathematics, as did mathematician–Magus John Dee (d. 1608) in his preface to the first English translation of Euclid. There was now a confident assertiveness about European learning. Francis Bacon reflected it in *Novum Organum* (1620)—a work as ambitious in its embrace as the Aristotelian learning it proposed to replace. As the 17th century began, Europe's economic and political center had moved north and the times were ripe for new exploration and conquest in mathematics as in colonization.

Suggestions for Further Reading

Carl B. Boyer, *A History of Mathematics* (New York: Wiley 1968).

Moritz Cantor, *Vorlesungen über die Geschichte der Mathematica.* 4 vols. (Leipzig: Teubner, 1894-1908). (Subsequent additions and corrections appear in the third series of the journal *Bibliotheca Mathematica.*)

Jerome Cardan, *The Book of My Life* (1575) trans. from the Latin by Jean Stoner (New York: Dover, 1962).

Marshall Clagett, *Archimedes in the Middle Ages.* 3 vols. (Madison: University of Wisconsin Press, 1964 and Philadelphia: American Philosophical Society 1976 and 1978).

Edward Grant (ed.), *A Source Book in Medieval Science* (Cambridge, Mass.: Harvard University Press, 1974).

A. P. Juschkewitsch, *Geschichte der Mathematik in Mittelalter* (Leipzig: Teubner 1964). This is a German translation made by Viktor Ziegler from the Russian.

Michael S. Mahoney, "Mathematics" in David C. Lindberg (ed.), *Science in the Middle Ages* (Chicago: University of Chicago Press, 1978).

Karl Menninger, *Number Words and Number Symbols: A Cultural History of Numbers* (Cambridge, Mass.: The M.I.T. Press, 1969). This is an English translation made by P. Broneer of the revised German edition.

John E. Murdoch, "The Medieval Euclid: Salient Aspects of the Translations of the *Elements* by Adelard of Bath and Campanus of Novara," *Revue de Synthèse* 89 (ser. 3, nos. 49-52, 1968): 67-94.

Paul Tannery, *Mémoires Scientifiques.* vols. 4 and 5 (Paris: L. Cerf 1922).

For information on the lives of Medieval and Renaissance mathematicians one should consult appropriate volumes in Charles Coulston Gillispie (ed.), *Dictionary of Scientific Biography,* 15 vols. (New York: Charles Scribner's Sons, 1970-78).

Chapter V
The Medieval–Renaissance–Reformation Periods in Europe

LEONARDO OF PISA (*c.* 1170–*c.* 1240)

Leonardo of Pisa (or Leonardo Fibonacci), the first great mathematician of the Latin West in the Middle Ages, was the son of Guilielmo Bonacci, a secretary of the Republic of Pisa. In 1192 the father became director of the Pisan trading colony in Bugia (now Bougie), Algeria. Expecting the son to become a merchant, the father brought Leonardo to Bugia to study calculation with an Arab master and to receive other excellent instruction *(ex mirabili magisterio).* As described later in the *Liber abbaci* (1202), he enjoyed learning "the art of nine Indian numerals." The father also sent Leonardo on business trips to Egypt, Syria, Greece (Byzantium), Sicily, and Provence. During these travels he studied other numerical systems and calculating methods but found none as satisfactory as the Hindu–Arabic. He also amassed a rich source of geometric and algebraic information from the ancient Greek and Arabic masters including Diophantus, al-Khwārizmī, and al-Khayyāmī. Around 1200 Leonardo returned to Pisa. During the next 25 years he wrote on mathematics and its practical commercial applications.

His books include *Liber abbaci* ("Book of the Abacus," 1202), *Practica geometriae* ("Practice of Geometry," 1220 or 1221), and his masterpiece *Liber quadratorum* ("Book of Square Numbers," 1225), which he devoted to solving second degree Diophantine or indeterminate equations. These books, especially the widely copied and imitated *Liber abbaci,* brought him to the attention of Holy Roman Emperor Frederick II, a patron of science who invited Leonardo to a meeting of his court in Pisa in the 1220s. For the emperor and his aide, John of Palermo, Leonardo solved Diophantine equations. When Leonardo revised the *Liber abbaci* in 1228, he dedicated it to Michael Scott, an astrologer and the emperor's chief scholar (a man whom Dante later condemned to hell in the *Inferno* because of his magical deceptions). We know nothing of the life of Leonardo after 1228, except that in 1240 the city of Pisa awarded him an annuity of 20 Pisan pounds for services rendered.

Leonardo decisively pioneered the revival of mathematics in the Latin West. No previous medieval scholar compares with him in gathering, giving fresh consideration to, and expanding upon ancient Greek, Hindu, and Arabic mathematical knowledge. At a time when most Europeans used Roman numerals and only a few European savants knew of Hindu–Arabic figures, he introduced the latter (*i.e.*, 9 8 7 6 5 4 3 2 1 and the sign 0) to the West in his influential *Liber abbaci.* After presenting the notation and explaining the principle of place value, he employed Hindu–Arabic numerals in arithmetical operations and then applied them to such commercial problems as interest, profit margins, money changing, and conversion of weights and measures. (Subsequently, many Pisan and Florentine merchants adopted his new methods often being more responsive than academic scholars.) The rabbit problem of the *Liber abbaci* is its most famous. It leads to the number sequence 1, 1, 2, 3, 5, 8, 13, 21, 34, in which each term is the sum of the prior two. This first recurrent number sequence known in Europe is today called the Fibonacci sequence.

While practical mathematics was his chief concern, Leonardo also contributed to theoretical or speculative mathematics. In the eighth chapter of *Liber abbaci,* which mainly examines the theory of proportion, he significantly presented complete proofs in the manner of Euclid. Beginning with Boethius in the 5th century, medieval geometers usually omitted proofs or gave only partial ones. Leonardo displayed a mastery of the Euclidean axiomatic method in *Practica geometriae,* which also examines Archimedes' determination of the value of π and extends Euclid's classification of irrationals as given in Book X of the *Elements.* His *Liber quadratorum* is a collection of theorems systematically arranged with proofs invented by the author. Besides providing general solutions of indeterminate equations of second degree, he treated certain cubic equations. Like al-Khayyāmī, he held that cubic equations could not be solved algebraically. Perhaps the most creative work in this book concerns congruent numbers, those which give the same remainder when divided by a given number. The *Liber quadratorum* shows Leonardo to be the major contributor to the origins of number theory between Diophantus and Fermat.

53. From *Liber abbaci**

(The Rabbit Problem)

LEONARDO OF PISA

How many pairs of rabbits can be bred from one pair in one year?

A man has one pair of rabbits at a certain place entirely surrounded by a wall. We wish to know how many pairs can be bred from it in one year, if the nature of these rabbits is such that they breed every month one other pair and begin to breed in the second month after their birth. Let the first pair breed a pair in the first month, then duplicate it and there will be 2 pairs in a month. From these pairs one, namely the first, breeds a pair in the second month, and thus there are 3 pairs in the second month. From these in one month two will become pregnant, so that in the third month 2 pairs of rabbits will be born. Thus there are 5 pairs in this month. From these in the same month 3 will be pregnant, so that in the fourth month there will be 8 pairs. From these pairs 5 will breed 5 other pairs, which added to the 8 pairs gives 13 pairs in the fifth month, from which 5 pairs (which were bred in that same month) will not conceive in that

*Source: Reprinted by permission of the publishers from *A Source Book in Mathematics, 1200-1800,* edited by D. J. Struik, Cambridge, Mass.: Harvard University Press, Copyright © 1969 by the President and Fellows of Harvard College.

month, but the other 8 will be pregnant. Thus there will be 21 pairs in the sixth month. When we add to these the 13 pairs that are bred in the 7th month, then there will be in that month 34 pairs . . . [and so on, 55, 89, 144, 233, 377, . . .]. Finally there will be 377. And this number of pairs has been born from the first-mentioned pair at the given place in one year. You can see [below] how we have done this, namely by combining the first number with the second, hence 1 and 2, and the second with the third, and the third with the fourth . . . At last we combine the 10th with the 11th, hence 144 with 233, and we have the sum of the above-mentioned rabbits, namely 377, and in this way you can do it for the case of infinite numbers of months.[1]

Month	**Pairs**
	1
first	2
second	3
third	5
fourth	8
fifth	13
sixth	21
seventh	34
eighth	55
ninth	89
tenth	144
eleventh	233
twelfth	377

NOTE

1. This sequence of numbers, 1, 2, 3, 5, 8, . . ., u_n, . . ., with the property that $u_n = u_{n-1} + u_{n-2}$, $u_0 = 1$, $u_1 = 1$, is called a *Fibonacci series.* It has been the subject of many investigations, and is closely connected with the golden section, that is, the division of a line segment *AB* by a point *P* such that *AP* : *AB* = *PB* : *AP*. See, for example, R. C. Archibald, "The golden section," *American Mathematical Monthly 25* (1918), 232-238; D'Arcy W. Thompson, *On growth and form* (Cambridge University Press, New York, 1942), 912-933; H. S. M. Coxeter, "The golden section, phyllotaxis, and Wythoff's game," *Scripta Mathematica 19* (1950), 135-143; E. B. Dynkin and W. A. Uspenski, *Mathematische Unterhaltungen, II* (Deutscher Verlag der Wissenchaften, Berlin, 1956); and N. N. Vorob'ev, *Fibonacci numbers,* trans. by H. Mors (Blaisdell, New York, London, 1961).

Chapter V
The Medieval–Renaissance–Reformation Periods in Europe

NICOLE ORESME (*c.* 1320-1382)

The French scholastic Nicole Oresme was a Norman who was probably born near Caen. Almost nothing is known about his family and early life. Presumably he studied the liberal arts in the 1340s at the University of Paris where one of his teachers was the celebrated master Jean Buridan. It is plausible that he took his main interest in natural philosophy from Buridan. In 1348, Oresme was listed as a scholarship student in theology at the College of Navarre in the University of Paris and as a master of the Norman nation in that college. After teaching liberal arts and pursuing theological studies, he took his theological mastership in 1355 or 1356 and became Grand-Master of the College of Navarre in the latter year.

The friendship of Oresme with the dauphin (the future King Charles V the Wise) seems to have begun during the late 1350s and continued until Charles' death in 1380. Oresme listed himself as a secretary to the dauphin in 1359 and was sent to Rouen in 1360 to negotiate a loan for him.

Oresme declined an appointment as archdeacon of Bayeux in 1361 to stay in Navarre, but left in November 1362 when he was appointed canon at Rouen. A few months later he became canon at Sainte-Chapelle in Paris. He was appointed dean of the cathedral at Rouen in 1364—a dignity he held until 1377.

Charles V ascended the French throne in 1364, and Oresme probably served as the king's chaplain and counselor even though he appears to have resided regularly in Rouen until 1369 and in Paris only afterwards. From about 1369 to 1377, he skillfully translated a series of Aristotelian Latin texts (the *Ethics, Politics,* and *Economics*) into French and wrote commentaries on them at the behest of, and with stipends from, the king. Oresme's reward for this scholarly service was his appointment as bishop of Lisieux in 1377, but he probably did not move to that city until 1380.

Working within a basic Aristotelian framework, Oresme made impressive contributions to natural philosophy. In these endeavors he displayed a moderately skeptical temper coupled with a rational and naturalistic outlook. He vigorously opposed astrology, perhaps because he felt that the purveyors of occult arts exercised too much influence on King Charles. In treating celestial and terrestrial physics, he made subtle emendations and some relatively radical speculations, again within an Aristotelian orthodoxy. His application of the metaphor of the mechanical clock to the heavens, while seemingly a sharp departure, did not commit him to an umbrated version of modern celestial mechanics. The moving forces in his universe remained Aristotelian intelligences, not quantifiable physical forces. In terrestrial physics, he drew upon the kinematics developed at Merton College, Oxford, and gave a geometrical proof of the Merton mean-speed theorem.

Like other scholastics at Oxford and Paris, Oresme began to think about change and rates of change in quantitative terms. This approach culminated in his doctrine of the latitude of forms. His tract *De configurationibus* (1350s) made original contributions to this doctrine by the imaginative use of two dimensional geometrical figures to represent, or graph, intensities in qualities (*i.e.*, velocities changing with time). In his graphs, the base line represented the subject or time and the perpendicular, the velocity or degree of intensity. A rectangle represented a uniform motion (motion with constant velocity) over a given time, while a right triangle represented a uniformly difform motion (motion with constant acceleration) beginning at zero. In *Questiones super geometriam Euclidis,* Oresme proved that a right triangle representing constant acceleration equals a rectangle representing a uniform motion when at the velocity of the middle instant of acceleration—a proof used by Galileo in his work on the acceleration law in the *Discorsi* (1638).

Oresme proposed other germinal ideas in mathematics. In the tract *Algorismus proportionum* ("Algorism of Ratios"), he utilized a notation, where the fractional exponents (ratios) were expressed as pure numbers and gave rules for manipulating them. In modern symbols the square root of two was written as $(\frac{1}{2})2^{P}$, which implies $2^{\frac{1}{2}}$. Similarly $(1^{P}\frac{1}{2})4$ implies $4^{3/2}$, which he correctly stated equals 8. He noted that $3 = 3^{2/3} \cdot 3^{1/3}$. In his study of geometric infinite series in *De configurationibus,* he distinguished some convergent from some divergent series. He did not uncritically assume summation of series as did many contemporary thinkers. In chapter eight he proved that the series $1 + 2(\frac{1}{2}) + 3(\frac{1}{4}) + 4(\frac{1}{8}) + \ldots + n(\frac{1}{2})^{n-1} \ldots = 4$ by showing that it was equivalent to $1 + 1 + \{1 + \frac{1}{2} + \frac{1}{4} + \ldots + \frac{1}{2}n + \ldots\} = 4$, where the sum in brackets (Zeno's series) was known. His work on kinematics stimulated this research on series. Oresme also contributed to monetary theory by giving an early version of what became known as Gresham's law.

54. From *De configurationibus**

(The Latitude of Forms)

NICOLE ORESME

ON QUADRANGULAR QUALITY

A certain quality is imaginable by a rectangle—in fact by any such rectangle constructed on the same base—and by no other type of figure can it be designated. This last part is made clear by means of the prior statements in chapter six.

Source: From Marshall Clagett, trans. and ed., *Nicole Oresme and the Medieval Geometry of Qualities and Motions* (Madison: The University of Wisconsin Press; © 1968 by The Board of Regents of the University of Wisconsin System), 189, 191, 193, and 195. Professor Marshall Clagett, School of Historical Studies, Institute for Advanced Study, Princeton, New Jersey 08549. Letters in the legends beneath the diagrams are abbreviations for Clagett's manuscript sources.

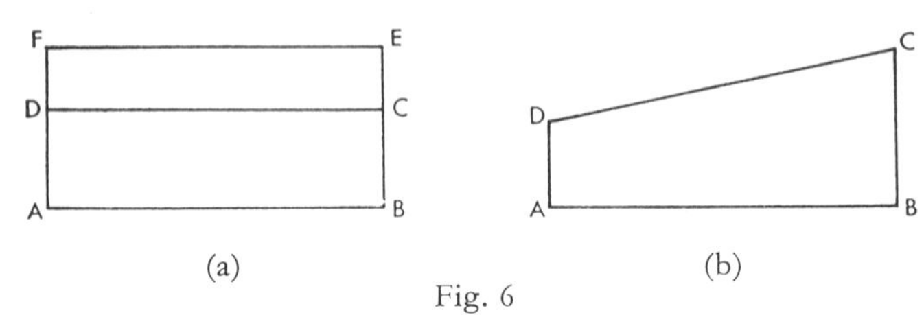

(a) (b)

Fig. 6

Figures in MSS *BLSCG*. Letters *C* and *D* in figure (a) are interchanged in MS *L*.

And so let there be rectangle *ABCD* [see Fig. 6(a)]. Therefore, it is possible that the quality of line *AB* be proportional in intensity to this rectangle in altitude. Therefore, it will be proportional to any rectangle constructed on *AB*, because all such rectangles are of proportional, although unequal, altitude. Therefore, by chapter seven, this quality is imaginable by rectangle *ABCD* and similarly by rectangle *ABEF* which is greater and also by one that is less. Moreover, any such quality is said to be "uniform" or "of equal intensity" in all of its parts.

Again it ought to be known that some quality is imaginable by a quadrangle having two right angles on the base and the other two angles unequal, e.g., by quadrangle *ABCD* [see Fig. 6(b)] and by every quadrangle constructed on base *AB* which is of proportional altitude, whether it be greater or less, as is clear in chapter seven. Moreover, any such quality is spoken of as "uniformly difform terminated in both extremes at some degree," so that the more intense extreme is designated in the acute angle *C* and the more remiss in the obtuse angle *D*. The superior line, e.g., line *CD*, is called "the line of summit," or in relation to quality it can be called "the line of intensity" because the intensity varies according to its variation.

And so every uniform quality is imagined by a rectangle and every quality uniformly difform terminated at no degree is imaginable by a right triangle. Further, every quality uniformly difform terminated in both extremes at some degree is to be imagined by a quadrangle having right angles on its base and the other two angles unequal. Now every other linear quality is said to be "difformly difform" and is imaginable by means of figures otherwise disposed according to manifold variation. Some modes of the "difformly difform" will be examined later. The aforesaid differences of intensities cannot be known any better, more clearly, or more easily than by such mental images and relations to figures, although certain other descriptions or points of knowledge could be given which also become known by imagining figures of this sort: as if it were said that a uniform quality is one which is equally intense in all parts of the subject, while a quality uniformly difform is one in which if any three points [of the subject line] are taken, the ratio of the distance between the first and the second to the distance between the second and the third is as the ratio of the excess in intensity of the first point over that of the second point to the excess of that of the second point over that of the third point, calling the first of those three points the one of greatest intensity.

Let us clarify this first with respect to a quality uniformly difform which is terminated at no degree and which is designated or imagined by ΔABC [see Fig. 7(a)]. With the three perpendicular lines *BC*, *FG*, and *DE* erected, then let *HE* be drawn parallel to line *DF* and similarly *GK* parallel to line *FB*. Therefore, the two small triangles *CKG* and *GHE* are formed and they are equiangular. Hence, by [proposition] VI.4 of [the *Elements* of] Euclid,[1] $GK/EH = CK/GH$,

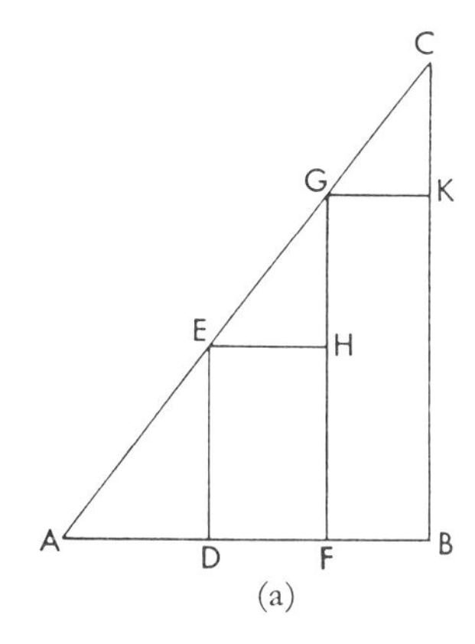

(a)

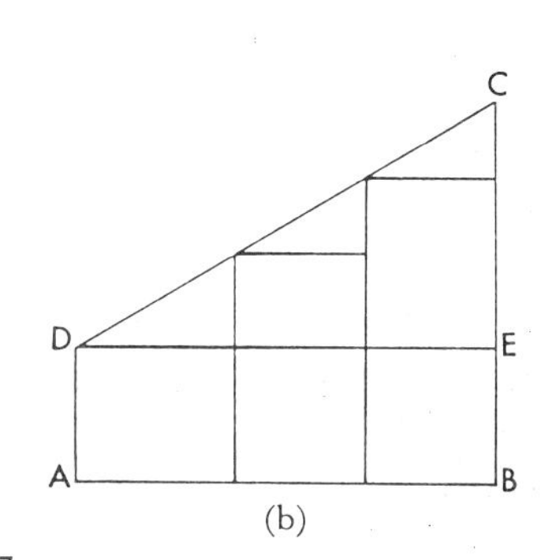

(b)

Fig. 7
Figure (a) in MSS *BLEDGSC*, with letters *B* and *C* interchanged in MS *B*. Figure (a) is reversed in orientation in MS *E*. Figure (b) in MSS *BLDCSG*.

CK and *GH* being excesses. And since *GK* = *FB* and similarly *EH* = *DF*, so *FB*/*DF* = *CK*/*GH*, *FB* and *DF* being the distances on the base of the three points and *CK* and *GH* being the excesses of altitude proportional to the intensity of these same points. Since, therefore, the quality of line *AB* is such that the ratio of the intensities of the points of the line is as the ratio of the altitudes of the lines perpendicularly erected on those same points, that which has been proposed is evidently clear, namely that the ratio of the excess in intensity of the first point over the second to the excess of the second over the third is the same as the ratio of the distance between the first and second points to the distance between the second and the third, and similarly for any other three points. Hence what we have premised in regard to a quality difform in this way is quite fitting, and so it (this quality) was well designated by such a triangle.

By the same method the aforesaid description of property can be demonstrated for a quality uniformly difform terminated in both extremes at [some] degree, and thus for one which we let be imagined by quadrangle *ABCD* in which line *DE* is drawn parallel to base *AB* forming Δ*DEC* [see Fig. 7(b)]. Then let lines of altitude be drawn in the quadrangle and also transversals parallel to the base in this triangle, thus forming small triangles. And then one can easily argue concerning the excesses and the distances in this triangle just as was argued in the other one. This will be easily apparent to one who is observant.

Further, every quality which is disposed in [any] other way than those described earlier is said to be "difformly difform." It can be described negatively as a quality which is not equally intense in all parts of the subject nor in which, when any three points of it are taken, the ratio of the excess of the first over the second to the excess of the second over the third is equal to the ratio of their distances.

NOTE

1. See the Commentary, I. viii, lines 25, 27.

55. From *Questiones super geometriam Euclidis*[1*]

(The Latitude of Forms)

NICOLE ORESME

QUESTION 10

Consequently it is sought whether some quadrangular surface is uniformly difform in altitude.

It is argued in the negative: for no altitude is difformly uniform, therefore no altitude is uniformly difform. The consequence holds by analogy. The antecedent is evident, for in that which is uniform or equal there is no difformity or inequality.

The opposite is argued: there is some uniform altitude, therefore there is some uniformly difform altitude.

In the first place we must consider the question under inquiry. Then secondly we must apply it to the matter as concerned with mean qualities.

In connection with the first, it is to be known that the altitude of a surface is measured by a perpendicular line lying directly upon the base, as can be evident in a figure [see Fig. 1]. Secondly, it is to be noted that a surface is said to be uniformly and equally high when all the lines by which the altitude is measured are equal; it is said to be difformly high when they are unequal and they rise to a line which is not parallel to the base. Thirdly, it is to be noted that an altitude is said to be uniformly difform when any three or more of the lines which are at equal distances apart exceed one another according to arithmetic proportion, *i.e.*, by the amount that one line exceeds the second, so the second exceeds the third [see Fig. 2]. From this it is evident that the upper line limiting them [*i.e.*, the perpendiculars] is a straight line not parallel to the base. Fourthly, it is to be noted that an altitude is said to be difformly difform when the [perpendicular] lines do not

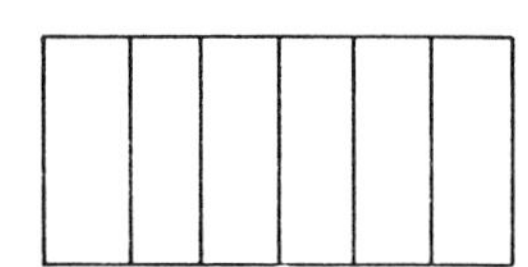

Fig. 1
In *c* here and in *s* on f. 107, c. 2 and labelled there *uniformiter uniffornis*.

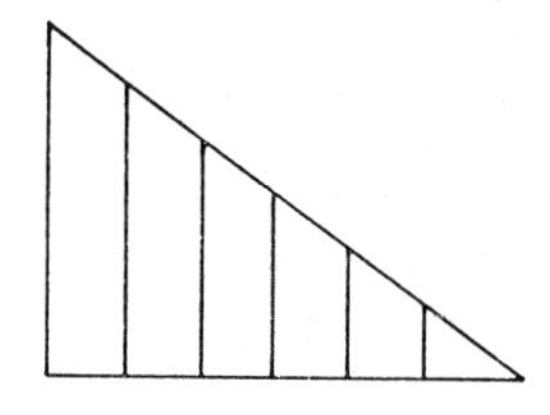

Fig. 2
In *c* here and in *s* on f. 107, c. 2 and labelled there *uniformiter difformis*.

Source: From Marshall Clagett, trans. and ed., *Nicole Oresme and the Medieval Geometry of Qualities and Motions* (Madison: The University of Wisconsin Press; © 1968 by The Board of Regents of the University of Wisconsin System), 527, 529, 531, 533, and 535. Professor Marshall Clagett, School of Historical Studies, Institute for Advanced Study, Princeton, New Jersey 08549.

exceed one another in this manner; and in such a case the line crossing through their summits is not a straight line [see Fig. 3]. And the difformity in altitude varies according to the variation of such a [summit] line.

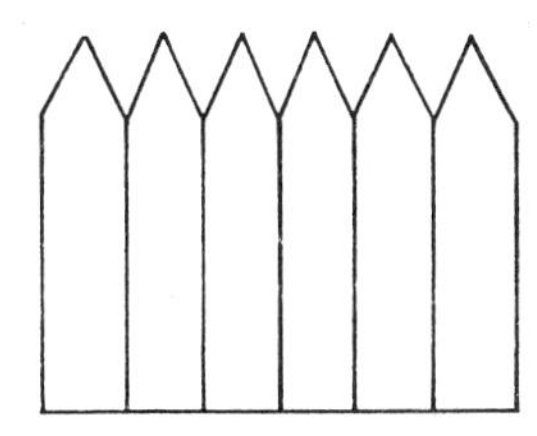

Fig. 3
In *s* on f. 107r, c. 2 and labelled *difformiter difformis*.

As for the second part, namely the mathematical mean which is in qualities and velocities, it is to be noted firstly that in quality two things are to be imagined, namely intensity according to degrees and extension through the subject; and therefore such a quality is imagined to have two dimensions.[2] Accordingly we sometimes say that it has "latitude," understanding by this, "intensity," on the ground that we understand its "extension" by the term "longitude." [Hence every latitude presupposes longitude.][3] Secondly it is to be noted that a quality can be imagined to reside in a point, or in an indivisible subject like a soul. It can also be imagined to be in a line, as well as in a surface or in a body.

1. Hence let this be the first conclusion, that the quality of a point or an indivisible subject is to be imagined as a line, for it has only one dimension, namely intensity. From this it follows that such a quality, like knowledge or virtue, ought not to be described as either "uniform" or "difform," just as a line is not properly said to be "uniform" or "difform." It follows also that one speaks improperly of a latitude of knowledge or virtue since no longitude is to be imagined there and every latitude presupposes longitude.

2. The second conclusion is that the quality of a line is to be imagined as a surface whose longitude is the rectilinear extension of the subject and whose latitude is its intensity which is imagined by lines perpendicular to the line which is the subject.

3. The third conclusion is that the quality of a surface is to be imagined, using a similar imagery, by means of a body whose longitude and latitude constitute the extension of the subject and whose depth is the intensity of the quality. And by like reasoning the quality of a whole body would have to be imagined as a body whose longitude and latitude would be the extent of the whole body and the depth its intensity.[4] But someone may raise a doubt: if the quality of a point is imagined as a line, the quality of a line as a surface, and that of a surface as a body having three dimensions, therefore the quality of a body will be imagined to have four dimensions and be in another genus of quantity. I answer that such is not necessary, for just as a flowing point imaginatively produces a line, a line a surface, a surface a body, so if a body were imagined to flow it is not necessary for it to produce a fourth kind of quantity but in fact only a body. And it is on this account that Aristotle says in the first book of the *On the Heaven*[5] that from this, *i.e.*, from a body, no passage to another genus of quantity takes place by this method of imagining. One ought to speak similarly in the matter at hand. Hence one ought to speak [thus] of the quality of this line and similarly of the quality of a surface and of a body.

4. The fourth conclusion is that a uniform linear quality is to be imagined by a rectangle that is uniformly high, so that the extension is imagined by the base[6] and the intensity is measured by a [summit] line parallel to it, as is evident in the figure [see Fig. 4] [and it is obvious, for just as any line which would be erected on the given [base] line would be equal to another, so any point there would be imagined as equally intense].

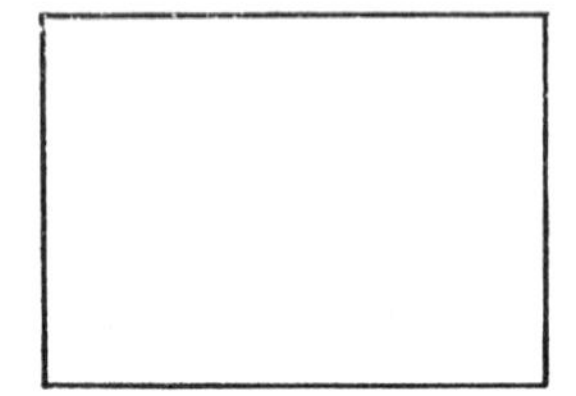

Fig. 4
In *c* and *s* (in *s* the figure is that already given earlier with altitude lines drawn).

But a quality uniformly difform is to be imagined by a surface which would be uniformly difformly high, so that the line of altitude [*i.e.*, the summit line,] would not be parallel to the base, as is evident in the figure [see Fig. 5]; still it would be a straight line. This can be proved.

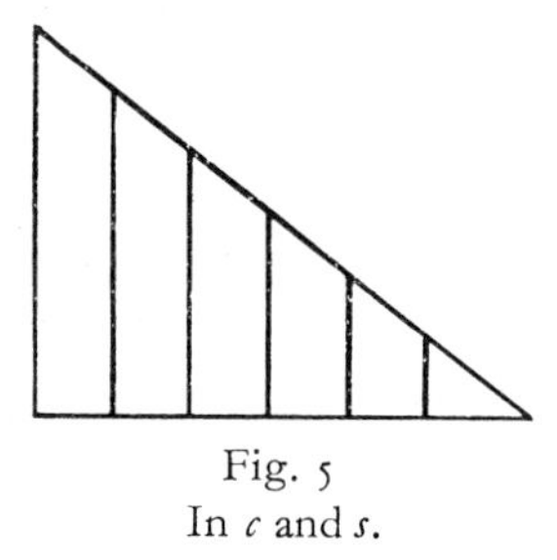

Fig. 5
In *c* and *s*.

The ratio in intensity of [any] points would be as the ratio in altitude of the perpendicular lines on these points. And this can be in two ways just as a surface uniformly difform in altitude can exist in two ways. In one way such a quality is terminated at no degree [*i.e.*, zero] and then it is like a surface uniformly difformly high [terminated in one extreme] at no degree, *i.e.*, like a triangle. Or [it is terminated] on both sides at a degree; in this case it is like a quadrangle whose [line of] altitude [*i.e.*, summit line] would be a straight line not parallel to the base [see Fig. 6].

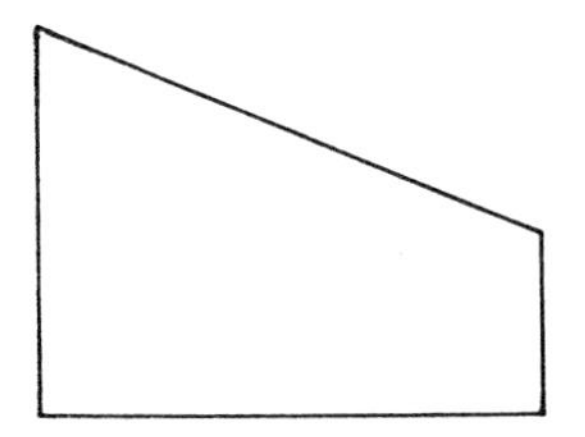

Fig. 6
Not in *s* or *c*. I have added the figure.

5. The penultimate [conclusion] is that from this latter together with the aforesaid it can be proved that a quality uniformly difform is equal to the middle degree, *i.e.*, that it would be just as great in quantity as if it were uniform at the middle degree. And this can be proved as for a surface [see Fig. 7].[7]

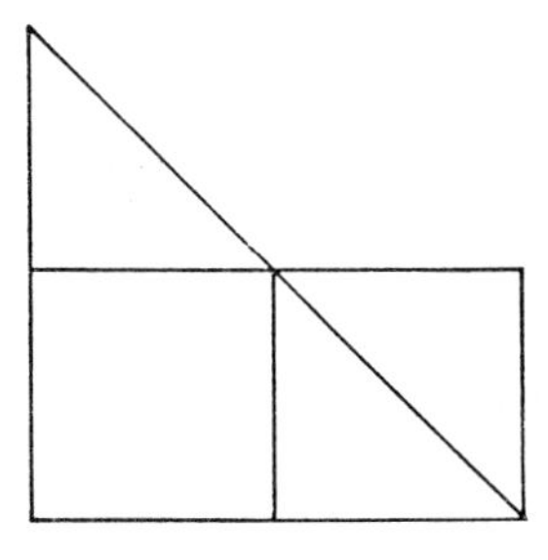

Fig. 7
Neither *c* nor *s* has it here, but *s* has it for essentially the same proposition in Q. 15, *q.v.*; *c* does have a right triangle and below a confused figure which perhaps had its source in a figure like this.

6. The last conclusion is that a quality difformly difform is to be imagined as a surface whose subject line would be the base and whose altitude [*i.e.*, summit line,] would be a line which is neither straight nor parallel to the base. From this it is evident that such difformity can be imagined in almost an infinitude of ways according as this line of altitude [*i.e.*, summit line,] can be multiply varied, as is evident in the figure [see Fig. 8].

But one might say: "Master, it is not necessary for it to be so imagined." I answer that the imagination [*i.e.*, imagery,] is a good one. This is evident by Aristotle who imagines time by means

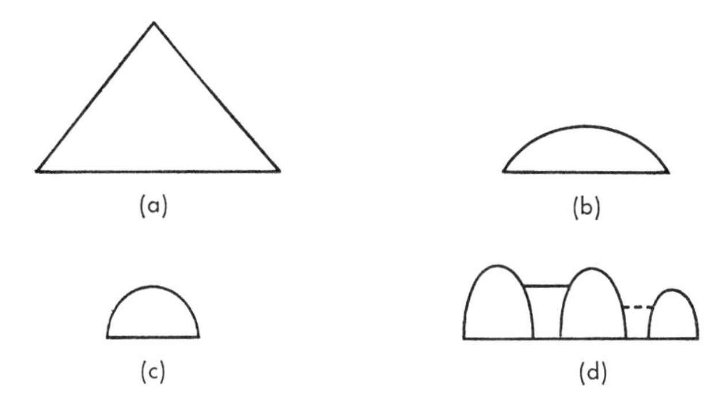

Fig. 8
(a), (b), and (c) are in *s* (along with three rectangular figures). (d) is in *c* (along with a rectangular figure).

of a line. Similarly in perspective it is expressly imagined that active force is to be imagined by means of triangular surfaces. Further, following this imagination I can more easily understand those things which are said about qualities uniformly difform and so on. Therefore, I say that the imagination is a good one.

NOTES

1. The full title of the work as given by Busard in his edition is: *Quaestiones super geometriam Euclidis per Magistrum Nicholaum Oresme Probum Philosophum et solemnem disputate Parisius.*

2. This description of qualities as having two dimensions should be compared with Oresme's brief statement in the *Questiones super libros de generatione et corruptione* quoted above in Introduction II.A, fn. 18.

3. The bracketed phrase added from *s* seems to complete the thought of the preceding phrase, but it may have crept in here from its clearly genuine place in line 40.

4. Cf. *De configurationibus,* I.iv, where this is treated much more clearly. Here in the *Questiones* he seems to say that the volume of the subject can somehow be reduced to the longitude and latitude of the imaginative body used to represent the quality of the original body, thus leaving the depth of the imaginative body to represent intensity. In the *De configurationibus,* it is clear that each of the infinite parallel planes making up the body can be imagined to have a body erected on it, so that we have a forest of interlacing bodies that represent the quantity of the quality. Here again we see evidence of a more mature consideration of a subject in the *De configurationibus.*

5. *De caelo,* Bk. I, 268a 30—b 3. In the Moerbeke translation accompanying Thomas Aquinas' *Expositio* (Turin, Rome, 1952), 8, Text No. 4, the passage reads:

> Sed illud quidem palam, quoniam non est in aliud genus transitio, quemadmodum ex longitudine in superficiem, in corpus autem ex superficie: non enim adhuc talis perfecta erit magnitudo.

Thomas' commentary is even more instructive for Oresme's treatment (*ibid.,* 11):

> Tertium est manifestum ex praemissis, scilicet quod non fit transitus a corpore in aliud genus magnitudinis, sicut fit transitus ex longitudine in superficiem, et ex superficie in corpus. Et utitur modo loquendi quo utuntur geometrae, imaginantes quod punctus motus facit lineam, linea vero mota facit superficiem, superficies autem corpus. A corpore autem non fit transitus ad aliam magnitudinem: quia talis exitus, sive processus, ad aliud genus magnitudinis, est secundum defectum eius a quo transitur (unde etiam motus naturalis est actus imperfecti). Non est autem possibile quod corpus, quod est perfecta magnitudo, deficiat secundum hanc rationem, quia est continuum secundum omnem modum: et ideo non potest fieri transitus a corpore in aliud genus magnitudinis.

6. In Introduction II, I have already remarked that for Oresme the base line, as well as the intensity line, was an *ymaginatio,* but,

of course, the base line as an extension is abstracted from the extension of the subject, while the intensity perpendicular as a line is not abstracted from the intensity of the quality since intensity is not essentially extended. This is why Oresme often calls the base line the subject, as if the two were identical. . . .

7. Note that a corrupt proof appears in manuscript *s* and is included in the variant readings. This proof was not, I believe, in the original text, which seems to have said that the proposition "can (*i.e.*, will be able to) be proved as of a surface." Although the addition in *s* appears hopelessly corrupt, I believe that it means to divide the combined triangle and rectangle into five equal triangles so that the four small triangles of the right triangle are equal to the four small triangles of the rectangle, as in the accompanying figure.

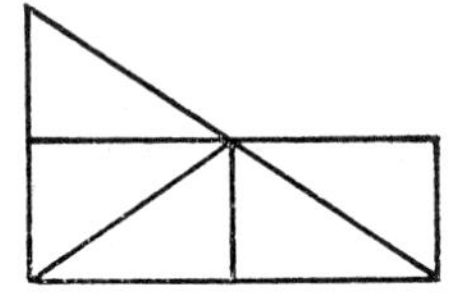

Chapter V
The Medieval–Renaissance–Reformation Periods in Europe

GIROLAMO CARDANO (1501-1576)

Cardano, a complex man of the Italian Renaissance who possessed a powerful intellect, childlike credulity, and strong superstition, was also a central figure in the early phase of the Scientific Revolution. Indeed, the significance of his contributions to mathematics suggests that he ranks with Copernicus (astronomy), Vesalius (anatomy), and Paracelus (alchemy and medicine) as one of the founders of that revolution.

Cardano was the illegitimate son of Fazio Cardano, a learned jurist, and the ignorant and irascible Chiara Mischeri. He was often ill and mistreated by his mother, but his father (a friend of Leonardo da Vinci) urged him to study the classics, mathematics, and astrology. In 1520 he began higher studies at the University of Pavia but completed them in 1526 at Padua, where he graduated in medicine.

Beginning in poverty, Cardano rose to fame only after many difficulties. Throughout his career he suffered from delusions augmented by public condemnation (including torture). He first practiced medicine in Saccolongo (near Padua) from 1526 to 1532. Despite continuing economic hardship, he later described this six-year period as the happiest time of his life. In 1531 he married Lucia Bandareni; they had two sons and a daughter.

Through the intercession of the Milanese senator Sfondrato, whose son he cured, Cardano was allowed to practice medicine in Milan from 1534 to 1543, eventually becoming the city's most famous physician. By the late 1530s, he ranked second only to Vesalius as an authority in theoretical medicine in Europe, and this made many of his Milanese colleagues envious. Besides practicing medicine, Cardano taught mathematics, astronomy, dialectics and Greek, at the "piattine" schools in Milan. Clearly, mathematics was not yet a profession that could provide employment for its practitioners. He displayed a mastery of calculation and a confidence in dealing with algebraic equations in his *Practica arithmetice* ("Practice of Mathematics," 1539). In what became a celebrated affair, Cardano enticed Niccolò Tartaglia to divulge his newly found general solution for cubic equations. Though Tartaglia swore him to secrecy, Cardano published the solution in his monumental book *Artis magnae* (1545), which credited Tartaglia and Ludovico Ferrari with the solution. In a time of widespread intrigue when secrecy was the norm regarding new mathematical discoveries, Tartaglia accused Cardano of pirating his ideas.

Cardano was appointed professor of medicine at the University of Pavia in 1543, and his massive *De Subtilitate* ("The Subtlety of Things"), with sections adumbrating natural science, was published in 1550. An animistic account of the organic and inorganic realms including cosmology, technology, astrology, alchemy, the occult, and natural magic, this book enhanced his reputation beyond the bounds of medicine. The author's assertion that mathematics had only a limited range of application in the natural world provoked a debate with the humanist Julius Caesar Scaliger, who claimed a more general application and more subtle nature for mathematics.

Although technically a professor at Pavia until 1560, Cardano traveled about Western Europe from 1552 to 1559, because many of its crowned heads and nobility solicited his medical services. In 1552, he went to Scotland to treat the asthma of Archbishop John Hamilton of St. Andrews. He journeyed on to the court of Edward VI of England and then to Paris, where he defended the "heresies" of Vesalius in discussions with the university's medical faculty. In 1560, at the height of his fame and prosperity, he was banished from Pavia following a series of personal calamities. The worst involved his elder and favorite son, Giovanni, who was tried and executed for poisoning his wife. Cardano was crushed. His medical practice declined, and his rivals criticized his work. He now turned to gambling (in the forms of dice, chess, and cards) and imagining terrible punishments for his critics.

In 1562 Cardano found it advisable to accept a professorship at the University of Bologna, where he lived in moderate comfort until 1570. In that year the Inquisition imprisoned him on a charge of heresy stemming from his casting a horoscope of Jesus. This, together with his bent for other modes of free thought, barred him from further teaching or publishing. Through the intercession of several influential churchmen, he was released by the Inquisition and went to Rome, where Pope Pius V granted him a small pension for his services as an astrologer. He spent his final years writing commentaries and preparing his autobiography, *De Vita Propria* ("The Book of My Life"). This was to be the last of over 200 treatises on medicine, mathematics, physics, philosophy, religion, and music.

Cardano's reputation as the leading mathematician of his time rests chiefly on his contributions to algebra and probability. Algebra then was sometimes referred to as *ars magna* ("the great art") in contradistinction to the lesser art of arithmetic. This accounts for the title of his major book, *Artis magnae*, which contained the first published account of general solutions for cubic and quartic equations by means of radical expressions. (His disciple and son-in-law, Ludovico Ferrari, was credited with having discovered the latter solution.) Cardano also advanced novel ideas about relations between the roots and coefficients of equations. In solving for roots, he followed custom by accepting positive rational and irrational numbers as real and considering negative numbers to be fictitious. In addition, he calculated formally with imaginary numbers but did not accept them. The *Artis magnae* substantially influenced the rapid growth of algebra in Europe.

Cardano's *Liber de ludo alaea* ("Book on Games of Chance") was a pioneering attempt to establish a theory of probability a century before similar investigations by Fermat and Pascal. Since it was only published posthumously in 1663 (after their work), it had little influence. In the book Cardano stated the law of large numbers and gave the so-called "power law"—that is, if p is the probability of an event, then the probability of the event being repeated n times is p^n. In mechanics, he fervently admired Archimedes, studied the lever and inclined plane, and affirmed the impossibility of perpetual motion except in the heavens.

56. From the *Artis Magnae**

GIROLAMO CARDANO

CHAPTER XI ON THE CUBE AND FIRST POWER EQUAL TO THE NUMBER

Scipio Ferro of Bologna well-nigh thirty years ago discovered this rule and handed it on to Antonio Maria Fior of Venice, whose contest with Niccolò Tartaglia of Brescia gave Niccolò occasion to discover it. He [Tartaglia] gave it to me in response to my entreaties, though withholding the demonstration.† Armed with this assistance, I sought out its demonstration in [various] forms. This was very difficult. My version of it follows.

DEMONSTRATION

For example, let GH^3 plus six times its side GH equal 20, and let AE and CL be two cubes the difference between which is 20 and such that the product of AC, the side [of one], and CK, the side [of the other], is 2, namely one-third the coefficient of x. Marking off BC equal to CK, I say that, if this is done, the remaining line AB is equal to GH and is, therefore, the value of x, for GH has already been given as [equal to x].

†In 1539, Tartaglia gave Cardano his method in obscure verses that begin as follows:

Quando che'l cubo con le case appresso
Se ogguaglia a qualche numero discreto
Trovan dui altri, differenti in esso . . .

When x^3 together with px
Are equal to a q
Then take u and v, $u \neq v$

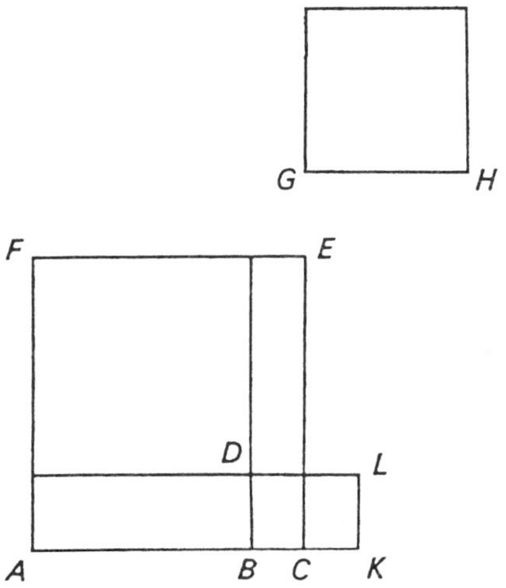

In accordance with the first proposition of the sixth chapter of this book, I complete the bodies DA, DC, DE, and DF; and as DC represents BC^3, so DF represents AB^3, DA represents $3(BC \times AB^2)$ and DE represents $3(AB \times BC^2)$.[1] Since, therefore, $AC \times CK$ *equals 2*, $AC \times 3CK$ will equal 6, the coefficient of x; therefore $AB \times 3(AC \times CK)$ makes $6x$ or $6AB$,[2] wherefore three times the product of AB, BC, and AC is $6AB$. Now the difference between AC^3 and CK^3—manifesting itself as BC^3, which is equal to this by supposition—is 20, and from the first proposition of the sixth chapter is the sum of the bodies DA, DE, and DF. Therefore these three bodies equal 20.

**Source:* From Girolamo Cardano, *The Great Art or Rules of Algebra* trans. by T. Richard Witmer (1968), 96-99 and 218-220. Reprinted by permission of the M.I.T. Press. Footnotes renumbered.

Now assume that BC is negative:

$$AB^3 = AC^3 + 3(AC \times CB^2) + (-BC^3) + 3(-BC \times AC^2),$$

by that demonstration. The difference between $3(BC \times AC^2)$ and $3(AC \times BC^2)$, however, is [three times] the product of AB, BC, and AC. Therefore, since this, as was demonstrated, is equal to $6AB$, add $6AB$ to the product of $3(AC \times BC^2)$, making $3(BC \times AC^2)$. But since BC is negative, it is now clear that $3(BC \times AC^2)$ is negative and the remainder which is equal to it is positive. Therefore,

$$3(CB \times AB^2) + 3(AC \times BC^2) + 6AB = 0.$$[3]

It will be seen, therefore, that as much as is the difference between AC^3 and BC^3, so much is the sum of

$$AC^3 + 3(AC \times CB^2) + 3(-CB \times AC^2) + (-BC^3) + 6AB.$$

This, therefore, is 20 and, since the difference between AC^3 and BC^3 is 20, then, by the second proposition of the sixth chapter, assuming BC to be negative,

$$AB^3 = AC^3 + 3(AC \times BC^2) + (-BC^3) + 3(-BC \times AC^2).$$

Therefore since we now agree that

$$AB^3 + 6AB\text{[4]} = AC^3 + 3(AC \times BC^2) + 3(-BC \times AC^2\text{[5]}) + (-BC^3) + 6AB,$$

which equals 20, as has been proved, they [*i.e.*, $AB^3 + 6AB$] will equal 20. Since, therefore,

$$AB^3 + 6AB = 20,$$

and since

$$GH^3 + 6GH = 20,$$

it will be seen at once and from what is said in I, 35 and XI, 31 of the *Elements* that GH will equal AB. Therefore GH is the difference between AC and CB. AC and CB, or AC and CK, the coefficients, however, are lines containing a surface equal to one-third the coefficient of x and their cubes differ by the constant of the equation. Whence we have the rule:

RULE

Cube one-third the coefficient of x; add to it the square of one-half the constant of the equation; and take the square root of the whole. You will duplicate[6] this, and to one of the two you add one-half the number you have already squared and from the other you subtract one-half the same. You will then have a *binomium* and its *apotome*. Then, subtracting the cube root of the *apotome* from the cube root of the *binomium*, the remainder [or] that which is left is the value of x.[7]

For example,

$$x^3 + 6x = 20.$$

Cube 2, one-third of 6, making 8; square 10, one-half the constant; 100 results. Add 100 and 8, making 108, the square root of which is $\sqrt{108}$. This you will duplicate: to one add 10, one-half the constant, and from the other subtract the same. Thus you will obtain the *binomium* $\sqrt{108} + 10$ and its *apotome* $\sqrt{108} - 10$. Take the cube roots of these. Subtract [the cube root of the] *apotome* from that of the *binomium* and you will have the value of x:

$$\sqrt[3]{\sqrt{108} + 10}\text{[8]} - \sqrt[3]{\sqrt{108} - 10}$$

In modern notation the chapters beginning with eleven deal with the following problems:

11. $x^3 + ax = b$
12. $x^3 = ax + b$
13. $x^3 + a = bx$
14. $x^3 = ax^2 + b$
15. $x^3 + ax^2 = b$
16. $x^3 + a = bx^2$
17. $x^3 + ax^2 + bx = c$
18. $x^3 + ax = bx^2 + c$
19. $x^3 + ax^2 = bx + c$
20. $x^3 = ax^2 + bx + c$
21. $x^3 + a = bx^2 + cx$
22. $x^3 + ax + b = cx^2$
23. $x^3 + ax^2 + b = cx$

24. On 44 derivative equations, including $x^6 + 6x^4 = 100$
25. On imperfect and particular rules

Chapter 26 and subsequent chapters also examine biquadratic equations.

The following example, Problem III in Chapter 37, involves imaginary numbers. Modern notation has been substituted into the problem.

CHAPTER XXXVII

PROBLEM III

Likewise, if I say I have 12 *aurei* more than Francis and the square of mine is 128 more than the cube of Francis' *aurei,* we let Francis have $-x$ and I have 12 *aurei* minus x. The square of mine is $144 + x^2 - 24x$, and this is equal to $-x^3 + 128$. Therefore

$$16 + x^2 + x^3 = 24x,$$

wherefore x is 4, and so much in the negative is what Francis lacks. I have 8 *aurei* of my own.

RULE II

The second species of negative assumption involves the square root of a negative. I will give an example: If it should be said, Divide 10 into two parts the product of which is 30 or 40, it is clear that this case is impossible. Nevertheless, we will work thus: We divide 10 into two equal parts, making each 5. These we square, making 25. Subtract 40, if you will, from the 25 thus produced, as I showed you in the chapter on operations in the sixth[9] book, leaving a remainder of -15, the square root of which added to or subtracted from 5 gives parts the product of which is 40. These will be $5 + \sqrt{-15}$ and $5 - \sqrt{-15}$.

DEMONSTRATION

In order that a true understanding of this rule may appear, let *AB* be a line which we will say is 10 and which is divided into two parts, the rectangle based on which must be 40. Forty, however, is four times 10; wherefore we wish to

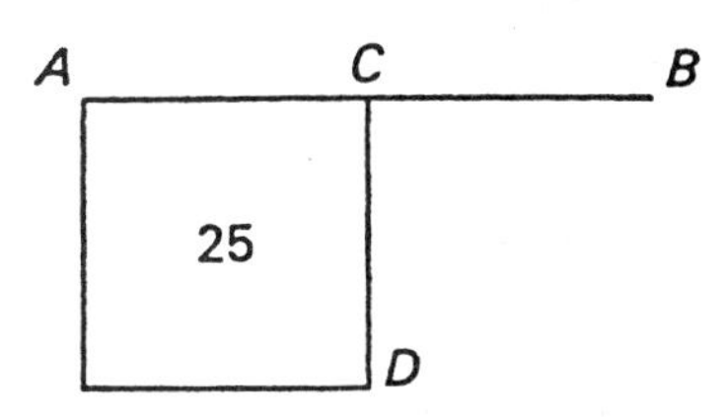

quadruple the whole of *AB*. Now let *AD* be the square of *AC*, one-half of *AB*, and from *AD* subtract 4*AB*, ignoring the number.[10] The square root of the remainder, then—if anything remains—added to or subtracted from *AC* shows the parts. But since such a remainder is negative, you will have to imagine $\sqrt{-15}$—that is, the difference between *AD* and 4*AB*—which you add to or subtract from *AC*, and you will have that which you seek, namely $5 + \sqrt{25 - 40}$ and $5 - \sqrt{25 - 40}$, or $5 + \sqrt{-15}$ and $5 - \sqrt{-15}$. Putting aside the mental tortures involved,[11] multiply $5 + \sqrt{-15}$ by $5 - \sqrt{-15}$, making $25 - (-15)$ which is $+15$. Hence this product is 40. Yet the nature of *AD* is not the same as that of 40 or of *AB*, since a surface is far from the nature of a number and from that of a line, though somewhat closer to the latter. This truly is sophisticated,[12] since with it one cannot carry out the operations one can in the case of a pure negative and other [numbers]. [Likewise,] one cannot determine what it [the solution] is by adding the square of one-half the [given] number to the number to be produced and to or from the square root of this sum adding and subtracting half that which is to be divided.[13] For example, in this case you could divide 10 into two parts whose product is 40; add 25, the square of one-half of 10, to 40, making 65; from the square root of this subtract 5 and also add 5 to it; you then have parts with the likeness of $\sqrt{65} + 5$ and $\sqrt{65} - 5$. [But] while these numbers differ by 10, their sum is $\sqrt{260}$, not 10. So progresses arithmetic subtlety the end of which, as is said, is as refined as it is useless.

NOTES

1. 1570 and 1663 vary considerably from here on to the end of the demonstration. They read:

We will have, therefore, four propositions, two of which have already been mentioned—namely, that $AC \times CK$ or CB is 2 and that the difference between AC^3 and CB^3 is 20. The third can be deduced from these and is that, since the product of $3AB \times BC \times AC$ is equal to the sum of [the text has, "the difference between"] DE and DA and that $3AB \times AC \times BC$ is $6AB$ for, from the first proposition, the product of AC and CB is 2, therefore three times this is 6, and this product times AB is $6AB$. This, however, is the sum of [the text has, "the difference between"] DE and DA. The fourth [proposition], which derives from the second and third corollaries in the sixth chapter, [is] that DF [*i.e.*, AB^3] is the difference between $AC^3 + 3(AC \times CB^2)$ and $CB^3 + 3(CB \times AC^2)$. Let, therefore, α be AC^3, β BC^3 [the text has ABC^3], γ $3CB \times AC^2$, δ $3AC \times CB^2$, ϵ the difference between α and β, ζ the difference between γ and δ, and η [1570 has β; 1663's character is illegible] the difference between $\alpha + \delta$ and $\beta + \gamma$. Therefore [there is what appears to be a superfluous *cum* inserted at this point] ϵ is composed of $\zeta + \eta$ [1570 again has a β and again 1663's character is illegible], as can readily be demonstrated numerically and by example as shown in the margin. But ϵ is 20, from the second assumption, ζ is $6AB$, and η [1570's character is illegible and 1663 has θ] is AB^3. Therefore, $AB^3 + 6AB$—that is, plus $6x$, for AB is the root of its cube—is equal to 20. Therefore, since $GH^3 + 6GH$ [the text has $BH^3 + 6BH$ here and the next place it occurs] is equal to 20, $GH^3 + 6GH$ will be equal to $AB^3 + 6AB$. Hence AB is x and this is the difference between two sides the product of which is 2 and the cubes of which differ by 20, which was to be demonstrated. From this we construct the rule.

$$20\left|\begin{matrix}24 & 1\\ & \\ 4\,. & 14\end{matrix}\right| 13 \left|\begin{matrix}25\\ \\ 18\end{matrix}\right| 7$$

As it is obvious from the preceding the text of 1570 and 1663 is quite corrupt. These items are especially bothersome:

(1) the *differentiae DE et DA* which occurs twice. To make sense, we have either to assume, as I have here, that *differentia* is an error for *aggregatum* or that, in the earlier definitions of DA as *triplus CB in quadratum AB* and of DE *as triplus, AB in quadratum BC, AB* should be replaced by AC. Either is consistent with the later development of the argument.

(2) the *cubus abc* which I read as a typographical error for *cubus bc*.

(3) the confusion between eta, beta, and theta at various points.

(4) the misprinting of BH for GH at four places.

2. *fiunt 6 res AB, seu sexcuplum AB.*

3. *faciunt nihil.*

4. The text has a spare *cum* at this point, as though something more were to be added.

5. 1545 has AB^2.

6. 1545 has *seminabis;* 1570 and 1663 have *servabis*. The former, corrected to read *geminabis* in accord with later passages, is followed here.

7. *I.e.*, if $x^3 + ax = N$, $x = \sqrt[3]{\sqrt{(a/3)^3 + (N/2)^2} + N/2} - \sqrt[3]{\sqrt{(a/3)^3 + (N/2)^2} - N/2}$.

8. 1570 and 1663 have ℞ *b: cub:* ℞ 108 *p:* 10, the *b:* being a misprint for *v:.*

9. 1570 and 1663 have "fourth."

10. *absque numero.*

11. *dimissis incruciationibus.* We may perhaps suspect Cardano of indulging in a play on words here, for this can also be translated "the cross-multiples having canceled out," with the whole sentence then reading, "Multiply $5 + \sqrt{-15}$ by $5 - \sqrt{-15}$ and, the cross-multiples having canceled out, the result is $25 - (-15)$, which is $+15$." *Cf.* the translation of this passage by Professor Vera Sanford in David Eugene Smith, *A Source Book in Mathematics* (New York, 1959 reprint), I, 202.

12. In the 38th problem of the *Ars Magna Arithmeticae,* Cardano remarks: "Note that $\sqrt{9}$ is either a $+3$ or -3, for a plus [times a plus] or a minus times a minus yields a plus. Therefore $\sqrt{-9}$ is neither $+3$ nor -3 but is some recondite third sort of thing" *(quaedam tertia natura abscondita).*

13. *quae vere est sophistica, quoniam per eam, non ut in puro m: nec in aliis, operationes exercere licet, nec venari quid sit* [1663 at this point puts a period and inserts *Modus*] *est ut addas quadratum medietatis numeri numero producendo, et a* ℞ *aggregati minuas ac addas dimidium dividendi.*

Chapter V

The Medieval–Renaissance–Reformation Periods in Europe

FRANÇOIS VIÈTE (also called Franciscus Vieta; 1540–1603)

Viète, the son of a lawyer and notary, studied law at the University of Poitiers. After receiving a bachelor's degree in 1560, he began a legal practice in his native city of Fontenay. From 1564 to 1570 he tutored Catherine de Parthenay, the daughter of Antoinette d'Anberterre, who remained a lifelong confidante. In 1571 he went to Paris, where Charles IX appointed him legal advocate to the city's *parlement*. From 1573 to 1579, he was legal counselor to the *parlement* of Brittany at Rennes. Henry III, who put him in charge of special missions in 1576, named him master of request in Paris and a member of the royal privy council in 1579.

A Huguenot sympathizer in a time of religious persecution, Viète was the victim of political enemies who succeeded in having him banished from court from 1584 to 1589. Henry III recalled him to serve at Tours in 1589. He continued in that role after the accession of Henry IV.

As his wealth increased, Viète spent more of his leisure hours studying cryptography and mathematics. During the war with Spain in 1589 and 1590, he successfully decoded for Henry IV intercepted letters containing a complex cipher of over 400 characters. Philip II of Spain, convinced that this work could not be done by natural means, complained to the pope that the French were using black magic against him. Viète lived in Paris from 1594 to 1597, in Fontenay from 1597 to 1599, and again in Paris from 1599 to 1603. He requested release from service to the king because of poor health in 1602.

Viète engaged in two major scientific polemics. In the first of these disputes, he opposed the new Gregorian calendar (1582) prepared by Christof Clavius, suggesting instead a different method based directly on the Julian calendar and publishing libelous, vehement, and unjustified attacks against Clavius. In the second dispute, he was more successful. In 1592 the Italo-French humanist Joseph Justus Scaliger proffered a solution to the quadrature of the circle, the trisection of an angle, and the construction of two mean proportions between two given line segments by means of ruler and compass only. Viète proved that Scaliger's assertions were erroneous.

Viète had two highly fruitful periods of leisure and research—the first between 1564 and 1568 and the second between 1584 and 1589. His writings from these periods contain his greatest discoveries. In the first period, he completed *Harmonicum coeleste* (available in manuscript in Paris and Florence), a work that applied trigonometry to astronomy, and commenced composing his *Canon mathematicus seu ad triangula* ("Mathematical Laws Applied to

Triangles," 1579). In both, mathematics is closely connected with astronomical and cosmological research. Interestingly, Viète accepted Ptolemaic rather than Copernican astronomy because he believed that the latter was not valid geometrically. His *Canon* was probably the first text in Western Europe to develop systematically plane and spherical triangle solutions using all six trigonometric lines, which later became functions. Viète based his trigonometric tables largely on the chords and half chords of Ptolemy's *Almagest* and al-Zarqālī's *Canones.* He divided the radius of his unit circle into 60 parts. He also presented decimal fractions independently of Stevin and helped make them standard in western Europe.

During the second period, Viète prepared most of the *In artem analyticem isagoge* ("Introduction to the Analytical Art," 1591). In this—his most important book—he brought together the ancient geometrical methods of Euclid, Archimedes, Apollonius, and Pappus with the numerical algebra of Diophantus, Cardano, Tartaglia, Bombelli, and Stifel. Viète's *Isagoge,* the first book on symbolic algebra in Western Europe, resembles a modern algebra text. It uses consonants to denote known quantities, vowels to denote unknowns, and has a syncopated form for powers, where A quadratus, A cubus, . . . represents A^2, A^3, ... For his pioneering work on symbolism Viète is often called the father of modern algebraic notation.

Viète made other contributions to algebra. In 1615, he contributed to the theory of equations in the treatise *De aequationum recognitione et emendatione* ("Concerning the Recognition and Emandation of Equations"). In 1646, Franz van Schooten published Viète's collected works, but they were already relegated to a secondary status in mathematics because of the appearance of Cartesian (analytic) geometry with its more felicitous symbols. All of Viète's books except this last one were printed and distributed at his expense.

57. From *In artem analyticem isagoge**

(The New Algebra)

FRANCQIS VIÈTE

CHAPTER I. ON THE DEFINITION AND PARTITION OF ANALYSIS, AND ON THOSE THINGS WHICH ARE OF USE TO ZETETICS

In this chapter Viète refers to Pappus's distinction between analysis and synthesis, and between zetetic and poristic analysis, referring also to Euclid and Theon. There also should be, he writes, a third kind of analysis, the rhetic or exegetic,

so that there is a zetetic art by which is found the equation[2] or proportion between the magnitude that is being sought and the given things; a poristic art by which from the equation or proportion the truth of the required theorem is investigated, and an exegetic

*Source: Reprinted by permission of the publishers from *A Source Book in Mathematics, 1200-1800,* edited by D. J. Struik, Cambridge, Mass.: Harvard University Press, Copyright © 1969 by the President and Fellows of Harvard College. Regular text is Viète's; bold face text is Struik's. Headings are from Viète.

art by which from the constructed equation or proportion there is produced the magnitude itself that is being sought. And the whole threefold analytical art may be defined as the science of finding the truth in mathematics. But what truly belongs to the zetetic art is established by the art of logic through syllogisms and enthymemes,[3] of which the foundations are those very symbols[4] by which equations and proportions are obtained . . . The zetetic art, however, has its own form of proceeding, since it applies its logic not to numbers—which was the boring habit of the ancient analysis—but through a logistic which in a new way has to do with species.[5] This logistic is much more successful and powerful than the numerical one for comparing magnitudes with one another in equations, once the law of homogeneity has been established and there has been constructed, for that purpose, a traditional series or scale of magnitudes ascending or descending by their own nature from genus to genus, by which scale the degrees and genera of magnitudes in equations may be designated and distinguished.

CHAPTER II. ON THE SYMBOLS FOR EQUATIONS AND PROPORTIONS

Here Viète takes a number of postulates and propositions from Euclid, such as:

1. The whole is equal to the sum of its parts;
2. Things that are equal to the same thing are equal among themselves;
3. If equals are added to equals, the sums are equal; . . .

8. If like proportionals are added to like proportionals, then the sums are proportional;[6] . . .

15. If there are three or four magnitudes, and the product of the extreme terms is equal to either that of the middle one by itself or that of the middle terms, then these magnitudes are proportional.[7]

CHAPTER III. ON THE LAW OF HOMOGENEOUS QUANTITIES, AND THE DEGREES AND GENERA OF THE MAGNITUDES THAT ARE COMPARED

The first and supreme law of equations or of proportions, which is called the law of homogeneity, since it is concerned with homogeneous quantities, is as follows:

1. Homogeneous quantities must be compared to homogeneous quantities [*Homogenea homogeneis comparari*].

Indeed, it cannot be known how heterogeneous quantities can be affected among themselves, as Adrastus says.[8] Hence:

If a magnitude is added [*additur*] to a magnitude, it is homogeneous with it.

If a magnitude is subtracted [*subdicitur*] from a magnitude, it is homogeneous with it.

If a magnitude is multiplied [*ducitur*] by a magnitude, the result is heterogeneous with both.

Since they did not, these ancient Analysts, attend to this, the result was much obscurity and blindness.

2. Magnitudes which by their own nature ascend or descend proportionally from genus to genus are called scalars.[9]

The first of the scalar magnitudes is side or root [*latus seu radix*].[10]

The second is square [*quadratum*].

The third is cube.

The fourth is squared square [*quadrati-quadratum*].

The fifth is squared-cube . . .

The ninth is cubed-cubed-cube.

And the further ones can from here be named by this series and method . . .

The genera of the magnitudes that we have to compare so that they may be named in the order of the scalars are:

(1) Length and breadth [*longitudo, latitudo*],

(2) Plane,

(3) Solid,

(4) Plane-plane,

(5) Plane-solid . . .

(9) Solid-solid-solid.

And the further ones can be named from here by this series and method.

[*text omitted*]

5. In a series of scalars, the degree in which the magnitude stands compared to the side is called the power [*potestas*]. The other inferior scalars are called parodic[11] grades to this power.

6. The power is pure when it is free from affection. By affection is meant that a homogeneous magnitude is mixed with a magnitude of lower power together with a coefficient.[12]

To this Van Schooten adds: "A pure power is a square, cube . . . But an affected power is in the second grade: a square together with a plane composed of a side and a length or breadth;[13] in the third grade: a cube together with a solid composed of a square and a length or latitude."

7. Adjunct magnitudes which multiply scalars lower in relation to a certain power and thus produce homogeneous magnitudes are call subgradual.

Van Schooten adds: "Subgraduals are length, breadth, plane, solid, plane-plane, etc. Thus if there be a squared square with which is mixed a plane-plane which is the side multiplied with a solid, then the solid will be a subgradual magnitude, and in relation to the squared square the side will be a lower scalar."

CHAPTER IV. ON THE RULES FOR THE CALCULATION BY SPECIES [*logistica speciosa*]

Numerical calculation [*logistica numerosa*] proceeds by means of numbers, reckoning by species by means of species or forms of things, as, for instance, the letters of the alphabet.

Van Schooten adds: "Diophantus operates with numerical calculation in the thirteen books of his *Arithmetica*, of which only the first six are extant, and are now available in Greek and Latin, illustrated by the commentaries of the very erudite Claude Bachet.[14] But the calculation by species has been explained by Viète in the five books of his *Zetetics*,[15] which he has chiefly arranged from selected questions of Diophantus, some of which he solves by his own peculiar method. Wherefore, if you wish to understand with profit the distinction between the two logistics, you must consult Diophantus and Viète together." He then compares specifically certain problems of Diophantus with his and with Viète's solutions.

There are four canonical rules for the calculation by species.

RULE I

To add a magnitude to a magnitude. Take two magnitudes A and *B*. We wish to add the one to the other. But, since homogeneous magnitudes cannot be affected to heterogeneous ones, those which we wish to add must be homogeneous magnitudes. That one is greater than the other does not constitute diversity of genus. Therefore, they may be fittingly added by means of a coupling or addition; and the aggregate will be A plus B, if they are simple lengths or breadths. But if they stand higher in the scale, or if they share in genus with those that stand higher, they will be denoted in the appropriate way, say A square plus B plane, or A cube plus B solid, and similarly in further cases.

However, the Analysts are accustomed to indicate the affection of summation by the symbol $+$.

RULE II

To subtract a magnitude from a magnitude. **This leads in an analogous way to $A - B$, A square $- B$ square, A is larger than B, also to rules such as $A - (B + D) = A - B - D$; Viète writes $=$ instead of our $-$.**

RULE III

To multiply a magnitude by a magnitude. Take two magnitudes A and B. We wish to multiply the one by the other.

Since then a magnitude has to be multiplied by a magnitude they will by their multiplication produce a magnitude heterogeneous with respect to each of them; their product will rightly be designated by the word "in" or "under" [*sub*], e.g., *A* in *B*, which will mean that the one has been multiplied by the other, or, as others say, under *A* and *B*, and this simply when *A* and *B* are simple lengths or breadths.[16]

But if the magnitudes stand higher in the scale, or if they share in genus with these magnitudes, then it is convenient to add the names themselves, e.g., *A* square in *B*, or *A* square in *B* plane solid, and similarly in other cases.

If, however, among magnitudes that have to be multiplied, two or more are of different names, then nothing happens in the operation. Since the whole is equal to its parts, the products under the segments of some magnitude are equal to the product under the whole. And when the positive name [*nomen adfirmatum*] of a magnitude is multiplied by a magnitude also of positive name, the product will be positive, and negative [*negatum*] when it is negative.[17]

From which precept it follows that by the multiplication of negative names the product is positive, as when $A - B$ is multiplied by $D - G$; since the product of the positive *A* and the negative *G* is negative, which means that too much is taken away [and similarly negative *B* into positive]. Therefore, in compensation, when the negative *B* is multiplied by the negative *G* the product is positive.[18]

The denominations of the factors that ascend proportionally from genus to genus in magnitude behave, therefore, in the following way:

> A side multiplied by a side produces a square,
> A side multiplied by a square produces a cube . . .
> And conversely, a square multiplied by a side produces a cube . . .
> A solid multiplied by a solid-solid produces a solid-solid-solid,
> And conversely, and so on in that order.

RULE IV

To divide a magnitude by a magnitude. This leads in an analogous way to such expressions as $\frac{B \text{ plane}}{A}$, $\frac{B \text{ cube}}{A \text{ plane}}$ and so forth. Furthermore to add $\frac{Z \text{ plane}}{G}$ to $\frac{A \text{ plane}}{B}$; the sum will be

$$\frac{G \text{ in } A \text{ plane} + B \text{ in } Z \text{ plane}}{B \text{ in } G}.$$

To multiply $\frac{A \text{ plane}}{B}$ by Z; the result will be $\frac{A \text{ plane in } Z}{B}$.

CHAPTER V. CONCERNING THE LAWS OF ZETETICS

The way to do Zetetics is, in general, directed by the following laws:

> 1. If we ask for a length, but the equation or proportion is hidden under the cover of the data of the problem, let the unknown to be found be a side.
> 2. If we ask for a plane . . . let the unknown to be found be a square.
>
> [*text omitted*]
>
> 9. If the element that is homogeneous under a given measure happens to be combined with the element that is homogeneous in conjunction, there will be antithesis.

These laws amount to introducing (1) x, (2) x^2, (3) x^3, (4) the law of homogeneity, as in $x = ab$; and to (5) denoting the unknown by vowels *A, E,* . . . and the given magnitudes by consonants, *B, G, D,* . . . , (6) constructing $x^2 = ab + cd$, or, as Viète writes it: A square equal to *B* in *C* + *D* in *F*; (7) forming $ax \pm bx$ ("homogeneous in conjunction"); (8) forming $x^3 + ax^2 - bx^2 = c^2d + e^3$; (9) passing from $x^3 + ax^2 + bx^2 - c^2d + e^2f = g^3$ to $x^3 + ax^2 - bx^2 = c^2d - e^2f + g^3$ ("antithesis"). Then Viète continues with Propositions marked (10) to (12), which state that an equation is not changed by antithesis, by hypobibasm, and by parabolism. Hypobibasm means dividing

by the unknown, as passing from $x^3 + ax^2 = b^2x$ to $x^2 + ax = b^2$, parabolism is dividing by a known magnitude. Nos. (13) and (14) deal with the relation of equations to proportions.

These are the titles of the next chapters:

VI. Concerning the examination of theorems by means of the poristic art.

VII. Concerning the function of the rhetic art.

VIII. The notation of equations and the epilogue to the art.[19]

This chapter ends as follows:

29. Finally, the analytic art, now having been cast into the threefold form of zetetic, poristic, and exegetic, appropriates to itself by right the proud problem of problems, which is

THERE IS NO PROBLEM THAT CANNOT BE SOLVED.[20]

NOTES

1. Footnote omitted.

2. Viète writes *aequalitas,* equality, but the term "equation," now used, seems to fit the meaning better. The stress on proportion is due to the respect in which Book V of Euclid's *Elements* was held as a model whereby the contradiction between arithmetic and geometry could be overcome by rigorous mathematical reasoning.

3. An enthymeme is a syllogism incompletely stated, perhaps by leaving out the major or the minor premise; for example, in "John is a liar, therefore he is a coward," the premise, "every liar is a coward," is omitted.

4. Symbols, *symbola,* had here more the meaning of typical rules or stipulations.

5. Hence the name "logistica speciosa" for Viète's new type of calculation. The term *logistike* was used by the Greeks for the art of calculation, in contrast to *arithmetike,* number theory. Viète's term "species" is probably the translation of Diophantus' *eidos,* the term in a particular expression, primarily in reference to the specific power of the unknown it contains. See further the J. Winfree Smith translation of the *Isagoge,* pp. 21-22.

6. If $a : b = c : d$, then $(a + c) : (b + d) = a : b = c : d$.

7. If a, b, c, d are such that either $ac = b^2$ or $ad = bc$, then either $a : b = b : c$ or $a : b = c : d$.

8. Reference to a reference in Theon: "For Adrastus says that it is impossible to know how heterogeneous magnitudes may be in a ratio to one another." Who Adrastus was does not seem to be known.

9. *Scalares* means "ladder magnitudes," literally, steps or rungs of a ladder. Viète follows Diophantus in the naming of the powers. The term *scalar,* of vector-analysis fame, is due to W. R. Hamilton (1853).

10. This is the *cosa,* or *res,* of the cossists, hence x in our notation. The next scalars are x^2 (square), x^3, x^4, and so forth. In Viète these quantities have dimensions.

11. *Parodic* is from Greek *para, hodas,* on the way, coming up.

12. x^5 is pure, $x^5 + ax^4$ is affected.

13. x^2 is pure, $x^2 + ax$ is affected.

14. Bachet's edition of Diophantus is of 1621, and was the inspiration of Fermat's work on numbers (see selection I.6).

15. In this work of 1593 Viète gives many examples of his *logistica speciosa.*

16. In arithmetic the custom was to use "in": *ducta in;* in geometry, "under": a rectangle is "under" its sides.

17. + in + is +; + in − is −.

18. $(A - B)(D - G) = AD - AG - BD + BG$.

19. Chapter VI mentions the retracing of the zetetic process by synthesis; Chapter VII the special application of the analytic art, after solution, to special arithmetic and geometric problems. Here Viète speaks of the "exegetic art." Chapter VIII is the discussion of different possible expressions and equations, stressing homogeneity. There are 29 rules.

20. Quod est, Nullum non problema solvere.

Chapter V
The Medieval–Renaissance–Reformation Periods in Europe

SIMON STEVIN (1548-1620)

We know little of the life of the Flemish mathematician and engineer Simon Stevin, who helped to make standard the use of decimal fractions in Western Europe and contributed notably to statics and hydrostatics. He was the natural son of Antheunis Stevin and Cathelijne van de Poort, two wealthy citizens of Bruges (in what is today Belgium). As a young man he worked briefly as a merchant's clerk in Bruges and Antwerp before traveling to Poland, Prussia, Denmark and Norway sometime between 1571 and 1577.

In 1581, Stevin was in Leiden, Holland, and entered its famous university two years later. His reasons for leaving the southern Netherlands for the new Republic of Holland in the north are not known. Perhaps the Spanish occupation of the south and the accompanying persecution caused the move. In any case, Stevin actively participated in the economic and cultural renaissance of Holland. He was an engineer until 1604, when he became quartermaster-general of the Dutch Army, serving under Maurice of Nassau, Prince of Orange. At the same time, he served Maurice as science and mathematics tutor and advised him on many occasions on navigational and military matters.

For the seafaring Dutch Republic, navigation was crucially important. Stevin had studied astronomy and was among the first to accept unconditionally the Copernican system in 1608. He gave a theory of tides based heavily on empirical data and recommended a world-wide survey of geomagnetic deviation to provide a basis for determining longitude. (The subsequent invention of the ship's chronometer solved the problem more simply.)

Stevin wrote widely on military subjects. In those writings he systematically studied camps, military equipment, and sieges as well as recommending improvements in the art of fortification. He wanted strongholds defended with modern artillery rather than the traditional small firearms. As a starting point for these studies, he drew upon the ancient Roman methods of fortification as recounted by Polybius. Stevin further helped to devise new types of sluices and locks for flooding areas and driving off invaders, a technique so important in the defense of Holland.

Stevin organized a school for engineers in Leiden and administered Prince Maurice's estate. About 1600, he constructed a 26-passenger wind-powered, sail carriage that moved along the seashore at a speed faster than that possible with horses. He married Catherine Cray in 1610, and they had four children.

The commercial and industrial prosperity of the cities of northern Italy and the Netherlands, together with the discovery of major writings of ancient science (especially those of Archimedes, Apollonius, and Diophantus), underlay the early Scientific Revolution. Stevin responded to both of these developments. For example, he incorporated commercial applications in his research, as his first book on interest tables (1582) demonstrates. The book first made public rules for single and compound interest and offered tables for the rapid computation of annuities and discounts. These tables, which big banking houses had previously kept secret, soon came into common use in Holland. Stevin also examined ancient sources, considering them a part of an age of wisdom. As the Dutch jurist Hugo Grotius commented, he wanted to bring about a second age of wisdom.

The chief achievements of Stevin in mathematics date from 1585, when he completed the pamphlet *De Thiende (The Tenth)* and its French translation *La Disme* as well as the book *L'arithmetique*. *De Thiende* introduced decimal fractions into Western Europe for general use and demonstrated that mathematical operations (addition, subtraction, multiplication, and division) could be performed easily with them. Stevin's notation, however, was unwieldy, for example $23^{57}/_{100}$ would be 23 ⓪ 5 ① 7 ②. Largely through the efforts of Viète and Napier, who improved Stevin's notation, the decimal system of numeration was widely adopted within 50 years. Although Stevin declared that the general adoption of the decimal division of coinage, weights and measures, and degrees of arc was only a matter of time, its systematic introduction for weights and measures in continental Europe did not occur until the French Revolution (1789). Stevin's *L'arithmetique,* which treats arithmetic and algebra, advanced the interesting view that all numbers (including negatives and irrationals) were of the same nature—a view that his contemporaries did not accept but later algebraists vindicated. Also about this time Stevin wrote a treatise on perspective that contains what amounts to an early, cumbrous attempt at integration.

In 1586, Stevin published his major book on mechanics, *De Behinselen der Weeghconst* ("Statics and Hydrostatics"), which built upon and went beyond the work of Archimedes. It includes the theory of the lever, a new proof for the law of the inclined plane, the determination of the center of gravity, and what, in principle, amounts to the parallelogram diagram of forces. (It gave a new impetus to the study of statics in Western Europe.) Below the diagram demonstrating the law of the inclined plane, the author inscribed the cherished maxim: "Wonder en is gheen wonder"—"What appears a wonder is not a wonder." Here was a confident expression of the new science. Stevin was saying that many matters, believed by people in the Middle Ages and Renaissance to be part of the supernatural realm, were actually part of the natural world and could be scientifically explained.

58. From *De Thiende**

(Decimal Fractions)

SIMON STEVIN

THE PREFACE OF SIMON STEVIN

To Astronomers, Land-meters, Measurers of Tapestry, Gaugers, Stereometers in general, Money-Masters, and to all Merchants, Simon Stevin wishes health.

Many, seeing the smallness of this book and considering your worthiness, to whom it is dedicated, may perchance esteem this our conceit absurd. But if the proportion be considered, the small quantity hereof compared to human imbecility, and the great utility unto high and ingenious intendments, it will be found to have made comparison of the extreme terms, which permit not any conversion of proportion. But what of that? Is this an admirable invention? No certainly: for it is so mean as that it scant deserves the name of an invention, for as the countryman by chance sometime finds a great treasure, without any use of skill or cunning, so hath it happened herein. Therefore, if any will think that I vaunt myself of my knowledge, because of the explicitation of these utilities, out of doubt he shows himself to have neither judgment, understanding, nor knowledge, to discern simple things from ingenious inventions, but he (rather) seems envious of the common benefit; yet howsoever, it were not fit to omit the benefit hereof for the inconvenience of such calumny. But as the mariner, having by hap found a certain unknown island, spares not to declare to his Prince the riches and profits thereof, as the fair fruits, precious minerals, pleasant champion,[1] etc., and that without imputation of self-glorification, even so shall we speak freely of the great use of this invention; I call it great, being greater than any of you expect to come from me. Seeing then that the matter of this Dime (the cause of the name whereof shall be declared by the first definition following) is number, the use and effects of which yourselves shall sufficiently witness by your continual experiences, therefore it were not necessary to use many words thereof, for the astrologer knows that the world is become by computation astronomical (seeing it teaches the pilot the elevation of the equator and of the pole, by means of the declination of the sun, to describe the true longitudes, latitudes, situations and distance of places, etc.) a paradise, abounding in some places with such things as the earth cannot bring forth in other. But as the sweet is never without the sour, so the travail in such computations cannot be unto him hidden, namely in the busy multiplications and divisions which proceed of the 60th progression of degrees, min-

Source: This English translation of *De Thiende* is based on one made by Richard Norton in 1608. It is given, except for the introduction, in D. J. Struik (ed.), *A Source in Mathematics 1200-1800* (1969), 7-11, and is reprinted by permission of Harvard University Press, For further information on *De Thiende,* see Dirk Struik (ed.), *The Principal Works of Simon Stevin,* vol. II A (1958), 373-385. Another English translation, prepared by Vera Sanford, is given in David Eugene Smith (ed.), *A Source Book in Mathematics* (1929), 20-34.

utes, seconds, thirds, etc. And the surveyor or land-meter knows what great benefit the world receives from his science, by which many dissensions and difficulties are avoided which otherwise would arise by reason of the unknown capacity of land; besides, he is not ignorant (especially whose business and employment is great) of the troublesome multiplications of rods, feet, and oftentimes of inches, the one by the other, which not only molests, but also (though he be very well experienced) causes error, tending to the damage of both parties, as also to the discredit of landmeter or surveyor, and so for the money-masters, merchants, and each one in his business. Therefore how much they are more worthy, and the means to attain them the more laborious, so much the greater and better is this Dime, taking away those difficulties. But how? It teaches (to speak in a word) the easy performance of all reckonings, computations, & accounts, without broken numbers, which can happen in man's business, in such sort as that the four principles of arithmetic, namely addition, subtraction, multiplication, & division, by whole numbers may satisfy these effects, affording the like facility unto those that use counters. Now if by those means we gain the time which is precious, if hereby that be saved which otherwise should be lost, if so the pains, controversy, error, damage, and other inconveniences commonly happening therein be eased, or taken away, then I leave it willingly unto your judgment to be censured; and for that, that some may say that certain inventions at the first seem good, which when they come to be practised effect nothing of worth, as it often happens to the searchers of strong moving,[2] which seem good in small proofs and models, when in great, or coming to the effect, they are not worth a button: whereto we answer that herein is no such doubt, for experience daily shows the same, namely by the practice of divers expert land-meters of Holland, unto whom we have shown it, who (laying aside that which each of them had, according to his own manner, invented to lessen their pains in their computations) do use the same to their great contentment, and by such fruit as the nature of it witnesses the due effect necessarily follows. The like shall also happen to each of yourselves using the same as they do. Meanwhile live in all felicity.

THE ARGUMENT

The dime has two parts, that is definitions & operations. By the first definition is declared what *dime* is, by the second, third, and fourth what *commencement, prime, second,* etc. and *dime numbers* are. The operation is declared by four propositions: the addition, subtraction, multiplication, and division of dime numbers. The order whereof may be successively represented by [the following statement].

The dime has two parts: (1) definitions, as what is dime, commencement, prime, second, etc., and dime number, and (2) operations or practice of the addition, subtraction, multiplication, and division.

And to the end the premises may the better be explained, there shall be hereunto an appendix adjoined, declaring the use of the dime in many things by certain examples, and also definitions and operations, to teach such as do not already know the use and practice of numeration, and the four principles of common arithmetic in whole numbers, namely addition, subtraction, multiplication, & division, together with the Golden Rule, sufficient to instruct the most ignorant in the usual practice of this art of dime or decimal arithmetic. [*Editor's note:* The appendix describes how to apply the decimal method of counting to surveying, wine gauging, the measuring of cloth, and other trades and professions.]

THE FIRST PART: OF THE DEFINITIONS OF THE DIME

THE FIRST DEFINITION

Dime is a kind of arithmetic, invented by the tenth progression, consisting in characters of ciphers, whereby a certain number is described and by which also all accounts which happen in human affairs are dispatched by whole numbers, without fractions or broken numbers.

Explication. Let the certain number be one thousand one hundred and eleven, described by the characters of ciphers thus 1111, in which it appears that each 1 is the 10th part of his precedent character 1; likewise in 2378 each unity of 8 is the tenth of each unity of 7, and so of all the others. But because it is convenient that the things whereof we would speak have names, and that this manner of computation is found by the consideration of such tenth or dime progression, that is that it consists therein entirely, as shall hereafter appear, we call this treatise fitly by the name of *Dime,* whereby all accounts happening in the affairs of man may be wrought and effected without fractions or broken numbers, as hereafter appears.

THE SECOND DEFINITION

Every number propounded is called COMMENCEMENT, whose sign is thus ⓪.

Explication. By example, a certain number is propounded of three hundred sixty-four: we call them the 364 *commencements,* described thus 364 ⓪, and so of all other like.

THE THIRD DEFINITION

And each tenth part of the unity of the COMMENCEMENT we call the PRIME, whose sign is thus ①, *and each tenth part of the unity of the prime we call the SECOND, whose sign is* ②, *and so of the other: each tenth part of the unity of the precedent sign, always in order one further.*

Explication. As 3 ① 7 ② 5 ③ 9 ④, that is to say: 3 *primes,* 7 *seconds,* 5 *thirds,* 9 *fourths,* and so proceeding infinitely, but to speak of their value, you may note that according to this definition the said numbers are 3/10, 7/100, 5/1000, 9/10000, together 3759/10000, and likewise 8 ⓪ 9 ① 3 ② 7 ③ are worth 8, 9/10, 3/100, 7/1000, together 8 937/1000, and so of other like. Also you may understand that in this *dime* we use no fractions, and that the multitude of signs, except ⓪, never exceed 9, as for example not 7 ① 12 ②, but in their place 8 ① 2 ②, for they value as much.

THE FOURTH DEFINITION

The numbers of the second and third definitions beforegoing are generally called DIME NUMBERS.

THE SECOND PART OF THE DIME: OF THE OPERATION OR PRACTICE

THE FIRST PROPOSITION: OF ADDITION

Dime numbers being given, how to add them to find their sum.

The Explication Propounded: There are 3 orders of dime numbers given, of which the first 27 ⓪, 8 ①, 4 ②, 7 ③, the second 37 ⓪, 6 ①, 7 ②, 5 ③, the third 875 ⓪, 7 ①, 8 ②, 2 ③.

The Explication Required: We must find their total sum.

Construction. The numbers given must be placed in order as here adjoining, adding them in the vulgar manner of adding of whole numbers in this manner. The sum (by the first problem of our French Arithmetic[3]) is 941304, which are (that which the signs above the numbers do show) 941 ⓪ 3 ① 0 ② 4 ③. I say they are the sum required.

		⓪	①	②	③
	2	7	8	4	7
	3	7	6	7	5
8	7	5	7	8	2
9	4	1	3	0	4

Demonstration. The 27 ⓪ 8 ① 4 ② 7 ③ given make by the 3rd definition before 27, $^{8}/_{10}$, $^{4}/_{100}$, $^{7}/_{1000}$, together $27^{847}/_{1000}$ and by the same reason the 37 ⓪ 6 ① 7 ② 5 ③ shall make $37^{675}/_{1000}$ and the 875 ⓪ 7 ① 8 ② 2 ③ will make $875^{782}/_{1000}$, which three numbers make by common addition of vulgar arithmetic $941^{304}/_{1000}$. But so much is the sum 941 ⓪ 3 ① 0 ② 4 ③ ; therefore it is the true sum to be demonstrated. Conclusion: Then dime numbers being given to be added, we have found their sum, which is the thing required.

Note that if in the number given there want some signs of their natural order, the place of the defectant shall be filled. As for example, let the numbers given be 8 ⓪ 5 ① 6 ② and 5 ⓪ 7 ② , in which the latter wanted the sign of ① ; in the place thereof shall 0 ① be put. Take then for that latter number given 5 ⓪ 0 ① 7 ② , adding them in this sort.

	⓪	①	②
	8	5	6
	5	0	7
1	3	6	3

This advertisement shall also serve in the three following propositions, wherein the order of the defailing figures must be supplied, as was done in the former example.

THE SECOND PROPOSITION: OF SUBTRACTION

A dime number being given to subtract, another less dime number given: out of the same to find their rest.

Explication Propounded: Be the numbers given 237 ⓪ 5 ① 7 ② 8 ③ and 59 ⓪ 7 ① 3 ② 9 ③ .

The Explication Required: To find their rest.

Construction. The numbers given shall be placed in this sort, subtracting according to vulgar manner of subtraction of whole numbers, thus.

		⓪	①	②	③
2	3	7	5	7	8
	5	9	7	3	9
1	7	7	8	3	9

The rest is 177839, which values as the signs over them do denote 177 ⓪ 8 ① 3 ② 9 ③ , I affirm the same to be the rest required.

Demonstration. The 237 ⓪ 5 ① 7 ② 8 ③ make (by the third definition of this Dime) $237^{5}/_{10}$, $^{7}/_{100}$, $^{8}/_{1000}$, together $237^{578}/_{1000}$, and by the same reason the 59 ⓪ 7 ① 3 ② 9 ③ value $59^{739}/_{1000}$, which subtracted from $237^{578}/_{1000}$, there rests $177^{839}/_{1000}$, but so much doth 177 ⓪ 8 ① 3 ② 9 ③ value; that is then the true rest which should be made manifest.

Conclusion. A dime being given, to subtract it out of another dime number, and to know the rest, which we have found.

THE THIRD PROPOSITION: OF MULTIPLICATION

A dime number being given to be multiplied, and a multiplicator given: to find their product.

The Explication Propounded: Be the number to be multiplied 32 ⓪ 5 ① 7 ② , and the multiplicator 89 ⓪ 4 ① 6 ② .

The Explication Required: To find the product.

Construction. The given numbers are to be placed as here is shown, multiplying according to the vulgar manner of multiplication by whole numbers, in this manner, giving the product 29137122. Now to know how much they value, join the two last signs together as the one ② and the other ② also, which together make ④ , and say that the last sign of the product shall be ④ , which being known, all the rest are also known by their continued order. So that the product required is 2913 ⓪ 7 ① 1 ② 2 ③ 2 ④ .

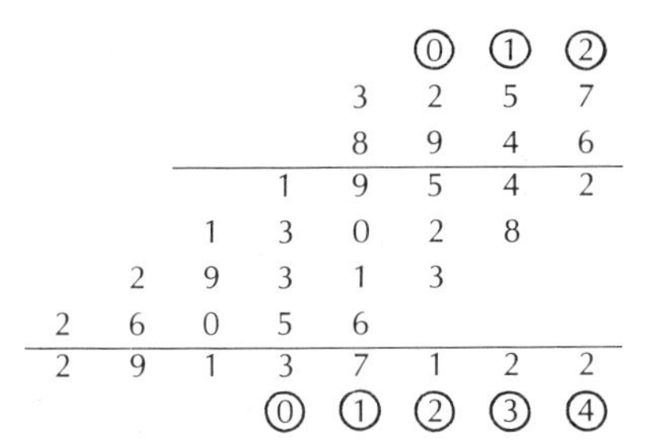

Demonstration. The number given to be multiplied, 32 ⓪ 5 ① 7 ② (as appears by the third definition of this Dime), $32\frac{5}{10}$, $\frac{7}{100}$, together $32\frac{57}{100}$; and by the same reason the multiplicator 89 ⓪ 4 ① 6 ② value $89\frac{46}{100}$ by the same, the said $32\frac{57}{100}$ multiplied gives the product $2913\frac{7122}{10000}$. But it also values 2913 ⓪ 7 ① 1 ② 2 ③ 2 ④.

It is then the true product, which we were to demonstrate. But to show why ② multiplied by ② gives the product ④, which is the sum of their numbers, also why ④ by ⑤ produces ⑨, and why ⓪ by ③ produces ③, etc., let us take $\frac{2}{10}$ and $\frac{3}{100}$, which (by the third definition of this Dime) are 2 ① 3 ②, their product is $\frac{6}{1000}$, which value by the said third definition 6 ③; multiplying then ① by ②, the product is ③, namely a sign compounded of the sum of the numbers of the signs given.

Conclusion. A dime number to multiply and to be multiplied being given, we have found the product, as we ought.

Note: If the latter sign of the number to be multiplied be unequal to the latter sign of the multiplicator, as, for example, the one 3 ④ 7 ⑤ 8 ⑥, the other 5 ① 4 ②, they shall be handled as aforesaid, and the disposition thereof shall be thus.

	④	⑤	⑥		
	3	7	8		
		5	4	②	
	1	5	1	2	
1	8	9	0		
2	0	4	1	2	
④	⑤	⑥	⑦	⑧	

THE FOURTH PROPOSITION: OF DIVISION

A dime number for the dividend and divisor being given: to find the quotient.

Explication Proposed: Let the number for the dividend be 3 ⓪ 4 ① 4 ② 3 ③ 5 ④ 2 ⑤ and the divisor 9 ① 6 ②.

Explication Required: To find their quotient.

Construction. The numbers given divided (omitting the signs) according to the vulgar manner of dividing of whole numbers, gives the quotient 3587; now to know what they value, the latter sign of the divisor ② must be subtracted from the latter sign of the dividend, which is ⑤, rests ③ for the latter are also manifest by their continued order, thus 3 ⓪ 5 ① 8 ② 7 ③ are the quotient required.

```
  1
 18
3164
7617        ⓪  ①  ②  ③
344352      (3  5  8  7
 96666
  999
```

Demonstration. The number dividend given 3 ⓪ 4 ① 4 ② 3 ③ 5 ④ 2 ⑤ makes (by the third definition of this Dime) 3, $\frac{4}{10}$, $\frac{4}{100}$, $\frac{3}{1000}$, $\frac{5}{10000}$, $\frac{2}{100000}$, together $3\frac{44352}{100000}$, and by the same reason the divisor 9 ① 6 ② values $\frac{96}{100}$, by which $3\frac{44352}{100000}$, being divided, gives the quotient $3\frac{587}{1000}$; but the said quotient values 3 ⓪ 5 ① 8 ② 7 ③, therefore it is the true quotient to be demonstrated.

Conclusion. A dime number being given for the dividend and divisor, we have found the quotient required.

Note: If the divisor's signs be higher than the signs of the dividend, there may be as many such ciphers 0 joined to the dividend as you will, or as many as shall be necessary: as for example, 7 ② are to be divided by 4 ⑤, I place after the 7 certain ⓪, thus 7000, dividing them as aforesaid, and in this sort it gives for the quotient 1750 ⓪.

3̸ 2̸
7̸ 0̸ 0̸ 0̸ (1 7 5 0 ⓪
4̸ 4̸ 4̸ 4̸

It happens also sometimes that the quotient cannot be expressed by whole numbers, as 4 ① divided by 3 ② in this sort, whereby appears that there will infinitely come 3's, and in such a case you may come so near as the thing requires, omitting the remainder. It is true, that 13 ⓪ 3 ① 3⅓ ②, or 13 ⓪ 3 ① 3 ② 3⅓ ③ etc. shall be the perfect quotient required. But our invention in this Dime is to work all by whole numbers. For seeing that in any affairs men reckon not of the thousandth part of a mite, es, grain, etc., as the like is also used of the principal geometers and astronomers in computations of great consequence, as Ptolemy and Johannes Montaregio,[4] have not described their tables of arcs, chords, or sines in extreme perfection (as possibly they might have done by multinomial numbers), because that imperfection (considering the scope and end of those tables) is more convenient than such perfection.

1̸ 1̸ 1̸ (1 ⓪①②
4̸ 0̸ 0̸ 0̸ 0 0 0 (1 3 3 3
3̸ 3̸ 3̸ 3̸

Note 2. The extraction of all kinds of roots may also be made by these dime numbers; as, for example, to extract the square root of 5 ② 2 ③ 9 ④, which is performed in the vulgar manner of extraction in this sort, and the root shall be 2 ① 3 ②, for the moiety or half of the latter sign of the numbers given is always the latter sign of the root; wherefore, if the latter sign given were of a number impair, the sign of the next following shall be added, and then it shall be a number pair; and then extract the root as before. Likewise in the extraction of the cubic root, the third part of the latter sign given shall be always the sign of the root; and so of all other kinds of roots.

1̸
5̸ 2̸ 9̸
2̸ 3̸

4̸

NOTES

1. Champion, comp. French "champagne," field, landscape. Comp. e.g. Deut. XI, 30, author. transl. of 1611: "the Canaanites which dwell in the campions."

2. This is a translation of the Dutch "roersouckers," after Stevin's French version: "chercheurs de fort mouvements." It probably stands for people who start moving things, take initiative, comp. the archaic Dutch expressions "roermaker," "roerstichter" (information from Prof. Dr. C. G. N. De Vooys). The Dutch has "vonden der roersouckers," where "vonden" stands for "findings, inventions," and the whole expression for something like "widely proclaimed innovations."

3. *L'Arithmétique de Simon Stevin de Bruges* (Leyden, 1585); see Stevin, *The principal works* (Swets-Zeitlinger, Amsterdam), vol. IIB (1958). Problem I (p. 81) is: "Given two arithmetical integer numbers. Find their sum."

4. Johannes Montaregio (1436-1476) is best known under his latinized name, Johannes Regiomontanus. This craftsman, humanist, astronomer, and mathematician of Nuremberg influenced the development of trigonometry by means of his widely used book *De triangulis omnimodis* (written c. 1464, printed in Nuremberg in 1533). The sines, for Regiomontanus as well as for Stevin, were half chords; see our note to the following text on Napier and Selection III.2.

Chapter V
The Medieval–Renaissance–Reformation Periods in Europe

JOHN NAPIER (1550-1617)

The son of Sir Archibald Napier and his first wife Janet Bothwell, John Napier was the eighth laird of Merchiston, an estate near Edinburgh, Scotland. He matriculated at St. Andrews (1563), where he pursued theological studies, and traveled abroad after graduating, probably to continue his learning. Napier returned to Scotland in 1571 and married Elizabeth Stirling in the next year. They had two children and resided at Gartner Castle. After Elizabeth died in 1579, he married Agnes Chisholm; they had ten children. Following his father's death in 1608, he moved to Merchiston Castle where he lived for the remainder of his life.

Napier was a fervent Scottish Protestant and landowner in an age of tumult in religion and politics. He staunchly opposed "popery" urging the Scottish king, James VI (the future James I of England), to "purge his house and court of all Papists, Atheists, and Newtrals." As a landowner, Napier concentrated on improving his crops and cattle. He and his sons experimented with various manures, discovered how to use common salt to enrich the soil. His family was granted a monopoly for his new method of tillage. A man of many talents, he invented a hydraulic screw and revolving axle for controlling the water level in coal pits (1579) and devised military plans for armored chariots and burning mirrors to destroy enemy ships. Except for his authority as a divine, his ingenuity might have gotten him persecuted as a warlock. This was not merely a case of folk backwardness. James VI had written a book on demonology, and in this period mathematicians were suspected of being in league with the powers of darkness.

Logarithms were Napier's major contribution to mathematics. He invented them after 20 years of pondering the sequences of numbers obtained by repeated multiplication leading to a primitive recognition of exponentiation. One step in his preparation was an examination of Tycho Brahe's method of computation using Ibn Yunus' identity—$2 \cos \alpha \cos \beta = \cos(\alpha + \beta) + \cos(\alpha - \beta)$—which had reduced multiplication to addition. Initially, Napier called logarithms "artificial numbers" but later coined the term logarithm meaning "number of the ratio." (In 1728, Euler defined the logarithm of a number as the exponent to which a certain number, called the base, must be raised to give the original number.) Although Napier did not work in terms of base, he used $(1 - 1/10^7)^{10^7}$, which is nearly $\lim_{n\to\infty} (1 - 1/n)^n$. Virtually he had base $1/e$. He described his invention in two Latin treatises entitled *Mirifici logarithmorum canonis descriptio* ("The Description of the Wonderful Canon of Logarithms," 1614)* and *Mirifici logarithmorum canonis constructio* ("The Construction of the Wonderful Canon of Logarithms," published posthumously in 1619). His logarithms quickly became a powerful too' in computations, especially in astronomy, which in turn was important to navigation. They were enthusiastically advanced by Henry Briggs, who used base 10, as well as by Jobst Bürgi, William Oughtred, and René Descartes.

Napier contributed to other areas of mathematics. He was among the first to use effectively Stevin's decimal notation. In addition, he investigated imaginary roots of equations and developed "Napier's Analogies" mnemomic devices in spherical trigonometry that proved helpful in navigation. Finally, he invented a calculating device known as "Napier's Bones," an ancestor of the slide rule.

*The opening verse of the *Descriptio* follows:

Hic liber est minimus, si spectes verba, sed usum
Sic spectes, Lector, maximus hic liber est
Disce, scies parvo tantum debere libello
Te, quantum magnis mille voluminibus.

which Struik freely translates as follows:

The use of this book is quite large, my dear friend,
No matter how modest it looks,
You study it carefully and find that it gives
As much as a thousand big books.

59. From *Mirifici logarithmorum canonis constructio**

(Logarithms)

JOHN NAPIER

1. A logarithmic table [*tabula artificialis*] is a small table by the use of which we can obtain a knowledge of all geometrical dimensions and motions in space, by a very easy calculation . . . It is picked out from numbers progressing in continuous proportion.

2. Of continuous progressions, an arithmetical is one which proceeds by equal intervals; a geometrical, one which advances by unequal and proportionally increasing or decreasing intervals . . .

3. In these progressions we require accuracy and ease in working. Accuracy is obtained by taking large numbers for a basis; but large numbers are most easily made from small by adding cyphers.[1]

Thus instead of 1000000, which the less experienced make the greatest sine,[2] the more learned put 10000000, whereby the difference of all sines is better expressed. Wherefore also we use the same for radius and for the greatest of our geometrical proportionals.

*Source: This English translation is taken from W. R. Maconald, *The construction of the wonderful canon of logarithms* (1889). It appears with notes and explanations in D. J. Struik (ed.), *A Source Book,* 12-21, and is reprinted by permission of Harvard University Press, For further information on Napier and his logarithms, see C. G. Knott (ed.), *Napier Tercentenary Memorial Volume* (1915).

4. In computing tables, these large numbers may again be made still larger by placing a period after the number and adding cyphers . . .

5. In numbers distinguished thus by a period in their midst, whatever is written after the period is a fraction [*quicquid post periodam notatur fractio*], the denominator of which is unity with as many cyphers after it as there are figures after the period.[3]

Thus 10000000.04 is the same as $10000000^{4}/_{100}$; also 25.803 is the same as $25^{803}/_{1000}$; also 9999998.0005021 is the same as $9999998^{5021}/_{10000000}$, and so of others.

6. When the tables are computed, the fractions following the period may then be rejected without any sensible error. For in our large numbers, an error which does not exceed unity is insensible and as if it were none . . .

Then follow in Arts. 7-15 some rules for accurate counting with large numbers.

16. Hence, if from the radius with seven cyphers added you subtract its 10000000th part, and from the number thence arising its 10000000th part, and so on, a hundred numbers may very easily be continued geometrically in the proportion subsisting between the radius and the sine less than it by unity, namely between 10000000 and 9999999; and this series of proportionals we name the First table.

First table

10000000.0000000
1.0000000
9999999.0000000
.9999999
9999998.0000001
.9999998
9999997.0000003
to be continued up to
9999900.0004950

Thus from radius, with seven cyphers added for greater accuracy, namely,

10000000.0000000

subtract

1.0000000

you get

9999999.0000000;

from this subtract

.9999999,

you get

9999998.0000001;

and proceed in this way . . . until you create a hundred proportionals, the last of which, if you have computed rightly, will be 9999900.0004950.

17. The Second table proceeds from radius with six cyphers added, through fifty other numbers decreasing proportionally in the proportion which is easiest, and as near as possible to that subsisting between the first and last numbers of the First table.

Second table

10000000.000000
100.000000
9999900.000000
99.999000
9999800.001000
to be continued up to
9995001.222927

Thus the first and last numbers of the First table are 10000000.0000000 and 9999900.0004950, in which proportion it is difficult to form fifty proportional numbers. A near and at the same time an easy proposition is 100000 to 99999, which may be continued with sufficient exactness by adding six cyphers to radius and continually subtracting from each number its own 100000th part . . . and this table contains, besides radius which is the first, fifty other proportional numbers, the last of which, if you have not erred, you will find to be 9995001.222927.[4]

Article 18 has a Third table of 69 columns, from 10^{12} down by 2000th parts to 9900473.57808.

19. The first numbers of all the columns must proceed from radius with four cyphers added, in the proportion easiest and nearest to that subsisting between the first and the last numbers of the first column.

As the first and the last numbers of the first column are 10000000.0000 and 9900473.5780, the easiest proportion very near to this is 100 to 99. Accordingly sixty-eight numbers are to be continued from radius in the ratio of 100 to 99 by subtracting from each one of them its hundredth part.

20. In the same proportion a progression is to be made from the second number of the first column through the second numbers in all the columns, and from the third through the third, and from the fourth through the fourth, and from the others respectively through the others.

Thus from any number in one column, by subtracting its hundredth part, the number of the same rank in the following column is made, and the numbers should be placed in order as follows.

Here follows a table of "Proportionals of the Third Table," with 69 columns, the last number in the sixty-ninth column being 4998609.4034, roughly half the original number 10000000.0000.

21. Thus, in the Third table, between radius and half radius, you have sixty-eight numbers interpolated, in the proportion of 100 to 99, and between each two of these you have twenty numbers interpolated in the proportion of 10000 to 9995; and again, in the Second table, between the first two of these namely between 10000000 and 9995000, you have fifty numbers interpolated in the proportion of 100000 to 99999; and finally, in the First table, between the latter, you have a hundred numbers interpolated in the proportion of radius or 10000000 to 9999999; and since the difference of these is never more than unity, there is no need to divide it more minutely by interpolating means, whence these three tables, after they have been completed, will suffice for computing a Logarithmic table.

Hitherto we have explained how we may most easily place in tables sines or natural numbers progressing in geometrical proportion.

22. It remains, in the Third table at least, to place beside the sines or natural numbers decreasing geometrically their logarithms or artificial numbers increasing arithmetically.

Articles 23 and 24 represent arithmetic increase and geometric decrease by points on a line.

25. Whence a geometrically moving point approaching a fixed one has its velocities proportionate to its distances from the fixed one.

Thus referring to the preceding figure [Fig. 1], I say that when the geometrically moving point *G* is at *T*, its velocity is as the distance *TS*, and when *G* is at 1 its velocity is as 1*S*, and when at 2 its velocity is as 2*S*, and so of the others. Hence, whatever be the proportion of the distances *TS*, 1*S*, 2*S*, 3*S*, 4*S*, etc., to each other, that of the velocities of *G* at the points *T*, 1, 2, 3, 4, etc., to one another, will be the same.

For we observe that a moving point is declared more or less swift, according as it is seen to be borne over a greater or less space in equal times. Hence the ratio of the spaces traversed is necessarily the same as that of the velocities. But the ratio of the spaces traversed in equal times, *T*1, 12, 23, 34, 45, etc., is that of the distances *TS*, 1*S*, 2*S*, 3*S*, 4*S*, etc. Hence it follows that the ratio to one another of the distances of *G* from *S*, namely *TS*, 1*S*, 2*S*, 3*S*, 4*S*, etc., is the same as that of the velocities of *G* at the points *T*, 1, 2, 3, 4, etc., respectively.

26. The logarithm of a given sine is that number which has increased arithmetically with the same velocity throughout as that with which radius began to decrease geometrically, and in the same time as radius has decreased to the given sine.

Let the line *TS* [Fig. 2] be the radius,

Fig. 1
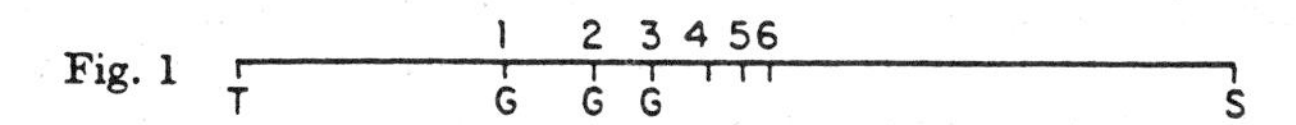

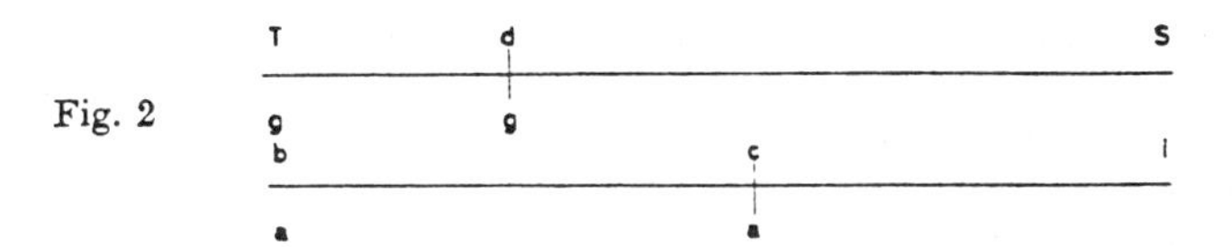

Fig. 2

and *dS* a given sine in the same line; let g move geometrically from *T* to *d* in certain determinate moments of time. Again, let *bi* be another line, infinite towards *i*, along which, from *b*, let *a* move arithmetically with the same velocity as g had at first when at *T*; and from the fixed point *b* in the direction of *i* let *a* advance in just the same moments of time up to the point *c*. The number measuring the line *bc* is called the logarithm of the given sine *dS*.[5]

27. Whence nothing is the logarithm of radius [*Unde sinus totius nihil est pro artificiali*] . . .

28. Whence also it follows that the logarithm of any given sine is greater than the difference between radius and the given sine, and less than the difference between radius and the quantity which exceeds it in the ratio of radius to the given sine. And these differences are therefore called the limits of the logarithm.

Thus, the preceding figure being repeated [Fig. 3], and *ST* being produced beyond *T* to *o*, so that *oS* is to *TS* as *TS* to *dS*, I say that *bc*, the logarithm of the sine *dS*, is greater than *Td* and less than *oT*. For in the same time that g is borne from *o* to *T*, g is borne from *T* to *d*, because (by 24) *oT* is such a part of *oS* as *Td* is of *TS*, and in the same time (by the definition of a logarithm) is *a* borne from *b* to *c*; so that *oT*, *Td*, and *bc* are distances traversed in equal times. But since g when moving between *T* and *o* is swifter than at *T*, and between *T* and *d* slower, but at *T* is equally swift with *a* (by 26); it follows that *oT* the distance traversed by g moving swiftly is greater, and *Td* the distance traversed by g moving slowly is less, than *bc* the distance traversed by the point *a* with its medium motion, in just the same moments of time; the latter is, consequently, a certain mean between the two former. Therefore *oT* is called the greater limit, and *Td* the less limit of the logarithm which *bc* represents.

29. Therefore to find the limits of the logarithm of a given sine.

By the preceding it is proved that the given sine being subtracted from radius the less limit remains, and that radius being multiplied into the less limit and the product divided by the given sine, the greater limit is produced, as in the following example.

30. Whence the first proportional of the First table, which is 9999999, has its logarithm between the limits 1.0000001 and 1.0000000 . . .

31. The limits themselves differing insensibly, they or anything between them may be taken as the true logarithm . . .

32. There being any number of sines decreasing from radius in geometrical proportion, of one of which the logarithm or its limits is given, to find those of the others.

This necessarily follows from the definitions of arithmetical increase, of geometrical decrease, and of a logarithm . . . So that, if the first

Fig. 3

logarithm corresponding to the first sine after radius be given, the second logarithm will be double of it, the third triple, and so of the others; until the logarithms of all the sines be known . . .

33. Hence the logarithms of all the proportional sines of the First table may be included between near limits, and consequently given with sufficient exactness . . .

34. The difference of the logarithms of radius and a given sine is the logarithm of the given sine itself . . .

35. The difference of the logarithms of two sines must be added to the logarithm of the greater that you may have the logarithm of the less, and subtracted from the logarithm of the less that you may have the logarithm of the greater . . .

36. The logarithms of similarly proportioned sines are equidifferent.

This necessarily follows from the definitions of a logarithm and of the two motions . . . Also there is the same ratio of equality between the differences of the respective limits of the logarithms, namely as the differences of the less among themselves, so also of the greater among themselves, of which logarithms the sines are similarly proportioned.

37. Of three sines continued in geometrical proportion, as the square of the mean equals the product of the extremes, so of their logarithms the double of the mean equals the sum of the extremes. Whence any two of these logarithms being given, the third becomes known . . .

38. Of four geometrical proportionals, as the product of the means is equal to the product of the extremes; so of their logarithms, the sum of the means is equal to the sum of the extremes. Whence any three of these logarithms being given, the fourth becomes known . . .[6]

39. The difference of the logarithms of two sines lies between two limits; the greater limit being to the radius as the difference of the sines to the less sine, and the less limit being to radius as the difference of the sines to the greater sine . . .[7]

Articles 40-46 show how to find logarithms.

47. In the Third table, beside the natural numbers, are to be written their logarithms; so that the Third table, which after this we shall always call the Radical table, may be made complete and perfect . . .

48. The Radical table being now completed, we take the numbers for the logarithmic table from it alone.

For as the first two tables were of service in the formation of the third, so this third Radical table serves for the construction of the principal Logarithmic table, with great ease and no sensible error.

49. To find most easily the logarithms of sines greater than 9996700.

This is done simply by the subtraction of the given sine from radius. For (by 29) the logarithm of the sine 9996700 lies between the limits 3300 and 3301; and these limits, since they differ from each other by unity only, cannot differ from their true logarithm by any sensible error, that is to say, by an error greater than unity. Whence 3300, the less limit, which we obtain simply by subtraction, may be taken for the true logarithm. The method is necessarily the same for all sines greater than this.

50. To find the logarithms of all sines

The Radical Table

First Column		Second Column			69th Column	
Natural Numbers	Logarithms	Natural Numbers	Logarithms		Natural Numbers	Logarithms
10000000.0000	.0	9900000.0000	100503.3		5048858.8900	6834225.8
9995000.0000	5001.2	9895050.0000	105504.6	. . .	5046334.4605	6839227.1
9990002.50000	10002.5	9890102.4750	110505.8		5043011.2932	6844228.3
⋮	⋮	⋮	⋮	. . .	⋮	⋮
9900473.5700	100025.0	9801468.8423	200528.2		4998609.4034	6934250.8

embraced within the limits of the Radical table.

Multiply the difference of the given sine and table sine nearest it by radius. Divide the product by the easiest divisor, which may be either the given sine or the table sine nearest it, or a sine between both, however placed. By 39 there will be produced either the greater or less limit of the difference of the logarithms, or else something intermediate, no one of which will differ by a sensible error from the true difference of the logarithms on account of the nearness of the numbers in the table. Wherefore (by 35), add the result, whatever it may be, to the logarithm of the table sine, if the given sine be less than the table sine; if not, subtract the result from the logarithm of the table sine, and there will be produced the required logarithm of the given sine.

Two examples are given. In the first the given sine is 7489557, the table sine of which nearest to it is 7490786.6119. The computation gives 2890752 for the logarithm.

51. All sines in the proportion of two to one have 6931469.22 for the difference of their logarithms [because the number is the logarithm of sine 5000000].

52. All sines in the proportion of ten to one have 23025842.34 for the difference of their logarithms.

Article 53 contains a short table of given proportions of sines and corresponding differences of logarithms; Art. 54 deals with the logarithms of all sines outside the limits of the Radical table.

55. As half radius is to the sine of half a given arc, so is the sine of the complement of the half arc to the sine of the whole arc . . .[8]

56. Double the logarithm of an arc of 45 degrees is the logarithm of half radius . . .

57. The sum of the logarithms of half radius and any given arc is equal to the sum of the logarithms of half the arc and the complement of the half arc. Whence the logarithm of the half arc may be found if the logarithms of the other three be given . . .

Article 58 deals with the logarithms of all arcs not less than 45 degrees.

59. To form a logarithmic table.

Here follows a description of the construction of a table of 45 pages, each page devoted to one degree divided into minutes.

Napier's table is constructed in quite the same form as that used at present, except that the second (sixth) column gives sines for the number of degrees indicated at the top (bottom) and of minutes in the first (seventh) column, the third (fifth) column gives the corresponding logarithm, and the fourth column gives the *differentiae* between the logarithms in the third and fifth columns, these being therefore essentially logarithmic tangents or cotangents. A few entries follow.

0° min	sines	logarithm	+/− differentiae	logarithm	sines	
0	0	infinitum	infinitum	0	10000000	69
1	2909	81425681	81425680	1	10000000	59
2	5818	74494213	74494211	2	9999998	58
3	8727	70439560	70439560	4	9999998	57
			. . . .			
30° min	sines	logarithm	+/− differentiae	logarithm	sines	
0	5000000	6931469	5493059	1483410	8660254	60
1	5002519	6926432	5486342	1440090	8658799	59
2	5005038	6921399	5479628	1441771	8657344	58
			. . . .			
44° min						
59	7069011	3468645	5818	3462827	7071068	1
60	7071068	3465735	0	3465735	7071068	0
						min 45°

Hence log sin 3′ = log 8727 = 70439560, log sin 30°1′ = log 5002519 = 6926432, log sin 45° = log 7071068 = 3465735; (half of log sin 30°, Art. 56), also log sin 90° = log 10000000 = 0.

NOTES

1. Cyphers = zeros (see Selection I.1, Leonardo of Pisa).

2. Sin 90° = 100.0000, hence the radius R of the circle is 10^6. The sine of an angle was always defined as half the chord belonging to the double angle, hence $\sin \alpha = CB = \frac{1}{2}$ chord $2\alpha = \frac{1}{2}CD$ [Fig. 4]. The numerical values of the sines therefore depended on the choice of R. Euler introduced dimensionless sines and tangents by consistently writing $R = 1$ (1748, see our extract of the *Introductio*, Selection V.15).

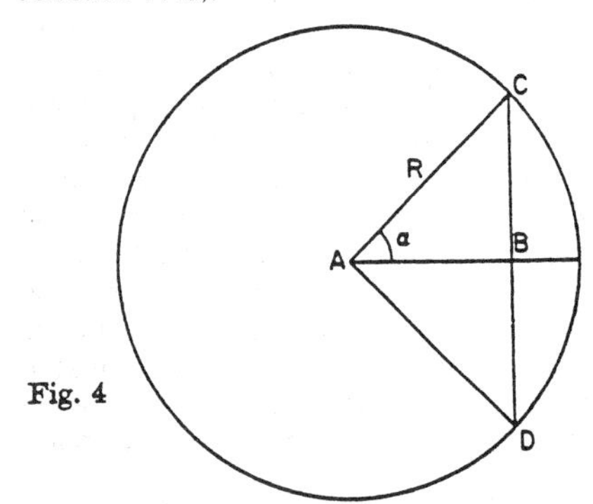

Fig. 4

3. The clumsy notation for decimal fractions of Stevin is here replaced by the method of the decimal point. Napier's authority made this method generally accepted.

4. This should be 9995001.224804.

5. In the language of the calculus: let $TS = a$ ($= 10^7$), $dS = y$; then the initial velocity ($t = c$) at g is a (see Art. 25), hence the velocity of g at d is $(d/dt)(a - y) = -dy/dt = y$, hence $y = a\,e^{-t}$. When $bc = x$, then $x = at =$ Nap log y. Hence Nap log $y = a \ln a/y$, so that (by Art. 27) for $y = a$, Nap log $a = 0$, where $\ln = \log_e$, the natural logarithm. The familiar rules for logarithmic computation do not apply:

$$\text{Nap log } xy = a(\ln a - \ln x - \ln y).$$

We should not be confused by the terms "radius" and "sine"; what is meant is a line segment TS and a section $dS \leqq TS$. When $a = 1$ the Nap log and the ln differ only in sign; this may have caused the confusion in some textbooks, which insist on calling the natural logarithms Napierian or Neperian logarithms.

6. The modern theorem for the logarithm of a product does not hold, since the logarithm of unity is not zero. Hence Arts. 37 and 38, to express special cases.

7. This is proved by the principle of proportion and of Article 36. This rule is used first in Arts. 40 and 41 as an illustration to find the logarithm of 9999975.5 from that of the nearest sine in the First table, 9999975.0000300, noting that the limits of the logarithms of the latter number are 25.0000025 and 25.000000, that the difference of the logarithms of the two numbers by the rule just given is .4999712, and that the limits for the logarithm of 9999975.5 are therefore 24.5000313 and 24.5000288, whence Napier lists the logarithm as 24.5000300.

Articles 41 to 45 illustrate the fact that one may now calculate the logarithms of all the "proportionals" in the First, Second, and Third tables, as well as of the sines or natural numbers not proportionals in these tables but near or between them.

8. Only here does Napier begin to introduce angles into the construction of his tables. Napier proves Arts. 55-57 by geometric principles and the preceding theorems concerning logarithms. He then often speaks of the logarithms of the arcs, meaning logarithms of the corresponding sines.

Chapter VI

The Scientific Revolution at its Zenith (1620-1720)

Introduction. From 1620 to 1720 the uses, influence, and serious pursuit of mathematics markedly expanded in Europe. A central feature of that expansion was the profound effect that mathematics had upon the form and content of astronomy and the mechanical philosophy of nature (essentially physics) as the Scientific Revolution moved toward its zenith. This revolution was in turn a central component in a "century of genius"[1] that saw an uneasy tension develop between the domains of deductive reason and religious faith. The tension deepened as an *esprit géomètrique* permeated Europe's new intellectual culture. Exemplifying the new spirit, the French savant Bernard Fontenelle (1657–1757) claimed that any book, whether on natural philosophy or even theology, ethics, or history would be better written by a geometer. Baruch Spinoza carried the axiomatic method in to his masterpiece, the *Ethics* (1677), as did Isaac Newton in *Principia Mathematica* (1687). By the mid-seventeenth century a new type of savant included substantive mathematics among his accomplishments. Many artists, men of letters, physicians, and lawyers shared this appreciation. They considered mathematics in conjunction with new instruments, like the barometer and telescope, to be the key that would unlock the secrets of nature. Mathematics had displaced theology as queen of the sciences, though not without difficulties.

A succession of extraordinary achievements unparalleled since early Hellenistic times occurred in European mathematics. One of them was analytic geometry, which was created by the imaginative linking of geometry with the theory of equations and trigonometry. Formative number theory and probability also made substantial advances. Most fruitful of all was the invention of two different forms of the infinitesimal calculus. Internally, a more critical study of ancient Greek, Arabic, medieval and Renaissance texts provided a source for these developments. For example, the masterworks of Euclid and Archimedes, while extant earlier, were only now sufficiently mastered and disseminated to support new advances. Externally, geometers and calculators responded to new needs and practices of society and commercial capitalism.

Mathematicians during the seventeenth and eighteenth centuries had increased opportunities to address socioeconomic concerns, many of which fostered a vigorous study of nature. Exploration of new continents, expanding commerce, and changing forms of warfare posed new mathematical problems: navigators required improved trigonometric tables, star charts, and more reliable maps; field commanders wanted better guns and cannons; mine operators desired improved pumps; while gamblers and even magistrates in the courts appealed to tables of expectations from a nascent probability theory. Success stimulated investment, and a share of the new wealth was spent to solve these problems.

Even more crucial for mathematics than technological advances in cartography, ballistics, and hydraulics, was its becoming the language of astronomy and natural philosophy. This meant not merely more complicated calculations but the acceptance only of descriptions of natural phenomena formulated in terms of general mathematical laws. These laws supported a shift from the conception of nature as an active principle, as the Aristotelians held, to an ordered assembly of phenomena existing in a grand mechanism. Kepler's three laws of planetary motion[2] and Newton's law of attraction illustrate a continuing mathematization of astronomy, while Boyle's law (that volume varies inversely with pressure) carried the procedure to pneumatics. Huygens' principle of the conservation of *vis viva* (roughly kinetic energy) applied to collisions. In the circumstances presented above, mathematics came to be centered in the leading commercial and university cities of western and central Europe and at royal courts: Florence and Padua; Paris; London, Oxford, and Cambridge; Leiden and Utrecht in Holland; Brunswick in Hanover; and Basel.

During the second half of the seventeenth century an evolving institutional framework for the emergent natural sciences—in England and France especially—and a greatly improved network for transmitting the ideas and methods of leading thinkers benefited and, to an extent, reoriented the approach to mathematical inquiry. Fundamental changes were underway in the organizational support of mathematics, even though scientific activity and learned journals were still chiefly financed by noble, royal, and bourgeois patrons. To be sure, these patrons continued to support individuals and small groups such as the short-lived Florentine *Accademia del Cimento* (Academy of Experiments)[3] in the earlier mode. In addition, monarchs now chartered two durable, national scientific societies—the Royal Society of London in 1662 and the Paris *Académie Royale des Sciences* in 1666. These institutions grew out of informal gatherings of "worthy persons inquisitive into natural philosophy."[4] The Anglo-Irish chemist Robert Boyle called the London predecessor the "Invisible College." Both institutions, unlike their predecessors, were professionalized having bureaucratized organizational structures, strict membership criteria, and controls over publications. Ostensibly, they were established to harness scientific studies to practical ends for "the service of the king and the state." More importantly they fostered collaborative projects and evaluations among men of science. Furthermore, leading members did not restrict themselves to immediate practical ventures but produced ambitious theoretical work, as epitomized in the treatises of Christiaan Huygens and the Cassinis in Paris and Isaac Newton and Edmond Halley in England.

Conservative universities offered no comparable support for the new natural sciences. The supremacy of humane studies continued unchallenged and the older sciences were relegated to a secondary place. Nevertheless, a few notable professorial chairs were established in mathematics and natural philosophy. There had been distinguished chairs in these subjects at Padau, Wittenberg,[5] and the *College Royale* in Paris, and Gresham College had supported geometers and astronomers in London since late Elizabethan times. Oxford now added the Savilian chair, Cambridge the Lucasian chair, and Basel established the professorship filled successively by Jakob and Johann Bernoulli.

The spread of print technology led to the publication of learned journals, which allowed for a wider and more rapid circulation of mathematical ideas toward the end of the seventeenth century. The secrecy that prevailed in mathematics during the sixteenth century slowly began to disappear, as did the practice of deciding priority of discovery through public debates and contests. These debates were a relic of Scholastic disputes. Already at the mid-seventeenth century the French priest Marin Mersenne and the German-born scientific intelligencer Henry Oldenburg, the future secretary of the Royal Society, had enlisted correspondents from across Europe and the Levant. In the absence of general postal services within kingdoms, their efforts were indispensable to the regular exchange of scientific ideas and led to a crucial process of review and rebuttal to elicit and certify objective natural truths. Mersenne and Oldenburg did not limit the geographical bounds of scientific inquiry to the emerging nation-states; their circles of correspondents gave it an international scope. The two major scientific societies helped to found scientific journalism with the publication of voluminous communications. In the pages of the *Philosophical Transactions*[6] Fellows of the Royal Society addressed communications and announced discoveries to the whole world, and in the *Mémoires* Paris Academicians did the same. The articles in these journals ranged widely over natural history, including the investigation of minerals, flora and fauna, and the heavens; physical experiments relating to dynamics, optics, and acoustics; medical findings; and wonders and prodigies.

Mathematicians had two other specialized journals in which to publish—the *Journal des Sçavans*, a weekly founded in Paris in 1665, and *Acta Eruditorum*, the first scientific periodical in Germany, cofounded in Leipzig (1682) by Otto Mencke and Gottfried Leibniz. These journals were extremely important at a time when publishers hesitated to publish mathematical books because they were not very profitable.

By 1720, therefore, mathematics was gradually becoming professionalized, a status not fully attained until the nineteenth century. To become a profession the discipline required permanent employment for its members in universities, government, commerce and industry, its own scholarly organizations, vigorous schools of thought in mathematical education, and journals solely devoted to the discipline.

In Europe the century after 1620 was a time of crisis leading to a rise of absolutism with profound changes in politics, religion and economics.[7] Disintegrating empires together with a technological innovation played a decisive role in the political change. Cannon and gunpowder, "the weapon of cowards," had become widely available in western and central Europe since their introduction in the late fifteenth century. The new weaponry undermined the political stability of already weak regimes and strengthened a patchwork of rival dynastic states. Fortresses already vulnerable to cannons and mortars were further weakened by advances in military engineering applied to siege warfare. Monarchs employing the new technology, now gained absolute power over kingdoms at the expense of local princes and nobles. Only Muscovy in the east escaped the derogation of empires. The jockeying for wealth, power, and glory among competing absolute monarchs led to brutal wars, depicted as "the sport of kings." Frequent wars, famines and especially the worst plagues in three hundred years more than doubled mortality rates. Despite retarded and irregular growth Europe's population still rose from about 95 millions in 1600 to about 120 millions in 1700.[8] The Hapsburgs of Austria and Spain fought against the Bourbons of France in the two major wars that frame the era—the Thirty Years' War (1618–48) and the War of the Spanish Succession (1700–13).

Besides dynastic pluralism and the use of separate vernacular languages over a common Latin at royal courts, religious disunity furthered separatism in Europe. New Protestant faiths contended vigorously with the Catholic branch of Christianity. Religious strife persisted through 1720.

The Thirty Years' War linked political and religious passions so intensely that warfare reached a new brutality. In the eastern portion of the Holy Empire,[9] European Christians fought essentially a civil war. Its immediate causes were twofold: (1) Hapsburg rulers in Vienna struggled to impose absolute political authority over separatist Protestant nobility in Bohemia and (2) the same Catholic Hapsburgs attempted to restore traditional Catholicism to Protestant regions of Bohemia, Poland, and Austria. The war quickly destroyed most of the Bohemian high culture. With the introduction into the conflict of Swedish forces under Gustavus Adolphus and (after 1635) of French forces and financial subsidies, the devastation mounted. Marauding soldiers destroyed many towns and industries in the south German states and caused the death of an estimated one-third of their population. The lack of tight discipline in the military increased the carnage. The war was the worst European catastrophe since the Black Death three centuries earlier. One casualty was the social and economic structure that had supported Tycho Brahe and Johann Kepler in Prague under Rudolf II. That structure could not be rebuilt for over a half century.

The Treaty of Westphalia ended hostilities in the Holy Roman Empire in 1648. This settlement reasserted religious stability on the hoary principle of the Peace of Augsburg (1550) that each ruler determine the religion of subjects within his land. An added provision recognized Calvinism in Germany thus officially acknowledging the split in Christianity. The political price for a religious diversity guaranteeing the freedom of Calvinists as well as that of Lutherans and Catholics was the prolongation of a fragmented Germany consisting of more than 300 political subdivisions. The treaty, which diminished the authority of the Hapsburg rulers in Germanic Europe, placed dynastic interests above religious allegiance and thus is often called "a secular charter for Europe."

A reversal in geopolitical roles in economics also undermined Hapsburg fortunes. The Spanish had acquired economic hegemony in the early sixteenth century, when Italian cities and sea power declined. By the mid-seventeenth century, however, Spanish finances were crippled, partly because of Dutch and English piracy. The once formidable army was no better than the fleet in protecting Spanish interests. Seeking to break Hapsburg encirclement, France had declared war on Spain in 1635 as part of the Thirty Years' War and badly defeated its army at the Battle of Rocroi (1643). The two continued their conflict until 1659, when French victories forced the Spanish to sign the humiliating Treaty of the Pyrenees. By that time economic dominance had shifted decisively from the Mediterranean lands to those on the Atlantic and Baltic trade routes. Spain's sixteenth century enemies—Great Britain, France, and briefly the Netherlands—were now in the forefront.

The struggle among preindustrial capitalist powers extended quickly to colonial trade. In the commercial rivalry between Spain, France and Great Britain, Spain was clearly on the defensive by the early eighteenth century. This rivalry marks a second era in European overseas expansion that had begun with discovery, exploration, and emigrations. It centered on North and South America, the West Indies, and India. According to the prevailing economic doctrine of mercantilism, scarce resources required colonial powers to forge tight systems of national commerce and to encourage domestic manufacturing for export. Colonies were to supply natural resources to the mother country and could not trade with other countries. Navigation laws and high tariffs enforced these restrictions. This policy of national monopoly promoted the accumulation of gold and silver by the mother country. Early eighteenth century western Europeans imported coffee, furs, tobacco, corn, potatoes and tea. They also engaged in the profitable traffic in African slaves who ultimately numbered in the millions. The slave trade often followed a triangular pattern—slaves from Africa were traded for sugar in the West Indies, sugar for rum in New England, and rum for more slaves in Africa.

Even with expanding commercial capitalism, it would be a mistake to view the century after 1620 as one of widespread prosperity. A continuing pattern of inflation dating from the 1550s and known as the "Price Revolution" caused real wages to fall. The multicauses of that spectacular price rise probably included the influx of bullion from America, the growth of populations competing for goods, and pure greed.

The emergence of Britain and France as the ascendant powers in western Europe came only after several decades of internal strife in both countries during the mid-seventeenth century. In England the Stuart monarchs' struggle with Parliament provoked a constitutional crisis in which religious issues were the most prominent. By imposing forced loans, collecting illegal taxes (such as ship money) and billeting troops in private English homes, Charles I (1625–1649) prompted the House of Commons to issue the Petition of Right (1628) forbidding these practices without its consent. Charles' unyielding support of the *Book of Common Prayer* and the episcopal system also antagonized powerful English Puritans. His actions led to the Civil War in 1642. Oliver Cromwell and the New Model Army not only defeated the king and his forces, the Cavaliers, but also executed Charles in 1649. Cromwell thereupon ruled a Puritan Republic as Lord Protector until his death in 1658. During the Puritan Commonwealth and Protectorate London reached a population of almost a half million and began to surpass Amsterdam whose Bourse was the chief financial center of Europe. British economic strength lay in wool manufacture and exports, extensive coal-mining, iron production and, above all, in trade (shipping).

The power of Parliament increased after the Stuart restoration in 1660. The early reign of Charles II (1660–1685) saw public anxiety not over politics but the revived bubonic plague (1664–1665) and the Great Fire in London (1665–1666). Parliament confirmed the supremacy of law over the arbitrary rule of the king when it replaced the Catholic James II with his Protestant daughter Mary and her husband William III of Orange during the bloodless "Glorious Revolution" of 1688–1689. The House of Commons issued a famous Bill of Rights, and Britain ended the century as a model of limited monarchy, parliamentary government, and measured religious toleration for Protestant Nonconformists but not Catholics.

If the sixteenth century belonged to Spain, the seventeenth and early eighteenth centuries belonged to France. In France, an absolute monarchy crushed the possibilities of representative government and religious pluralism. Cardinal Richelieu (1585–1642), chief minister to Louis XIII, was the man most responsible for building absolutism and with it the modern centralized state. Using Machiavelli's principle of *raison d'etat* ("reasons of state"), Richelieu gave unrestricted political power to the state in its struggle against the nobility. Resentment against these policies led after his death to a civil war known as the Fronde (1649–1652) between nobles and king. The Fronde produced near anarchy during the early minority of Louis XIV when Cardinal Mazarin held power. During the seventy-two year reign of Louis XIV (1643–1715), France with the largest population in Europe, a rich soil, and strong traditions based on a common language attained military hegemony and cultural primacy in Europe.

Louis, the "Sun King," skillfully pursued governmental centralization and embarked on expansionist wars. His Jesuit tutor Bishop Bossuet had inculcated in the young monarch the doctrine of the "divine right of kings." Answerable only to God, the King was recognized as the embodiment of the state. This idea set the style for courtly life at the Palace of Versailles, which was built on the outskirts of

Paris. Amid architectural grandeur, wigged courtiers followed catechized etiquette which (together with the size of their wigs) prescribed a hierarchical status. Notwithstanding the pomp, they suffered physical torment by staying in damp, freezing apartments and, perhaps, psychological ones as well by having to offer constant flattery to the King and his retainers.[10] Still, Louis XIV successfully drew thousands of nobles to lavish Versailles thereby domesticating the aristocracy. He further weakened the influence of the nobility by installing members of the privileged middle class as ministers of his Council and *intendants* (administrators) of the royal provinces. Financially the new French absolutism was based on a mercantilist state capitalism created by Jean-Baptiste Colbert (1619–1683). To support an ambitious foreign policy, the war minister Louvois, known as the king's "evil genius," transformed a poorly organized army into the most powerful and best administered in Europe.[11]

Besides a centralized monarchy Louis also came to desire religious conformity. Influenced by Jesuit confessors, he suppressed the Jansenists, Catholic Puritans with Augustinian views. A secret marriage to the pious Madame de Maintenon lessened his philandering and made him more determined to unify France religiously. In 1685 he revoked the Edict of Nantes (1598) that had granted relative freedom of worship to Huguenots, French Calvinists comprising nearly 10% (or 1.75 million) of the population. Repressive laws accompanied the revocation, and dragoons (mounted infantry) were used to force conversions. This rigid intolerance forced the emigration of about 100,000 Huguenots, many of whom were businessmen and skilled artisans.

Economic surpluses accumulated in the 1670s largely from exports of linen, cotton, woolen cloth, and luxury goods, especially silk, were spent on war. Intent on glory, the Sun King's court embarked on a series of large-scale aggressions to achieve natural and defensible frontiers to the north and east—War of Devolution (1667–1668), Dutch War (1672–1678),[12] and War of the League of Augsburg (1688–1697). This could be attempted because the French nation–state enjoyed hegemony; no single power could withstand it. Fearing French domination other states formed defensive alliances to establish the balance of power. During the War of the Spanish Succession (1702–1713), which came at the end of Louis' reign, a Grand Alliance of armies defeated the French outside France. The war was fought not only in Europe but in America, Asia, and wherever the powers had colonies and trading stations. The costs almost bankrupted the French economy. While the Treaties of Utrecht-Rastadt (1712–1714) recognized Philip V, a Bourbon and Louis' grandson, as king of Spain, they also acknowledged Britain's emerging primacy as a sea power and granted her dominance in the slave trade.

Central and eastern Europe were also changing. More agrarian and with large estates populated by serfs, the states of the region had fewer cities and no overseas empires. Monarchical authority and state power were prominent issues. After Westphalia the Austrian Hapsburgs had reduced their participation in north German affairs and consolidated their influence in Bohemia, Austria, the Tyrol, Croatia, and Hungary. At the time, two new powers, Brandenburg-Prussia and Russia, were emerging. Hohenzollern rulers merged the scattered and vulnerable territories of Brandenburg, an area poor in natural resources and with a barren soil, into a state that grew in size. Frederick William (1640–1688), the Great Elector, began the consolidation by creating a standing army to ruthlessly crush independent towns and territories of noblemen. His efficient bureaucracy, especially the "general war commissariat," amassed a full treasury by following the Cameral doctrines of rational economies. After defeating rebellious Junkers (rural nobility), Frederick William recruited their sons to lead the army and bureaucracy, beginning a longstanding tradition. He also attracted skilled immigrants, including 20,000 French Huguenots to Berlin after 1685. Below the aristocracy the populace was obedient and, in the case of most of the peasantry, servile.

Previously remote and isolated on the periphery of Europe, Muscovy emerged as a military presence of the first order. After a "Time of Troubles" in the early seventeenth century, the new Romanov tsars gradually subdued the boyars (nobles), attempted to control the Russian Orthodox Church, and crushed continuing peasant uprisings. Peter the Great (ruled 1682–1725) introduced a forced program of modernization. His government was centralized by introducing a series of "colleges" or bureaus headed by ministers and by establishing a nine-man Senate (1711) to guide the bureaucracy. The main task of the bureaucracy was to collect taxes for an enlarged army. Peter opened new schools, acquired craftsmen from the West, founded an iron industry in the Urals, and built St. Petersburg on the Gulf of Finland. The Ottoman Empire (now badly decayed) and Sweden (a northern outpost making its first sortie into European power politics) had hemmed in Russia in the past. Petrine Russia went to war and defeated both of them. With Sweden's defeat in the Great Northern War (1700–21), Russia became the leading power on the Baltic.

Europe's turbulent "century of genius" embraced the visual arts, literature, and thought, as well as mathematics. For some scholars this coexistence of high creativity in culture and high disorder in society is paradoxical; to others it indicates the tenacity of Europe's republic of letters, expanded patronage from wealthy bourgeoisie, and the robust vitality of the age. The visual arts, which were generally suffused with a refined geometric taste, continued to flourish amid change. In painting, the Baroque style with dramatic scenes and vivid colors replaced the Classic. Leadership in painting, etching, and drawing shifted from Italy to Flanders and Holland. The greatest early Baroque masters were the learned, multilingual Peter Paul Rubens (d. 1640) in Antwerp and Rembrandt van Rijn (d. 1669) in Amsterdam. Rubens, whose stylistic authority equalled that of Michelangelo a century before, was known for his rounded, fleshy female nudes and poignant crucifixions. Rembrandt developed the concept of art as a form of reasoned organization that paralleled Descartes' rational cosmology and Spinoza's mathematical ethics and that appealed to the prosperous Dutch bourgeoisie with its order. Known for naturalism, effective lighting, and honest detail, Rembrandt gave a special place to etchings. His *Three Trees* suggests the vastness and power of nature. Major figures elsewhere were Spain's court painter Diego Velázquez (d. 1660), who created sweeping impressionistic scenes (such as the surrender of Breda), and Prague's Karl Skreta (d. 1674), who synthesized Roman and Bolognese naturalism in his warm chiaroscuroes. The early eighteenth century saw the French painter Jean Antoine Watteau (d. 1721) depart from historic themes to portray in a graceful and sensual manner the *fêtes gallantes*—secular fantasies—of Louis XIV's court and its imitators.

Italy generally retained leadership in architecture and sculpture. At the Papal Court in Rome, Gianlorenzo Bernini (d. 1680) constructed his famous spiraled columned baldacchino for St. Peter's and the spectacular "Fountain of the Four Rivers," while in Piedmont the Theatine priest Guarini (d. 1683) demonstrated a distinctive geometric taste for square designs beneath convex surfaces and for receding tunnels in steeped arches. He took the latter idea from Islamic flying arches and superimposed it to form the faleric of the dome of the Cathedral of Turin. Also of note, the English architect Christopher Wren (d. 1723), a Fellow of the Royal Society, rebuilt much of London, including St. Paul's Cathedral, after the Great Fire.

Reserving their foremost loyalty for religion, many men of letters continued to appeal to Biblical authority, but a critical rational spirit especially evident in investigations of the physical world led some to skepticism. Literature in England and

France expressed the competing themes. In England the new sciences figured in the writings of the brooding clergyman John Donne (d. 1631) and the poet John Milton (d. 1674), whose majestic epic *Paradise Lost* (1667) deals with evil, pride, and man's place in the cosmos. Donne condemned Copernicus and his views in *Ignatius his Conclave*, while Milton moved away from Ptolemaic toward Copernican astronomy in his later poetry. (Addressing a major issue of the Puritan Revolution and the Restoration Milton wrote *Areopagitica* (1664), an eloquent defense of freedom of the press.) Milton and many Puritan reformers, who saw scientific activity as a cognate to their scheme for change,[13] supported the new sciences. Many other English authors did not. Walter Charleton spoke of "the dark lanthorne of reason," while others portrayed God's plan and interventions in the universe as inaccessible to human reason, a view suggested by John Bunyan (d. 1688) in *Grace Abounding* (1666). Bunyan also wrote *Pilgrim's Progress* (1684) tracing a Puritan guest for salvation.

Drama and the essay dominated French literature. Blaise Pascal (d. 1662), an austere Christian moralist and mathematician, created modern French prose. His *Provincial Letters* (1656–1657) was a sustained assault on the casuistry of the Jesuits. In Paris the plays of comedian Jean-Baptise Poquelin (d. 1673), whose adopted name was Moliére, ridiculed pretentiousness (*La Misanthrope*, 1661). The poetic dramatist Pierre Corneille (d. 1684) probed the tensions between Christian morals and vengeful ambition, and the master tragedian Jean Racine (d. 1699) examined forbidden passions.[14]

As early as the mid-seventeenth century, a separation developed between compliant religion and Cartesian rationalism in French intellectual circles. This distressed Pascal, who believed that the immensity and wonder of nature exceeded the bounds of mathematical thought. His *Pensées* censured those who rely solely on deductive reasoning. Pascal, like Galileo, believed that the underlying structure of the material world was given in mathematical handwriting of the three branches of geometry: mechanics (motion), arithmetic (number) and geometry (space). To attempt to fathom the inexhaustible diversity of the firmament beyond the intelligible underlying structure, however, one must depend on faith. "The God of the Christians is not simply the author of geometrical truth," he wrote. "I cannot forgive Descartes" for eliminating the analogy between God and the world—that is, the "*imagines*" and "*vestigia Deo in mundo*"—and making God into simply a watchmaker.

At the close of the century, the Huguenot refugee Pierre Bayle (d. 1706) turned critical rationalism toward Biblical criticism. A genial Pyrrhonist (after the ancient Greek skeptic), Bayle attacked superstition, a credulous acceptance of miracles, and a literal interpretation of the Bible in his *Historical and Critical Dictionary* (1697). The French Encyclopedists of the mid-18th century adopted his sometimes oblique methods of criticism of the misuse of Biblical authority.

Studies in the law and political theory addressed urgent issues for the centralized state and absolute monarchy. The Dutch diplomat Hugo Grotius (d. 1645), a disciple of Cicero and Seneca, denounced aggressive wars in *De jure belli ac pacis* ("Rights of War and Peace," 1625) and persuasively argued for an international law based upon man's nature and his place in the universe. The German jurists Samuel Pufendorf (d. 1672) and Christian Thomasius (d. 1728) further developed the concept of natural law. Political theorists attempted to answer questions of political obligation, primarily those relating to the necessity of obedience and the right of rebellion. Thomas Hobbes (d. 1679), an Englishman, was the most original. In *Leviathan* (1651) he offered a purely intellectual ground for absolutism rather than advancing the traditional appeal to divine right as Bossuet had. Contrary to Aristotle and Aquinas, he asserted that the natural state of man was "solitary, poor, nasty, brutish, and short." Man must form a "great Leviathan," or central government, whose ruler was absolute. Geometry, "the only science . . . God (hath) bestow(ed) on mankind," provided a model for civil order because it began with agreed upon definitions and principles.[15] Moral philosophers were berated for not discharging the duty to their science as well as geometers had to theirs.[16] Although Hobbes did not desire tyranny, he indicted revolution for the anarchy it brought in the form of civil war. The harshness of his explanation of human psychology, religious skepticism, and materialism made him the *bête noir* of political theorists. His countryman John Locke (d. 1704) opposed absolutism in the celebrated *Two Treatises of Government* (1690). A puritan who provided the intellectual justification of the Glorious Revolution, Locke argued that monarchs must preserve the natural law and not transgress it. Expressing the view of the commercial bourgeoisie, he spoke for a representative government of which the king was protector and guarantor.

By definition the Scientific Revolution brought vast and irreversible changes in the scientific view of the physical universe and, more generally, in Western thought. The late Herbert Butterfield compared its importance in the West to that of the rise of Christianity. One of its achievements, the change from a geostatic model of a small, immutable, and hierarchically-ordered universe to a heliocentric, changing, and infinite one, displaced humanity from the center of the cosmos and gradually fostered a spectacular secularization of thought. It undermined and eventually displaced hallowed traditions in morals, psychology and religion.

The Scientific Revolution, so named by Alexandre Koyré, transpired over the course of two centuries and was the handiwork of a small minority of several hundred men and a few women.[17] New ideas were advanced and then abandoned. Others that proved useful and increasingly precise were partially or wholly accepted, elaborated, and consolidated into larger and more elegant conceptions. Understanding grew. Slowly the work accumulated in the studies, libraries, observatories, academies, and crude laboratories of Europe. The most notable developments occurred in Italy, Poland, the German states, Bohemia, the Netherlands, France and Great Britain.

Beginning in the mid-sixteenth century amid the revival of learning known as Renaissance humanism, it embraced anatomy, alchemy, astronomy and mathematics. The Flemish Andreas Vesalius (d. 1564), one of four major founders, corrected the anatomical work of Galen on the basis of careful dissections of human cadavers. Human dissections had long been severely limited in Christian Europe and were usually performed in medical classes by untutored barber surgeons. Vesalius' accurate findings given in *De humani corporis fabrica* ("On the Fabric of the Human Body," 1543) and studies of the heart and blood movement at Padua prepared the way for the discovery of the circulation of the blood by William Harvey (d. 1657). Harvey, physician to Charles I, detailed his discovery in *De motu cordis* (1628).

Alchemy and medical chemistry were pursued as avidly. The irascible Paracelsus, another and more inflammatory founder, dramatically rejected ancient authority by openly burning books by Aristotle, Galen, and Avicenna. Recommending that new observations be made in the medical sciences, he introduced departures from the past by replacing Galenic herbal medicine with chemical cures using diluted mercury. He also rejected the Peripatetic doctrine of four elements—earth, air, fire, and water—proposing instead the three principles—ideal forms of salt, sulphur and mercury. Primal substance or, more narrowly, the theory of matter occupied many seventeenth century thinkers. Pierre Gassendi (d. 1655) revived the Democritan-Epicurean theory of atomism, which most theologians opposed because of its materialism, while Robert Boyle attempted to define the term "element" in *The Scyptical Chymist* (1661). In addition, the German physician Georg Stahl (d. 1734) elaborated the phlogiston theory of combustion in chemical science. Alchemy covered more than these pursuits. A complex segment of natural philosophy, radically different from mechanics, it had strong roots in the Hermetic, Cabalistic, and Natural Magic traditions. The Hermetic tradition derived from the teachings of the ancient magi, while the Cabalistic stemmed from number mysticism in Roman Alexandria.[18] These traditions also contributed to the new scientific activity in astronomy and mathematics. No less than Isaac Newton spent decades on an extended program of study of alchemy while at the peak of his career.

Neither medicine nor alchemy equalled the advances made in the "high sciences"—mathematical astronomy, the mechanical philosophy of nature, optics and mathematics. Their often interrelated development proceeded through three phases beginning with Copernicus. In *De revolutionibus orbium coelestium* ("On the Revolution of the Heavenly Spheres," 1543), Copernicus set forth a heliostatic theory of the heavens in place of the Ptolemaic geostatic model. Copernican astronomy was not entirely novel: his universe, while larger than Aristotle's, was finite (bounded by the fixed stars), and celestial motions were circular and uniform as in Aristotle and Ptolemy. Nevertheless, there was strong criticism. Catholic and Protestant religious critics railed at a Copernican astronomy which seemingly contradicted the divine testimony of Scripture in the Old Testament,[19] while Aristotelians correctly observed that it lacked a supporting physics. The Danish astronomer Tycho Brahe (d. 1601) attempted to mediate the models of Copernicus and Ptolemy by placing the earth at the center of the universe with the moon and sun circling it. The Tychonic system had all other planets circle the sun. The Italian cosmologist Giardano Bruno (d. 1600) enthusiastically accepted Copernican astronomy and speculated that the universe was heliocentric and infinite. For his view on the material nature of the soul and his praise of ancient Egyptian religion, Bruno was burned at the stake.[20] New ideas were not without danger, especially in an age that associated them with the dark arts.

A second phase of advance that included a telling attack upon Aristotelian physics and increasing attention to mathematization began with the mature work of Johann Kepler (d. 1631) and Galileo Galilei (d. 1641). Kepler's demonstration of the elliptical orbits of planets based on a painstaking analysis of the path of Mars sounded the death knell for Aristotelian celestial physics with its circular motion. Kepler's three planetary laws, so fundamental to Newton's grand synthesis, went largely unappreciated by contemporaries. Although Kepler's Latin satisfied academic fashion, his treatises were too mathematical to be grasped by most scholars of his day. Kepler recognized his isolation when he wrote in the preface to *Astronomia nova* (1609): "It is a hard task to write mathematical and above all astronomical books for few can understand them. . . . I myself, who am considered a mathematician become tired when reading my own work." His studies of numerous curves and their application to astronomy reflected the deep belief of a mystical visionary in Pythagorean and Neoplatonic perfection.

Galileo, a disciple of "divine Archimedes," rejected the Aristotelians' radical separation of terrestrial from celestial physics. Galileo's deduction of the law of freely falling bodies and the parabolic paths of projectiles broke with the tenets of Aristotelian terrestrial motion, while his telescopic discovery of the rugged surface of the moon contested the Aristotelian notion of perfect celestial bodies and perfect crystalline spheres. Galileo went far toward the portrayal of a rational universe with regularity of behavior throughout—a universe better described by mathematical laws, numerical relationships, and measured quantities than by Scholastic logic.

Seventeenth century thinkers examined two books: the Book of Nature and the Scriptures. In *Saggiatore* (1623) Galileo wrote: "The Book (of nature) is . . . written in mathematical language, . . . without whose help it is impossible to comprehend a single word of it; without which one wanders in vain through a dark labyrinth."[21] Here was a confession of faith in mathematics not theology. No longer could well-founded mathematical models be easily dismissed as mere hypotheses or artful devices that "need not be true or even probable," as Osiander had stated in the introduction to Copernicus' *De revolutionibus*. Galileo's advocacy of the authority of mathematics in describing the physical universe and his continuing defense of Copernican astronomy in *Dialogue of the Two Chief World Systems* (1632) led to his famous trial the next year. This trial and his recantation under threat of torture became the *cause célèbre* of the period. Galileo's most brilliant mathematical book, *Two New Sciences* (1638) advanced kinematics. A new epoch was beginning, where mathematics was used more extensively in physics.

By the 1620s the ferment and new discoveries in the high sciences fostered refinements in method and a new systematic treatment. The two methods were induction and an expansion of deduction, the first propounded by the English stateman Francis Bacon (d. 1626), and the second by the French thinker René Descartes (d. 1650). Taking a confident "ancient versus moderns" approach, Bacon struck out against a reverence for old authority in *Novum Organum* (1620), a part of his *Great Instauration* intended to surpass Aristotle's *Organon*. He found Scholastic logic with its syllogisms acceptable for civil business and opinion but "not nearly subtle enough to deal with nature." Bacon conceived of the emerging natural sciences as knowledge grounded in sensual experience. His universe was mechanistic and filled with the actions of corpuscles, essentially atoms. To investigate, discover, and explain the structure of the physical world, he proposed a new form of induction leading to the "invention not of arguments but of arts. . . ." His induction aimed to reach operation. He used a striking analogy from the world of insects: Induction was not simply a process of collection of physical facts (resembling the work of the ants), nor was it speculating in abstraction (resembling the spider that spins its web from itself); it was more closely joined like the activity of the bee that "gathers its materials from the flowers of the garden and the field, but transforms and digests it by a power of its own." To make correct inductive inferences, investigators were to purge from the intellect four prejudicing idols that belonged to the tribe (excessive anthropomorphism), cave (personal prejudice), market place (social conventions), and theater (ambiguity of language).

Bacon's interest in the new natural sciences went beyond establishing the inductive method in his theory of knowledge. He wanted them to be public, cooperative, and cumulative. He was a propagandist for the new sciences and a canvasser of their achievements. Unfortunately, Bacon knew little about their actual substance, however. He treated mathematics simply as a tool, rejected the "absurd opinions" of Copernican theory, and misunderstood Galileo.

Descartes, who epitomized the geometric spirit of the age, set forth a new method of deduction and formulated a comprehensive scientific system. He divided the world into two parts: mind (unextended thinking substance) and matter (extended substance). He, like Galileo, thought that mathematical laws describe the material world, and these laws could be grasped by the speculative use of God-given, human reason. His guide to deductive reason in the *Discourse on Method* (1637) was just as revolutionary as Bacon's induction, containing optimism about new learning and spleen against Scholasticism. To reach his method Descartes had begun by doubting everything, including his own existence. He found he could not doubt the existence of doubt itself and concluded that certainty was to be found in the act of pure thinking. Descartes' ruthless rationalism pursued the ideal of mathematical certainty and accepted as true only "clear and distinct" ideas. It progressed through logical steps from the simplest to the most complex issues. Interwoven in his reasoning was the empirical criterion of a crucial experiment to confirm theory. In general, his deductive method reversed the Baconian order of inquiry by beginning with the simple and general, developing a chain of causation, and ending with specifics. Descartes' rationalism included conjectures or hypotheses as a basic tool of research. Although he sincerely denied that his absolute separation of mind and matter (dualism) endangered established religion, its application led others to skepticism and materialism.

Descartes explained the material world in an all-embracing corpuscular mechanism in *Principles of Philosophy* (1644). He banished the spirits, sympathies, and occult qualities of the Scholastics from the material world and considered only mechanical causes and effects. Matter in motion was the fundamental, palpable reality existing within a clockwise universe. His corpuscles of matter resembled atoms but yet retained continuous divisibility. Reduced to its simplest form, Cartesian mechanics was one of impacting billiard balls. Bodies acted on each other in a plenum (Descartes admitted no vacuum), where action-by-contact (impulsion) was the only admissible force. Under the Cartesian law of inertia motion acquired a status equal to that of rest, and the status distinction between rectilinear and circular motion vanished. These were two major departures from Aristotle. Descartes criticized Galileo for not concentrating on general cases but rather on particular ones, and for granting existence to a state of nothingness, the vacuum. An invariant "quantity of motion" (magnitude times speed) was conserved in Cartesian mechanics. Celestial motions, which maintained the Aristotelian separation from terrestrial mechanics, depended on theoretical vortices.

Although Descartes wished to geometrize physics, his system requiring continuous mechanical contacts was almost entirely speculative. It lacked a mathematical technique, such as the calculus, to make highly accurate calculations describing or predicting physical phenomena. The notion of "more or less" was not yet expunged from physics. Still, Cartesian science proved compelling. It supplanted Aristotelian physics and enjoyed primacy in continental Europe until the late 1730s. Jacques Rohault synthesized it in his famous textbook, *Treatise on Physics* (1671), and Malebranche's circle in Paris extensively developed it. Its acme in influence came with the publication of Fontenelle's *Essay on the Plurality of Worlds* in 1686. The next year Newton published his *Principia*.

The Scientific Revolution culminated in the late seventeenth century in spectacular breakthroughs in the mechanical philosophy of nature, optics, and the theory of matter. Isaac Newton in England made the greatest headway, presenting the core of his ideas in *Principia Mathematica* and *Opticks* (1704). In these Newton prepared a grand synthesis of the prevailing corpuscular philosophy, mechanics, and mathematics. Building upon ancient Greek and Gassendan atomism, he portrayed the physical universe as composed of impenetrable particles acted upon by forces, like magnetism. These forces were real and prior to particles, a novel view differing from Descartes, who had believed that forces were the result of a movement of particles. In the *Principia* Newton enunciated sound definitions and systematized dynamics. His laws of motion demonstrated that the force of attraction inversely proportional to the square of the distance accurately describes planetary motions according to Kepler's laws and the motion of freely falling and projected bodies on earth according to Galileo's laws. Thus, celestial and terrestrial mechanics were united for the first time, and Copernican astronomy justified. The law of inertia was fundamental to Newton's dynamics, and space and time were absolute. The *Prin-*

cipia included a consistent system of equations dealing primarily with two bodies influenced by mutual gravitational forces. These equations had a restricted field of application.

Newton's second seminal book was the *Opticks.* It presented his corpuscular theory of light and the nut-shell theory that matter occupies but a small portion of a nearly vacuous universe. In arriving at conclusions he followed a critical empirical methodology. In the tradition of Bacon and Locke, Newton required that theory be anchored in experience. Later editions of the *Opticks* conclude with 31 queries that go beyond physical optics and raise questions about magnetism, electricity, and the ether.

On the Continent, scholars made major advances generally within the framework of Cartesian physics. Most accepted its mechanical causes and impulsions as sound, for example, and believed that only some details needed correcting. The Cartesians also attacked Newton's attraction, which had no mechanical cause, as an occult quality. Huygens, a disciple of Archimedes and Galileo, rejected one element of the Cartesian system, the principle of the conservation of "quantity of motion." He proposed instead the conservation of the magnitude of a body times the square of its velocity *(vis viva)* in collisions. Huygens, who invented the pendulum clock, discussed centripetal and centrifugal accelerations in his magnum opus, *Horologium Oscillatorium* (1673). He also proposed an undulatory theory of light in his *Treatise on Light* (1690). His student Leibniz modified the conservation of *vis viva* and used it to unify his rudimentary analytical mechanics. Further Leibniz had animate monads as the primal substance of the universe. These metaphysical points of force (energy) differed from each other only in degrees of awareness. His universe was everywhere alive; it was a natural, not a passive machine. The Bernoulli brothers, who corresponded with Leibniz, investigated mechanics and hydraulics. They made the principle of sufficient reason, so fundamental to Leibniz's metaphysics (or rational theology), the basis of Classical physics. It states: "If all events from now to eternity were constantly observed, it would be found that everything occurs for a definite reason."

Mathematics too had enjoyed spectacular development since the mid-sixteenth century. Algebra and arithmetical computations particularly had flourished. Highlighting the work in the elementary theory of equations was Girolamo Cardano's solutions for the general cubic and quartic and Francois Viète's attempt to join Euclidean geometry with numerical algebra. Viète improved algebraic symbolism as well. In computations, Stevin invented decimal fractions. Extensive calculations, particularly those required in astronomy and navigation, culminated in John Napier's logarithms, which reduced multiplication and division to addition and subtraction.

After 1620 there was a broadening of creative mathematical activity. Geometry, having lagged behind algebra, now revived with results that were unmatched since antiquity. Renaissance painters had given new directions to the science of perspective—or projections. The German engraver Albrecht Dürer (d. 1528) viewed subjects through a grid and the cartographer Mercator (d. 1594) extended three dimensional projection techniques to his map-making. The study of the Latin translation of the surviving first four books of Apollonius' *Conic Sections* by Frederigo Commandino (d. 1575), of new curves in astronomy, and of optics, especially the design of lenses for the telescope, had increased attention to the geometry of projected shapes. Seeking to rationalize and unify a range of contemporary *ad hoc* graphical techniques, to improve upon them, and to integrate projective methods with mathematics, the French architect Girard Desargues (d. 1661) invented a new branch of mathematics—projective geometry. His *Brouillon project . . .* (1639) on conic sections concentrated on projective transformations and mathematical invariance, those properties of figures remaining unchanged through various transformations.

The *Brouillon project's* projective geometry had little impact upon the learned scientific community. Desargues printed only fifty copies, sending them to colleagues. Perhaps only the young Blaise Pascal fully recognized the importance of Desargues' work. Pascal's *Essay on Conics* (1640) contained an important new principle: stated in modern terms, If a hexagon is inscribed in a conic, the three points of intersection of the pairs of opposite sides are collinear. Pascal's principle gave Pappus' theorem of three points on each of two lines as a special case. The *Brouillon project* was also communicated to Fermat and Descartes, who discerned the ability of the author. Descartes called it a "beautiful invention" and "ingenious" in its simplicity. Both considered the scope of projective geometry too limited to bear comparison with their algebraic geometry, however. Desargues' highly original vocabulary and difficult symbolism may have prevented less gifted readers from appreciating it. As the vogue for Cartesian analytic geometry and the infinitesimal calculus grew, Desargues' new geometry was largely forgotten. The Frenchman Philippe de La Hire's contributions to polar theory were an exception, and so was Leibniz's study of the invariant properties of central projections. Projective geometry was not resumed fully until the nineteenth century when mathematical invariance began to dominate physics. Two decades after Jean Victor Poncelet began to establish the field in the 1820s, Michael Chasles rediscovered Desargues' earlier contributions.

With projective geometry Desargues had wanted to introduce exactitude and efficiency into the practical graphical techniques of sculptors, artists, draftsmen, engravers, artisans, and stonecutters. His prolix attempts to explain himself were largely rebuffed by the people he was seeking to teach. Offensive to those committed to older techniques and, above all, to the powerful craft guides, they generated sharp opposition. After several harsh polemics in the 1640s, Desargues demonstrated the efficacy of his graphical techniques by collaborating in designing houses and mansions in Paris, distinguished in several cases by complex, beautiful staircases, and public buildings in Lyons with architecturally delicate rooms.

Analytic geometry was the chief achievement in seventeenth century geometry. Whereas projective geometry was simply a branch of the discipline, analytic geometry offered a novel and powerful method. The way had been well-prepared for its origination. As part of the humanist revival of ancient wisdom in the sixteenth century, geometers had translated and mastered ancient Greek mathematical classics, especially those of Apollonius and Archimedes. These studies showed the stringency required in proofs. The Renaissance mathematicians sought more; they also wanted a heuristic tool for invention, missing in Aristotle and also unavailable in many ancient geometric classics. They began to seek clues to a true analytic method of discovery in Archimedes' *Quadrature of the Parabola,* Apollonius' *Conic Sections* and *On Contacts,* Diophantus' *Arithmetica,* the writings of Heron and Proclus, and above all in Pappus' *Mathematical Collection.* The search resulted in Viète's symbolic algebra that was substituted for ancient geometrical algebra. An important feature of Viète's *Introduction to the Analytic Art* (1591) was its operational, though not modern, notation. His notation systematically used vowels to represent variables and consonants for parameters and employed abbreviations to designate algebraic operations and exponents. Thus, Viète might write B cub + C in B quad aequatur C plano in B + D solido for our $B^3 + B^2C = BC + D$. His notation clarified the roles of variables and parameters in equations, focused attention on solution procedures, and shifted emphasis from a particular solution to the general—an important advance toward the algorithmic approach of the calculus. Finally, Viète connected algebra and geometry through his determination of equations that correspond to various geometrical constructions. He did this only for determinate equations in one unknown, however. The next step to indeterminate equations with continuous variables in two unknowns was taken almost simultaneously by Pierre Fermat (d. 1665) and René Descartes (d. 1650).

Fermat and Descartes took complementary approaches to founding essentially the same analytic geometries. According to the fundamental principle of the discipline, every equation—in modern notation $f(x,y) = 0$—defines a curve. Fermat usually began with an algebraic equation and derived the geometric locus of the corresponding curve. His sources were the writings of Viète, his reconstruction of Apollonius' lost treatise *Plane Loci,* and Archimedean locus problems to determine fulcrum points and centers of gravity along an ideal balance.[22] Fermat set forth his ideas in the brief paper "Ad locos et solidos isagoge" (1636), which he circulated in manuscript form. The "Isagoge" not only considered the straight line, circles, and conic sections, it also proposed new mechanical curves. Although Fermat's work was in many ways more systematic than Descartes, it suffered by comparison for its adherence to Viète's outdated symbolism. For example, Fermat used A and E for x and y. The "Isagoge" was first printed posthumously in 1679 robbing Fermat of recognition. Fermat never wrote for publication.

Descartes began analytic geometry with the geometric construction of curves and then derived their algebraic equations. He believed that these equations lacked the clear and distinct ideas of geometry. His sources were apparently Ramist methods of invention, a Renaissance ordering system for mnemonics, and Hermetic efforts to construct knowledge from a set of mental loci as well as algebra and geometry. His *Géométrie* (1637) was outstanding for its notation. It introduced the system of designating exponents by Hindu-Arabic numerals and using the first letters of the alphabet for parameters and those near the end for variables. Neither he nor Fermat had a coordinate framework for curves. They both used a single axis with moving ordinates (the verticals). Descartes studied geometric curves and kinematically constructed (mechanical) curves, which he erroneously said the ancient Greeks ignored. He also presented a general method for finding the normal to a curve at any point—calling this "the most useful and most general problem . . . in geometry." The tangent is perpendicular to the normal. The construction of tangent lines is in turn, a corollary of determining maximum and minimum values. Descartes' circle method of constructing tangent lines was prohibitively tedious when the equation of a curve is not a simple algebraic one.[23]

In 1638 he engaged in an acrid debate with Fermat over methods of maxima and minima, its corollary method of tangents, and analytic geometry. He called Fermat's ideas paralogistic and limited but soon saw his error. The debate subsided but not before it poisoned his relations with Fermat's spokesman, Roberval (d. 1675).

Analytic geometry gradually gained acceptance. Descartes' choice to publish in French rather than Latin and deliberately to leave gaps in his presentation slowed the process. The second enlarged edition of the Latin translation of the *Géométrie* (1659–1661) made by Franz van Schooten and containing Jan Hudde's rule for finding double roots of equations gave analytic geometry tremendous influence. Hudde's rule simplified Descartes' circle method for tangents. By this time the contributions of Fermat were largely forgotten. Analytic geometry allowed European mathematicians to generate a plethora of curves from equations, a sharp contrast to the ancient Greeks who had few curves. Next to conic sections, the cycloid was the most studied curve. (The cycloid is the curve traced out by a fixed point on a circle rolling along a straight line.) Roberval carefully examined it in the 1630s and claimed to have rectified it by reducing the problem to one of integrating the sine. Constructing a straight line segment equal in length to a given curve previously had been thought impossible for algebraic curves. Torricelli and Wren later made the same rectification. Because the cycloid was the subject of or proved the solution to a number of hotly disputed challenge problems, it became known as the "Helen of Geometry." It was, for example, the solution of the Bernoullis' brachistochrone problem.

Number theory and a calculus of probabilities were also developed. Fermat contributed to both. Adopting a Janus-like approach that looked to the past and the future, he first meticulously examined Diophantus' *Arithmetica* for guidance in number theory. He broke with the past, however, seeking integral solutions rather than the Diophantine rational ones. Fermat especially investigated divisibility in the domain of integers (lesser theorem) and Pythagorean triples (last theorem). (These theorems are discussed in Fermat's biography on this anthology.) Fermat sought strict methods for number theory and separated it from algebra. He worked alone in the field and could not convince his contemporaries of its importance.

Probability emerged from a different intellectual context than algebra, geometry, or number theory. These last three were part of "high science," *scientia,* with its exact demonstrable knowledge. Probability belonged to the "low sciences," such as Paracelsian alchemy, with their signs or indicators and with counts of the frequency of their correctness. It fell in the domain of *opiniones,* which consisted of a body of convictions that were only likely to be true *(versimiles).* The social roots of probability lay in risks associated with extensive gaming activity and with the contingencies covered by the contract law of the latter 16th century commercial capitalism. Grotius wrote that contracts promote "beneficial intercourse among men." Latin cannon law included aleatory contracts that dealt with risks and annuities. One might enter into an aleatory contract with only a slight possibility of success in a venture. Rules were established to ascertain the fairness or equity of the agreement and the prudent possibilities for profit or loss.

Mathematical probability dates formally from the mid-seventeenth century, when chance and expectation came to be associated with *probabilis.* The algebra of Viète and Descartes had led to a widespread belief in the efficiency of mathematics. As a new vogue for gaming swept the French court, Pascal exchanged letters with Fermat in the spring and summer of 1654 on the problem of points—how to divide fairly the stakes from an unfinished game of chance. They began to establish the conceptual foundations of probability theory, starting with denoting the chance of success as equal to favorable outcome compared to all possible outcomes. The correspondence ended when Fermat returned to number theory. Pascal was associated with the Port Royal *Logic* (1662), which attempted to quantify the probability of the occurrence of a good or evil or of improbable events like miracles. Pascal's wager on the existence of God in the *Pensées* reflected a prudential approach to expectation.[24] While in Paris in 1655, Huygens learned of the Pascal-Fermat correspondence and prepared the first treatise on probability, *Calculation on Hazard.* This seminal book, which was written in Dutch, computed and catalogued possible outcomes of fair games. Van Schooten's Latin translation, *De ratiociniis in Ludo aleae* (1657), brought out the term *expectatio,* or expectation, for the payoff table. Jakob Bernoulli included this Latin treatise in his *Ars conjectandi,* 1713, which gave a coherent mathematical treatment to chance proportions, moral (that is, social and psychological) probability, and the simplest form of the law of large numbers. Bernoulli's book made probability a subfield of mathematics. Probability had growing applications in the gathering of demographic data and in insurance calculations during a time of intensive warfare.

Extensive groundwork preceded the inventions of the infinitesimal calculus. Two major sources were the formulation of several area and tangent methods, from which the basic algorithms of the calculus were taken, and the development of powerful infinitesimal series techniques. Kepler's study of curves associated with planetary motion and Galileo's investigation of constant acceleration had increased awareness of the significance of the area under a curve (quadrature) and the slope of a curve. Descartes, Fermat, Roberval, and Torricelli on the Continent and Barrow in England devised methods for constructing tangents and its kindred method of finding maxima and minima, that is extreme values, of a quantity. On a plotted curve the latter involved finding slope changes near maximum and minimum and points of inflection. Fermat's method of maxima and minima nearly created a differential calculus of infinitesimals. Attempting to find two line segments, B and a part of it A, such that the rectangle ($AB - A^2$) contained on them is a maximum, he proceeded as follows: Substitute A + E for A. This gives the pseudo-equality

$$(A + E)B - (A + E)^2 \approx AB - A^2$$

$$AB + EB - A^2 - 2AE - E^2 \approx AB - A^2$$

(subtract common terms from both sides)

$$EB - 2AE - E^2 \approx 0$$

(Divide by E)

$$B - 2A - E \approx 0$$

(discard the E term and change to equality)

$B = 2A$. The result is a square.

Fermat's method posed thorny problems for contemporary mathematicians and has been interpreted in various ways. Dividing by E and then setting E equal to zero was troublesome. Fermat could not prove his method. To him, however, his method's working in practice was more important. Descartes severely criticized this approach. It is tempting today to read too much into Fermat's method by setting $E = x$ and $A = f(x)$. However, Fermat did not treat quantity A as a function, nor E as an infinitesimal, and the method does not involve the concept of limits. It is an algebraic process.

Investigations of the method of exhaustion and the arithmetic of the infinite produced rudimentary integration techniques. Kepler began by computing exactly and approximately the volumes of ninety wine casks in *Nova stereometria* . . . (1615). He summed solid "indivisible" components. Kepler's intuitive considerations and Galileo's study of the infinite led Bonaventura Cavalieri (d. 1647) to attempt to legitimize Kepler's techniques and to modify and extend the Archimedean method of exhaustion. To these ends he systematically employed infinitesimal techniques in his *Geometria indivisibilibus* ("Geometry of Indivisibles," 1635) and *Exercitationes geometricae sex* ("Six Geometrical Exercises," 1647). Where Kepler divided geometrical figures into indivisibles of the same dimension, Cavalieri had indivisibles of a lower dimension make up a figure. An area was composed of equidistant and parallel line segments and a volume of equidistant and parallel plane sections. Cavalieri's method of indivisibles compared two geometrical figures by distributive or collective means. It was based on a principle named after him.

> If two solids have equal altitudes, and if sections made by planes parallel to the bases and at equal distance from them are always a given ratio, then the volumes of the solids are in this ratio.

While Cavalieri's principle obscures the limit process in his computations, with it he could obtain many results more simply and quickly than with the method of exhaustion. His finding of the area of the ellipse and volume of a sphere show this. He also found what in modern notation is equivalent to

$$\int_0^a x^n dx = \frac{a^{n+1}}{n+1}, \text{ for any natural number } n.$$

Torricelli and John Wallis also advanced infinitesimal analysis. Torricelli recognized the dangers of paradoxes in the doctrine of indivisibles and ongoing debates about the nature of indivisibles and the structure of the continuum. Precise meanings for infinitesimal magnitude and infinite sum were not yet known. They, nevertheless, remained useful in providing new mathematical results. In *De motu gravium* (1644) Torricelli discovered the inverse nature of integration and differentiation. Like Isaac Barrow, who made the same discovery in 1670, he failed to realize its importance, however.

Wallis came close to integration in his *Arithmetica infinitorum* (1656). Through a daring sequence of interpolations by analogy he arrived at the famous infinite product: $\frac{4}{\pi} = \frac{3}{2} \cdot \frac{3}{4} \cdot \frac{5}{4} \cdot \frac{5}{6} \cdot \frac{7}{6}$. This was part of his attempt to develop a method to compute by means of arithmetical indivisibles the area of a quadrant of a unit circle. He was searching, therefore, for the series expansion of $\frac{\pi}{4}$ by integrating from 0 to 1—(in modern terms) $\frac{\pi}{4} = \int_0^1 (1-x^2)^{1/2}\, dx$. Others had found the area under x^m. Wallis formalized the rule for the general indefinite integral as the equivalent of $\int x^m dx = \frac{x^{m+1}}{m+1}$. His "functional thinking" relied on sequences of numerical expressions rather than on geometric curves. Because he used several unproved assumptions and analogies, such as those leading to the infinite product for $\frac{4}{\pi}$, Hobbes harshly portrayed the *Arithmetica infinitorum* as a "scab of symbols" and Fermat criticized it as "incomplete" induction.

Newton and Leibniz independently invented the infinitesimal calculus.[25] They were the first to offer a logically coherent and general body of procedures for the subject and a symbolism. The style and content of their inventions differed, however. Newton prepared the fluxional calculus and Leibniz a type of calculus based on differentials and integrals. Newton's formative discoveries date from his final undergraduate years at Cambridge, when he pored over and taught himself the substance of such writings as van Schooten's translation of Descartes' *Géométrie,* Wallis' *Arithmetica Infinitorum,* and Nicolas Mercator's *Logarithmotechnia* (1668), a book that summed the area of a hyperbolic segment in the manner of Cavalieri's indivisibles. Wallis had reviewed the Mercator book for the *Philosophical Transactions.* By the end of 1665 Newton generalized the binomial theorem $(a + b)^n$ allowing series expansions for fractional and negative powers. Previously these were known for positive integral powers only. He recognized the coefficients {1}, {1,1}, {1,2,1}, {1,3,3,1}, . . . as those in Pascal's triangle. By Wallisian interpolation, that is to say, a process of shrewd guessing by analogy, he found expressions for series coefficients for fractional values of n. He could, therefore, find the integral of $(1 - x^2)^{1/2}$.

During the three decades after 1665 Newton developed his fluxional calculus. His results appeared in letters and three tracts *De analysi per aequationes numero terminorum infinitas* ("On Analysis by Means of Equations with an Infinite Number of Terms," written 1669, published 1711), *De quadratura curvarum* ("On the Quadrature of Curves," completed 1693, published 1704), and the more extensive *Methodus fluxionum,* ("Method of Fluxions," published posthumously in 1736). His *Principia* gave the first public indication of the fluxional calculus, although it is not mentioned by name. Besides a recognition of the importance of the inverse relationship of differentiation and integration, its central themes were power series expansions, algorithms, concepts of variables as moving continuously in time, and the doctrine of prime and ultimate ratios. In the fluxional calculus the variable quantities of analytic geometry became smoothly flowing quantities, or fluents. Newton's algorithm for determining the fluxion, or time rate of change, of any variables depended upon x becoming $x + o$, where $\dot{x}o$ is a discrete infinitesimal increment, or moment of x. The concept of moments long troubled Newton and, as indicated in the problem below, was not satisfactorily resolved. The fluxion, $\dot{x}$, equals in Leibniz's notation dx/dt. To find the slope of curve $x^2 = by + c$, Newton proceeded as follows:

$$(x + \dot{x}o)^2 = b(y + \dot{y}o) + c$$

$$x^2 + 2x\dot{x}o + (\dot{x}o)^2 + by + b\dot{y}o + c,$$ (by neglecting terms in o^2 and subtracting equals from both sides, he found)

$$2x\dot{x}o = b\dot{y}o$$

$$\frac{2x}{b} = \frac{\dot{y}}{\dot{x}} .$$

The ratio of fluxions, or "ultimate ratio of evanescent quantities," is the derivative. Newton's integral was the indefinite integral; he worked with the inverse rate of change to solve area and volume problems.

Leibniz's studies leading to the calculus began under the guidance of Huygens in Paris between 1673 and 1676. His work rests on at least three basic components. Leibniz sought a *characteristica generalis,* a symbolic language, for translating mathematical methods and statements into algorithms and formulas. Second, he investigated sequences of the differences and sums of numbers, such as in the harmonic triangle, and found the two sequences were formed by mutually inverse operations—a critical insight when transposed to quadrature and tangent problems in geometry. Third, he discovered the "characteristic triangle"—dx, dy, and ds—in Pascal's *Letters de A. Dettonville.* "A light suddenly burst upon him." Applying the characteristic triangle within his "transmutation theorem" yielded transformations of the quadrature of one curve into that of another, with the second curve related to the first by tangency. His successful application of the transmutation rule to general parabolas and hyperbolas convinced him of the existence of the calculus.

An application of the transmutation theorem to the "arithmetical quadrature of the circle" produced the infinite series $\pi/4 = 1 - 1/3 + 1/5 - 1/7 + \ldots$ now bearing Leibniz's name. His method was simpler than Newton's numerous extrapolations. After dividing the geometric series $\frac{1}{1 + z^2} = 1 - z^2 + z^4 - \ldots,$ he employed term-wise integration. Leibniz presented these results given in a 1675 manuscript—results later checked by using derivatives of trigonometric functions.[26] That manuscript contained a beginning of his facile symbolism, the extended "$\int$" and "d," as well as a discussion of their inverse relationship and rules for their use, including the correct chain rule for differentiation—$d(xy) = xdy + ydx$.

Leibniz published a series of articles on the calculus in *Acta Eruditorum* between 1684 and 1695. His papers of 1684 and 1686 were the first two publications on the subject. The 1684 paper contains the dx and dy symbols for differentials and the chain rule. These articles demonstrate that the differential, not the derivative, was fundamental to his version of the calculus, that he handled integration as summation, and that his calculus with its stress on notation was more analytical than Newton's. Numerous misprints and a deliberate attempt to be obscure detracted from these articles.

After 1690 the calculus evolved rapidly in Europe amid a controversy over its origins. Jakob and Johann Bernoulli mastered the Leibnizian version from *Acta* articles and correspondence with Leibniz. Flaws in the *Acta* articles prohibited scholars of lesser caliber from doing likewise. They learned Leibnizian differential calculus from l'Hôpital's *Analyse des infiniment petits pour l'intelligence des lignes courbes* ("Analysis of Infinitely Small Quantities for the Understanding of Curved Lines," 1696). Shortly after 1700 the Paris Academicians accepted it. (The calculus was about to displace geometry as the premier field in mathematics.) Since Newton did not publish on the fluxional calculus until 1704, there was little competition from that quarter.

A nasty quarrel over the exact origin of the calculus erupted during the first decades of the eighteenth century. Long simmering, it effectively became a public matter in 1708, when John Keill declared at the Royal Society that Leibniz was a plagiarist. Newton had recruited Keill to be his spokesman. When Leibniz sought a remedy from the charges of the upstart *(homo novus)* Keill, the Royal Society appointed a commission to look into the matter and report back in 1711. The dispute put a strain on Newton, who presided over the Royal Society and wrote the commission report, *Commercium Epistolicum.* That report absolved Keill of doing any injury, confirmed Newton as "first inventor" of the calculus and censured Leibniz for allegedly concealing knowledge of the calculus gained from correspondence with Collins and Oldenburg of the Royal Society in 1676. This defamation of Leibniz had at least two unfortunate consequences. The *Commercium* settled nothing; the dispute continued. As it did, the British cut themselves off from Leibniz's differential calculus and its versatile notation. On a personal level, the report sullied Leibniz's intellectual reputation for over two centuries in the English-speaking world.

NOTES

1. Alfred North Whitehead first made this famous characterization of the seventeenth century in print in Chapter III of *Science and the Modern World* (1925).

2. Kepler's first two laws describe properties of the orbit of a particular planet, namely its shape and variation in speed during the complete orbit. By contrast, the third law presents a general relationship between motions of different planets. These laws follow:

First Law: Each planet moves around the sun in an elliptical orbit with the sun located at one focus of the ellipse.

Second Law: The radius vector drawn from the sun to any planet sweeps out equal area in equal times as the planet moves along its orbit.

Third Law: The ratio of the square of the time period of revolution (T) of any planet to the cube of its semimajor axis (A) is the same constant for all planets in our solar system. In symbols, $\frac{T^2}{R^3} = k$.

3. For the development of scientific activity around the *Academia* see Eric Cochrane, *Florence in the Forgotten Centuries: 1527–1800* (Chicago: University of Chicago Press, 1973).

4. For information on these groups of intellectuals meeting for informal discussions see Joseph Ben-David, *The Scientist's Role in Society* (Englewood Cliffs, N.J.: Prentice-Hall, 1971), pp. 63ff.

5. Mathematics and astronomy flourished at Wittenberg in the latter sixteenth century. This happened after the Protestant educational reformer Philipp Melanchthon (1497–1560) emphasized the study of mathematical disciplines in the curricula of German universities. The French pedagogue Peter Ramus compared Melanchthon's role in supporting mathematics in the German states to Plato's in ancient Greece.

6. The English zoologist and paleontologist Thomas Henry Huxley wrote in the later nineteenth century: "If all the books in the world except the *Philosophical Transactions* were destroyed, it is safe to say that the foundations of physical science would remain unshaken." This statement, while nationalistic and extravagant, has much truth in it.

7. While chronological demarcations are useful as guides, they should not be taken to mark sharp discontinuities. One reason for choosing 1620 besides the political, military, and intellectual developments given in the introduction is the voyage of the Mayflower, which marks the first of many migrations from Northern Europe to the New World.

8. Pockets of more efficient agriculture and distribution of nutrition partly account for this rise.

9. It was Voltaire who called the Holy Roman Empire neither holy nor Roman nor an empire. The Empire was a nominal political organization of some 360 principalities comprising the German nation.

10. Kings and courtiers across Europe imitated Versailles. One of its habits, the taking of coffee or tea in special houses or salons where gossip, news and literary works circulated, was quite important in French intellectual life during the Enlightenment.

11. Louvois adopted close order drills and strict discipline for troops, introduced military uniforms, and made it official policy to supply French armies with food while on the march. He established a professional officer corps with regular pay and promotions based largely on merit, and he reinforced a sense of solidarity with the king by parades and ritual functions that also included the troops and their commanders.

12. In the Dutch War Holland opened its dikes to stop the French armies. The French monarch was greatly offended by Netherlands cartoons showing the sun (Louis was the "Sun King") being eclipsed by a moon of Dutch cheese.

13. Charles Webster, *The Great Instauration* (London: Duckworth, 1975). For Milton's views supporting Copernican astronomy, see the seventh and eighth books of *Paradise Lost*.

14. Pursuing social diversions and the exotic, the reading public in Louis XIV's court equally prized with this exemplary drama, books like *The Miraculous History of a 380-Year-Old Man in the Portuguese Indies, Eight Times Married, Whose Teeth Have Twice Fallen Out and Grown in Again*.

15. *Leviathan,* chapter 4, in *The English Works of Thomas Hobbes* ed. W. Molesworth (London, 1839–1845), vol. III, p. 23. He traced the universal acceptance of Euclid's axioms to their not being "contrary to any man's right of dominion, or the interest of men that have dominions." *Leviathan,* chapter (*English Works,* III, p. 91).

16. Dedicatory letter to *De cive* (*English Works,* vol. III, p. iv). Hobbes, who tutored the young, exiled Charles II in geometry, enthusiastically studied Euclid's *Elements* and was a member of Mersenne's circle. He criticized Descartes' analytic geometry and, after claiming to have squared the circle, engaged in a vitriolic debate with John Wallis, who knew better.

17. For the contributions of women, see Carolyn Merchant, *The Death of Nature: A Feminist Reappraisal of the Scientific Revolution.* (New York: Harper & Row, 1978).

18. For information on these traditions, see Allen G. Debus, *Man and Nature in the Renaissance* (Cambridge: Cambridge University Press, 1980).

19. For a different response by a prominent Lutheran group, see Robert S. Westman, "The Melanchthon Circle, Rheticus, and the Wittenberg Interpretation of the Copernican Theory," *Isis,* 66 (1975), 165-194.

20. See Frances A. Yates, *Giardano Bruno and the Hermetic Tradition* (Chicago: University of Chicago Press, 1964).

21. Translation from *Opere,* 4, 171, as given in Morris Kline, *Mathematical Thought from Ancient to Modern Times* (New York: Oxford University Press, 1972), p. 329.

22. See Michael S. Mahoney, *The Mathematical Career of Pierre de Fermat (1601–1665)* (Princeton: Princeton University Press, 1973).

23. His circle method was algebraic. To find the tangent line to a curve at a given point, in modern notation $y = f(x)$ at $P(x,y)$, he first located point $C(v,0)$, where the normal to curve P intersects the x-axis and then drew the perpendicular to the normal line. Generally, the curve is intersected in a second point by the circle with center $C(v,0)$ and radius $r = vP$. The "double point" intersection means finding double roots for the equation $y^2 + (v-x) = r^2$.

24. Pascal (1641) and Leibniz (1673–75) also developed calculating machines. Their technological development and broad use in society lay nearly 300 years in the future.

25. See their biographies for their many other contributions to mathematics.

26. Consider $\frac{\pi}{4} = \int_0^1 \frac{dx}{1+x^2}$. For $1 + x^2$ substitute $1 + \tan^2\alpha$.
By the chain rule we know that

$$\frac{d\tan\alpha}{d\alpha} = \frac{d}{d\alpha}\left(\frac{\sin\alpha}{\cos\alpha}\right) = \frac{\sin\alpha}{\cos^2\alpha}\frac{d\cos\alpha}{d\alpha} + \frac{1}{\cos\alpha}\frac{d\sin\alpha}{d\alpha}$$

$$= \frac{\sin^2\alpha + \cos^2\alpha}{\cos^2\alpha} = \frac{1}{\cos^2\alpha} = \sec^2\alpha.$$

Further $1 + \tan^2\alpha = \sec^2\alpha$. Thus

$$\int_0^1 \frac{dx}{1+x^2} = \int_0^{\frac{\pi}{4}} \frac{d\tan\alpha}{\sec^2\alpha\, d\alpha} = \int_0^{\frac{\pi}{4}} \frac{\sec^2\alpha}{\sec^2\alpha}\, d\alpha = \int_0^{\frac{\pi}{4}} d\alpha = \pi/4.$$

[Both Newton and Leibniz knew the series expansions for sines and cosines. For Euler's work on this problem, see selection 85 in this anthology.]

SUGGESTIONS FOR FURTHER READING

E. J. Aiton, "Kepler's Second Law of Planetary Motion," *Isis* 60 (1969), 75-90.

Angus Armitage, *Copernicus: the Founder of Modern Astronomy*. New York: A. S. Barnes and Company, Inc., 1957.

_________, *John Kepler*. London: Faber and Faber, 1966.

M. E. Baron, *The Origins of the Infinitesimal Calculus*. London: Pergamon, 1969.

Isaac Barrow, *Geometrical Lectures*, ed. by J. M. Child. Chicago: Open Court, 1916.

A. E. Bell, *Christiaan Huygens and the Development of Science in the Seventeenth Century*.

N. L. Biggs, "The Roots of Combinatorics," *Historia Mathematica* 6 (1979), 109-136.

H. J. M. Bos, "Differentials, Higher Order Differentials, and the Derivative in the Leibnizian Calculus," *Archive for History of Exact Science*, 14 (1974–75), 1-90.

Carl Boyer, *History of Analytic Geometry*. New York: Scripta Mathematica, 1956.

_________, *History of the Calculus*. New York: Dover, 1959.

E. M. Bruins, "Computation of Logarithms by Huygens," *Janus* 65 (1978), 97-104.

Florian Cajori, *A History of Mathematical Notations*. Chicago: Open Court, 1928–1929, 2 vols.

Ronald Calinger, *Gottfried Wilhelm Leibniz*. Troy, N.Y.: Allen Memorial, 1976.

I. B. Cohen, *The Newtonian Revolution*. Cambridge: Cambridge University Press, 1980.

Cynthia Conwell Cook, *Western Mathematics Comes of Age*. Garden City, N.Y.: Doubleday & Co., Inc., 1977.

J. L. Coolidge, "The Story of Tangents," *American Mathematical Monthly*, 58 (1951) 449-462.

Allen Debus, *Man and Nature in the Renaissance*. Cambridge: Cambridge University Press, 1978.

E. J. Dijksterhius, *The Mechanization of the World Picture*. New York: Oxford University Press, 1961.

J. Earman, "Infinities, Infinitesimals, and Indivisibles: the Leibnizian Labyrinth," *Studia Leibnitiana* 7 (1975), 236-251.

Charles Henry Edwards, *The Historical Development of the Calculus*. New York: Springer-Verlag, 1979.

Emil A. Fellmann, *G. W. Leibniz Marginalia in Newtoni Principia Mathematica*. Paris: Vrin, 1973.

Eric Forbes, "Descartes and the Birth of Analytic Geometry," *Historia Mathematica* 4 (1977), 141-151.

J. T. Fraser and N. Lawrence, eds., *The Study of Time*. New York: Springer-Verlag, 1975.

Hans Freudenthal, "Huygens' Foundations of Probability," *Historia Mathematica* 7 (1980), 113-117.

Owen Gingerich, ed., *The Nature of Scientific Discovery*. Washington, D.C.: The Smithsonian Institution, 1975.

H. H. Goldstine, *A History of Numerical Analysis From the 16th Through the 19th Century*. New York: Springer, 1977.

I. Grattan-Guinness, ed., *From the Calculus to Set Theory, 1630–1910: An Introductory History*. London: Duckworth, 1980.

N. T. Gridgeman, "John Napier and the History of Logarithms," *Scripta Mathematica* 29 (1973), 49-65.

Ian Hacking, *The Emergence of Probability*. Cambridge: Cambridge University Press, 1975.

Roger Hahn, *The Anatomy of a Scientific Institution: the Paris Academy of Sciences.* Berkeley: University of California Press, 1971.

A. Rupert Hall, *Philosophers at War: The Quarrel Between Newton and Leibniz.* Cambridge: Cambridge University Press, 1980.

———— and Laura Tilling, eds., *The Correspondence of Isaac Newton.* Cambridge: Cambridge University Press, 1959– , 7 vols.

Marie Boas Hall, "Leibniz and the Royal Society, 1670–1676," *Studia Leibnitiana Supplementa,* vol. XVII, Tome I (1978), 171-182.

Joseph E. Hofmann, "Leibniz in Paris." Cambridge: Cambridge University Press, 1974.

C. Jensen, "Pierre Fermat's Method of Determining Tangents of Curves and Its Application to the Conchoid and the Quadratrix," *Centaurus* 14 (1969), 72-85.

H. W. Jones, "A Seventeenth-Century Debate," *Annals of Science* 31 (1974), 307-333.

P. Kitcher, "Fluxions, Limits, and Infinite Littleness: A Study of Newton's Presentation of the Calculus," *Isis* 64 (1973), 33-49.

Morris Kline, *Mathematical Thought from Ancient to Modern Times.* New York: Oxford University Press, 1977.

Thomas S. Kuhn, *The Copernican Revolution.* New York: Vintage Books, 1962.

Timothy Lenoir, "Descartes and the Geometrization of Thought: The Methodological Background of Descartes' Géometrie," *Historia Mathematica* 6 (1979), 355-379.

J. A. Lohne, "Essays on Thomas Harriot," *Archive for History of Exact Sciences* 20 (1979), 189-312.

Michael S. Mahoney, *The Mathematical Career of Pierre de Fermat, 1601–1665.* Princeton, N.J.: Princeton University Press, 1973.

Carolyn Iltis Merchant, *The Death of Nature: A Feminist Reappraisal of the Scientific Revolution.* New York: Harper & Row, 1978.

J. F. Montucla, *Histoire des Mathematiques.* Paris: Albert Blanchard (reprint), 1960, 3 vols.

Ernest Mortimer, *Blaise Pascal: The Life and Work of a Realist.* New York: Harper and Brothers, 1959.

R. H. Naylor, "Galileo's Theory of Projectile Motion," *Isis* 71 (1980), 550-571.

Karl Pearson, *The History of Statistics in the 17th and 18th Centuries Against the Changing Background of Intellectual, Scientific, and Religious Thought,* edited by E. S. Pearson. New York: Macmillan, 1978.

P. Ribenboim, "The Early History of Fermat's Last Theorem," *The Mathematical Intelligencer* 11 (1976), 7-21.

J. F. Scott, *A History of Mathematics.* London: Taylor and Francis, 1960 (2nd ed., 1969).

————, *The Scientific Work of René Descartes.* London: Taylor and Francis, 1952 (reprinted 1976).

C. J. Scriba, "The Inverse Method of Tangents: A Dialogue Between Leibniz and Newton," *Archive for History of Exact Sciences* 2 (1962), 113-137.

D. E. Smith and M. L. Latham, *The Geometry of René Descartes.* Chicago: Open Court, 1925.

Mark Smith, "Galileo's Theory of Indivisibles: Revolution or Compromise," *Journal for the History of Ideas* 37 (1976), 571-588.

P. Stromholm, "Fermat's Methods of Maxima and Minima and of Tangents, A Reconstruction," *Archive for History of Exact Sciences* 5 (1968), 47-69.

N. I. Stayakhin, *History of Mathematical Logic from Leibniz to Peano*. Cambridge, Mass.: MIT Press, 1969.

René Taton, *The Beginnings of Modern Science*, trans. by A. J. Pomerans. London: Thames and Hudson, 1964.

E. G. R. Taylor, *The Mathematical Practitioners of Tudor and Stuart England, 1485–1714*. Cambridge: Cambridge University Press, 1954.

O. Toeplitz, *The Calculus: A Genetic Approach*. Chicago: University of Chicago Press, 1963.

Jack R. Vrooman, *René Descartes: A Biography*. New York: G. P. Putnam's Sons, 1970.

Richard S. Westfall, *Never at Rest: A Biography of Isaac Newton*. Cambridge: Cambridge University Press, 1980.

_________, "Newton's Marvelous Years of Discovery and their Aftermath: Myth versus Manuscript," *Isis* 256 (1980), 109-121.

D. T. Whiteside, ed., *The Mathematical Papers of Isaac Newton*. Cambridge: Cambridge University Press, 1967–1976, 7 vols.

_________, "The Mathematical Principles Underlying Newton's *Principia Mathematica*," *Journal of the History of Astronomy*, I (1970), 116-138.

_________, "Patterns of Mathematical Thought in the Later 17th Century," *Archive for History of Exact Science* I (1960–1962), 179-388.

A. P. Yushkevich, "Comparaisons des conceptions de Leibniz et de Newton sur le calcul infinitesimal," *Studia Leibnitiana Supplementa*, vol XVII, Tomes I (1978), 69-80.

Chapter VI
The Scientific Revolution at Its Zenith (1620-1720)

Section A
Algebra, Analytic Geometry, and Arithmetic

RENÉ DESCARTES (1596-1650)

The reputed founder of modern philosophy was the third child of Joachim Descartes, counselor of the *parlement* of Rennes, and Jeanne Brochard. The father was, in modern terms, a lawyer and judge. Both parents belonged to the *noblesse de robe,* a social class of lawyers and their families whose status fell below the nobility but above the bourgeoisie. When the mother died a year after Descartes' birth, it seemed the sickly and pale child would not survive. When the father remarried in 1600, the young boy was raised by his maternal grandmother and a dedicated nurse.

At the age of eight, Descartes was sent to the Royal College at La Flèche where the Jesuits taught. The rector, Father Charlet, a distant relative, permitted the frail boy to spend mornings in bed meditating. While Descartes praised the teachers, he later severely criticized the syllabus of studies in the humanities and philosophy for its lack of unity and of "assurances" for establishing the truth of given knowledge. Philosophic studies in particular did not satisfy him, in part because they were based on the Aristotelian syllogism and a Scholastic logic that often led to quibbling. Moreover, they were primarily oriented toward theology. He was drawn to poetry, with its "enthusiasm and power of imagination," but only mathematics, with the self-evidence of its reasoning, offered some consolation in his search for fundamental truths. At La Flèche, however, mathematics was principally oriented toward practical applications and its role in the military arts—important concerns for young nobles.

After leaving La Flèche in 1614, Descartes traveled first to Paris and then to Poitiers, where he enrolled in the University of Poitiers as a law student. Philosophy and the pleasures of the world competed for the attention of the young gentleman who was making major decisions about his life. In 1615 and 1616, he freed himself of friends and family to pursue the study of mathematics. He went to Paris to meet Claude Mydorge, a leading French geometer, and had continuing contacts with the polymath, Father Marin Mersenne, a former pupil at La Flèche. These meetings proved helpful to his mathematical studies. In 1616 he received a degree in law from Poitiers.

In 1618, the 22-year-old Descartes joined the Dutch army of Prince Maurice of Nassau in Holland as an unpaid officer. Careers in the army and the church were the two chief occupations of young gentlemen in the 17th century, and it was probably the father's wish that his son serve the prince as a soldier. However, Descartes was the restless student, and the "idleness and dissipation" of camp life bored him. While at Breda with Prince Maurice, he met Isaac Beekman, for whom he solved a challenge problem in mathematics. Beekman and Descartes discussed recent developments in mathematics, including the work of Viète. In 1619 Descartes reported to Beekman of catching a glimpse of "an entirely new science"—the first inkling of analytic geometry. Later that year, after travels to Denmark and Germany, Descartes joined the Bavarian army at the beginning of the Thirty Years' War. On November 10, while near Ulm, there occurred the famous *poêle* (stove) episode of three consecutive dreams following what may have been either an illness or mystical crisis. Descartes interpreted the dreams to mean that he begin his studies with systematic doubt about everything and then proceed to build a universal science based on self-evident, deductive principles—an ambitious program that he made his life's mission nine years later. This was the core of his search for truths and certain knowledge. He left the military after serving in the Imperial Army in Hungary in 1621.

Descartes' wandering continued from 1622 to 1628—the final years of his formative intellectual life. Visits to his family were infrequent, but during one in 1623 he settled his affairs by selling the small estate at Perron inherited from his mother. The investment of the money he received permitted him to lead the life of a gentleman with simple tastes and needs. After travels to Italy from 1623 to 1625, he returned to Paris to participate in meetings with savants, especially of Mersenne's circle, and to study intensely the telescope and construction of optical instruments.

In the autumn of 1628 Descartes moved to Holland, where he lived except for short absences until 1649. The religious toleration, internal peace, new universities, and commercial prosperity of Holland appealed to him. Seeking solitude and tranquility, he usually avoided cities and moved 18 times to prevent distractions. The striking ideas he presented in voluminous correspondence and publications were to provoke bitter opposition both in Dutch universities and among ecclesiastics who accused the devout Catholic of materialism and atheism.

The years in Holland were highly productive. In 1633, displaying a mixture of caution and boldness, Descartes withheld the publication of his treatise supporting Copernican astronomy, *Le Monde . . .* ("The World"), after learning of Galileo's condemnation by the Inquisition. His most famous work, *Discourse de la Méthode* ("Discourse on Method," 1637), which was written in elegant French, was a guide for "correctly conducting the reason and seeking truth in the sciences." It began with his incontrovertible truth—*cogito, ergo sum* ("I think, therefore I am")—and set forth a deductive method that contrasted with Francis Bacon's largely inductive approach. The *Discourse* presented his dualism in ontology—the world is composed of mind (unextended thinking substance) and matter (extended substance)—and had three appendices. The first of which was *Géométrie.*

Descartes' studies were interrupted in 1640 when the death of his five-year-old daughter Francine (whose mother was probably his servant Hélene) left him desolate. After recovering, he had his *Meditations* (1641) published and completed the magisterial *Principia Philosophiae* ("Principles of Philosophy," 1644), a comprehensive account of the origin and nature of the universe that profoundly influenced the learned world. The *Principia* gave a mechanistic cosmology (clock model of the universe) and explained motion as action-by-contact. It rejected the void and depicted the universe as a plenum, explained celestial motions with a vortex theory (whirlpools of ether), and proposed the law of the conservation of momentum (*mv*). Besides his theoretical work in physics, Descartes made extensive anatomical dissections on animals and numerous observations in optics. He succeeded in providing a complete scientific system that his disciples, such as Malebranche, refined and extended. Cartesian science marks a transition between ancient and modern scientific theories—that is to say, it quickly superceded Aristotelian ideas and dominated European thought in natural philosophy until Newtonian dynamics and optics displaced it a century later.

By 1647 Descartes contemplated leaving Holland and returning to France. The attacks on his ideas by Protestant divines in Leiden may have prompted this consideration. He met Pascal in 1647 and perhaps suggested barometric experiments to him. The next year he decided against returning to a France that was experiencing chronically poor harvests, a fiscal crisis, and was on the verge of a civil war known as the Fronde. In 1649, after much hesitation, Descartes accepted Queen Christina's invitation to come to Stockholm to be her philosopher. Required to rise early in the bitter cold to tutor her, he contracted pneumonia in 1650 and died.

The chief contribution of Descartes to mathematics was his creation of coordinate or analytic geometry, which associates algebraic equations with curves and surfaces. The ancient geometry of Apollonius and Pappus together with the algebra of Diophantus, Ramus, Cardano and Viète were sources for Descartes' new creation, which appeared in the appendix to the *Discourse* entitled *Géométrie*. The coordinates with which he plotted equations are still called "Cartesian coordinates" in his honor. Analytic geometry aided his project to "carry geometry into physics"—that is, to geometrize physics. This was most important. Analytic geometry was, Descartes held, a constructive science that applied to the natural world and not a purely contemplative discipline. He considered "mathematics for its own sake" to be an idle searching or a vain play of the mind.

Descartes made other contributions to mathematics and occasionally engaged in heated polemics to defend his viewpoint. He stated Euler's theorem for polyhedra ($V + F = E + 2$) and discovered the correct law of refraction of light independently of Willebrord Snel. His quarrels largely involved matters relating to the origins of the differential and integral calculus. He rejected Pierre Fermat's method of maxima and minima and rule of tangents because they were not the result of strict *a priori* deduction. He proved an irreconcilable adversary of Gilles Roberval, a leading proponent of the geometry of indivisibles taken directly from Archimedes. Descartes excluded infinitesimals from his own research because they lacked intuitive clarity.

60. From the *Regulae**

(Rule IV: "In search for the truth of things a method is indispensable")

RENÉ DESCARTES

For the human mind has in it a something divine, wherein are scattered the first seeds of useful modes of knowledge. Consequently it often happens that, however neglected and however stifled by distracting studies, they spontaneously bear fruit. Arithmetic and geometry, the easiest of the sciences, are instances of this. We have sufficient evidence that the ancient geometers made use of a certain "analysis" which they applied in the resolution of all their problems, although, as we find, they grudged to their successors knowledge of this method. There is now flourishing a certain kind of arithmetic, called algebra, which endeavours to determine in regard to numbers what the ancients achieved in respect of geometrical shapes. These two sciences are no other than the spontaneous fruits above mentioned; they are products of the innate principles of the method here in question; and I do not wonder that these sciences, dealing as they do with such objects, the simplest of all objects, should have yielded a harvest so much more rewarding than the other sciences in which greater obstructions tend to choke all growth. But certainly these other sciences, if only we cultivate them with due care, can also be brought to full maturity.

This, indeed, is what I have chiefly had in view in preparing this treatise. For I should not ascribe much value to these rules, if they sufficed only for the solution of those unprofitable problems with which logicians and geometers are wont to beguile their leisure. I should then only be claiming for myself the power to argue about trifles more subtly than others. Though I shall often have to speak about geometrical shapes and numbers, that is only because no other disciplines can furnish us with examples so evident and so certain. Whoever, therefore, pays due regard to what I have in view will easily see that nothing is less intended by me than ordinary mathematics, and that I am indeed engaged in expounding quite another discipline, of which these examples are rather the outer covering than the constituents. For this discipline claims to contain the primary rudiments of human reason, and to extend to the eliciting of truths in every field whatsoever. To speak freely, I am convinced that it is a more powerful method of knowing than any handed down to us by human agency, and that it is indeed the source from which those older disciplines have sprung. In dwelling, as I have done, on the outer [mathematical] integument, my purpose has not been to cover over and conceal this discipline, with a view to warding off the vulgar, but rather to clothe and adorn it, that it may thereby be the more suitably conformed to our human powers.

When first I applied my mind to the mathematical disciplines, I read through most of what writers on these subjects

*Source: From Norman Kemp Smith (ed. and trans.), *Descartes' Philosophical Writings* (1952), 17-21. Reprinted by permission of Macmillan, London and Basingstroke. Footnote omitted.

are wont to teach; and I paid special attention to arithmetic and geometry, because they were said to be the simplest, and as it were a path leading to all the rest. But in neither field did I meet with authors who fully satisfied me. I did indeed read in them many things regarding numbers which on calculation I found to be true; and in respect of geometrical shapes they exhibited things in a certain manner to the eyes, and from them drew consequent conclusions. But to the mind itself they failed to submit evidence why these things are so or how they have been discovered. Accordingly I am not surprised that many people, even among the talented and learned, on sampling these sciences, very soon set them aside as being idle and puerile; or else, judging them to be exceedingly difficult and intricate, they have stopped short at the very threshold. For truly there is nothing more futile than to occupy ourselves with bare numbers and imaginary shapes, as if we could be content to rest in the knowledge of such trifles. Nothing, too, is more futile than to accustom ourselves to those superficial demonstrations which are discovered more often by chance than by skill, and which address themselves so much more to the eyes and to the imagination than to the understanding that we in a manner disaccustom ourselves to the use of our reason. Moreover, there can be no more perplexing task than to tackle by any such manner of proof new difficulties bearing on unordered numbers. When, however, I afterwards bethought myself how it could be that the first practitioners of philosophy refused to admit to the study of wisdom anyone not previously versed in mathematics, *i.e.* viewed mathematics as being the easiest of all disciplines, and as altogether indispensable for training our human powers and for preparing them to lay hold of the other more important sciences, I could not but suspect that they were acquainted with a mathematics very different from that which is commonly cultivated in our day. Not that I imagined that they had full knowledge of it. Their extravagant exultations, and the sacrifices they offered for what are minor discoveries, suffice to show how rudimentary their knowledge must have been. Nor am I shaken in this opinion by those machines of theirs, which historians have eulogised. The machines may have been quite simple and may well have been lauded as miraculous by the ignorant and wonder-loving multitude. I am convinced that certain primary seeds of truth implanted by nature in our human minds—seeds which in us are stifled owing to our reading and hearing, day by day, so many diverse errors—had such vitality in that rude and unsophisticated ancient world, that the mental light by which they discerned virtue to be preferable to pleasure and honour to utility, although they knew not why this should be so, likewise enabled them to recognise true ideas in philosophy and mathematics, although they were not yet able to obtain complete mastery of them. Certain vestiges of this true mathematics I seem to find in Pappus and Diophantus, who, though not belonging to that first age, yet lived many centuries before our time. These writers, I am inclined to believe, by a certain baneful craftiness, kept the secrets of this mathematics to themselves. Acting as many inventors are known to have done in the case of their discoveries, they have perhaps feared that their method being so very easy and simple, would, if made public, diminish, not increase the public esteem. Instead they have chosen to propound, as being the fruits of their skill, a number of sterile truths, deductively demonstrated with great show of logical subtlety, with a view to winning an amazed admiration, thus dwelling indeed on the results obtained by way of their method, but without disclosing the method itself—a disclosure which would have completely undermined that amazement. Lastly, in the present age there have been certain very able men who have attempted to revive this mathematics. For it seems to be no other than this very science which has been given the barbarous name, algebra—provided, that is to say, that it can be extricated from the tortuous array of numbers and from the complicated geometrical shapes by which it is overwhelmed, and that it be no longer lacking in the transparency and unsurpassable clarity which, in our view, are proper to a rightly ordered mathematics.

These were the thoughts which recalled me from the particular studies of arithmetic and geometry to a general investigation of mathematics; and my first inquiry was as to what precisely has been intended by this widely used name, and why not only the sciences above mentioned, but also astronomy, music, optics, mechanics and the several other sciences are spoken of as being parts of mathematics. It does not here suffice to consider the etymology of the word; for since the term mathematics simply signifies [in Latin] *disciplina* [*i.e.,* science], all the other sciences can, with as much right as geometry, be so entitled. Yet, as we see, there is almost no one, with the least tincture of letters, who does not easily distinguish, in the matters under question, between what relates to mathematics and what relates to the other disciplines. What, on more attentive consideration, I at length came to see is that those things only were referred to mathematics in which order or measure is examined, and that in respect of measure it makes no difference whether it be in numbers, shapes, stars, sounds or any other object that such measure is sought, and that there must therefore be some general science which explains all that can be inquired into respecting order and measure, without application to any one special subject-matter, and that this is what is called "universal mathematics"—no specially devised designation, but one already of long standing, and of current use as covering everything on account of which the other sciences are called parts of mathematics. How greatly it excels in utility and power the sciences which depend on it, is evident from this, that it can extend to all the things of which those other sciences take cognisance, and many more besides. Such difficulties as it may contain are also found in those other sciences, while these, on their part, owing to the particularity of their objects, exhibit yet other difficulties peculiar to themselves. Everyone knows the name of this discipline, and understands with what it deals, however inattentive he may be to it. How then is it that so many men laboriously pursue those other disciplines, and yet that no one is concerned to inform himself regarding this universal discipline upon which they depend? Assuredly I should marvel, were I not well aware that everyone thinks it to be the easiest of all disciplines, and had I not long since observed that we leave aside what we consider ourselves easily able to comprehend, hastily reaching out to what is new and more imposing.

61. From *Discours de la Méthode**

(Four Fundamental Rules of Logic)

RENÉ DESCARTES

Along with other philosophical disciplines I had, in my early youth, made some little study of logic, and, in the mathematical field, of geometrical analysis and of algebra—three arts or sciences, which, it seemed to me, ought to be in some way helpful towards what I had in view. But on looking into them I found that in the case of logic, its syllogisms and the greater part of its other precepts are serviceable more for the explaining to others the things we know (or even, as in the art of Lully, for speaking without judgment of the things of which we are ignorant) than for the discovery of them; and that while it does indeed yield us many precepts which are very good and true, there are so many others, either harmful or superfluous, mingled with them, that to separate out what is good and true is almost as difficult as to extract a Diana or a Minerva from a rough unshaped marble block. As to the analysis of the ancients and the algebra of the moderns, besides extending only to what is highly abstract and seemingly of no real use, the former is so confined to the treatment of shapes that it cannot exercise the understanding without greatly fatiguing the imagination, and the latter is in such subjection to certain rules and other requirements that out of it they have made an obscure and difficult art, which encumbers the mind, not a science helpful in improving it. I was thus led to think that I must search for some other method which will comprise all that is advantageous in these three disciplines, while yet remaining exempt from their defects. A multiplicity of laws often furnishes the vicious with excuses for their evil-doing, and a community is much the better governed if, with only a very few laws, it insists on a quite strict observance of them. So, in like manner, in place of the numerous precepts which have gone to constitute logic, I came to believe that the four following rules would be found sufficient, always provided I took the firm and unswerving resolve never in a single instance to fail in observing them.

The first was to accept nothing as true which I did not evidently know to be such, that is to say, scrupulously to avoid precipitance and prejudice, and in the judgments I passed to include nothing additional to what had presented itself to my mind so clearly and so distinctly that I could have no occasion for doubting it.

The second, to divide each of the difficulties I examined into as many parts as may be required for its adequate solution.

The third, to arrange my thoughts in order, beginning with things the simplest and easiest to know, so that I may then ascend little by little, as it were step by step, to the knowledge of the more complex, and, in doing so, to

*Source: From Norman Kemp Smith (ed. and trans.), *Descartes' Philosophical Writings* (1952), 128–131. Reprinted by permission of Macmillan, London and Basingstoke. Footnotes omitted.

assign an order of thought even to those objects which are not of themselves in any such order of precedence.

And the last, in all cases to make enumerations so complete, and reviews so general, that I should be assured of omitting nothing.

Those long chains of reasonings, each step simple and easy, which geometers are wont to employ in arriving even at the most difficult of their demonstrations, have led me to surmise that all the things we human beings are competent to know are interconnected in the same manner, and that none are so remote as to be beyond our reach or so hidden that we cannot discover them—that is, provided we abstain from accepting as true what is not thus related, *i.e.,* keep always to the order required for their deduction one from another. And I had no great difficulty in determining what the objects are with which I should begin, for that I already knew, viz. that it was with the simplest and easiest. Bearing in mind, too, that of all those who in time past have sought for truth in the sciences, the mathematicians alone have been able to find any demonstrations, that is to say, any reasons which are certain and evident, I had no doubt that it must have been by a procedure of this kind that they had obtained them. In thus starting from what is simplest and easiest I did not as yet anticipate any other advantage than that of accustoming my mind to pasture itself on truths, and to cease from contenting itself with reasons that are false. Nor while doing so, had I any intention of endeavouring to master all the various sciences which are commonly entitled mathematical. Having observed that however different their objects, all agree in considering only the diverse relations or proportions to be found as holding between them, I thought it best to treat only of these proportions, taking them in a quite general manner, and without ascribing to them any other objects than those which might serve to facilitate the knowing of them (though without in any way restricting them to these objects), so that afterwards I might be the better able to transfer them to all the other things to which they may apply. Then, noting that to obtain knowledge of these proportions I should sometimes have to consider them one by one, and sometimes to retain them in memory, or to embrace several together, I decided that for the better apprehending of each singly, I should view it as holding between lines, there being nothing simpler and nothing that I can represent more distinctly by way of my imagination and senses; and that for the retaining of several in the memory, or for embracing several things simultaneously, I should express them by certain symbols [*i.e.,* numbers or letters] as briefly as possible. In this way, I should be borrowing all that is best in geometry and algebra, and should be correcting all the defects of the one by help of the other.

62. From *Géométrie**

(Theory of Equations)[1]

RENÉ DESCARTES

It is necessary that I make some general statements concerning the nature of equations, that is, of sums composed of several terms, in part known, in part unknown, of which some are equal to the others, or rather, all of which considered together are equal to zero, because this is often the best way to consider them.

Know then that in every equation there are as many distinct roots, that is, values of the unknown quantity, as is the number of dimensions of the unknown quantity.[2]

Suppose, for example, $x = 2$ or $x - 2 = 0$,[3] and again, $x = 3$, or $x - 3 = 0$. Multiplying together the two equations $x - 2 = 0$ and $x - 3 = 0$, we have $xx - 5x + 6 = 0$, or $xx = 5x - 6$. This is an equation in which x has the value 2 and at the same time x has the value 3. If we next make $x - 4 = 0$ and multiply this by $xx - 5x + 6 = 0$, we have $x^3 - 9xx + 26x - 24 = 0$, another equation, in which x, having three dimensions, has also three values, namely, 2, 3, and 4.

It often happens, however, that some of the roots are false, or less than nothing. Thus, if we suppose x to represent the defect[4] of a quantity 5, we have $x + 5 = 0$ which, multiplied by $x^3 - 9xx + 26x - 24 = 0$, gives $x^4 - 4x^3 - 19xx + 106x - 120 = 0$, as an equation having four roots, namely three true roots, 2, 3, and 4, and one false root, 5.

It is evident from this that the sum of an equation containing several roots is always divisible by a binomial consisting of the unknown quantity diminished by the value of one of the true roots, or plus the value of one of the false roots. In this way, the dimension of an equation can be lowered.

On the other hand, if the sum of an equation is not divisible by a binomial consisting of the unknown quantity plus or minus some other quantity, then this latter quantity is not a root of the equation. Thus the last equation $x^4 - 4x^3 - 19xx + 106x - 120 = 0$ is divisible by $x - 2$, $x - 3$, $x - 4$, and $x + 5$, but is not divisible by x plus or minus any other quantity, which shows that the equation can have only the four roots, 2, 3, 4, and 5.

We can determine from this also the number of true and false roots that any equation can have, as follows: An equation can have as many true roots as it contains changes of sign, from + to − or from − to +; and as many false roots as the number of times two + signs or two − signs are found in succession.[5] Thus, in the last equation, since $+x^4$ is followed by $-4x^3$, giving a change of sign from + to −, and $-19xx$ is followed by $+106x$ and $+106x$ by -120, giving two more changes, we know

*Source: This translation is based on *The Geometry of René Descartes,* trans. by D. E. Smith and M. L. Latham (Chicago: Open Court, 1925). It appears in D. J. Struik (ed.), *A Source Book in Mathematics, 1200-1800* (1969), 90-93 and is reprinted by permission of Harvard University Press,

there are three true roots; and since $-4x^3$ is followed by $-19xx$ there is one false root.

It is also easy to transform an equation so that all the roots that were false shall become true roots, and all those that were true shall become false. This is done by changing the sign of the second, fourth, sixth, and all even terms, leaving unchanged the signs of the first, third, fifth, and other odd terms. Thus, if instead of

$$+x^4 - 4x^3 - 19xx + 106x - 120 = 0$$

we write

$$+x^4 + 4x^3 - 19xx - 106x - 120 = 0,$$

we get an equation having only one true root, 5, and three false roots, 2, 3, and 4.

If the roots of an equation are unknown and it be desired to increase or diminish each of these roots by some known number, we must substitute for the unknown quantity another quantity greater or less by the given number.[6] Thus, if it be desired to increase by 3 the value of each root of the equation

$$x^4 - 4x^3 - 19xx + 106x - 120 = 0,$$

put y in the place of x, and let y exceed x by 3, so that $y - 3 = x$. Then for xx put the square of $y - 3$, or $yy - 6y + 9$; for x^3 put its cube, $y^3 - 9yy + 27y - 27$; and for x^4 put its fourth power [*quarré de quarré*], or $y^4 - 12y^3 + 54yy - 108y + 81$. Substituting these values in the above equation, and combining, we have

$$\begin{array}{lrrrrr}
 & y^4 + 12y^3 & + 54yy & - 108y & + 81 & \\
 & + 4y^3 & - 36yy & + 108y & - 108 & \\
 & & - 19yy & + 114y & - 171 & \\
 & & & - 106y & + 318 & \\
 & & & & - 120 & \\
\hline
 & y^4 - 8y^3 & - 1yy & + 8y & & = 0, \\
\text{or} & y^3 - 8yy & - 1y & + 8 & & = 0,
\end{array}$$

whose true root is now 8 instead of 5, since it has been increased by 3.

If, on the other hand, it is desired to diminish by 3 the roots of the same equation, we must make $y + 3 = x$ and $yy + by + 9 = xx$, and so on, so that instead of

$$x^4 + 4x^3 - 19xx - 106x - 120 = 0$$

we have

$$\begin{array}{rrrrr}
y^4 + 12y^3 & + 54yy & + 108y & + 81 & \\
+ 4y^3 & + 36yy & + 108y & + 108 & \\
 & - 19yy & - 114y & - 171 & \\
 & & - 106y & - 318 & \\
 & & & - 120 & \\
\hline
y^4 + 16y^3 & + 71yy & - 4y & - 420 & = 0.
\end{array}$$

It should be observed that increasing the true roots of an equation diminishes the false roots by the same amount; and on the contrary diminishing the true roots increases the false roots; while diminishing either a true or a false root by a quantity equal to it makes the root zero; and diminishing it by a quantity greater than the root renders a true root false or a false root true. Thus by increasing the true root 5 by 3, we diminish each of the false roots, so that the root previously 4 is now only 1, the root previously 3 is zero, and the root previously 2 is now a true root, equal to 1, since $-2 + 3 = +1$. This explains why the equation $y^3 - 8yy - y + 8 = 0$ has only three roots, two of them, 1 and 8, being true roots, and the third, also 1, being false; while the other equation $y^4 - 16y^3 + 71yy - 4y - 420 = 0$ has only one true root, 2, since $+5 - 3 = +2$, and three false roots, 5, 6, and 7.

NOTES

1. *(Editor's Note).* In this translation Struik leaves the terms "dimension" for degree, "true" for positive roots, and "false" for negative ones. Please note that the terms "real" and "imaginary" begin in their modern sense in Descartes' text.
2. This is Descartes's formulation of the fundamental theorem of algebra. . . .
3. Descartes writes: "x equal to 2 or $x - 2$ equal to nothing, and again $x \propto 3$, or $x - 3 \propto 0$." He does not use the sign $=$, already introduced by Recorde and Harriot; see note 1.
4. A defect [*défaut*] is the negative of a positive number; thus -5 is the defect of 5, that is, the remainder when 5 is subtracted from zero.

5. This is the sign rule as stated by Descartes. It was formulated in a more precise manner by Isaac Newton in his *Arithmetica universalis* (Cambridge, 1707) and by C. F. Gauss in "Beweis eines algebraischen Lehrsatzes," *Crelle's Journal für die reine und angewandte Mathematik 3* (1828), 1-4; *Werke*, III, 65-70. We now can express it in the following way: If $f(x) = 0$ is an equation of degree n with real coefficients, where $f(x) = a_0x^n + a_1x^{n-1} + \ldots + a_{n-1}x + a_n$, then the number of positive roots is equal to or an even number less than the number of variations in the signs of successive terms. Multiple roots have to be counted in accordance with their multiplicity, and zero is not a positive root.

6. This change of variable of an equation by means of substitution of the type $y = x + a$ is not new, and is . . . one of the substitutions used by Cardan in his *Ars magna* (1545). Descartes here shows its use in the search for positive and negative roots. The notation, as elsewhere in Descartes's writings, strikes us as quite modern.

63. From *Géométrie* (1637)*

(The Principle of Nonhomogeneity)

RENÉ DESCARTES

All problems in geometry can easily be reduced to such terms that a knowledge of the lengths of certain straight lines is sufficient for their construction.

Just as arithmetic consists of only four or five operations, namely, addition, subtraction, multiplication, division, and the extraction of roots, which may be considered a kind of division, so in geometry, to find required lines it is merely necessary to add or subtract other lines; or else, taking one line which I shall call the unit in order to relate it as closely as possible to numbers, and which can in general be chosen arbitrarily, and having given two other lines, to find a fourth line which shall be to one of the given lines as the other is to the unit (which is the same as multiplication); or, again, to find a fourth line which is to one of the given lines as the unit is to the other (which is equivalent to division); or, finally, to find one, two, or several mean proportionals between the unit and some other line (which is the same as extracting the square root, cube root, etc., of the given line). And I shall not fear to introduce these arithmetical terms into geometry, for the sake of greater clarity.

For example, let *AB* [Fig. 1] be taken as the unit, and let it be required to multiply *BD* by *BC*. I have only to join the points *A* and *C*, and draw *DE* parallel to *CA*; then BE is the product of *BD* and *BC*.

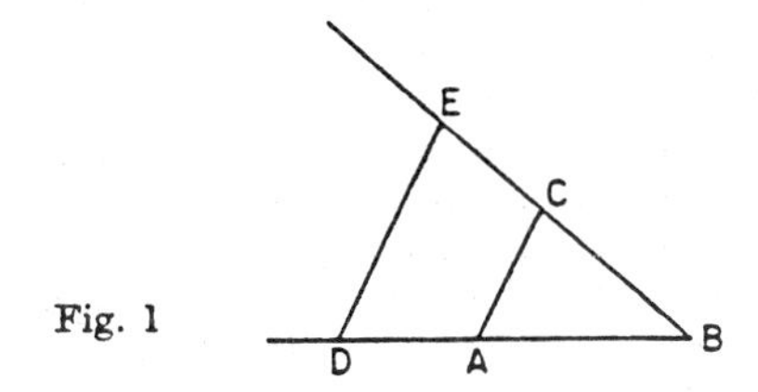

Fig. 1

If it be required to divide *BE* by *BD*, I join *E* and *D*, and draw *AC* parallel to *DE*; then *BC* is the result of the division.

*Source: This translation is based on *The Geometry of René Descartes*, trans. by D. E. Smith and M. L. Latham (Chicago: Open Court, 1925). It appears in D. J. Struik (ed.), *A Source Book in Mathematics, 1200-1800* (1969), 150-157 and is reprinted by permission of Harvard University Press, Copyright © 1969 by the President and Fellows of Harvard College. Footnotes renumbered.

Or, if the square root of *GH* [Fig. 2] is desired, I add, along the same straight line, *FG* equal to the unit; then, bisecting *FH* at *K*, I describe the circle *FIH* about *K* as a center, and draw from *G* a perpendicular and extend it to *I*, and *GI* is the required root. I do not speak here of cube root, or other roots, since I shall speak more conveniently of them later.

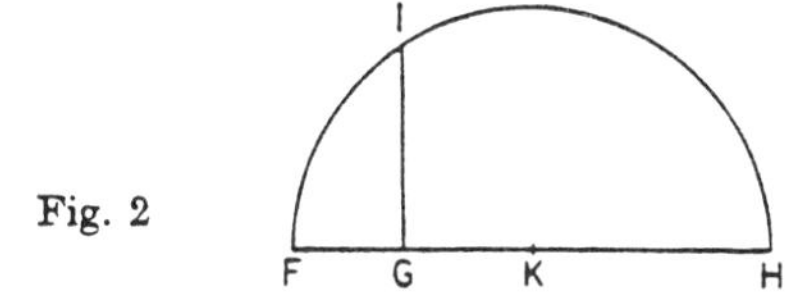

Fig. 2

Often it is not necessary thus to draw the lines on paper, but it is sufficient to designate each by a single letter. Thus, to add the lines *BD* and *GH*, I call one *a* and the other *b*, and write $a + b$. Then $a - b$ will indicate that *b* is subtracted from *a*; *ab* that *a* is multiplied by *b*; a/b[1] that *a* is divided by *b*; *aa* or a^2 that *a* is multiplied by itself; a^3 that this result is multiplied by *a*, and so on, indefinitely. Again, if I wish to extract the square root of $a^2 + b^2$, I write $\sqrt{a^2 + b^2}$; if I wish to extract the cube root of $a^3 - b^3 + abb$, I write $\sqrt[3]{a^3 - b^3 + abb}$,[2] and similarly for other roots. Here it must be observed that by a^2, b^3, and similar expressions, I ordinarily mean only simple lines, which, however, I name squares, cubes, etc., so that I make use of the terms employed in algebra.

It should also be noted that all parts of a single line should as a rule be expressed by the same number of dimensions, when the unit is not determined in the problem. Thus, a^3 contains as many dimensions as *abb* or b^3, these being the component parts of the line which I have called $\sqrt[3]{a^3 - b^3 + abb}$. It is not, however, the same thing when the unit is determined, because it can always be understood, even where there are too many or too few dimensions; thus, if it be required to extract the cube root of $a^2b^2 - b$, we must consider the quantity a^2b^2 divided once by the unit, and the quantity *b* multiplied twice by the unit.

Finally, so that we may be sure to remember the names of these lines, a separate list should always be made as often as names are assigned or changed. For example, we may write $AB = 1$, that is *AB* is equal to 1; $GH = a$, $BD = b$, and so on.[3]

If, then, we wish to solve any problem, we first suppose the solution already effected,[4] give names to all the lines that seem needful for its construction—to those that are unknown as well as to those that are known. Then, making no distinction between known and unknown lines, we must unravel the difficulty in any way that shows most naturally the relations between these lines, until we find it possible to express a single quantity in two ways. This will constitute what we call an equation, since the terms of one of these two expressions are equal to those of the other. And we find as many such equations as there are supposed to be unknown lines; but if, after considering everything involved, so many cannot be found, it is evident that the question is not entirely determined. In such a case we may choose arbitrarily lines of known length for each unknown line to which there corresponds no equation.

If there are several equations, we must use each in order, either considering it alone or comparing it with the others, so as to obtain a value for each of the unknown lines; and so we must combine them until there remains a single unknown line which is equal to some known line, or whose square, cube, fourth power, fifth power, sixth power, etc., is equal to the sum or difference of two or more quantities, one of which is known, while the others consist of mean proportionals between unity and this square, or cube, or fourth power, etc., multiplied by other known lines. I express this as follows:

$$z = b,$$

or

$$z^2 = -az + bb,$$

or

$$z^3 = az^2 + bbz - c^3,$$

or

$$z^4 = az^3 - c^3z + d^4, \text{ etc.}$$

That is, z, which I take for the unknown quantity, is equal to b; or, the square of z is equal to the square of b diminished by the square of z, plus the square of b multiplied by z, diminished by the cube of c; and similarly for the others.

Thus, all the unknown quantities can be expressed in terms of a single quantity, whenever the problem can be constructed by means of circles and straight lines, or also by conic sections, or even by some other curve only one or two degrees more composed.

But I shall not stop to explain this in more detail, because I should deprive you of the pleasure of mastering it yourself, as well as of the advantage of training your mind by working over it, which is in my opinion the principal benefit to be derived from this science. Because I find nothing here so difficult that it cannot be worked out by any one at all familiar with ordinary geometry and with algebra, who will consider carefully all that is set forth in this treatise.

That is why I shall content myself with the statement that if the student, in solving these equations, does not fail to make use of division wherever possible, he will surely reach the simplest terms to which the problem can be reduced.

And if it can be solved by ordinary geometry, that is, by the use of straight lines and circles traced on a plane surface, when the last equation shall have been entirely solved there will remain at most only the square of an unknown quantity, equal to the product of its root by some known quantity, increased or diminished by some other quantity also known.[5] Then this root or unknown line can easily be found. For example, if I have $z^2 = az + bb$, I construct [Fig. 3] a right triangle NLM with one side LM, equal to b, the square root of the known quantity bb, and the other side, LN, equal to $\frac{1}{2}a$, that is to half the other known quantity which was multiplied by z, which I suppose to be the unknown line. Then prolonging MN, the hypotenuse [*la baze*] of this triangle, to O, so that NO is equal to NL, the whole line OM is the required line z. It is expressed in the following way:

$$z = \tfrac{1}{2}a = \sqrt{\tfrac{1}{4}aa + bb}.$$

But if I have $yy = -ay + bb$, where y is the quantity whose value is desired, I construct the same right triangle NLM, and on the hypotenuse MN lay off NP equal to NL, and the remainder PM is y, the desired root. Thus I have

$$y = -\tfrac{1}{2}a + \sqrt{\tfrac{1}{4}aa + bb}.$$

In the same way if I had

$$x^4 = -ax^2 + b^2,$$

PM *would be* x^2 and I should have

$$x = \sqrt{-\tfrac{1}{2}a + \sqrt{\tfrac{1}{4}aa + bb}},$$

and so for other cases.

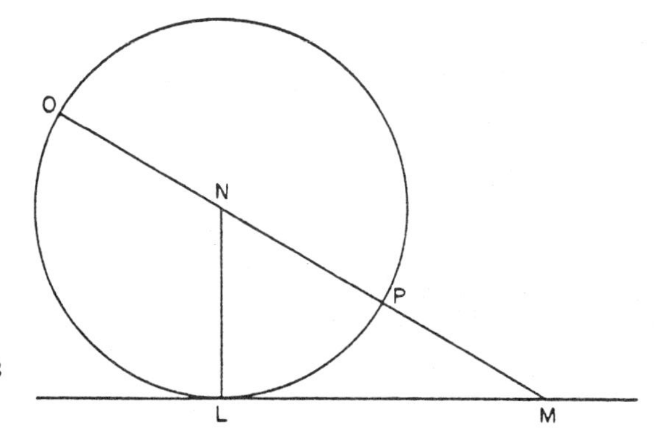

Fig. 3

Finally, if I have $z^2 = az - bb$, I make NL equal to $\frac{1}{2}a$ and LM equal to b as before; then [*Fig. 4*], instead of joining the points M and N, I draw MGR parallel to LN, and with N as a center describe a circle through L cutting MGR in the points G and R; then z, the line sought, is either MG or MR, for in this case it can be expressed in two ways, namely,

$$z = \tfrac{1}{2}a + \sqrt{\tfrac{1}{4}aa - bb}$$

and

$$z = \tfrac{1}{2}a - \sqrt{\tfrac{1}{4}aa - bb}.$$

And if the circle described about N and passing through L neither cuts nor touches the line MGR, the equation has no root, so that we may say that the construction of the problem is impossible.[6]

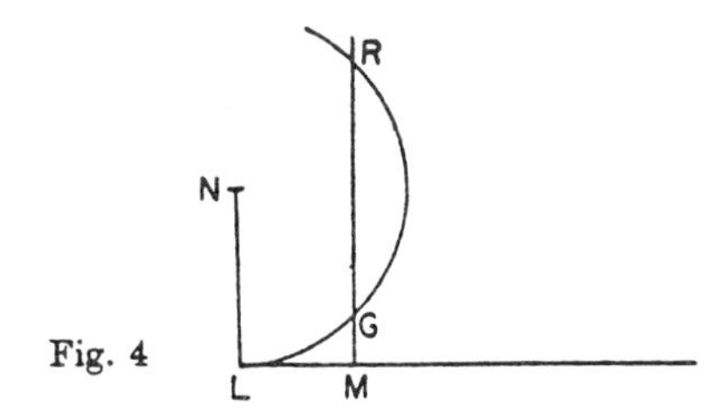

Fig. 4

These same roots can be found by many other methods. I have given these very simple ones to show that it is possible to construct all the problems of ordinary geometry by doing no more than the little covered in the four figures that I have explained. This is one thing which I believe the ancient mathematicians did not observe, for otherwise they would not have put so much labor into writing so many books in which the very sequence of the propositions shows that they did not have a sure method of finding them all, but rather gathered those propositions on which they had happened by accident.

(Descartes' Method of Obtaining the Equation of a Curve)

I wish to know the *genre*[7] of the curve EC [Fig. 1], which I imagine to be described by the intersection of the ruler GL and the rectilinear plane figure $CNKL$, whose side KN is produced indefinitely in the direction of C, and which, being moved in the same plane in such a way that its side KL always coincides with some part of the line BA (produced in both directions), imparts to the ruler GL a rotary motion about G (the ruler being so connected to the figure $CNKL$ that it always passes through L).[8] If I wish to find out to what *genre* this curve belongs, I choose a straight line, as AB, to which to refer all its points, and in AB I choose a point like A at which to begin the calculation. I say that I choose the one and the other, because we are free to choose them as we like, for while it is necessary to use care in the choice in order to make the equa-

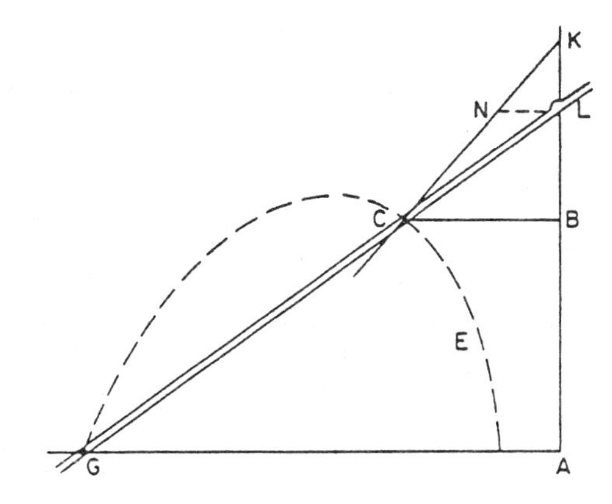

Fig. 1

tion as short and simple as possible, yet no matter what line I should take instead of AB the curve would always prove to be of the same *genre,* a fact easily demonstrated.

Then I take on the curve an arbitrary point, as C, at which I will suppose that the instrument to describe the curve is applied. Then I draw through C the line CB parallel to GA. Since CB and BA are unknown and indeterminate quantities, I shall call one of them y and the other x. But in order to find the relation between these quantities I consider also the known quantities which determine the description of the curve, as GA, which I shall call a; KL, which I shall call b; and NL, parallel to GA, which I shall call c. Then I say that as NL is to LK, or as c is to b, so CB, or y, is to BK, which is therefore equal to $\frac{b}{c}y$. Then BL is equal to $\frac{b}{c}y - b$, and AL is equal to $x + \frac{b}{c}y - b$. Moreover, as CB is to LB, that is, as y is to $\frac{b}{c}y - b$, so AG or a is to LA or $x + \frac{b}{c}y - b$. Multiplying the second by the third, we get $\frac{ab}{c}y - ab$ equal to

$$xy + \frac{b}{c}yy - by,$$

which is obtained by multiplying the first by the last. Therefore, the required equation is

$$yy = cy - \frac{cx}{b}y + ay - ac.$$

From this equation we see that the curve EC belongs to the first *genre,* it being, in fact, a hyperbola.

If in the instrument used to describe the curve we substitute for the straight line CNK this hyperbola or some other curve of the first *genre* lying in the plane $CNKL$, the intersection of this curve with the ruler GL will describe, instead of the hyperbola EC, another curve, which will be of the second *genre.*

Thus, if CNK be a circle having its center at L, then we shall describe the first Conchoid of the Ancients,[9] while if we use a parabola having KB as diameter we shall describe the curve which, as I have already said, is the first and simplest of the curves required in the problem of Pappus, that is, the one which furnishes the solution when five lines are given in position.[10]

Then Descartes continues with his solution of the problem of Pappus, which leads him to the consideration of conic sections and other curves with several types of equations, such as

$$y^2 = 2y - xy + rx - x^2,$$

$$y^3 - 2ay^2 - a^2y + 2a^3 = axy,$$

$$x^2 = ry - \frac{r}{q}y^3,$$

$$y^2 - by^2 - cdy + bcd + dxy = 0.$$

Here also is Descartes' method of finding the equation of a normal to a curve. This method was in a sense opposed to that of Fermat, whose method was based on finding first the equation of a tangent to a curve . . . and thus came close to the idea of a derivative.

NOTES

1. Descartes writes $\frac{a}{b}$.
2. Descartes writes $\sqrt{C \cdot a^3 - b^3 + abb}$.
3. Descartes writes $AB \propto 1$, etc.; see Selection II.8, note 3.
4. This is the "analysis" of Pappus, see Selection II.5.
5. Hence $z^2 = az \pm b$.
6. Descartes here follows the common practice of his day, which considered only the types $z^2 + az - b^2 = 0$, $x^2 - az - b^2 = 0$, and $z^2 - az + b^2 = 0$ of quadratic equations, ignoring the type $z^2 + az + b^2 = 0$, since it has no positive roots (a is a segment, hence positive). Only much later (Newton) did mathematicians begin to associate coordinates with negative numbers. All coordinates in Descartes are positive. The name

"coordinate" does not appear in Descartes; this term is due to Leibniz.

7. Earlier in Book II, Descartes has defined the *genre* of a curve. In our terms: If an algebraic curve has degree $2n - 1$ or $2n$, its *genre* is n. This terminology may have been inspired by the problem of Pappus. Newton translates *genre* by *genus*.

8. The instrument thus consists of three parts: (1) a ruler AK of indefinite length, fixed in the plane; (2) a ruler GL, also of indefinite length, passing through a pivot G in this plane (but not on AK); and (3) a triangle LNK, KN indefinitely extended toward KC, to which the ruler GL is connected at L so as to make the triangle slide with its side KL along AB.

9. Pappus mentions four types of conchoid (shell curves); the first is the one we still call a conchoid, in polar coordinates $r = a + b \sec \theta$. It is a curve of the third degree, therefore of the second *genre* of Descartes.

10. This is also a curve of the second *genre*.

Chapter VI
The Scientific Revolution at Its Zenith (1620–1720)

Section A
Algebra, Analytic Geometry, and Arithmetic

PIERRE DE FERMAT (1601-1665)

Pierre Fermat came from a prosperous bourgeois family. His father, Dominique, was the proprietor of a successful leather business; his mother, Claire de Long, was from a family of jurists. Pierre loved classical studies and proved remarkably adept in them. A master of Greek and Latin, he studied classical literature, dabbled in philosophy, and composed Latin poetry. He also studied two modern languages, Italian and Spanish. He probably attended the University of Bordeaux, before receiving a degree in Civil Laws from the University of Orleans in 1631. In the same year, he married his cousin Louise de Long. They had five children.

Because mathematics did not become a profession with employment opportunities and scholarly organizations for another two centuries, Fermat followed mathematics only as an avocation and made his living as a jurist. In 1631, he purchased the offices of counselor and master of requests at the Toulouse *parlement* (law court). At that time judicial offices were sold in France. Principally because of a high rate of mortality among his colleagues, he rose rapidly in the legal profession. In 1638, he became *conseiller aux enquêtes* in Toulouse, and four years later he was named to the highest councils of the Toulouse *parlement*—the Criminal Court and then the Grand Chamber (acting as the latter's chief spokesman in 1648). However, certain critics viewed his legal work as occasionally less than satisfactory. In 1664, the Languedoc *intendant* (local royal administrative official) described him in deprecating terms to Jean-Baptiste Colbert, the finance minister to Louis XIV. Probably by reason of seniority, Fermat, a staunch Catholic, presided over the Chambre de l'Edit, which had jurisdiction in suits between Huguenots and Catholics.

Throughout his life Fermat proved an extraordinary mathematical genius, especially in the origins of the infinitesimal calculus and in the origins of the theory of numbers, where he was without a peer for more than a century. He presented his important conceptions and discoveries in correspondence with his friends (primarily Bernard Frénicle de Bessy, Pierre Carcavi, and Marin Mersenne) rather than in publications. He enjoyed good relations with several other savants in Paris among whom he circulated handwritten manuscripts, but copies of these are less faithful to Fermat's ideas than his correspondence. In 1670, his son, Samuel, edited and enriched his father's notes on Diophantus, which contain many of his discoveries in the theory of numbers. In 1679, Samuel edited *Varia Opera,* a collection of his father's writings.

Fermat independently discovered coordinate (or analytic) geometry and an early stage of the calculus as well as contributing significantly to the theory of numbers, probability, and optics. A disciple of Francois Viète since his days in Bordeaux, he early pursued symbolic algebra and defended his master's algebraic notation, even when Descartes rendered it obsolete. His study of ancient sources—including the extant writings of Archimedes, Apollonius, Euclid, Pappus, and above all the *Arithmetica* of Diophantus (which was published in a Latin translation in 1621)—also influenced his research. By 1636, Fermat had devised a system of analytic geometry that he recorded in his brief book, *Ad Locos Planos et Solidos Isagoge* ("Introduction to Plane and Solid Loci," publ. posth. 1679). It was nearly identical to Descartes' version in *Géométrie,* an appendix to the celebrated *Discours de la méthode* (1637). Neither man plotted curves in a coordinate system but generated them from a single axis with a moving ordinate. Fermat's method of maxima and minima evolved into an algorithm (rule of mathematical procedure) that was equivalent to differentiation. With this procedure he was able to find the equations of tangents. Viewing Fermat's methods as rivals to his own, Descartes asserted in 1638 that they were paralogistic in their reasoning and had only limited application. These assertions initiated an acrid priority dispute in analytic geometry. As a result of this analysis of curves by Fermat and his building upon the work of Archimedes on summation processes for quadrature, including upper and lower limits, he came very near to discovering the calculus.

Today Fermat is probably best known for his pioneering work in his favorite field, the theory of numbers, where he concentrated on primeness and divisibility. Based on his finding that numbers of the form $2^x + 1$, where $x = 2^n$ and n is an integer, are prime for $n = 1$ through 4, he conjectured that all such numbers are prime. (In the 18th century Leonhard Euler showed this to be false for $n = 5$. Numbers of this form are not prime for n from 5 through 16.) Fermat also stated the equivalent of his lesser theorem: $(a^p - a)$ is divisible by p (*i.e.,* $a^{p-1} \equiv 1 \mod p$), when p is prime and a, p are relatively prime. His "last" or "great" theorem asserts that there are no positive integers x, y, and z such that $x^n + y^n = z^n$, for $n > 2$. Referring to it in 1637, Fermat wrote in his copy of Claude-Gaspard Bachet's book, *Diophanti,* "I have discovered a truly remarkable proof but this margin is too small to contain it." Subsequent historical research suggests that this statement is not correct. The last theorem is still unproven.

Through correspondence with Blaise Pascal, Fermat helped to create the foundation for the theory of probability. The two co-founders proceeded differently. Fermat utilized a superior approach that relied on direct computation rather than general mathematical formulas. Nevertheless, Christian Huygens soon superceded his probabilistic methods with more sophisticated ones in *De ludo aleae* (1657). Fermat vainly tried to interest Pascal in the theory of numbers. In the 1640s, some commentators associated Fermat's study of different forms of numbers with the vogue in Paris for magic and astrology. In the late 1650s and 1660s, Fermat investigated the laws of refraction in optics. His belief that light does not travel faster in denser media led to a polemic with the Cartesians. He postulated *Fermat's principle of least time* in optics: a ray of light in traveling between two points always takes the path requiring least time. With this principle he made early usage of the calculus of variations, as Euler and Lagrange illustrated in the 18th century.

64. From a letter to Bernard Frénicle de Bessy (October 10, 1640)*

(Are numbers of the form $2^n + 1$ prime when $n = 2^t$?)[1]

PIERRE FERMAT

It seems to me after this that it is important to tell you on what foundation I construct the demonstrations of all that concerns the geometrical progressions, which is as follows:

Every prime number is always a factor [*mesure infailliblement*] of one of the powers of any progression minus 1, and the exponent [*exposant*] of this power is a divisor of the prime number minus 1. After one has found the first power that satisfies the proposition, all those powers of which the exponents are multiples of the exponent of the first power also satisfy the proposition.

Example: Let the given progression be

1	2	3	4	5	6	
3	9	27	81	243	729	etc.

with its exponents written on top.

Now take, for instance, the prime number 13. It is a factor of the third power minus 1, of which 3 is the exponent and a divisor of 12, which is one less than the number 13, and because the exponent of 729, which is 6, is a multiple of the first exponent, which is 3, it follows that 13 is also a factor of this power 729 minus 1.

And this proposition is generally true for all progressions and for all prime numbers, of which I would send you the proof if I were not afraid to be too long.

But it is true that every prime number is a factor of a power plus 1 in any kind of progression; for, if the first power minus 1 of which the said prime number is a factor has for exponent an odd number, then in this case there exists no power plus 1 in the whole progression of which this prime number is a factor.

Example: Because in the progression of 2 the number 23 is a factor of the power minus 1 which has 11 for exponent, the said number 23 will not be a factor of any power plus 1 of the said progression to infinity.

If the first power minus 1 of which the given prime number is a factor has an even number for exponent, then in this case the power plus 1 which has an exponent equal to half this first exponent will have the given prime as a factor.

The whole difficulty consists in finding the prime numbers which are not factors of any power plus 1 in a given progression, for this, for instance, is useful for finding which of the prime numbers are factors of the radicals of the perfect numbers, and to find a thousand other things as, for example, why it is that the 37th power minus 1 in the progression of 2 has the factor 223. In one

**Source:* The Latin texts of Fermat's extant writings together with a French translation have been published in *Oeuvres de Fermat* (4 volumes, 1891-1912). This English translation of two of his letters with notes added is taken from D. J. Struik (ed.), *A Source Book in Mathematics, 1200-1800,* 28-31 and is reprinted by permission of Harvard University Press, Copyright © 1969 by the President and Fellows of Harvard College.

word, we must determine which are the prime numbers that factor their first power minus 1 in such a way that the exponent of the said power be an odd number—which I think very difficult [*fort malaisé*].

Fermat then continues with other striking properties of powers, also of numbers of the form $2^n + 1$, which, he believed, are all prime if n is a power of 2.[2]

NOTES

1. (Editor's note). Leonhard Euler showed that $2^n + 1$ is not prime for $n = 32$ in the *Commentarii I* of the St. Petersburg Academy of Sciences. His proof appeared in the *Commentarii I* (1732/33), published 1738), 20-48.
2. These numbers $2^n + 1$, $n = 2^t$, when prime, are known as Fermat numbers. See O. Ore, *Number theory and its history* (McGraw-Hill, New York, 1948).

65. From two letters of February 1657*

(Challenge to Mathematicians: Find an Infinity of Integer Solutions for the "Pell" Equation, that is $x^2 - Ay^2 = 1$, where A may be any Nonsquare Integer)[1]

PIERRE FERMAT

Fermat, after observing that "Arithmetic has a domain of its own, the theory of integral numbers," defines his problem as follows:

Given any number not a square, then there are an infinite number of squares which, when multiplied by the given number, make a square when unity is added.

Example.—Given 3, a nonsquare number; this number multiplied by the square number 1, and 1 being added, produces 4, which is a square.

Moreover, the same 3 multiplied by the square 16, with 1 added makes 49, which is a square.

And instead of 1 and 16, an infinite number of squares may be found showing the same property; I demand, however, a general rule, any number being given which is not a square.

It is sought, for example, to find a square which when multiplied into 149, 109, 433, etc., becomes a square when unity is added.

In the same month (February 1657), Fermat, in a letter to Frénicle, suggests the same problem, and expressly states the condition, implied in the foregoing, that the solution be in integers:

Every nonsquare is of such a nature that one can find an infinite number of squares by which if you multiply the number given and if you add unity to the product, it becomes a square.

Example.—3 is a nonsquare number, which multiplied by 1, which is a square, makes 3, and by adding unity makes 4, which is a square.

The same 3, multiplied by 16, which is a square, makes 48, and with unity added makes 49, which is a square.

There is an infinity of such squares which when multiplied by 3 with unity added likewise make a square number.

I demand a general rule,—given a

*Source: The Latin texts of Fermat's extant writings together with a French translation have been published in *Oeuvres de Fermat* (4 volumes, 1891-1912). This English translation of two of his letters with notes added is taken from D. J. Struik (ed.), *A Source Book in Mathematics, 1200-1800,* 28-31 and is reprinted by permission of Harvard University Press,

nonsquare number, find squares which multiplied by the given number, and with unity added, make squares.

What is for example the smallest square which, multiplied by 61 with unity added, makes a square?

Moreover, what is the smallest square which, when multiplied by 109 and with unity added, makes a square?

If you do not give me the general solution, then give the particular solution for these two numbers, which I have chosen small in order not to give too much difficulty.

After I have received your reply, I will propose another matter. It goes without saying that my proposition is to find integers which satisfy the question, for in the case of fractions the lowest type of arithmetician could find the solution.

Connected with this problem are a number of others, assembled by Fermat's friend Jacques de Billy (1602-1669), a Jesuit teacher of mathematics in Dijon, in his *Doctrinae analyticae inventum novum* (ed. S. Fermat; Toulouse 1670), translated in Fermat, *Oeuvres,* III, 325-398. They begin with the Diophantine problem (called a double equation), to make both $2x + 12$ and $2x + 5$ squares (answer $x = 2$). Part III (p. 376) begins (we change to modern notation):

On the procedure for obtaining an infinite number of solutions which give square or cubic values to expressions in which enter more than three terms of different degrees.

1. I shall discuss here in particular expressions which contain the five terms in x^4, x^3, x^2, x, and the constant, but I also wish to discuss expressions with four terms which may be all positive [true], or mixed with negative [false] terms. We wish to give these expressions square values (in the case of five terms), or cubic ones (in the case of four terms), and this in an infinity of ways. In general we must say that for the square value at least the coefficient of the term in x^4 or the constant term must be a square; as to the cubic values, the coefficient of x^3 or the constant term must be a cube.

Applied to making $x^4 + 4x^3 + 6x^2 + 2x + 7$ a square, De Billy writes $(x^2 + 2x + 1)^2 = x^4 + 4x^3 + 6x^2 + 4x + 1$, which, set equal to the given form, gives $x = 3$.

In the case of $x^4 + 4x^3 + 10x^2 + 20x + 1$ De Billy equates this to $(1 + 10x - 45x^2)^2$, and gets $x = {}^{113}/_{253}$, then he equates it to $(x^2 + 2x - 1)^2$, and gets $x = -3$, and so on.

Then, by substituting for x the value $x + x_0$, where x is a "primitive" solution for example $x_0 = -3$, or $x_0 = -4$, and repeating the process, he obtains new solutions. For $x \to x - 3$ he requires that $x^4 - 8x^3 + 28x^2 - 40x + 4$ be a square, which gives $x = {}^7/_2$; hence $x = {}^1/_2$ is a solution of the orginal equation. Here he turned a "false" solution into a "true" one. This process can be repeated.

It was from these problems by Fermat that Euler, in the paper of 1732/33, started his research on the "Pell" equation.

NOTE

1. (Editor's Note). The roots of the "Pell" equation may be traced to antiquity. The *Cattle Problem,* which Archimedes studied, leads to such an equation, wherein $A = 4{,}729{,}494 = 2 \cdot 3 \cdot 7 \cdot 11 \cdot 29 \cdot 353$. Several early Indian mathematicians also examined these equations.

Chapter VI
The Scientific Revolution at Its Zenith (1620–1720)

Section A
Algebra, Analytic Geometry, and Arithmetic

BLAISE PASCAL (1623-1662)

Pascal, a profound moralist, geometer, polemicist, and master of French prose, was born in Clermont Ferrand, France. He was the second son of Étienne Pascal, a lawyer, and the pious Antoinette Begon. Theirs was a devout Christian home. His mother died in 1626 when Pascal was three, so his father raised the children—Gilberte, Blaise, and Jacqueline—and guided their early education. Étienne moved the family to Paris in 1631. In teaching Blaise, the father encouraged him to master Greek and Latin and to think for himself. After preparatory studies, Blaise began at age 12 to study geometry on his own. The father soon noticed the son's precocity when Blaise reportedly reinvented the equivalent of Euclid's first 32 theorems.

In 1635 the elder Pascal, who associated with eminent artists and men of science, introduced his son to Father Marin Mersenne's "Academie Parisienne," where he met Desargues, Fermat, and Roberval. The young Pascal actively participated in its research and discussions. He was the only other member of the circle to support Girard Desargues' *Brouillon project,* which laid the foundations of synthetic projective geometry and a unified theory of conic sections. In 1640, Pascal wrote "Essais pour les coniques," which built upon the work of Apollonius so skillfully that it won the praise of René Descartes.

In 1640, as the intensity of the geometrical and other intellectual labors began to affect seriously the health of the frail teenager, Pascal went to Rouen to join his father who had been appointed an *intendant* (local administrator) there in 1639. From 1640 to 1647 Blaise only occasionally returned to the intense intellectual stimulation of Paris. Even so, his health further declined in 1641, and he briefly reduced his geometrical studies sharply. He had already achieved a reputation in mathematics that was enhanced in 1642, when he designed an "arithmetic machine" to help his father with computations involved in tax collecting. The machine—essentially the first digital calculator—employed the movement of gears to add and subtract. Pascal supervised a team of workers who produced a working model in 1645. He sold the calculating machine, but few could afford it because of its high price. His contemporaries considered the calculator his greatest achievement.

The Pascal family had respected religious practice, but religion became especially important in their lives only in 1646 when they were introduced to Jansenism. The Jansenist movement in the Catholic Church revived Augustinian views by repudiating free will and accepting predestination. The 23-year-old Blaise became a convert to the austere doctrines through the Abbé of Saint-Cyran and probably influenced his entire family to do likewise. His sister, Jacqueline, subsequently entered the Jansenist convent at Port Royal. Later, Pascal became involved in several polemics with the Jesuits, the enemies of the Jansenists.

Although Pascal turned his attention to religious and theological questions after 1646, this did not end his intense research in natural philosophy and mathematics. In 1646 and 1647 he defended the possibility of a vacuum against Descartes and his followers, who accepted the Aristotelian *horror vacui*. As part of his defense, Pascal conducted experiments at high altitudes that supported theories on air pressure by Galileo and Evangelista Torricelli, who discovered the principle of the barometer. When his health worsened in 1647, Pascal returned to Paris, where he maintained contact with secular intellectuals and the Port Royal Jansenists. When the controversy over the vacuum continued, he conducted a famous barometric experiment while ascending to the summit of Puy de Dome (1648), a mountain overlooking Clermont Ferrand. These studies in turn led him to investigate the action of fluids under the pressure of air. By the early 1650s, he had probably derived the basic principles of hydrostatics, and his results appeared in the book, *Great Experiment Concerning the Equilibrium of Fluids* (1663).

On the advice of his physician, Pascal put aside some of his research in 1651 and instead frequented polite society with his friends, the young Duc de Roannez and the Chevalier de Méré. With them he read Epictetus and Montaigne, and a problem posed by de Méré about the division of stakes in a game of chance led Pascal to study the theory of probability. In his correspondence with Pierre Fermat in 1654 and the *Traité du triangle arithmétique* (completed in the same year), he laid the foundations for the calculus of probability.

By late 1654, Pascal developed "an extreme aversion for the beguilements of the world," and he experienced a second conversion in the "nuit de feu" on November 23. He resolved to retire from the political world, to submit himself totally to Jesus, and to devote himself henceforth to meditation and religious matters. The Jansenists soon employed Pascal's talents; his Port Royal friends, Antoine Arnauld and Pierre Nichole, suggested that he write the *Lettres provinciales* (1656-57). This he did under the pseudonym Louis de Montalte, and the style and polemic of the *Lettres* captivated Paris.

Despite the immersion in religion and his ill health, which became acute in 1658, Pascal did not forget mathematics. To distract his mind from a persistent toothache in 1658, he provided a complete solution to the problem of the cycloid. His contemporaries, including John Wallis, had failed to reach this. He also prepared a manual for teaching geometry at Port Royal. Geometry remained for him the "highest exercise of the mind," but as he wrote to Fermat with mystical enthusiasm, "it is only a trade, . . . and I am steeped in studies so far from that mentality that scarcely do I remember there is any such."

After helping to plan a public transportation project in Paris in 1659, Pascal spent his final years planning a book on Christian apologetics. Fragments he wrote were discovered at his death by the Jansenists and published by them under the inappropriate title *Pensées* in 1670.

Pascal's substantial contributions to mathematics went beyond refining projective geometry, inventing his digital calculating machine, and pioneering in the theory of probability. In his treatise on the arithmetical triangle, an idea known earlier to the 13th-century Chinese mathematician, Yang Hui, and the 15th-century Persian mathematician, al-Kāshī, he provided a method for finding binomial coefficients, stressed reasoning by recursion, and correctly stated a formula for finding the number of combination of n things taken r at a time, *i.e.*, in modern form $\frac{n!}{r!\,(n-r)!}$. In other work, Pascal believed that he had perfected Bonaventura Cavalieri's "theory of indivisibles" and engaged in contests with Wallis, Sluse, Huygens, and others on the subject (1658). His research on this theory, his study of the cycloid, and his early steps toward a concept of a definite integral show that he came very near to the integral calculus. His *Lettres de A. Dettonville . . .* (1659) later influenced Gottfried Leibniz in the discovery of the infinitesimal calculus.

66. From *Traité du triangle arithmétique . . .**

(The Pascal Triangle)

BLAISE PASCAL

I designate as the arithmetic triangle a figure of which the construction is as follows [Fig. 1]. Through an arbitrary point G, I draw 2 lines perpendicular to each other, GV and $G\zeta$, on each of which I take as many equal and continuous parts as I like, beginning at G, which I call 1, 2, 3, 4, etc., and these numbers are the indices [*exposans*] of the divisions of the lines.

Then I join the points of the first division, which are on each of the two lines, by another line that forms a triangle of which this line is the *base*.

I also join the two points of the second division by another line that forms a second triangle of which this line is the *base*.

And joining in this way all the division points which have the same indices I form with them as many *triangles* and *bases*.

I draw through every one of the division points lines parallel to the sides, and these by their intersections form small squares which I call cells [*cellules*].

And the cells that are between two parallels that run from left to right are called *cells of the same parallel rank,* such as the cells G, σ, π, etc., or ϕ, ψ, θ, etc.

And those that are between two lines that run from the top downward are called *cells of the same perpendicular rank,* such as the cells G, ϕ, A, D, etc. and these: σ, ψ, B, etc.

And those that the same base traverses diagonally are called *cells of the same base,* such as the following: D, B, θ, λ, and these: A, ψ, π.

The cells of the same base that are equally distant from their extremities are called *reciprocal,* such as these: E, R

**Source:* The French original appears in L. Brunschvicg and P. Bourtroux (eds.), *Oeuvres de Pascal* (1909), 405 seq. This English translation with notes added is taken from D. J. Struik (ed.), *A Source Book in Mathematics, 1200-1800,* 21-26 and is reprinted by permission of Harvard University Press, Copyright © 1969 by the President and Fellows of Harvard College. Footnotes renumbered.

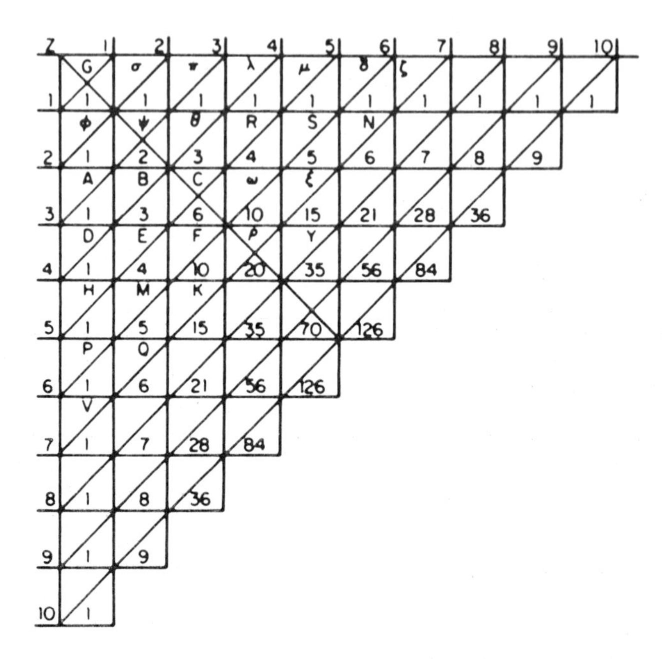

Fig. 1

and B, θ, because the index of the parallel rank of the one is the same as the index of the perpendicular rank of the other, as appears in the example, where E is in the second perpendicular and in the fourth parallel rank, and R is in the second parallel and in the fourth perpendicular rank, reciprocally. It is easy enough to show that those which have their indices reciprocally equal are in the same base and equally distant from its extremities.

It is also quite easy to demonstrate that the index of the perpendicular rank of any cell whatsoever, added to the index of its parallel rank, exceeds the index of its base by unity.

For example, the cell F is in the third perpendicular rank and in the fourth parallel one, and in the sixth base, and its two indices of the ranks $3 + 4$ exceed the index 6 of the base by unity, which results from the fact that the two sides of the triangle are divided into an equal number of parts, but that is rather understood than demonstrated.

The following remark is of the same nature: that every base contains one cell more than the preceding one, and every one contains as many cells as its index has units; the second base $\phi\sigma$, for instance, has two cells, the third $A\psi\pi$ has three of them, etc.

We now place numbers in each cell and this is done in the following way: the number of the first cell which is in the right angle is arbitrary, but once it has been placed all the other numbers are determined, and for this reason it is called the *generator* of the triangle. And every one of the other numbers is specified by this sole rule:

The number of each cell is equal to that of the cell preceding it in its perpendicular rank plus that of the cell which precedes it in its parallel rank. For instance, the cell F, that is, the number of the cell F, is equal to cell C plus cell E, and so the others.

From this many consequences can be drawn. Here are the most important ones, where I consider the triangles whose generator is unity, but what can be said about them will also apply to the others.

FIRST CONSEQUENCE

In every arithmetic triangle all the cells of the first parallel rank and of the first perpendicular rank are equal to the generator.

Indeed, by the construction of the triangle, every cell is equal to the cell which precedes it in its perpendicular rank plus the cell that precedes it in its parallel rank. Now, the cells of the first parallel rank have no cells which precede them in their perpendicular ranks, nor have those of the first perpendicular rank any in their parallel ranks: hence they are all equal to each other and to the generating first number.

And so ϕ is equal to G + zero, that is, ϕ is equal to G.

And so A is equal to ϕ + zero, that is, ϕ.

And so σ is equal to G + zero, and π equal to σ + zero.

And so the others.

Using a more modern notation, in which we call P_l^k the cell of parallel rank l and vertical rank k, so that

$$P_l^k = \frac{(k + l - 2)!}{(k - 1)!(l - 1)!},$$

we can write the next "consequences" as follows:

2. $$P_l^k = \sum_{i=1}^{k} P_{l-1}^i; \quad \text{e.g., } \omega = R + \theta + \psi + \varphi;$$

3. $$P_l^k = \sum_{i=1}^{k} P_i^{k-1}; \quad \text{e.g., } C = B + \psi + \sigma;$$

4. $$P_l^k - 1 = \sum_{i=1}^{k-1} \sum_{j=1}^{l-1} P_j^i; \quad \text{e.g., } \xi - g = R + \theta + \psi + \varphi + \lambda + \pi + \sigma + G,$$

where $g = 1$, the generator;

5. $$P_l^k = P_k^l; \quad \text{e.g., } \varphi = \sigma = G, \pi = A = G, D = \lambda = G.$$

6. All $P_l^k = P_k^l$, k fixed; e.g., $\sigma\psi BEM\varphi$ is equal to $\varphi\psi\theta RSN$;

7. $$\sum_{l,k=1,\ldots,n} P_l^k = 2 \sum_{i,j=1,\ldots,n-1} P_j^i, \quad k + l = \text{fixed number} = a, \quad i + j = a - 1;$$

e.g., $D + \lambda + B + \theta = 2A + 2\psi + 2\pi$;

8. $$\sum_{l,k=1,\ldots,n} P_l^k = 2^{n-2}, \quad k + l = n;$$

9. $$1 + 2 + \cdots + 2^n = 2^{n+1} - 1;$$

10. $$\sum_{l=n}^{p} P_l^k = 2 \sum_{i=n-1}^{p-1} P_j^i + P_{n-1}^p \quad [\text{e.g., } P_4^1 + P_3^2 + P_2^3 = 2(P_3^1 + P_2^2) + P_4^3].$$

$k + l = n, \quad i + j = n - 1, \quad p = n - 2;$ e.g., $D + B + \theta = 2A + 2\varphi + \pi;$

11. $$P_l^l = 2P_l^{l-1} = 2P_{l-1}^l; \quad \text{e.g., } C = \theta + B = 2B.$$

TWELFTH CONSEQUENCE

In every arithmetic triangle, if two cells are contiguous in the same base, the upper is to the lower as the number of cells from the upper to the top of the base is to the number of those from the lower to the bottom, inclusive.

Let the two contiguous cells, arbitrarily chosen on the same base, be E, C; then I say that

$\underbrace{E}_{\text{lower one}}$ is to $\underbrace{C}_{\text{upper one}}$ as

$\underbrace{2}_{\text{because there are two cells between } E \text{ and the first, namely } E, H;}$ is to $\underbrace{3}_{\text{because there are three cells between } C \text{ and the top, namely } C, R, \mu.}$

Although this proposition has an infinite number of cases I shall give for it a very short demonstration by supposing two lemmas:

The first one, evident in itself, is that this proposition occurs in the second base; because it is clear enough that ϕ is to σ as 1 is to 1.

The second one is that if this proposition is true in an arbitrary base, it will necessarily be true in the next base. From which it is clear that it will necessarily be true in all bases, because it is true in the second base because of the first lemma; hence by means of the second lemma it is true in the third base, hence in the fourth base, and so on to infinity.[1]

It is therefore necessary to demonstrate only the second lemma, and this can be done in the following way. Let this proposition be true in an arbitrary base, as in the fourth one D, that is, if D is to B as 1 is to 3, and B to θ as 2 to 2, and θ to λ as 3 to 1, etc., then I say that the same proportion will be true in the next base, $H\mu$, and that, for example, E is to C as 2 is to 3.

Indeed, D is to B as 1 is to 3, by hypothesis.

Hence $\underbrace{D + B}_{E}$ is to B as $\underbrace{1 + 3}_{4}$ is to 3.

E is to B as 4 is to 3.

In the same way: B is to θ as 2 is to 2, by hypothesis.

Hence $\underbrace{B + \theta}_{C}$ is to B as $\underbrace{2 + 2}_{4}$ is to 2.

C is to B as 4 is to 2.

But B is to E as 3 is to 4.

Hence, by the double proportion,[2] C is to E as 3 is to 2. Q.E.D.

The proof can be given in the same way in all the other cases, since this proof is founded only on the fact that this proportion is true in the preceding base, and that every cell is equal to its preceding one plus the one above it, which is true in all cases.[3]

There follow more "consequences," numbered 13-19.[4] The article ends with a "Problem":

Given the indices of the perpendicular and of the parallel rank of a cell, to find the number of the cell, without using the arithmetic triangle.

For example, let it be proposed to

find the number of the cell ξ of the fifth perpendicular rank and of the third parallel rank.

Having taken all the numbers that precede the index of the perpendicular rank 5, that is, 1, 2, 3, 4, take as many natural numbers beginning with the index of the parallel rank 3, that is, 3, 4, 5, 6.

Now multiply the first numbers into each other, and let the product be 24. Multiply the other numbers into each other, and let the product be 360, which divided by the other product 24, gives 15 as the quotient. This quotient is the desired number.

Indeed, ξ is to the first number of its base V in composed ratio of all the ratios of the cells among themselves, that is,

ξ is to V in composed ratio of

$$\underbrace{\xi \text{ to } \rho}_{3 \text{ to } 4} + \underbrace{\rho \text{ to } K}_{4 \text{ to } 3} + \underbrace{K \text{ to } Q}_{5 \text{ to } 2} + \underbrace{Q \text{ to } V,}_{6 \text{ to } 1,}$$

or by the twelfth consequence:

ξ is to V as 3 into 4 into 4 into 5 into 6 into 4 into 3 into 2 into 1,

But V is unity; hence ξ is the quotient of the division of the product of 3 into 4 into 5 into 6 by the product of 4 into 3 into 2 into 1.

Note. If the generator were not unity we should have to multiply the quotient by the generator.

This paper is followed by several others, in which the Pascal triangle is applied. First it is used to sum the arithmetical sequences of different orders 1, 2, 3, 4, etc.; 1, 3, 6, 10, etc., 1, 4, 10, 20, . . . (these sequences are called "numbers of the first, second, etc. order" [*ordres numériques*], then to the solution of certain games of chance, to the finding of combinations, to the raising of binomials to different powers, to the summation of the squares, cubes, etc., of the terms of an arithmetical series, etc., and to the proof that (in our present notation)

$$\int_0^a x^p\,dx = \frac{a^{p+1}}{p+1},\ p \text{ a positive integer.}$$

NOTES

1. This seems to be the first satisfactory statement of the principle of complete induction. See H. Freudenthal, "Zur Geschichte der vollständigen Induktion," *Archives Internationales des Sciences 22* (1953), 17-37.

2. The text has "proportion troublée," probably a misprint for "proportion doublée."

3. The meaning of this is as follows. Given

$$P_k^l : P_{k-1}^{l+1} = \frac{l}{k-1} \text{ (in base } k + l - 1).$$

But

$$P_k^l + P_{k+1}^{l+1} = P_{k-1}^{l+1} \text{ (rule of formation of the triangle);}$$

hence

$$P_k^{l+1} : P_{k-1}^{l+1} = \frac{l+k-1}{k-1},$$

$$P_{k-1}^{l+1} : P_{k-2}^{l+2} = \frac{l+1}{k-2},$$

$$P_{k-1}^{l+2} : P_{k-1}^{l+1} = \frac{l+k-1}{l+1};$$

hence

$$P_k^{l+1} : P_{k-1}^{l+2} = \frac{l+1}{k-1} \text{ (in base } k + l).$$

4. For example, consequence 17 states that

$$\sum_{i=1}^{k} P_i^l : \sum_{j=1}^{l} P_k^j = k : l,$$

e.g., $(B + \psi + \sigma) : (B + A) = 3{:}2$.

These consequences can all be found in the translation of Pascal's paper in Smith, *Source book,* pp. 74-75.

Chapter VI
The Scientific Revolution at Its Zenith (1620–1720)

Section B
Origins of the Infinitesimal Calculus

JOHANN KEPLER (1571-1630)

Johann Kepler, the German astronomer, cosmologist, and natural philosopher, was a major, transitional figure in the early Scientific Revolution. He is perhaps best known for discovering the three laws of planetary motion—planar elliptical planetary orbits, equality of areas, and the 3/2 ratio. With these and other findings Kepler helped to discredit the clumsy Ptolemaic geocentric model of the heavens and to establish the Copernican heliostatic theory, which he improved into a far more accurate heliocentric theory. The first notable Copernican after Copernicus (except for Rheticus), Kepler was also a founder of modern geometrical optics, providing the first correct explanation of how humans see and coming close to formulating Snel's sine law of refraction. His wearing of eyeglasses (in an age when it was uncommon) served as a stimulus to his study of the formation of images with lenses. Bold in his scientific ideas and attracted to mystical speculation, he was congenial in his dealings with others. He was known for his modest manner, ready wit, scrupulous honesty, and wealth of knowledge.

Kepler grew up in unsettled times and unsettled surroundings. His father, Heinrich, was a mercenary soldier; his mother, Katarina Guldenmann, was the quarrelsome daughter of an innkeeper. Physically, Kepler was short, frail, nearsighted, in poor health, and plagued by fears of fevers, stomach ailments, and impotency. Since his family was impoverished, only the enlightened scholarship policy of the dukes of Württemberg made it possible for him to attend the University of Tübingen (1589-94), where he was a "straight A" student and was profoundly influenced by Michael Maestlin, a cautious Copernican. After receiving the M.A. from Tübingen in 1591, Kepler entered theological studies to prepare for the Lutheran ministry. However, he changed the direction of his career in 1594, when he accepted the position of teacher of mathematics at Graz in southern Austria. Throughout his life he maintained strong theological and mystical interests which he consciously strove to incorporate into his scientific work.

Kepler resided at Graz from 1594 until 1598, where he taught arithmetic, astronomy, the poetry of Vergil, and rhetoric. An impecunious teacher, he also compiled careful almanacs and skillfully cast horoscopes, even though he considered astrology to be the foolish little daughter of respectable astronomy. In the summer of 1595, he came up with the seminal idea of assigning the orbits of the six known planets to spheres circumscribed around the five regular polyhedra or "platonic bodies," which were nested one within another. Although the origin of this idea was not Baconian–scientific—it derived from an aesthetic sense combined with a Platonic and Pythagorean vision of celestial harmony—his geometrical schema did, in fact, provide the correct relations for the distances between the planets and the sun. He explained the nesting idea in *Mysterium Cosmographicum* ("Cosmographic Mystery," 1596). In 1597 he married Barbara Müller who died in 1611. He had to leave Graz in 1598, when Catholic authorities ordered the expulsion of all Protestant teachers from the town.

After an unsuccessful attempt to join the Tübingen faculty, Kepler became assistant to Tycho Brahe at the court of Rudolph II in Prague in 1599. He was part of a research group at Benatky Observatory. When Tycho died in 1601, Kepler succeeded him as imperial mathematician and gained access to Tycho's incomparable register of astronomical observations. Using these he began to construct the *Rudolphine Tables*, which provided far more accurate planetary positions than those of earlier calculational astronomers. His were within 10′ of actual position compared to earlier errors of 5°.

The Prague years (1601 to 1612) were the most creative of Kepler's life. In completing a painstaking study of the orbit of Mars begun for Tycho, he discovered that its orbit was an ellipse—with the sun at one focus—rather than a circle. This was a devastating blow to the Aristotelian cosmology of rotating physical spheres as a source of planetary motions. That planets move in elliptical orbits was his first planetary law. The first and second laws appeared in his book *Astronomia Nova* ("New Astronomy," 1609), in which he also gave a novel animistic–mechanical concept of "inertia," if his word *soul* is interpreted as acting as a physical force. The *New Astronomy* shows its author to be moving from the ancient geometrical description of the heavens to dynamical astronomy. In 1610 Kepler praised Galileo's astronomical *Siderius Nuncius* (which many ridiculed) and used a telescope for the first time. Although increasing religious turmoil in Bohemia forced Rudolph to abdicate in May 1611, Kepler was required to remain in Prague until Rudolph's death in 1612.

From 1612 through 1626, Kepler was imperial mathematician in Linz, where local Calvinists considered him a renegade and Catholics tried to make him a convert. In 1613, he married Susanna Reuttinger; they had seven children. Continuing his research and writing, Kepler completed the *Harmonice mundi* ("Harmonies of the World," 1618), which contained his third law. In this book, he maintained that the archetypal principles of the universe were based on geometry and musical harmonies—a view that differed sharply from Robert Fludd's more qualitative and alchemical view of the universe. He also wrote a five-volume introduction to Keplerian astronomy, entitled *Epitome Astronomiae Copernicae* ("Epitome of Copernican Astronomy," 1618–21), whose very title emphasized the espousal of a new cosmology. After the poet John Donne visited him in 1617, Kepler was offered a position in England in 1620, which he declined. After skillfully intervening in that year to win acquittal for his mother, who was being tried for witchcraft, he concentrated on attempting to publish the *Rudolphine Tables*. However, these efforts were hindered by a lack of funds (his salary was greatly in arrears) and by the ravages of the Thirty Years' War (Lutheran printers around Linz had their houses and presses burned). The printing of the *Tables* was finally completed at Ulm, Germany, in 1627.

After a siege of Linz in 1626, Kepler

gained permission to move to Regensburg (also known as Ratisbon), and the next year he moved to Sagan in Silesia with the imperial commander-in-chief, Albrecht von Wallenstein, as his patron. Still, he was nearly penniless without back pay. Leaving his family behind he traveled in 1630 to Regensburg to try to collect salaries owed him but died after contracting an acute fever.

In addition to making his primary contributions to astronomy, cosmology, and speculative physics, Kepler advanced mathematics. After refusing to confine himself to classical Archimedean procedures for gauging volumes, he presented a new, nonrigorous, but direct method in his treatise *Nova stereometria doliorum vinariorum* ("New Solid Geometry of Wine Barrels," 1615), which measured the actual volumes of wine casks by combining an infinite number of thin circular laminae or other cross sections. His method, which extended the range of Archimedes' results, represented a step toward the integral calculus. Also, by the 1620s he was among the first to use Napier's logarithms.

67. From *Nova steriometria doliorum vinariorum* (1615)*

(Integration Methods)

JOHANN KEPLER

PART I. THE SOLID GEOMETRY OF REGULAR BODIES

Theorem I. We first need the knowledge of the ratio between circumference and diameter. Archimedes taught:

The ratio of circumference to diameter is about 22:7. To prove it we use figures inscribed in and circumscribed about the circle. Since there is an infinite number of such figures, we shall, for the sake of convenience, use the hexagon [Fig. 1]. Let a regular hexagon *CDB* be inscribed in the circle; let its angles be *C, D, B,* its side *DB,* and *F* the point of intersection of the two tangents at *D* and *B* respectively. The line *AF* connects the center *A* with *F,* and intersects the line *DB* at *G,* the curve *DB* at *E*. But as *DGB* is a straight line, it is the shortest distance between *D* and *B*.

DEB, on the other hand, being a curve, is not the shortest distance between *D* and *B*. Hence *DEB* is longer than *DGB*. On the other hand, *BF* is

Fig. 1

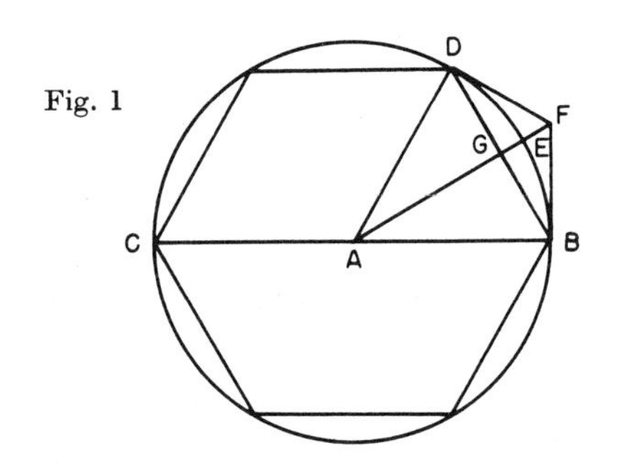

**Source:* These selections may be found in M. Caspar (ed.), *Gesammelte Werke* (1960), IX, 13-16 and 47-49. The English translation with notes added is taken from D. J. Struik (ed.), *A Source Book in Mathematics 1200-1800,* 192-197 and is reprinted by permission of Harvard University Press,

tangent to the circle and therefore all parts of the curve *EB* are between *FB* and *GB;* therefore, if *EB* were straight, it would altogether be shorter than *FB*. For *AEB, FEB* are equivalent to a right angle, and, as *EFB* is an acute angle, *EB,* opposite the smaller angle *EFB,* must be smaller than *FB,* since this is opposite the larger angle. And we can consider *EB* a straight line, because in the course of the proof the circle is cut into very small arcs, which appear to be equal to straight lines.[1]

Now since, as can be observed, the curve *DEB* is contained in the triangle *DBF,* it must be smaller than the lines *DF, FB,* since it bends toward the angle *DFB,* and still has not the slightest part outside the lines *DF, FB;* but the containing, according to common sense, is greater than the contained.[2] This would be different, were the curve *DEB* winding and irregular.

But as *DB* is a side of the inscribed hexagon, and *DF, FB* are two halves of the circumscribed hexagon, arc *DEB* must be a sixth of the circle, since it was greater than *DB* and smaller than *DF, FB;* 6 *DB* is smaller than the circumference of the circle and 12 *DF* (or *FB*) is greater than the circumference.

But the side *DB* of the regular hexagon is equal to the radius *AB*. Therefore 6 radii *AB,* that is, three diameters *CB* or (if the diameter is divided by 7) $^{21}/_{7}$ *CB* are shorter than the circumference.

And again, since *DG, GB* are equal, *GB* is half of *AB*. The square of *AB,* however, is equal to the sum of the squares of *AG* and *GB* and is the quadruple of the square of *GB*. Therefore the square of *AG* is three times the square of *GB*. The ratio therefore of the squares of *AB* and *AG* is $^{4}/_{3}$ of the lines, therefore the ratio *AB:AG* is $\sqrt{^{4}/_{3}}$, that is, the ratio of the numbers 100,000 : 86,603.[3] But as *AG:AB* = *GB:BF,* then also *BF:GB* is $\sqrt{^{4}/_{3}}$ and as *GB* is half of *AB,* for example, 50,000, *BF* must have about 57,737 of such parts. Twelvefold this total number therefore will be greater than the circumference of the circle. Computation gives the number 477,974 for those circles which have 200,000 for diameter. And those of diameter 7 have for twelve times *BF* the value 24 minus $^{1}/_{10}$. But this number is greater than the circumference itself; on the other hand the number 21 is smaller than the said circumference. And it is obvious that the curve *BE* is nearer to *BG* than the line *BF*. The circumference therefore is nearer the number 21 than 24 − $^{1}/_{10}$.[4] We suppose it differs by 1 from 21, from the other by 2 − $^{1}/_{10}$, and that it therefore doubtless is 22. This, however, Archimedes shows much more accurately by means of multisided figures of 12, 24, 48 sides; there it also becomes apparent how little the difference of the circumference from 22 is. Adrianus Romanus proved by the same method that when the diameter is divided into 20,000,000,000,000,000 parts, then about 62,831,853,071,-795,862 of those parts make up the circumference.[5]

Remark [*Episagma*]. Of the three conical lines, which are called parabola, hyperbola, and ellipse, the ellipse is similar to the circle, and I showed in the *Commentary on the motions of Mars* that the ratio of the length of the elliptic line to the arithmetic mean of its two diameters (which are called the right and transversal axes) is about equal to 22:7.[6]

Theorem II. *The area of a circle compared with the area of the square erected on the diameter has about the ratio* 11:14.

Archimedes uses an indirect proof in which he concludes that if the area exceeds this ratio it is too large. The meaning of it seems to be this [Fig. 2].

The circumference of the circle *BG* has as many parts as points, namely, an infinite number; each of these can be regarded as the base of an isosceles triangle with equal sides *AB,* so that there are an infinite number of triangles in the area of the circle. They all converge with their vertices in the center *A*.

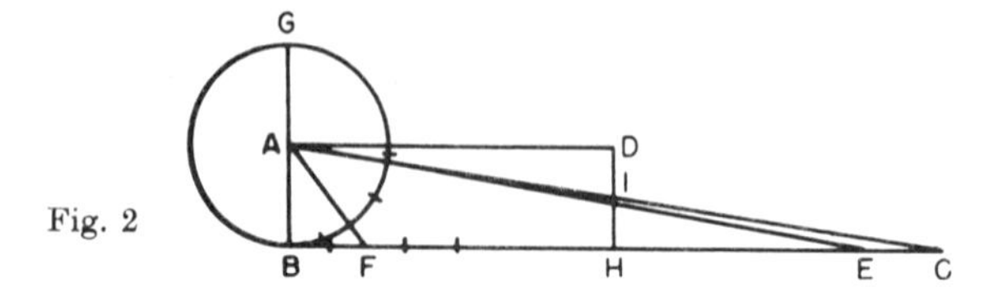

Fig. 2

We now straighten the circumference of circle *BG* out into the line *BC*, equal to it. The bases of these infinite triangles or sectors are therefore all supposed to be on the straight line *BC*, arranged one next to the other. Let *BF* be one of these bases, and *CE* any other, equal to it, and let the points *F, E, C* be connected with *A*. Since there are as many triangles *ABF, AEC* over the line *BC* as there are sectors in the area of the circle, and the bases *BF, EC* are equal, and all have the altitude *BA* in common (which is also one of the sectors), the triangles *EAC, BAF* will be equal, and equal to one of the circle sectors. As they all have their bases on *BC*, the triangle *BAC*, consisting of all those triangles, will be equal to all the sectors of the circle and therefore equal to the area of the circle which consists of all of them. This is equivalent to Archimedes' conclusion by means of an absurdity.

If now we divide *BC* in half at *H*, then *ABHD* forms a parallelogram. Let *DH* intersect *AC* in *I*. This parallelogram is equal to the circle in area. Indeed, *CB* is to its half *CH* as *AB* (that is, *DH*) is to its half *IH*. Therefore *IH* = *ID* and *HC* = *DA* (equal to *BH*). The angles at *I* are equal, and those at *D* and *H* are right angles. The triangle *ICH*, which is outside the parallelogram, is equal to triangle *IAD* by which the parallelogram exceeds the trapezoid *AIHB*.

If now the diameter *GB* is 7 parts, then its square will be 49. And since the circumference consists of 22 such parts—hence also *BC*—its half *BH* will consist of 11, hardly more or less. Multiply it by the semidiameter 3½, which is *AB*, and we get for the rectangle *AH* 38½ [38 *semis*]. Therefore, if the square of the diameter is 49, the area of the circle is as twice 49 or 98 to 77. Dividing by 7 we obtain 14 to 11, Q.E.D.

Corollary 1. The area of the sector of a circle (consisting of straight lines from the center intersecting the arc) is equal to the rectangle over the radius and half the arc.

Corollary 2 deals with the area of a segment of a circle.

The next theorems deal with the cone, cylinder, and sphere. In a supplement Kepler introduces conic sections and solids generated by these curves. Among the solids he discusses we find the torus, which he calls a ring [*annulus*].

Theorem XVIII. *Any ring with circular or elliptic cross section is equal to a cylinder whose altitude equals the length of the circumference which the center of the rotated figure describes, and whose base is the same as the cross section of the ring.*[7]

By cross section is meant the intersection of a plane through the center of the ring-shaped space and perpendicular to the ring-shaped surface. The proof of this theorem follows partly from theorem XVI[8] and can be established by the same means by which Archimedes taught as the principles of solid geometry.

Indeed, if we cut the ring *GCD* [Fig. 3] from its center *A* into an infinite number of very thin disks, any one of them will be the thinner toward the center *A*, the nearer its part, such as *E*, lies to the center *A* than to *F* and the normal through *F* erected in the intersecting plane to the line *ED*. It also will be the thicker the nearer it is to the point *D*. At such two extreme points,

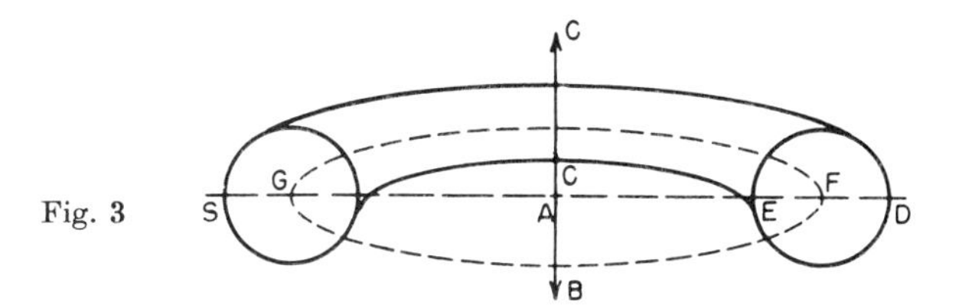

Fig. 3

such as *D* and *E,* the sum of the two thicknesses will be twice the one in the middle of the disk.

This consideration would not be valid if the parts at *E* and *D* of the disk on either side of the circumference *FG* and the perpendiculars through *F* and *G* were not equal and equally situated.

Corollary. This mode of measuring is valid for circular and for elliptical rings as well, high, narrow, or reclining, for open and closed rings alike, as indeed even for all rings whatever shape their cross section may have (instead of the circle *ED*)—so long as in the plane through *AD* perpendicular to the ring the parts on either side of *F* are equal and equally situated. We shall explore this in the case of a square section. Let the ring be of square shape and assume the square to be on *ED*. This ring can also be measured in another way. For it is the outer part of a cylinder whose base is a circle with AD as radius and whose height is DE. From this cylinder, according to Theorem XVI, the middle part has to be subtracted, that is, the cylinder whose base is the circle of radius *AE* and whose height is *ED*. The product, therefore, of *ED* and the circular area *AD* minus the circular area *AE* is equal to the volume of the ring with a square as cross section. And if *ED* is multiplied by the difference of the squares of *AD* and *AE,* then the ratio of this body to the fourth part of the ring would be as the square to the circle, therefore as 14 to 11. Let *AE* be equal to 2, *AD* equal to 4, then its square is 16; but the square of *AE* is 4, therefore the difference of the squares is 12; this number multiplied by the altitude 2 gives the volume as 24, of which the quadruple is 96. Since 14 is to 11 as 96:$75^3/_7$, the volume of the square ring is $75^3/_7$. This is according to the computation of Theorem XVI. And according to the preceding method, if *AF* is 3, *FG* is 6. Since 7 is to 22 as 6 is to 19 minus $^1/_7$, this therefore will be the length of the circumference *FG,* the altitude of the cylinder. And since $ED = 2$, its square is 4. To obtain the base of the cylinder, multiply therefore 4 by $(19 - {}^1/_7)$. In this way also we see the truth of the theorem.

Theorem XIX *and Analogy. A closed ring is equal to a cylinder whose base is the circle of the cross section and whose height equals the circumference of the circle described by its center.*

As this method is valid for every ring, whatever the ratio of *AE* and *AF* may be, therefore, it also holds for a closed ring, in which the center *F* of the circle *ED* describes the circle *FG,* where *FG* is equal to the rotated *AD* itself. This is because such a closed ring is intersected from *A* in disks that have no thickness at *A* and at *D* twice the thickness of that at *F*. Hence the circle through *D* is twice that through *F*.

Corollary. The cylindric body that is created by rotation of *MIKN* [Fig. 4a], the four-sided figure of straight and curved lines, is according to the same consideration equal to a column with this figure as base and the length of the circle *FG* as height. But the outer fringe *IKD* that surrounds the cylindric body—as a wooden hoop surrounds a barrel—clearly does not yield to this theorem, and must be computed by other means.

Analogy. Moreover, this method is valid for all cylindric bodies or parts of apples (or figs), no matter how slender,

Fig. 4

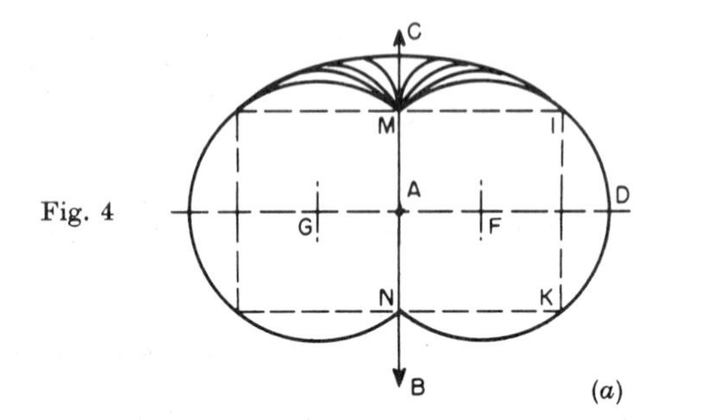

(a)

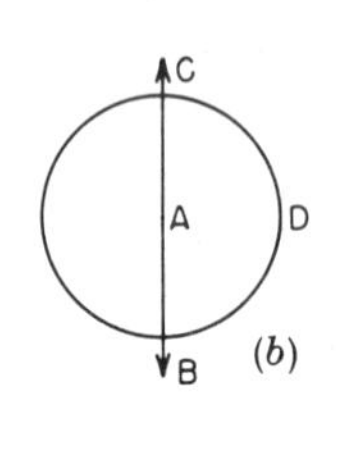

(b)

until *I, K* coincide with *M, N,* which happens in the formation of the sphere [Fig. 4*b*], where instead of the two lines *MN* and *IK* there exists only one, namely, *BC*. For this body the demonstration and use of this theorem fail for the first time.

Corollary. The ratio of the sphere to the closed ring created by the same circle is 7 to 33, since one-third of the radius multiplied by four times the area of the greatest circle, or two-thirds of the diameter multiplied by the area of the greatest circle, produce a cylinder equal to the sphere.[9] And a cylinder equal to the closed ring has the same base, and its altitude is the circumference [formed by the center]. Therefore as the circumference is to two-thirds of the diameter, that is, 33:7,[10] so is the ring to the sphere.[11]

NOTES

1. This statement of Kepler's was attacked by Paul Guldin, in his *Centrobaryca seu de centro gravitatis* (2 vols.; Vienna, 1635, 1641). There exists no geometric proof whatever, wrote Guldin, that a circular arc, be it as small as you like, may be equated to a straight line. Guldin (1577-1645), a Swiss-born Jesuit mathematician who taught in Rome, Vienna, and Graz, was critical not only of the methods of Kepler, but also of those of Cavalieri. His book also contains the "rules of Guldin"; see note 7.

2. Here Guldin criticized again: if this were evident, then Archimedes would not have found it necessary (in *De sphaero et cylindro*) to prove that the circumference of a polygon circumscribed about a circle is larger than that of a circle: "In geometry we should not trust too much in what is evident."

3. Kepler wrote "one-half of $^4/_3$" [*semisesquitertia*], expressing the square root by "one-half," a mode of expression that goes back to Boethius (sixth century A.D.) and even to Euclid. . . . This mode of expression, with its "logarithmic" flavor, has a relation to the ancient theory of music.

4. The actual value is $12 \cdot {}^7/_2 \tan 30° = 24.25$. Kepler writes "24, minus decima."

5. Adriaen Van Roomen (1561-1615) had published this in his *Ideae mathematicae* (Louvain, 1593).

6. For this approximation of the circumference C of the ellipse of semiaxes a and b as $C = {}^{22}/_7(a + b)$, see Kepler's *Astronomia nova* (Heidelberg, 1609), *Gesammelte Werke*, ed. Caspar, III, 368. This statement of Kepler's was also criticized by Guldin. However, for planetary orbits, with small eccentricity, Kepler's approximation was not bad; if developed up to fourth powers of the eccentricity e, it is only $({}^1/_{32})\pi e^4$ greater than the circumference C (*ibid.*, III, 484; IV, 480).

7. In Theorem I and Theorem II Kepler had replaced Archimedes' *reductio ad absurdum* with a more direct proof, and in a vague way identified the points on the circumference with very small segments. His reasoning reminds us of Antiphon; see T. L. Heath, *Manual of Greek mathematics* (Clarendon Press, Oxford, 1931), 140. In Theorems XVIII and XIX we find the solid divided into very small disks. These theorems are special cases of the so-called Guldin or Pappus theorem, which in the version of Pappus runs as follows: "The ratio of two

perfect [complete] surfaces of rotation is composed of the ratio of the rotated areas and of that of the straight lines drawn perpendicularly to the axes of rotation from the centers of gravity of the rotated areas of the axes"; Pappus, *Mathematical collection*, Book VII, trans. Ver Eecke (see Selection III.3, note 1), pp. 510-511. It should be pointed out that some scholars believe that this theorem is a later insertion. Kepler and Guldin probably found their theorems independently of Pappus.

The special case of the torus, which interested Kepler here, can be found in Heron's *Metrica* (c. A.D. 100), where it is attributed to a certain Dionysodoros (probably second century B.C.), author of a lost treatise on the torus (Heath, *Manual of Greek mathematics*, p. 385). Guldin, referring to Kepler's theorems on figures of rotation, stated his own rules and pointed out that Kepler had almost found them.

8. Theorem XVI deals with the ratio of conical segments of the same height and different bases.

9. The text writes "cube."

10. The ratio is $3\pi:2$.

11. On Kepler there exists in English a symposium of the History of Science Society, *Johann Kepler 1571-1630, a tercentenary commemoration of his work* (Williams and Wilkins, Baltimore, 1931). [See also Angus Armitage, *John Kepler* (Faber and Faber, London, 1966).]

Chapter VI
The Scientific Revolution at Its Zenith (1620–1720)

Section B
Origins of the Infinitesimal Calculus

GALILEO GALILEI (1564-1642)

The name of Galileo Galilei—Italian mathematician, physicist, and astronomer—is inextricably linked with the Scientific Revolution. During the second phase of that Revolution, roughly from 1615 to 1660, he clashed with the Catholic Church over Copernican astronomy and attempted to extend the mathematical treatment of natural philosophy (physics) from statics to kinematics and the strength of materials. Moreover, in natural philosophy he informally stated the essence of Newton's first two laws of motion and insisted that a panmathematical rationalism yielded a more correct and fruitful understanding of nature than the Scholastic logico-verbal approach. He combined the panmathematical rationalism with experimental methods that improved upon those of Renaissance technicians. For these achievements he is often called the founder of modern mechanics and experimental physics.

Galileo, the eldest of seven children, was born at Pisa in 1564, the year Michelangelo died and Shakespeare was born. His father, Vincenzio Galilei, a lutenist and musical theorist, appears to have taught him a contempt for dogmatic authority. About 1575 Galileo began studies in logic at the Monastery of Santa Maria at Vallombrosa (near Florence), but three years later his father withdrew him for a time when the son apparently wished to enter the order of monks as a novice. Galileo returned to school and enrolled as a medical student at the University of Pisa (at age 17). There he studied medicine, natural philosophy, and cosmology under Aristotelian professors.

In 1582 his observation of a swinging lamp in the cathedral at Pisa led to his discovery that a pendulum of a given length swings at a constant frequency regardless of the amplitude. Later, during a vacation period in 1583, Galileo met Ostilio Ricci, a family friend and teacher at the Tuscan court, who introduced him to the study of mathematics, a subject for which he quickly developed a passion. Despite a rapid progress in studies, Galileo was denied a scholarship, perhaps because of his self-assured, polemical nature. Lacking funds, he left the University of Pisa without a degree in 1585. Returning to Florence to study by himself, he consulted the works of Euclid and Archimedes, whose mathematical method he adopted. Although translated into Latin in the 13th century, the major works of Archimedes had not been assimilated. Now, new translations were available, in part by Ricci's teacher, the algebraist Nicolò Tartaglia.

From 1585 to 1589, Galileo was with his family in Florence. He had now grown into a ginger-haired, short necked, squarely built young man of lively disposition. A stare that suggested an opthalmological problem detracted from his appearance. He gave instruction in the mathematical sciences and actively studied the classics (especially his favorites Ovid, Vergil, and Seneca). His reputation as a man of learning grew. Following an Archimedean inspiration, he invented a hydrostatic balance in 1586. He traveled to Rome in the following year to meet the eminent Jesuit astronomer, Christopher Clavius, as well as other scholars. In 1588 the Florentine Academy invited him to lecture on the topography of Dante's *Inferno*.

Galileo came to the attention of influential men, including Cardinal del Monte, who successfully recommended that the Medici duke of Tuscany appoint him lecturer in mathematics at the University of Pisa. In 1589 Galileo began his academic career by returning to that university as a teacher. He moved to the celebrated University of Padua near Venice in 1592. His 18 years at Padua were the most creative of his life. In dynamics, he arrived at the law of free fall by 1604 and studied motion on inclined planes. In astronomy, he came to reject the Ptolemaic geocentric theory (and its underlying Aristotelian physics) and to accept Copernican theory. The appearance of a supernova in 1604 suggested to him that the heavens were not immutable as Aristotelian physics held. On hearing of Lippershey's telescope in 1609, he invented his own and quickly increased its power from 9 to 32. He published his observations in his first book, *Siderius nuncius* (*The Starry Messenger*, 1610), which brought him sudden fame. Contrary to Aristotelian cosmology, he revealed that the moon was mountainous (and thus not a smooth crystalline sphere), that the Milky Way contained a congeries of stars, and that Jupiter had four moons (which he opportunistically called the Medicean stars).

In 1610 Galileo returned to Florence to serve as "first philosopher and mathematician" to Cosimo II de'Medici, the grand duke of Tuscany—a post that allowed more time for research. He was not yet a full supporter of Copernican theory; his first decisive advocacy of it came in *Letters on Sunspots* in 1613. At the universities, Aristotelian professors united against him, while ecclesiastical authorities (especially the Dominicans), fulminated against his "blasphemous utterances." Appealing to a literal interpretation of the Old Testament, the Dominicans claimed that the earth could not move. They charged the abrasive scientist with raising "a new Pythagoreanism," an evil in their sight. In 1615 Galileo wrote an eloquent *Letter to the Grand Duchess* delineating the spheres of religious and scientific inquiry and pleading solemnly for freedom of thought in science. In February 1616 he was summoned by Cardinal Robert Bellarmine and the Roman Inquisition. Galileo was warned not to "hold or defend" Copernican astronomy, although he might discuss it as a "mathematical supposition." A month later Copernicus' *De Revolutionibus* was placed on the *Index* of forbidden books.

The years from 1616 to 1624 were spent in studies. With an acerbic wit and an irony that was ill-advised, Galileo demolished the ideas of a notable Jesuit astronomer in *Il saggiatore* (The Assayer, 1623), where he stated that the "Book of Nature is . . . written in mathematical characters" and can only be deciphered by those who are well-versed in mathematics. Shortly before the *Saggiatore* was published, Maffeo Barberini became Pope Urban VIII. The two men were on good terms, and he was once more allowed to write about the "systems of the world" (but only in a noncommittal way that recognized the primacy of religious authority). Galileo went well beyond the spirit and the letter of that injunction. His *Dialogo sopra due massimi sistemi del mondo* ("Dialogue on the Two Chief World Systems," 1632), a literary masterpiece, was a powerful Copernican tract that prompted his famous trial before the Roman Inquisition in 1633. Urban VIII was especially angered that his views had been put in the mouth of a simple-minded Aristotelian in the *Dialogue*. Under threat of torture, the aged and failing Galileo "abjured, cursed and detested" past Copernican errors.

Galileo lived his last eight years confined to his estate in Arcetri (outside Florence), where he could receive visitors, including the English poet John Milton. Although the Congregation of the *Index* forbade publication of his books, he was able to smuggle to Holland the *Discorsi e dimostrazioni matematiche intorno a du nuove scienze* ("Discourses and Mathematical Demonstrations Concerning Two New Sciences"). By the time this book appeared in print in 1638, Galileo was completely blind.

The *Two New Sciences*, which underlies much of modern physics, principally deals with the mathematical science of kinematics and the engineering science of strength of materials. It also examines uniform and accelerated motion, discusses parabolic trajectory, and studies the nature of sound and the speed of light. In pure mathematics, it stresses the paradoxes encountered since antiquity in handling the concepts of the infinite and infinitesimal *(parti non quanta)*. Galileo clearly differentiated between actual and potential infinity and set the integers in a one-to-one correspondence with their squares. In their subsequent elaboration of his ideas on the infinite and the infinitesimal, Galileo's pupils, Cavalieri and Torricelli, provided new integration methods shortly before the discovery of the calculus.

68. From *Two New Sciences* (1638)*

(Paradoxes of Infinity: The Relationship between Points and Lines, the Order of an Infinity, Infinitesimals, the Concept of a Continuum)

GALILEO GALILEI

Salviati. . . . Well, since paradoxes are at hand, let us see how it might be demonstrated that in a finite continuous extension it is not impossible for infinitely many voids to be found. At the same time we shall see, if nothing else, at least a solution of the most admirable problem put by Aristotle among those that he himself called admirable; I mean among his *Mechanical Questions*. And its solution may perhaps be no less enlightening and conclusive than that

**Source:* From Stillman Drake, trans. and ed., *Galileo Galilei: Two New Sciences* (Madison: The University of Wisconsin Press; © 1974 by The Board of Regents of the University of Wisconsin System), 28-40 and 55-57. Professor Stillman Drake, 219 Glen Road, Toronto, Ontario, M4W 2X2, Canada. The interlocuters are Salviati, Sagredo, and Simplicio. Footnotes renumbered.

which he himself alleges, and yet different from that which the learned Monsignor di Guevara very acutely considers.[1]

But first it is necessary to explain a proposition not touched on by others, upon which the solution of this question depends; and if I am not mistaken, this [proposition] will later entail other new and admirable things. To understand this, let us draw the diagram with attention. We are to think of an equilateral and equiangular polygon of any number of sides described around the center *G*. For the present, let this be a hexagon *ABCDEF*, similar to and concentric with which we shall draw a smaller hexagon marked *HIKLMN*, and extend one side of the larger, *AB*, indefinitely in the direction *S*. The corresponding side of the smaller, *HI*, is extended in the same direction by line *HT* parallel to *AS*, and through the center we draw *GV* parallel to both these.

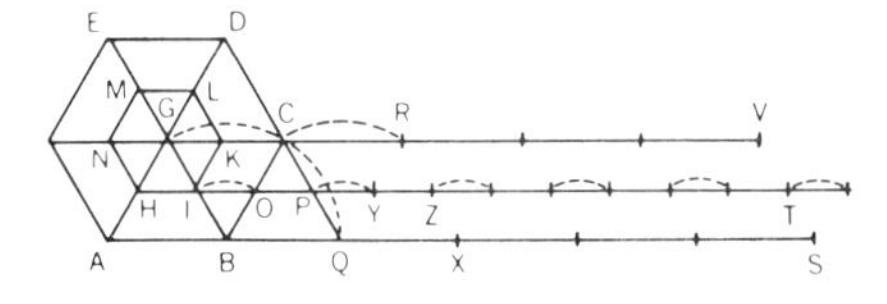

This done, we suppose the larger polygon to rotate along the line *AS*, carrying with it the smaller polygon. It is clear that the point *B*, one end of side *AB*, remains fixed. When revolution begins, the corner *A* rises and the point *C* drops, describing the arc *CQ*, so that side *BC* fits the equal line *BQ*. In this revolution, the corner *I* of the smaller polygon is lifted above line *IT*, because *IB* is oblique to *AS*; and point *I* does not return to the parallel *IT* until point *C* gets to *Q*, when point *I* will have dropped to *O* after describing arc *IO*, outside the line *HT*, the side *IK* having then passed to *OP*. During all this time, the center *G* will have been moving along outside the line *GV*, to which it does not return until it has described the entire arc *GC*.

This first step having been taken, the larger polygon is now situated with its side *BC* on line *BQ;* side *IK* of the smaller one is on line *OP*, having jumped over the part *IO* without touching it; and the center *G* has come to *C*, tracing its whole path outside the parallel *GV*. The entire figure is again at a place similar to its first position. Commencing the second turn and coming to the second place, side *DC* of the larger polygon will fit on the part *QX; KL* of the smaller, having first skipped the arc *PY*, falls on *YZ;* and the center, still moving outside *GV*, falls on it only at *R* after the big jump *CR*. And eventually, when one entire revolution has been made, the larger polygon will have touched, along *AS*, six lines equal [in all] to its perimeter, with nothing interposed [between them]; the smaller polygon will likewise have impressed six lines equal to its circumference but interrupted [*discontinuate*] by the interposition of five arcs, under which there are stretches which are parts of *HT* not touched by this polygon; and the center *G* has never met the parallel *GV* except at six points. From this, it is understood that the space passed over by the smaller polygon is almost equal to that passed by the larger one; that is, line *HT* [nearly equals] *AS*, being smaller only by the chord of one of these arcs, if we understand line *HT* to include the spaces of the five [skipped] arcs.[2]

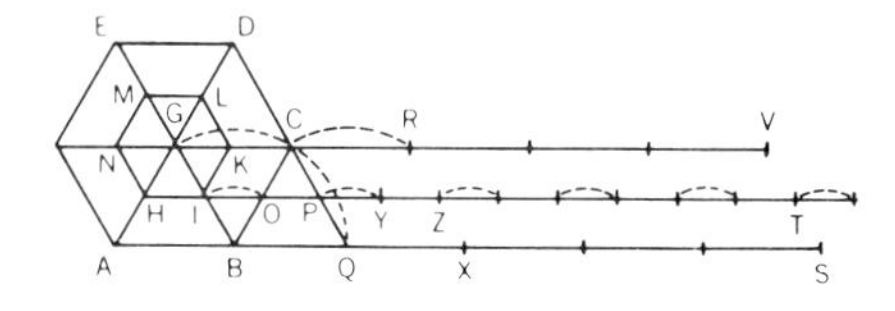

Now, what I have here set forth and explained by the example of these hexagons, I wish to be understood as happening with all other polygons, of as many sides as you please, provided that they are similar, concentric, and joined so that the turning of the larger governs that of the smaller, no matter how much smaller it may be. Understand, I say, that the lines passed over by these are approximately equal, when we count as space passed over by the smaller those intervals under the little arcs, which are not touched by any part of the perimeter of this smaller polygon. Therefore a larger polygon having a thousand sides passes over and measures a straight line equal to its perimeter, while at the same time the smaller one passes an approximately equal line, but one interruptedly composed of a thousand little particles equal to its thousand sides with a thousand void spaces interposed—for we may call these "void" in relation to the thousand linelets touched by the sides of the polygon. And what has been said thus far presents no difficulty or question.

But now tell me: if around some center, say this point *A*, we describe two concentric, joined circles, and from the points *C* and *B* on their radii we draw the tangents *CE* and *BF*, with the parallel *AD* to these [passing] through the center *A*; and if we suppose the

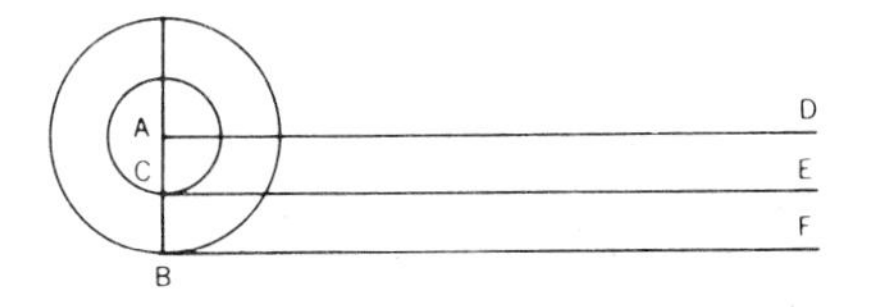

greater circle to be turned on the line *BF*, equal to its circumference as are likewise lines *CE* and *AD*; then, when the greater circle has completed one revolution, what will the smaller circle have done, and the center? The center will certainly have run over and touched the whole line *AD*; and the circumference of the smaller [circle] will with its contact have measured the whole of *CE*, behaving like the polygons considered above. The only difference is that there, the line *HT* was not touched in all its parts by the perimeter of the smaller polygon, for it left untouched, by the interposition of the voids skipped over, as many parts as those touched by the sides.[3] But here, in the circles, the circumference of the smaller circle is never separated from the line *CE* in such a way that any part of *CE* is not touched; nor is that in this circumference which touches ever less than that which is touched in the straight line [*CE*]. How then, without skipping, can the smaller circle run through a line so much longer than its circumference?

Sagredo. I was wondering whether one might say that just as the center of the circle, all alone, being but a single point drawn along on *AD*, touches the whole of that line, so the points of the smaller circumference, driven by the larger circumference, might be dragged through some particles of the line *CE*.

Salviati. This cannot be, for two reasons. First, because there would be no more reason that some of the contacts analogous to *C*, rather than others, should be dragged along some parts of the line *CE*. If this were the case, and such contacts being infinitely many by reason of their being points, the draggings along *CE* would be infinitely many; and being quantified [*quanti*], these would form an infinite line; but *CE* is finite. The second reason is that since the larger circle in its revolution continually changes its [point of] contact, the smaller circle cannot avoid likewise [continually] changing its contact, as it is only through the point *B* that a line can be drawn to the center *A* and still pass through the point *C*. So whenever the larger circle changes contact, the smaller does also; nor does any point of the smaller [circle] touch more than one point of the straight line *CE*.

Besides, even in the revolution of the polygons, no point of the perimeter of the smaller is fitted to more than one point of the line that is measured by that same perimeter. This may easily be understood by considering the line *IK* as parallel to *BC*, so that until *BC* falls on *BQ*, *IK* remains lifted above *IP*, nor does it fall [flat] before that very instant in which *BC* is united with *BQ*. But at that instant *IK* as a whole unites with *OP*, and later on it is just as suddenly lifted above it.

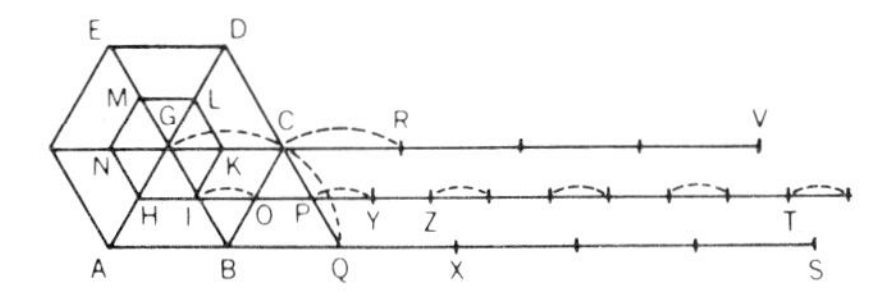

Sagredo. This business is truly very intricate, and no solution at all occurs to me; so tell us what occurs to you.

Salviati. I return to the consideration of the polygons discussed earlier, the effect of which is intelligible and already understood. I say that in polygons of one hundred thousand sides, the line passed over and measured by the perimeter of the larger—that is, by the hundred thousand sides extended [straight and] continuously—is equal to that measured by the hundred thousand sides of the smaller, but with the interposition [among these] of one hundred thousand void spaces.[4] And just so, I shall say, in the circles (which are polygons of infinitely many sides), the line passed over by the infinitely many sides of the large circle, arranged continuously [in a straight line], is equal in length to the line passed over by the infinitely many sides of the smaller, but in the latter case with the interposition of as many voids between them. And just as the "sides" [of circles] are not quantified, but are infinitely many, so the interposed voids are not quantified, but are infinitely many; that is, for the former [line touched by the larger circle there are] infinitely many points, all filled [*tutti pieni*], and for the latter [line touched by the smaller circle], infinitely many points, part of them filled points and part voids.

Here I want you to note how, if a line is resolved and divided into parts that are quantified and consequently numbered [*numerate*], we cannot then arrange these into a greater extension than that which they occupied when they were continuous and joined, without the interposition of as many void [finite] spaces. But imagining the line resolved into unquantifiable parts—that is, into its infinitely many indivisibles—we can conceive it immensely expanded without the interposition of any quantified void spaces, though not without infinitely many indivisible voids.

What is thus said of simple lines is to be understood also of surfaces and of solid bodies, considering those as composed of infinitely many unquantifiable atoms; for when we wish to divide them into quantifiable parts, doubtless we cannot arrange those in a larger space than that originally occupied by the solid unless quantified voids are interposed—void, I mean, at least of the material of the solid. But if we take the highest and ultimate resolution [of surfaces and bodies] into the prime components, unquantifiable and infinitely many, then we can conceive such components as being expanded into immense space without the interposition of any quantified void spaces, but only of infinitely many unquantifiable voids. In this way there would be no contradiction in expanding, for instance, a little globe of gold into a very great space without introducing quantifiable void spaces—provided, however, that gold is assumed to be composed of infinitely many indivisibles.[5]

Simplicio. It seems to me that you are traveling along the road of those voids scattered around by a certain ancient philosopher.[6]

Salviati. At least you do not add, "who denied Divine Providence," as in a similar instance a certain antagonist of our Academician very inappropriately did add.[7]

Simplicio. Indeed I perceived, not without disgust, the hatred in that malicious opponent; yet I shall not touch on that, not only by reason of the bounds of good taste, but because I know how far such ideas are from the temperate and orderly mind of such a man as you, who are not only religious and pious, but Catholic and devout.

Getting back to the point, I feel many

difficulties that are born of the reasoning just heard; doubts from which I really don't know how to free myself. For one, I advance this: if the circumferences of the two circles are equal to the two straight lines *CE* and *BF*, the latter taken as continuous and the former with the interposition of infinitely many void points, how can *AD*, described by a center that is one point only, be called equal to this point, of which [entities] it contains infinitely many? Also, this composing the line of points, the divisible of indivisibles, the quantified of unquantifiables—these reefs seem to me to be hard to pass. And not absent from my difficulties is the necessity of assuming the void, so conclusively refuted by Aristotle.

Salviati. There are these [difficulties] indeed, and others; but let us remember that we are among infinites and indivisibles, the former incomprehensible to our finite understanding by reason of their largeness, and the latter by their smallness. Yet we see that human reason does not want to abstain from giddying itself about them. Taking some liberties on that account, I am going to produce a fantastic idea of mine which, if it concludes nothing necessarily, will at least by its novelty occasion some wonder. Or perhaps it will seem to you inopportune to digress at length from the road that we started on, and hence will be distasteful.

Sagredo. Please let us enjoy the benefit and privilege that comes from speaking with the living and among friends, about things of our own choice and not by necessity, which is very different from dealing with dead books that excite a thousand doubts and resolve none of them. So make us partners in whatever reflections suggest themselves to you in the course of our discussions. We do not lack time to continue and resolve the other matters we have undertaken, thanks to our present freedom from necessary occupations. In particular, the doubts raised by Simplicio are by no means to be skipped over.

Salviati. Be it so, since that is the way you wish it. Let us begin from the first—how a single point can ever be understood to be equal to a line. The most that can be done at present is for me to try to put at rest, or at any rate to moderate, this improbability with an equal or greater one, as a marvel is sometimes put to rest by a miracle. I shall do this by showing you two equal surfaces, and two bodies, also equal, with the said surfaces as their bases. These will [all] go continually and equally diminishing during the same time, their remaining parts always being equal, until finally the surfaces and the solids terminate their preceding perceptual equality by one solid and one surface becoming a very long line, while the other solid and the other surface become a single point; that is, the latter two become a single point, and the former two, infinitely many points.

Sagredo. This seems to me a truly remarkable proposal; let me hear its explanation and demonstration.

Salviati. We must draw a diagram for it, since the proof is purely geometrical.[8]

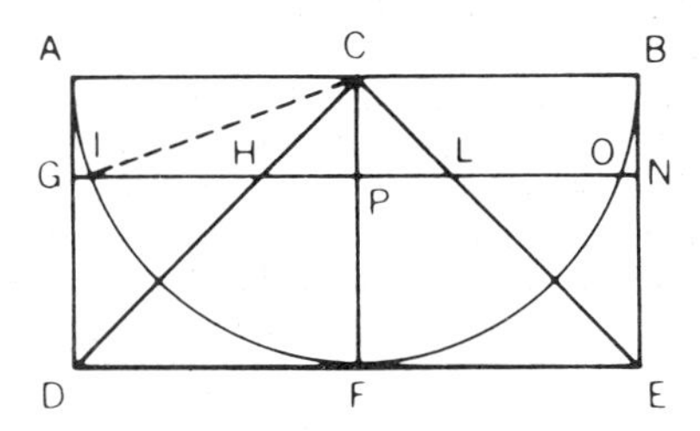

Take the semicircle *AFB* whose center is *C*, and around it the rectangular parallelogram *ADEB*; from the center to points *D* and *E*, draw the straight lines *CD* and *CE*. Next imagine the whole figure rotated around the fixed radius *CF*, perpendicular to the straight lines *AB* and *DE*. It is manifest that a cylinder will be described by the rectangle *ADEB*, a hemisphere by the semicircle *AFB*, and a cone by the triangle *CDE*. We now suppose the hemisphere removed, leaving [intact] the cone and those remains of the cylinder which in shape resemble a soupdish, for which reason we shall call it by that name.

First, we shall prove this soupdish to be equal [in volume] to the cone. Then, drawing a plane parallel to the circle at the base of the soupdish, of diameter *DE* and with center *F,* we shall prove that this plane, passing for example through the line *GN,* and cutting the soupdish at points *G, I, O* and *N,* and the cone at points *H* and *L,* leaves the part of the cone *CHL* always equal to the part of the soupdish whose cross section is represented by the "triangles" *GAI* and *BON*. Moreover, we shall prove that any base of the cone, say the circle whose diameter is *HL,* is equal to that circular surface which is the base for that part of the soupdish; this is, as it were, a washer [*nastro,* ribbon] of breadth *GI*.

Note here what sort of things mathematical definitions are; that is, the mere imposition of names, or we might say abbreviations of speech, arranged and introduced in order to remove the tedious drudgery that you and I felt before we agreed to call one surface the "washer," and presently feel until we call the [upper section of the] soupdish the "cylindrical razor." Now, call these what you will, it suffices to understand that the plane at any level, provided that it is parallel to the base, or circle of diameter *DE,* always makes the two solids equal; that is, the part of the cone *CHL,* and the upper part of the soupdish [*i.e.,* the cylindrical razor]. Likewise it makes equal the two surfaces that are the bases of those solids; that is, the washer and the circle *HL*.

From this follows the marvel previously mentioned; namely, that if we understand the cutting plane to be gradually raised toward the line *AB,* the parts of the solids it cuts are always equal, as likewise are the surfaces that form their bases. Lifting it more and more, the two always-equal solids, as well as their always-equal bases, finally vanish—the one pair in the circumference of a circle, and the other pair in a single point, such being the upper rim of the soupdish and the summit of the cone. Now, during the diminution of the two solids, their equality was maintained right up to the end; hence it seems consistent to say that the highest and last boundaries of the reductions are still equal, rather than that one is infinitely greater than the other, and so it appears that the circumference of an immense circle may be called equal to a single point!

What happens in the solids likewise happens in the surfaces that are their bases. These also maintain equality throughout the diminution in which they share; and at the end, in the instant of their ultimate diminution, the washer reaches its limit in the circumference of a circle, and the base of the cone in a single point. Now, why should these not be called equal, if they are the last remnants and vestiges left by equal magnitudes?[9]

Note next that if these vessels were as large as the immense celestial hemispheres, and the ultimate edges and the points of the contained cones always preserved their equality, those edges would terminate in circumferences equal to great circles of the celestial orbs, and the cones [would terminate] in single points. Hence, along the line in which such speculations lead us, the circumferences of all circles, however unequal [in size], may be called equal to one another, and each of them [may be called] equal to a single point!

Sagredo. The speculation appears to me so delicate and wonderful that I should not oppose it even if I could. To me it would seem a sort of sacrilege to

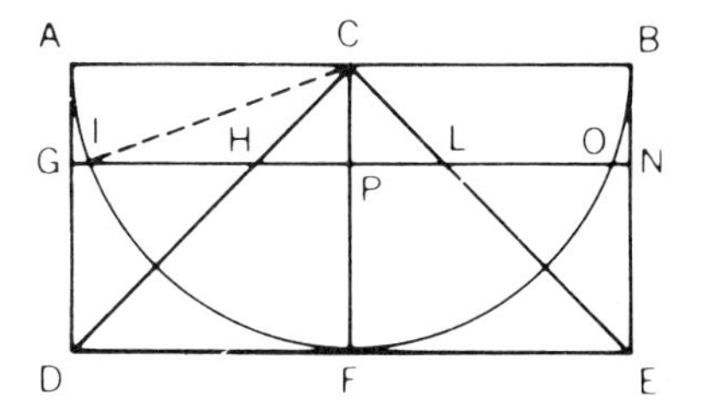

mar so fine a structure, trampling on it with some pedantic attack. Still, for our full satisfaction, let us have that proof, which you call geometrical, of the constant maintenance of equality between those solids, and between their bases. I think this must be very clever, since the philosophical meditation stemming from this conclusion is so subtle.

Salviati. The demonstration is also brief and easy. In the diagram drawn, angle *IPC* being a right angle, the square of the radius *IC* is equal to the two squares of the sides *IP* and *PC*. But the radius *IC* is equal to *AC*, and this to *GP*; and *CP* is equal to *PH*. Therefore the square of the line *GP* is equal to the two squares on *IP* and *PH*, and four times the former equals four times the sum of the latter; that is, the square of the diameter *GN* is equal to the two squares *IO* and *HL*. And since circles are to each other as the squares of their diameters, the circle of diameter *GN* will be equal to the two circles of diameters *IO* and *HL*; hence removing the common circle whose diameter is *IO*, the remaining circle *GN* will be equal to the circle whose diameter is *HL*.

So much for the first part [areas]. As to the other part [volumes], let us skip that proof for the present; if we wish to see it, we shall find it in the twelfth proposition of the second book of *De centro gravitatis solidorum* by Signor Luca Valerio, the new Archimedes of our age, who makes use of it for another proposition of his.[10] [We omit the proof] also because in our case it is enough to have seen how the two surfaces described are always equal, and that in

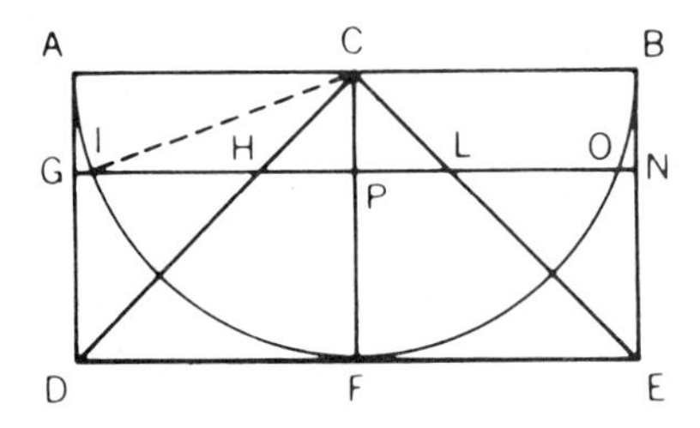

diminishing always equally, they tend to end, the one in a single point, and the other in the circumference of a circle of any size whatever; for our marvel turns on this consequence alone.

Sagredo. The proof is as ingenious as the reflection based on it is remarkable. Now let us hear something about the second difficulty advanced by Simplicio, if you have anything new to say about it, which I believe may not be the case, since the controversy has been so widely agitated.

Salviati. I shall give you my own special thought on it, first repeating what I said a while ago; that is, that the infinite is inherently incomprehensible to us, as indivisibles are likewise; so just think what they will be when taken together!

If we want to compose a line of indivisible points, we shall have to make these infinitely many, and so it is necessary [here] to understand simultaneously the infinite and the indivisible. Many indeed are the things I have on many occasions turned over in my mind on this matter. Some of them, perhaps the most important, I may not recall offhand; but in the progress of the argument I may happen to awaken objections and difficulties in you, and especially in Simplicio, in meeting which I shall remember things that without such stimulus would remain asleep in my imagination. So, with our customary freedom, let it be agreed that we bring in our human caprices, as we may well call them in contrast with those theological [*sopranaturale*] doctrines that are the only true and sure judges of our controversies and the unerring guides through our obscure and dubious, or rather labyrinthine, opinions.[11]

One of the first objections usually produced against those who compound the continuum out of indivisibles is that one indivisible joined to another indivisible does not produce a divisible thing, since if it did, it would follow that even the indivisible was divisible; because if two indivisibles, say two points, made a quantity when joined, which would be a divisible line, then this would be even better composed of three, or five, or seven, or some other odd number [of indivisibles]. But these lines would then be capable of bisection, making the middle indivisible capable of being cut. In this, and other objections of the kind, satisfaction is given to its partisans by telling them that not only two indivisibles, but ten, or a hundred, or a thousand do not compose a divisible and quantifiable magnitude; yet infinitely many may do so.[12]

Simplicio. From this immediately arises a doubt that seems to me unresolvable. It is that we certainly do find lines of which one may say that one is greater than another; whence, if both contained infinitely many points, there would have to be admitted to be found in the same category a thing greater than an infinite, since the infinitude of points of the greater line will exceed the infinitude of points of the lesser. Now, the occurrence of an infinite greater than the infinite seems to me a concept not to be understood in any sense.

Salviati. These are some of those difficulties that derive from reasoning about infinites with our finite understanding, giving to them those attributes that we give to finite and bounded things. This, I think, is inconsistent, for I consider that the attributes of greater, lesser, and equal do not suit infinities, of which it cannot be said that one is greater, or less than, or equal to, another.[13] In proof of this a certain argument once occurred to me, which for clearer explanation I shall propound by interrogating Simplicio, who raised the difficulty. I assume that you know quite well which are square numbers, and which are not squares.

Simplicio. I know well enough that a square number is that which comes from the multiplication of a number into itself; thus four and nine and so on are square numbers, the first arising from two, and the second from three, each multiplied by itself.

Salviati. Very good. And you must also know that just as these products are called squares, those which thus produce them (that is, those which are multiplied) are called sides, or roots. And others [numbers] that do not arise from numbers multiplied by themselves are not squares at all. Whence if I say that all numbers, including squares and non-squares, are more [numerous] than the squares alone, I shall be saying a perfectly true proposition; is that not so?

Simplicio. One cannot say otherwise.

Salviati. Next, I ask how many are the square numbers; and it may be truly answered that they are just as many as are their own roots, since every square has its root, and every root its square; nor is there any square that has more than just one root, or any root that has more than just one square.[14]

Simplicio. Precisely so.

Salviati. But if I were to ask how many *roots* there are, it could not be denied that those are as numerous as all the numbers, because there is no number that is not the root of some square. That being the case, it must be said that square numbers are as numerous as all numbers, because they are as many as their roots, and all numbers are roots. Yet at the outset we said that all the numbers were many *more* than all the squares, the majority being non-squares. Indeed, the multitude of squares diminishes in ever-greater ratio as one moves on to greater numbers, for up to one hundred there are ten squares, which is to say that one-tenth are squares; in ten thousand, only one one-hundredth part are squares; in one million, only one one-thousandth. Yet in the infinite number, if one can conceive that, it must be said that there are as many squares as all numbers together.

[*Text omitted here.*]

Salviati. These are among the marvels that surpass the bounds of our imagination, and that must warn us how gravely one errs in trying to reason about infinites by using the same attributes that we apply to finites; for the natures of

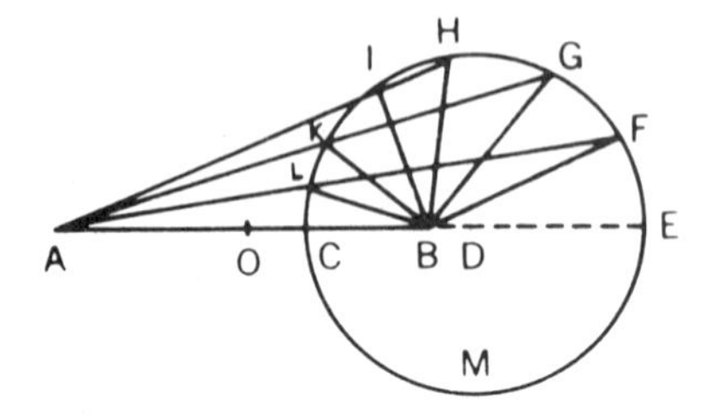

these have no necessary relation [*convenienza*] between them. Apropos of this, I do not wish to pass by in silence a remarkable event that just now occurs to me, illuminating the infinite difference and even the repugnance and contrariety of nature encountered by a bounded quantity in passing over to the infinite. Let us take this straight line *AB*, of any length whatever, and in it take some point *C* that divides it into unequal parts. I say that pairs of lines leaving from the points *A* and *B*, and preserving between themselves the same ratio as that of the parts *AC* and *BC*, will intersect in points that all fall on the circumference of the same circle. For example, *AL* and *BL*, coming from points *A* and *B* and having the same ratio as parts *AC* and *BC*, meet in a point *L*; with the same ratio, another pair *AK* and *BK* meet in K; others [are] *AI* and *BI*, *AH* and *HB*, *AG* and *GB*, *AF* and *FB*, *AE* and *EB*. I say that the meeting-points *L*, *K*, *I*, *H*, *G*, *F*, and *E* all fall on the circumference of the same circle. Thus if we imagine the point *C* moving continuously, under the rule that the lines produced from it to the fixed limits *A* and *B* shall maintain always the same ratio as that of the original parts *AC* and *CB*, then that point *C* will describe the circumference of a circle, as I shall next prove. And the circle described in this way will be ever greater, infinitely, according as the point *C* is taken closer to the midpoint *O* [of *AB*], while the circle will be smaller [which is] described by a point closer to the end *B*. Thus, following the above rule, circles will be described by motion of the infinitely many points that can be taken in the line *OB*, and the circles are of every size—less than the pupil of a flea's eye, or even greater than the equator of the celestial sphere [*primo mobile*].

[*Text omitted here. The discussion of the paradoxes of infinity continues with an examination of the concept of the continuum.*]

I don't know, Simplicio, whether the learned Peripatetics, to whom I grant as quite true the concept that the continuum is divisible into ever-divisibles in such a way that in continuing such division and subdivision one would never reach an end, will be willing to concede to me that none of their divisions is the last—as indeed none is, since there always remains another—and yet that there indeed exists a last and highest, and it is that which resolves the line into infinitely many indivisibles. I admit that one will never arrive at this by successively dividing [the line] into a greater and greater multitude of parts. But by employing the method I propose, that of distinguishing and resolving the whole infinitude at one fell swoop—an artifice that should not be denied to me—I believe that they should be satisfied, and should allow this composition of the continuum out of absolutely indivisible atoms. Especially since this is a road that is perhaps more direct than any other in extricating ourselves from many intricate labyrinths. One such, in addition to that already mentioned of the [problem of the] coherence of the parts of solids, is the understanding of rarefaction and condensation, without our stumbling into the inconsistency of being forced by the former [rarefaction] to admit void spaces,[15] and by the latter [condensation, to admit] the [inter]penetration of bodies, both these involving contradictions that seem to me to be cleverly avoided by assuming the said composition of indivisibles.

Simplicio. I don't know what the Peripatetics would say, inasmuch as the considerations you have set forth would strike them, I believe, for the most part as novelties, and as such they would need to be examined. It may be that the Peripatetics would find replies and solutions capable of untying those knots that I, from the shortness of time and the frailty of my intellect, cannot at present resolve. So leaving aside for now that [Peripatetic] faction, I should indeed like to hear how the introduction of these indivisibles facilitates the comprehension of condensation and rarefaction, while at the same time it circumvents [both] the void and the [inter]penetration of bodies.

Sagredo. I too will hear this with pleasure, as it is still obscure to my mind. Provided, that is, that I shall not be defrauded of hearing, in accordance with what you said to Simplicio a short time ago, the reasonings of Aristotle in refuting the void, and then the solutions thereof at which you arrive, as is only fitting if you assume that which he denies.

Salviati. Both shall be done. As to the first, it is necessary that just as we shall make use, in regard to rarefaction, of the line described by the smaller circle when that is driven by the revolution of the larger, which line is longer than its own circumference; so, for an understanding of condensation, we must show how the larger [circle, driven] by the revolution of the lesser, describes a straight line shorter than its own circumference. For a clear explanation of this, let us consider what happens with the polygons.

In a diagram similar to the previous one, let there be two hexagons, *ABC* and *HIK*, around the common center *L*,

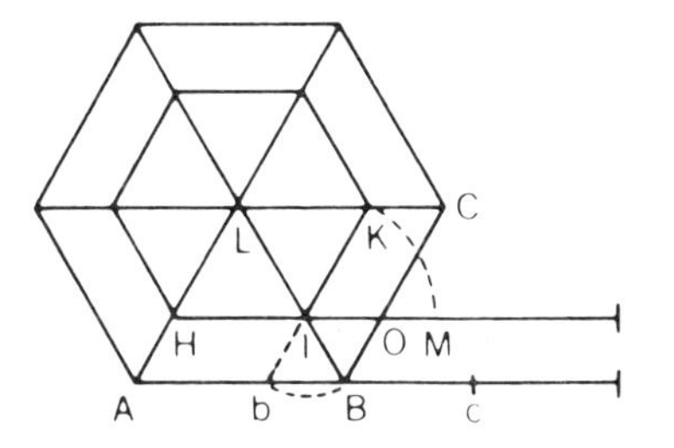

and the parallel lines *HOM* and *ABc* on which they must be revolved. Let the corner *I* of the smaller polygon be fixed, and turn this polygon until the side *IK* falls on the parallel [*MH*]. In this motion, point *K* will describe arc *KM*, and side *KI* will unite with part *IM*. Let us see what side *CB* of the larger polygon will do. Since the revolution is made about the point *I*, the line *IB* with the end *B* will go back, describing the arc *Bb* below the parallel *cA*, so that when side *KI* is joined with line *MI*, side *BC* will unite with line *bc*, going forward only as much as the part *Bc*, and leaving behind the part subtended by the arc *Bb*, which comes to be superimposed on the line *BA*. Assuming the rotation driven by the smaller polygon to continue in this way, it will trace and cover on its own parallel a line equal to its perimeter. But the larger [hexagon] will pass over a line shorter than its perimeter by one less line [of length] *bB* than [the number of] its sides, and this line will be approximately equal to that described by the lesser polygon, which it will exceed by only the length *bB*. Here, then, without any contradiction [*repugnanza*], is revealed the reason why the sides of the larger polygon, when driven by the smaller, do not cover a line greater than that traveled by the smaller; for a part of each side is superimposed on that which precedes and is adjacent to it.

Now consider the two circles around center *A*, placed on their parallels so that the smaller touches one of these at point *B*, and the larger [touches] the other at point *C*. The smaller [circle] commencing to roll, its point *B* will not remain motionless for any time while the [imaginary] line *BC* goes backward carrying point *C*, as happened in the polygons, where point *I* remained fixed until side *KI* fell on line *IM*. There, line *IB* did carry *B* (one end of side *CB*) backward to *b* so that side *BC* fell on *bc*, superimposing part *Bb* on line *BA*, and advancing only by the part *Bc* equal to *IM*, or to one side of the

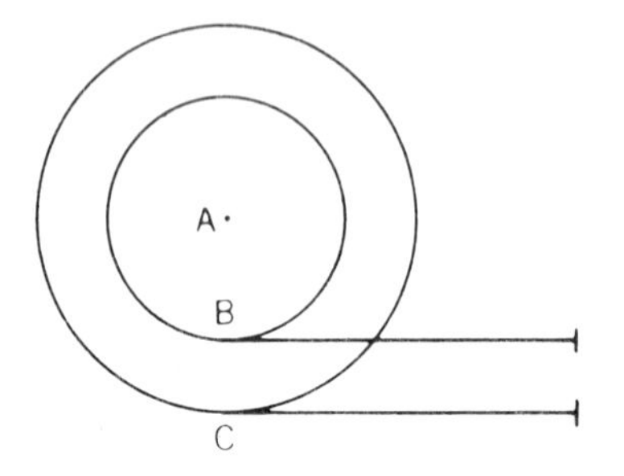

smaller polygon. On account of these superpositions, equal to the excesses of the larger sides over the smaller, the residual advances made, equal to the sides of the lesser polygon, come to compose in one entire revolution the straight line equal to that marked and measured by the smaller polygon.

If we were to apply similar reasoning to the case of circles, we should have to say that where the sides of any polygon are contained within some number, the sides of any circle are infinitely many; the former are quantified and divisible, the latter unquantifiable and indivisible; either end of each side of the revolving polygon stays fixed for a time (that is, that fraction of the time of an entire revolution, which the side is of the entire perimeter), whereas in circles the delays of the ends of their infinitely many sides are momentary, because an instant in a finite time is a point in a line that contains infinitely many [points]. The backward turns made by sides of the larger polygon [each] advancing as far as the side of the lesser [polygon] are not those of a whole side, but only of its excess over a side of the smaller; in circles, the point (or "side") *C*, during the instantaneous rest of end *B*, moves back as much as its excess over the "side" *B*, and advances by as much as [point] *B*.[16] To sum up, the infinitely many indivisible sides of the greater circle, with their infinitely many indivisible retrogressions, made in the infinitely many instantaneous rests of the infinitely many ends of the infinitely many sides of the lesser circle, together with their infinitely many advances, equal to the infinitely many sides of the lesser circle, compose and describe [*disegnano*, mark] a line, equal to that described by the lesser circle, which contains in itself infinitely many unquantifiable superpositions, making a compacting and condensation without any [inter]penetration of quantified parts.[17]

This is not to be understood as happening in the line divided into quantified parts, as in the perimeter of any polygon, which extended into a straight line cannot be compressed into [any] shorter length except by superposition and interpenetration of its sides. The compacting of infinitely many unquantifiable parts without interpenetration of quantified parts, and the previously explained expansion of infinitely many indivisibles with the interposition of indivisible voids, I believe to be the most that can be said to explain the condensation and rarefaction of bodies without the necessity of introducing interpenetration of bodies and [appealing to] quantified void spaces. If anything in it pleases you, make capital of that; if not, ignore this as idle, and my reasoning along with it, and go search for some other explanation that will bring you more peace of mind. I repeat only this: we are among infinites and indivisibles.

NOTES

1. Giovanni di Guevara (1561-1641), Bishop of Teano, had discussed this problem with Galileo but took a different approach in his *In Aristotelis Mechanicas comentarii* (Rome, 1627).

2. Whether there are exactly five skipped spaces, or something more than that, is crucial to the paradox. This difficulty does not really vanish when Galileo passes to the circle "at one fell swoop" (pp. 92-93), any more than n ever becomes $n + 1$ or some intermediate quantity; see also notes 3 and 4, below.

3. Here Galileo says that for the hexagon there would be not five, but six skipped spaces; cf. note 2, above.

4. Strictly speaking, it is necessary to add ". . . of which 99,999 are each equal to a side of the smaller polygon, while one is the excess over that of a side of the larger polygon."

5. The reference here seems to be to point-atoms of gold, since *minima naturalia,* indivisible physically but divisible mathematically, could not be infinitely many in a finite bulk.

6. The reference is probably to Epicurus (341-270 B.C.) as expounded by Lucretius (98-55 B.C.), since the scattered voids are interstitial, as well as for reasons implied in note 7, below.

7. Orazio Grassi (1583-1654), in his *Ratio ponderum librae et simbellae* (Paris, 1626), which was an attack on Galileo's *Il Saggiatore* of 1623. Cf. *Opere,* VI, 475-76, where it is the Epicureans that are named though Democritus (460-357 B.C.), is usually considered as the chief atomist of antiquity. Cf. note 6, above.

8. The ensuing paradox had been hinted at in the *Dialogue,* p. 247 (*Opere,* VII, 271-72). Galileo had previously sent it to Buonaventura Cavalieri (1598?-1647) to caution him regarding the perils of the "method of indivisibles" in geometry. The paradox has a double purpose here: to illustrate the nature of mathematical definitions, and to show the pitfalls of analogy in transferring the word "equal" from entities of n dimensions to their supposed counterparts of $n - 1$ dimensions. Cf. notes 9 and 12, below.

9. Use of the word "equal" in this way violates Berkeley's axiom that conclusions reached from the behavior of given entities and dependent on their existence cannot be rigorously applied to other entities deprived of them when they vanish. Sagredo's reply bears out the view of Abraham Kästner (1719-1800) and Bernard Bolzano (1781-1848) that Galileo inserted this paradox not as a conclusion to be accepted, but solely to stimulate careful thought.

10. Luca Valerio (1552-1618) met Galileo at Pisa about 1590 and later corresponded with him at Padua. The book cited was first published at Rome in 1603/04; Bk. II, Prop. XII, includes a demonstration from which Galileo derived the foregoing paradox.

11. The ensuing argument required very tactful treatment, since the proposition that a line might be composed of indivisibles, strongly opposed by Aristotle, had been condemned as heretical in 1415 by the Council of Constance. John Wyclif was exhumed and his body burned for this and other Epicurean doctrines. Cf. note 7, above, and see Aristotle, *Physica,* Bk. VI; *De caelo,* 299a.10 ff., as well as the pseudo-Aristotelian treatise *On Indivisible Lines.*

12. The meaning here is not that the adversaries are literally satisfied, but that this is the proper reply to them. Galileo's position is quite unrelated to the older discussions cited in note 11, above, or to that of Thomas Bradwardine (1290?-1349) in his *De continuo,* all of which debates concerned "indivisibles" that differed only in size from whatever they were supposed to compose. Galileo speaks here of elements having one less dimension than the aggregates in which they are supposed to exist; it was of such "indivisibles" that Cavalieri (note 8, above) made use in his geometry. In evaluating the role of Cavalieri's work in the development of the calculus, his "indivisibles" have frequently been confused by historians with infinitesimal magnitudes having the same dimensionality as the continuum to be analyzed, an approach studiously avoided by Cavalieri himself.

13. Having previously warned against the dangers in applying the word "equal" to infinites in the same sense as to finite magnitudes (notes 8 and 9, above), Galileo next turns to the positive integers to introduce the idea of one-to-one correspondence. His conclusions are valid and consistent, though by a much later extension of the concept of number we are now permitted to speak of different orders of infinite aggregates.

14. Negative roots were excluded under the Euclidean definition of number.

15. Empty spaces of finite dimensions would create physical problems, since in Galileo's view natural effects would continually destroy them (nature's horror of a void). Empty points in the mathematical sense raised no physical problem; hence Galileo employed them to attract physically adjacent particles and to keep them joined up to the limit of resistance to fracture. . . .

16. According to Berkeley's axiom (note 9, above), Galileo could not move from very small sides to point-sides for circles without losing the right to speak of any excess analogous to that existing between the sides of polygons of no matter how many sides. Being unable to argue consistently with his own views that points in one circle are somehow greater than those in another, he was left with no proper basis for the analogy here.

17. The purpose of this clearly deliberate complication in an ostensible summing up was to discourage all attempts to simplify paradoxes which in Galileo's opinion were genuine and needed to be thought about and thought through again and again.

69. From "On the Transformation and Simplification of the Equations of Loci" (*c.* 1640)[1]*

(Integration)

PIERRE FERMAT

APPLICATIONS OF THE GEOMETRIC PROGRESSION TO THE QUADRATURE[2] OF PARABOLAS AND INFINITE HYPERBOLAS

Archimedes did not employ geometric progressions except for the quadrature of the parabola; in comparing various quantities he restricted himself to arithmetic progressions. Was this because he found that the geometric progression was less suitable for the quadrature? Was it because the particular device that he used to square the parabola by this progression can only with difficulty be applied to other cases? Whatever the reason may be, I have recognized and proved that this progression is very useful for quadratures, and I am willing to present to modern mathematicians my invention which permits us to square, by a method absolutely similar, parabolas as well as hyperbolas.

The entire method is based on a well-known property of the geometric progression, namely the following theorem:

Given a geometric progression the terms of which decrease indefinitely, the difference between two consecutive terms of this progression is to the smaller of them as the greater one is to the sum of all following terms.[3]

This established, let us discuss first the quadrature of hyperbolas:

I define hyperbolas as curves going to infinity, which, like *DSEF* [Fig. 1], have the following property. Let *RA* and *AC* be asymptotes which may be extended indefinitely; let us draw parallel to the asymptotes any lines *EG, HI, NO, MP, RS,* etc. We shall then always have the same ratio between a given power of *AH* and the same power of *AG* on one side, and a power of *EG* (the same as or different from the preceding) and the

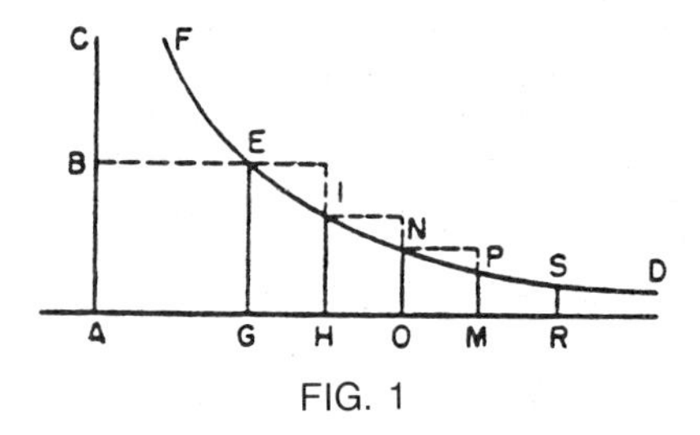

FIG. 1

**Source:* This translation made from the *Oeuvres de Fermat,* edited by P. Tannery and C. Henry, I (1891), 255-259 and III (1896), 216-219, appears in D. J. Struik (ed.), *A Source Book in Mathematics, 1200-1800* (1969), 219-222. It is reprinted by permission of Harvard University Press, Copyright © 1969 by the President and Fellows of Harvard College. Footnotes renumbered. The biography of Pierre Fermat precedes selection 64.

same power of HI on the other. I mean by powers not only squares, cubes, fourth powers, etc., the exponents of which are 2, 3, 4, etc., but also simple roots the exponent of which is unity.[4]

I say that all these infinite hyperbolas except the one of Apollonius,[5] or the first, may be squared by the method of geometric progression according to a uniform and general procedure.

Let us consider, for example, the hyperbolas the property of which is defined by the relations $AH^2/AG^2 = EG/HI$ and $AO^2/AH^2 = HI/NO$, etc. I say that the indefinite area which has for base EG and which is bounded on the one side by the curve ES and on the other side by the infinite asymptote GOR is equal to a certain rectilinear area.

Let us consider the terms of an indefinitely decreasing geometric progression; let AG be the first term, AH the second, AO the third, etc. Let us suppose that those terms are close enough to each other that following the method of Archimedes we could adequate [*adégaler*] according to Diophantus,[6] that is, equate approximately the rectilinear parallelogram $GE \times GH$ and the general quadrilateral $GHIE$; in addition we shall suppose that the first intervals GH, HO, OM, etc. of the consecutive terms are sufficiently equal that we can easily employ Archimedes' method of exhaustion by circumscribed and inscribed polygons. It is enough to make this remark once and we do not need to repeat it and insist constantly upon a device well known to mathematicians.

Now, since $AG/AH = AH/AO = AO/AM$, we have also $AG/AH = GH/HO = HO/OM$, for the intervals. But for the parallelograms,

$$\frac{EG \times GH}{HI \times HO} = \frac{HI \times HO}{ON \times OM}.$$

Indeed, the ratio $EG \times GH/HI \times HO$ of the parallelograms consists of the ratios EG/HI and GH/HO; but, as indicated, $GH/HO = AG/AH$; therefore, the ratio $EG \times GH/HI \times HO$ can be decomposed into the ratios EG/HI and AG/AH. On the other hand, by construction, $EG/HI = AH^2/AG^2$ or AO/AG, because of the proportionality of the terms; therefore, the ratio $GE \times GH/HI \times HO$ is decomposed into the ratios AO/AG and AG/GH; now AO/AH is decomposed into the same ratios; we find consequently for the ratio of the parallelograms: $EG \times GH/HI \times HO = AO/AH = AH/AG$.

Similarly we prove that $HI \times HO/NO \times MO = AO/AH$.

But the lines AO, AH, AG, which form the ratios of the parallelograms, define by their construction a geometric progression; hence the infinitely many parallelograms $EG \times GH$, $HI \times HO$, $NO \times OM$, etc., will form a geometric progression, the ratio of which will be AH/AG. Consequently, according to the basic theorem of our method, GH, the difference of two consecutive terms, will be to the smaller term AG as the first term of the progression, namely, the parallelogram $GE \times GH$, to the sum of all the other parallelograms in infinite number. According to the adequation of Archimedes, this sum is the infinite figure bounded by HI, the asymptote HR, and the infinitely extended curve IND.

Now if we multiply the two terms by EG we obtain $GH/AG = EG \times GH/EG \times AG$; here $EG \times GH$ is to the infinite area the base of which is HI as $EG \times GH$ is to $EG \times AG$. Therefore, the parallelogram $EG \times AG$, which is a given rectilinear area, is adequated to the said figure; if we add on both sides the parallelogram $EG \times GH$, which, because of infinite subdivisions, will vanish and will be reduced to nothing, we reach a conclusion that would be easy to confirm by a more lengthy proof carried out in the manner of Archimedes, namely, that for this kind of hyperbola the parallelogram AE is equivalent to the area bounded by the base EG, the asymptote GR, and the curve ED infinitely extended.

It is not difficult to extend this idea to all the hyperbolas defined above except the one that has been indicated.

COMMENTARY BY D. J. STRUIK

Fermat then extends his method to parabolas. His reasoning can be translated as follows.

Divide the interval $0 \leqslant x < a$ into parts by the points $x_1 = a$, $x_2 = ar$, $x_3 = ar^2, \ldots, r < 1$, which are separated by the intervals $l_1 = a(1 - r)$, $l_2 = ar(1 - r)$, $l_3 = ar^2(e - r), \ldots$. If $y = x^n$ ($n = p/q$, $p, q \lesseqgtr 0$) is the equation of the "hyperbola" or "parabola," then the values of y corresponding to $x_1, x_2, x_3, \ldots$ are $y_1 = a^n$, $y_2 = a^n r^n$, $y_3 = a^n r^{2n}, \ldots$. Then the sum S of the rectangles $l_1 x_1 + l_2 x_2 + l_3 x_3 + \cdots$ is

$$S = a(1-r)a^n + ar(1-r)a^n r^n + ar^2(1-r)a^n r^{2n} + \cdots$$

$$= (1-r)a^{n+1}(1 + r^{n+1} + r^{2n+2} + \cdots)$$

$$= \frac{1-r}{1-r^{n+1}}a^{n+1}.$$

When $r = s^q$ ($s < 1$) and $n \neq -1$, then

$$\int_0^a x^n\,dx = a^{n+1} \lim \frac{1-r}{1-r^{n+1}} = a^{n+1} \lim \frac{1-s^q}{1-s^{p+q}}$$

$$= \frac{qa^{n+1}}{p+q} = \frac{a^{n+1}}{n+1}.$$

As we see, this procedure holds for n positive and negative, but it fails for $n = -1$.

This method approaches our modern method of limits; it uses the concept of the limit of an infinite geometric series.

NOTES

1. *Editor's note.* This paper generalizes Bonaventura Cavalieri's integral, which in modern symbols is $\int_0^a x^n dx = \frac{a^{n+1}}{n+1}$ where n is positive whole number, to n fractional or negative.

2. Fermat uses the Greek term *tetragonizein* for "to perform a quadrature," a practice not uncommon in the seventeenth century.

3. This is Fermat's way of expressing that the sum of a convergent series $a + ar + ar^2 + \cdots + ar^n + \cdots = a/(1 - r)$.

4. This may mean "exponents that are unit fractions."

5. The hyperbola of Apollonius is the ordinary hyperbola, of which, if its equation is $xy = a^2$, the integral $\int_0^\infty y\,dx$ diverges.

6. The term *adequatio* is a Latin translation of the Greek term *parisótēs,* by which Diophantus denoted an approximation to a certain number as closely as possible. See T. L. Heath, *Manual of Greek Mathematics* (Clarendon Press, Oxford, 1931), 493. Fermat uses the term to denote what we call a limiting process.

70. From "On a Method for the Evaluation of Maxima and Minima"[1]*

(Fermat obtained a general method to find the extrema of a given function. His algorithm was subsequently developed into the method of the "characteristic triangle," dx, dy, and ds.)

PIERRE FERMAT

The whole theory of evaluation of maxima and minima presupposes two unknown quantities and the following rule:

Let a be any unknown of the problem (which is in one, two, or three dimensions, depending on the formulation of the problem). Let us indicate the maximum or minimum by a in terms which could be of any degree. We shall now replace the original unknown a by $a + e$ and we shall express thus the maximum or minimum quantity in terms of a and e involving any degree. We shall adequate [*adégaler*], to use Diophantus' term,[2] the two expressions of the maximum or minimum quantity and we shall take out their common terms. Now it turns out that both sides will contain terms in e or its powers. We shall divide all terms by e, or by a higher power of e, so that e will be completely removed from at least one of the terms. We suppress then all the terms in which e or one of its powers will still appear, and we shall equate the others; or, if one of the expressions vanishes, we shall equate, which is the same thing, the positive and negative terms. The solution of this last equation will yield the value of a, which will lead to the maximum or minimum, by using again the original expression.

Here is an example:

To divide the segment AC [Fig. 1] *at E so that AE × EC may be a maximum.*

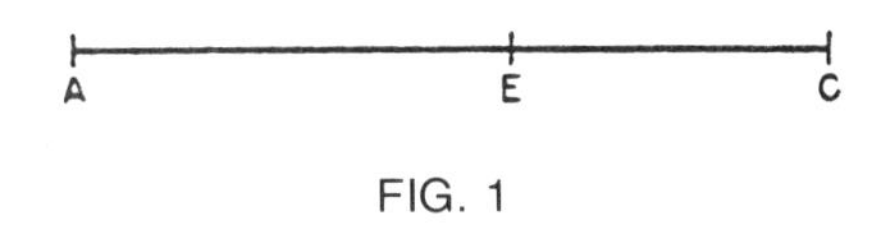

FIG. 1

We write $AC = b$; let a be one of the segments, so that the other will be $b - a$, and the product, the maximum of which is to be found, will be $ba - a^2$. Let now $a + e$ be the first segment of b; the second will be $b - a - e$, and the product of the segments, $ba - a^2 + be - 2ae - e^2$; this must be adequated with the preceding: $ba - a^2$. Suppressing common terms: $be \sim 2ae + e$. Suppressing e: $b = 2a$.[3] To solve the problem we must consequently take the half of b.

We can hardly expect a more general method.

ON THE TANGENTS OF CURVES

We use the preceding method in order to find the tangent at a given point of a curve.

Let us consider, for example, the parabola BDN [Fig. 2] with vertex D and of diameter DC; let B be a point on

*Source: This translation from the *Oeuvres de Fermat,* edited by P. Tannery and C. Henry, appears in D. J. Struik (ed.), *A Source Book in Mathematics, 1200-1800* (1969), 223–227. It is reprinted by permission of Harvard University Press,

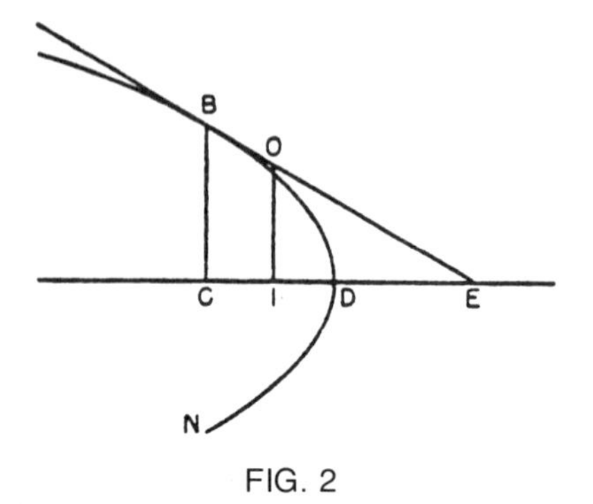

FIG. 2

it at which the line BE is to be drawn tangent to the parabola and intersecting the diameter at E.

We choose on the segment BE a point O at which we draw the ordinate OI; also we construct the ordinate BC of the point B. We have then: $CD/DI > BC^2/OI^2$, since the point O is exterior to the parabola. But $BC^2/OI^2 = CE^2/IE^2$, in view of the similarity of triangles. Hence $CD/DI > CE^2/IE^2$.

Now the point B is given, consequently the ordinate BC, consequently the point C, hence also CD. Let $CD = d$ be this given quantity. Put $CE = a$ and $CI = e$; we obtain

$$\frac{d}{d-e} > \frac{a^2}{a^2+e^2-2ae}.$$ [4]

Removing the fractions:

$$da^2 + de^2 - 2dae > da^2 - a^2e.$$

Let us then adequate, following the preceding method; by taking out the common terms we find:

$$de^2 - 2dae \sim -a^2e,$$

or, which is the same,

$$de^2 + a^2e \sim 2dae.$$

Let us divide all terms by e:

$$de + a^2 \sim 2da.$$

On taking out de, there remains $a^2 = 2da$, consequently $a = 2d$.

Thus we have proved that CE is the double of CD—which is the result.

This method never fails and could be extended to a number of beautiful problems; with its aid, we have found the centers of gravity of figures bounded by straight lines or curves, as well as those of solids, and a number of other results which we may treat elsewhere if we have time to do so.

I have previously discussed at length with M. de Roberval[5] the quadrature of areas bounded by curves and straight lines as well as the ratio that the solids which they generate have to the cones of the same base and the same height.

Now follows the second illustration of Fermat's "e-method," where Fermat's e = Newton's o = Leibniz' dx.[6]

CENTER OF GRAVITY OF PARABOLOID OF REVOLUTION, USING THE SAME METHOD[7]

Let $CBAV$ (Fig. 3) be a paraboloid of revolution, having for its axis IA and for its base a circle of diameter CIV. Let us find its center of gravity by using the same method which we applied for maxima and minima and for the tangents of curves; let us illustrate, with new examples and with new and brilliant applications of this method, how wrong those are who believe that it may fail.

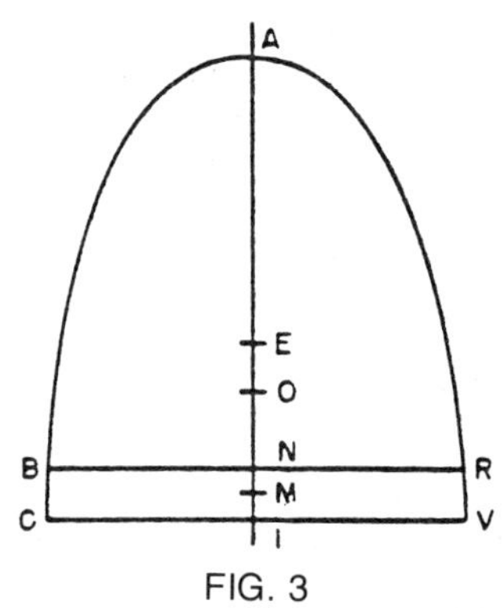

FIG. 3

In order to carry out this analysis, we write $IA = b$. Let O be the center of gravity, and a the unknown length of the segment AO; we intersect the axis IA by any plane BN and put $IN = e$, so that $NA = b - e$.

It is clear that in this figure and in similar ones (parabolas and paraboloids) the centers of gravity of segments cut off by parallels to the base divide the axis in a constant proportion (indeed, the argument of Archimedes can be extended by similar reasoning from the case of a parabola to all parabolas and paraboloids of revolution[8]). Then the center of gravity of the segment of which NA is the axis and BN the radius of the base will divide AN at a point E such that $NA/AE = IA/AO$, or, in formula, $b/a = (b - e)/AE$.

The portion of the axis will then be $AE = (ba - ae)/b$ and the interval between the two centers of gravity, $OE = ae/b$.

Let M be the center of gravity of the remaining part $CBRV$; it must necessarily fall between the points N, I, inside the figure, in view of Archimedes' postulate 9 in *On the equilibrium of planes*, since $CBRV$ is a figure completely concave in the same direction.[9]

But

$$\frac{\text{Part } CBRV}{\text{Part } BAR} = \frac{OE}{OM},$$

since O is the center of gravity of the whole figure CAV and E and M are those of the parts.

Now in the paraboloid of Archimedes,

$$\frac{\text{Part } CAV}{\text{Part } BAR} = \frac{IA^2}{NA^2} = \frac{b^2}{b^2 + e^2 - 2be};$$

hence by dividing,

$$\frac{\text{Part } CBRV}{\text{Part } BAR} = \frac{2be - e^2}{b^2 + e^2 - 2be}.$$

But we have proved that

$$\frac{\text{Part } CBRV}{\text{Part } BAR} = \frac{OE}{OM}.$$

Then in formulas,

$$\frac{2be - e^2}{b^2 + e^2 - 2be} = \frac{OE\ (= ae/b)}{OM};$$

hence

$$OM = \frac{b^2ae + ae^3 - 2bae^2}{2b^2a - be^2}.$$

From what has been established we see that the point M falls between points N and I; thus $OM < OI$; now, in formula, $OI = b - a$. The question is then prepared from our method, and we may write

$$b - a \sim \frac{b^2ae + ae^3 - 2bae^2}{2b^2e - be^2}.$$

Multiplying both sides by the denominator and dividing by e:

$$2b^3 - 2b^2a - b^2e + bae \sim b^2a + ae^2 - 2bae.$$

Since there are no common terms, let us take out those in which e occurs and let us equate the others:

$$2b^3 - 2b^2a = b^2a, \text{ hence } 3a = 2b.$$

Consequently

$$\frac{IA}{AO} = \frac{3}{2}, \text{ and } \frac{AO}{OI} = \frac{2}{1},$$

and this was to be proved.[10]

The same method applies to the centers of gravity of all the parabolas ad infinitum as well as those of paraboloids of revolution. I do not have time to indicate, for example, how to look for the center of gravity in our paraboloid obtained by revolution about the ordinate;[11] it will be sufficient to say that, in this conoid, the center of gravity divides the axis into two segments in the ratio 11/5.

NOTES

1. This paper was sent by Fermat to Father Marin Mersenne, who forwarded it to Descartes. Descartes received it in January 1638. It became the subject of a polemic discussion between him and Fermat (*Oeuvres*, I, 133). On Mersenne, see Selection I.6, note 1, of Struik.

2. See Selection IV.7, note 5, of Struik. where we have written (following the French translation in *Oeuvres*, III, 122) $be \sim 2ae + e^2$, Fermat wrote: *B* in *E* adaequabitur *A* in *E* bis + *Eq* (*Eq* standing for *E* quadratum). The symbol $\sim$ is used for "adequates."

4. Fermat wrote: *D* ad *D* − *E* habebit majorem proportionem quam *Aq.* ad *Aq.* +

Eq. − *A* in *E* bis (*D* will have to *D* − *E* a larger ratio than A^2 to $A^2 + E^2 - 2AE$).

5. See the letters from Fermat to Roberval, written in 1636 (*Oeuvres*, III, 292-294, 296-297).

6. The gist of this method is that we change the variable x in $f(x)$ to $x + e$, e small. Since $f(x)$ is stationary near a maximum or minimum (Kepler's remark), $f(x+e) - f(x)$ goes to zero faster than e does. Hence, if we divide by e, we obtain an expression that yields the required values for x if we let e be zero. The legitimacy of this procedure remained, as we shall see, a subject of sharp controversy for many years. Now we see in it a first approach to the modern formula: $f'(x) = \lim_{e\to 0} \frac{f(x+e) - f(x)}{e}$, introduced by Cauchy (1820-21).

7. This paper seems to have been sent in a letter to Mersenne written in April 1638, for transmission to Roberval. Mersenne reported its contents to Descartes. Fermat used the term "parabolic conoid" for what we call "paraboloid of revolution."

8. "All parabolas" means "parabolas of higher order," $y = kx^n$, $n > 2$. The reference is to Archimedes' *On floating bodies*, II, Prop. 2 and following; see T. L. Heath, *The works of Archimedes* (Cambridge University Press, Cambridge, England, 1897; reprint, Dover, New York), 264ff.

9. This is postulate 7 in the modern Heiberg edition, and is translated in Heath, p. 190, as follows: "In any figure whose perimeter is concave in (one and) the same direction the center of gravity must be within the figure." (On the term "concave in the same direction," see Heath, p. 2.)

10. These relations were known to Archimedes (see note 8). But Fermat solved this problem on centers of gravity, hence a problem in the integral calculus, with what we might call an application of the principle of virtual variations.

11. Here *ACI* of Fig. 3 is rotated about *CI*.

71. From "On the Sines of a Quadrant of a Circle" (1659)*

(In the mid-17th century, French mathematician Gilles Roberval proposed the cycloid, or roulette as he called it, as a test curve for different methods relating to infinitesimals. The cycloid became the "apple of discord" among geometers. Roberval also introduced its so-called companion, that is, the sine curve. He influenced Pascal, who integrated $\sin^n x$, $n = 1, 2, 3, 4$, by means of an early form of a characteristic triangle, that is, *dx*, *dy*, and *ds*. Pascal's paper given here partially rejects indivisibles and presages the indefinite integral.)

BLAISE PASCAL

Let *ABC* [Fig. 1] be a quadrant of a circle of which the radius *AB* will be considered the axis and the perpendicular radius *AC* the base; let *D* be any point on the arc from which the sine *DI* will be drawn to the radius *AC*; and let *DE* be the tangent on which we choose the points *E* arbitrarily, and from these

**Source:* This translation of *Traité des sinus du quart de cercle* is made from the *Oeuvres de Pascal,* edited by L. Brunschwicg and P. Boutroux, and appears in D. J. Struik (ed.), *A Source Book in Mathematics, 1200-1800* (1969), 239-241. It is reprinted by permission of Harvard University Press, The biography of Blaise Pascal precedes selection 66.

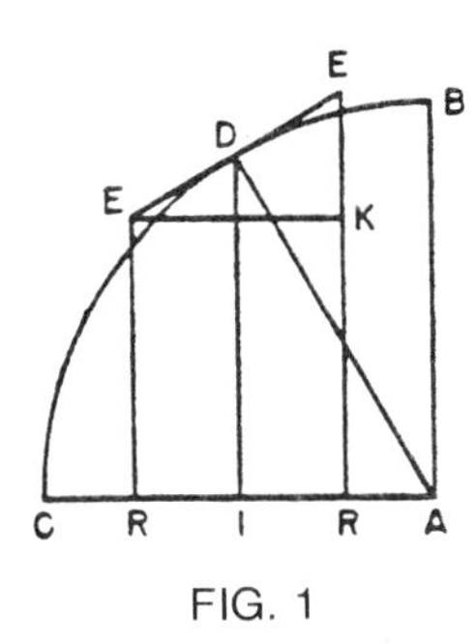

FIG. 1

points we draw the perpendiculars *ER* to the radius *AC*.[1]

I say that the rectangle formed by the sine[2] *DI* and the tangent *EE* is equal to the rectangle formed by a portion of the base (enclosed between the parallels) and the radius *AB*.

For the radius *AD* is to the sine *DI* as *EE* is to *RR*, or to *EK*, which is clear because of the similarity of the right-angled triangles *DIA*, *EKE*, the angle *EEK* or *EDI* being equal to the angle *DAI*.

Proposition I. The sum of the sines of any arc of a quadrant is equal to the portion of the base between the extreme sines, multiplied by the radius.[3]

Proposition II. The sum of the squares of those sines is equal to the sum of the ordinates[4] of the quadrant that lie between the extreme sines, multiplied by the radius.[5]

Proposition III. The sum of the cubes of the same sines is equal to the sum of the squares of the same ordinates between the extreme sines, multiplied by the radius.[6]

Proposition IV. The sum of the fourth powers of the same sines is equal to the sum of the cubes of the same ordinates between the extreme sines, multiplied by the radius.

And so on to infinity.

Preparation for the proof. Let any arc *BP* be divided into an infinite number of parts by the points *D* [Fig. 3] from which we draw the sines *PO*, *DI*, etc. . . . ; let us take in the other quadrant of the circle the segment *AQ*, equal to *AO* (which measures the distance between the extreme sines of the arc, *BA*, *PO*); let *AQ* be divided into an infinite number of equal parts by the points *H*, at which the ordinates *HL* will be drawn.

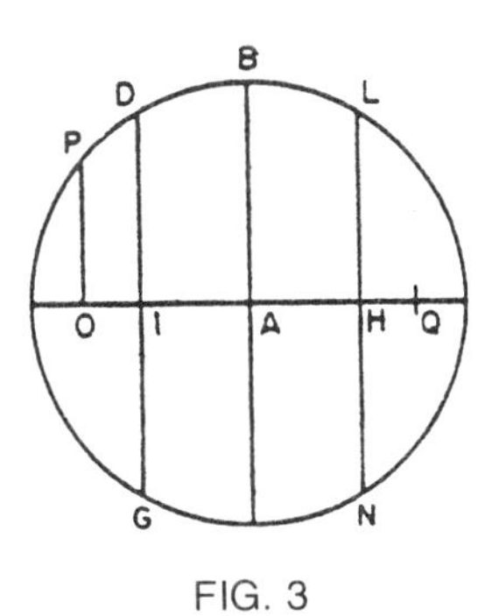

FIG. 3

Proof of Proposition I. I say that the sum of the sines *DI* (each of them multiplied of course by one of the equal small arcs *DD*) is equal to the segment *AO* multiplied by the radius *AB*.

Indeed, let us draw at all the points *D* the tangents *DE* [Fig. 1], each of which intersects its neighbor at the points *E*; if we drop the perpendiculars *ER* it is clear that each sine *DI* multiplied by the tangent *EE* is equal to each distance *RR* multiplied by the radius *AB*. Therefore, all the quadrilaterals formed by the sines *DI* and their tangents *EE* (which are all equal to each other) are equal to all the quadrilaterals formed by all the portions *RR* with the radius *AB*; that is (since one of the tangents *EE* multiplies each of the sines, and since the radius *AB* multiplies each of the distances), the sum of the sines *DI*, each of them multiplied by one of the tangents *EE*, is equal to the sum of the distances *RR*, each multiplied by *AB*. But each tangent *EE* is equal to each one of the equal arcs *DD*. Therefore the sum of the sines multiplied by one of the equal small arcs is equal to the distance *AO* [Fig. 3] multiplied by the radius.

Note. It should not cause surprise when I say that all the distances *RR* are equal to *AO* and likewise that each tangent *EE* is equal to each of the small arcs *DD*, since it is well known that, even though this equality is not true when the number of the sines is finite, nevertheless the equality is true when the number is infinite; because then the sum of all the equal tangents *EE* differs from the entire arc *BD*, or from the sum of all the equal arcs *DD*, by less than any given quantity: similarly the sum of the *RR* from the entire *AO*.

Proof of Proposition II. I say that the sum of the squares of the sines *DI* (each of them multiplied by one of the equal small arcs *DD*) is equal to the sum of the *HL*, or to the area *BHQL*, multiplied by the radius *AB*.

For if the sines *DI* as well as the ordinates *HL* are extended to the circumference on the other side of the base, intersecting them at the points *G* and *N*, it is clear that each *DI* will be equal to each *IG*, and *HN* to *HL*.

In order to prove the proposition that all the squares of the *DI* times *DD* are equal to all the *HL* times *AB*, it is enough to prove that the sum of all the *HL* times *AB*, or all the *HN* times *AB*, or the area *QNN* multiplied by *AB*, is equal to all the *GI* times *ID* times *EE*, or to all the *GI* times *RR* times *AB* (since *ID* times *EE* is equal to each *RR* times *AB*). Then, by taking out the common quantity *AB* we have to prove that the area *AQNN* is equal to the sum of the rectangles *GI* times *RR*: this is clear, since the sum of the rectangles formed by each *GI* and each *RR* differs only by less than any given quantity from the area *AOGN* or, what is the same, *AQNN*, since the segment *AQ* was constructed to be equal to *AO*; and this was to be proved.

NOTES

1. The triangle *EEK* of this figure led Leibniz to his early researches into the calculus; it gave him the idea of the "characteristic" triangle, when *EE* is small. The segment *EE* is a tangent in Pascal's essay. With Leibniz it became a chord. For the different forms of this triangle see D. Mahnke, "Neue Einblicke in die Entdeckungsgeschichte der höheren Analysis," *Abhandlungen der preussischen Akademie der Wissenschaften, Kl. Math. Phys. 1* (1925), 1-64.

2. As with all authors up to the eighteenth century, Pascal's sine of an angle ϕ is a line, and not a ratio. It is what we now write $R \sin \phi$, R being the radius of the circle.

3. This is equivalent to our formula $\int_{\phi_0}^{\phi_1} \sin \phi \, d\phi = \cos \phi_0 - \cos \phi_1$.

4. To understand the difference between "ordinates" and "sines" we must take Pascal's *triligne* ("trilinear figure"), formed by two perpendicular lines *AB* and *AC* and a (convex) curve *BLC* (Fig. 2); see *Oeuvres*, VIII, 369). The points *D* divide *AB* into equal parts; the points *E* do the same for *AC*, and the points *L* for the arc *BLC* (every arc is of the same length). Then: *(a)* the perpendiculars to *AB*, from *D* to the curve, are the "ordinates to the axis" *(ordonnées à l'axe); (b)* the perpendiculars to *AC*, from *E* to the curve, are the "ordinates of the base" *(ordonnées à la base); (c)* the perpendiculars from *L* to *AC* are the "sines on the base"; *(d)* the perpendiculars from *L* to *AB* are the "sines on the axis." The difference between "ordinates" and "sines" is expressed at present by the change in the variable of integration.

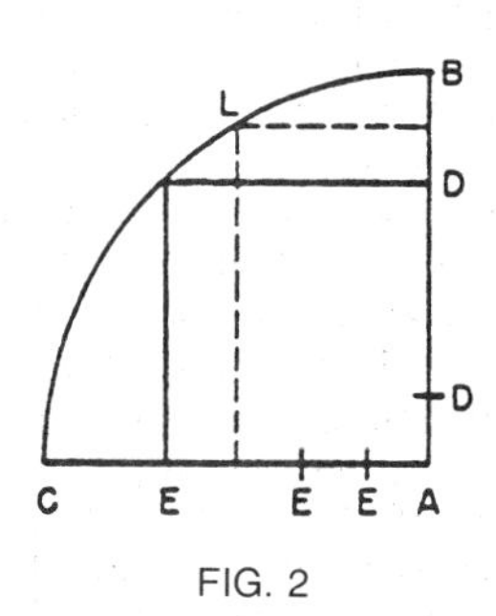

FIG. 2

5. Proposition II is equivalent to our formula $\int_{\phi_0}^{\phi_1} \sin^2 \phi \, d\phi = -\int_{\phi_0}^{\phi_1} \sin \phi \, d(\cos \phi)$, and therefore expresses a change of variable. The area is equal to $\frac{1}{2}(\phi_1 - \phi_0) - \frac{1}{2}(\sin \phi_1 \cos \phi_1) + \frac{1}{2}(\sin \phi_0 \cos \phi_0)$.

6. Hence $\int_{\phi_0}^{\phi_1} \sin^3 \phi \, d\phi = -\int_{\phi_0}^{\phi_1} \sin^2 \phi \, d(\cos \phi)$.

Chapter VI
The Scientific Revolution at Its Zenith (1620–1720)

Section C
The Discovery of the Differential and Integral Calculus

GOTTFRIED WILHELM LEIBNIZ (1646–1716)

The universal genius Gottfried Leibniz was a progenitor of German idealism, a pioneer of the Continental Enlightenment, and a principal architect of modern science. His philosophical, logical, and scientific ideas (as systematized and modified by his disciple Christian Wolff) dominated the intellectual life of Germany until displaced by Kant's Critical Philosophy in the 1780s. Today, Leibniz is perhaps best known for his law of metaphysical optimism—the world in its entirety is the best of all possible worlds and improves with each passing instant. (It was this law and its author that the French author Voltaire was attacking in his famous satire *Candide* (1759).)

Leibniz was born into a pious Lutheran family of Slavic origin in Leipzig, two years before the close of the devastating Thirty Years' War (1618–1648). His father, Frederick Leibniz (a notary, jurist, and moral philosopher), died in 1652, and his mother, Catherine Schmuck (Frederick's third wife), raised her son alone. The precocious young Leibniz attended Nicholai school but was largely self-taught. Studying in his father's library, he had, by the age of 12, taught himself Latin and some Greek in order to be able to read the literature of the classical Greeks, ancient Romans, and the early Church fathers.

In 1661 Leibniz entered the University of Leipzig to study law. At the university Jacob Thomasius taught him Aristotelian and Scholastic philosophy, as well as the modern thinking of Bacon, Hobbes, Grotius, Galileo and Descartes. He began to think of reconciling Aristotelian and modern thought. After completing his baccalaureate in 1664, he worked on a master's degree at Leipzig. In 1666 he applied for the degree of doctor of law but was refused admission to the program, probably because he had not completed a five-year law internship. He consequently left Leipzig (forever as it turned out) and enrolled at the University of Altdorf, near Nürnberg (Nuremberg), which almost at once granted him the doctorate in jurisprudence. Following the brilliant defense of his doctoral dissertation, he was offered a professorship at Altdorf but declined it because German universities were then in low esteem and because he had devised major legal and political reforms that, in turn, led him to seek the patronage of kings and princes.

Following the model of Cicero and Francis Bacon, Leibniz pursued the active life of a courtier, at first for the elector–archbishop of Mainz. He labored to codify Roman laws and apply them to German areas, and he also began to introduce natural law in jurisprudence. In 1672 the young jurist was sent to Paris to try to dissuade Louis XIV from attacking German lands, but his diversionary Egyptian plan, which called for building a Suez Canal, did not impress Louis. Nevertheless, Leibniz resided in Paris until 1676, practicing law, mastering the French language, and examining Cartesian thought firsthand with Pierre Malebranche and others. He improved his debating style under the tutelage of Antoine Arnauld and studied mathematics and physics under Christian Huygens. The studies with Huygens led him to discover an early stage of the differential calculus by the end of 1675. He also built a calculating machine that improved upon Blaise Pascal's, and he became a Fellow of the Royal Society of London in 1673. However, there was no academic or income-producing governmental position for Leibniz in Catholic France in 1675 or 1676, probably because he was a Lutheran.

By the summer of 1676, Leibniz had completed his preparatory studies and was about to embark on a new phase. He would develop his mature ideas in the following years and engage in heated polemics to defend them. To support himself, he accepted a position in October 1676 as librarian, judge, and minister-without-portfolio from John Frederick, duke of Hanover. He was to spend the rest of his career in Hanoverian service. On the way to Hanover, he stopped in Holland to meet the microscopist Antoni van Leeuwenhoek and the philosopher Baruch Spinoza, whom he criticized. Following his arrival in Brunswick, he launched a movement to reunify the Christian churches (Irenicism) and assisted in the effort to raise Hanover from ducal status to that of an electoral member of the Holy Roman Empire. In 1682 he helped found a learned journal, *Acta Eruditorum,* and served as its editor. Ernst August named him historian for the House of Brunswick in 1685, a position he held for the rest of his life. In history, Leibniz appealed to the critical study of primary sources and pioneered in historicism by assuming that history yields important knowledge and recognizing the importance of studying every people, not just the ancients.

The shy and bookish Leibniz did not limit himself to history after 1685. He also labored to construct a strong institutional and conceptual framework for the sciences in central Europe and Russia. Largely through his efforts, the Berlin Academy of Sciences was founded in 1700, and his suggestion to Peter the Great for a Russian academy became a reality in St. Petersburg in 1725. His ideas were expounded in the *Discourse on Metaphysics* (1686), the *Theodicy* (1710), and in a voluminous correspondence. In methodology, he set forth an organic rationalism based largely on the principle of sufficient reason. Although inspired by Cartesian rationalism, Leibniz (like Huygens) rejected the Cartesian principle of the conservation of momentum *(mv)* in dynamics. Instead he accepted the principle of the conservation of *vis viva* (mv^2) as the unifying principle for his theory of dynamics. The primal substance of his plenum universe was animate, percipient monads that were metaphysical points of energy. In epistemology, he opposed John Locke's belief that the mind is a *tabula rasa* ("blank tablet") at birth and that man learns only through his senses. He believed that the distinct mind and matter operate like two perfectly synchronized clocks (the pre-established harmony).

In his last years, Leibniz became embroiled in controversies with the British Newtonians. The first of these involved a dispute about the discovery of the calculus. Who was justified in claiming priority—Leibniz or Newton? In 1712–13 the Royal Society, presided over by Newton himself, condemned Leibniz for plagiarism—a judgment which has not been borne out by the historical record. In 1714 and 1715 Leibniz corresponded with Newton's spokesman Samuel Clarke defending the view that space, time, and motion are relative. Two years before his death, Leibniz was in Hanover in near isolation and ill health. Elector George Lewis, who had become George I of Great Britain, ordered him to complete the family history, *Brunswick Annals*. When Leibniz died, only his secretary followed the coffin to the cemetery.

The two chief achievements of Leibniz in mathematics were his discovery of an early stage of the differential and integral calculus independently of Newton and his pioneering work in symbolic logic. His first published paper on the differential calculus—a term that he coined—appeared in 1684 in *Acta Eruditorum*. This paper gave the general rules for differentiation, used differentials rather than derivatives, and offered the $d \cdot$ notation, dx and dy, for differentials. Two years later he published a paper on the inverse tangent problem or integral calculus—a term that he and Johann Bernoulli introduced. This paper contained the capital *S* symbol for integration. In a series of early papers dating to 1695, Leibniz obtained the result $\pi/4 = 1 - 1/3 + 1/5 - \ldots$ (Leibniz series) and recognized that integration and differentiation are inverse operations—the fundamental theorem of the calculus. His work was not without flaws. Although he based the calculus on differentials, Leibniz offered neither a consistent nor satisfactory definition of them. Rigorous foundations for the calculus lay far in the future.

The contributions of Leibniz to the calculus and symbolic logic proved very influential. During the 18th century, the Bernoulli brothers, Leonhard Euler, Louis Lagrange, and other leading Continental mathematicians adopted Leibniz's felicitous notation and incorporated his results into their extensive elaboration of the calculus. In symbolic logic, Leibniz had attempted to create more than a *calculus ratiocinator,* that is an abstract logic of formulas. He had pursued a *characteristica universalis,* which could express current logic through written signs more precisely and more clearly than through words. Late in the 19th century his symbolic logic bore fruit in the work of Gottlob Frege, Guiseppe Peano, and Bertrand Russell.

Leibniz advanced many other areas of mathematics. By 1674 he produced a working model of a computing machine that could add, subtract, divide, multiply, and extract roots. It was superior to any other in use during his lifetime. He also worked on the origins of complex numbers, used binary numbers, presented determinants as well as an early version of topology *(analysis situs),* and developed combinatorics.

72. From "A New Method for Maxima and Minima as Well as Tangents, Which is Impeded Neither by Fractional nor by Irrational Quantities, and a Remarkable Type of Calculus for This" (1684)*

(Differential Calculus)

GOTTFRIED WILHELM LEIBNIZ

Let an axis *AX* [Fig. 1; simplified from Leibniz's figure] and several curves such as *VV, WW, YY, ZZ* be given, of which the ordinates *VX, WX, YX, ZX*, perpendicular to the axis, are called *v, w, y, z* respectively. The segment *AX*, cut off from the axis [*abscissa ab axe*[1]] is called *x*. Let the tangents be *VB, WC, YD, ZE,*

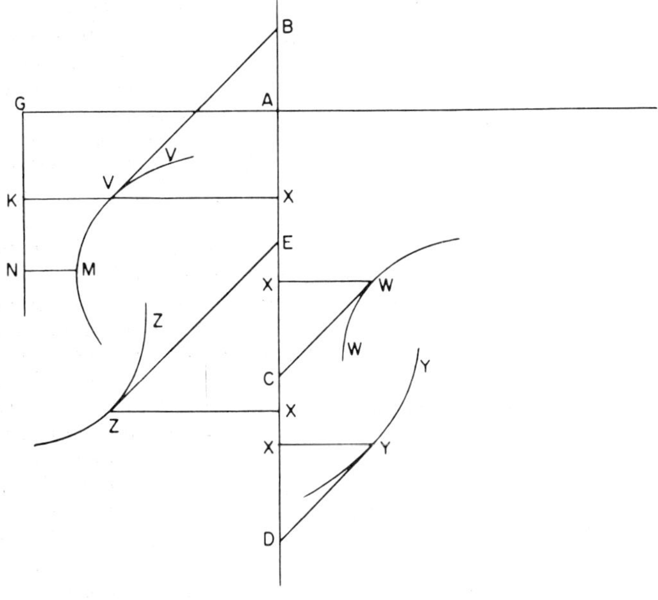

FIG. 1

*Source: This translation of "Nova methodus pro maximis et minimis, itemque tangentibus, quae nec fractas nec irrationales quantitates moratur, et singulare pro illi calculi genus" is taken from D. J. Struik (ed.), *A Source Book in Mathematics, 1200-1800* (1969), 272-280, and is reprinted by permission of Harvard University Press,

intersecting the axis respectively at B, C, D, E. Now some straight line selected arbitrarily is called dx, and the line which is to dx as v (or w, or y, or z) is to XB (or XC, or XC, or XE) is called dv (or dw, or dy, or dz),[2] or the difference[3] of these v (or w, or y, or z). Under these assumptions we have the following rules of the calculus.

If a is a given constant, then $da = 0$, and $d(ax) = a\,dx$. If $y = v$ (that is, if the ordinate of any curve YY is equal to any corresponding ordinate of the curve VV), then $dy = dv$. Now *addition* and *subtraction:* if $z - y + w + x = v$, then $d(z - y + w + x) = dv = dz - dy + dw + dx$. Multiplication: $d(xv) = x\,dv + v\,dx$, or, setting $y = xv$, $dy = x\,dv + v\,dx$. It is indifferent whether we take a formula such as xv or its replacing letter such as y. It is to be noted that x and dx are treated in this calculus in the same way as y and dy, or any other indeterminate letter with its difference. It is also to be noted that we cannot always move backward from a differential equation without some caution, something which we shall discuss elsewhere.

Now *division:*

$$d\frac{v}{y} \text{ or } \left(\text{if } z = \frac{v}{y}\right) dz = \frac{\pm v\,dy \mp y\,dv}{yy}.$$

The following should be kept well in mind about the *signs.*[4] When in the calculus for a letter simply its differential is substituted, then the signs are preserved; for z we write dz, for $-z$ we write $-dz$, as appears from the previously given rule for addition and subtraction. However, when it comes to an explanation of the values, that is, when the relation of z to x is considered, then we can decide whether dz is a positive quantity or less than zero (or negative). When the latter occurs, then the tangent ZE is not directed toward A, but in the opposite direction, down from X. This happens when the ordinates z decrease with increasing x. And since the ordinates v sometimes increase and sometimes decrease, dv will sometimes be positive and sometimes be negative; in the first case the tangent VB is directed toward A, in the latter it is directed in the opposite sense. None of these cases happens in the intermediate position at M, at the moment when v neither increases nor decreases, but is stationary. Then $dv = 0$, and it does not matter whether the quantity is positive or negative, since $+0 = -0$. At this place v, that is, the ordinate LM, is *maximum* (or, when the convexity is turned to the axis, *minimum),* and the tangent to the curve at M is directed neither in the direction from X up to A, to approach the axis, nor down to the other side, but is parallel to the axis. When dv is infinite with respect to dx, then the tangent is perpendicular to the axis, that is, it is the ordinate itself. When $dv = dx$, then the tangent makes half a right angle with the axis. When with increasing ordinates v its increments or differences dv also increase (that is, when dv is positive, $d\,dv$, the difference of the differences, is also positive, and when dv is negative, $d\,dv$ is also negative), then the curve turns toward the axis its *concavity,* in the other case its *convexity.*[5] Where the increment is maximum or minimum, or where the increments from decreasing turn into increasing, or the opposite, there is a *point of inflection.*[6] Here concavity and convexity are interchanged, provided the ordinates too do not turn from increasing into decreasing or the opposite, because then the concavity or convexity would remain. However, it is impossible that the increments continue to increase or decrease, but the ordinates turn from increasing into decreasing, or the opposite.[7] Hence a point of inflection occurs when $d\,dv = 0$ while neither v nor $dv = 0$. The problem of finding inflection therefore has not, like that of finding a maximum, two equal roots, but three. This all depends on the correct use of the signs.

Sometimes it is better to use *ambiguous signs,* as we have done with the division, before it is determined what the precise sign is. When with increasing

$x\,v/y$ increases (or decreases), then the ambiguous signs in $d\dfrac{v}{y} = \dfrac{\pm v\,dy \mp y\,dv}{yy}$ must be determined in such a way that this fraction is a positive (or negative) quantity. But $\mp$ means the opposite of $\pm$, so that when one is $+$ the other is $-$ or vice versa. There also may be several ambiguities in the same computation, which I distinguish by parentheses. For example, let $\dfrac{v}{y} + \dfrac{y}{z} + \dfrac{x}{v} = w$; then we must write

$$\frac{\pm v\,dy \mp y\,dv}{yy} + \frac{(\pm)y\,dz\,(\mp)z\,dy}{zz} + \frac{((\pm))x\,dv\,((\mp))v\,dx}{vv} = dw,$$

so that the ambiguities in the different terms may not be confused. We must take notice that an ambiguous sign with itself gives $+$, with its opposite gives $-$, while with another ambiguous sign it forms a new ambiguity depending on both.

Powers. $dx^a = ax^{a-1}\,dx$; for example, $dx^3 = 3x^2\,dx$. $d\dfrac{1}{x^a} = -\dfrac{a\,dx}{x^{a+1}}$; for example, if $w = \dfrac{1}{x^3}$, then $dw = -\dfrac{3\,dx}{x^4}$.

Roots. $d\sqrt[b]{x^a} = \dfrac{a}{b}\,dx\sqrt[b]{x^{a-b}}$ (hence $d\sqrt[2]{y} = \dfrac{dy}{2\sqrt[2]{y}}$, for in this case $a = 1$, $b = 2$), therefore $\dfrac{a}{b}\sqrt[b]{x^{a-b}} = \tfrac{1}{2}\sqrt[2]{y^{-1}}$, but y^{-1} is the same as $\dfrac{1}{y}$; from the nature of the exponents in a geometric progression, and $\sqrt[2]{\dfrac{1}{y}} = \dfrac{1}{\sqrt[2]{y}}$, $d\dfrac{1}{\sqrt[b]{x^a}} = \dfrac{-a\,dx}{b\sqrt[b]{x^{a+b}}}$.

The law for integral powers would have been sufficient to cover the case of fractions as well as roots, for a power becomes a fraction when the exponent is negative, and changes into a root when the exponent is fractional. However, I prefer to draw these conclusions myself rather than relegate their deduction to others, since they are quite general and occur often. In a matter that is already complicated in itself it is preferable to facilitate the operations.

Knowing thus the *Algorithm* (as I may say) of this calculus, which I call *differential calculus,* all other differential equations can be solved by a common method. We can find maxima and minima as well as tangents without the necessity of removing fractions, irrationals, and other restrictions, as had to be done according to the methods that have been published hitherto. The demonstration of all this will be easy to one who is experienced in these matters and who considers the fact, until now not sufficiently explored, that dx, dy, dv, dw, dz can be taken proportional to the momentary differences, that is, increments or decrements, of the corresponding x, y, v, w, z. To any given equation we can thus write its differential equation. This can be done by simply substituting for each *term* (that is, any part which through addition or subtraction contributes to the equation) its differential quantity. For any other quantity (not itself a term, but contributing to the formation of the term) we use its differential quantity, to form the differential quantity of the term itself, not by simple substitution, but according to the prescribed Algorithm. The methods published before have no such transition. They mostly use a line such as DX or of similar kind, but not the line dy which is the fourth proportional to DX, DY, dx—something quite confusing. From there they go on removing fractions and irrationals (in which undetermined quantities occur). It is clear that our method also covers transcendental[8] curves—those that cannot be reduced by algebraic computation, or have no particular degree—and thus holds in a most general way without any particular and not always satisfied assumptions.

We have only to keep in mind that to find a *tangent* means to draw a line that connects two points of the curve at an infinitely small distance, or the continued side of a polygon with an infinite number of angles, which for us takes the place of the *curve*. This infinitely small distance can always be expressed by a known differential like dv, or by a relation to it, that is, by some known tangent. In particular, if y were a transcendental quantity, for instance the ordinate of a cycloid, and it entered into a computation in which z, the ordinate of another curve, were determined, and if we desired to know dz or by means of dz the tangent of this latter curve, then we should by all means determine dz by means of dy, since we have the tangent of the cycloid. The tangent to the cycloid itself, if we assume that we do not yet have it, could be found in a similar way from the given property of the tangent to the circle.

Now I shall propose an example of the calculus, in which I shall indicate division by $x:y$, which means the same as x divided by y, or $\frac{x}{y}$.[9] Let the *first* or given equation be[10] $x:y + (a + bx)(c - xx):(ex + fxx)^2 + ax\sqrt{gg + yy} + yy:\sqrt{hh + lx + mxx} = 0$. It expresses the relation between x and y or between AX and XY, where a, b, c, e, f, g, h are given. We wish to draw from a point Y the line YD tangent to the curve, or to find the ratio of the line DX to the given line XY. We shall write for short $n = a + bx$, $p = c - xx$, $q = ex + fxx$, $r = gg + yy$, and $s = hh + lx + mxx$. We obtain $x:y + np: qq + ax\sqrt{r} + yy:\sqrt{s} = 0$, which we call the *second* equation. From our calculus it follows that

$$d(x:y) = (\pm\, x\, dy \mp y\, dx):yy,^{11}$$

and equally that

$$d(np:qq) = [(\pm)2np\, dq\, (\mp)q(n\, dp + p\, dn)]:q^3,$$

$$d(ax\sqrt{r}) = +ax\, dr:2\sqrt{r} + a\, dx\sqrt{r},$$

$$d(yy:\sqrt{s}) = ((\pm))yy\, ds\, ((\mp))\, 4ys\, dy:2s\sqrt{s}.$$

All these differential quantities from $d(x:y)$ to $d(yy:\sqrt{s})$ added together give 0, and thus produce a *third* equation, obtained from the terms of the second equation by substituting their differential quantities. Now $dn = b\, dx$ and $dp = -2x\, dx$, $d = e\, dx + 2fx\, dx$, $dr = 2y\, dy$, and $ds = l\, dx + 2mx\, dx$. When we substitute these values into the third equation we obtain a *fourth equation, in which the only remaining differential quantities, namely* dx, dy, are all outside of the denominators and without restrictions. Each term is multiplied either by dx or by dy, so that the law of homogeneity always holds with respect to these two quantities, however complicated the computation may be. From this we can always obtain the value of $dx:dy$, the ratio of dx to dy, or the ratio of the required DX to the given XY. In our case this ratio will be (if the fourth equation is changed into a proportionality):

$$\mp x:yy - axy:\sqrt{r}\ (\mp)\ 2y:\sqrt{s}$$

divided by

$$\mp 1:y\ (\pm)\ (2npe + 2fx):q^3\ (\mp)\ (-2nx + pb)\ :qq + a\sqrt{r}((\pm))yy(l + 2mx)\ :2s\sqrt{s}.$$

Now x and y are given since point Y is given. Also given are the values of n, p, q, r, s expressed in x and y, which we wrote down above. Hence we have obtained what we required. Although this example is rather complicated we have presented it to show how the above mentioned rules can be used even in a more difficult computation. Now it remains to show their use in cases easier to grasp.

Let two points C and E [Fig. 2] be given and a line SS in the same plane. It is required to find a point F on SS such that when E and C are connected with F the sum of the rectangle of CF and a

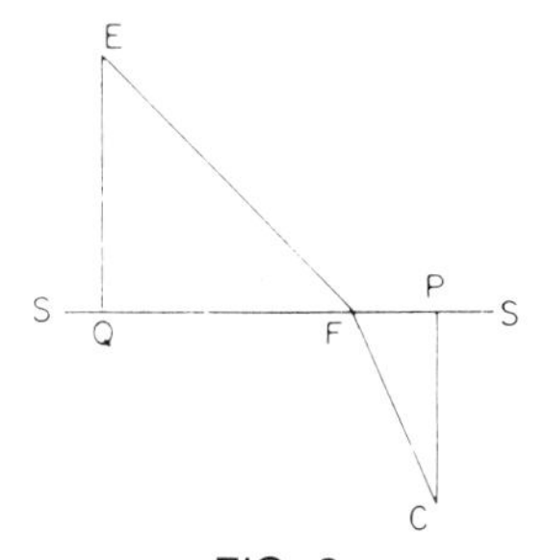

FIG. 2

given line h and the rectangle of FE and a given line r are as small as possible.[12] In other words, if SS is a line separating two media, and h represents the density of the medium on the side of C (say water), r that of the medium on the side of E (say air), then we ask for the point F such that the path from C to E via F is the shortest possible. Let us assume that all such possible sums of rectangles, or all possible paths, are represented by the ordinates KV of curve VV perpendicular to the line GK [Fig. 1]. We shall call these ordinates w. Then it is required to find their minimum NM. Since C and E [Fig. 2] are given, their perpendiculars to SS are also given, namely CP (which we call c) and EQ (which we call e); moreover PQ (which we call p) is given. We denote $QF = GN$ (or AX) by x, CF by f, and EF by g. Then $FP = p - x$, $f = \sqrt{cc + pp - 2px + xx}$ or $= \sqrt{l}$ for short; $g = \sqrt{ee + xx}$ or $= \sqrt{m}$ for short. Hence

$$w = h\sqrt{l} + r\sqrt{m}.$$

The differential equation (since $dw = 0$ in the case of a minimum) is, according to our calculus,

$$0 = +h\,dl : 2\sqrt{l} + r\,dm : 2\sqrt{m}.$$

But $dl = -2(p - x)\,dx$, $dm = 2x\,dx$; hence

$$h(p - x) : f = rx : g.$$

When we now apply this to dioptrics, and take f and g, that is, CF and EF, equal to each other (since the refraction at the point F is the same no matter how long the line CF may be), then $h(p - x) = rx$ or $h : r = x : (p - x)$, or $h : r = QF : FP$; hence the sines of the angles of incidence and of refraction, FP and QF, are in inverse ratio to r and h, the densities of the media in which the incidence and the refraction take place. However, this density is not to be understood with respect to us, but to the resistance which the light rays meet. Thus we have a demonstration of the computation exhibited elsewhere in these *Acta* [1682], where we presented a general foundation of optics, catoptrics, and dioptrics.[13] Other very learned men have sought in many devious ways what someone versed in this calculus can accomplish in these lines as by magic.

This I shall explain by still another example. Let *13* [Fig. 3] be a curve of such a nature that, if we draw from one of its points, such as *3*, six lines *34, 35, 36, 37, 38, 39* to six fixed points *4, 5, 6, 7, 8, 9* on the axis, then their sum is equal to a given line. Let *T14526789* be the axis, *12* the abscissa, *23* the ordinate, and let the tangent *3T* be required.

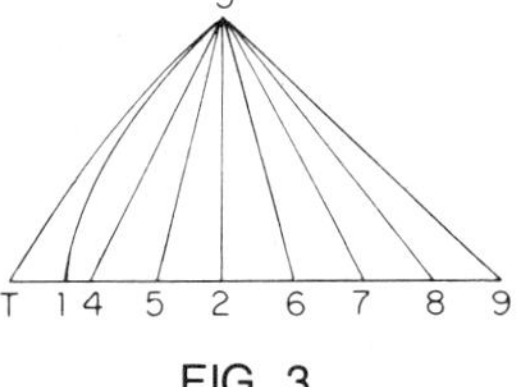

FIG. 3

Then I claim that *T2* is to *23* as $\frac{23}{34} + \frac{23}{35} + \frac{23}{36} + \frac{23}{37} + \frac{23}{38} + \frac{23}{39}$ to $-\frac{24}{34} - \frac{25}{35} + \frac{26}{36} + \frac{27}{37} + \frac{28}{38} + \frac{29}{39}$. The same rule will hold if we increase the number of terms, taking not six but ten or more fixed points. If we wanted to solve this problem by the existing tangent methods, removing irrationals, then it would be a most tedious and sometimes insuperable task; in this case we would have to set up the condition that the rectangular planes and solids which can be constructed by means of all possible combinations of two or three of these lines are equal to a given quantity.[14] In all these cases and even in more complicated ones our methods are of astonishing and unequaled facility.

And this is only the beginning of much more sublime Geometry, pertaining to even the most difficult and most beautiful problems of applied mathematics, which without our differential calculus or something similar no one could attack with any such ease. We shall add as appendix the solution of the problem which De Beaune proposed to Descartes and which he tried to solve in Vol. 3 of the *Letters,* but without success.[15] It is required to find a curve WW such that, its tangent WC being drawn to the axis, XC is always equal to a given constant line a. Then XW or w is to XC or a as dw is to dx. If dx (which can be chosen arbitrarily) is taken constant, hence always equal to, say, b, that is, x or AX increases uniformly, then $w = \frac{a}{b}\,dw$. Those ordinates w are therefore proportional to their dw, their increments or differences, and this means that if the x form an arithmetic progression, then the w form a geometric progression. In other words, if the w are numbers, the x will be logarithms, so that the curve WW is logarithmic.

NOTES

1. Note the Latin term *abscissa.* This term, which was not new in Leibniz's day, was made by him into a standard term, as were so many other technical terms. In the article "De linea ex lineis numero infinitis ordinatim ductis inter se concurrentibus formata . . . ," *Acta Eruditorum 11* (1692), 168-171 (Leibniz, *Mathematische Schriften,* Abth. 2, Band I (1858), 266-269), in which Leibniz discusses evolutes, he presents a collection of technical terms. Here we find *ordinata, evolutio, differentiare, parameter, differentiabilis, functio,* and *ordinata* and *abscissa* together designated as *coordinatae.* Here he also points out that ordinates may be given not only along straight but also along curved lines. The term *ordinate* is derived from *rectae ordinatim applicatae,* "straight lines designated in order," such as parallel lines. The term *functio* appears in the sentence: "the tangent and some other functions depending on it, such as perpendiculars from the axis conducted to the tangent."

2. When the subtangent—a term Leibniz used in a paper in the *Acta Eruditorum* (1694; *Mathematische Schriften,* Abth. 2, Band I, 306), though it may be older—is denoted by δ, Leibniz defines $dy : dx = y : \delta$, or $\delta = y : dy/dx$. We may express this by saying that Leibniz takes the derivative (geometrically, in the form of the tangent) without further definition, and defines the differentials in terms of the derivative.

3. Leibniz uses the term *differentia* and conceives it as a finite line segment. What we now call *differential* would long after Leibniz often be called *difference.* Leibniz also uses other terms. As to the meaning of the differentials, see the end of this selection.

4. The ambiguity in signs is due to the fact that s is taken positive. Systematic discrimination between positive and negative senses in analytic geometry came only with Monge and Möbius in the early nineteenth century.

5. Leibniz has "concavity" and "convexity" interchanged.

6. Leibniz' term is *punctum flexii contrarii* (point of opposite flection). On this term see T. F. Mulcrone, *The Mathematics Teacher 61* (1968), 475-478.

7. There seems to be something wrong here: when $y = x^2$, $dy = 2x\,dx$; then, when x passes from negative to positive ($dx > 0$), dy increases while y first decreases and then increases. However, see note 4.

8. This may be the first time that the term "transcendental" in the sense of "nonalgebraic" occurs in print.

9. From this suggestion by Leibniz dates the general adoption of this notation; see J. Tropfke, *Geschichte,* 3rd ed., II (1933), 30. See also the reference to Mengoli in G. Castelnuovo, *Le origini del calcolo infinitesimale nell'era moderna,* 153.

10. We have retained Leibniz' notation : but substituted parentheses for superscript bars: Leibniz writes

$x : y + \overline{\overline{a + bx}\ \overline{c\ xx}}$: quadrat. $ex + fxx + ax\sqrt{gg + yy} + yy : \sqrt{hh + lx + mxx}$ aequ. 0.

11. Leibniz writes $d, x : y$.

12. For this problem, due to Fermat (*Oeuvres,* II (1844), 457), see note 13.

13. In this paper, "Unicum opticae, catoptricae et dioptricae principium," *Acta Eruditorum 1* (1683), 186-190, dealing with the laws of refraction and reflection, Leibniz makes known for the first time in print that he has his own *methodus de maximis et minimis.*

14. If the coordinates of point i are a_i, $i = 4, 5, \ldots,$ and those of point 3 are x, y, then

this result can immediately be obtained by differentiating $\Sigma_i \sqrt{(x-a_i)^2+y^2}$. Leibniz writes -24, -25 because his segments are all positive.

15. This is an inverse-tangent problem; Leibniz quotes *Les lettres de René Descartes* (3 vols.; Paris, ed. C. de Clerselier, 1657-1667). The problem is part of a long series of investigations that begins with the invention of logarithms by Napier by comparing an arithmetic and a geometric series and leads up to the full recognition of the inverse relation of the two functions $y = \log x$ and $x = e^y$ by Euler. Florimond De Beaune (1601-1652), a jurist at Blois, had written to Descartes about some curves; Descartes's answer of 1639 exists (Descartes, *Oeuvres*, II, 510-519); it was printed in the above-mentioned seventeenth-century edition of Descartes's letter and was studied by Leibniz. One of the curves was defined by a geometric description equivalent to the equation $dy/dx = (x-y)/b$. By means of the substitution $x' = b - x + y$, $y' = y$ this equation is transformed into $dy'/dx' = -y'/b$, the differential equation of what we now call the logarithmic curve. Descartes comes to the equivalent of this result; without mentioning logarithms he derives an inequality that can be written in our notation

$$\frac{1}{n}+\frac{1}{n+1}+\cdots+\frac{1}{m-1}>\log\frac{m}{n}>\frac{1}{n+1}+\frac{1}{n+2}+\cdots+\frac{1}{m}$$

($n < m - 1$; m, n positive integers); see C. J. Scriba, "Zur Lösung des 2. Debeauneschen Problems durch Descartes," *Archive for History of Exact Sciences 1* (1961), 406-419. Descartes, like Napier, lets the logarithms grow when the argument decreases, while Briggs, who introduces 10 as base, lets argument and function grow at the same time. The next important steps, known to Leibniz, were Grégoire De Saint Vincent's determination of the area enclosed by a hyperbola, two ordinates, and an asymptote (1647), which Alfons Anton De Sarasa (1649) interpreted with the aid of logarithms, and Nikolaus Mercator's series (1667) for this area of the hyperbola: $a - \frac{a^2}{2} + \frac{a^3}{3} - \frac{a^4}{4} + \cdots$.

73. From "Supplementum geometriae dimensoriae . . ."[1] in *Acta Eruditorum* (1693)*

(The Fundamental Theorem of the Calculus)

GOTTFRIED WILHELM LEIBNIZ

I shall now show *that the general problem of quadratures can be reduced to the finding of a line that has a given law of tangency (declivitas),* that is, for which the sides of the characteristic triangle have a given mutual relation. Then I shall show how this line can be described by a motion that I have invented. For this purpose [Fig. 1] I assume for every curve $C(C')$ a *double characteristic triangle,*[2] one, TBC, that is assignable, and one, GLC, that is inassignable,[3] and these two are similar. The inassignable triangle consists of the parts GL, LC, with the elements of the coordinates CF, CB as sides, and GC, the element of arc, as the base or hypotenuse. But the assignable triangle TBC consists of the axis, the ordinate, and the tangent, and therefore contains

*Source: This translation, made from Leibniz's *Mathematische Schriften*, Abth. 2, Band I, 294-301, appears in D. J. Struik (ed.), *A Source Book in Mathematics, 1200-1800* (1969), 282-284. It is reprinted by permission of Harvard University Press,

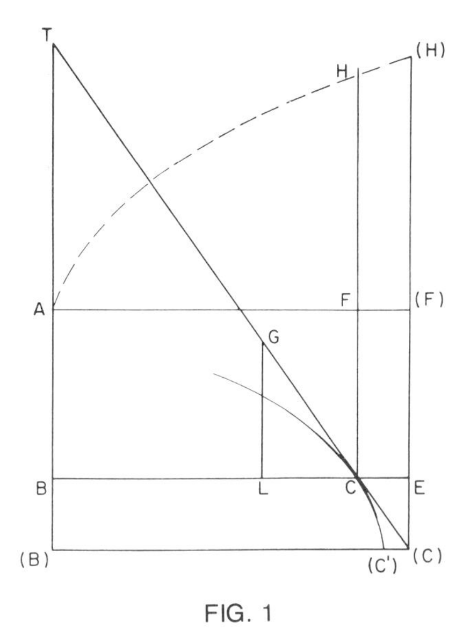

FIG. 1

the angle between the direction of the curve (or its tangent) and the axis or base, that is, the inclination of the curve at the given point C. Now let $F(H)$, the region of which the area has to be squared,[4] be enclosed between the curve $H(H)$, the parallel lines FH and $(F)(H)$, and the axis $F(F)$; on that axis let A be a fixed point, and let a line AB, the conjugate axis, be drawn through A perpendicular to AF. We assume that point C lies on HF (continued if necessary); this gives a new curve $C(C')$ with the property that, if from point C to the conjugate axis AB (continued if necessary) both its ordinate CB (equal to AF) and tangent CT are drawn, the part TB of the axis between them is to BC as HF to a constant [segment] a, or a times BT is equal to the rectangle AFH (circumscribed about the trilinear figure $AFHA$).[5] This being established, I claim that the rectangle on a and $E(C)$ (we must discriminate between the ordinates FC and $(F)(C)$ of the curve) is equal to the region $F(H)$. When therefore I continue line $H(H)$ to A, the trilinear figure $AFHA$ of the figure to be squared is equal to the rectangle with the constant a and the ordinate FC of the squaring curve as sides. This follows immediately from our calculus. Let $AF = y$, $FH = z$, $BT = t$, and $FC = x$; then $t = zy:a$, according to our assumption; on the other hand, $t = y\,dx:dy$ because of the property of the tangents expressed in our calculus. Hence $a\,dx = z\,dy$ and therefore $ax = \int z\,dy = AFHA$. Hence the curve $C(C')$ is the quadratrix with respect to the curve $H(H)$, while the ordinate FC of $C(C')$, multiplied by the constant a, makes the rectangle equal to the area, or the sum of the ordinates $H(H)$ corresponding to the corresponding abscissas AF. Therefore, since $BT : AF = FH : a$ (by assumption), and the relation of this FH to AF (which expresses the nature of the figure to be squared) is given, the relation of BT to FH or to BC, as well as that of BT to TC, will be given, that is, the relation between the sides of triangle TBC.[6] Hence, all that is needed to be able to perform the quadratures and measurements is to be able to describe the curve $C(C')$ (which, as we have shown, is the quadratrix), when the relation between the sides of the assignable characteristic triangle TBC (that is, the law of inclination of the curve) is given.

Leibniz continues by describing an instrument that can perform this construction.

NOTES

1. Leibniz distinguishes here between *geometria dimensoria,* which deals with quadratures and *geometria determmatrix,* which can be reduced to algebraic equations.

2. In Fig. 1 Leibniz assigns the symbol (C) to two points which we denote by (C) and (C'). If, with Leibniz, we write $CF = x$, $BC = y$, $HF = z$, then $E(C) = dx$, $CE = F(F) = dy$, and $H(H)(F)F = z\,dy$. First Leibniz introduces curve $C(C')$ with its characteristic triangle, and then later reintroduces it as the squaring curve [*curva quadratrix*] of curve $AH(H)$.

3. For want of anything better we use Leibniz's terms *assignabilis* and *inassign-*

abilis. G. Kowalewski, *Leibniz über die Analysis des Unendlichen,* Ostwald's *Klassiker,* No. 162 (Engelmann, Leipzig, 1908), 30, uses the German *angebbar* and *unangebbar,* "indicable" and "unindicable." For "differential" Leibniz in our text uses the term "element." Observe also the use of the term "coordinates" (Latin *coordinatae*).

4. The Latin is here a little more expressive than the English. From the Latin *quadrare* we can derive *quadrans, quadrandus, quadratrix, quadratura,* which can be translated by "to square," "squaring," "to be squared," "squaring curve" or "quadratrix," and "quadrature."

5. This is Pascal's expression; see Selections IV.11, 12 [in Struik].

6. This reasoning is still very much like that of Barrow, Gregory, and Torricelli, but because Leibniz possesses the converse relation $a\,dx = x\,dy - \int a\,dx = \int x\,dy$ he needs only one demonstration, where Barrow needed two (Lecture X, 11; XI, 19; Selection IV.14 [in Struik]).

Chapter VI
The Scientific Revolution at Its Zenith (1620–1720)

Section C
The Discovery of the Differential and Integral Calculus

ISAAC NEWTON (1642–1727)

Isaac Newton was born on Christmas, the posthumous child of a yeoman father and Hannah (née Ayscough) Newton. At birth he was a physical weakling who, it was said, could have fit into a quart mug. He grew up in his father's house near the hamlet of Woolsthorpe. When his mother remarried, he was placed in the care of his maternal grandmother for eight years. The English Civil War had begun, and Cromwell was rising to power. Raids by armed men were frequent, and even in places where there was no immediate danger people lived in fear. Isaac, a solitary child without playmates, turned to meditation. He began to construct mechanical contrivances—fiery kites, lanterns, and models of mills. When her second husband died, his mother returned to Woolsthorpe in 1653 intending to make a farmer of Isaac. However, his uncle and the master of Grantham school convinced her that the boy was unsuited for such work and should be sent to a university.

In 1661, Newton was admitted to Trinity College, Cambridge, where he received the Bachelor of Arts degree in 1665, when the great plague was just beginning. As the disease spread across the country, universities were closed. From June 1665 through 1666, he stayed at home in Woolsthorpe but made at least one visit to Cambridge. In his musings as an elderly man, Newton remembered this period as an *annus mirabilis* during which he laid the foundations for his monumental scientific achievement. In mathematics, he discovered by induction the general binomial theorem and invented an early stage of the calculus. In optics, he decomposed white light with his prism, and, in mechanics, he enunciated the inverse-square law of attraction. His extraordinary scientific creativity was underway. His studies of the calculus and spectroscopy were in embryo, while in mechanics he continued to examine the Cartesian vortex theory, which he later rejected accepting attraction instead. After returning to Trinity College in 1667, he completed the M.A. degree in 1668, whereupon he was appointed Lucasian Professor of Mathematics, a position he held until 1701. Presumably, the first incumbent, Isaac Barrow, recognized Newton as a prodigy and resigned so that he might have this prestigious chair. Few students attended Newton's lectures in algebra and dynamics, fewer still understood them. His teaching departed from the dominant Cartesian physics presented in the school texts. He was a quirky, retiring scholar who was known to eat sparingly, skip meals, and be careless of outward appearances, sometimes wearing unkempt clothing. But, more importantly, he was compulsively precise in his research and had an amazing ability to "think on a problem" to the exclusion of all else.

In the two decades after 1668, Newton began to win a reputation in learned circles as the scientific genius of Britain, a reputation recognized throughout Britain by the 1710s. During the early 1670s, he concentrated on optical research. After inventing a reflecting telescope, he was elected a Fellow of the Royal Society (1672), the highest scientific honor in Britain. Shortly thereafter published a new theory of white light and colors, which he based upon his discovery of the chromatic composition of white light. His new theory founded the science of spectroscopy. A controversy arose when Robert Hooke accused Newton of stealing optical ideas from him, leading Newton to decide to publish no further works on optics until after Hooke's death in 1702. After the episode with Hooke, he grew to dislike debate. In 1684 came the famous visit of the astronomer, Edmund Halley, who not only urged Newton to write the classic *Principia Mathematica* (1687) but also agreed to pay for its printing. In the *Principia,* Newton incorporated Kepler's laws of planetary motion and Galileo's law of free fall into a general dynamics. Newton's theory united celestial and terrestrial dynamics under the law of gravitation and justified Copernican astronomy.

After the publication of the *Principia,* Newton's activity was no longer confined to the cloistered, academic world of Cambridge. He was elected a Whig member of Parliament in 1689. In London, he met John Locke and Richard Bentley, with whom he discussed theological and biblical questions. In autumn 1693, he suffered from insomnia and an attack of depression that amounted to a psychic crisis. He made unjustified accusations against his friends, John Locke and Samuel Pepys, accusing the former of attempting to "embroil him with women." Locke remained a friend and did not respond and Newton, who never married, soon recovered his equilibrium. He was appointed warden of the Mint in 1696, was reelected M.P. in 1701, but was defeated in a third election try in 1706. In 1703 he was elected president of the Royal Society, which he ruled over with an iron hand. In 1705 Queen Anne knighted him.

In his final years Newton was occupied with a number of scientific, mystical, and humanistic concerns. His second major book, *Opticks* (1704), presented a corpuscular theory of light, the nut-shell theory of matter, and the "method of fluxions" in an appendix entitled, *Tractatus de quadratura curvarum* ("Treatise on the quadrature of curves," 1704). From 1711 through 1712, he guided the deliberations of the committee of the Royal Society in investigating the question of whether he or Leibniz deserved priority and originality in the invention of the calculus. It was Newton, not the committee, who wrote the *Commercium epistolicum* (1712) according Newton priority. In his last years, Newton continued studies of alchemy and hermetic philosophy that he had pursued since the 1670s, as well as of Biblical chronology and history. He died in 1727, following a lengthy struggle with gout and inflamed lungs, and was buried in Westminster Abbey, an honor reserved previously for royalty.

Newton contributed to many branches of mathematics, including algebra and number theory, Euclidean and analytic geometry, classification of cubic curves, finite differences, methods of approximation and probability. Some of his writings on mathematics are *Arithmetica Universalis* (1707), *Methodus Differentialis* (1711), and *The Method of Fluxions and Infinite Series* (publ. posth., 1736). According to Newton, his chief influence on mathematics comes from his theory of infinite series and the "method of fluxions," an early stage of the calculus. Building upon John Wallis' methods of interpolation and extrapolation, he generalized the binomial theorem, that is, the expansion of $(a+b)^n$. Previously it had been known for positive integral exponents; in 1665, Newton produced binomial expansions for negative and fractional n as well. However, he did not supply a rigorous proof. This was later offered by Euler for all real n and by Abel for all n of complex value. Even though Newton's binomial expansions were obtained without rigorous proofs, they were correct and put emphasis on infinite series, whose study was to dominate the calculus.

It took time and intense creativity for Newton to discover and begin to systematize the calculus. In 1664, his last undergraduate year at Cambridge, he began copious studies of Cartesian coordinate geometry and prepared extensive notes on the infinitesimal analysis found in Wallis' *Arithmetica Infinitorum* (1656) and Barrow's *Lectiones* (1664). His research into the interlocking structure of these two branches of mathematics laid the foundations for his discovery in 1666 that integration and differentiation are inverse operations in his "method of fluxions." A fluxion is a derivative taken with respect to time, expressed in his dot notation as $\dot{x} = dx/dt$.

Newton did not publish on his new fluxional analysis for some time. His first efforts to publish on it were thwarted by the unsaleability of technical treatises and the severe depression in the book trade following the Great Fire of London (1666). Neither his fluxional tract of October 1666 nor his famous "De Analysi per Aequationes Infinitas" ("On Analysis by Infinite Equations," *c.* 1669) appeared in print before 1700. Moreover, the hail of criticism that greeted his novel, corpuscular theory of light in the early 1670s made him hesitant to publish his new scientific ideas. The *Principia* was Newton's first publication to involve the "method of fluxions," although it made no explicit reference to fluxions. In Book One, he spoke of prime and ultimate ratios, essentially conceiving of the derivative of an algebraic function as $\lim_{h \to o} \left\{ \frac{f(x+h) - f(x)}{h} \right\}$. In Book Two, he referred to an infinitesimal or evanescent increase in x as the "moment" of x and denoted it by $\dot{x}o$. Newton soon grew uneasy with infinitesimals, or differentials, that could "vanish." Thus, his first full exposition of the fluxional calculus, the *Tractatus de quadratura curvarum*, written in 1693, hardly mentions them. It abandons static aggregates of infinitesimals as a basis for the calculus and instead relies on variable quantities known as fluents that are generated by the continuous motion of points, lines, and planes, their rates of change, the fluxion, and prime and ultimate ratios. Again, this exposition was not published until 1704.

As one would expect, Newton did not complete the development of the calculus. While he was aware of the importance of convergence in infinite series, others had to develop convergency tests. Likewise, his concept of limit was hazy. The solid foundations for the calculus were not established until over a century later by Augustin Cauchy and Karl Weierstrass. Significantly, in providing the beginning stage of the calculus, Newton went beyond classical Greek geometry with an independent science that could handle a vastly expanded range of problems in physics.

74. From *Specimens of a Universal* [*System of*] *Mathematics* (written c. 1684)*

ISAAC NEWTON

A certain method of resolving problems by convergent series devised by me about eighteen years ago[1] had, by my very honest friend[2] Mr. John Collins, around that time been announced to Mr. James Gregory, the renowned Professor of Mathematics at Edinburgh University in Scotland, as being in my possession. This method Mr. Gregory learnt—a measure of the power of his intellect—from but a single series of this type which had been passed on to him, but not much aliorum afterwards he was snatched away by an untimely death.[3] From his papers, extant in which were certain calculations though without a description of the method, his celebrated successor in the Mathematical Chair, David Gregory, also learnt this method of calculation and developed it in a neat and stimulating tract: in this he revealed not only his expertise in mathematical topics but also the utmost probity, candidly acknowledging what he himself had taken from his predecessor and what his predecessor had received from Collins.[4] While still reading it[5] I began at once to reflect that, since I had often been asked to publish something, I had now less excuse to resist the entreaties of my friends and thwart the expectation of others,[6] and that I were better advised to be swiftly acquiescent rather than have to submit with annoyance at a later, less opportune time. Through the agency[7] of Mr. H. O. certain letters regarding series of this sort once passed between me and Mr. Leibniz, now the Duke of Hanover's eminent minister in public affairs. By publishing these I shall certainly oblige the reader more than if (after what Mr. Gregory has done) I should write up the whole subject afresh, and especially so since in them is contained Leibniz' extremely elegant method, far different from mine, of attaining the same series—one about which it would be dishonest to remain silent while publishing my own. Mine shall, indeed, be described in such a way that Mr. Gregory's book, with its more lavish explanation of points which are here touched upon piecemeal and omission of others which are here more copiously described, may profitably fill the rôle of an introduction. The letters, however, are to this effect.[8]

NOTES

1. That is, in 1665-6. Newton here seems to refer to his discovery of the general binomial expansion early in 1665 and perhaps the "mechanical" extraction of the root of an algebraic equation as an infinite series according to "Vieta's Analyticall resolution of powers" adumbrated in his October 1666 tract but not, of course, developed at length till he came to write his 1671 treatise.

2. This replaces the more fulsome phrase "rerum Mathematicarum cultor eximius" (that outstanding cultivator of things

*Source: This translation of *Matheseos Universalis Specimena* is from D. T. Whiteside (ed.), *The Mathematical Papers of Isaac Newton*, vol. IV, 1674-1684 (1971), 527-531. It is reprinted by permission of Cambridge University Press. Footnotes are renumbered.

mathematical). Newton knew very well the moderate limits of Collins' mathematical talents and not even the latter's recent death (in November 1683) would provoke him more than momentarily to an undeserved epithet.

3. James Gregory died in late October 1675, when he was not quite 37 years old and still in the prime of life, within a few hours of having had a severe stroke accompanied by blindness and paralysis. Newton's source of information is David Gregory who wrote in his *Exercitatio Geometrica*. . . .

4. Compare David Gregory's *Exercitatio Geometrica*: 4. An overwhelming number of the examples to which David applies his rediscovered Gregorian method in the body of his book come in fact (though this is not always made clear in the case of still unpublished *adversaria*) from the papers and printed works of his uncle James, though he at one point refers to Nicolaus Mercator's *Logarithmotechnia* for the series expansion of log (1+x) and twice cites René-François de Sluse's *Miscellanea* (Liège, 1668), once as source for the general Slusian "pearl," once as inventor of the Slusian conchoid.

5. The subjunctive mood is manifestly a slip on Newton's part: the act of reading is coterminous with the reaction it inspired.

6. In the present mathematical context Newton would probably rank Barrow and Collins (now both dead) as friends, Oldenburg (also deceased seven years before) and John Wallis as "other" acquaintances!

7. "mediante" (through the mediation) is cancelled. "H.O." is, of course, "H[enrico] O[ldenburgo]" (Henry Oldenburg). The designation of Leibniz—after the spring of 1676 Librarian (and unofficial legal adviser) to the Duke of Hanover—as his "minister in negotijs publicis" (minister of state?) probably points to Newton's lack of awareness of contemporary political realities at this period.

8. In sequel Newton indicates extracts relating to infinite series from the five letters which passed (by way of Oldenburg) between himself and Leibniz during the period June 1676 to July 1677. We relate these brief citations of terminal phrases (quoted by Newton, in the case of the three Leibniz letters, from copies furnished him by Oldenburg and Collins) to the reproductions of the originals as sent, given by H. W. Turnbull in his edition of *The Correspondence of Isaac Newton,* 2, 1960. While rightly choosing not to tamper with the text of his *epistola prior* of 13 June 1676—nor, of course, with Leibniz' words—Newton has slightly revised and clarified passages in his *epistola posterior* of the following 24 October.

75. From a Letter to Henry Oldenburg on the Binomial Series (June 13, 1676)*

ISAAC NEWTON

Though the modesty of Mr. Leibniz, in the extracts from his letter which you have lately sent me, pays great tribute to our countrymen for a certain theory of infinite series, about which there now begins to be some talk, yet I have no doubt that he has discovered not only a method for reducing any quantities whatever to such series, as he asserts, but also various shortened forms, perhaps like our own, if not even better. Since, however, he very much wants to know what has been discovered in this subject by the English, and since I myself fell upon this theory some years ago, I have sent you some of those things which occurred to me in order to satisfy his wishes, at any rate in part.

Fractions are reduced to infinite series by division; and radical quantities by

Source: This translation from H. W. Turnbull, F.R.S. (ed.), *The Correspondence of Isaac Newton,* vol. II (1960), 32-33. Reprinted by permission of Cambridge University Press. Henry Oldenburg was secretary of the Royal Society.

extraction of the roots, by carrying out those operations in the symbols just as they are commonly carried out in decimal numbers. These are the foundations of these reductions: but extractions of roots are shortened by this theorem,

$$(P + PQ)^{m/n} = P^{m/n} + \frac{m}{n} AQ + \frac{m-n}{2n} BQ + \frac{m-2n}{3n} CQ + \frac{m-3n}{4n} DQ + \text{etc.}$$

where $P + PQ$ signifies the quantity whose root or even any power, or the root of a power, is to be found; P signifies the first term of that quantity, Q the remaining terms divided by the first, and m/n the numerical index of the power of $P + PQ$, whether that power is integral or (so to speak) fractional, whether positive or negative. For as analysts, instead of aa, aaa, etc., are accustomed to write a^2, a^3, etc., so instead of $\sqrt{a}$, $\sqrt{a^3}$, $\sqrt{c{:}a^5}$, etc. I write $a^{1/2}$, $a^{3/2}$, $a^{5/3}$, and instead of $1/a$, $1/aa$, $1/a^3$, I write a^{-1}, a^{-2}, a^{-3}. And so for

$$\frac{aa}{\sqrt{c : (a^3 + bbx)}}$$

I write $aa(a^3 + bbx)^{-1/3}$, and for

$$\frac{aab}{\sqrt{c{:}\{(a^3 + bbx)\,(a^3 + bbx)\}}}$$

I write $aab(a^3 + bbx)^{-2/3}$: in which last case, if $(a^3 + bbx)^{-2/3}$ is supposed to be $(P + PQ)^{m/n}$ in the Rule, then P will be equal to a^3, Q to bbx/a^3, m to -2, and n to 3. Finally, for the terms found in the quotient in the course of the working I employ A, B, C, D, etc., namely, A for the first term, $P^{m/n}$; B for the second term, $m/n\ AQ$; and so on. For the rest, the use of the rule will appear from the examples.

Example 1

$$\sqrt{(c^2+x^2)} \text{ or } (c^2+x^2)^{1/2} = c + \frac{x^2}{2c} - \frac{x^4}{8c^3} + \frac{x^6}{16c^5} - \frac{5x^8}{128c^7} + \frac{7x^{10}}{256[c]^9} + \text{etc.}$$

For in this case, $P = c^2$, $Q = x^2/c^2$, $m = 1$, $n = 2$,

$$A(= P^{m/n} = (cc)^{1/2}) = c,\ B(= (m/n)AQ) = x^2/2c,$$

$$C\left(= \frac{m-n}{2n} BQ\right) = -\frac{x^4}{8c^3};$$

and so on.

76. From Letter to Henry Oldenburg on General Method for Finding Quadratures[1] (October 24, 1676)*

ISAAC NEWTON

I can hardly tell with what pleasure I have read the letters of those very distinguished men Leibniz and Tschirnhaus. Leibniz's method for obtaining convergent series is certainly very elegant, and it would have sufficiently revealed the genius of its author, even if he had written nothing else. But what he has scattered elsewhere throughout his letter is most worthy of his reputation—it leads us also to hope for very great things from him. The variety of ways by which the same goal is approached has given me the greater pleasure, because three methods of arriving at series of that kind had already become known to me, so that I could scarcely expect a new one to be communicated to us. One of mine I have described before; I now add another, namely, that by which I first chanced on these series—for I chanced on them before I knew the divisions and extractions of roots which I now use. And an explanation of this will serve to lay bare, what Leibniz desires from me, the basis of the theorem set forth near the beginning of the former letter.

At the beginning of my mathematical studies, when I had met with the works of our celebrated Wallis, on considering the series by the intercalation of which he himself exhibits the area of the circle and the hyperbola, the fact that, in the series of curves whose common base or axis is x and the ordinates

$$(1-x^2)^{0/2},\ (1-x^2)^{1/2},\ (1-x^2)^{2/2},$$

$$(1-x^2)^{3/2},\ (1-x^2)^{4/2},\ (1-x^2)^{5/2},\ \text{etc.},$$

if the areas of every other of them, namely

$$x,\ x - \tfrac{1}{3}x^3,\ x - \tfrac{2}{3}x^3 + \tfrac{1}{5}x^5,$$

$$x - \tfrac{3}{3}x^3 + \tfrac{3}{5}x^5 - \tfrac{1}{7}x^7,\ \text{etc.}$$

could be interpolated, we should have the areas of the intermediate ones, of which the first $(1-x^2)^{1/2}$ is the circle: in order to interpolate these series I noted that in all of them the first term was x and that the second terms $\tfrac{0}{3}x^3$, $\tfrac{1}{3}x^3$, $\tfrac{2}{3}x^3$, $\tfrac{3}{3}x^3$, etc., were in arithmetical progression, and hence that the first two terms of the series to be intercalated ought to be $x-\tfrac{1}{3}(\tfrac{1}{2}x^3)$, $x-\tfrac{1}{3}(\tfrac{3}{2}x^3)$, $x-\tfrac{1}{3}(\tfrac{5}{2}x^3)$, etc. To intercalate the rest I began to reflect that the denominators 1, 3, 5, 7, etc. were in arithmetical progression, so that the numerical coefficients of the numerators only were still in need of investigation. But in the alternately given areas these were the figures of powers of the number 11, namely of these, 11^0, 11^1, 11^2, 11^3, 11^4, that is, first 1; then 1, 1; thirdly, 1, 2, 1; fourthly 1, 3, 3, 1; fifthly 1, 4, 6, 4, 1, etc. And so I began to inquire how the remaining figures in these series could

*Source: This translation from H. W. Turnbull, F.R.S. (ed.), *The Correspondence of Isaac Newton*, vol. II (1960), 130–134 and 148–149. Reprinted by permission of Cambridge University Press.

be derived from the first two given figures, and I found that on putting m for the second figure, the rest would be produced by continual multiplication of the terms of this series,

$$\frac{m-0}{1} \times \frac{m-1}{2} \times \frac{m-2}{3} \times \frac{m-3}{4} \times \frac{m-4}{5}, \text{ etc.}$$

For example, let $m=4$, and $4 \times \frac{1}{2}(m-1)$, that is 6 will be the third term, and $6 \times \frac{1}{3}(m-2)$, that is 4 the fourth, and $4 \times \frac{1}{4}(m-3)$, that is 1 the fifth, and $1 \times \frac{1}{5}(m-4)$, that is 0 the sixth, at which term in this case the series stops. Accordingly, I applied this rule for interposing series among series, and since, for the circle, the second term was $\frac{1}{3}(\frac{1}{2}x^3)$, I put $m=\frac{1}{2}$, and the terms arising were

$$\frac{1}{2} \times \frac{\frac{1}{2}-1}{2} \text{ or } -\frac{1}{8},$$

$$-\frac{1}{8} \times \frac{\frac{1}{2}-2}{3} \text{ or } +\frac{1}{16},$$

$$\frac{1}{16} \times \frac{\frac{1}{2}-3}{4} \text{ or } -\frac{5}{128},$$

and so to infinity. Whence I came to understand that the area of the circular segment which I wanted was

$$x - \frac{\frac{1}{2}x^3}{3} - \frac{\frac{1}{8}x^5}{5} - \frac{\frac{1}{16}x^7}{7} - \frac{\frac{5}{128}x^9}{9} \text{ etc.}$$

And by the same reasoning the areas of the remaining curves, which were to be inserted, were likewise obtained: as also the area of the hyperbola and of the other alternate curves in this series $(1+x^2)^{0/2}$, $(1+x^2)^{1/2}$, $(1+x^2)^{2/2}$, $(1+x^2)^{3/2}$, etc. And the same theory serves to intercalate other series, and that through intervals of two or more terms when they are absent at the same time. This was my first entry upon these studies, and it had certainly escaped my memory, had I not a few weeks ago cast my eye back on some notes.

But when I had learnt this, I immediately began to consider that the terms

$$(1-x^2)^{0/2}, \quad (1-x^2)^{2/2}, \quad (1-x^2)^{4/2}, \quad (1-x^2)^{6/2}, \quad \text{etc.,}$$

that is to say,

$$1, \quad 1-x^2, \quad 1-2x^2+x^4, \quad 1-3x^2+3x^4-x^6, \quad \text{etc.}$$

could be interpolated in the same way as the areas generated by them: and that nothing else was required for this purpose but to omit the denominators 1, 3, 5, 7, etc., which are in the terms expressing the areas; this means that the coefficients of the terms of the quantity to be intercalated $(1-x^2)^{1/2}$, or $(1-x^2)^{3/2}$, or in general $(1-x^2)^m$, arise by the continued multiplication of the terms of this series

$$m \times \frac{m-1}{2} \times \frac{m-2}{3} \times \frac{m-3}{4}, \text{ etc.,}$$

so that (for example)

$(1-x^2)^{1/2}$ was the value of $1-\frac{1}{2}x^2-\frac{1}{8}x^4-\frac{1}{16}x^6$ etc.,

$(1-x^2)^{3/2}$ of $1-\frac{3}{2}x^2+\frac{3}{8}x^4+\frac{1}{16}x^6$, etc.,

and

$(1-x^2)^{1/3}$ of $1-\frac{1}{3}x^2-\frac{1}{9}x^4-\frac{5}{81}x^6$, etc.

So then the general reduction of radicals into infinite series by that rule, which I laid down at the beginning of my earlier letter became known to me, and that before I was acquainted with the extraction of roots. But once this was known, that other could not long remain hidden from me. For in order to test these processes, I multiplied

$$1-\tfrac{1}{2}x^2-\tfrac{1}{8}x^4-\tfrac{1}{16}x^6, \text{ etc.}$$

into itself; and it became $1-x^2$, the remaining terms vanishing by the continuation of the series to infinity. And even so $1-\frac{1}{3}x^2-\frac{1}{9}x^4-\frac{5}{81}x^6$, etc. multiplied twice into itself also produced $1-x^2$. And as this was not only sure proof of these conclusions so too it guided me to try whether, conversely, these series, which it thus affirmed to be roots of the quantity $1-x^2$, might not be extracted out of it in an arithmetical manner. And the matter turned out well. This was the form of the working in square roots.

$$
\begin{array}{l}
1-x^2(1-{}^1/_2\,x^2-{}^1/_8\,x^4-{}^1/_{16}\,x^6, \text{ etc.} \\
1 \\
\hline
0-x^2 \\
\quad -x^2+{}^1/_4\,x^4 \\
\hline
\qquad -{}^1/_4\,x^4 \\
\qquad -{}^1/_4\,x^4+{}^1/_8\,x^6+{}^1/_{64}\,x^8 \\
\hline
\qquad\quad 0 \quad -{}^1/_8\,x^6-{}^1/_{64}\,x^8.
\end{array}
$$

After getting this clear I have quite given up the interpolation of series, and have made use of these operations only, as giving more natural foundations. Nor was there any secret about reduction by division, an easier affair in any case. But soon I attacked the resolution of affected equations and obtained it. Whence the ordinates, the segments of the axes and any other right lines at once became known from the areas or arcs of the curves being given. For the return to them needed nothing beyond the solution of the equations by which the areas or arcs were given in terms of the given right lines.

At that time the plague breaking out forced me to flee hence and think about other things. Yet, soon after, I added a certain way of finding logarithms from the area of an hyperbola, which I here append. Let *dFD* be an hyperbola, *C* its centre, *F* its vertex, and let *CAFE* = 1 be an inscribed square. In *CA* take *AB*, *Ab*, on this side and that, equal to ${}^1/_{10}$ or 0·1. Then, the perpendiculars *BD*, *bd* being erected to terminate on the

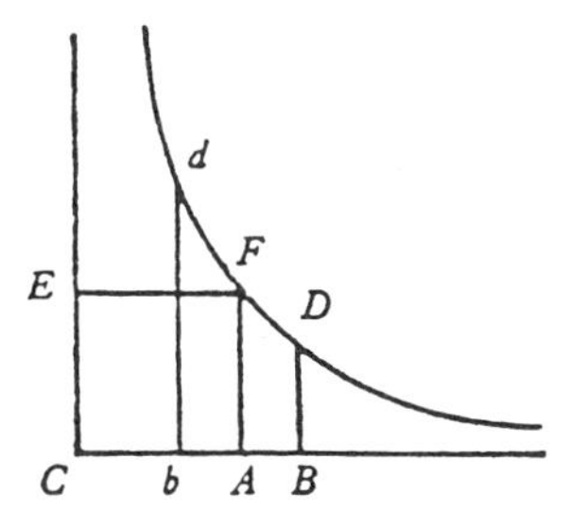

hyperbola, the semi-sum of the areas *AD* and *Ad* will be

$$=0{\cdot}1+\frac{0{\cdot}001}{3}+\frac{0{\cdot}00001}{5}+\frac{0{\cdot}0000001}{7}$$

and the semi-difference

$$=\frac{0{\cdot}01}{2}+\frac{0{\cdot}0001}{4}+\frac{0{\cdot}000001}{6}+\frac{0{\cdot}00000001}{8}, \text{ etc.}$$

which give on reduction

0·1000000000000	0·0050000000000
3333333333	250000000
20000000	1666666
142857	12500
1111	100
9	1
0·1003353477310	0·0050251679267

The sum of these, 0·1053605156577, is *Ad*, and the difference, 0·0953101798043, is *AD*. And in the same way, if *AB*, *Ab* are taken on this side and that, equal to 0·2, the result *Ad* = 0·2231435513142 and *AD* = 0·1823215567939 will be had. Thus, having obtained the hyperbolic logarithms of the four decimal numbers 0·8, 0·9, 1·1 and 1·2, since (1·2/0·8) × (1·2/0·9) = 2, and 0·8 and 0·9 are less than unity, add their logarithms to twice the logarithm of 1·2 and you will have 0·6931471805597 for the hyperbolic logarithm of the number 2. To the triple of this add log 0·8 (since (2×2×2)/0·8=10) and you will have 2·3025850929933 for the logarithm of the number 10. Thence, by addition the logarithms of the numbers 9 and 11 follow at once; so that the logarithms of all the primes 2, 3, 5, 11 are in readiness. In addition, merely by lowering the numbers in the above calculation by decimal places, and by addition, the logarithms of the decimals 0·98, 0·99, 1·01, 1·02 are obtained; as also of 0·998, 0·999, 1·001, 1·002. And then by addition and subtraction the logarithms of the primes 7, 13, 17, 37, etc., emerge. And these, combined with the above and divided by the logarithm of the number 10, become true logarithms for inserting in the table. But afterwards I have obtained them more closely.

I am ashamed to tell to how many places I carried these computations, having no other business at that time: for then I took really too much delight in these inventions. But when there appeared that ingenious work, the *Logarithmotechnia* of Nicolas Mercator (whom I suppose to have made his discoveries first), I began to pay less attention to these things, suspecting that either he knew the extraction of roots as well as division of fractions, or at least that others upon the discovery of division would find out the rest before I could reach a ripe age for writing. Yet at the very time when this book appeared, a certain compendium of the method of these series was communicated by Mr. Barrow (then professor of mathematics) to Mr. Collins; in which I had indicated the areas and lengths of all curves, and the surfaces and volumes of solids from given right lines, and that conversely from these as given the right lines could be determined; and the method there disclosed I had illustrated by various series. When afterwards a regular correspondence developed between us, Collins, a man born to promote the art of mathematics, did not cease to suggest that I should make these things public. And five years ago when, urged by my friends, I had planned to publish a treatise on the refraction of light and on colours, which I then had in readiness, I began again to think about these series and I compiled a treatise on them too, with a view to publishing both at the same time. But on the occasion of the Reflecting Telescope, when I had sent you a letter in which I briefly explained my ideas of the nature of light, something unexpected caused me to feel that it was my business to write to you in haste about the printing of that letter. Then frequent interruptions that immediately arose from the letters of various persons (full of objections and of other matters) quite deterred me from the design and caused me to accuse myself of imprudence, because, in hunting for a shadow hitherto, I had sacrificed my peace, a matter of real substance.

About that time, from just a single one of my series which Collins had sent him, Gregory, after much reflection, as he wrote back to Collins, arrived at the same method, and he left a treatise on it which we hope is going to be published by his friends. Indeed, with his strong understanding he could not fail to add many discoveries of his own, and it is in the interest of mathematics that these should not be lost. Moreover, I myself had not completely finished my treatise when I desisted from the proposal, nor has my mind to this day returned to the task of adding the rest. In fact there was wanting that part in which I had decided to explain the mode of solving problems which cannot be reduced to squarings; although I had done something to lay its foundations.

But in that treatise infinite series played no great part. Not a few other things I brought together, among them the method of drawing tangents which the very skillful Sluse communicated to you two or three years ago, about which you wrote back [to him] (on the suggestion of Collins) that the same method had been known to me also. We happened on it by different reasoning: for, as I work it, the matter needs no proof. Nobody, if he possessed my basis, could draw tangents any other way, unless he were deliberately wandering from the straight path. Indeed we do not here stick at equations in radicals involving one or each indefinite quantity, however complicated they may be; but without any reduction of such equations (which would generally render the work endless) the tangent is drawn directly. And the same is true in questions of maxima and minima, and in some others too, of which I am not now speaking. The foundation of these operations is evident enough, in fact; but because I cannot proceed with the explanation of it now, I have preferred to conceal it thus: *6accdæ 13eff7i3l9n 4o4qrr4s8t12vx.*

[*Portion of letter omitted.*]

THE ANAGRAM

This inverse problem of tangents, when the tangent between the point of contact and the axis of the figure is of given length, does not demand these methods. Yet it is that mechanical curve the determination of which depends on the area of an hyperbola. The problem is also of the same kind, when the part of the axis between the tangent and the ordinate is given in length. But I should scarcely have reckoned these cases among the sports of nature. For when in the right-angled triangle, which is formed by that part of the axis, the tangent and the ordinate, the relation of any two sides is defined by any equation, the problem can be solved apart from my general method. But when a part of the axis ending at some point given in position enters the bracket, then the question is apt to work out differently.

The communication of the solution of affected equations by the method of Leibniz will be very agreeable; so too an explanation how he comports himself when the indices of the powers are fractional, as in this equation.

$$20 + x^{3/7} - x^{6/5}y^{2/3} - y^{7/11} = 0,$$

or surds, as in

$$(x^{\sqrt{2}} + x^{\sqrt{7}})^{\sqrt[3]{2/3}} = y,$$

where $\sqrt{2}$ and $\sqrt{7}$ do not mean coefficients of x, but indices of powers or dignities of it, and $\sqrt[3]{2/3}$ means the power of the binomial $x^{\sqrt{2}} + x^{\sqrt{7}}$. The point, I think, is clear by my method, otherwise I should have described it. But a term must at last be set to this wordy letter. The letter of the most excellent Leibniz fully deserved of course that I should give it this more extended reply. And this time I wanted to write in greater detail because I did not believe that your more engaging pursuits should often be interrupted by me with this rather austere kind of writing.

1. *Turnbull's Note.* Oldenburg transmitted Newton's letter of June 13, 1676 to Leibniz, who responded in a letter (August 17, 1676), revealing his results in finding quadratures and hinting that he had a general method. Leibniz's letter interested Newton, who wrote a second letter to Oldenburg dated October 24, 1676. His second letter guardedly presented by means of an anagram a general method for finding quadratures and its inverse problem of tangents.

77. From *Principia Mathematica* (1687)*

(Prime and Ultimate Ratios: The Theory of Limits)

ISAAC NEWTON

BOOK ONE
THE MOTION OF BODIES
SECTION I

The method of first and last ratios of quantities, by the help of which we demonstrate the propositions that follow.

LEMMA I

Quantities, and the ratios of quantities, which in any finite time converge continually to equality, and before the end of that time approach nearer to each other than by any given difference, become ultimately equal.

If you deny it, suppose them to be ultimately unequal, and let D be their ultimate difference. Therefore they cannot approach nearer to equality than by that given difference D; which is contrary to the supposition.

LEMMA II

If in any figure AacE, *terminated by the right lines* Aa, AE, *and the curve* acE, *there be inscribed any number of parallelograms* Ab, Bc, Cd, *etc., comprehended under equal bases* AB, BC, CD, *etc., and the sides, Bb, Cc, Dd, etc., parallel to one side* Aa *of the figure; and the parallelograms* aKbl, bLcm, cMdn, *etc., are completed: then if the breadth of those parallelograms be supposed to be diminished, and their number to be augmented* in infinitum, *I say, that the ultimate ratios which the inscribed figure* AKbLcMdD, *the circumscribed figure* AalbmcndoE, *and curvilinear figure* AabcdE, *will have to one another, are ratios of equality.*

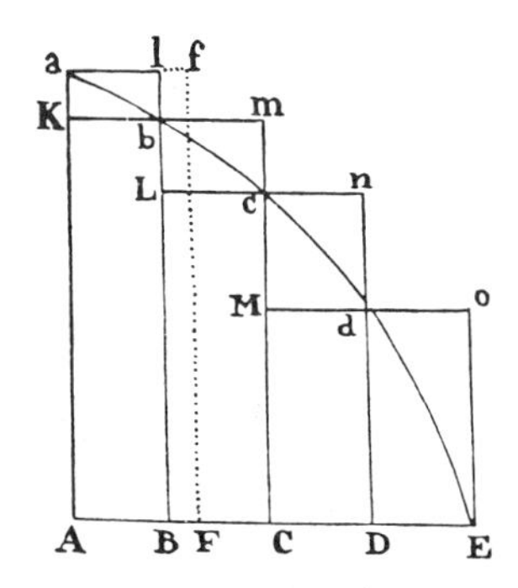

For the difference of the inscribed and circumscribed figures is the sum of the parallelograms K*l*, L*m*, M*n*, D*o*, that is (from the equality of all their bases), the rectangle under one of their bases K*b* and the sum of their altitudes A*a*, that is, the rectangle AB*la*. But this rectangle, because its breadth AB is supposed diminished *in infinitum,* becomes less than any given space. And therefore (by Lem. I) the figures inscribed and circumscribed become ultimately equal one to the other; and much more will the intermediate curvilinear figure be ultimately equal to either. Q.E.D.

Source: From *Sir Isaac Newton's Mathematical Principles of Natural Philosophy and His System of the World,* trans. by Andrew Motte (1729) and revised by Florian Cajori (1946), 29-39 and 249-251. This translation is reprinted by permission of the University of California Press.

LEMMA III

The same ultimate ratios are also ratios of equality, when the breadths AB, BC, DC, *etc., of the parallelograms are unequal, and are all diminished* in infinitum.

For suppose AF equal to the greatest breadth, and complete the parallelogram FA*af*. This parallelogram will be greater than the difference of the inscribed and circumscribed figures; but, because its breadth AF is diminished *in infinitum,* it will become less than any given rectangle. Q.E.D.

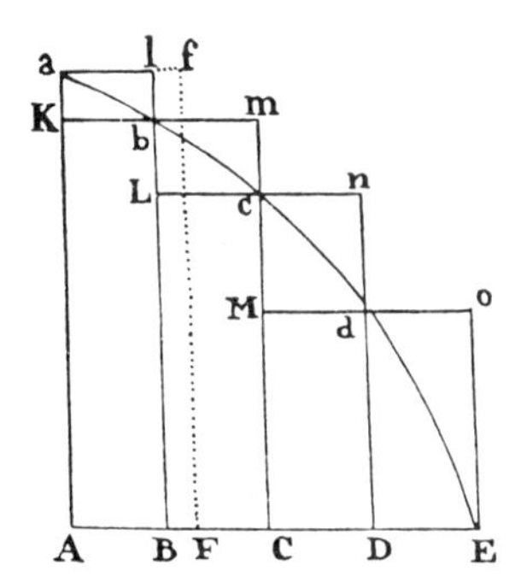

Cor. I. Hence the ultimate sum of those evanescent parallelograms will in all parts coincide with the curvilinear figure.

Cor. II. Much more will the rectilinear figure comprehended under the chords of the evanescent arcs *ab, bc, cd,* etc., ultimately coincide with the curvilinear figure.

Cor. III. And also the circumscribed rectilinear figure comprehended under the tangents of the same arcs.

Cor. IV. And therefore these ultimate figures (as to their perimeters *ac*E) are not rectilinear, but curvilinear limits of rectilinear figures.

LEMMA IV

If in two figures AacE, PprT, *there are inscribed (as before) two series of parallelograms, an equal number in each series, and, their breadths being diminished* in infinitum, *if the ultimate ratios of the parallelograms in one figure to those in the other, each to each respectively, are the same: I say, that those two figures,* AacE, PprT, *are to each other in that same ratio.*

For as the parallelograms in the one are severally to the parallelograms in the other, so (by composition) is the sum of all in the one to the sum of all in the other; and so is the one figure to the other; because (by Lem. III) the former figure to the former sum, and the latter figure to the latter sum, are both in the ratio of equality. Q.E.D.

Cor. Hence if two quantities of any kind are divided in any manner into an equal number of parts, and those parts, when their number is augmented, and their magnitude diminished *in infinitum,* have a given ratio to each other, the first to the first, the second to the second, and so on in order, all of them taken together will be to each other in that

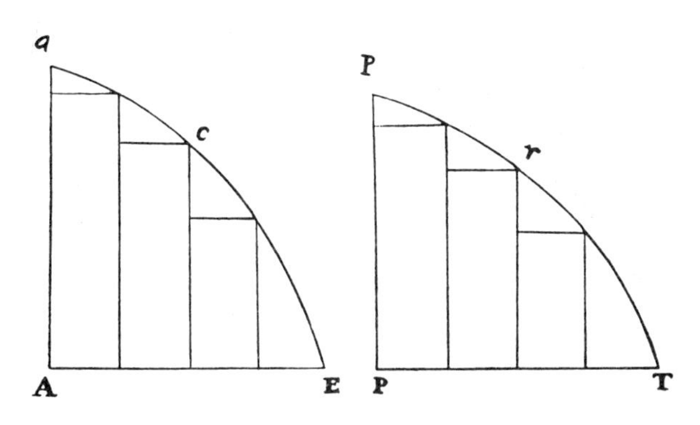

same given ratio. For if, in the figures of this Lemma, the parallelograms are taken to each other in the ratio of the parts, the sum of the parts will always be as the sum of the parallelograms; and therefore supposing the number of the parallelograms and parts to be augmented, and their magnitudes diminished *in infinitum,* those sums will be in the ultimate ratio of the parallelogram in the one figure to the correspondent parallelogram in the other; that is (by the supposition), in the ultimate ratio of any part of the one quantity to the correspondent part of the other.

LEMMA V

All homologous sides of similar figures, whether curvilinear or rectilinear, are proportional; and the areas are as the squares of the homologous sides.

LEMMA VI

If any arc ACB, *given in position, is subtended by its chord* AB, *and in any point* A, *in the middle of the continued curvature, is touched by a right line* AD, *produced both ways; then if the points* A *and* B *approach one another and meet, I say, the angle* BAD, *contained between the chord and the tangent, will be diminished* in infinitum, *and ultimately will vanish.*

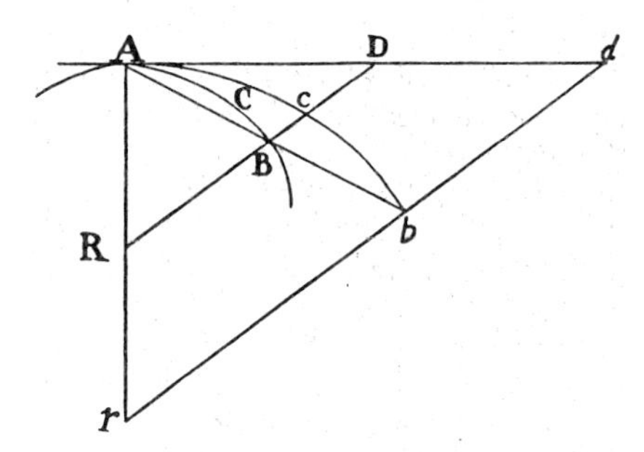

For if that angle does not vanish, the arc ACB will contain with the tangent AD an angle equal to a rectilinear angle; and therefore the curvature at the point A will not be continued, which is against the supposition.

LEMMA VII

The same things being supposed, I say that the ultimate ratio of the arc, chord, and tangent, any one to any other, is the ratio of equality.

For while the point B approaches towards the point A, consider always AB and AD as produced to the remote points *b* and *d;* and parallel to the secant BD draw *bd;* and let the arc A*cb* be always similar to the arc ACB. Then, supposing the points A and B to coincide, the angle *d*A*b* will vanish, by the preceding Lemma; and therefore the right lines A*b*, A*d* (which are always finite), and the intermediate arc A*cb,* will coincide, and become equal among themselves. Wherefore, the right lines AB, AD, and the intermediate arc ACB (which are always proportional to the former), will vanish, and ultimately acquire the ratio of equality. Q.E.D.

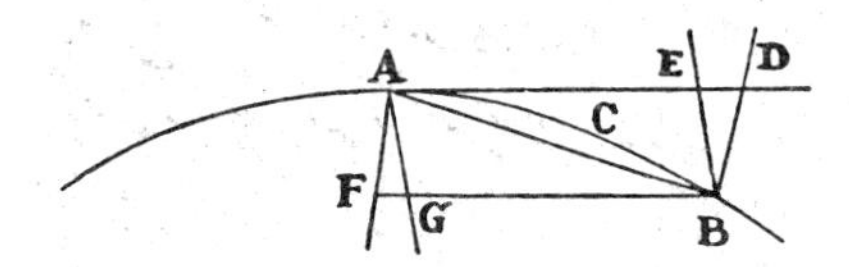

Cor. I. Whence if through B we draw BF parallel to the tangent, always cutting any right line AF passing through A in F, this line BF will be ultimately in the ratio of equality with the evanescent arc ACB; because, completing the parallelogram AFBD, it is always in a ratio of equality with AD.

Cor. II. And if through B and A more right lines are drawn, as BE, BD, AF, AG, cutting the tangent AD and its parallel BF; the ultimate ratio of all the abscissas AD, AE, BF, BG, and of the chord and arc AB, any one to any other, will be the ratio of equality.

Cor. III. And therefore in all our reasoning about ultimate ratios, we may freely use any one of those lines for any other.

LEMMA VIII

If the right lines AR, BR, *with the arc* ACB, *the chord* AB, *and the tangent* AD, *constitute three triangles* RAB, RACB, RAD, *and the points* A *and* B *approach and meet: I say, that the ultimate form of these evanescent triangles is that of similitude, and their ultimate ratio that of equality.*

[*Editor's Note:* This proof omitted.]

LEMMA IX

If a right line AE, *and a curved line* ABC, *both given by position, cut each other in a given angle,* A; *and to that right line, in another given angle,* BD, CE *are ordinately applied, meeting the curve in* B, C; *and the points* B *and* C *together approach towards and meet in the point* A: *I say, that the areas of the triangles* ABD, ACE, *will ultimately be to each other as the squares of homologous sides.*

For while the points B, C, approach towards the point A, suppose always AD to be produced to the remote points *d* and *e*, so as A*d*, A*e* may be proportional to AD, AE; and the ordinates *db*, *ec*, to be drawn parallel to the ordinates DB and EC, and meeting AB and AC produced in *b* and *c*. Let the curve A*bc* be similar to the curve ABC, and draw the right line A*g* so as to touch both curves in A, and cut the ordinates DB, EC, *db*, *ec*, in F, G, *f*, *g*. Then, supposing the length A*e* to remain the same, let the points B and C meet in the point A; and the angle *c*A*g* vanishing, the curvilinear areas A*bd*, A*ce* will coincide

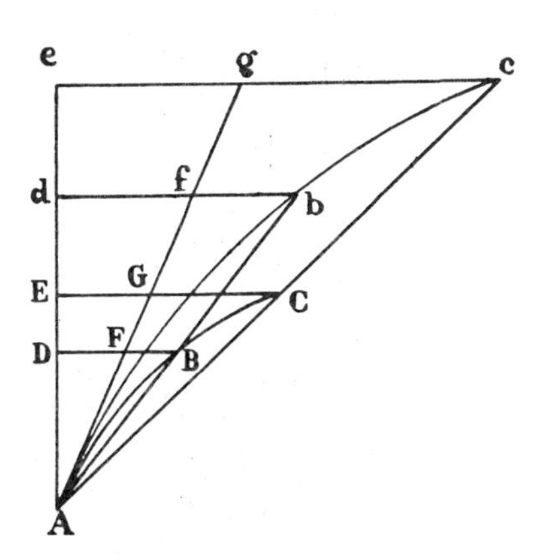

with the rectilinear areas A*fd*, A*ge*; and therefore (by Lem. V) will be one to the other in the duplicate ratio of the sides A*d*, A*e*. But the areas ABD, ACE are always proportional to these areas; and so the sides AD, AE are to these sides. And therefore the areas ABD, ACE are ultimately to each other as the squares of the sides AD, AE. Q.E.D.

LEMMA X

The spaces which a body describes by any finite force urging it, whether that force is determined and immutable, or is continually augmented or continually diminished, are in the very beginning of the motion to each other as the squares of the times.

[*Editor's Note.* The proof of Lemma X and its corollaries are omitted.]

Scholium. If in comparing with each other indeterminate quantities of different sorts, any one is said to be directly or inversely as any other, the meaning is, that the former is augmented or diminished in the same ratio as the latter, or as its reciprocal. And if any one is said to be as any other two or more, directly or inversely, the meaning is, that the first is augmented or diminished in the ratio compounded of the ratios in which the others, or the reciprocals of the others, are augmented or diminished. Thus, if A is said to be as B directly, and C directly, and D inversely, the meaning is, that A is augmented or diminished in the same ratio as $B \cdot C \cdot {}^{1}/_{D}$, that is to say, that A and ${}^{BC}/_{D}$ are to each other in a given ratio.

LEMMA XI

The evanescent subtense of the angle of contact, in all curves which at the point of contact have a finite curvature, is ultimately as the square of the subtense of the conterminous arc.[1]

Case I. Let AB be that arc, AD its tangent, BD the subtense of the angle of contact perpendicular on the tangent, AB the subtense of the arc. Draw BG perpendicular to the subtense AB, and AG perpendicular to the tangent AD,

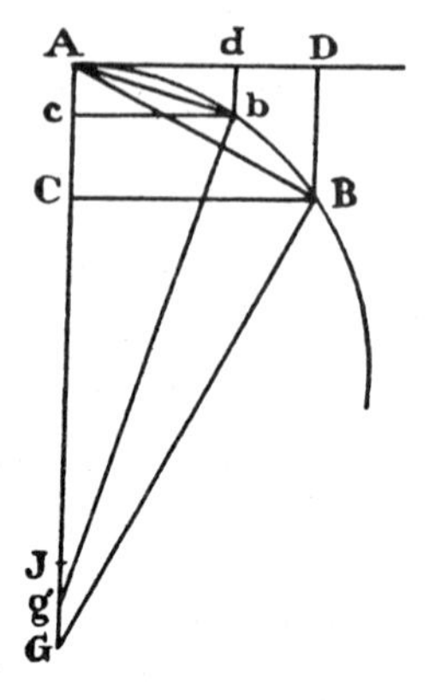

meeting in G; then let the points D, B, and G approach to the points *d, b,* and *g,* and suppose J to be the ultimate intersection of the lines BG, AG, when the points D, B have come to A. It is evident that the distance GJ may be less than any assignable distance. But (from the nature of the circles passing through the points A, B, G, and through A, *b, g*),

$$AB^2 = AG \cdot BD, \text{ and}$$
$$Ab^2 = Ag \cdot bd.$$

But because GJ may be assumed of less length than any assignable, the ratio of AG to Ag may be such as to differ from unity by less than any assignable difference; and therefore the ratio of AB^2 to Ab^2 may be such as to differ from the ratio of BD to *bd* by less than any assignable difference. Therefore, by Lem. I, ultimately,

$$AB^2 : Ab^2 = BD : bd. \qquad \text{Q.E.D.}$$

Case 2. Now let BD be inclined to AD in any given angle, and the ultimate ratio of BD to *bd* will always be the same as before, and therefore the same with the ratio of AB^2 to Ab^2. Q.E.D.

Case 3. And if we suppose the angle D not to be given, but that the right line BD converges to a given point, or is determined by any other condition whatever; nevertheless the angles D, *d,* being determined by the same law, will always draw nearer to equality, and approach nearer to each other than by any assigned difference, and therefore, by Lem. I, will at last be equal; and therefore the lines BD, *bd* are in the same ratio to each other as before. Q.E.D.

Cor. I. Therefore since the tangents AD, A*d,* the arcs AB, A*b,* and their sines, BC, *bc,* become ultimately equal to the chords AB, A*b,* their squares will ultimately become as the subtenses BD, *bd.*

Cor. II. Their squares are also ultimately as the versed sines of the arcs, bisecting the chords, and converging to a given point. For those versed sines are as the subtenses BD, *bd.*

Cor. III. And therefore the versed sine is as the square of the time in which a body will describe the arc with a given velocity.

Cor. IV. The ultimate proportion,

$$\Delta ADB : \Delta Adb = AD^3 : Ad^3 = DB^{3/2} : db^{3/2},$$

is derived from

$$\Delta ADB : \Delta Adb = AD \cdot DB : Ad \cdot db$$

and from the ultimate proportion

$$AD^2 : Ad^2 = DB : db.$$

So also is obtained ultimately

$$\Delta ABC : \Delta Abc = BC^3 : bc^3.$$

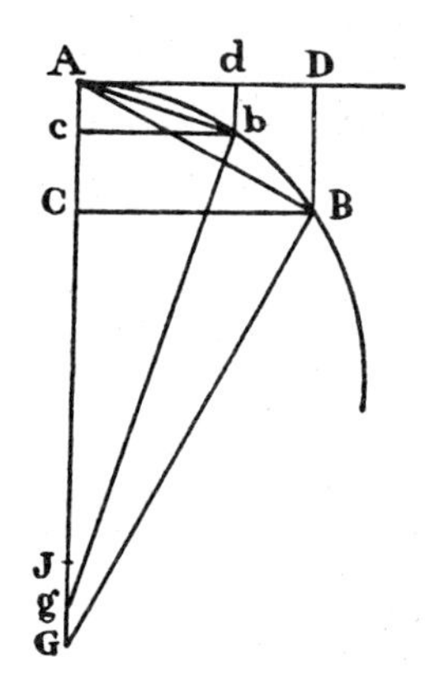

Cor. V. And because DB, *db* are ultimately parallel and as the squares of the lines AD, A*d,* the ultimate curvilinear areas ADB, A*db* will be (by the nature of the parabola) two-thirds of the rectilinear triangles ADB, A*db,* and the segments AB, A*b* will be one-third of the same triangles. And thence those areas and those segments will be as the squares of the tangents AD, A*d*, and also of the chords and arcs AB, AB.

Scholium. But we have all along supposed the angle of contact to be neither infinitely greater nor infinitely less than the angles of contact made by circles and their tangents; that is, that the curvature at the point A is neither infinitely small nor infinitely great, and that the interval AJ is of a finite magnitude.

[Text omitted here.]

These Lemmas are premised to avoid the tediousness of deducing involved demonstrations *ad absurdum,* according to the method of the ancient geometers. For demonstrations are shorter by the method of indivisibles; but because the hypothesis of indivisibles seems somewhat harsh, and therefore that method is reckoned less geometrical, I chose rather to reduce the demonstrations of the following Propositions to the first and last sums and ratios of nascent and evanescent quantities, that is, to the limits of those sums and ratios, and so to premise, as short as I could, the demonstrations of those limits. For hereby the same thing is performed as by the method of indivisibles; and now those principles being demonstrated, we may use them with greater safety. Therefore if hereafter I should happen to consider quantities as made up of particles, or should use little curved lines for right ones, I would not be understood to mean indivisibles, but evanescent divisible quantities; not the sums and ratios of determinate parts, but always the limits of sums and ratios; and that the force of such demonstrations always depends on the method laid down in the foregoing Lemmas.

Perhaps it may be objected, that there is no ultimate proportion of evanescent quantities; because the proportion, before the quantities have vanished, is not the ultimate, and when they are vanished, is none.[2] But by the same argument it may be alleged that a body arriving at a certain place, and there stopping, has no ultimate velocity; because the velocity, before the body comes to the place, is not its ultimate velocity; when it has arrived, there is none. But the answer is easy; for by the ultimate velocity is meant that with which the body is moved, neither before it arrives at its last place and the motion ceases, not after, but at the very instant it arrives; that is, that velocity with which the body arrives at its last place, and with which the motion ceases. And in like manner, by the ultimate ratio of evanescent quantities is to be understood the ratio of the quantities not before they vanish, nor afterwards, but with which they vanish. In like manner the first ratio of nascent quantities is that with which they begin to be. And the first or last sum is that with which they begin and cease to be (or to be augmented or diminished). There is a limit which the velocity at the end of the motion may attain, but not exceed. This is the ultimate velocity. And there is the like limit in all quantities and proportions that begin and cease to be. And since such limits are certain and definite, to determine the same is a problem strictly geometrical. But whatever is geometrical we may use in determining and demonstrating any other thing that is also geometrical.

It may also be objected, that if the ultimate ratios of evanescent quantities are given, their ultimate magnitudes will be also given: and so all quantities will consist of indivisibles, which is contrary to what Euclid has demonstrated concerning incommensurables, in the tenth Book of his *Elements.* But this objection is founded on a false supposition. For those ultimate ratios with which quantities vanish are not truly the ratios of ultimate quantities, but limits towards which the ratios of quantities decreasing without limit do always converge; and to which they approach nearer than by any given difference, but never go beyond, nor in effect attain to, till the quantities are diminished *in infinitum.* This thing will appear more evident in quantities infinitely great. If two quantities, whose difference is given, be augmented *in infinitum,* the ultimate ratio of these quantities will be given, namely, the ratio of equality; but it does not from thence follow, that the ultimate or greatest quantities themselves, whose ratio that is, will be given. Therefore if in what follows, for the sake of being more easily understood, I should happen to mention quantities as least, or evanescent, or ultimate, you are not to suppose that quantities of any determinate magnitude are meant, but such as are conceived to be always diminished without end.

BOOK TWO
MOMENTS

LEMMA II

The moment of any genitum *is equal to the moments of each of the generating sides multiplied by the indices of the powers of those sides, and by their coefficients continually.*

I call any quantity a *genitum* which is not made by addition or subtraction of divers parts, but is generated or produced in arithmetic by the multiplication, division, or extraction of the root of any terms whatsoever; in geometry by the finding of contents and sides, or of the extremes and means of proportionals. Quantities of this kind are products, quotients, roots, rectangles, squares, cubes, square and cubic sides, and the like. These quantities I here consider as variable and indetermined, and increasing or decreasing, as it were, by a continual motion or flux; and I understand their momentary increments or decrements by the name of moments; so that the increments may be esteemed as added or affirmative moments; and the decrements as subtracted or negative ones. But take care not to look upon finite particles as such. Finite particles are not moments, but the very quantities generated by the moments. We are to conceive them as the just nascent principles of finite magnitudes. Nor do we in this Lemma regard the magnitude of the moments, but their first proportion, as nascent. It will be the same thing, if, instead of moments, we use either the velocities of the increments and decrements (which may also be called the motions, mutations, and fluxions of quantities), or any finite quantities proportional to those velocities. The coefficient of any generating side is the quantity which arises by applying the genitum to that side.

Wherefore the sense of the Lemma is, that if the moments of any quantities A, B, C, etc., increasing or decreasing by a continual flux, or the velocities of the mutations which are proportional to them, be called *a, b, c,* etc., the moment or mutation of the generated rectangle AB will be $aB+bA$; the moment of the generated content ABC will be $aBC+bAC+cAB$; and the moments of the generated powers A^2, A^3, A^4, $A^{1/2}$, $A^{3/2}$, $A^{1/3}$, $A^{2/3}$, A^{-1}, A^{-2}, $A^{-1/2}$ will be $2aA$, $3aA^2$, $4aA^3$, $1/2aA^{-1/2}$, $3/2aA^{1/2}$, $1/3aA^{-2/3}$, $2/3aA^{-1/3}$, $-aA^{-2}$, $-2aA^{-3}$, $-1/2aA^{-3/2}$ respectively; and, in general, that the moment of any power $A^{n/m}$ will be $\frac{n}{m}aA^{\frac{n-m}{m}}$. Also, that the moment of the generated quantity A^2B will be $2aAB + bA^2$; the moment of the generated quantity $A^3B^4C^2$ will be $3aA^2B^4C^2 + 4bA^3B^3C^2 + 2cA^3B^4C$; and the moment of the generated quantity $\frac{A^3}{B^2}$ or A^3B^{-2} will be $3aA^2B^{-2} - 2bA^3B^{-3}$; and so on. The Lemma is thus demonstrated.

Case 1. Any rectangle, as AB, augmented by a continual flux, when, as yet, there wanted of the sides A and B half their moments $\frac{1}{2}a$ and $\frac{1}{2}b$, was $A-\frac{1}{2}a$ into $B-\frac{1}{2}b$, or $AB-\frac{1}{2}a\ B-\frac{1}{2}b\ A+\frac{1}{4}ab$; but as soon as the sides A and B are augmented by the other half-moments, the rectangle becomes $A+\frac{1}{2}a$ into $B+\frac{1}{2}b$, or $AB+\frac{1}{2}a\ B+\frac{1}{2}b\ A+\frac{1}{4}ab$. From this rectangle subtract the former rectangle, and there will remain the excess $aB+bA$. Therefore with the whole increments *a* and *b* of the sides, the increment $aB+bA$ of the rectangle is generated. Q.E.D.

Case 2. Suppose AB always equal to G, and then the moment of the content ABC or GC (by Case I) will be $gC+cG$, that is (putting AB and $aB+bA$ for G and g), $aBC+bAC+cAB$. And the reasoning is the same for contents under ever so many sides. Q.E.D.

Case 3. Suppose the sides A, B, and C, to be always equal among themselves; and the moment $aB+bA$, of A^2, that is, of the rectangle AB, will be $2aA$; and the moment $aBC+bAC+cAB$ of A^3, that is, of the content ABC, will be $3aA^2$. And by the same reasoning the moment of any power A^n is naA^{n-1}. Q.E.D.

Case 4. Therefore since $\frac{1}{A}$ into A is I, the moment of $\frac{1}{A}$ multiplied by A, together with $\frac{1}{A}$ multiplied by a, will be the moment of I, that is, nothing. Therefore the moment of $\frac{1}{A}$, or of A^{-1}, is $\frac{-a}{A^2}$. And generally since $\frac{1}{A^n}$ into A^n is I, the moment of $\frac{1}{A^n}$ multiplied by A^n together with $\frac{1}{A^n}$ into naA^{n-1} will be nothing. And, therefore, the moment of $\frac{1}{A^n}$ or A^{-n} will be $-\frac{na}{A^{n+1}}$. Q.E.D.

Case 5. And since $A^{1/2}$ into $A^{1/2}$ is A, the moment of $A^{1/2}$ multiplied by $2A^{1/2}$ will be a (by Case 3); and, therefore, the moment of $A^{1/2}$ will be $\frac{a}{2A^{1/2}}$ or $\frac{1}{2}aA^{-1/2}$. And generally, putting $A^{m/n}$ equal to B, then A^m will be equal to B^n, and therefore maA^{m-1} equal to nbB^{n-1}, and maA^{-1} equal to nbB^{-1}, or $nbA^{-m/n}$; and therefore $\frac{m}{n}aA^{\frac{n-m}{n}}$ is equal to b, that is, equal to the moment of $A^{m/n}$. Q.E.D.

Case 6. Therefore the moment of any generated quantity A^mB^n is the moment of A^m multiplied by B^n, together with the moment of B^n multiplied by A^m, that is, $maA^{m-1}B^n + nbB^{n-1}A^m$; and that whether the indices m and n of the powers be whole numbers or fractions, affirmative or negative. And the reasoning is the same for higher powers. Q.E.D.

Cor. I. Hence in quantities continually proportional, if one term is given, the moments of the rest of the terms will be as the same terms multiplied by the number of intervals between them and the given term. Let A, B, C, D, E, F be continually proportional; then if the term C is given, the moments of the rest of the terms will be among themselves as $-2A$, $-B$, D, 2E, 3F.

Cor. II. And if in four proportionals the two means are given, the moments of the extremes will be as those extremes. The same is to be understood of the sides of any given rectangle.

Cor. III. And if the sum or difference of two squares is given, the moments of the sides will be inversely as the sides.

NOTES

1. If AD = x and BD = y, then the equation of the curve near A may be written as $y = \alpha x^2 + \beta x^3 + \ldots$.

2. Newton is wrestling with the concept of "ultimate ratio of evanescent quantities" and Zeno's paradoxes. Berkeley later wrote about this type of argument in the *Analyst*.

78. From the Introduction to the *Tractatus de quadratura curvarum* (1704)*

ISAAC NEWTON

1. I consider mathematical quantities in this place not as consisting of very small parts; but as described by a continued motion. Lines are described, and thereby generated not by the apposition of parts, but by the continued motion of points; superficies by the motion of lines; solids by the motion of superficies; angles by the rotation of the sides; portions of time by a continual flux: and so in other quantities. These geneses really take place in the nature of things, and are daily seen in the motion of bodies. And after this manner the ancients, by drawing moveable right lines along immoveable right lines, taught the genesis of rectangles.

2. Therefore considering that quantities, which increase in equal times, and by increasing are generated, become greater or less according to the greater or less velocity with which they increase and are generated; I sought a method of determining quantities from the velocities of the motions or increments, with which they are generated; and calling these velocities of the motions or increments *fluxions,* and the generated quantities *fluents.* I fell by degrees upon the method of fluxions, which I have made use of here in the quadrature of curves, in the years 1665 and 1666.

3. Fluxions are very nearly as the augments of the fluents generated in equal but very small particles of time, and, to speak accurately, they are in the *first ratio* of the nascent augments; but they may be expounded by any lines which are proportional to them.

4. Thus if the areas *ABC, ABDG* [Fig. 1] be described by the ordinates *BC, BD* moving along the base *AB* with an uniform motion, the fluxions of these areas shall be to one another as the describing ordinates *BC* and *BD,* and may be expounded by these ordinates, because that these ordinates are as the nascent augments of the areas.

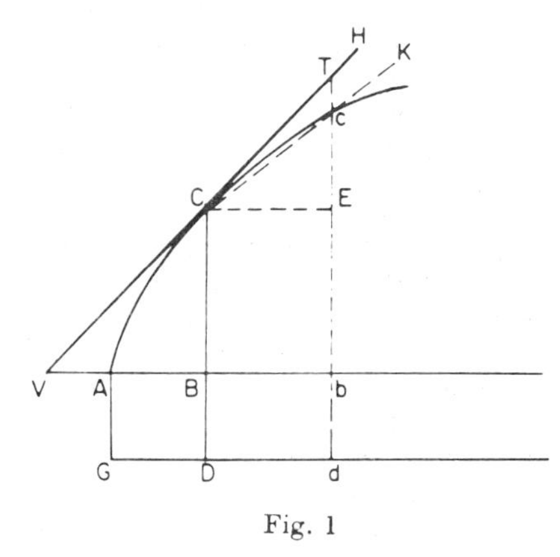

Fig. 1

5. Let the ordinate *BC* advance from its place into any new place *bc*. Complete the parallelogram *BCEb,* and draw the right line *VTH* touching the curve in *C,* and meeting the two lines *bc* and *BA*

*Source: This translation by J. Stewart (1745) appears in D. J. Struik (ed.), *A Source Book in Mathematics, 1200-1800* (1969), 303-308. It is reprinted by permission of Harvard University Press, Copyright © 1969 by the President and Fellows of Harvard College. In this introduction Newton presents the characteristic triangle in number 5 and defines a tangent as the limit of a secant.

produced in T and V: and Bb, Ec, and Cc will be the augments now generated of the abscissa AB, the ordinate BC and the curve line ACc; and the sides of the triangle CET are in the *first ratio* of these augments considered as nascent, therefore the fluxions of AB, BC, and AC are as the sides CE, ET, and CT of that triangle CET, and may be expounded by these same sides, or, which is the same thing, by the sides of the triangle VBC, which is similar to the triangle CET.

6. It comes to the same purpose to take the fluxions in the *ultimate ratio* of the evanescent parts. Draw the right line Cc, and produce it to K. Let the ordinate bc return into its former place BC, and when the points C and c coalesce, the right line CK will coincide with the tangent CH, and the evanescent triangle CEc in its ultimate form will become similar to the triangle CET, and its evanescent sides CE, Ec, and Cc will be *ultimately* among themselves as the sides CE, ET, and CT of the other triangle CET are, and therefore the fluxions of the lines AB, BC, and AC are in this same ratio. If the points C and c are distant from one another by any small distance, the right line CK will likewise be distant from the tangent CH by a small distance. That the right line CK may coincide with the tangent CH, and the ultimate ratios of the lines CE, Ec, and Cc may be found, the points C and c ought to coalesce and exactly coincide. The very smallest errors in mathematical matters are not to be neglected.

7. By the like way of reasoning, if a circle described with the center B and radius BC be drawn at right angles along the absciss AB, with an uniform motion, the fluxion of the generated solid ABC will be as that generating circle, and the fluxion of its superficies will be as the perimeter of that circle and the fluxion of the curve line AC jointly. For in whatever time the solid ABC is generated by drawing that circle along the length of the absciss, in the same time its superficies is generated by drawing the perimeter of that circle along the length of the curve AC. You may likewise take the following examples of this method.

8. *Let the right line PB [Fig. 2], revolving about the given pole P, cut another right line AB given in position: it is required to find the proportion of the fluxions of these right lines AB and PB.*

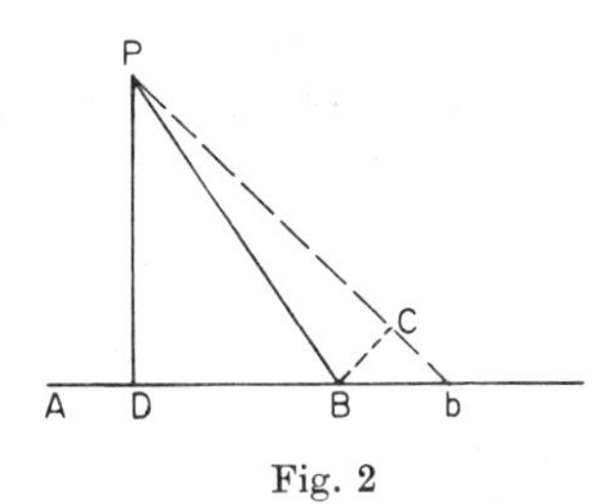

Fig. 2

Let the line PB move forward from its place PB into the new place Pb. In Pb take PC equal to PB, and draw PD to AB in such manner that the angle bPD may be equal to the angle bBC; and because the triangles bBC, bPD are similar, the augment Bb will be to the augment Cb as Pb to Db. Now let Pb return into its former place PB, that these augments may evanish, then the ultimate ratio of these evanescent augments, that is the ultimate ratio of Pb to Db, shall be the same with that of PB to DB, PDB being then a right angle, and therefore the fluxion of AB is to the fluxion of PB in that same ratio.

9. *Let the right line PB, revolving about the given pole P, cut other two right lines given in position,* viz. *AB and AE in B and E: the proportion of the fluxions of these right lines AB and AE is sought.*

Let the revolving right line PB [Fig. 3] move forward from its place PB into the new place Pb, so as to cut the lines AB, AE in the points b and e: and draw BC parallel to AE meeting Pb in C, and it will be $Bb:BC::Ab:Ae$, and $BC:Ee::PB:PE$, and by joining the ratios, $Bb:Ee::Ab \times PB:Ae \times PE$. Now let Pb return into its former place PB, and the

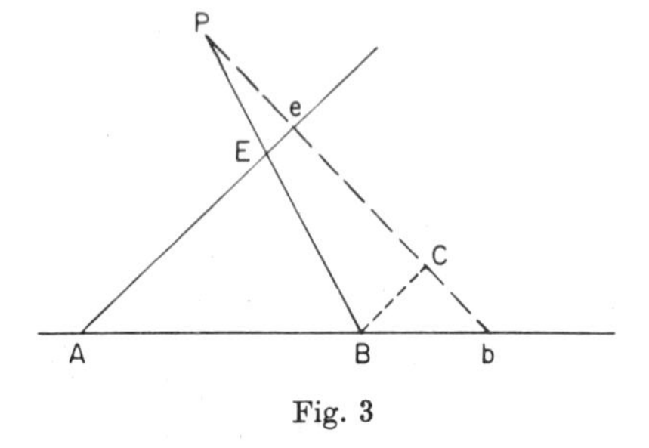

Fig. 3

evanescent augment Bb will be to the evanescent augment Ee as $AB \times PB$ to $AE \times PE$; and therefore the fluxion of the right line AB is to the fluxion of the right line AE in the same ratio.

10. Hence if the revolving right line PB cut any curve lines given in position in the points B and E, and the right lines AB, AE now becoming moveable, touch these curves in the points of section B and E: the fluxion of the curve, which the right line AB touches, shall be to the fluxion of the curve, which the right line AE touches, as $AB \times PB$ to $AE \times PE$. The same thing would happen if the right line PB perpetually touched any curve given in position in the moveable point P.

11. *Let the quantity x flow uniformly, and let it be proposed to find the fluxion of* x^n.

In the same time that the quantity x, by flowing, becomes $x + o$, the quantity x^n will become $(x + o)^n$, that is, by the method of infinite series, $x^n + nox^{n-1} + \frac{n^2-n}{2} oox^{n-2} +$ etc. And the augments o and $nox^{n-1} + \frac{n^2-n}{2} oox^{n-2} +$ etc. are to one another as 1 and $nx^{n-1} + \frac{n^2-n}{2} ox^{n-2} +$ etc.

Now let these augments vanish, and their ultimate ratio will be 1 to nx^{n-1}.

12. By like ways of reasoning, the fluxions of lines, whether right or curve in all cases, as likewise the fluxions of superficies, angles, and other quantities, may be collected by the method of *prime* and *ultimate* ratios. Now to institute an analysis after this manner in finite quantities and investigate the *prime* or *ultimate* ratios of these finite quantities when in their nascent or evanescent state, is consonant to the geometry of the ancients: and I was willing to show that, in the method of fluxions, there is no necessity of introducing figures infinitely small into geometry. Yet the analysis may be performed in any kind of figures, whether finite or infinitely small, which are imagined similar to the evanescent figures; as likewise in these figures, which, by the method of indivisibles, use to be reckoned as infinitely small, provided you proceed with due caution.

From the fluxions to find the fluents, is a much more difficult problem, and the first step of the solution is equivalent to the quadrature of curves; concerning which I wrote what follows some considerable time ago.

13. In what follows I consider indeterminate quantities as increasing or decreasing by a continued motion, that is, as flowing forwards, or backwards, and I design them by the letters z, y, x, v, and their fluxions or celerities of increasing I denote by the same letters pointed $\dot{z}$, $\dot{y}$, $\dot{x}$, $\dot{v}$. There are likewise fluxions or mutations more or less swift of these fluxions, which may be called the second fluxions of the same quantities z, y, x, v, and may be thus designed $\ddot{z}$, $\ddot{y}$, $\ddot{x}$, $\ddot{v}$: and the first fluxions of these last, or the third fluxions of z, y, x, v, are thus denoted $\overset{\therefore}{z}$, $\overset{\therefore}{y}$, $\overset{\therefore}{x}$, $\overset{\therefore}{v}$: and the fourth fluxions thus $\overset{::}{z}$, $\overset{::}{y}$, $\overset{::}{x}$, $\overset{::}{v}$. And after the same manner that $\overset{\therefore}{z}$, $\overset{\therefore}{y}$, $\overset{\therefore}{x}$, $\overset{\therefore}{v}$ are the fluxions of the quantities $\ddot{z}$, $\ddot{y}$, $\ddot{x}$, $\ddot{v}$, and these the fluxions of the quantities $\dot{z}$, $\dot{y}$, $\dot{x}$, $\dot{v}$; and these last fluxions of the quantities z, y, x, v: so the quantities z, y, x, v may be considered as the fluxions of others, which I shall design thus $\overset{\prime}{z}$, $\overset{\prime}{y}$, $\overset{\prime}{x}$, $\overset{\prime}{v}$; and these as the fluxions of others $\overset{\prime\prime}{z}$, $\overset{\prime\prime}{y}$, $\overset{\prime\prime}{x}$, $\overset{\prime\prime}{v}$; and these last still as the fluxions of others z, $\overset{\prime\prime\prime}{y}$, $\overset{\prime\prime\prime}{x}$, $\overset{\prime\prime\prime}{v}$. Therefore $\overset{\prime\prime}{z}$, $\overset{\prime}{z}$, z, $\dot{z}$, $\ddot{z}$, $\overset{\therefore}{z}$, $\overset{::}{z}$, $\overset{\cdot\cdot}{\overset{\therefore}{z}}$, etc. design a series of quan-

tities whereof every one that follows is the fluxion of the one immediately preceding, and every one that goes before, is a flowing quantity having that which immediately succeeds, for its fluxion. The like is the series $\sqrt{az \overset{\prime\prime}{-} zz}$, $\sqrt{az \overset{\prime}{-} zz}$, $\sqrt{az - zz}$, $\sqrt{az \overset{\cdot}{-} zz}$, $\sqrt{az \overset{\cdot\cdot}{-} zz}$, $\sqrt{az \overset{\cdot\cdot\cdot}{-} zz}$; as likewise the series $\frac{az+zz}{a \overset{\prime\prime}{-} z}$, $\frac{az+zz}{a \overset{\prime}{-} z}$, $\frac{az+zz}{a-z}$, $\frac{az+zz}{a \overset{\cdot}{-} z}$, $\frac{az+zz}{a \overset{\cdot\cdot}{-} z}$, $\frac{az+zz}{a \overset{\cdot\cdot\cdot}{-} z}$, etc.

14. And it is to be remarked that any preceding quantity in these series is as the area of curvilinear figure of which the succeeding is the rectangular ordinate, and the absciss is z: as $\sqrt{ax \overset{\prime}{-} zz}$ the area of a curve, whose ordinate is $\sqrt{az-zz}$, and absciss z. The design of all these things will appear in the following propositions.

PROPOSITION I
PROBLEM I

15. *An equation being given involving any number of flowing quantities, to find the fluxions.*[1]

Solution. Let every term of the equation be multiplied by the index of the power[2] of every flowing quantity that it involves, and in every multiplication change the side or root of the power into its fluxion, and the aggregate of all the products with their proper signs, will be the new equation.

16. *Explication.* Let a, b, c, d, etc. be determinate and invariable quantities, and let any equation be proposed involving the flowing quantities z, y, x, etc. as $x^3 - xy^2 + a^2z - b^3 = 0$. Let the terms be first multiplied by the indexes of the powers of x, and in every multiplication for the root, or x of one dimension write $\dot{x}$, and the sum of the factors will be $3\dot{x}x^2 - \dot{x}y^2$. Do the same in y, and there arises $-2xy\dot{y}$. Do the same in z, and there arises $aa\dot{z}$. Let the sum of these products be put equal to nothing, and you'll have the equation $3\dot{x}x^2 - \dot{x}y^2 - 2xy\dot{y} + aa\dot{z} = 0$. I say the relation of the fluxions is defined by this equation.

17. *Demonstration.* For let o be a very small quantity, and let $o\dot{z}$, $o\dot{y}$, $o\dot{x}$ be the moments, that is the momentaneous synchronal increments of the quantities z, y, x. And if the flowing quantities are just now z, y, x, then after a moment of time, being increased by their increments $o\dot{z}$, $o\dot{y}$, $o\dot{x}$, these quantities shall become $z + o\dot{z}$, $y + o\dot{y}$, $x + o\dot{x}$: which being wrote in the first equation for z, y, and x, give this equation $x^3 + 3x^2o\dot{x} + 3xoo\dot{x}\dot{x} + o^3\dot{x}^3 - xy^2 - o\dot{x}y^2 - 2xo\dot{y}y - 2\dot{x}o^2\dot{y}y - xo^2\dot{y}\dot{y} - \dot{x}o^3\dot{y}\dot{y} + a^2z + a^2o\dot{z} - b^3 = 0$.

Subtract the former equation from the latter, divide the remaining equation by o, and it will be $3\dot{x}x^2 + 3\dot{x}\dot{x}ox + \dot{x}^3o^2 - \dot{x}y^2 - 2x\dot{y}y - 2\dot{x}o\dot{y}y - xo\dot{y}\dot{y} - \dot{x}o^2\dot{y}\dot{y} + a^2\dot{z} = 0$. Let the quantity o be diminished infinitely, and neglecting the terms which vanish, there will remain $3\dot{x}x^2 - \dot{x}y^2 - 2x\dot{y}y + a^2\dot{z} = 0$. Q.E.D.

18. *A fuller explication.* After the same manner if the equation were $x^3 - xy^2 + aa\sqrt{ax-y^2} - b^3 = 0$, thence would be produced $3x^2\dot{x} - \dot{x}y^2 - 2xy\dot{y} + aa\overset{\cdot}{\sqrt{ax-y^2}} = 0$. Where if you would take away the fluxion $\overset{\cdot}{\sqrt{ax-y^2}}$, put $\sqrt{ax-y^2} = z$, and it will be $ax - y^2 = z^2$, and by this proposition $a\dot{x} - 2\dot{y}y = 2\dot{z}z$, or $\frac{a\dot{x}-2\dot{y}y}{2z} = \dot{z}$, that is $\frac{a\dot{x}-2\dot{y}y}{2\sqrt{ax-yy}} = \overset{\cdot}{\sqrt{ax-yy}}$. And thence $3x^2\dot{x} - \dot{x}y^2 - 2xy\dot{y} + \frac{a^3\dot{x}-2a^2\dot{y}y}{2\sqrt{ax-yy}} = 0$.

19. And by repeating the operation, you proceed to second, third, and subsequent fluxions. Let $zy^3 - z^4 + a^4 = 0$ be an equation proposed, and by the first operation it becomes $\dot{z}y^3 + 3z\dot{y}y^2 - 4\dot{z}z^3 = 0$; by the second $\ddot{z}y^3 + 6\dot{z}\dot{y}y^2 + 3z\ddot{y}y^2 + 6z\dot{y}^2y - 4\ddot{z}z^3 - 12\dot{z}^2z^2 = 0$, by the third, $\dddot{z}y^3 + 9\ddot{z}\dot{y}y^2 + 18\dot{z}\dot{y}^2y + 3z\dddot{y}y^2 + 18z\ddot{y}\dot{y}y + 6z\dot{y}^3 - 4\dddot{z}z^3 - 36\ddot{z}\dot{z}z^2 - 24\dot{z}^3z = 0$.

20. But when one proceeds thus to second, third, and following fluxions, it is proper to consider some quantity as flowing uniformly, and for its first fluxion to write unity, for the second and subsequent ones, nothing. Let there be given the equation $zy^3 - z^4 + a^4 = 0$,

as above; and let z flow uniformly, and let its fluxion be unity: then by the first operation it shall be $y^3 + 3z\dot{y}y^2 - 4z^3 = 0$; by the second $6\dot{y}y^2 + 3z\ddot{y}y^2 + 6z\dot{y}^2y - 12z^2 = 0$; by the third $9\ddot{y}y^2 + 18\dot{y}^2y + 3z\dddot{y}y^2 + 18z\ddot{y}\dot{y}y + 6z\dot{y}^3 - 24z = 0$.

But in equations of this kind it must be conceived that the fluxions in all the terms are of the same order, i.e., either all of the first order $\dot{y}$, $\dot{z}$; or all of the second $\ddot{y}$, $\dot{y}^2$, $\dot{y}\dot{z}$, $\dot{z}^2$; or all of the third $\dddot{y}$, $\ddot{y}\dot{y}$, $\ddot{y}\dot{z}$, $\dot{y}^3$, $\dot{y}^2\dot{z}$, $\dot{y}\dot{z}^2$, $\dot{z}^3$, etc. And where the case is otherwise the order is to be completed by means of the fluxions of a quantity that flows uniformly, which fluxions are understood. Thus the last equation, by completing the third order, becomes $9\dot{z}\ddot{y}y^2 + 18\dot{z}\dot{y}^2y + 3z\dddot{y}y^2 + 18z\ddot{y}\dot{y}y + 6z\dot{y}^3 - 24z\dot{z}^3 = 0$.[3]

NOTES

1. Newton prefers to differentiate equations, but later also differentiates functions, often given as areas.

2. [*Footnote by the translator, Stewart*] The word translated here *power* is *dignitas*, dignity, by which must be understood not only perfect, but also imperfect powers or surd roots, which are expressed in the manner of perfect powers, as is well known, by fractional indexes. In which sense $x^{1/2}$, $x^{2/3}$, etc. are powers; ½ and ⅔ their indexes, and x the side or root. I use the word power, because dignity is seldom used in English in this sense.

3. Newton insists on homogeneity, which requires that each term of the equation has the same number of "pricks."

Chapter VI
The Scientific Revolution at Its Zenith (1620–1720)

Section D
The Bernoullis

JAKOB BERNOULLI (1654–1705)

The Bernoulli family is the most famous in the history of mathematics. From the late 17th century to the present time it has contributed distinguished, and sometimes eminent, men of learning. The reputation of the Bernoulli's began with the careers of the brothers Jakob and Johann.

Jakob Bernoulli came from a thriving mercantile family in Basel, Switzerland. His father Nikolaus was a druggist and town magistrate; his mother Margaretha Schönauer was the daughter of a banker. They were a Protestant family whose ancestors had fled Antwerp in 1583 to escape the Catholic persecution of the Huguenots. Following the wishes of his father who wanted him to become a Protestant pastor, Jakob received a master of arts degree in philosophy from the University of Basel in 1671 and a licentiate in theology in 1676. However, he had other interests. As he stated in his motto *Invito patre sidera verso* ("against my father's will I study the stars"), he investigated astronomy and mathematics on his own.

The father's efforts to make Jakob Bernoulli a cleric were futile; a career in higher education was his goal upon graduation from the university. He began as a tutor in Geneva in late 1676 and then spent the next two years in France, familiarizing himself with the newly dominant Cartesian science, including the work of Nicolas Malebranche. Seeking more first-hand information on recent advances in the sciences, he took a second educational trip in 1681–1682. At this time he met the mathematician Jan Hudde in the Netherlands as well as the natural philosophers Robert Boyle and Robert Hooke in England. The main results of Bernoulli's early research were a theory of comets that later proved inadequate and a theory of gravity that his contemporaries regarded highly.

From 1683 on Jakob Bernoulli taught at the University of Basel and devoted his time to research in mathematics, astronomy, and mechanics. His careful study of the second edition of Franz Schooten's Latin translation of Descartes' *Géométrie* (1659–1661), John Wallis' *Arithmetica Infinitorum* ("The Arithmetic of Infinitesimals," 1656), and Isaac Barrow's *Lectiones Geometricae* ("Geometrical Lectures," 1664–1670) led him to the problem of infinitesimal geometry. In 1687 he was named professor of mathematics at the University of Basel. Already in 1683 his younger brother Johann had come to live with him and pursue university studies. Thereafter, the careers of the two men were closely linked, not always with happy results. To keep abreast of recent developments in the sciences, they read the *Journal des Sçavans* and *Acta Eruditorum*. In the late 1680s, they concentrated on articles by Leibniz and Walther von Tschirnhaus on the calculus and its application to mechanics. To gain more information they corresponded with Leibniz and the Dutch physicist Christiaan Huygens. By 1689, after much effort, Jakob and Johann mastered Leibniz's calculus, although they mistakenly believed it to be simply a computational formalism of Barrow's ideas. Jakob then began to study probability theory and to stress induction in the sciences.

After 1690, the two brothers continued their research even though an antagonism was growing between them. Both were self-willed, aggressive, and irritable men who had an exaggerated need for recognition. Johann, who had greater intuitive power and descriptive ability, was the more acute in working on mathematical formulations. Jakob possessed a deeper intellect but took longer to reach his results. Jakob came to resent being eclipsed by his gifted younger brother. Johann's boast in 1691 that he had first solved the velaria problem, which involves the shape of a sail under the pressure of wind, annoyed Jakob. In his research of the early 1690s, Jakob studied the logarithmic spiral, determined evolutes (envelopes of the normals to a curve), and introduced polar coordinates. He also solved problems in the differential calculus, examined elasticity, and became convinced that continuity exists in the processes of nature. He agreed with the Latin proverb *natura non fecit saltum* ("nature does not leap"). In response to Leibniz's proposal of finding the curve of constant descent in a gravitational field, Jakob presented an analysis that first employed the term "integral" in the modern mathematical sense (1690). He also posed as a counterproblem finding the shape of catenaries—the curves formed by cords suspended between two end points. His exhaustive studies of these later found applications to suspension bridges and high voltage transmission lines.

Work on the birth of the calculus of variations led to a complete break between Jakob and Johann Bernoulli. In 1696, Johann proposed determining the "curve of quickest descent" between two points—the brachistochrone problem—and set a six-month limit for its solution. The two Bernoullis, Leibniz, and Newton found the solution, a cycloid. Although Johann ingeniously reduced the mechanical problem to an optical one and resolved it by using Fermat's principle of least time, he failed to recognize that variational problems involve finding a function to make an integral an extreme. Jakob saw this and posed for solution the so-called isoperimetric problem. Johann's solution based on a differential equation of second degree was incorrect. Jakob believed in 1697 that a differential equation of third degree was required, something he proved in 1701. The two made ugly critical remarks about each other. Johann, who had resided in Holland since 1695, refused to return to Switzerland as long as his brother lived, and he kept his word. Leibniz tried to mediate between the two men, but Jakob believed that he favored Johann.

In his final years, Jakob Bernoulli completed the five dissertations in his *Theory of Series* (1682–1704) and worked on, but never completed, his most original book, the *Ars Conjectandi* ("The Art of Conjecturing," publ. posth. 1713). His long-held interest in mysticism increased as he grew older. The recurrence of a similar spiral in the logarithmic spiral—$\log r = a\theta$—fascinated him. He directed that the spiral be inscribed on his tombstone with the words *Eadem mutata resurgo* ("Though changed I shall rise the same").

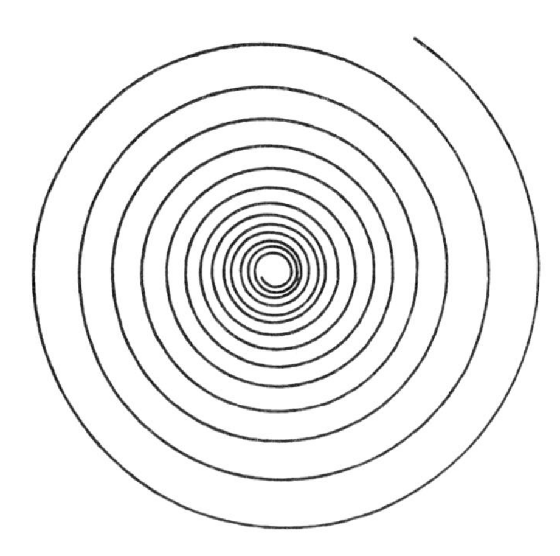

Eadem mutata resurgo

Jakob Bernoulli substantially advanced the calculus in its early stage, the theory of probability, and mechanics. He was probably the first student of the calculus to focus on the question of divergence and convergence of infinite series. In the *Theory of Series,* he compared series according to how rapidly they diverged. He knew that the harmonic series diverges (Proposition 16). His attempts to sum $\sum_{k=1}^{\infty} k^{-2}$ were unsuccessful. This series was first summed in closed form by Euler and subsequently in a more general form by Riemann. The *Ars Conjectandi* was a significant book on probability. Even before the time of Cardano, the gambling of the nobility and the attendant problem of the division of stakes had proved a powerful impetus to such study. In his research, Bernoulli had carefully examined the *Aleae Ludo* (1657), a tract on games of chance by Christiaan Huygens. The *Ars Conjectandi* extended the theory of combinations and permutations, developed the so-called Bernoulli numbers with which its author derived the exponential series (x^y), and used the formula for combinations to prove the binomial theorem for the case of n a positive integral. Its treatment of mathematical and moral probability included the law of large numbers. The results presented in *Ars Conjectandi* are basic to modern sampling theory.

The law of large numbers was an important finding. It allowed mathematicians to calculate the probability of certain occurrences happening. In fair coin tosses, for example, one can find the probability that $|h(n)/n - \frac{1}{2}| > \epsilon$, where $h(n)$ = heads and n = tosses. As n approaches infinity, the probability is equal to zero. This law led Bernoulli to restate Leibniz's principle of sufficient reason as the influential law of determinism in classical mechanics: "If all events from now to eternity were continually observed . . . , it would be found that everything in the world occurs for definite reasons."

79. From *Ars Conjectandi* (1713)*

(The Law of Large Numbers)[1]

JAKOB BERNOULLI

We have now reached the point where it seems that, to make a correct conjecture about any event whatever, it is necessary only to calculate exactly the number of possible cases,[2] and then to determine how much more likely it is that one case will occur than another. But here at once our main difficulty arises, for this procedure is applicable to only a very few phenomena, indeed almost exclusively to those connected with games of chance. The original inventors of these games designed them so that all the players would have equal prospects of winning, fixing the number of cases that would result in gain or loss and letting them be known beforehand, and also arranging matters so that each case would be equally likely. But this is by no means the situation as regards the great majority of the other phenomena that are governed by the laws of nature or the will of man. In the game of dice, for instance, the number of possible cases [or throws] is known, since there are as many throws for each individual die as it has faces; moreover all these cases are equally likely when each face of the die has the same form and the weight of the die is uniformly distributed. (There is no reason why one face should come up more readily than any other, as would happen if the faces were of different shapes or part of the die were made of heavier material than the rest.) Similarly, the number of possible cases is known in drawing a white or a black ball from an urn, and one can assert that any ball is equally likely to be drawn: for it is known how many balls of each kind are in the jar, and there is no reason why this or that ball should be drawn more readily than any other. But what mortal, I ask, could ascertain the number of diseases, counting all possible cases, that afflict the human body in every one of its many parts and at every age, and say how much more likely one disease is to be fatal than another—plague than dropsy, for instance, or dropsy than fever—and on the basis make a prediction about the relationship between life and death in future generations? Or who could enumerate the countless changes that the atmosphere undergoes every day, and from that predict today what the weather will be a month or even a year from now? Or again, who can pretend to have penetrated so deeply into the nature of the human mind or the wonderful structure of the body that in games which depend wholly or partly on the mental acuteness or the physical agility of the players he would venture to predict when this or that player would win or lose? These and similar forecasts depend on factors that are completely obscure, and which constantly deceive our senses by the endless complexity of their interrelationships, so that it would be quite pointless to attempt to proceed along this road.

There is, however, another way that will lead us to what we are looking for and enable us at least to ascertain *a pos-*

*Source: This selection appears in James R. Newman (ed.), *The World of Mathematics* (1956), vol. 3, 1452–1454. It is reprinted by permission of Simon and Schuster, Inc.

teriori what we cannot determine *a priori,* that is, to ascertain it from the results observed in numerous similar instances. It must be assumed in this connection that, under similar conditions, the occurrence (or nonoccurrence) of an event in the future will follow the same pattern as was observed for like events in the past. For example, if we have observed that out of 300 persons of the same age and with the same constitution as a certain *Titius,* 200 died within ten years while the rest survived, we can with reasonable certainty conclude that there are twice as many chances that Titius also will have to pay his debt to nature within the ensuing decade as there are chances that he will live beyond that time. Similarly, if anyone has observed the weather over a period of years and has noted how often it was fair and how often rainy, or has repeatedly watched two players and seen how often one or the other was the winner, then on the basis of those observations alone he can determine in what ratio the same result will or will not occur in the future, assuming the same conditions as in the past.

This empirical process of determining the number of cases by observation is neither new nor unusual; in chapter 12 and following of *L'art de penser*[3] the author, a clever and talented man, describes a procedure that is similar, and in our daily lives we can all see the same principle at work. It is also obvious to everyone that it is not sufficient to take any single observation as a basis for prediction about some [future] event, but that a large number of observations are required. There have been instances where a person with no education and without any previous instruction has by some natural instinct discovered—quite remarkably—that the larger the number of pertinent observations available, the smaller the risk of falling into error. But though we all recognize this to be the case from the very nature of the matter, the scientific proof of this principle is not at all simple, and it is therefore incumbent on me to present it here. To be sure I would feel that I were doing too little if I were to limit myself to proving this one point with which everyone is familiar. Instead there is something more that must be taken into consideration—something that has perhaps not yet occurred to anyone. *What is still to be investigated is whether by increasing the number of observations we thereby also keep increasing the probability that the recorded proportion of favorable to unfavorable instances will approach the true ratio, so that this probability will finally exceed any desired degree of certainty,* or whether the problem has, as it were, an asymptote. This would imply that there exists a particular degree of certainty that the true ratio has been found which can never be exceeded by an increase in the number of observations: thus, for example, we could never be more than one-half, two-thirds, or three-fourths certain that we had determined the true ratio of the cases. The following illustration will make clear what I mean: We have a jar containing 3,000 small white pebbles and 2,000 black ones, and we wish to determine empirically the ratio of white pebbles to the black—something we do not know—by drawing one pebble after another out of the jar, and recording how often a white pebble is drawn and often a black. (I remind you that an important requirement of this process is that you put back each pebble, after noting its color, before drawing the next one, so that the number of pebbles in the urn remains constant.) Now we ask, is it possible by indefinitely extending the trials to make it 10, 100, 1,000, etc., times more probable (and ultimately "morally certain") that the ratio of the number of drawings of a white pebble to the number of drawings of a black pebble will take on the same value (3:2) as the actual ratio of white to black pebbles in the urn, than that the ratio of the drawings will take on a different value? If the answer is no, then I admit that we are likely to fail in the attempt to ascertain the number of instances of each case [*i.e.*, the number of white and of black pebbles] by observation. But if it is true that we can finally attain moral certainty by this method[4] . . . then we can determine the number of instances *a posteriori* with almost as great accuracy as if they were known to us *a priori*. Axiom 9 [presented in an earlier chapter] shows that in our everyday lives, where moral certainty is regarded as absolute certainty, this consideration enables us to make a prediction about any event involving chance that will be no less scientific than the predictions made in games of chance. If, instead of the jar, for instance, we take the atmosphere or the human body, which conceal within themselves a multitude of the most varied processes or diseases, just as the jar conceals the pebbles, then for these also we shall be able to determine by observation how much more frequently one event will occur than another.

Lest this matter be imperfectly understood, it should be noted that the ratio reflecting the actual relationship between the numbers of the cases—the ratio we are seeking to determine through observation—can never be obtained with absolute accuracy; for if this were possible, the ruling principle would be opposite to what I have asserted: that is, the more observations were made, the *smaller* the probability that we had found the correct ratio. The ratio we arrive at is only approximate: it must be defined by two limits, but these limits can be made to approach each other as closely as we wish. In the example of the jar and the pebbles, if we take two ratios, 301/200 and 299/200, 3001/2000 and 2999/2000, or any two similar ratios of which one is slightly less than 1½ and the other slightly more, it is evident that we can attain any desired degree of probability that the ratio found by our many repeated observations will lie between these limits of the ratio 1½, rather than outside them.

It is this problem that I decided to publish here, after having meditated on it for twenty years. . . .

. . . If all events from now through eternity were continually observed (whereby probability would ultimately become certainty), it would be found that everything in the world occurs for definite reasons and in definite conformity with law, and that hence we are constrained, even for things that may seem quite accidental, to assume a certain necessity and, as it were, fatefulness. For all I know that is what Plato had in mind when, in the doctrine of the universal cycle, he maintained that after the passage of countless centuries everything would return to its original state.

NOTES

1. Translated from *Klassische Stücke der Mathematik,* selected by A. Speiser (Zürich, 1925), pp. 90-95. The selection is from the German translation of the *Ars Conjectandi* by R. Haussner in Ostwald's *Klassiker der exakten Wissenschaften,* Leipzig, 1899, nr. 108.

2. For "case," the correct translation of the German, one may read *result* or *outcome.*

3. *La logique, ou L'art de penser,* by Antoine Arnauld and Pierre Nicole, 1662. (Makes use of Pascal, Fragment no. 14.) There are in fact *two* authors but Bernoulli makes it appear there is only one.

4. Bernoulli demonstrates that this is true in his next chapter.

Chapter VI
The Scientific Revolution at Its Zenith (1620–1720)

Section D
The Bernoullis

JOHANN BERNOULLI (1667–1748)

Johann Bernoulli was the tenth child in his family. His father, Nikolaus, attempted to draw him into the family business. After an unsuccessful apprenticeship as a salesman, however, he received permission from his father to enroll at the University of Basel in 1683. He resided with his brother Jakob. At age 18, Johann received the master of arts degree. At his father's urging, he took up the study of medicine but privately studied mathematics and experimental physics with his brother. He went to Paris in 1691, where he participated in the mathematical circle of Nicolas Malebranche (then the foremost Cartesian). In their discussions Bernoulli disseminated Leibniz's calculus. He also gave calculus lessons to Guillaume-François-Antoine de L'Hospital—lessons which became the basis for the first textbook on the differential calculus, L'Hospital's *Analyse des infiniment petits* ("Analysis of the Infinitely Small," 1696). This text contained the method for evaluating the indeterminate form 0/0, which is incorrectly known today as L'Hospital's rule. In 1693, Bernoulli began an extensive correspondence with Leibniz.

Through the intervention of Christiaan Huygens, Johann Bernoulli was offered the professorial chair in mathematics at Gröningen (Holland) in 1695. He accepted because his quarrels with his brother were growing and because he could not hope to obtain the mathematics professorship at Basel as long as Jakob lived. In September 1695 he, his wife Dorothea Falkner, and their seven-month-old son Nikolaus left for Holland. Two sons were born later—Daniel, the most famous of the Bernoullis, and Johann II. While at Gröningen he did not curb his "Flemish pugnacity." The theologians with whom he argued about natural philosophy charged him with Spinozism, a late-17th-century word for atheism.

Upon the death of Jakob in 1705, Johann succeeded him at the University of Basel. Johann would have preferred to accept other offers extended to him by the Universities of Leiden and Utrecht, but family concerns drew him to his native city where he spent the rest of his life. He was the most distinguished member of the university faculty. In the early 1720s, he taught his greatest student, Leonard Euler. Euler was to be one of his two heroes; the other was Leibniz. His activities were not limited to university affairs, as a member of the Basel school board, he worked to reform its humanistic Gymnasium.

Influential far beyond Switzerland, Johann took part in two major continental scientific quarrels. The Royal Society of London had proclaimed Leibniz a plagiarist in 1712, and Bernoulli was outraged when this news reached Switzerland. Already a strong supporter of Leibniz, he now became his outspoken champion against British critics. Neither Brook Taylor nor Colin Maclaurin, leading British Newtonians, were fully Bernoulli's equal in debates. Bernoulli attacked Newton's method of fluxions and in challenge problems showed the superiority of Leibniz's differential calculus.

On the Continent, the developing Newtonian dynamics and optics began to supplant the dominant Cartesian science during the 1720s and 1730s. The Paris Academy of Sciences was the focal point for the struggle between the two. Once again Bernoulli was an adversary of Newton's ideas, which he considered unduly narrow, especially in comparison to Leibniz's. The chief controversy was over the Cartesian vortex theory and its explanation of celestial motions by whirlpools of the ether—a theory Newton had criticized in Book II of the *Principia.* Bernoulli advocated a revised vortex theory in the papers he submitted to the Paris Academy for its prestigious, biennial prize competition. He won three times—for papers on the transmission of momentum (1727), the motions of the planets in aphelion (1730), and the inclination of the planetary orbits toward the solar equator (1734)—and shared the 1734 prize with his son Daniel who took a Newtonian position. It is said that the quarrelsome and obstinate Johann so begrudged Daniel his position and share of success that he ordered him out of the family house. Johann's work delayed but did not prevent the triumph of Newtonian science that occurred at the Paris Academy between 1734 and 1740.

The relations between Johann Bernoulli and his son Daniel did not improve greatly after 1734; their work in physics was a continuing source of antagonism. Johann advanced continuum mechanics and wrote *Hydraulica* (1738), which made some progress in the study of the internal pressure exerted by fluids on tubes. His book, however, was overshadowed by Daniel's classic *Hydrodynamica* (1738)—a term the son introduced.

The most significant contributions of Johann Bernoulli were to mathematics. He at first worked closely with his brother on probability and the differential calculus. Johann's results appeared in memoirs in *Acta Eruditorum* and the *Journal des Sçavans. Acta Eruditorum* for 1694 contained his discovery of the series now known as the Taylor series; the 1697 volume had his skillful paper on the calculus of variations. He was the first to realize that variational problems involve making a given integral a maximum or minimum. His pioneering efforts in the calculus of variations were to be superceded by his pupil Euler. In elaborating the calculus, Johann Bernoulli concentrated on the integration of differential equations; indeed, he defined the integral. From his study of the exponential function, x^y, he derived the natural exponential function, e^x, which is equal to its derivative. Subsequently, Euler made e the base for natural logarithms. Johann and his sons extended the calculus to cover two and three independent variables and provided a flexible formalism to handle the additional degrees of freedom.

80. From "The Curvature of a Ray in Nonuniform Media" (1697)*

(The Brachistochrone)

JOHANN BERNOULLI

The curvature of a ray in nonuniform media, and the solution of the problem to find the brachystochrone, that is, the curve on which a heavy point falls from a given position to another given position in the shortest time, as well as on the construction of the synchrone or the wave of the rays.

. . . We have a just admiration for Huygens, because he was the first to discover that a heavy point on an ordinary *Cycloid* falls in the same time [*tautochronos*], whatever the position from which the motion begins.[1] But the reader will be greatly amazed [*an non obstupescus plane*], when I say that exactly this *Cycloid*, or *Tautochrone of Huygens*, is our required *Brachystochrone*. I reached this understanding in two ways, one indirect and one direct. When I pursued the first, I discovered a wondrous agreement between the curved path of a light ray in a continuously varying medium and our *Brachystochrone*. I also found other rather mysterious things [*in quibus nescio quid arcani subest*] which might be useful in dioptric investigations. It is therefore true, as I claimed when I proposed the problem, that *it is not just naked speculation*, but *also very useful for other branches of knowledge*, namely, for dioptrics. But in order to confirm my words by the deed, let me here give the first mode of proof!

Fermat, in a letter to De la Chambre,[2] has shown that a light ray passing from a thin to a more dense medium, is bent toward the perpendicular in such a way that, under the supposition that the ray moves continuously from the light to the illuminated point, it follows the path that requires the shortest time. With the aid of these principles he showed that the sine of the angle of incidence and the sine of the angle of refraction are in inverse proportion to the densities of the media, hence directly as the velocities with which the light ray penetrates these media. Later Leibniz, in the *Acta Eruditorum*, 1682, pp. 185 sequ., and soon afterward the famous Huygens in his *Treatise on Light*, p. 40,[3] have demonstrated this more comprehensively and, by most valid arguments, have established the physical, or better the metaphysical, principle which Fermat seems to have abandoned at the insistence of Clerselier, remaining satisfied with his geometric proof and giving up his rights all too lightly.

Now we shall consider a medium that is not homogeneously dense, but consists of purely parallel horizontally superimposed layers, of which each consists of diaphanous matter of a certain density decreasing or increasing according to a certain law. It is then manifest that a ray which we consider as a particle will not be propagated in a

*Source: This translation of "Curvatura radii in diaphanis non uniformibus" is taken from D. J. Struik (ed.), *A Source Book in Mathematics, 1200-1800* (1969), 392-396.

straight line, but in a curved path. This has already been considered by Huygens in his above-mentioned *Treatise on Light,* but he did not determine the nature of this minimizing curve such that the particle, whose velocity increases and decreases depending on the density of the medium, will pass from point to point in the shortest time. We know that the sines of the angles of refraction at the separate points are to each other inversely as the densities of the media or directly as the velocities of the particles, so that the brachystochrone curve has the property that the sines of its angles of inclination with respect to the vertical are everywhere proportional to the velocities. But now we see immediately that the brachystochrone is the curve that a light ray would follow on its way through a medium whose density is inversely proportional to the velocity that a heavy body acquires during its fall. Indeed, whether the increase of the velocity depends on the constitution of a more or less resisting medium, or whether we forget about the medium and suppose that the acceleration is generated by another cause according to the same law as that of gravity, in both cases the curve is traversed in the shortest time. Who prohibits us from replacing the one by the other?

In this way we can solve the problem of an arbitrary law of acceleration, since it is reduced to the determination of the path of a light ray through a medium of arbitrarily varying density. Hence let FGD [Fig. 1] be the medium bounded by the horizontal line FG on which the luminous point A is situated. Let the curve AHE, with vertical axis AD, be given, its ordinates HC determining the densities of the medium at altitude AC or the velocities of the light rays or particles at M. Let the curved line of that light ray, which we wish to determine, be ABM. Let us write for AC, x; for CH, t; for CM, y; and for the differentials Cc, dx; diff. $mn = dy$; diff. $Mm = dz$, finally, let a be an arbitrary constant. Then Mm is the total sine, mn the sine of the angle of refraction or the angle of inclination of the curve with respect to the vertical. As we have said before, the ratio of mn to CH is constant, hence

$$dy : t = dz : a,$$

so that

$$a\,dy = t\,dz,$$

or

$$aa\,dy^2 = tt\,dz^2 = tt\,dx^2 + tt\,dy^2.$$

This gives a general differential equation for the required curve ABM:

$$dy = t\,dx : \sqrt{(aa - tt)}.$$

In this way I have solved at one stroke two important problems—an optical and a mechanical one—and have achieved more than I have demanded from others: I have shown that two problems, taken from entirely separate fields of mathematics, have the same character.

Now let us take a special case, namely the common hypothesis first in-

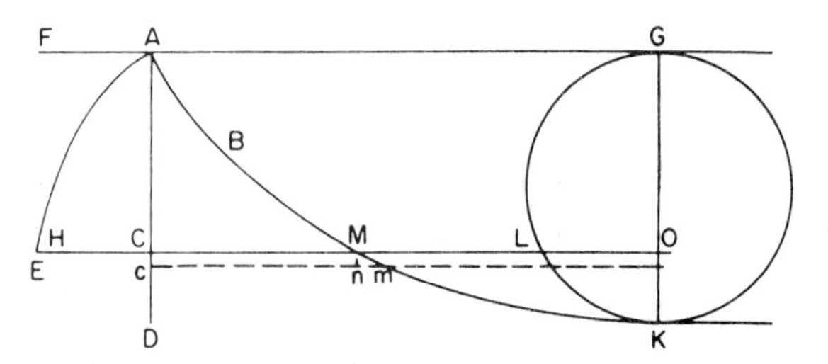

FIG. 1

troduced by Galilei, who proved that the velocities of falling bodies are to each other as the square roots [*in ratione subduplicata*] of the altitudes traversed—then this is really the given problem. Under this assumption the given curve *AHE* is a parabola $tt = ax$, hence $t = \sqrt{ax}$. If this value is substituted in the original equation, we obtain

$$dy = dx\sqrt{\frac{x}{a-x}},$$

from which I conclude that the *Brachystochrone* is the ordinary *Cycloid*. For when the circle *GLK* of radius *a* rolls on *AG* and the rolling starts at *A*, the point *K* describes a cycloid, of which the differential equation is exactly

$$dy = dx\sqrt{\frac{x}{a-x}},$$

if $AC = x$, $CM = y$.

Bernoulli then shows this analytically by writing

$$dx\sqrt{\frac{x}{a-x}} =$$

$$\frac{1}{2}\frac{a\,dx}{\sqrt{ax-x^2}} - \frac{1}{2}\frac{a\,dx-2x\,dx}{\sqrt{ax-x^2}},$$

which integrated gives

$$CM = \text{arc } GL - LO,$$

from which, since $MO = CO - \text{arc } GL + LO = \text{arc } LK + LO$, **it follows that** $ML = \text{arc } LK$.[4]

To solve the problem completely he then shows that from a given point as vertex a cycloid can be described that passes through a second given point.

Before I end I must voice once more the admiration that I feel for the unexpected identity of Huygens' tautochrone and my brachystochrone. I consider it especially remarkable that this coincidence can take place only under the hypothesis of Galilei, so that we even obtain from this a proof of its correctness. Nature always tends to act in the simplest way, and so it here lets one curve serve two different functions, while under any other hypothesis we should need two curves, one for tautochronic oscillations, the other for the most rapid fall. If, for example, the velocities were as the altitudes, then both curves would be algebraic, the one a circle, the other one a straight line.[5]

Bernoulli then introduces the *synchrone:* the curve *PB* [Fig. 2] in a vertical plane such that a heavy body falling from *A* along this curve reaches the points *B* in the same time as a heavy body falling on the cycloid *AB*. Referring to Huygens, he concludes that *PB* is also a cycloid intersecting all cycloids with initial point *A* at a right angle. He ends by suggesting that other orthogonal trajectories of given families of curves be found.[6]

NOTES

1. Huygens, *Horologium oscillatorium* (Paris, 1673), Proposition XXV: In a cycloid with vertical axis and with its vertex down,

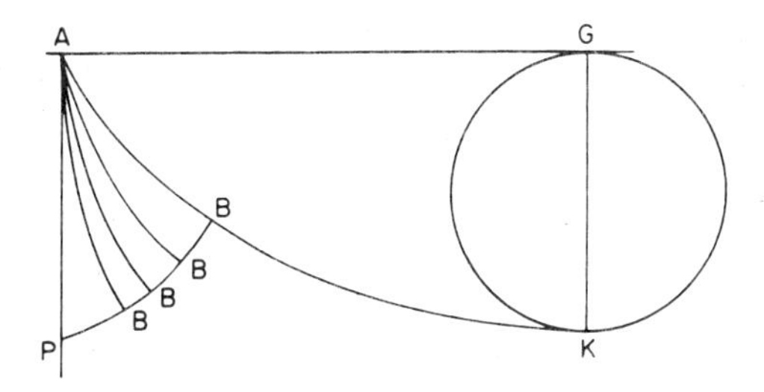

FIG. 2

the times of descent in which a mobile particle, starting from rest at an arbitrary point of the curve, reaches the lowest point are equal among themselves, and have to the time of the vertical fall along the total axis of the cycloid a ratio equal to that of the semicircumference of a circle to its diameter. *Oeuvres complètes,* XVIII (1934), 185.

2. Fermat's letters to Martin Cureau de la Chambre are of 1657 and 1662 (*Oeuvres,* II, 354-359, 457-463). The law of refraction was published by Descartes in his *Dioptrique* (1637). Fermat first opposed it, but then reestablished it by a maximum-minimum principle.

3. Huygens, *Traité de la lumière* (Leiden, 1690), 40; *Oeuvres complètes,* XIX (1737), 489.

4. This gives the equation of the cycloid in the form

$$x = \frac{a}{2}(1 - \cos t), \quad y = \frac{a}{2}(t - \sin t),$$

$$t = \pi - \phi, \quad \text{arc } LK = a\phi.$$

The differential equation can already be found in Leibniz's first paper on the integral calculus of 1686.

5. The cases mentioned are $t = ax$ and $t = ax^{1/3}$.

6. Johann Bernoulli does not yet use the term "orthogonal trajectories." The concept played an important role in the work of Leibniz and Bernoulli in those days. The connection with Huygens's theory of light was clear. The term "trajectory" dates from an article by Johann Bernoulli in the *Acta Eruditorum* of 1698 (*Opera omnia,* I, 266).

Chapter VII

The Age of Enlightenment and the French Revolution (1720-1800)

Introduction. The spectacular development of European mathematics in the hands of a few masters continued throughout the 18th century. Above all, the methods of the differential and the fluxional calculus were combined to form the unified mathematical discipline of analysis, which was then greatly extended. This was largely the result of the work of the Swiss-born Leonhard Euler, who dominated mathematics. These inventions widened the scope of mathematics and intensified its relations with rational mechanics[1] and astronomy, that is, celestial mechanics, the exact science in which the calculus was applied with the most dramatic results. Attempts to solve special, previously unassailable problems encouraged growth. Studies of the shape of vibrating strings fixed at their end points and of elastic beams under tension led to solutions requiring new principles and analytical methods, and so did successful efforts to describe more accurately lunar motion and the flow of water through pipes. Mathematical analysis was employed with extraordinary success in problem-solving by Daniel Bernoulli, Alexis Clairaut, Jean d'Alembert, Leonhard Euler, Louis Lagrange, and others. Three of its component parts—differential equations,[2] infinite series, and the calculus of variations—quickly developed into distinct areas of inquiry. Mathematical analysis itself not only gained autonomy as a branch of mathematics, forming a triumvirate with geometry and algebra, but also displaced geometry (both Euclidean and more recently Cartesian) from the primacy in mathematics that it had held for two millennia.

Of course 18th-century mathematics had other significant developments. It consisted of far more than the creation and exploitation of analysis. The term "mathematics" was applied to an array of subjects. D'Alembert's "Detailed System of Human Knowledge" given at the conclusion of his *Preliminary Discourse to the Encyclopedia* (1751) accurately maps these. "Pure mathematics" consisted of arithmetic (which included algebra and the calculus) and geometry, while the larger contiguous realm of "mixed mathematics" included the exact sciences of geometric astronomy, optics, acoustics, pneumatics, and rational mechanics, which, in turn, covered the more technological fields of ballistics, navigation and shipbuilding.

The principal achievements in the branches of mathematics outside of analysis were less original and largely motivated by it. In algebra improved methods of solution were devised for polynomial equations and the theory of numbers began to be transformed from a collection of disconnected results into a systematic science. Conventional probability theory, which was still in a formative stage, received a more analytical expression. In addition, analytic geometry matured, and in the latter part of the 18th century the application of analysis to geometry led to the founding of the differential geometry[3] of curves and spaces. Differential geometry, in turn, made possible a critical examination of Euclidean concepts. This examination together with studies in traditional geometry of the problem of parallels, beginning with the work of Girolamo Saccheri, were to culminate in the invention of non-Euclidean geometries in the nineteenth century. Eighteenth-century geometric studies thus set the stage for a revival of geometry which did not, however, regain its former preeminence.

The appearance of modern professional trappings for mathematics begun in the latter part of the seventeenth century was accelerated in the next century. Institutions lent more support to mathematicians and mathematicians found new opportunities to publish their work in a growing number of learned journals. Because the term "mathematician" still connoted practitioners of such fields as astrology and number mysticism (a connotation acquired in late Alexandrian times), those we call mathematicians by present-day definitions were still generally called geometers or algebraists in the 1700s. Leading geometers and algebraists mostly belonged to either the Royal Society of London, the Paris Academy of Sciences, or to the two newer national science academies in Berlin and St. Petersburg. These institutions were the chief centers of substantial collaborative research in the sciences. Although the Royal Society lacked royal patronage and did not provide pensioned positions together with equipment and social privileges, its counterparts on the Continent did. Their members were assigned projects leaving ample time for independent research, a fortunate stimulus for new, imaginative investigations.

While the Royal Society played a relatively minor role in the mathematical sciences during the Enlightenment, the Continental academies vigorously promoted them. The Royal Society suffered from a poor financial situation and its separation from Leibnizian mathematics.[4] Without royal patronage, buildings and sufficient equipment, it resorted to selecting wealthy, as well as simply able, members. Consequently, from the 1730s the Royal Society began more and more to resemble a mere gentlemen's club. With regular royal financing, a succession of distinguished members and a location at the center of a flourishing, receptive intellectual culture, the Paris Academy now dominated the natural sciences and was particularly outstanding late in the century. The St. Petersburg Academicians included Daniel Bernoulli and Jakob Hermann for brief periods and Leonhard Euler for nearly three decades. In Prussia, Frederick the Great rejuvenated the Brandenburg Society and renamed it the Berlin Academy. The Frederician Academicians included Pierre Maupertuis, Johann Castillon, and Johann Lambert as well as Euler and Lagrange, who successively supervised its mathematical–physical section. At mid-century the St. Petersburg *Commentarii* and the Berlin *Histoire* were perhaps as important for the mathematical sciences as the Paris *Mémoires*. Although articles on mathematics appeared in as many as 210 new periodicals during the 18th century, according to one estimate, few if any of these periodicals had standards comparable to the academy journals, and there was no notable journal devoted to mathematics exclusively. (The latter is not surprising in an age when no other science had a disciplinary journal and when a generalist spirit of knowledge persisted in the face of a growing trend toward specialization.)

Universities had a lesser influence upon the development of mathematics. Indeed, the few universities which were prominent in the sciences a century earlier waned or barely held their own, especially in England. Cambridge and Oxford fared ill. Increasing fees excluded students from all but the prosperous gentry and wealthy bourgeoisie. Among these students there was little interest in mathematics, except for military applications, and no employment existed for professional mathematicians. The new analytical mathematics, which departed from traditions of the past and which was fostered by socially dissident classes, was considered "unintellectual" in comparison to the dominant classical studies, philosophy, and rhetoric.[5] Scientific chairs, increasingly filled by theologians and classical scholars, were all too often rewards for past services to the king.

The importance of mathematics to the new sciences, to the development of method in logical reasoning, and to a more precise awareness of economic factors in national life was taken up by a few universities on the Continent. These institutions, which were centers of research and distinguished teaching in at least a few subjects, deeply influenced the intellectual life of the Enlightenment. However, they did not escape from disruptive theological disputes, especially (in Catholic countries) between Jansenists and Jesuits and from strong opposition to the curricular reform that recognized the importance of new disciplines. Chief among the nearly twenty French universities teaching the new mathematical sciences was the *Collège de France.* The renowned University of Paris, guardian of spiritual orthodoxy, was deeply shaken by religious quarrels and contributed little. It did not include a faculty of sciences until the Napoleonic reförms in 1808. The *École Polytechnique* was founded only in 1795. On occasion, there were outstanding scholars at the Universities of Leiden and Utrecht in Holland. Professors Johann and Daniel Bernoulli graced the University of Basel in Switzerland. Italian universities added almost nothing to mathematics, in sharp contrast to their role a century earlier. The dampening effect stemming from the trial of Galileo, the continuing influence of the Inquisition, the power of entrenched Aristotelians and the eroding economic status of Italian cities had their effect. The retiring Maria Agnesi (1718–99) did advance original ideas in the calculus in her *Instituzioni analytiche* (1748), while only an honorary member of the faculty at Bologna, a university at which she never taught.

In the German states Halle (in Prussia) and Göttingen (founded in 1737 in Hanover) were vital centers, and in Sweden Uppsala flourished briefly. German intellectual life at mid-century was dominated by the Leibniz–Wolffian philosophy, the ideas of Leibniz as organized and modified by the great philosopher's chief continuator, Christian Wolff (1679–1754). Kant's observation that it was Wolff who introduced rigor *(Gründlichkeit)* into German philosophy is an evidence of the esteem that was once paid this now largely-forgotten figure. Wolff, whose writings included seven volumes of *Vernünftige Gedanken,* stressed mathematical method formalized by syllogistic logic as the best way to obtain the certain knowledge of *Wissenschaft.* From Halle (and for a time Marburg)[6] his ideas spread across Europe to France, East Prussia (Königsberg), Russia (St. Petersburg) and Sweden (Uppsala). At Uppsala Anders Celsius, who improved the teaching of astronomy, praised Wolff's method and the clarity of its axioms, definitions and assumptions, while Samuel Klingenstierna made original contributions to geometrical optics. Late in the century Georg Klügel (1739–1812) joined Halle's faculty. He questioned the parallel postulate and contributed to analytic trigonometry by refining Euler's ratio concept of trigonometric functions and establishing that the six basic formulas are valid for a right spherical triangle. Klügel also applied mathematics to optics, dynamics, and astronomy.

More important than Halle was Göttingen. Stressing natural science and possessing the finest university library in Europe, Göttingen prepared the groundwork for the eminence that German universities were to attain in mathematics during the nineteenth century. Abraham Kästner (1719–1800) of its faculty was an important expositor of mathematical ideas. Young Carl Gauss reportedly shunned Kästner's lectures because they were too elementary, but subsequently followed Kästner's thought on parallelism and on the concept of actual infinity (which Kästner opposed). The science academies and these universities provided the principal institutional setting in which the conceptual development of mathematics occurred.

In the seventy years after Leibniz and Newton first wrote on the calculus, mathematicians developed it from a loose collection of methods for solving problems concerning curves into the unified mathematical discipline of analysis. Following the quarrel over priority for the invention of the calculus, there were two camps: the British and the Continentals. Stubbornly adhering to Newton's methods and less facile notation, the British essentially isolated themselves from the new directions taken by Continental mathematicians and made only a few advances. Brook Taylor (d. 1731) and Colin Maclaurin (d. 1746) provided most of these. Taylor had proposed his famous power series expansion formula by 1715. In his *Treatise of Fluxions* (1742), Maclaurin named and utilized this formula and gave the integral test for the convergence of infinite series. Mathematicians, however, did not yet recognize the importance of the Taylor series; that came only with the work of Lagrange after the further development of interpolation methods. Maclaurin also unsuccessfully attempted to find satisfactory foundations for the calculus based on Archimedean geometric methods. This reinforced an emphasis on geometry in Britain. As pointed out earlier, neither the Royal Society nor British universities strongly supported mathematics during the Enlightenment, and after mid-century Britain had no one comparable to the Continental masters.

On the Continent mathematicians made rapid progress with the calculus. In Basel Johann Bernoulli discovered the exponential function, e^x, by studying the inverse of the (natural) logarithm[7] and first published in 1742 his lectures to l'Hopital on the method of integrals. This publication complemented his lectures on differential equations that were available in l'Hôpital's *Analyse*. Bernoulli's definition of the integral as the inverse of the differential was widely accepted over Leibniz's, which referred to the summation of infinitely small quantities. He and his son Daniel also extended the calculus to handle two or three independent variables by the use of partial derivatives. In Paris the Academy of Sciences emphasized the calculus rather than geometry and Alexis Clairaut and Jean d'Alembert developed differential equations. The shaping of analysis through many discoveries and methods together with the codification of the discipline, however, were chiefly the work of Euler, the brilliant student of Johann Bernoulli. Euler accomplished this in a trilogy of magisterial textbooks: *Introduction to the Analysis of Infinities* (2 vols., 1748), *Textbooks on the Differential Calculus* (1755), and *Textbooks on the Integral Calculus* (3 vols., 1768–70).

In the *Introduction* Euler fashioned analysis into a major branch of mathematics separate from geometry and algebra. He accomplished this by first making clear its content and methods. For Newton variable (kinematic) geometric quantities had been the principal object of study in the calculus, but as these grew more intricate and remote Continental masters shifted emphasis to associated formulas and their formal manipulations. Euler completed that transition by explicitly basing analysis upon analytic expressions, especially functions, and infinite processes. He believed that all functions could be expanded as infinite or generalized power series using binomial expansion, long division, and other methods. Following Johann Bernoulli (1718), he wrote in paragraph four of the *Introduction:*

> A function of a variable quantity is an analytic expression composed in any way from this variable quantity and from numbers or constant quantities.

After finding arbitrary functions (that were not analytic) in his study of the problem of the vibrating string, he broadened his definition in the preface to the *Differential Calculus:*

> If some quantities so depend on other quantities that if the latter are changed the former undergo change, then the former quantities are called functions of the latter.[8]

Euler's identification of the centrality of functions rather than curves not only separated analysis from traditional geometry, it also allowed the arithmetization of geometry.

In the first part of the *Introduction* Euler inventoried and classified functions in the manner employed today. Logarithmic, exponential, trigonometric and their inverses, and elementary transcendental functions were covered. Since this text was to be a survey of concepts and methods of analytic geometry and analysis preliminary to the study of the calculus, Euler introduced these functions without recourse to the integral calculus, which was no mean feat. In chapter VII he interpreted logarithms as exponents, presenting the first systematic treatment of them in history with the base a for natural logarithms. Subsequently he denoted this base with the letter e. In paragraph 123 Euler gave its decimal expansion to 23 places ($e = 1 + \frac{1}{1!} + \frac{1}{2!} + \frac{1}{3!} + \ldots = 2.718281828459\ldots$). Next, he standardized trigonometric functions as ratios which supplanted the previously dominant view of them as half chords of a central angle 2α. After deriving the infinite series for each, he deduced in a simple, nonrigorous fashion what is usually called De Moivre's identity, that is, $(\cos x + i \sin x)^n = \cos nx \pm i \sin nx$ and discovered the famous relation between exponential and trigonometric functions, $e^{\pm ix} = \cos x \pm \sin nx$, where $i = \sqrt{-1}$. Complex numbers were admitted equally with real numbers for independent variables of functions, which was a significant step.

Euler greatly enriched the differential and integral calculus. The first two chapters of the *Differential Calculus* introduced the subject as a limiting case of a calculus of infinitely small differences, an approach akin to Leibniz's. Euler rejected the infinitesimal and old view of the differential as quantities less than any given magnitude and yet not equal to zero, however. The derivative (dy/dx), he argued, had a finite ratio, even though dy and dx were in fact equal to zero. That is to say, the latter quantities were diminished until they vanished completely or disappeared. To justify retaining dx while dropping $(dx)^2$, he said that $(dx)^2$ vanished before dx. His version of the calculus based on the quotient of "qualitative zeroes," while not contradictory, did not suffice to handle many of the new problems that subsequently arose. His lack of hesitation to operate with sums that did not converge also led to difficulties, especially when followed by lesser mathematicians. His *Integral Calculus* thorougly treated integration expressible by algebraic and elementary functions. It investigated many definite integrals, including the Eulerian integrals, now called beta and gamma functions following Legendre's designation. Euler also contributed to the theory of elliptic integrals and introduced double integrals. He revealed numerous methods for solving ordinary and partial differential equations that were important in mechanics. With its clear and systematic treatment of the calculus, Euler's trilogy established the scope and style of the subject for the next half century.

Histories of the calculus during the 18th century customarily pay close attention to the way rigor approximated present standards or failed to reach them. Unless treated with care, the problem of *rigor* can be counterproductive. It tends to obscure the fact that mathematicians as a rule concentrate on solving problems, not on pondering reflexively the fundamental concepts and methods used to obtain solutions. Nor does an emphasis on the search for critical techniques bring out that most 18th-century problems concerning the physical world were non-pathological. Their solution could be tested by empirical rather than epistemological criteria; they did not require a stringent formalism. Despite the absence of satisfactory, logical foundations, the computational tools of the differential calculus rapidly developed, as did integration regarded as the inverse of differentiation. Leading mathematicians recognized the need for sound foundations or what they called the metaphysics of the calculus before 1750, however. Euler drew graphs as a guide to indicate continuity or discontinuity and convergence or divergence, and he trusted to his intuition. The shift to formalism and logical foundations did not occur until after a period of free creation of a sufficient body of materials. A Scriptural verse expresses well the situation with regard to rigor in the eighteenth century.

> Be therefore not anxious about tomorrow;
> for tomorrow will be anxious for the things of itself.
> Sufficient unto the day is the evil thereof. (Matthew 6:34)

The pursuit of rigor and fundamental concepts in the calculus emerged gradually during the 18th century. Bishop Berkeley (d. 1753) brought these needs into sharp focus in his critique of the calculus, entitled *The Analyst* (1734). While praising the calculus as "the general key by help whereof the modern mathematicians unlock the secrets of Geometry, and consequently of Nature," he showed that powerful conclusions rested on logical inconsistencies and ambiguous concepts. The ratios of both Newton's fluxions and Leibniz's differentials posed difficulties. When defined as ultimate ratios of evanescent increments, they were not precisely finite nor yet zero. Seizing on that imprecision Berkeley called these increments "the ghosts of departed quantities." He also criticized the Newtonian extraction of the derivative nx^{n-1} of x^n from the increment $(x + o)^n - x^n$. First one divided the expansion by o, supposing o does not equal zero, and then sets o equal to zero to eliminate terms. This seemed "a most inconsistent way of arguing" that "would not be allowed if in Divinity." Berkeley ingeniously proposed that mathematicians arrived at correct results and great success by a chain of mutually compensatory errors implicit in the rules of the calculus. Several British mathematicians made spirited rejoinders to Berkeley in adamant defense of Newton's work. Of these only Maclaurin's *Treatise of Fluxions* displayed an understanding of both Berkeley's criticisms and Newton's fluxional calculus. Given no more than the contemporary stage of development of the calculus, Maclaurin could hardly be expected to provide the missing logical foundations. Benjamin Robins (d. 1751) subsequently argued that the concept of limits was crucial to resolving foundational questions in his *Mathematical Tracts* (1761, see vol. 2, p. 49).

On the Continent advances were made in foundations during the second half of the century through the work of d'Alembert and Lagrange as well as by two prize competitions at the Berlin and St. Petersburg Academies. D'Alembert focussed on the limit concept. In the article "Differential" in volume four of the *Encyclopédie* (1754), he defined the derivative as a limit of the ratio of increments and wrote:

> [Newton] never considered the differential calculus to be the study of infinitesimals, but the method of prime and ultimate ratios, that is to say, the method of finding the limits of these ratios.[9]

D'Alembert asserted further in the article "Limite" (1765) that "the theory of limits is the true metaphysics of the calculus." The limit was defined as follows:

> One magnitude is said to be the limit of another magnitude when the second may approach the first within any given magnitude, however small, though the second magnitude may never exceed the magnitude it approaches.

However, as the late historian Carl Boyer noted, d'Alembert was almost alone in considering the limit of a function to be the key concept around which the calculus was to be unified; most mathematicians still looked to the differential. His adherence to a geometric approach prevented him from putting the limit concept into a purely algorithmic form and thus kept it vague.

Lagrange had different ideas for making the foundations more sound. Although Berkeley's "compensation of errors" thesis impressed him in the late 1750s, he had by 1772 proposed that the differential calculus was legitimately based on the Taylor power series expansions of all functions. In his scheme functions retained the central role that Euler had assigned them. With the definability and generality of functions still hotly disputed and the emphasis on Taylor's series questioned, few major mathematicians accepted Lagrange's viewpoint at this time. Even Lagrange probably had some reservations.

The Berlin Academy, whose mathematical–physical section Lagrange now directed, fostered a search for a more conclusive explanation. In 1784 it announced a prize competition on the development of "a clear and precise theory of what is called the infinite in mathematics." Simon L'Huillier (d. 1840) won the prize in 1786 for a memoir that improved upon d'Alembert's discussion of limits. Proceeding with exemplary rigor, L'Huillier presented the derivative as the limit of the difference quotient and, in the modern style, considered $\frac{dy}{dx}$ to be a single sign rather than a ratio. He also introduced "lim" as a notation. L'Huillier's memoir was not widely read, perhaps because of its many verbose passages and its general lack of clarity.

In 1787 the St. Petersburg Academy opened a prize competition seeking to resolve a dispute over the nature of functions that had arisen between Euler and d'Alembert during their study of the vibrating string problem. Did arbitrary functions introduced by the integration of differential equations with three or more variables "represent any curves or surfaces whatsoever, either algebraic or transcendental, either mechanical, discontinuous, or produced by a simple movement of the hand?"—as Euler had argued—or did "these functions comprise only continuous curves represented by an algebraic or transcendental equation?"—as d'Alembert had maintained. The Alsatian Louis Arbogast (d. 1803), a partisan of Lagrange's latest views on foundations, submitted the prize essay. Nearly all functions employed in the eighteenth century were continuous, meaning constancy of analytic expression over an interval rather than the modern preference for connectedness of a graph. In a bold extension of Euler's ideas, Arbogast distinguished two ways that continuity could be broken. One type of discontinuous functions was defined by different laws (*i.e.*, those today referred to as piecewise smooth); whereas the second type was altogether intractable to definition. In addition, Arbogast introduced the modern sense of discontinuous, that is discontiguity with jumps. By assuming a geometric approach, he extended continuity to include contiguity, or connectedness.

The 18th-century search for rigorous foundations for the differential calculus culminated in Lagrange's *Theory of Analytic Functions* (1797). According to the subtitle of this comprehensive textbook, Lagrange wanted to expunge all considerations of the infinitely small (including the absolute zeroes of Euler), vanishing quantities, and fluxions. This segment of his ambitious program was ultimately unsuccessful. In a refinement of his earlier viewpoint, he based the calculus upon the safe analytic functions, rather than arbitrary ones, and gave a proof that all analytic functions can be expanded into Taylor series. Augustin Cauchy, however, soon discovered that some simple functions are not analytic, e.g., $f(x) = e^{-1/x^2}$. In addition, Lagrange's proof lacked the assumption of termwise differentiation necessary to complete it. To prove power series expansions required first accepting the existence of derivatives, a step he sought to avoid. Lagrange's lasting significance in foundations stems from his derivation for the first time of the Taylor series with remainder and from his thorough focus on investigation of the basic properties of continuous curves.

The calculus of variations and differential equations arose jointly from the study of various special problems and the effort to mathematicize mechanics. The calculus of variations, which deals with extreme value problems, requires maximizing or minimizing an integral. Its immediate origins lay in the Bernoullis' brachistochrone problem, in which an extreme value depends upon the form of a curve. The Bernoullis also proposed more difficult problems of the same type called isoperimetric problems. Euler generalized the evolving methods of solution for a large variety of problems in *Methodus Inveniendi Lineas Curvas Maximi Minimive Proprietate Gaudentes* ("The Art of Finding Curved Lines Which Enjoy Some Maximum or Minimum," 1744). With that work the calculus of variations emerged as a distinct branch of analysis. Euler's combination of geometric and analytical arguments based on successive differences, infinite series, and finite sums was often complicated, so that he recognized the need for a more satisfactory general method. In the appendix of the *Methodus* he solved two famous problems; the shape of an elastic beam under such tension that its stored potential energy is a minimum was obtained by an elliptic integral, and a special case of the principle of least action was presented.

Concern to exploit this principle substantially motivated mathematicians who subsequently developed the calculus of variations. Enlightenment men of science generally believed for theological, philosophical, or aesthetic reasons that nature is economical in operation. Pierre Maupertuis (d. 1759), who became Euler's official superior as president of the Berlin Academy, developed the principle of least action as a capstone of Newtonian dynamics. Impressed by the work of Euler, the young Lagrange propounded the first general formal statement of this principle for mass-point systems. From the "beautiful theorem" at the close of the *Methodus* Lagrange also developed a purely analytical procedure for the calculus of variations. That theorem held that a material point under the influence of central forces describes the same trajectory as the element of the curve multiplied by the integral of the velocity, assuming the latter product is a maximum or minimum. He skillfully employed variational principles in his *Mécanique analytique* (1788).

Leading Continental geometers expanded the field of differential equations as they further mathematicized mechanics. This mathematicization transformed mechanics into a thorough analytical science, whose methods, in turn, were unified by formulating basic laws as mathematical formulas, especially differential equations. Since Newton had written the *Principia* with a geometric format, setting and solving the differential equations for it was a central concern. Euler set a framework for analytic mechanics in *Mechanica* (1736), and definitively formulated the differential equations for Newton's laws (1750). These equations, now known as "Newton's equations," are $F_x = Ma_x$, $F_y = Ma_y$ and $F_z = Ma_z$. Euler continued to lay the foundations for analytic mechanics in *Theoria Motus Corporum Solidorum Seu Rigidorum* (1765). The Euler equations of rigid body motion simplified the complex set of differential equations for describing such motion. They involve torques on the body and rotation, usually about the center of mass. Lagrange's classic text on analytic mechanics and Laplace's on celestial mechanics soon followed. Laplace gave the first comprehensive principles for solving ordinary differential equations.

Clairaut, d'Alembert, Daniel Bernoulli, Euler, and Lagrange devised new differential equations as they competed with each other in attempts to solve physical problems. They met with varying degrees of success in their efforts to solve the famous three-body problem, especially the mutually attracting earth, sun, and moon. Its solution was critical for navigators anxious to develop moon tables accurate enough to permit a reliable determination of position at sea. Reliable tables based on Euler's theory and Tobias Mayer's observational astronomy appeared in the 1760s. Lagrange provided the most accurate procedure for approximating perturbations with a system of 12 differential equations in the 1770s, but the three-body problem has defied exact solution to the present. Partial differential equations were discovered to solve the vibrating string problem and major problems in hydrodynamics, such as the efflux of fluid from a vessel and determining the shape of the earth. Daniel Bernoulli's *Hydrodynamica* (1733–38), Clairaut's *Théorie de la figure de la Terre* (1743), and Euler's 1755 paper with the equations for the motion of compressible fluids are basic works in hydrodynamics.

Enlightenment developments in the theory of numbers and algebra were only minor. Objection to negative numbers persisted, and debates over complex numbers grew more strident. In a prize paper of 1747 d'Alembert asserted that complex numbers had the form $a + b\sqrt{-1}$. However, in the *Encyclopédie* he remained discretely silent on these numbers. Complex numbers were occasionally employed, but their position was not strengthened appreciably until the first proof of the fundamental theorem of algebra by Gauss (1799). The major figure in number theory between Fermat and Gauss was Euler. Building on Diophantus and Fermat, he contributed to the theory of divisibility, in part by giving three proofs of Fermat's lesser theorem. Euler worked directly or through correspondence with his best friend in St. Petersburg, Christian Goldbach (d. 1764). A discussion with Goldbach about Fermat's assertion that all numbers $2^{2^n} + 1$ are prime motivated his proof that it was invalid for $n = 5$. Euler also studied the zeta-function ($\zeta(s) = \sum_{n=1}^{\infty} \frac{1}{n^s}$). Other findings of Euler and Lagrange in the theory of numbers are given in their biographies. Euler's Berlin colleague Lambert used his expression of the number e as a continued fraction in proofs that e^x (for x any positive integer) and π are irrational.

In Paris Adrien-Marie Legendre (d. 1833) took up the work of Euler, Lambert, and Lagrange in the theory of numbers. In 1785 he independently formulated and named the law of biquadratic reciprocity and gave an incomplete proof for it. Legendre published his findings and many of those of his predecessors in *Essai sur la théorie des nombres* (1798, 3rd ed., 1830). It contains his law for the distribution of prime numbers (if m signifies the number of primes that are less than n, then $m = \frac{n}{\log n - 1.08366}$). This collection suffers from its lack of general conceptions. A new stage in the theory of numbers was about to open with the work of Gauss.

The names of the major figures in 18th-century algebra are familiar ones. The period had two major textbooks. Clairaut wrote *Élémens d'algèbre*. Like his text on geometry, it follows a pseudo-historical approach. Its section on elementary problems appeals only to common sense. As the reader works up to the solution of fourth degree equations, difficulty gradually increases and symbolism is introduced. Euler's *Vollständige Anleitung zur Algebra* ("Complete Introduction to Algebra," 1770) is today considered the best algebra text of the century. A contemporary might not have expected this after Lagrange's comment to d'Alembert in 1770: "It contains nothing of interest except for a treatise on the Diophantine equations, which is, in truth, excellent." Lagrange added brief notes to the French translation

(1773). Euler's search for solutions by radical of the quintic equation and his approximating methods of solution for equations brought out the paradox of higher degree equations, which Abel and Galois were to settle.

Lagrange and Alexandre-Theophile Vandermonde (d. 1796) published the most subtle studies concerning the resolution of equations by radicals. Lagrange's critical-historical investigation appeared in two massive memoirs at the Berlin Academy (1771 and 1773); Vandermonde's analogous but independent study was published in the *Mémoires* of the Paris Academy (1774). Examining the conditions for solvability of equations, Lagrange provided a general method which could be applied successfully to second, third, and fourth degree equations but not to those of the fifth degree. He abandoned the search for a solution of the quintic by algebraic operations, concluding that it was likely impossible—an insight on which Abel later capitalized. Lagrange also introduced the theory of permutation or substitution groups upon which Galois expanded.

In probability theory Enlightenment mathematicians developed many analytical results and broadened the fundamental concept of expectation. Abraham de Moivre (d. 1754) wrote *The Doctrine of Chances* (1718), *Annuities on Lives* (1724), and *Miscellanea Analytica* (1730). *The Doctrine of Chances* was a masterpiece. In it de Moivre proved the weak law of large numbers by following upon the work of Jakob Bernoulli and introduced the concepts of statistical independence and conditional probability, *i.e.*, the probability of A conditioned on the event A_1 occurring. There is statistical independence if and only if the conditional probability of two events is equal. De Moivre further first utilized the probability integral $\int_0^\infty e^{-x^2}\, dx = \frac{\pi}{2}$ and originated Stirling's formula, *i.e.*, $n! = \sqrt{2\pi n}e^{-n}n^n$. Using Stirling's formula, he approximated the binomial probability distribution which proved very fruitful as the Gaussian or normal frequency curve. Statistical inference evolved slowly. The Scottish cleric John Arbuthnot (d. 1735) had introduced it in a paper concerning the regularity of births of both sexes (1710). The English Nonconformist Thomas Bayes (d. 1761) established a mathematical basis for statistical inference by using probability inductively and by analogy when no prior probability of a statistical hypothesis existed. With this technique, he could find the probability even when no antecedent events were known. The technique prevailed until challenged by George Boole in *Laws of Thought* (1854).

Mathematicians heatedly debated the definition of probabilistic expectation until the end of the 18th century.[10] In attempting to enlarge the domain of probability, the inherited concept of expectation with its equitable and prudential aspects was altered to embrace the whole of rational decision making. Economic, legal, and psychological considerations were taken into account. The conduct of "reasonable man" was the fixed standard. In an article for the St. Petersburg *Commentarii* (1738) Daniel Bernoulli attempted to resolve the St. Petersburg paradox which concerned the number of heads or tails in a coin toss. Its importance was that no reasonable man would wager based solely on its mathematical analysis. This reluctance seemed an affront to common sense. Bernoulli's resolution made a distinction between mathematical and moral expectation. The latter signified a mean economic utility or "the power . . . to procure . . . felicity." D'Alembert, who believed Bernoulli oversimplifed economic risk taking, strongly criticized probability theory in several *Encyclopédie* articles. Imbued with the sensationalist views of Locke and Condillac, he castigated conventional probability theory because its practitioners neglected experience as they emphasized the pursuit of abstract truth.

These arguments had their effect later in the century. Georges Buffon (d. 1788) distinguished three types of certainty—mathematical, physical, and moral—and attempted to quantify them. Where Bernoulli's view turned on economic factors and d'Alembert's on physical experience, Buffon added a psychological standard of reasonable belief. Marquis de Condorcet (d. 1794) clearly distinguished between absolute and subjective probability in an effort to justify the classical definition of expectation as an average. Unlike his colleagues, Condorcet applied statistics to political economy and voting processes rather than to natural science. Proceeding in the spirit of d'Alembert, Laplace formulated the Bayes-Laplace theorem on inverse probabilities and wrote *Théorie analytique des probabilités* (1812), which systematized classical probability. It is discussed in the biography of Laplace in this anthology.

Enlightenment geometers examined the "problem of parallels," criticized Euclid, and created differential geometry. The problem of parallels refers to the necessity of Euclid's fifth postulate in Book I of the *Elements*. The Jesuit Girolamo Saccheri (d. 1733) attempted to prove its necessity in *Euclides ab omni vindicatus* ("Euclid Vindicated from All Faults," 1733). His work was an advance over past efforts addressing the same topic. He explicitly formulated for the first time three hypotheses: the sum of the angles in a triangle is either equal to, always greater than, or always less than two right angles. Then by *reductio ad absurdum* he attempted to eliminate the obtuse and acute angle hypotheses as self-contradictory leaving the first Euclidean hypothesis as the true alternative. Saccheri believed that he had disproved the hypothesis of obtuse angle but could not satisfactorily eliminate that of the acute angle. He finally rejected the latter only because he found its consequence of asymptotic straight lines "repugnant to the nature of the straight line."[11] Although Saccheri's proferred confirmation of Euclidean geometry was to prove to be inconclusive, his argument concerning the consequences of denying the parallel postulate had great merit. Since his book was not widely read, it had a limited influence. The problem of parallels continued to taunt geometers.

Clairaut and Euler contributed significantly to geometry during the mid-18th century. Influenced by anti-Euclidean ideas of Petrus Ramus and Blaise Pascal, Clairaut rose to the attack in his didactic work *Elemens de geometrie* (1741). He found the first propositions of geometry in land measurement and advanced the subject by a natural or pseudo-historical path to self-evident truths in more abstract areas. Clairaut sought "only to avoid any false steps that . . . [geometry's first inventors] might have had to take." His goal was to encourage beginners to rediscover geometry by themselves. He and Euler also introduced powerful analytic techniques into geometry by founding three-dimensional differential geometry. Clairaut introduced the theory of space curves, that is curves of double curvature, in the 1730s and treated geodesics of surfaces of revolution at length in his classic treatise on the shape of the earth (1743), but he soon afterward turned to other subjects. Following a proposal of Johann Bernoulli Euler had given differential equations for geodesics on surfaces as early as 1728, and he subsequently pioneered the study of topological surfaces. Euler's chief contribution to differential geometry was the theory of surfaces, which he established in a landmark paper of 1760.[12] In that paper he determined "the radius of curvature of any plane section of a surface" and established "a proper idea of the curvature of surfaces." Euler still assigned to Euclidean, three-space the primary role in creating geometry and endowed surfaces with an induced geometry.

Differential geometry was not generally recognized and developed until Gaspard Monge (d. 1818) renewed its study by expanding upon the work of Clairaut and Euler. In the 1770s and 1780s he completed Euler's study of developable surfaces and introduced the notions of focal surface, normal, congruence of lines, and line of curvature to the theory of surfaces. Through his rich lectures at the Ecole Polytechnique at the turn of the century and his text *Application de l'analyse a la géométrie* (1807), Monge influenced many students to investigate differential geometry.

Imaginative investigations of the problem of parallels conducted in the latter part of the eighteenth century continued to suggest that Euclidean geometry might be flawed. At the Berlin Academy Johann Heinrich Lambert (d. 1777) wrote *Theorie der Parallelinien.* He left this manuscript aside in 1766 and it was not published until 1786. Although Saccheri's name is not mentioned, Lambert expanded upon his Italian predecessor's hypotheses. Noticing that the second, or obtuse, hypothesis describes angles in spherical geometry he argued by analogy that the acute hypothesis might occur on an imaginary sphere. In either case one might dispense with the parallel postulate. Unfortunately he considered this thought to be only a clever trick and did not make a connection between it and his study of hyperbolic functions (that extended Euler's work). For Lambert the problem of parallels was simply foundational—a matter of consistency and necessity—and not a matter of the nature of space. His interest in Euclid waned soon after he began corresponding with Kant (1765–70). His influence on the great philosopher probably pertained more to the subjects of deductive reason, space, time, matter, and force. Lambert's *Theore der Parallelinien* may have had some influence on Gauss, who borrowed it at the University of Göttingen library in 1795 and 1797. At the University of Edinburgh John Playfair (d. 1819) considered Euclid's treatment of linear parallelism unsatisfactory. Following an idea suggested by Proclus, he gave the parallel postulate in a modern form in his *Elements of Geometry* (1795). Today known as Playfair's postulate, it states (in one current form): Given a line L and a point p not on L there exists one and only one line M containing p and no points of T.

These studies did not soften the staunch defense of Euclid given by Kant and Legendre who asserted that it was both true for our physical world and unique. Early in his career Kant had believed that alternative consistent geometries were possible in principle. Afterwards, he characterized geometry as "the most faithful interpretor of all phenomena" in his *Inaugural Dissertation* (1770). By the time he wrote *Critique of Pure Reason* (1781), he was asserting that non-Euclidean geometries could be invented for purely imaginary space, but Euclidean geometry alone describes the actual physical universe or even highly probable space. The synthetic a priori proofs of Euclidean geometry provided authoritative knowledge about the physical universe. Euclidean space retained its privileged status. It is highly ironic that Kant, who defined the limits of pure reason, stopped short of recognizing the closure property in geometry. Euclidean geometry applies to local space and non-Euclidean geometry to large, curved space.

More important for what has been described as "the return to Euclid" was Legendre's *Elements de géométrie* (1794). It dominated elementary instruction in the subject through its many editions and translations for nearly a century. The tireless Legendre wanted a return to the rigor of Euclid in instruction; he opposed the lesser formalism of Clairaut and the Encyclopedists in didactics. He mistakenly thought that he had disproved the hypothesis of the obtuse angle and confirmed the parallel postulate. A disciple of Newton, Legendre accepted absolute Euclidean space. Not even his studies in spherical geometry and spherical trigonometry could change his mind. Legendre's *Elements* subsequently made it difficult for the inventors and early advocates of non-Euclidean geometries to have their ideas accepted.

Brief Survey of the Enlightenment. European mathematics developed within the social and intellectual context of the Enlightenment from 1720 until 1794, when a revulsion set in against its methods and ideas in the midst of the French Revolution.

Immanuel Kant coined the German word "Aufklärung,"[13] from which our word Enlightenment comes. This period is also called *siècle des lumières* in French or simply the Age of Reason. The term "Reason" does not mean sheer speculation about grandiose systems—this feature of seventeenth century thinking was criticized. Rather it refers to scientific method anchored in experience. In Voltaire's words there had to be an appeal *au fait*—to the facts!—rather than traditional methods of theology and metaphysics. Unless used with caution, the term "Reason" can obscure the full range of the age's rich and diverse developments, such as its deep sentimentality.

Every period is one of contrast and historic change. The Enlightenment was unusually so based on its sharp growth in agricultural productivity, population, business activity, industrial technologies, paths of commerce, and Empire. The Enlightenment was also an age embracing reforms in the society of the Ancien Regime,[14] changing contours of religion, cosmopolitan culture, and powerful, new thought, especially in economics, history, and the sciences. It was dynamic; "Everything that was alive and most active in [its] life was of a new order, indeed not merely new but outspokenly hostile to the past."[15] "The age," stated Samuel Johnson in 1783, "is running mad after innovation."

The pace of change had quickened during the mature Enlightenment after the 1740s. The Philosophes, a group of thinkers and critics, and the ideas that they championed, such as toleration, humane behavior, and often the diffusion of the sciences, became fashionable. Their diverse ideas and programs began to influence energetic and paternalistic monarchs known as enlightened despots who ruled in central and eastern Europe. The enlightened despots mixed these ideas with *raison d'état* in their pursuit of stronger, more effective monarchical government. Patronage of Philosophes, savants, and men of science increased as monarchs competed with each other for their services. The mature Enlightenment spread from Paris across Europe (and even across the Atlantic to North America), although its progress was not unopposed.

Central to the changing order was the continuing sharp tension between religion based upon faith in revelation and discursive reason of the republic of letters and the exact sciences.[16] The triumph of reason provoked a "crisis of conscience." The success of Newtonian dynamics, the epitome of the new sciences, was decisive, however. The precision of Newtonian science and its promise of control over nature were not only impressive by contemporary standards but also appealed to the quantitative sense and the commitment to practicality of the Enlightenment mind. While astronomers of the previous century worked with errors of over one degree of arc, Newtonian theory developed an accuracy to one-thousandth of a percent. Religious authority retreated before an advancing spirit of secularism or, what was worse, witnessed the transmutation of orthodox faith into Deism, a more deliberately rational doctrine than Christian theism. There was a parallel tendency to replace religious allegiance as the primary means for achieving human betterment and happiness. Materialism, skepticism and doctrinal quarrels also shook Orthodox Christianity.

A qualified, rational optimism in the inevitability of material, intellectual, and moral progress spread sparingly among literate Europeans. The optimistic temper varied across Europe, usually taking deeper root in western Europe and the English colonies in America. Since literary men soberly recognized the immense difficulties facing society, only a few, like Helvetius, Turgot, and Condorcet actually embraced the theory of progress. The metaphysical optimism of Leibniz, embodied in the "best of all possible worlds" doctrine, was its major intellectual source. By the latter part of the eighteenth century it was neither immune from criticism nor was it facile and naïve as Romantic critics subsequently charged. Voltaire scathed an exaggeration of Leibnizian optimism in his satire *Candide* (1759), written shortly after the catastrophe of the Lisbon earthquake. The epitaph for the eighteenth-century theory of progress—the Philosophe Condorcet's rosy *Esquisse d'un tableau progrès de l'esprit humain* ("Sketch of the Progress of the Human Mind," 1794)—recognized, to be sure, an inherent and inescapable improvement in the world through the application of reason. At the same time Condorcet noted the darkness, ignorance, superstition, and suffering that still existed in the world.

The optimistic temper was also encouraged by the conviction that human suffering could be reduced. The first stirrings of modern humanitarianism came with the rising criticism of dreadful forms of suffering by advocates now known as partisans of humanity. As historian Michel Foucault has argued, the use of torture and mutilation ended rather suddenly in the decades after the 1760s and was replaced by confinement and surveillance in prisons.[17] Mass public hangings of men, women, and children were disappearing. The Inquisition had less effect in Catholic lands, although people were still broken on the wheel in *autos de fe*. Despite the humanitarian trend, the slave trade did reach huge proportions; about six million black Africans survived the Atlantic crossing. Slaves worked in the expanding plantation system that brought great profits. Still reformers criticized slavery with some effect. It was monstrous and degrading to master and slave alike. To keep the level of adversity in perspective, one should bear in mind that by twentieth century standards (except for the prosperous aristocracy and comfortable bourgeoisie) life was then very harsh.

Eighteenth-century society continued to be basically agrarian with its traditional hierarchical class structure and aristocratic privilege. Over 75 percent of the people lived in the country. The tillers of the soil west of the Elbe River were generally peasants subject to feudal dues, while east of the Elbe in Prussia, Austria, and Russia they were serfs legally bound to a particular plot of land and lord. Russian serfs were economic commodities who were little more than chattel slaves. In the west the peasantry were poor and tended to live in wretched housing on manorial estates, where they worked for the substantial aristocracy and the higher clergy. As a group the aristocracy stood atop the social pyramid of the time. In a few instances the peasantry worked on the estates of the wealthy bourgeoisie. Since the landowners usually found ways to escape most taxes, the burden of state taxation fell on the peasantry. In this society change came slowly.

Nevertheless at least four fundamental changes affected the conditions of European life. The first was a momentous increase in food production. A new husbandry was available to replace the two-field system of cultivation employed since the high Middle Ages. Yet that system and its three-field variant for more fertile lands were used in only half the arable land in 1700. Fallowing was necessary to prevent exhaustion of the soil but under the three-field system two-thirds of the land rather than one-half could be tilled. Concomitant gradual changes transformed densely populated and prosperous Holland into more efficient high farming land. High farming replaced the traditional and slow method of fallowing with the rotation of selected legumes, such as clover and turnips, with Europe's principal grains, wheat and rye. The Dutch and Flemish also fenced in animals, which allowed them not only to save fields for tillage but also to collect manure as fertilizer and to breed animals selectively. Consequently, supplies of meat and milk increased.[18] In England Jethro Tull (d. 1741) used iron plows to turn the soil more deeply and improved the "horse hoe" to remove weeds from between crops planted in rows following the new French model, while Charles Townsend (d. 1738) pioneered the use of the turnip to enrich the nitrogen content of the soil. Perhaps the most important development in agronomy was the spread of potato cultivation. This humble immigrant from the New World has a fine supply of carbohydrates and calories and, if the skins are eaten, vitamins A and C as well. An acre of potatoes produces enough nutrition in different soils to feed a small family all winter. The potato was at first considered fit only for animals. By the 1770s, however, it was an important dietary supplement for humans and was well on its way to becoming a staple. These improvements in agriculture began in western Europe and spread slowly eastward.

Better nutrition did not eliminate the enforced social and economic dependency of the peasants but under the new agriculture landlords sought handsome profits from their crops. Peasant society clung to the goal of ensuring a steady quantity and quality of local food supplies by proven, traditional methods. Dark bread made from a mixture of ground wheat and rye was quite literally the staff of life. Traveling east, one generally found that the food supply became more precarious and peasants often resisted agricultural innovation. In Britain the gentry consolidated or enclosed their lands in search of larger commercial profits. The enclosure movement forced poor cottagers and smaller farmers (who needed common pasturage) from the land and led to misery and riots. However, agriculture became more productive. Across the Continent peasant uprisings, known as Jacqueries (from the French surname), were the response to agricultural innovation, unfair pricing, new rents or dues and harsh treatment. These uprisings often sought to preserve or restore tradition. The most massive revolt, the Pugachev Rebellion (1773–1775), occurred against Catherine the Great in Russia. Eager for new taxes and dependent on the nobility, European governments used their armies and militia to smash the peasants. In the Pugachev revolt the survival of the monarchy and aristocratic privilege were at stake.

The second change related to improved nutrition was a sharp upsurge in population. Europe (excluding most of the Balkans) went from an estimated 100 to 120 million people in 1700 to about 190 million in 1800. About two-thirds of the demographic growth came after 1750 and represents the beginning of the modern population explosion.[19] There was growing pressure on food supplies in the country as well as in the towns and cities. Cities expanded their walls and acted, in the words of the historian Fernand Braudel, as "electric transformers" serving to "increase tension, accelerate the rhythm of exchange and ceaselessly stir up men's lives."[20] The number of towns with populations over 20,000 increased considerably. Broadly speaking, there were three types of cities: market centers, commercial and financial centers, and great capitals into which vast wealth flowed. London approached one million people by 1800, Paris exceeded a half million, Berlin 170,000, and St. Petersburg 250,000. As the most dynamic element in the urban society, the middle class was on a collision course with the entrenched aristocracy.

Demographers evaluate the causes for the sharp population increase differently. The chief factor allowing for sustained growth was probably a larger and more varied food supply. Population growth was undoubtedly also a cause of agricultural improvements. The weather, far better than in the previous century, also added to the size of the harvest. Terrible famines had occurred in 1693–94 and 1709–10, when as many as one million people died of starvation. Although there was a grain shortage in 1769, the next grain crisis did not occur until the drought and hard freeze of 1787–88. Besides better nutrition there were improvements in medicine and public health leading to a reduced infant mortality and the effective disappearance of bubonic plague. Other diseases, such as smallpox, dysentery, and cholera, still acted as controls, however. Better hygiene may have promoted the population increase but only in a minor way. In the more populous cities the contrast between noble townhouses and the wretched quarters of the poor was stark indeed. Wells were polluted, the air fetid, and general sanitation standards low. Finally, changes in warfare fostered a larger European population. The 18th century had two prolonged periods of peace from 1720 to 1734 and from 1763 to 1778. Armies came under stricter control and generals avoided pitched battles. Soldiers were clothed, fed, and housed by the state so that looting was not required for subsistence. In addition, soldiers were recruited from among criminals, vagrants, and debtors, leaving local areas safer and more productive.

The birth of the Industrial Revolution was the third major change. The clustering of population in cities and the growing agricultural productivity nearby provided concentrations of potential workers, inexpensive basic sustenance and a potential market. Although James Watt's patenting of the steam engine in the 1760s accelerated it, this revolution had been brewing for some time. It had other important components. A new form of industrial organization, the factory system, proved superior in some respects to the prevailing "domestic system." Under the domestic system raw products, like yarns, simple machines like handlooms, and parts for mirrors were distributed to peasant weavers and artisans who did the work in their cottages. The finished product was collected by the owner or his agent. Factories were originally an expedient to introduce large machinery and later to allow the introduction of steam power. They served to bring people together to work under supervision and in time to put a premium on "industrial discipline." As factories spread, there was tenacious resistance by guilds of handloom weavers and artisans.

The Industrial Revolution spread slowly from west to east. One nation, Britain, and one industry, textiles, were acquiring industrial dominance by the 1790s. The British possessed three crucial elements for success in industry—a relatively well-educated segment of the population, plentiful supplies of capital and natural resources (coal and iron ore), and superior textile technologies. John Kay's flying shuttle (1733) halved the number of weavers needed to work looms; James Hargreaves' spinning jenny (1770) allowed the simultaneous working of 16 spindles; while Richard Arkwright's water frame (1769) and Samuel Crompton's spinning mule (1779) allowed fine and coarse yarns to be spun in unprecedented quantities. On the Continent, France, which emphasized luxury items like tapestries, remained the industrial leader until the Revolution. Bohemia was also important in industrial production. Sweden and Russia increased their output of iron, so important for military purposes; but in the west there was a shortage of timber for smelting purposes.

The mid-18th century saw an intense rivalry in colonial trade between France and Britain world-wide as well as renewed conflict on the Continent, chiefly between Prussia and Austria for hegemony in central Europe. Russia was drawn into the Continental wars. Commercial competition and central European warfare were separate but not unrelated; the struggle among these five powers was the major feature of Enlightenment politics. Although Spain retained its sprawling colonial system and the Netherlands was important in transit trade, neither could compete with France and Britain, whose centralized national governments, effective systems of water transport (at home and abroad), and advanced naval technology (ordnance) placed them far ahead of others. The French overseas trade increased fourfold, while the carrying capacity of the British merchant fleet increased fivefold. Enhanced trade with the New World, India and the Baltic put pressure on Europe's market economy, spurred industrial growth among leading colonial powers, and led to new forms of smuggling.[21]

The flash points in the duel between France and Britain were in North America and the West Indies. The West Indies, the "jewels of empire," produced cotton, coffee, tobacco, indigo, and above all sugar. Sugar became a basic consumption in Europe used in coffee, tea, and cocoa, in making candy and preserving fruits, and in brewing. With planters unable to attract free labor, planting and indeed much of the prosperity of the transatlantic trade rested on slavery. Britain with her vocal, well organized pressure groups of shippers, merchants and bankers gained a world empire. This occurred as two decades of war on the Continent eroded France's position as a colonial competitor.

Shortly after he ascended the throne in 1740, the 27-year-old Frederick II of Prussia had his troops seize the prosperous Hapsburg province of Silesia, over which the untested 23-year-old Maria Theresa ruled. This ungallant act upset the continental balance of power and began the War of the Austrian Succession (1740–48). The Peace of Aix-la-Chapelle—little more than a truce—recognized the Prussian occupation of Silesia; but Maria Theresa had proven a formidable opponent. Led by able ministers, such as Prince Wenzel von Kaunitz, her government had gained the allegiance of its Hungarian subjects by guaranteeing their relative autonomy. It also established an effective bureaucracy, the Public and Cameral Directory, to administer Hapsburg lands and collect taxes under Viennese control rather than that of local aristocrats. During the war France's de facto prime minister, Cardinal Fleury (d. 1743), abandoned a planned attack on British trade to support Prussia's aggression against France's old enemy, Austria. It was a fateful decision for the future.

Other factors contributed to the Continental wars. In the "diplomatic revolution" of 1756 Britain allied with Prussia, while France joined Austria. This diplomatic realignment set the stage for the Seven Years' War that began when Frederick invaded Saxony and Austria, France, and Russia joined against him. Without massive financial aid from Britain and the death of his implacable foe Empress Elizabeth of Russia in 1762 coupled with the new tsar's withdrawal from the fray, he would have lost. Instead Frederick saved the integrity of his state and the peace in central Europe was based on the *status quo ante bellum*.

Nevertheless the Seven Years' War had far-reaching consequences. Prussia was recognized as a European power. More impressive to Europe was the outcome of the struggle between France and Britain outside the Continent. France suffered military losses to Britain in North America and India and her maritime trade fell from 30 million to 4 million livres. Large debts and reform efforts faced governments for the remainder of the century. Britain lost the 13 American colonies and saw demands for parliamentary reform persist until 1785, when the 22-year-old prime minister William Pitt the Younger, abandoned them. France was on the road to revolution.

A veritable explosion of business activity together with a new economic theory comprised the fourth fundamental change in Europe. The year 1720 was not an auspicious beginning. Excessive speculation led to a crash of London's South Sea Company involving trade with South America and Paris's Mississippi Company designed to exploit the wealth of Louisiana. While the bubble had burst on the speculative boom associated with these enterprises, ruining many fortunes and causing Bishop Berkeley to ponder the moral ruin of the old world, this experience did not stem business activity. It expanded even more rapidly, although with cyclic crashes. The new agricultural productivity, growing maritime trade, and the developing industrial revolution that employed machines in production for profit gave rise to a new stage of economic history in Europe—industrial and agricultural capitalism. The new entrepreneur hoping to make a profit in an uncertain market tended to be hostile to existing mercantilist theory, the economics of scarcity, based on monopolistic guilds and absolutist states. He sought freedom to make profits and endorsed those economic theories that proposed natural laws akin to Newtonian theory in dynamics.

Several new schools of thought broke sharply with the old mercantile tradition. In France the economic reformers were called Physiocrats. The word physiocracy literally means "the rule of nature." Their leading spokesman, the physician François Quesnay (d. 1774) believed that there was an analogy between William Harvey's circulation of the blood and the circulation of wealth in society. Quesnay wanted impediments to the "free" flow of wealth, originating in the land, removed. He and his disciples popularized the slogan *laissez-passer, laissez-faire.* Influenced by the physiocrats, the Scot Adam Smith (d. 1790) wrote *Inquiry into the Nature and Causes of the Wealth of Nations* (1776). He urged the abolition of the strictly regulated English mercantile system and advocated a new economy based on goods and services, not gold and silver. Once the mercantile regulations were abolished, Smith, a Philosophe, believed that an invisible hand would harmoniously and automatically direct the economy. Although this view seems theologically inspired, he was not dogmatic. While he urged a "hands off" policy by government in most areas, he argued for government to play a major role in maintaining transportation and opening dangerous new trade routes.

The Philosophes, the leading voices of the Enlightenment, championed change and reform. Active thinkers and propagandists, they set out to discredit blind authority, time honored myths, superstition, and outworn institutions, whether of a closed, intolerant church hierarchy or of absolute monarchs. They sought to create in their place a new society based on religious toleration, civil liberties, a justice more nearly equal, and representative government. They communicated these ideas to a wider audience than previously addressed through books, pamphlets, novels, encyclopedias, dictionaries, and magazines. In choosing to call themselves Philosophes, the men of the Enlightenment stressed the power of critical reason and common sense. The writings of Isaac Newton on physical science, John Locke on psychology, and Pierre Bayle on literary criticism chiefly inspired them to use rational criticism. By extending man's critical mind to human activity, they believed they could change tradition, practice, and deeply rooted customs of society.

Despite strenuous quarrels over many issues, the Philosophes shared as a goal the pursuit of human liberty and may be likened to a family. Most were Frenchmen who benefitted from a rich language. They and many aristocrats frequented fashionable Parisian salons of talented women, such as Madame Geoffrin, where they engaged in sparkling conversation and sharpened critical thinking. Other Philosophes were Italian (Cesare Beccaria), British (David Hume, Edward Gibbon, and Jeremy Bentham), German (Gotthold Lessing and Immanuel Kant), and American (Thomas Jefferson). In publishing their ambitious ideas and assumptions, their Continental members faced in many instances official censorship and imprisonment.

Historian Peter Gay has divided the Philosophes into three generations. The aristocrat Montesquieu (d. 1755) epitomizes the first. His influential book *The Spirit of the Laws* (1748) extolled representative government with checks and balances and critically examined all forms of government, including monarchy, as products of history and climate. Montesquieu established a basis for social criticism that the second generation extended. The second generation leaders, Voltaire (d. 1778) and Denis Diderot (d. 1784), were members of the middle class. Their generation had the widest, most profound influence of the Philosophes.

Voltaire, born François-Marie Arouet, praised religious diversity and free intellectual inquiry. Early in his career he pointed to the English model as a basis for these in *Philosophical Letters on the English* (1734). The premier literary figure of his century, Voltaire turned satire, ridicule, and invective against the Catholic Church, viewing it as a refuge of intolerance, and against all clergy, who supported persecution or exhorted their followers to wars over creeds. He found wars and militarism hateful. Voltaire's *Philosophical Dictionary* (1764) drew attention to the immoral acts of biblical heroes. His motto was *Écrasez l'infâme* ("Crush the Infamous Thing"), by which he meant repression, an intolerant Church, and prejudice. A tireless worker, who often spent a 20-hour day on study and writing, he also defended men whose chances for a fair trial were endangered by charges of heresy.

Diderot, the son of a cutler, was as indefatigable as Voltaire. His most brilliant achievement was editing the famous *Encyclopédie*—a task to which he devoted 25 years of his life. Although many contemporary encyclopedias are dull compendia, Diderot's work was revolutionary. It was an instrument to "change the general way of thinking" in the Ancien Regime. More than 100 Philosophes contributed; their articles scorned superstition, popularized the sciences and mathematics, and described crafts and industries. Official censorship was generally avoided by placing criticism of religion and social inequality in such unlikely places as articles on metallurgy and lace-making. Still the *Encyclopédie* experienced censorship and after 1759 suppression. Completed in 1772 (17 volumes of text and 11 volumes of illustration), it became a standard reference work.

If Voltaire and Diderot shaped in many ways the mind of Enlightenment Europe, Rousseau (d. 1778) won its heart. He appealed to the emotions (sensibility) rather than to deductive reason and urged people to "follow nature," but clearly he did not accept primitivism. He believed that artifices in advancing society corrupted humans. Rousseau urged a more open process for educating the young in *Émile* (1762). Among adults he called for an extreme form of democracy based on a social contract between rulers and ruled. Guidance was to be gained from a General Will within his collective society, a concept chiefly inspired by Plato and the idealized ancient Greek *polis*. Immanuel Kant, a third generation Philosophe, indicated Rousseau's influence in adding a more human perspective to Enlightenment thinking. After reading *Émile* Kant wrote:

> By inclination I am an inquirer. I feel a consuming thirst for knowledge, the unrest which goes with the desire to progress in it, and satisfaction at every advance in it. There was a time when I believed this constituted the honor of humanity, and I despised the people, who know nothing. Rousseau corrected me in this. This blinding prejudice disappeared. I learned to honor man.[22]

While their intellectual achievement was great, the Philosophes were equally concerned about applying their ideas. Most of them favored neither Montesquieu's conservative reforms nor Rousseau's contractual democracy but looked to monarchs to bring about change. The second generation succeeded in ameliorating many past evils by working with enlightened rulers, notably Frederick II of Prussia, Maria Theresa and Joseph II of Austria, and Catherine II of Russia. There followed a series of limited "enlightened" reforms, even as the royal reformers continued the usual policies of "irrational" militarism. Early in his reign the Francophile Frederick corresponded with Voltaire and hosted him in Berlin, "the Sparta of the North." Frederick turned the Berlin Academy into a vital intellectual center by recruiting Maupertuis, Euler, La Mattrie, and Lagrange; he allowed Catholics and Jews to settle in his Lutheran country; and he ordered a codification of law to unify the Prussian system and reduce aristocratic influence.

In Austria the reforms began in Maria Theresa's court. Her ministers broke the power of the clergy at the University of Vienna and reformed its curriculum; they also reduced the *robot*, the services due the landlord by the serf, and they abolished torture for minor crimes. Maria Theresa's eldest son, Joseph II, who was "a good deal of a prig," perhaps carried enlightened despotism to its farthest extent. In the 1780s he granted toleration to all Christians against the wishes of the Catholic Church and abolished serfdom, a privilege of the nobility. These measures, which alienated all too many powerful groups, were revoked by his successor, Leopold II.

In imperial Russia Catherine II, who is known today for her many lovers and animal paramours, brought a limited imprint of the Enlightenment. Its effect extended to the nobility and extremely small resident intelligentsia. The nobility hired French tutors; the monarch consulted Diderot and Euler on educational reforms, and the *Encyclopédie* was allowed to circulate briefly. Catherine, who cultivated friendship with many Philosophes, also convened a legislative commission to revise the Russian laws and government based on instructions containing their ideas. These revisions were not realized during her reign.

Eighteenth-century Europe had a vital high culture. Literature and music flourished. To give but a few examples, England had the poetry of Alexander Pope (d. 1744) and the literary criticism of Dr. Samuel Johnson (d. 1784), who was a conservative opponent of the Philosophes. The urbane Dr. Johnson compiled the epoch-making *Dictionary of the English Language* (1755). The modern novel began with Samuel Richardson's sentimental *Pamela* (1740). From the ideas of the German *Sturm und Drang* Johann Wolfgang von Goethe (d. 1832) brilliantly developed the novel. In Central Europe the baroque masters Johann Sebastian Bach (d. 1750) and George Frederick Handel (d. 1759) composed elaborate church music and created new secular, instrumental music. Late in the century the Croatian Franz Joseph Haydn (d. 1809) and the Austrian Wolfgang Amadeus Mozart (d. 1791) perfected chamber and symphonic music in graceful classical forms.

Natural science and history were the two dominant disciplines of the Enlightenment. From them Philosophes and savants drew inspiration. Natural science offered precise methodologies and a new vision of the universe. History developed a critical and often elegant style that could be used to cut away pretense, ineptitude, and abuses in social and political affairs. That style appeared in the modern historical literature that began with Giambattista Vico's *The New Science* (1725), Voltaire's *Age of Louis XIV* (1751), and Gibbon's *Decline and Fall of the Roman Empire* (1766–88).

The diffusion and elaboration of Newtonian science was the most important development in 18th-century science. Newton had presented the core of his ideas on dynamics in the *Principia* (1687) and on the corpuscular theory of light in the *Opticks* (1704). Only those expert in mathematical science could substantially expand upon or modify his dynamics. On the Continent Newtonian science gradually supplanted the dominant Cartesian science. Precise empirical research, on the movement of tides, the orbits of comets, stellar motion, and the shape of the earth together with efforts of popularizers, like Algarotti and Voltaire, led to the Newtonian triumph. The University of Leiden faculty supported Newtonian ideas by the late 1720s. The principal center for the Newtonian-Cartesian polemic, however, was the Paris Academy. The Cartesian position became desparate in the 1730s: Maupertuis accepted Newton's dynamics in his *Discourse on the Shape of Stars* (1732); Voltaire praised Newton's ideas in *Philosophical Letters* (1734); Daniel Bernoulli won a prize for a paper supporting Newton's astronomy (1734); and Maupertuis and Clairaut confirmed that the earth was flattened at the poles on the Lapland expedition (1736–37) as Newton's theory held. The earth was not lemon-shaped as the Cartesians believed. Voltaire published *The Elements of Newton's Philosophy* (1740) dealing mainly with optics. Subsequently, Newtonian science was accepted by the University of Paris faculty in 1745 and by the Marquis d'Argens in the 1746 edition of his book *La philosophie du bons sens* (1737). In the 1750s d'Alembert expounded Newtonian ideas in the *Encyclopédie*.

After 1740 Newtonian science spread eastward and had dramatic confirmations. At the Berlin Academy Maupertuis and Euler extensively developed Newtonian dynamics and the calculus. Euler was selective. He rejected Newton's corpuscular optics for Huygens' wave theory. At Königsberg Kant went beyond Newton by proposing that attraction and Newtonian dynamics operated throughout the entire universe. The precision of Clairaut's prediction of the return of Halley's comet (1759) and Laplace's establishment of the long-term stability of the solar system further confirmed the power of Newtonian science. These Newtonian studies were decisive in convincing Europe's intelligentsia that there is a mathematical design in nature and that the laws of nature are mathematical.

There were many significant developments in science besides those in dynamics. In medicine Hermann Boerhaave (d. 1738) reformed clinical teaching and bedside observation at Leiden; Caspar Wolff (d. 1794) disproved preformationism and established epigenesis in embryology; and Edward Jenner (d. 1823) devised an effective smallpox vaccine. In botany Carl Linnaeus (d. 1778) established a new system of classification with binomial nomenclature. In studies of electricity the American Benjamin Franklin (d. 1790) formulated a theory of general electrical action. Finally, chemistry became a modern quantitative science as mechanics had a century earlier. Following criticisms of the phlogiston theory of combustion and the discovery of dephlogisticated air, Antoine Lavoisier (d. 1794) completed the revolution in chemistry. He discovered the role oxygen played in combustion, decomposed water into hydrogen and oxygen, stated the principle of conservation of matter in chemical reactions, carefully defined the term element, and listed 55.[23]

NOTES

1. "Rational mechanics" refers to the new mechanics beginning with Newton and the Bernoullis that was based on the mathematical method and guided by physical experience. The term distinguished the new mechanics from what might be called speculative Aristotelian physics and the investigation of machines.

2. When several variables describe the state of a physical system, differential equations prescribe rates of change of these state variables.

3. Luigi Bianchi first used the term "differential geometry" in 1894.

4. The origins of the separation are covered at the end of the introduction to chapter VI in the discussion of the quarrel over the question of priority in the discovery of the calculus. The consequences of the separation are mentioned below in the section on mathematical analysis.

5. See J. H. Plumb, *The Death of the Past* (Boston: Houghton Mifflin Company, 1971), pp. 54-58.

6. German universities were not without their religious quarrels. Wolff was dismissed from Halle in 1723 because of his praise of the non-Christian Chinese and his determinism; he returned in 1741, however.

7. In modern notation, if $z = \frac{1}{y}$, then one finds $z = \lim_{n\to\infty} \log_e (1 + \frac{z}{n})^n$.

8. As cited in C. H. Edwards, Jr., *The Historical Development of the Calculus* (New York: Springer-Verlag, 1979), 270.

9. Taken together his two articles (1754 and 1765) present the derivative as $\frac{dy}{dx} = \lim_{\Delta x\to 0} \frac{\Delta y}{\Delta x}$.

10. See Lorraine J. Daston, "Probabilistic Expectation and Rational in Classical Probability Theory," *Historia Mathematica* 7 (1980), 234-260.
11. As quoted in Jeremy Gray, "Non-Euclidean Geometry—A Re-Interpretation" *Historia Mathematica* 6 (1979), p. 248.
12. "Recherches sur la courbure des surfaces," *Opera Omnia* (1) 28, 1-22.
13. In his article entitled, "Answer to the Question: What is Enlightenment." (1784), Kant also offered a motto for the period: "*Sapere aude* (Dare to Know!)"—a motto taken from the Wolffians.
14. The term Ancien Regime (Old Regime) was introduced during the French Revolution. It generally designated the life and institutions of prerevolutionary Europe under absolute monarchs and a privileged aristocracy.
15. Alexis de Tocqueville gave this accurate portrayal of the Enlightenment in the nineteenth century.
16. There were also debates about the merits and limits of reason, whether deductive or inductive, among the intelligentsia. In comparisons of the cognate systems of the seventeenth and eighteenth centuries, historians often refer to the *esprit géomètrique* of the former and Condillac's *esprit systématique* of the latter.
17. I am indebted to Roderick Brumbaugh for pointing out precisely Foucault's findings.
18. Europeans did not often drink milk, believing (with some justification before pasteurization) that it caused a host of maladies. Primarily they used it to make cheese and butter.
19. One response was the grim *Essay on Population* (1798) by the Reverend Thomas Malthus (d. 1834).
20. Fernand Braudel, *Capitalism and Material Life 1400–1800* (New York: Harper Torchbooks, 1973), p. 73.
21. Improved sextants and chronometers facilitated expanding maritime trade and exploration. Exploration, such as that of Captain Cook in Australia, fired the imagination of Europeans.
22. *Gesammelte Schriften*, vol. II, (1942), p. 44.
23. For an account of the role of science in France in the late eighteenth century, see Charles C. Gillispie, *Science and Polity in France at the End of the Old Regime* (Princeton: Princeton University Press, 1981).

SUGGESTIONS FOR FURTHER READING

A. D. Aleksandrov, A. N. Kolmogorov, and M. A. Lavrent'ev, eds., *Mathematics: Its Content, Methods and Meaning*. Cambridge, Mass.: The MIT Press, 1969. 3 vols.

Keith Baker, *Condorcet: From Natural Philosophy to Social Mathematics*. Chicago: University of Chicago Press, 1975.

E. J. Barbeau, "Euler Subdues a Very Obstreperous Series," *The American Mathematical Monthly* 86, 356-372.

Robert J. Baum, *Philosophy and Mathematics: From Plato to the Present*. San Francisco: Freeman, 1974.

P. P. Bockstaele, "Mathematics in the Netherlands from 1750 to 1830," *Janus* 65 (1978), 67-95.

H. Behnke, et al., eds., *Fundamentals of Mathematics*. Cambridge, Mass.: The MIT Press, 1974. 3 vols.

Valentin Boss, *Newton & Russia: The Early Influences, 1698–1796*. Cambridge, Mass.: Harvard University Press, 1972.

Felix E. Browder and Saunders MacLane, "The Relevance of Mathematics," in Lynn Arthur Steen, ed., *Mathematics Today*. New York: Springer-Verlag, 1978, 323-351.

Carl Boyer, *The History of the Calculus and its Conceptual Development*. New York: Dover Publications, Inc., 1959.

Ronald Calinger, "Euler's *Letters to a Princess of Germany* as an Expression of His Mature Scientific Outlook," *Archive for History of Exact Science* 15 (1976), 210-233.

_________, "Kant and Newtonian Science: The Pre-Critical Period," *Isis* 70 (1979), 349-363.

I. B. Cohen, *The Newtonian Revolution*. Cambridge: Cambridge University Press, 1980.

J. L. Coolidge, "The Number e," *American Mathematical Monthly* 57 (1950), 591-602.

L. J. Daston, "D'Alembert's Critique of Probability Theory," *Historia Mathematica* 6 (1979), 259-279.

Philip J. Davis and Reuben Hersh, *The Mathematical Experience*. Boston: Birkhauser, 1981.

P. Dedron and J. Itard, *Mathematics and Mathematicians*. London: Transworld Publications, 1973.

L. E. Dickson, *History of the Theory of Numbers*. New York: Chelsea, 1952. 3 vols.

C. H. Edwards, Jr., *The Historical Development of the Calculus*. New York: Springer-Verlag, 1979.

S. B. Engelsman, "Lagrange's Early Contributions to the Theory of First Order Partial Differential Equations," *Historia Mathematica* 7 (1980), 7-23.

Leonhard Euler, *Opera Omnia*, Berlin: B. G. Teubner and Basel Birkhäuser, 1911— ; 4 series 78 vols.

Emil Alfred Fellman, "Leonhard Euler," in *Enzyklopädie Die Grossen des Weltgeschichte*. Zurich: Kindler Verlag, 1975. Vol. 6, 497-531.

Eric G. Forbes, *The Euler-Mayer Correspondence (1751–1755)*. New York: American Elsevier Publishing Company, Inc., 1971.

_________, *Tobias Mayer's Opera Inedita*. New York: American Elsevier Publishing Company, Inc., 1971.

Hans Freudenthal, *Mathematics as an Educational Task*. Dordrecht, Holland: D. Reidel Publishing Company, 1973.

H. H. Frisinger, "Mathematics and Our Founding Fathers," *Mathematics Teacher* 69 (1976), 301-307.

Charles C. Gillispie. *The Edge of Objectivity*. Princeton: Princeton University Press, 1960.

Arthur Gittleman, *History of Mathematics*. Columbus, Ohio: Charles E. Merrill Publishing Company, 1975.

H. H. Goldstine, *A History of Numerical Analysis from the 16th through the 19th Century*. New York: Springer-Verlag, 1977.

Judith V. Grabiner, *The Origins of Cauchy's Rigorous Calculus*. Cambridge, Mass.: The MIT Press, 1981.

I. Grattan-Guinness, ed., *From the Calculus to Set Theory, 1630–1910: An Introductory History*. London: Duckworth, 1980.

J. J. Gray and Laura Tilling, "Johann Heinrich Lambert, Mathematician and Scientist, 1728–1777," *Historia Mathematica* 5 (1978), 13-41.

R. R. Hamburg, "The Theory of Equations in the 18th Century: The Work of Joseph Lagrange," *Archive for History of Exact Science* 16 (1978), 17-36.

Thomas Hankins, *Jean d'Alembert.* Oxford: Clarendon Press, 1970.

John Heilbronn, *Electricity in the 17th and 18th Century.* Berkeley: University of California Press, 1979.

Mary Hesse, *Revolutions and Reconstructions in the Philosophy of Science.* Bloomington: Indiana University Press, 1980.

Douglas R. Hofstadter, *Gödel, Escher, Bach: an Eternal and Golden Braid.* New York: Basic Books, Inc., Publishers, 1979.

A. G. Howson, ed., *Developments in Mathematical Education.* Cambridge: Cambridge University Press, 1973.

Morris Kline, *Mathematics: An Introduction to Its Spirit and Use: Readings from Scientific American.* San Francisco: Freeman, 1978.

_________, *Mathematics: The Loss of Certainty.* New York: Oxford University Press, 1980.

Ya Khurgin, *Did You Say Mathematics?* Moscow: MIR, 1974.

Imre Lakatos, *Mathematics, Science and Epistemology.* Cambridge: Cambridge University Press, 1978.

Cornelius Lanczos, *Numbers Without End.* Edinburgh: Oliver & Boyd, 1968.

V. I. Lysenko, *Nikolai Ivanovich Fuss 1755–1826.* Moscow: Nauka, 1975.

Lang W. Miller, "Kant's Philosophy of Mathematics," *Kant-Studien* 66 (1975), 297-308.

James R. Newman, ed., *The World of Mathematics.* New York: Simon and Schuster, 1959–60. 4 vols.

George E. Owen, *The Universe of the Mind.* Baltimore: The Johns Hopkins University Press, 1971.

Charles B. Paul, *Science and Immortality: The Eloges of the Paris Academy of Sciences (1699–1791).* Berkeley: University of California Press, 1981.

Olaf Pederson, "The 'Philomaths' of 18th Century England," *Centaurus* 8 (1963), 238-262.

George Polya, *Mathematical Discovery.* New York: Wiley, 1962. 2 vols.

G. S. Rousseau and Roy Porter, eds., *The Ferment of Knowledge.* Cambridge: Cambrige University Press, 1980.

O. B. Sheynin, "P. S. Laplace's Work on Probability," *Archive for History of Exact Science* 16 (1977), 137-187.

G. C. Smith, "Thomas Bayes and Fluxions," *Historia Mathematica* 7 (1980), 379-388.

Stephen M. Stigler, "Napoleonic Statistics: the Work of Laplace," *Biometrika* 62 (1975), 503-517.

N. I. Styazhkin, *History of Mathematical Logic from Leibniz to Peano.* Cambridge, Mass.: The MIT Press, 1969.

René Taton, "Inventaire Chronologique de l'Oeuvre de Lagrange," *Review D'Histoire des Sciences* 27 (1974), 3-36.

Otto Toeplitz, *The Calculus, a Genetic Approach.* Chicago: University of Chicago Press, 1963.

Clifford Truesdell, *Essays in the History of Mechanics*. New York: Springer-Verlag, 1968.

________, "Leonard Euler, Supreme Geometer (1707–1783)," in Harold E. Pagliaro, ed., *Irrationalism in the Eighteenth Century*. Cleveland: The Press of Case Western Reserve University, 1972, 51-97.

________, "A Program Toward Rediscovering the Rational Mechanics of the Age of Reason," *Archive for History of Exact Sciences*, 1 (1960), 1-36.

Alexander Vucinich, *Science in Russian Culture: A History to 1860*. Stanford, Calif.: Stanford University Press, 1963.

P. J. Wallis, "British Philomaths—Mid-Eighteenth Century and Earlier," *Centaurus* 17 (1972), 301-314.

Richard S. Westfall, *The Construction of Modern Science*. Cambridge: Cambridge University Press, 1980.

Gerald J. Whitrow, *The Natural Philosophy of Time* (New York: Oxford University Press, 2nd ed., 1980).

A. P. Youschkevitch, "The Concept of Function Up To the Middle of the 19th Century," *Archive for History of Exact Sciences* 16 (1976), 37-85.

Chapter VII
The Enlightenment (1720–1800)

Section A
Elaboration and Criticism of Mathematical Analysis

BROOK TAYLOR (1685–1731)

During the early Enlightenment, Brook Taylor was one of the few English mathematicians who could hold his own in disputes with Continental rivals, especially with John Bernoulli. The eldest son of John and Olivia Bart Taylor, he grew up in a comfortable family of the minor nobility. In 1701 Brook entered St. John's College, Cambridge (where he studied under John Mackin and John Keill). He received the LL. B. (1709) and LL. D. (1714). Elected a Fellow of the Royal Society in 1712, he sat on the committee that decided whether Newton or Leibniz deserved priority for the invention of the calculus. He was elected secretary of the Royal Society in 1714 but resigned the position in 1718 because of ill health and possibly from lack of interest in that confining task.

Taylor's most productive mathematical period dates from 1713 to 1719. He published two books on mathematics in 1715, entitled *Methodus Incrementorum Directa et Inversa* ("Direct and Indirect Methods of Incrementation") and *Linear Perspective.* During the period 1712 to 1724, he also wrote 13 articles for the *Philosophical Transactions* of the Royal Society and visited France on several occasions for social enrichment and to improve his health. In France, he met the mathematician Pierre Rémond de Montmort, and the two men subsequently corresponded with each other on the subjects of religion, infinite series, and probability. In this correspondence Taylor sometimes acted as an intermediary between Montmort and Abraham de Moivre in their studies of probability and suggested problems that needed to be solved.

After 1720, Taylor concentrated on art, philosophy, religion, music, and family matters. A brief rift with his morose father occurred when Taylor married a woman who had no fortune. After she died during childbirth in 1723, Taylor returned to his parent's home at Bifrons, Kent, and then remarried in 1725 (this time with his father's approval). He and his second wife, Sabetta Sambridge, moved to Bifrons estate when he inherited it in 1729. The next year Sabetta died during the birth of their daughter, Elizabeth, who survived. Taylor's delicate health rapidly worsened.

Taylor contributed to the early development of the calculus. He is best known for deriving the powerful formula for expanding a function into an infinite series that is now known as Taylor's theorem. He first explicitly stated it as Proposition VII of Theorem III in his book *Methodus Incrementorum.* In modern notation, it is

$$f(x+h) = f(x) + \frac{f'(x)h}{1!} + \frac{f''(x)h^2}{2!} + \frac{f'''(x)h^3}{3!} + \cdots + \frac{f^{(n)}(x)h^n}{n!} + \cdots$$

Conversant with the work of his predecessors, especially the British ones, he freely admitted a debt in deriving the theorem to Newton, Mackin, and Halley as well as to the German Kepler. Indeed he praised Newton, used his dot notation, and built upon Lemma V of Book III and Corollary II to Theorem III of the *Principia*. However, he was not so forthright about his indebtedness to contemporary Central European mathematicians. Even though he knew of Leibniz's pioneering work on finite differences and John Bernoulli's independent discovery of Taylor's theorem, which was given in *Acta eruditorum* (1694), he mentioned neither. Moreover, he did not worry over his lack of rigor in deriving the theorem nor did he seem to grasp the important role assigned to it by Lagrange, who in 1772 proclaimed it to be "the fundamental principle of differential calculus."

Because of its arithmetical exposition, the *Methodus* had little immediate influence in Britain where the primary attempts were to link the calculus to geometry or the physical notion of velocity. Peano later wanted to give John Bernoulli priority for the "Taylor series," but historians agree that Taylor deserves priority for integration by parts. Taylor and Bernoulli each claimed priority and sharply disagreed in other matters as well.

In the *Methodus* Taylor also established what is now called the calculus of finite differences as a new branch of higher analysis. With it he was able to reduce the motion of a vibrating elastic string to mechanical principles and to study the associated second order differential equation. He solved the equation $a^2\ddot{x} = \dot{s}y\dot{y}$, where $s = \sqrt{\dot{x}^2 + \dot{y}^2}$ and the differentiation is with respect to time, and gave $y + A \sin(x/a)$ as the form of the string at any time. In his overly concise book on linear perspective, he developed a theory of perspective in a formal and rigorous manner and presented the first general treatment of vanishing points and vanishing lines.

81. From *Methodus Incrementorum Directa et Inversa* (1715)*

(The Taylor Series)

BROOK TAYLOR[1]

Proposition VII. *Theorem* III. Let z and x be two variable quantities, of which z increases uniformly with given increments Δz.[2] Let $n\Delta z = v$, $v - \Delta z = \dot{v}$, $\dot{v} - \Delta z = \ddot{v}$, etc. Then I say that when z grows into $z + v$, then x grows into

$$z + \Delta x \frac{v}{1.\Delta z} + \Delta^2 x \frac{v\dot{v}}{1.2(\Delta z)^2} + \Delta^3 x \frac{v\dot{v}\ddot{v}}{1.2.3(\Delta z)^3} + \cdots.$$

Source: This translation made from pages 21 through 23 of the *Methodus* is reprinted by permission of the publishers from *A Source Book in Mathematics, 1200-1800,* edited by D. J. Struik, Cambridge, Mass.: Harvard University Press,

Fig. 1

DEMONSTRATION

x	Δx	$\Delta^2 x$	$\Delta^3 x$	$\Delta^4 x$ etc.
$x + \Delta x$	$\Delta x + \Delta^2 x$	$\Delta^2 x + \Delta^3 x$	$\Delta^3 x + \Delta^4 x$	etc.
$x + 2\Delta x + \Delta^2 x$	$\Delta x + 2\Delta x + \Delta^3 x$	$\Delta^2 x + 2\Delta^3 x + \Delta^4 x$	etc.	
$x + 3\Delta x + 3\Delta^2 x + \Delta^3 x$	$\Delta x + 3\Delta^2 x + 3\Delta^2 x + \Delta^4 x$	etc.		
$x + 4\Delta x + 6\Delta^2 x + 4\Delta^3 x + \Delta^4 x$	etc.			

The successive values of x, collected by continued addition, are x, $x + \Delta x$, $x + 2\Delta x + \Delta^2 x$, $x + 3\Delta x + 3\Delta^2 x + \Delta^3 x$, etc., as we see from the operation expressed in the table. But the numerical coefficients of the terms x, Δx, $\Delta^2 x$, etc. for these values of x are formed in the same way as the coefficients of the corresponding terms in the binomial expansion [*in dignitate binomii*]. And if n is the exponent of the expansion [*dignitatis index*], then the coefficients (according to Newton's theorem) will be 1, $\frac{n}{1}$, $\frac{n}{1}\frac{n-1}{2}$, $\frac{n}{1}\frac{n-1}{2}\frac{n-2}{3}$, etc. When, therefore, z grows into $z + n\Delta z$, that is, into $z + v$, then x will be equal to the series

$$x + \frac{n}{1}\Delta x + \frac{n}{1}\frac{n-1}{2}\Delta^2 x + \frac{n}{1}\frac{n-1}{2}\frac{n-2}{3}\Delta^3 x + \text{etc.}$$

But

$$\frac{n}{1} = \left(\frac{n\Delta z}{\Delta z}\right) = \frac{v}{\Delta z}, \quad \frac{n-1}{2} = \left(\frac{n\Delta z - \Delta z}{2\Delta z}\right) = \frac{\dot v}{2\Delta z}, \quad \frac{n-2}{3} = \left(\frac{n\Delta z - 2\Delta z}{3\Delta z}\right) = \frac{\ddot v}{3\Delta z},$$

etc. Hence in the time that z grows into $z + v$, x grows into[3]

$$x = x + \Delta x\frac{v}{1.\Delta z} + \Delta^2 x\frac{v\dot v}{1.2(\Delta z)^2} + \Delta^3 x\frac{v\dot v\ddot v}{1.2.3(\Delta z)^3} + \text{etc.}$$

Corollary I. If the Δz, Δx, $\Delta^2 x$, $\Delta^3 x$ remain the same, but the sign of v is changed so that z decreases and becomes $z - v$, then x decreases at the same time and becomes

$$x - \Delta x\frac{v}{1.\Delta z} - \Delta^2 x\frac{v\dot v}{1.2(\Delta z)^2} - \Delta^3 x\frac{v\dot v\ddot v}{1.2.3(\Delta z)^3} - \text{etc.}$$

or[4]

$$x - \Delta x\frac{v}{1.\Delta z} + \Delta^2 x\frac{vv_1}{1.2(\Delta z)^2} - \Delta^3 x\frac{vv_1v_{11}}{1.2.3(\Delta z)^3} + \text{etc.}$$

with v, $\ddot v$, etc. converted into $-v_1$, $-v_{11}$, etc.

Corollary II. If we substitute for evanescent increments the fluxions proportional to them, then all $\ddot v$, $\dot v$, v, v_1, v_{11} become equal. When z flows uniformly into $z + v$, x becomes[5]

$$x + x\frac{v}{1.\dot z} + \ddot x\frac{v^2}{1.2\dot z^2} + \dddot x\frac{v^3}{1.2.3\dot z^3} + \text{etc.},$$

or with v changing its sign, when z decreases to $z - v$, x becomes

$$x - \dot{x}\frac{v}{1.\dot{z}} + \ddot{x}\frac{v^2}{1.2\dot{z}^2} - \dddot{x}\frac{v^3}{1.2.3\dot{z}^3} + \text{etc.},$$

[*Text omitted.*]

Here follows one of Taylor's applications of the theorem.

Proposition VIII. *Problem* V. Given an equation which contains, apart from a uniformly increasing z, a certain number of other variables x. To find the value of x from given z by a series of an infinite number of terms.

Find all increments, to infinity, of the proposed equation by means of Proposition I. If $\Delta^n x$ be the infinite increment of x in the proposed equation, then by means of these equations will be given all increments $\Delta^n x$ and those with higher n expressed by means of increments of lower n. Let a, c, c_1, c_2, c_3, etc. be certain arbitrary values corresponding to z and x, Δx, $\Delta^2 x$, $\Delta^3 x$, etc.; then by means of these equations all terms c_n, c_{n+1}, and the following can be expressed in terms of the terms preceding c_n. Hence if we write $a + v$ for z, then x will be given by means of

$$x = c + c_1\frac{v}{1.\Delta z} + c_2\frac{v\dot{v}}{1.2(\Delta z)^2} + c_3\frac{v\dot{v}\ddot{v}}{1.2.3(\Delta z)^3} + \text{etc.}$$

(according to Proposition VII). Here the coefficients c, c_1, c_2, etc. of the terms whose number is n are given by the same number of conditions imposed on the problem.[6]

NOTES

1. Taylor used a complicated notation with dots and primes *(lineolae)* used as superscripts and subscripts, and the primes in both the *accent grave* and *accent aigu* position. We have kept his notation, except that instead of the increments $z_{\prime}$, $z_{\prime\prime}$, etc. we have written Δz, $\Delta^2 z$, etc., and for the v with subscript *accents aigus* we have written v_1, v_{11}, ... Taylor's notation also has its advantage. In his notation x'', x, x', $\dot{x}$, $\ddot{x}$ represent a sequence of functions of which each is the fluxion of the previous one; whereby Taylor remarks that the *lineolae* in x', x'' can be regarded as negative dots—an anticipation, if we like, of our modern operational notation D^{-2}, D^{-1}, D^0, D^1, D^2, and for the same purpose.

2. Since z flows uniformly, Δz is constant, so that $\Delta^2 z$, $\Delta^3 z$, etc. are all zero. Here $\Delta^2 z$ is the increment of Δz, $\Delta^3 z$ that of $\Delta^2 z$, etc.

3. This is Newton's well-known interpolation formula (*Principia*, Book III, Lemma 5); see also Newton, *Methodus differentialis* (London, 1711); *James Gregory tercentenary memorial volume*, ed. H. W. Turnbull (Bell, London, 1939), 119; H. W. Turnbull, *The mathematical discoveries of Newton* (Blackie, Glasgow, 1945), 46.

4. The v_{11}, v_1, v, $\dot{v}$, $\ddot{v}$ form a sequence of increments, so that $v_1 - \Delta z = v$, $v_{11} - \Delta z = v_1$, or $v_1 = (n + 1)\Delta z$, $v_{11} = (n + 2)\Delta z$.

5. This is the classical Taylor series, since in the Leibniz notation $\dot{x}/\dot{z} = dx/dz$, $x/(\dot{z})^2 = d^2x/dz^2$, etc. Taylor therefore obtained his series from Newton's interpolation formula by taking $\Delta x = 0$, $n = \infty$. Felix Klein has called Taylor's step "a transition to the limit of extraordinary audacity"; see *Elementary mathematics from an advanced standpoint*, trans. E. R. Hedrick and C. A. Noble, I (Dover, New York, 1924), 233. Although we shall not belittle this statement we must also take into account that Taylor's theorem had been "in the air" ever since James Gregory had it in a manuscript of 1671 (*Gregory tercentenary memorial volume*, pp. 123, 173, 356). See also A. Pringsheim, "Zur Geschichte des Taylorschen Lehrsatzes," *Bibliotheca mathematica (3) 1* (1900), 433-479; G. Eneström, "Zur Vorgeschichte der Entdeckung des Taylorschen Lehrsatzes," *ibid., 12* (1911-12), 333-336.

6. This proposition and several others give information on the number of arbitrary constants in difference and differential equations. Taylor, on page 27 of his book, shows how the differential equation $\ddot{x} - \dot{x}z - 2x = 0$ can be solved by means of $x = A + Bz + Cz^2 + Dz^3 +$

$Ez^4 + \cdots$, and by substituting this series as well as $\dot{x} = B + 2Cz + \cdots$, $\ddot{x} = 2C + 6Dz + \cdots$ he gets a series of recursion equations, from which he derives the solution

$$x = A + Bz + Az^2 + \tfrac{1}{3}Bz^3 + \tfrac{1}{3}Az^4 + \cdots$$

with two arbitrary constants. These series are what our textbooks often call "Maclaurin" series.

Taylor has no discussion of convergence.

GEORGE BERKELEY (1685–1753)

George Berkeley, a distinguished critic of scientific, philosophical, and political thought, was born and raised in County Kilkenny, Ireland, but considered himself an Englishman like his ancestors. After proving precocious in early schooling, he was enrolled in 1696 in Kilkenny College. In 1700, he entered Trinity College, Dublin, where he earned his B.A. (1704) and M.A. (1707). Although he was influenced at college by the writings of Francis Bacon, Robert Boyle, René Descartes, and Isaac Newton, it was the critical empiricism of John Locke and the Continental skepticism of Pierre Bayle and Nicolas Malebranche that most impressed him.

Berkeley quickly established a reputation in the world of learning. After his election as fellow of Trinity College in 1707, he spent two years examining and revising his ideas. He proposed a new defense of immaterialism based on his principle—*esse est percipi*, ("to be means to be perceived"). His radical argument for immaterialism accepted spiritual substance but rejected the reality of the material. His views appeared in *An Essay Towards a New Theory of Vision* (1709), which was a contribution to psychology, and *Principles of Human Knowledge* (1710). In these two short tracts, he maintained that the source of visual ideas is in our minds rather than in extramental objects. Among those on the Continent who commented upon his immaterialism was Gottfried Leibniz. Berkeley was ordained an Anglican priest in 1710 and lectured at Trinity on divinity, Greek, and Hebrew.

After requesting a leave of absence in 1713, he spent the next eight years traveling. He went first to London, where he met the clerics Samuel Clarke and Jonathan Swift, the essayists Richard Steele and Joseph Addison, and the poet Alexander Pope. He was received well in London circles. At the request of Steele, he wrote papers against freethinkers for the *Guardian*. Pope believed that Berkeley possessed "ev'ry virtue under heav'n." Berkeley traveled to Sicily in 1713–14 as a chaplain to the embassy staff to Lord Peterborough. He returned to Italy, staying in Rome and Naples as a tutor from 1716 to 1720. He studied antiquities and natural phenomena at his leisure and returned to London in 1720. A year later he wrote a treatise entitled *De motu*, which rejected Newton's absolute space, time, and motion. This critique, together with his definition of causal action in *Principles*, may have later influenced David Hume and Immanuel Kant.

The return to London had been unsettling. Frenzied financial speculation and the bursting of the South Seas Bubble led Berkeley to lose faith in the moral health of the Old World and to look to the New. He believed that citizens of the New World required education and the Protestant faith to assure their future. To help meet this goal he planned to found a college in Bermuda. Until he could raise the necessary financing, he returned to Dublin. He was a teacher and administrator at Trinity College from autumn 1721 to 1724, when he was appointed to the rich and influential Anglican deanery of Derry.

Berkeley married Anne Forster in

1728 and soon sailed for America. They resided on a 96-acre farm in Newport, Rhode Island, while awaiting a promised grant of £20,000 from the British Parliament and its leader Robert Walpole for the Bermudan college. However, the grant was never paid, and Berkeley was compelled to give up his plan for the college. While in America, he founded a study group in Newport, was a benefactor to Yale University, and corresponded with Samuel Johnson, a future president of King's College (Columbia University). He also wrote *Alciphron: or, the Minute Philosopher* (1732), which was not published until he returned to London. This finely written dialogue massively defended theism and Christianity from criticisms by freethinkers and brought together his thought on creation as an ordered chain of being earlier presented in the *Essay* (1707) and *Principles* (1710).

From 1731 to spring 1734, Berkeley resided in London. His *Alciphron* quickly became popular (going through a second edition in 1732 and being translated into French in 1734). It provoked widespread criticism, including that of the satirist Bernard de Mandeville, author of *Fable of the Bees*. His critics were now calling Berkeley paradoxical and a visionary. Replying to an attack on the appendix to *Alciphron* on vision, he wrote *The Theory of Vision, or Visual Language . . . Vindicated and Explained* (1733). In 1734, he published *The Analyst,* a celebrated critique of the foundations of the calculus that is discussed below.

In 1734, Berkeley was consecrated Anglican bishop of Cloyne (County Cork) in Dublin. His 18 years' episcopate was uneventful. Together with his wife and six children, he lived at the see-house at Cloyne. Theirs was a cultured home that became a dispensary during epidemics. Responding to the poverty and sickness about him, he wrote *The Querist* (1735–37) on basic economics and his last major work, *Siris* (1744), which examines the medicinal virtues of tar water, attempts to incorporate selected Newtonian concepts into chemical science, and discusses again the metaphysical great chain of being.

Weakened by ill health, Berkeley went to live in Oxford in 1752, probably because he desired to be near his son. He died the next year.

Berkeley's chief contribution to mathematics was *The Analyst; or a Discourse Addressed to an Infidel Mathematician*. (The infidel mathematician was Edmond Halley.) The *Analyst* did not impugn the use of the calculus for practical purposes but properly deplored its faulty foundations. At the same time, its author presented an ad hominem argument for religion. He questioned Newton's fluxion as a ratio of evanescent increments (infinitesimal) denoted by a little o. He lampooned these ratios as "the ghosts of departed quantities," and stated, "Certainly . . . he who can digest a second or third fluxion . . . need not, methinks be squeamish about any point in divinity." Berkeley also criticized Leibniz's differentials but maintained that that theory had correct results because of compensating errors. The long and fruitful controversy engendered by his book (beginning with James Jurin, Benjamin Robbins, and Colin Maclaurin) led ultimately to the desired satisfactory foundations.

82. From *The Analyst* (1734)*

(Criticism of the Foundations of the Calculus)

GEORGE BERKELEY

3. The Method of Fluxions is the general key by help whereof the modern mathematicians unlock the secrets of Geometry, and consequently of Nature. And, as it is that which hath enabled them so remarkably to outgo the ancients in discovering theorems and solving problems, the exercise and application thereof is become the main if not sole employment of all those who in this age pass for profound geometers. But whether this method be clear or obscure, consistent or repugnant, demonstrative or precarious, as I shall inquire with the utmost impartiality, so I submit my inquiry to your own judgment, and that of every candid reader. Lines are supposed to be generated by the motion of points, planes by the motion of lines, and solids by the motion of planes. And whereas quantities generated in equal times are greater or lesser according to the greater or lesser velocity wherewith they increase and are generated, a method hath been found to determine quantities from the velocities of their generating motions. And such velocities are called fluxions: and the quantities generated are called flowing quantities. These fluxions are said to be nearly as the increments of the flowing quantities, generated in the least equal particles of time; and to be accurately in the first proportion of the nascent, or in the last of the evanescent increments. Sometimes, instead of velocities, the momentaneous increments or decrements of undetermined flowing quantities are considered, under the appellation of moments.

4. By moments we are not to understand finite particles. These are said not to be moments, but quantities generated from moments, which last are only the nascent principles of finite quantities. It is said that the minutest errors are not to be neglected in mathematics:[1] that the fluxions are celerities, not proportional to the finite increments, though ever so small; but only to the moments or nascent increments, whereof the proportion alone, and not the magnitude, is considered. And of the aforesaid fluxions there be other fluxions, which fluxions of fluxions are called second fluxions. And the fluxions of these second fluxions are called third fluxions: and so on, fourth, fifth, sixth, etc. *ad infinitum*. Now, as our sense is strained and puzzled with the perception of objects extremely minute, even so the imagination, which faculty derives from sense, is very much strained and puzzled to frame clear ideas of the least particles of time, or the least increments generated therein: and much more so to comprehend the moments, or those increments of the flowing quantities in *statu nascenti,* in their very first origin or beginning to exist, before they become finite particles. And it seems still more difficult to conceive the

**Source:* The full text of *The Analyst* may be found in *The Works of George Berkeley* (London: Nelson, 1951) vol. IV, 65-102. This selection is taken from D. J. Struik (ed.), *A Source Book in Mathematics, 1200-1800* (1969), 334-338, and is reprinted by permission of Harvard University Press,

abstracted velocities of such nascent imperfect entities. But the velocities of the velocities, the second, third, fourth, and fifth velocities, etc., exceed, if I mistake not, all human understanding. The further the mind analyseth and pursueth these fugitive ideas the more it is lost and bewildered; the objects, at first fleeting and minute, soon vanishing out of sight. Certainly in any sense, a second or third fluxion seems an obscure mystery. The incipient celerity of an incipient celerity, the nascent augment of a nascent augment, *i.e.,* of a thing which hath no magnitude: take it in what light you please, the clear conception of it will, if I mistake not, be found impossible; whether it be so or no I appeal to the trial of every thinking reader. And if a second fluxion be inconceivable, what are we to think of third, fourth, fifth fluxions, and so on without end?

5. The foreign mathematicians are supposed by some, even of our own, to proceed in a manner less accurate, perhaps, and geometrical, yet more intelligible. Instead of flowing quantities and their fluxions, they consider the variable finite quantities as increasing or diminishing by the continual addition or subduction of infinitely small quantities. Instead of the velocities wherewith increments are generated, they consider the increments or decrements themselves, which they call differences, and which are supposed to be infinitely small. The difference of a line is an infinitely little line; of a plane an infinitely little plane. They suppose finite quantities to consist of parts infinitely little, and curves to be polygons, whereof the sides are infinitely little, which by the angles they make one with another determine the curvity of the line. Now to conceive a quantity infinitely small, that is, infinitely less than any sensible or imaginable quantity, or than any the least finite magnitude is, I confess, above my capacity. But to conceive a part of such infinitely small quantity that shall be still infinitely less than it, and consequently though multiplied infinitely shall never equal the minutest finite quantity, is, I suspect, an infinite difficulty to any man whatsoever; and will be allowed such by those who candidly say what they think; provided they really think and reflect, and do not take things upon trust.

6. And yet in the *calculus differentialis,* which method serves to all the same intents and ends with that of fluxions, our modern analysts are not content to consider only the differences of finite quantities: they also consider the differences of those differences, and the differences of the differences of the first differences. And so on *ad infinitum.* That is, they consider quantities infinitely less than the least discernible quantity; and others infinitely less than those infinitely small ones; and still others infinitely less than the preceding infinitesimals, and so on without end or limit. Insomuch that we are to admit an infinite succession of infinitesimals, each infinitely less than the foregoing, and infinitely greater than the following. As there are first, second, third, fourth, fifth, etc. fluxions, so there are differences, first, second, third, fourth, etc., in an infinite progression towards nothing, which you still approach and never arrive at. And (which is most strange) although you should take a million of millions of these infinitesimals, each whereof is supposed infinitely greater than some other real magnitude, and add them to the least given quantity, it shall never be the bigger. For this is one of the modest *postulata* of our modern mathematicians, and is a corner-stone or ground-work of their speculations.

[Paragraphs 7 and 8 omitted.]

9. Having considered the object, I proceed to consider the principles of this new analysis by momentums, fluxions, or infinitesimals; wherein if it shall appear that your capital points, upon which the rest are supposed to depend, include error and false reasoning; it will then follow that you, who are at a loss to conduct your selves, cannot with any decency set up for guides to other men. The main point in the method of fluxions is to obtain the fluxion or momentum of the rectangle or product of two indeterminate quantities. Inasmuch as from thence are derived rules for obtaining the fluxions of all other products and powers; be the coefficients or the indexes what they will, integers or fractions, rational or surd. Now, this fundamental point one would think should be very clearly made out, considering how much is built upon it, and that its influence extends throughout the whole analysis. But let the reader judge. This is given for demonstration. Suppose the product or rectangle AB increased by continual motion: and that the momentaneous increments of the sides A and B are a and b. When the sides A and B were deficient, or lesser by one half of their moments, the rectangle was $\overline{A - \frac{1}{2}a} \times \overline{B - \frac{1}{2}b}$ i.e., $AB - \frac{1}{2}aB - \frac{1}{2}bA + \frac{1}{4}ab$. And as soon as the sides A and B are increased by the other two halves of their moments, the rectangle becomes $\overline{A + \frac{1}{2}a} \times \overline{B + \frac{1}{2}b}$ or $AB + \frac{1}{2}aB + \frac{1}{2}bA + \frac{1}{4}ab$. From the latter rectangle subduct the former, and the remaining difference will be $aB + bA$. Therefore the increment of the rectangle generated by the entire increments a and b is $aB + bA$. *Q.E.D.* But it is plain that the direct and true method to obtain the moment or increment of the rectangle AB, is to take the sides as increased by their whole increments, and so multiply them together, $A + a$ by $B + b$, the product whereof $AB + aB + bA + ab$ is the augmented rectangle; whence, if we subduct AB the remainder $aB + bA + ab$ will be the true increment of the rectangle, exceeding that which was obtained by the former illegitimate and indirect method by the quantity ab. And this holds universally be the quantities a and b what they will, big or little, finite or infinitesimal, increments, moments, or velocities. Nor will it avail to say that ab is a quantity exceeding small: since we are told that *in rebus mathematicis errores quam minimi non sunt contemnendi.*[2]

10. Such reasoning as this for demonstration, nothing but the obscurity of the subject could have encouraged or induced the great author of the fluxionary method to put upon his followers, and nothing but an implicit deference to authority could move them to admit. The case indeed is difficult. There can be nothing done till you have got rid of the quantity ab. In order to [do] this the notion of fluxions is shifted: It is placed in various lights: Points which should be clear as first principles are puzzled; and terms which should be steadily used are ambiguous. But notwithstanding all this address and skill the point of getting rid of ab cannot be obtained by legitimate reasoning. If a man, by methods not geometrical or demonstrative, shall have satisfied himself of the usefulness of certain rules; which he afterwards shall propose to his disciples for undoubted truths; which he undertakes to demonstrate in a subtile manner, and by the help of nice and intricate notions; it is not hard to conceive that such his disciples may, to save themselves the trouble of thinking, be inclined to confound the usefulness of a rule with the certainty of a truth, and accept the one for the other; especially if they are men accustomed rather to compute than to think; earnest rather to go on fast and far, than solicitous to set out warily and see their way distinctly.

[The subject of the next sections can be summed up in the following argument. If $(x + c)^n - x^n = nx^{n-1}0 + \frac{n(n-1)}{1.2}x^{n-2}0^2 + \cdots$, and we divide by 0, we can get nx^{n-1}, the fluxion of x^n, only by first supposing that $0 \neq$ zero, then $0 =$ zero. "All which seems a most inconsistent way of arguing, and such as would not be allowed of in Divinity" (Sec. 14). Then follows, somewhat later:]

35. I know not whether it be worth while to observe, that possibly some men may hope to operate by symbols and suppositions, in such sort as to avoid the use of fluxions, momentums, and infinitesimals, after the following manner. Suppose x to be an absciss of a curve, and z another absciss of the same curve. Suppose also that the respective areas are xxx and zzz: and that $z - x$ is the increment of the absciss, and $zzz - xxx$ the increment of the area, without considering how great or how small these increments may be. Divide now $zzz - xxx$ by $z - x$, and the quotient will be $zz + zx + xx$: and, supposing that z and x are equal, this same quotient will be $3xx$, which in that case is the ordinate, which therefore may be thus obtained independently of fluxions and infinitesimals. But herein is a direct fallacy: for, in the first place, it is supposed that the abscisses z and x are unequal, without which supposition no one step could have been made; and in the second place, it is supposed they are equal; which is a manifest inconsistency, and amounts to the same thing that hath been before considered. And there is indeed reason to apprehend that all attempts for setting the abstruse and fine geometry on a right foundation, and avoiding the doctrine of velocities, momentums, etc. will be found impracticable, till such time as the object and end of geometry are better understood than hitherto they seem to have been. The great author of the method of fluxions felt this difficulty, and therefore he gave into those nice abstractions and geometrical metaphysics without which he saw nothing could be done on the received principles; and what in the way of demonstration he hath done with them the reader will judge. It must, indeed, be acknowledged that he used fluxions, like the scaffold of a building, as things to be laid aside or got rid of as soon as finite lines were found proportional to them. But then these finite exponents are found by the help of fluxions. Whatever therefore is got by such exponents and proportions is to be ascribed to fluxions: which must therefore be previously understood. And what are these fluxions? The velocities of evanescent increments? And what are these same evanescent increments? They are neither finite quantities, nor quantities infinitely small, nor yet nothing. May we not call them the ghosts of departed quantities?

NOTES

1. We recognize the statement made by Newton in his *Quadratura curvarum* (Selection V.7).

2. Newton's statement again, this time in Latin.

COLIN MACLAURIN (1698–1746)

Colin Maclaurin, the son of a cleric, was the most prominent mathematician of Great Britain during the 18th century. A child prodigy, he matriculated at the University of Glasgow at age 11 (1709) and received the master of arts degree there (1715). Following a competitive examination, he was appointed professor of mathematics at Marischal College, Aberdeen, in 1717. Two years later he visited London, where he met Isaac Newton and was elected a fellow of the Royal Society. Maclaurin's principal book, *Geometrica organica* (1720; *Organic Geometry, with the Description of Universal Linear Curves*) was published with Newton's imprimatur. It developed some theorems similar to those in Newton's *Principia* and introduced what has become known as Maclaurin's method for generating conics (circle, ellipse, hyperbola, and parabola).

Maclaurin left Scotland in 1722 to tutor the eldest son of Lord Polwarth, British plenipoteniary at Cambrai. The two traveled to Paris and Lorraine, where they resided for some time. In 1724, Maclaurin won the prestigious prize of the Paris Academy of Sciences for his paper "On the Percussion of Bodies" and returned to Aberdeen after his pupil died suddenly. Because he was absent for three years, Aberdeen had declared his chair vacant, so he went to Edinburgh. With the strong backing of Newton, he succeeded James Gregory in the chair of mathematics in 1725 and held this position until his death. In 1733, he married Anne Stewart.

The multi-talented Maclaurin was a skilled experimenter who built mechanical devices. He also improved maps, made actuarial tables for insurance companies, and shared with Leonhard Euler and Daniel Bernoulli the prize of the Paris Academy in 1740 for an essay on tides. In 1745, when Jacobite rebels (supporters of James II) marched on Edinburgh, he helped direct the city's defense (including preparing trenches and barricades) and worked to the point of exhaustion. When the Jacobites briefly captured the city, he fled to York. The fatigue from defense preparations and the ordeal of escape ruined his delicate health. He died of dropsy shortly after his return to Edinburgh in 1746.

Maclaurin developed and extended Newton's work on the method of fluxions (an early stage of the calculus), on geometry, and on physics. His *Treatise of Fluxions* (2 vols., 1742) first systematically elaborated Newton's version of the calculus. As Maclaurin stated in the preface, the *Treatise* responded to Bishop George Berkeley's criticism of the calculus for its faulty reasoning, mystery, and general lack of rigorous foundations. In the *Analyst* (1734), Berkeley had referred to Newton's fluxions as "the ghosts of departed quantities." Maclaurin sought to counter Berkeley's criticism by grounding the method of fluxions on ancient Greek geometrical methods and Archimedes' version of the method of exhaustion. His search for rigorous foundations proved unsuccessful; it took another century to develop these. Moreover, his geometrical presentation contributed to the neglect of mathematical analysis in Britain. In the *Treatise*, Maclaurin gave the integral test for the convergence of an infinite series and what is now known as Maclaurin's theorem for the expansion of a function x. The theorem is a special case of Taylor's theorem for $a = 0$:

$$f(x) = f(0) + x f'(0) + \frac{x^2}{2!} f''(0) + \frac{x^3}{3!} f'''(0) + \ldots$$

Maclaurin also wrote *Account of Sir Isaac Newton's Philosophical Discoveries*, which he had begun in 1728 at the urging of Newton's nephew, John Conduitt, and *Treatise of Algebra*. Both were published posthumously in 1748.

83. From *Treatise of Fluxions* (1742)*

(On Series and Extremes)

COLIN MACLAURIN

751. The following theorem is likewise of great use in this doctrine. Suppose that y is any quantity that can be expressed by a series of this form $A + Bz + Cz^2 + Dz^3 +$ etc. where A, B, C, etc. represent invariable coefficients as usual, any of which may be supposed to vanish. When z vanishes, let E be the value of y, and let $\dot{E}$, $\ddot{E}$, $\dddot{E}$, etc. be then the respective values of $\dot{y}$, $\ddot{y}$, $\dddot{y}$, etc. z being supposed to flow uniformly. Then $y = E + \frac{\dot{E}z}{\dot{z}} + \frac{\ddot{E}z^2}{1 \times 2\dot{z}^2} + \frac{\dddot{E}z^3}{1 \times 2 \times 3\dot{z}^3} + \frac{\ddddot{E}z^4}{1 \times 2 \times 3 \times 4\dot{z}^4} +$ etc. the law of the continuation of which series is manifest.

For since $y = A + Bz + Cz^2 + Dz^3 +$ etc. it follows that when $z = o$, A is equal to y; but (by the supposition) E is then equal to y; consequently $A = E$. By taking the fluxions, and dividing by $\dot{z}$, $\frac{\dot{y}}{\dot{z}} = B + 2Cz + 3Dz^2 +$ etc. and when $z = o$, B is equal to $\frac{\dot{y}}{\dot{z}}$, that is to $\frac{\dot{E}}{\dot{z}}$. By taking the fluxions again, and dividing by $\dot{z}$, (which is supposed invariable) $\frac{\ddot{y}}{\dot{z}^2} = 2C + 6Dz +$ etc. Let $z = o$, and substituting $\ddot{E}$ for $\ddot{y}$, $\frac{\ddot{E}}{\dot{z}^2} = 2C$, or $C = \frac{\ddot{E}}{2\dot{z}^2}$. By taking the fluxions again, and dividing by $\dot{z}$, $\frac{\dddot{y}}{\dot{z}^3} = 6D +$ etc. and by supposing $z = o$, we have $D = \frac{\dddot{E}}{6\dot{z}^3}$. Thus it appears that $y = A + Bz + Cz^2 + Dz^3 +$ etc. $= E + \frac{\dot{E}z}{\dot{z}} + \frac{\ddot{E}z^2}{1 \times 2\dot{z}^2} + \frac{\dddot{E}z^3}{1 \times 2 \times 3\dot{z}^3} + \frac{\ddddot{E}z^4}{1 \times 2 \times 3 \times 4\dot{z}^4} +$ etc.

This proposition may be likewise deduced from the binomial theorem. Let BD [Fig. 1], the ordinate of the figure FDM at B, be equal to E, $BP = z$, $PM = y$, and this series will serve for resolving

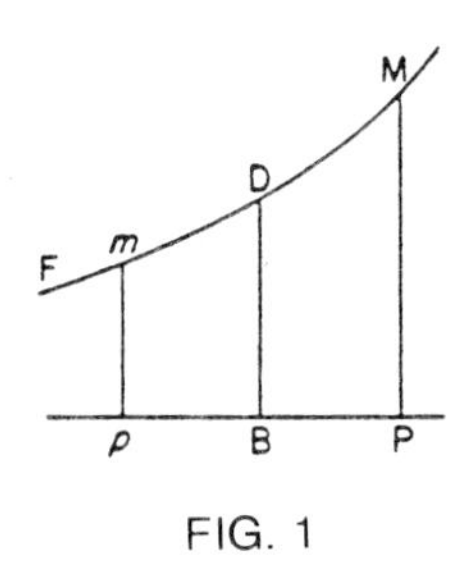

FIG. 1

*Source: Reprinted by permission of the publishers from *A Source Book in Mathematics, 1200-1800*, edited by D. J. Struik, Cambridge, Mass.: Harvard University Press, Copyright © 1969 by the President and Fellows of Harvard College.

the value of PM, or y, (some particular cases being excepted, as when any of the coefficients E, $\frac{\dot{E}}{\dot{z}}$, $\frac{\ddot{E}}{\dot{z}^2}$, etc. become infinite) into a series, not only in such cases as were described in the preceding articles, but likewise when the relation of y and z is determined by an affected equation, and in many cases when their relation is determined by a fluxional equation. This theorem was given by Dr. Taylor, *method. increm.* By supposing the fluxion of z to be represented by BP, or $\dot{z} = z$, we have $y = E + \dot{E} + \frac{\ddot{E}}{2} + \frac{\dddot{E}}{6} + \frac{\ddddot{E}}{24}$ + etc. (as was observed in Art. 255)[1] and hence it appears at what rate the fluxion of y of each order contributes to produce the increment or decrement of y, since $y - E = \dot{E} + \frac{\ddot{E}}{2} + \frac{\dddot{E}}{6} + \frac{\ddddot{E}}{24}$ + etc. If Bp be taken on the other side of B equal to BP, then $pm = A - Bz + Cz^2 - Dz^3$ + etc. = (the same quantities being represented by $\frac{\dot{E}}{\dot{z}}$, $\frac{\ddot{E}}{\dot{z}^2}$, etc., as before, or the base being supposed to flow the same way,) $E - \frac{\dot{E}z}{\dot{z}} + \frac{\ddot{E}z^2}{1 \times 2\dot{z}^2} - \frac{\dddot{E}z^3}{1 \times 2 \times 3\dot{z}^3} + \frac{\ddddot{E}z^4}{1 \times 2 \times 3 \times 4\dot{z}^4}$ − etc. consequently $PM + pm = 2E + \frac{2\ddot{E}z^2}{1 \times 2\dot{z}^2} + \frac{2\ddddot{E}z^4}{1 \times 2 \times 3 \times 4\dot{z}^4}$ + etc. . . .

[Then, in Arts. 858-861, Maclaurin gives his criterion for maxima and minima.]

858. When the first fluxion of the ordinate vanishes, if at the same time its second fluxion is positive, the ordinate is then a *minimum*, but is a *maximum* if its second fluxion is then negative; that is, it is less in the former, and greater in the latter case than the ordinates from the adjoining parts of that branch of the curve on either side. This follows from what was shown at great length in

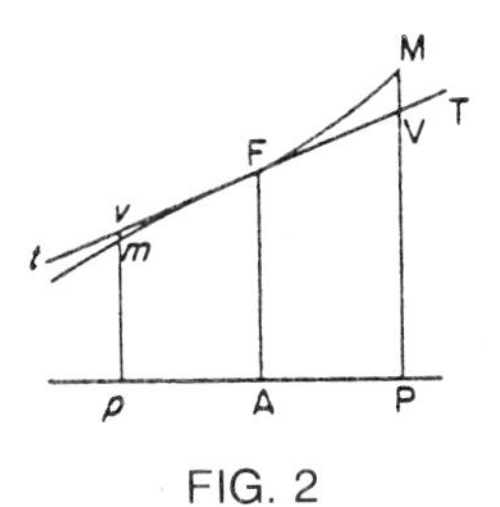

FIG. 2

Chap. 9. B. I, or may appear thus. Let the ordinate $AF = E$, $AP = x$ [Fig. 2], and the base being supposed to flow uniformly, the ordinate PM = (Art. 751) $E + \frac{\dot{E}x}{\dot{x}} + \frac{\ddot{E}x^2}{2\dot{x}^2} + \frac{\dddot{E}x^3}{6\dot{x}^3}$ + etc. Let Ap be taken on the other side of A equal to AP, then the ordinate $pm = E - \frac{\dot{E}x}{\dot{x}} + \frac{\ddot{E}x^2}{2\dot{x}^2} - \frac{\dddot{E}x^3}{6\dot{x}^3}$ + etc. Suppose now $\dot{E} = o$, then $PM = E * + \frac{\dot{E}x}{\dot{x}} + \frac{\ddot{E}x^2}{2\dot{x}^2}$ etc. and $pm = E * + \frac{\ddot{E}x^2}{2\dot{x}^2}$ − etc. Therefore if the distances AP and Ap be small enough, PM and pm will both exceed the ordinate AF when $\ddot{E}$ is positive; but will be both less than AF if $\ddot{E}$ be negative. But if $\ddot{E}$ vanish as well as $\dot{E}$, and $\dddot{E}$ does not vanish, one of the adjoining ordinates PM or pm shall be greater than AF, and the other less than it; so that in this case the ordinate is neither a *maximum* nor *minimum*. We always suppose the expression of the ordinate to be positive.

859. In general, if the first fluxion of the ordinate, with its fluxions of several subsequent orders, vanish, the ordinate is a *minimum* or *maximum*, when the number of all those fluxions that vanish is 1, 3, 5, or any odd number. The ordinate is a *minimum*, when the fluxion next to those that vanish is positive; but a *maximum* when this fluxion is negative. This appears from Art. 261, or by comparing the values of PM and pm in the last article. But if the number of all the fluxions of the ordinate of the first and subsequent successive orders that vanish be an even number, the ordinate is then neither a *maximum* nor *minimum*.

860. When the fluxion of the ordinate y is supposed equal to nothing, and an equation is thence derived for determining x, if the roots of this equation are all unequal, each gives a value of x that may correspond to a greatest or least ordinate. But if two, or any even number of these roots be equal, the ordinate that corresponds to them is neither a *maximum* nor *minimum*. If an odd number of these roots be equal, there is one *maximum* or *minimum* that corresponds to these roots, and one only. Thus if $\frac{\dot{y}}{\dot{x}} = x^4 + ax^3 + bx^2 + cx + d$, then supposing all the roots of the equation $x^4 + ax^3 + bx^2 + cx + d = 0$ to be real, if the four roots are equal there is no ordinate that is a *maximum* or *minimum;* if two or three of the roots only are equal, there are two ordinates that are *maxima* or *minima;* and if all the roots are unequal there are four such ordinates.

861. To give a few examples of the most simple cases. Let $y = a^2x - x^3$, then $\dot{y} = a^2\dot{x} - 3x^2\dot{x}$ and $\ddot{y} = -6x\dot{x}^2$. Suppose $\dot{y} = 0$, and $3x^2 = a^2$ or $x = \frac{a}{\sqrt{3}}$, in which case $\ddot{y} = \frac{-6a\dot{x}^2}{\sqrt{3}}$. Therefore $\ddot{y}$ being negative, y is a *maximum* when $x = \frac{a}{\sqrt{3}}$, and its greatest value is $\frac{2a^3}{3\sqrt{3}}$. If $y = aa + 2bx - xx$, then $\dot{y} = 2b\dot{x} - 2x\dot{x}$, and $\ddot{y} = -2\dot{x}^2$; consequently y is a *maximum* when $2b - 2x = 0$, or $x = b$. If $y = aa - 2bx + xx$ then $\dot{y} = 2b\dot{x} + 2x\dot{x}$, and $\ddot{y} = 2\dot{x}^2$; consequently y is now a *minimum* when $x = b$, if a be greater than b.

[Maclaurin also considers the cases in which $\ddot{y}$, $\dddot{y}$, $\ddddot{y}$, . . . vanish.]

NOTE

1. Maclaurin's book is divided into two parts. Book I is geometrical, Book II is computational. Our selection is from Book II. Articles 255 and 261 (to which he refers below) deal with the same matter in a geometrical way.

JEAN LE ROND D'ALEMBERT (1717–83)

The Frenchman Jean d'Alembert attained a reputation in the mathematical sciences before achieving fame as a *philosophe* during the Continental Enlightenment. In the sciences he substantially advanced mathematical analysis and rational mechanics, and he believed, like Locke and Condillac, that sense perception provides the basic evidence about the physical world. As a *philosophe,* he came to rank just below Voltaire and Denis Diderot, the general editor of the *Encyclopédie* (28 vols., 1751–72).

During the Enlightenment deductive reason was supplanting religious faith as the chief guide to social action among the educated public. D'Alembert maintained that the increased use of reason would lead to progress. He also advocated tolerance, free speech, and enlightened absolutism as well as criticizing established religion.

The illegitimate son of a salon hostess, Madame de Tencin, and a cavalry officer named Destouches-Canon, d'Alembert was abandoned on the steps of the Parisian church Saint Jean-Le-Rond by his mother who had just renounced her nun's vows and may have feared that civil authorities would forcibly return her to a convent if they learned of the birth. The father located the infant and found him a home with a humble glazier, named Rousseau, and his wife. They christened the child Jean le Rond for the church where he was found, and he lived with his adoptive parents until he was 47 years old. His natural father, though he did not reveal his identity, provided an annual annuity of 1200 livres and gained him admission to the prestigious College Mazarin, a Jansenist school which stressed classics and rhetoric. There d'Alembert developed an aversion for religious studies, and turned to law, becoming an advocate in 1738. He then briefly studied medicine before beginning work in the mathematical sciences, which he learned largely by himself. Later he would write that mathematics was "the only occupation which really interested me."

In 1739, d'Alembert submitted his first *memoir* to the Paris Academy of Sciences. During the next two years he submitted five more papers, which dealt with differential equations and with the motion of bodies in resisting media. He made himself familiar with the writings of Newton, L'Hôpital, the Bernoullis, and major contemporary geometers. Following several unsuccessful attempts to gain admittance to the Paris Academy, he was finally elected a member in 1741. After a two-year study of several problems in mechanics, he hastily published his most famous scientific work, *Traité de dynamique* (1743), which helped to formalize the new science of dynamics. The *Traité* contains "d'Alembert's principle," which maintains that Newton's third law of motion (every action has an equal and opposite reaction) holds for moving and rigidly fixed bodies. It also helped to resolve the controversy over the principle of the conservation of *vis viva* (mv^2). In this dispute the Newtonians and Cartesians asserted that the "quantity of motion" (mv) gave the correct measure of force in the study of collisions. The followers of Leibniz and Wolff disagreed; they claimed that mv^2 was the correct measure. Pointing out in the preface that Newton's force could be defined either as acting through space ($mv^2 = 2Fs$) or over time ($mv = mat = Ft$), d'Alembert declared this controversy over force measurement to be a false one—a quarrel of words.

By the middle of the 18th century, d'Alembert stood among the leading mathematicians and theoretical physicists in Europe. Three others were his French rival Alexis Clairaut, Daniel Bernoulli in Basel, and Leonhard Euler in Berlin and St. Petersburg, with Euler the most able of the group. In 1744, d'Alembert published a landmark treatise on fluid mechanics, which correctly established that if one assumes the earth to be a rotating fluid body, it must have an orange-like shape. Over the next three years he developed partial differential equations as a branch of the calculus and was the first to generally apply them to problems in physics, including that of the motion of vibrating chords. In 1749, his interest in the three-body problem in celestial mechanics led him to explain the precession of equinoxes—a gradual shift in the position of the earth's orbit—and the nutation or wobbling of the earth's axis. In his essay on hydrodynamics published in 1752, differential hydrodynamic equations were first expressed in terms of a field—a pioneering attempt in complex function theory—and the later discredited "d'Alembert's paradox" was introduced.

After 1750, d'Alembert turned increasingly to interests beyond the sciences, becoming associated with the *Encyclopédie*—the chief intellectual enterprise in Europe in the mid-18th century and the center of opposition to the *Ancien Regime*. He wrote the *Discourse préliminaire* (1751) to the *Encyclopédie* and served as its science editor for seven years. In 1756, he traveled to Geneva to enjoy a leisurely visit with Voltaire and to collect material for an article on the city. What he wrote was a tendentious four-page piece that appeared in the seventh volume of the *Encyclopédie*. In it d'Alembert claimed that some Genevan pastors "no longer believe in the divinity of Jesus Christ," and he praised them for their learning, their freedom from superstition, and their support of theatre. The publication of the article aroused a public furor in both Geneva and Paris, and d'Alembert prudently resigned the science editorship of the *Encyclopédie*. However, his action brought him strained relations with the shaken editor, Diderot, who considered him a deserter. The next year, after vehement public debate, the French government suspended the license of the *Encyclopédie*.

There were other tasks facing d'Alembert. The success of the *Discours préliminaire* and the intercession of Mme. du Deffand, whose home was a prominent salon for literary men and savants, had brought about his acceptance to the French Academy in 1754. He worked zealously to enhance its dignity and was made perceptual secretary in 1772. As his scientific and literary fame spread, foreign monarchs vied for his services. In 1764, he spent three months at Potsdam with Frederick the Great who wanted him to be president of the Berlin Academy. He refused the presidency, however, and recommended Euler for the position. His support for Euler healed a rift that had developed more than a decade earlier when d'Alembert believed that Euler had blocked his winning of a prize from the Berlin Academy for a paper on fluid mechanics. Refusing to leave Paris, the cultural capital of Europe, d'Alembert subsequently declined an offer from Catherine the Great who wanted him to improve the Russian educational system.

A small man with a highly pitched voice, d'Alembert was known in Parisian society for his gaiety, witty conversation, and talent for mimicry. He usually worked both in the morning and afternoon, spending his evenings in the salons where the cultivated public gathered. Practicing frugality, he was satisfied with his limited means. He enjoyed fair health until 1765 when he fell gravely ill. Although he never married, he moved at that time into the house of his lover, Mlle. de Lespinasse, and resided there until her death in 1776. He spent his last years in an apartment at the Louvre.

His contributions to mathematical analysis were extensive. Almost alone in this time he regarded the derivative as the limit of a quotient of increments, or what we now express as dy/dx. Eventually the calculus would be rationalized around the key concept of the limit, but d'Alembert was not able to put it in to a purely algorithmic form. He stressed the law of continuity in analysis and called equations with discontinuities impossible. His continuity requirement probably led him to the idea of a limit and made him examine the techniques for handling infinite series. In volume V of his *Opuscules mathématiques* (8 vols., 1761–80) he published d'Alembert's theorem (the ratio test for convergency). His theorem follows:

If $\lim \left[\frac{S_{n+1}}{S_n}\right] = r$ and $r < 1$, then the series $\sum_{n=1}^{\infty} S_n$ converges.

If $r > 1$, the series diverges; if $r = 1$, the test fails.

In mathematics, he also considered the parallel postulate in Euclidean geometry a "scandal" and worked on probability theory, applying it to games of chance and to determining life expectancy.

84. From "Differential," *Encyclopédie*, Vol. 4 (1754)*[1]

(On Limits)

JEAN d'ALEMBERT

. . . What concerns us most here is the metaphysics of the *differential* calculus.

This metaphysics, of which so much has been written, is even more important and perhaps more difficult to explain than the rules of this calculus themselves: various mathematicians, among them Rolle,[2] who were unable to accept the assumption concerning infinitely small quantities, have rejected it entirely, and have held that the principle was false and capable of leading to error. Yet in view of the fact that all results obtained by means of ordinary Geometry can be established similarly and much more easily by means of the *differential* calculus, one cannot help concluding that, since this calculus yields reliable, simple, and exact methods, the principles on which it depends must also be simple and certain.

Leibniz was embarrassed by the objections he felt to exist against infinitely small quantities, as they appear in the *differential* calculus; thus he preferred to reduce infinitely small to merely in-

*Source: Reprinted by permission of the publishers from *A Source Book in Mathematics, 1200-1800*, edited by D. J. Struik, Cambridge, Mass.: Harvard University Press,

comparable quantities. This, however, would ruin the geometric exactness of the calculations; is it possible, said Fontenelle,[3] that the authority of the inventor would outweigh the invention itself? Others, like Nieuwentijt,[4] admitted only *differentials* of the first order and rejected all others of higher order. This is impossible; indeed, considering an infinitely small chord of first order in a circle, the corresponding abscissa or versed sine is infinitely small[5] of second order; and if the chord is of the second order, the abscissa mentioned will be of the fourth order, etc. This is proved easily by elementary geometry, since the diameter of a circle (taken as a finite quantity) is always to the chord as the chord to the corresponding abscissa.[6] Thus, if one admits the infinitely small of the first order, one must admit all the others, though in the end one can rather easily dispense with all this metaphysics of the infinite in the *differential* calculus, as we shall see below.

Newton started out from another principle; and one can say that the metaphysics of this great mathematician on the calculus of fluxions is very exact and illuminating, even though he allowed us only an imperfect glimpse of his thoughts.

He never considered the *differential* calculus as the study of infinitely small quantities, but as the method of first and ultimate ratios, that is to say, the method of finding the limits of ratios. Thus this famous author has never differentiated quantities but only equations; in fact, every equation involves a relation between two variables and the differentiation of equations consists merely in finding the limit of the ratio of the finite differences of the two quantities contained in the equation. Let us illustrate this by an example which will yield the clearest idea as well as the most exact description of the method of the *differential* calculus.

Let AM [Fig. 1] be an ordinary parabola, the equation of which is $yy = ax$; here we assume that $AP = x$

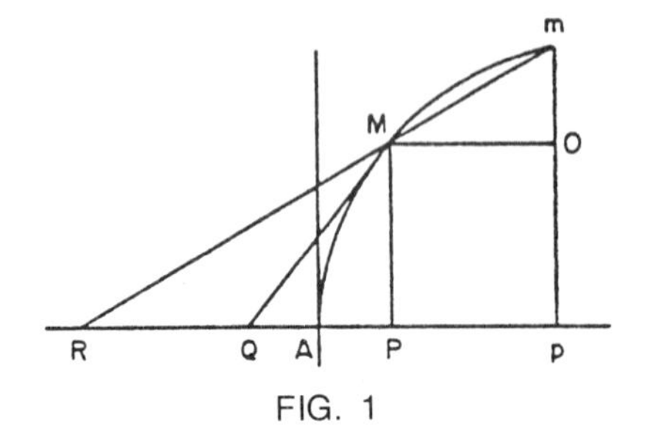

FIG. 1

and $PM = y$, and a is a parameter. Let us draw the tangent MQ to this parabola at the point M. Let us suppose that the problem is solved and let us take an ordinate pm at any finite distance from PM; furthermore, let us draw the line mMR through the points M, m. It is evident, *first,* that the ratio[7] MP/PQ of the ordinate to the subtangent is greater than the ratio MP/PR or mO/MO which is equal to it because of the similarity of the triangles MOm, MPR; *second,* that the closer the point m is to the point M, the closer will be the point R to the point Q, consequently the closer will be the ratio MP/PR or mO/MO to the ratio MP/PQ; finally, that the first of these ratios approaches the second one as closely as we please, since PR may differ as little as we please from PQ. Therefore, the ratio MP/PQ is the limit of the ratio of mO to OM. Thus, if we are able to represent the ratio mO/OM in algebraic form, then we shall have the algebraic expression of the ratio MP to PQ and consequently the algebraic representation of the ratio of the ordinate to the subtangent, which will enable us to find this subtangent. Let now $MO = u$, $Om = z$; we shall have $ax = yy$, and $ax + au = yy + 2yz + zz$. Then in view of $ax = yy$ it follows that $au = 2yx + zz$ and $z/u = a/(2y + z)$.

This value $a/(2y + z)$ is, therefore, in general the ratio of mO to OM, wherever one may choose the point m. This ratio is always smaller than $a/2y$; but the smaller z is, the greater the ratio will be and, since one may choose z as small as one pleases, the ratio $a/(2y + z)$ can be brought as close to the ratio $a/2y$ as we like. Consequently $a/2y$ is the limit of the ratio $a/(2y + z)$, that is to say, of the ratio mO/OM. Hence $a/2y$ is equal to the ratio MP/PQ, which we have found to be also the limit of the ratio of mO to Om, since two quantities that are the limits of the same quantity are necessarily equal to each other. To prove this, let X and Z be the limits of the same quantity Y. Then I say that $X = Z$; indeed, if they were to have the difference V, let $X = Z \pm V$: by hypothesis the quantity Y may approach X as closely as one may wish; that is to say, the difference between Y and X may be as small as one may wish. But, since Z differs from X by the quantity V, it follows that Y cannot approach Z closer than the quantity V and consequently Z would not be the limit of Y, which is contrary to the hypothesis.

From this it follows that MP/PQ is equal to $a/2y$. Hence $PQ = 2yy/a = 2x$. Now, according to the method of the *differential* calculus, the ratio of MP to PQ is equal to that of dy to dx; and the equation $ax = yy$ yields $a\,dx = 2y\,dy$ and $dy/dx = a/2y$. So dy/dx is the limit of the ratio of z to u, and this limit is found by making $z = 0$ in the fraction $a/(2y + z)$.

But, one may say, is it not necessary also to make $z = 0$ and $u = 0$ in the fraction $z/u = a/(2y + z)$, which would yield $0/0 = a/2y$? What does this mean? My answer is as follows. First, there is no absurdity involved; indeed $0/0$ may be equal to any quantity one may wish: thus it may be $= a/2y$. Secondly, although the limit of the ratio of z to u has been found when $z = 0$ and $u = 0$, this limit is in fact not the ratio of $z = 0$ to $u = 0$, because the latter one is not clearly defined; one does not know what is the ratio of two quantities that are both zero. This limit is the quantity to which the ratio z/u approaches more and more closely if we suppose z and u to be real and decreasing. Nothing is clearer than this; one may apply this idea to an infinity of other cases.[8]

Following the method of differentiation (which opens the treatise on the quadrature of curves by the great mathematician Newton), instead of the equation $ax + au = yy + 2yz + zz$ we might write $ax + a0 = yy + 2y0 + 00$, thus, so to speak, considering z and u equal to zero; this would have yielded $0/0 = a/2y$. What we have said above indicates both the advantage and the inconveniences of this notation: the advantage is that z, being equal to 0, disappears without any other assumption from the ratio $a/(2y + 0)$; the inconvenience is that the two terms of the ratio are supposed to be equal to zero, which at first glance does not present a very clear idea.

From all that has been said we see that the method of the *differential* calculus offers us exactly the same ratio that has been given by the preceding calculation. It will be the same with other more complicated examples. This should be sufficient to give beginners an understanding of the true metaphysics of the *differential* calculus. Once this is well understood, one will feel that the assumption made concerning infinitely small quantities serves only to abbreviate and simplify the reasoning; but that the *differential* calculus does not necessarily suppose the existence of those quantities; and that moreover this calculus merely consists in *algebraically determining the limit of a ratio, for which we already have the expression in terms of lines, and in equating those two expressions. This will provide us with one of the lines we are looking for.* This is perhaps the most precise and neatest possible definition of the *differential* calculus; but it can be understood only when one is well acquainted with this calculus, because often the true nature of a science can be understood only by those who have studied this science.

In the preceding example the known geometric limit of the ratio of z to u is the ratio of the ordinate to the subtangent; in the *differential* calculus we look for the algebraic limit of the ratio z to u and we find $a/2y$. Then, calling s the subtangent, one has $y/s = a/2y$; hence $s = 2yy/a = 2x$. This example is sufficient to understand the others. It will, therefore, be sufficient to make oneself familiar with the previous example concerning the tangents of the parabola, and, since the whole *differential* calculus can be reduced to the problem of the tangents, it follows that one could always apply the preceding principles to various problems of this calculus, for instance to find *maxima and minima*, points of inflection, cusps, etc. . . .[9]

What does it mean, in fact, to find a maximum or a minimum? It consists, it is said, in setting the difference[10] dy equal to zero or to infinity; but it is more precise to say that it means to look for the quantity dy/dx which expresses the limit of the ratio of finite dy to finite dx, and to make this quantity zero or infinite. In this way all the mystery is explained; it is not dy that one makes = to infinity: that would be absurd, since dy is taken as infinitely small and hence cannot be infinite; it is dy/dx: that is to say, one looks for the value of x that renders the limit of the ratio of finite dy to finite dx infinite.

We have seen above that in the *differential* calculus there are really no infinitely small quantities of the first order; that actually those quantities called u are supposed to be divided by other supposedly infinitely small quantities; in this state they do not denote either infinitely small quantities or quotients of infinitely small quantities; they are the limits of the ratio of two finite quantities. The same holds for the second-order differences and for those of higher order. There is actually no quantity in Geometry such as $d\,dy$; whenever $d\,dy$ occurs in an equation it is supposed to be divided by a quantity dx^2, or another of the same order. What now is $d\,dy/dx^2$? It is the limit of the ratio $d\,dy/dx$ divided by dx; or, what is still clearer, it is the limit of dz/dx, where $dy/dx = z$ is a finite quantity.

NOTES

1. (Editor's Note: For an *histoire du livre* of the *Encyclopédie*, one should consult Robert Darnton, *The Business of Enlightenment: A Publishing History of the Encyclopédie, 1775–1800* (Cambridge, Mass.: Harvard University Press, 1979).)

2. Michel Rolle (1652–1719), member of the Paris Academy, is best known for the theorem in the theory of equations called after him. In 1700 he took part in a debate in the Paris Academy on the principles of the calculus; see C. Boyer, *The History of the Calculus* (Dover, New York, 1949), 241.

3. Bernard le Bovier de Fontenelle (1657–1757) was a predecessor of d'Alembert as *secrétaire perpétuel* of the Academy. See Boyer, *History*, 241-242.

4. Bernard Nieuwentijt (1654–1718), a physician-burgomaster of Purmerend, near Amsterdam, opposed Leibniz's concept of the calculus.

5. Versed sin $\alpha = 1 - \cos \alpha = \alpha^2/2! - \alpha^4/4! + \cdots$ (d'Alembert still takes the dimension to be that of a chord, hence his vers α is really our R vers α).

6. $2R : 2R \sin \alpha/2 = 2R \sin \alpha/2 : R(1 - \cos \alpha)$.

7. D'Alembert writes $\frac{MP}{PQ}$.

8. Here d'Alembert refers to his articles on "Limit" and "Exhaustion" in the same *Encyclopédie*.

9. Here d'Alembert refers to his articles on these subjects.

10. D'Alembert makes little distinction between *différence* and *différentiel*.

LEONHARD EULER (1707–83)

Leonhard Euler was one of the two leading figures in the exact sciences in the 18th century. During that century, only the Savoyard Louis Lagrange compared with him in brilliance and achievement in mathematics and theoretical physics; no one in these fields compared with him in his prolific writing. Euler swiftly and clearly wrote over 866 books and articles, which constitute about one third of the entire corpus published between 1725 and 1800 on the subjects of mathematics, theoretical physics, and engineering mechanics. His publications fill 74 quarto volumes of 300 to 600 pages each. He also engaged in an extensive correspondence, exchanging as many as 5,000 letters with scientists, administrators, and savants across Europe. His letters, many of which are like articles in a modern research journal, cover a wide range of topics, including architecture, biology, chemical science, history, philosophy, religion, and technology.

Euler was born in Basel, Switzerland. His father, Paul, was a Zwinglian minister; his mother, Margaret Brucker, was the daughter of another minister. He grew up in the Swiss countryside in Riehen in a two-room parsonage with two younger sisters. At home, his mother instructed Leonhard in classical humanities, and his father, who had studied under Jakob Bernoulli, taught him mathematics and religion. As a child he developed the forthright disposition and deep religious conviction for which he was known lifelong.

In 1719, Euler was sent to Basel's humanistic Gymnasium for formal schooling and one year later enrolled at the University of Basel, where he displayed keen abilities and was graduated with first honors in 1722. He also revealed a phenomenal memory by reciting Vergil's *Aeneid* by heart. At his father's bidding in 1723, he began theological studies at the university in preparation for the ministry. However, he was already deeply interested in mathematics and, with effort, convinced the stern and difficult Johann Bernoulli to tutor him in mathematics and natural philosphy for one hour on Saturday afternoons. He read classics in these fields and presented problems that he could not solve. Bernoulli quickly recognized the boy's genius and helped to convince Paul Euler to allow his son to concentrate on the mathematical sciences.

In 1727, after failing to obtain a physics position in Basel, Euler joined the St. Petersburg Academy of Sciences. When the Russian government stopped its funds, he served as a medical lieutenant in the Russian navy from 1727 to 1730. He became professor of natural philosophy at the Academy in 1730 and first professor of mathematics (the premier post) in 1733 succeeding Daniel Bernoulli, who returned to Switzerland. Until then he had boarded at Daniel Bernoulli's home. Among the topics the two men discussed at dinner was Bernoulli's book *Hydrodynamica* (1738). In December 1733, Euler married Catherine Gsell, the daughter of a Dutch artist living in Russia. They had 13 children, five of whom survived childhood.

In the years from 1733 to 1741, Euler immersed himself in research on mathematics and phenomenology, setting a pattern that continued throughout his career. He carefully avoided the political dangers facing foreign scientists in St. Petersburg. The Holy Synod of the Russian Orthodox Church opposed Copernican astronomy and the new sciences in general, while the Russian nobility were hostile to German scholars. Through his articles and books his genius gradually became known to a wide audience, and he gained a European-wide reputation with his two-volume *Mechanica* (1736–38). This book first extensively applied the calculus to Newtonian dynamics. Following three years of intense work on astronomy and maps, he lost the sight of his right eye in 1738.

Amid the turmoil following the death of Empress Anna in Russia, Euler accepted in 1741 the invitation of Frederick II to join the Berlin Academy of Sciences. Prussia was becoming a great power in Europe, and Frederick wanted his academy to be a leading intellectual center. Euler was to serve as director of its mathematics section from 1744 to 1766. He resided on a farm outside Berlin (where he grew vegetables in the summer), and a small circle of colleagues gathered about him.

The picture of Euler that emerged in Berlin was that of a gracious, open, and self-assured man. In the early 1750s, he was drawn into a celebrated priority dispute. The Academy president Pierre Maupertuis believed that he had discovered the principle of least action, but Samuel König and the Wolffians claimed that Leibniz had done so earlier. Euler ably defended the priority of Maupertuis, and Voltaire, then an opponent of Maupertuis, criticized his stance. In the mid-1750s, Euler tutored Lagrange by correspondence and selflessly withheld from publication his work on the calculus of variations so that Lagrange might receive due credit for his contributions to that subject. In 1755, the Paris Academy of Sciences named him a foreign member; his winning of its prestigious biennial prize 12 times was a remarkable feat.

From 1760 to 1762, Euler completed the three-volume *Letters to a Princess of Germany* (1768–72), which represents his mature scientific outlook and shows him to be an original and eclectic thinker. Above all he supported Newtonian dynamics but opposed Newton's corpuscular theory of light, accepting instead Huygens' wave theory. He portrayed Cartesian science as an intermediary between Newtonian ideas and ancient thought. His chief criticism was reserved for the Leibniz–Wolffian philosophy, especially its monadic doctrine. Leibniz believed that animate, percipient, elastic monads were the primal substance of the universe; Euler maintained that passive, punctual, impenetrable elements of matter were. The *Letters* proved to be an extremely successful scientific popularization; it had been translated into 8 languages and had undergone over 40 editions by 1840.

After disagreeing with Frederick the Great on the matter of academic freedom, the presidency of the Academy, and the financial security for his children, Euler returned to Russia in 1766 when Catherine the Great made him a generous offer. Although a clumsy cataract operation in 1771 left Euler almost totally blind and in pain, this did not slow his research or writing. He dictated treatises on algebra, lunar motion, and optics to students and secretaries, doing calculations involving as many as 50 places in his head. These treatises have the same clarity that characterizes his other writings. In September 1783, he died of a brain hemorrhage.

Although Euler applied himself to all known areas of mathematics, he is generally best known for incorporating Leibniz's differential calculus and

Newton's method of fluxions into mathematical analysis—that is, the study of infinite series, a subject that he extensively developed. In his trilogy—*Introductio in analysin infinitorum* (2 vols., 1748), *Institutiones calculi differentialis* (1755), and *Institutiones calculi integralis* (3 vols., 1768–74)—he summarized the discoveries made in mathematical analysis during the mid-18th century. Probably on the basis of his study of trigonometry and logarithms, Euler made an early concept of function central to the calculus in his *Introductio,* but his improved definition of functions as uniquely paired values did not occur until the 1780s. In his trilogy and pertinent articles he recognized in essence the lack of rigorous foundations for analysis, which would preoccupy mathematicians nearly a century later. Euler's colleagues dubbed him "analysis incarnate" for his many contributions to analysis.

Among his major achievements in mathematics, Euler early in his career pioneered the theory of special functions, introducing the beta and gamma transcendental functions. He also represented trigonometric functions as ratios rather than Ptolemaic chords and used Taylor's theorem to deduce "Moivre's formula," $e^{\pm ix} = \cos x \pm \sin x$, the cardinal formula of analytical trigonometry. From this formula he derived the equation $e^{i\pi} + 1 = 0$. Euler brought order into the field of mathematical notation. Indeed, his notation (except for a few symbols) is our notation. As a result of their use in his *Introductio,* an impressive textbook from which generations learned analysis, the symbols cos, sin, *e,* the base for natural logarithms,[1] *i* for $\sqrt{-1}$ and π for the ratio of the circumference of a circle to its diameter gained general acceptance. He further introduced such notations as Σ for sum; $\int n$ for the sum of divisors of *n; A, B,* and *C* for the angles of a triangle and *a, b, c,* for the opposite sides; and little "*f*" for a function. Because of the number and importance of his writings together with the extensive use of his textbooks, his notation became conventional.

There was more. After reading Diophantus and Fermat, Euler began to make the theory of numbers into a science. Notably, he stated but could not prove the prime number theorem (1752) and the law of biquadratic reciprocity (1783). With the so-called Eulerian polyhedra formula (1750) connecting the vertices, edges, and faces of a closed convex polyhedron ($V - E + F = 2$), modern topology begins.

Physics also benefited greatly from Euler's attention. In addition to incorporating it into the mathematical sciences, he derived the so-called Newton's equations in his research in rational mechanics. Continuing the tradition of Johann Bernoulli, he elaborated continuum mechanics but at the same time set forth the kinetic theory of gases. His *Theoria Motus Corporum Solidorum seu Rigidorum,* ("Theory of the Motions of Rigid Bodies," 1765) laid the foundations for analytical mechanics. In celestial mechanics he advanced lunar theory and the three-body problem jointly with the French physicist Alexis Clairaut. Euler also conducted fundamental research in hydraulics, acoustics, optics, and elasticity.

---

[1]Euler had used the symbol *e* as early as 1727. It first appeared in print in the *Mechanica* (1736–1738). The symbol *e* denotes the sum of the series

$$1 + \frac{1}{1 \cdot 2} + \frac{1}{1 \cdot 2 \cdot 3} + \ldots .$$

85. From *Introductio in analysin infinitorum* (1748)*

(Trigonometry)

LEONHARD EULER

ON TRANSCENDENTAL QUANTITIES WHICH CAN BE OBTAINED FROM THE CIRCLE

§126. After logarithms and exponential quantities we shall investigate circular arcs and their sines and cosines, not only because they constitute another type of transcendental quantities, but also because they can be obtained from these very logarithms and exponentials when imaginary quantities are involved.

Let us therefore take the radius of the circle, or its sinus totus, = 1. Then it is obvious that the circumference of this circle cannot be exactly expressed in rational numbers; but it has been found that the semicircumference is by approximation

$$= 3.14159.26535.89793 \ldots$$

[127 decimal places are given[2]] for which number I would write for short

$$\pi,$$

so that π is the semicircumference of the circle of which the radius = 1, or π is the length of the arc of 180 degrees.[3]

§127. If we denote by z an arbitrary arc of this circle, of which I always assume the radius = 1, then we usually consider of this arc mainly the sine [*sinus*] and cosine [*cosinus*]. I shall denote the sine of the arc z in the future in this way

$$\sin. A.z, \quad \text{or only} \quad \sin. z$$

and the cosine accordingly

$$\cos. A.z, \quad \text{or only} \quad \cos. z.$$

Hence we shall have, since π is the arc of 180°,

$$\sin. 0 = 0, \quad \cos. 0 = 1$$

and

$$\sin. \tfrac{1}{2}\pi = 1, \quad \cos. \tfrac{1}{2}\pi = 0 \ldots$$

[Now follows a whole set of trigonometric formulas including the definitions $\text{tang. } z = \dfrac{\sin. z}{\cos. z}$, $\cot. z = \dfrac{\cos. z}{\sin. z}$, the addition formulas, and identities such as

$$\text{tang. } \frac{a+b}{2} = \frac{\sin. a + \sin. b}{\cos. a + \cos. b}.$$

Source: This translation of chapter eight of the *Introductio* is reprinted by permission of the publishers from *A Source Book in Mathematics, 1200-1800,* edited by D. J. Struik, Cambridge, Mass.: Harvard University Press,

Hereafter we omit the period after sin and cos and write i for $\sqrt{-1}$, as Euler also did in later work.[4]]

§132. Since

$$(\sin z)^2 + (\cos z)^2 = 1,$$

we shall have by factorization

$$(\cos z + i \sin z)(\cos z - i \sin z) = 1,$$

which factors, although imaginary [*etsi imaginarii*], still are of great use in combining and multiplying sines and cosines.

[Now comes De Moivre's theorem[5] (though the name is not mentioned), from which follows, in §133:]

$$\cos nz = \frac{(\cos z + i \sin z)^n + (\cos z - i \sin z)^n}{2}$$

and

$$\sin nz = \frac{(\cos z + i \sin z)^n - (\cos z - i \sin z)^n}{2i}.$$

When we develop these binomials in a series we shall get

$$\cos nz = (\cos z)^n - \frac{n(n-1)}{1.2}(\cos z)^{n-2}(\sin z)^2 + \text{etc.}$$

and

$$\sin nz = \frac{n}{1}(\cos z)^{n-1}\sin z - \frac{n(n-1)(n-2)}{1.2.3}(\cos z)^{n-3}(\sin)^2 + \text{etc.}$$

§134. Let the arc z be infinitely small; then we get $\sin z = z$ and $\cos z = 1$; let now n be an infinitely large number, while the arc nz is of finite magnitude. Take $nz = v$; then since $\sin z = z = v/n$ we shall have

$$\cos v = 1 - \frac{v^2}{1.2.3} + \frac{v^4}{1.2.3.4} - \ldots + \text{etc.}$$

and

$$\sin v = v - \frac{v^3}{1.2.3} + \frac{v^5}{1.2.3.4.5} - \ldots + \text{etc.}$$

[Then, by writing $v = \frac{m}{n}\cdot\frac{\pi}{2}$ Euler obtains a series for $\sin \frac{m}{n} 90°$ with terms up to $\frac{m^{29}}{n^{29}}$, and a series for $\cos \frac{m}{n} 90°$ with terms up to $\frac{m^{30}}{n^{30}}$, the coefficients given to 28 decimals; these are followed by series for the tangent and the cotangent. He shows that it is only necessary to know the numerical values of these quantities for the values from 0° to 30° to be able to find them all by identities such as $\sin(30 + z) = \cos z - \sin(30 - z)$. Here cosec. z and sec. z are introduced.]

§138. Let us now take in the formulas of §133 the arc z infinitely small and let n be an infinitely small number ϵ [Euler writes i] such that ϵz will take the finite value v. We thus have $\epsilon z = v$ and $z = v/\epsilon$, hence $\sin z = v/\epsilon$ and $\cos z = 1$. After substituting these values we find

$$\cos v = \frac{\left(1 + \frac{vi}{\epsilon}\right)^\epsilon + \left(1 - \frac{vi}{\epsilon}\right)^\epsilon}{2},$$

$$\sin v = \frac{\left(1 + \frac{vi}{\epsilon}\right)^{\epsilon} - \left(1 - \frac{vi}{\epsilon}\right)^{\epsilon}}{2i}.$$

In the previous chapter we have seen that

$$\left(1 + \frac{z}{\epsilon}\right)^{\epsilon} = e^{z},$$

where by e we denote the base of the hyperbolic logarithms; if we therefore write for z first iv, then $-iv$, we shall have

$$\cos v = \frac{e^{iv} - e^{-iv}}{2}$$

and

$$\sin v = \frac{e^{iv} - e^{-iv}}{2i}.$$

From these formulas we can see how the imaginary exponential quantities can be reduced to the sine and cosine of real arcs. Indeed, we have

$$e^{iv} = \cos v + i \sin v,$$

$$e^{-iv} = \cos v - i \sin v.$$

[Then follow in §139 some formulas for the logarithms leading up to

$$z = \frac{1}{2i}\, l\, \frac{\cos z + i \sin z}{\cos z - i \sin z},$$

where l indicates logarithm.]

§140. Since $\frac{\sin z}{\cos z} = \text{tang } z$, the arc z can be expressed by its tangent in such a way that we have

$$z = \frac{1}{2i}\, l\, \frac{1 + i \text{ tang } z}{1 - i \text{ tang } z}.$$

Now we have seen above (§123) that

$$l\,\frac{1 + x}{1 - x} = \frac{2x}{1} + \frac{2x^3}{3} + \frac{2x^5}{5} + \frac{2x^7}{7} + \text{etc.}$$

We now put $x = i$ tang z and shall obtain

$$z = \frac{\text{tang } z}{1} - \frac{(\text{tang } z)^3}{1} + \frac{(\text{tang } z)^5}{5} + \frac{(\text{tang } z)^7}{7} + \text{etc.}$$

If we therefore put tang $z = t$, so that z is the arc of which the tangent is t, which we shall indicate by $A.$ tang. t [our $\tan^{-1} t$], we shall have

$$z = A.\ \text{tang}.\ t.$$

Therefore, for known t, the corresponding arc will be

$$z = \frac{t}{1} - \frac{t^3}{3} + \frac{t^5}{5} - \frac{t^7}{7} + \frac{t^9}{9} - \text{etc.}$$

Therefore, if the tangent t is equal to the radius 1, the arc $z = 45°$ or $z = \pi/4$, and we shall have

$$\frac{\pi}{4} = 1 - \frac{1}{3} + \frac{1}{5} - \frac{1}{7} + \text{etc.},$$

which is the series first found by Leibniz to express the value of the circumference of the circle.

[The chapter ends with some other series for π that converge more rapidly.]

NOTES

1. [Does not appear in text reproduced here.]

2. Euler took this value from T. G. de Lagny, "Mémoire sur la quadrature du cercle," *Histoire de l'Académie Royale, Paris, 1719* (1727), 1e partie, 176–189, who computed π to 127 decimal places by means of a series for $\tan^{-1} 30°$.

3. The symbol π was never used in Antiquity; it seems first to have been used by William Jones (the editor of Newton's *Analysis per aequationes,* London, 1711) in his *Synopsis palmariorum matheseos* (London, 1706), p. 243. See D. E. Smith, *History of mathematics* (Ginn, New York, 1925), II, 312. Euler used π in his *Mechanica* (1736); see note 1 in his biography. See E. W. Hobson, *Squaring the circle* (Cambridge University Press, Cambridge, England, 1913). Euler, using the term *sinus totus* for the radius of the circle, adheres for the last time to the old terminology, in which the sine is a segment.

4. "In the following I shall denote the expression $\sqrt{-1}$ by the letter i so that $ii = -1$": Euler, *De formulis differentialibus angularibus,* presented to the Saint Petersburg Academy, 1777; published in the posthumous vol. IV of the *Institutiones calculi integralis* (1794), 183–194; *Opera omnia,* ser. I, vol. 19, 129–140, p. 130.

5. This theorem, now usually written $(\cos\phi + i\sin\phi)^n = \cos n\phi + i\sin n\phi$, appears at the opening of A. de Moivre, *Miscellanea analytica* (London, 1730), but in a different, more geometrical, form.

JOSEPH-LOUIS LAGRANGE (1736–1813)

Lagrange ranks with his intellectual mentor Leonhard Euler as one of the two leading mathematicians and theoretical physicists of the 18th century. He excelled in all branches of mathematical analysis, the theory of numbers, analytical mechanics, and celestial mechanics.

Lagrange was born in the Italian city of Turin into a family of French descent on his father's side. Though the father had a good position as treasurer to the Sardinian king, he lost his money in unsuccessful financial speculations, and the family lived modestly. Lagrange once commented, "If I had been rich, I probably would not have devoted myself to mathematics." His father wanted him to pursue the law, but in school he was drawn to the geometry of Euclid and Archimedes. An essay by Halley extolling the superiority of Newton's calculus over geometrical methods heightened his interest in mathematics.

His life divides naturally into three periods: the early years in Turin, from 1736 to 1766; the Berlin period, from 1766 to 1787; and the years in Paris, from 1787 to his death. In the mid-1750s, Lagrange began to establish his reputation; in a 1754 essay he devised a formal calculus by building upon the analogy between Newton's binomial theorem and Liebniz's rule for successive differentiation of two functions. To his chagrin he learned that Leibniz and Bernoulli had earlier made the same discovery and feared that he would be called a plagiarist, but he was not. About the same time he began corresponding with Euler and d'Alembert. Euler praised his work on the calculus of variations and also tutored him, while d'Alembert was his counsellor on political matters. Lagrange proved extremely diligent in his research. As a poorly paid professor at the Royal Artillery School in Turin from 1755 to 1766, he worked so relentlessly and ate so sparingly as to harm his health. Sustained by his association with Euler and d'Alembert, he brought the calculus of variations to maturity and applied it to mechanics; he also studied thoroughly the vibrating string problem.

Lagrange now began to win wide acclaim in the European scientific community and, to an extent, in the larger European intellectual community. He was elected an associate foreign member of the Berlin Academy of Sciences in 1756 and one year later took an important part in founding the Turin Academy of Sciences. In 1764, he won the prestigious biennial prize of the Paris Academy for an essay on the libration of the Moon and won that prize on three other occasions. Even so, his interests were never limited to the exact sciences. While returning from a trip to Paris and London (1763–65), he made a detour to Switzerland to visit Voltaire, "a character worth seeing."

Lagrange succeeded Euler as Director of the Mathematics Section of the Berlin Academy in 1766. Stung by the departure of a scientist of Euler's stature for St. Petersburg and aware of Lagrange's reputation, Frederick the Great now boasted that "the greatest king in Europe" must have "the greatest mathematician in Europe" at his court. The years in Berlin were extremely productive for Lagrange. In studying the consequences of Newton's law of gravitation for planetary and lunar motion, he derived a system of differential equations of the 12th order to solve in a general manner the three-body problem, *i.e.*, the mutual attraction of three large bodies. His analysis of the motions of Jupiter and Saturn also contributed a nascent explanation for the long-term stability explanation of the solar system. In mathematics he was at first primarily concerned with number theory and algebra before turning to infinitesimal analysis. He was able to solve a Fermat equation by skillfully applying the algorithm of continued fractions. He also investigated the properties of prime numbers, the arithmetic theory of quadratic forms, and the equation that Euler had mistakenly called Pell's. He also improved upon Euler's work on Diophantine analysis. A 1770 *memoir* opened a new era in algebra, which later inspired Galois to pursue group theory. Thus, he did much to live up to the emperor's boast.

When Frederick the Great died, Lagrange decided to accept an invitation (made through the intermediary Mirabeau) from Louis XVI to join the Paris Academy of Sciences. Lagrange arrived in Paris in 1787, where he was given apartments in the Louvre. A year later he published his classic *Mécanique analytique*, which extended and synthesized the work of Newton on mechanics, as well as that of the Bernoullis, Maupertuis, and Euler. His book based the entire science of mechanics on the principle of virtual velocities and was analytic throughout containing no geometrical figures. The treatise is most significant for reducing mechanics to the art of solving problems by means of theory and use of ordinary and partial differential equations, especially the so-called Lagrange equations.

Shy, diplomatic, and amenable, Lagrange not only survived the French Revolution (unlike his colleagues Condorcet and Lavoisier) but was treated throughout with honor and respect. In 1790, he served on the committee to standardize weights and measures, which proposed the adoption of the metric system for commerce. He experienced a state of lassitude during his first years in Paris that resulted collectively from his having left familiar surroundings in Berlin, a malaise following his publication of the *Mécanique analytique*, and the sudden changes of the early Revolution. The lassitude ended with his marriage to the 17-year-old Renée Le Monnier in 1792; she brought gaiety back into his life. However, the excesses of the Terror in France during the next two years disturbed him. When the great chemist Antoine Lavoisier was guillotined, he remarked, "it took only a moment to sever that head, and perhaps a century will not be sufficient to produce another like it."

Lagrange took an active part in improving university education under the two governments following the Terror (*i.e.*, the Revolutionary Convention of late 1794 and the Directory which assumed power the next year). In 1794, he helped to found a school that a year later became the École Polytechnique, where he and Gaspard Monge were the principal professors of mathematics. In 1795, he also taught elementary mathematics at the École Normale (with Laplace as his assistant) and was elected a member of the newly-founded Institut de France, the successor to the Paris Academy. Lagrange—now nearing sixty—continued to be honored under Napoleon, who came into full power soon after 1796. In 1808, he was named to the Legion of Honor and became Count of the Empire. When he died in April 1813, his body was interred in the Pantheon.

The last great mathematician of the 18th century, Lagrange opened the way for the abstract mathematics of the 19th century. Today he is probably best known not for his work on the calculus of variations, the theory of numbers, or algebra, but for his search for sound foundations for the calculus. In his *Theory of Analytical Functions* (1797), which contained his lectures given at the École Polytechnique, he attempted to demonstrate that Taylor's power series expansions alone were sufficient to provide the sought after satisfactory foundations for the calculus. In a supplementary article entitled "Lessons on the Calculus of Functions" (1801), he introduced a new symbolism for first derivative f', for second derivative f'', and so on. In the 19th century, Ampère, Cauchy, Weierstrass, and others successfully extended his search for sound foundations.

86. From "Attempt at a New Method for Determining the Maxima and Minima of Indefinite Integral Formulas" (1760–61)*

(The Calculus of Variations)

JOSEPH-LOUIS LAGRANGE

The first problem of this kind solved by the geometers is that of the *Brachystochrone,* or line of most rapid descent, which Mr. Jean Bernoulli proposed toward the end of the last century. It was solved only for particular cases, and it was not until some time later, on the occasion of the investigations on *Isoperimetrics,* that the great geometer whom we mentioned and his illustrious brother Mr. Jacques Bernoulli gave some general rules for solving several other problems of the same kind. But since these rules were not general enough, all these investigations were reduced by the famous Mr. Euler to a general method, in a work entitled *Methodus inveniendi . . . ,* an original work which everywhere radiates a deep knowledge of the calculus. But, however ingenious and fertile his method may be, we must recognize that it does not have all the simplicity that might be desired in a subject of pure analysis. The author has made us aware of this in Article 39 of Chapter II of his book, by the words, "A method free from a geometric solution is therefore required . . ."

Now here is a method that demands only a very simple application of the principles of the differential and integral calculus, but first of all I must warn you that, since this method demands that the same quantities vary in two different manners, I have, in order not to confuse these variations, introduced into my calculations a new characteristic δ. Thus δZ will express a difference of Z that will not be the same as dZ, but that nevertheless, will be formed by means of the same rules; so

*Source: The French original of this paper, "Essai d'une nouvelle méthode pour déterminer les maxima et les minima des formules intégrales indéfinies," is in the *Oeuvres des Lagrange,* I (1867), 355-362. This English translation of parts of it is taken from D. J. Struik (ed.), *A Source Book in Mathematics, 1200-1800* (1969), 407-410 and 412-413.

that when we have an equation $dZ = m\,dx$ we might just as well have $\delta Z = m\delta x$, and other expressions in the same way.

This being settled, I come first to the following problem.

I

Problem I. Given an indefinite integral expression represented by $\int Z$, where Z indicates a given arbitrary function of the variables x, y, z and their differentials [*différences*] dx, dy, dz, d^2x, d^2y, d^2z, . . . , to find the relation among these variables so that the formula $\int Z$ become a maximum or a minimum.

Solution. According to the known method *de maximis et minimis* we shall have to differentiate the proposed $\int Z$, and, regarding the quantities x y, z, dx, dy, dz, d^2x, d^2y, d^2z, . . . as variables, make the resulting differential [*différentielle*] equal to zero. When, therefore, we indicate these variations by δ, we shall have first, for the equation of the maximum or minimum,

$$\delta \int dZ = 0,$$

or, what is equivalent to it,

$$d \int \delta Z = 0.$$

Now, let Z be such that

$$\begin{aligned}\delta Z = {} & n\delta x + p\delta\, dx + q\delta\, d^2x + r\delta\, d^3x + \cdots \\ & + N\delta y + P\delta\, dy + Q\delta\, d^2y + R\delta\, d^3y + \cdots \\ & + \nu\delta z + \pi\delta\, dz + \chi\delta\, d^2z + \rho\delta\, d^3z + \cdots;\end{aligned}$$

then we obtain from it the equation

$$\begin{aligned}& \int n\delta x + \int p\delta\, dx + \int q\delta\, d^2x + \int r\delta\, d^3x + \cdots \\ & + \int N\delta y + \int P\delta\, dy + \int Q\delta\, d^2y + \int R\delta\, d^3y + \cdots \\ & + \int \nu\delta z + \int \pi\delta\, dx + \int \chi\delta\, d^2z + \int \rho\delta\, d^3z + \cdots = 0,\end{aligned}$$

but it is easily understood that

$$\delta\, dx = d\delta\, x, \quad \delta\, d^2x = d^2\delta x,$$

and the others in the same way; moreover, we find by the method of integration by parts,

$$\begin{aligned}& \int p\, d\delta x = p\delta x - \int dp\delta x, \\ & \int q\, d^2\delta x = q\, d\delta x - dq\delta x + \int d^2q\delta x, \\ & \int r\, d^3\delta x = r\, d^2\delta x - dr\, d\delta x + d^2r\delta x - \int d^3z\delta x,\end{aligned}$$

and the others in a similar way. The preceding equation will therefore be changed into the following:

$$\begin{aligned}& \int (n - dp + d^2q - d^3r + \cdots)\delta x \\ & + \int (N - dP + d^2Q - d^3R + \cdots)\delta y \\ & + \int (\nu - d\pi + d^2\chi - d^3\rho + \cdots)\delta z \\ & + (p - dq + d^2r - \cdots)\delta x + (q - dr + \cdots)d\delta x \\ \text{(A)} \qquad & \qquad + (r - \cdots)d^2\delta x + \cdots \\ & + (P - dQ + d^2R - \cdots)\delta y + (Q - dR + \cdots)d\delta y \\ & \qquad + (R - \cdots)d^2\delta y + \cdots \\ & + (\pi - d\chi + d^2\rho - \cdots)\delta z + (\chi - d\rho + \cdots)\, d\delta z \\ & \qquad + (\rho - \cdots)d^2\delta z + \cdots = 0,\end{aligned}$$

from which we obtain first the indefinite equation

$$(n - dp + d^2q - d^3r + \cdots)\delta x + (N - dP + d^2Q - d^3R + \cdots)\delta y - + (\nu - d\pi + d^2\chi - d^3\rho + \cdots)\delta z = 0, \tag{B}$$

and then the determinate equation

$$(p - dp + d^2r - \cdots)\delta x + (q - dr + \cdots)d\delta x + (r - \cdots)d^2\delta x + \cdots + (P - dQ + d^2R - \cdots)\delta y + (Q - dR + \cdots)d\delta y + (R - \cdots)d^2\delta y + \cdots + (\pi - d\chi + d^2\rho - \cdots)\delta z + (\chi - d\rho + \cdots)d\delta z + (\rho - \cdots)d^2\delta z + \cdots = 0. \tag{C}$$

This equation refers to the last part of the integral $\int Z$; but we must observe that, since each of its terms, such as $p\delta x$, depends on an integration by parts of the formula $\int p\, d\delta x$, we may add to or subtract from it a constant quantity. The condition by which this constant must be determined is that $p\delta x$ must vanish at the point where the integral $\int p\, d\delta x$ begins; we must therefore take away from $p\delta x$ its value at this point. From this we obtain the following rule. Let us express the first part of equation (C) generally by M, and let the value of M at the point where the integral $\int Z$ begins be indicated by $'M$, and at the point where this integral ends, by M'; then we have $M' - 'M = 0$ for the complete expression of equation (C). Now, in order to free the equations obtained from the undetermined differentials δx, δy, δz, $d\delta x$, $d\delta y$, . . . , we must first examine whether, by the nature of the problem, there exists some given relation among them, and then, having reduced them to the smallest number possible, we must equate to zero the coefficient of each of those that remain. If they are absolutely independent of each other, then equation (B) will give us immediately the three following:

$$n - dp + d^2q - d^3r + \cdots = 0,$$
$$N - dP + d^2Q - d^3R + \cdots = 0,$$
$$\nu - d\pi + d^2\chi - d^3\rho + \cdots = 0.$$

[Next follows the example

$$\int \frac{\sqrt{dx^2 + dy^2 + dx^2}}{\sqrt{x}},$$

which is the brachystochrone in empty space and leads *(a)* to the result that the curve is plane, and *(b)* to $dt = \sqrt{x}\, dx/\sqrt{c - x}$. The case of the brachystochrone on a surface is also discussed; here the relation $\delta z = p\delta x + q\delta y$ has to be taken into consideration. Lagrange takes the cases in which the end points are fixed, as well as those in which they are subjected to certain other conditions. This, says Lagrange, makes his method more general than that of Euler, since Euler keeps the end points fixed; moreover, he lets only y vary in Z. . . .]

Problem III. To find the equation of the maximum or the minimum of the formula $\int Z$, if Z is simply given by a differential equation that does not contain other differentials of Z than the first.

[This is the case in which we can write

$$\delta\, dZ + T\delta Z = n\delta x + p\delta\, dx + \cdots + N\delta y + P\delta\, dy + \cdots + \nu\delta z + \pi\delta\, dz,$$

which is then solved as a linear differential equation in δZ, taking $\delta\, dZ = d\delta Z$.

There are two appendices. In the first we find (a) the problem of the surface of least area among all surfaces with the same given perimeter:

$$\delta \int \int dx\, dy \sqrt{1 + p^2 + q^2} = 0, \qquad p = \left(\frac{dz}{dx}\right), \qquad q = \left(\frac{dz}{dy}\right),$$

which leads to the condition that both $p\, dx + q\, dy$ and $\dfrac{p\, dy - q\, dx}{\sqrt{1 + p^2 + q^2}}$ have to be exact differentials,[1] and *(b)* the problem of the surface of least area among all surfaces of equal volume:

$$\delta \left(\int \int z\, dx\, dy\right) = 0, \quad \delta \left(\int \int dx\, dy \sqrt{1 + p^2 + q^2}\right) = 0,$$

which leads to the condition that both $p\, dx + q\, dy$ and $\dfrac{p\, dy - q\, dx}{\sqrt{1 + p^2 + q^2}} + kx\, dy$ (k an arbitrary coefficient) must be exact differentials. This is verified for the sphere.

In the second appendix we find the problem of the polygon of largest area among all polygons of the same given number of sides. It is shown that this polygon is inscribed in a circle, a theorem proved geometrically by Cramer (*Histoire de l'Académie Royale, Berlin,* 1752). If only the sum of the sides is given, the polygon is regular.

Lagrange's paper was followed in the same number of the *Miscellanea Taurinensis,* pp. 196-298, by a longer one: "Application de différents problèmes de dynamique" (*Oeuvres,* I, 365-468).]

NOTE

1. Two examples of these "minimal surfaces," the catenoid and the right helicoid, were found by Jean-Baptiste Meusnier, a pupil of Monge's, in the *Mémoires des savants étrangers de l'Académie 10* (Paris, 1785). He also interpreted here Lagrange's analytic condition geometrically as indicating that the mean curvature is zero. The catenoid had already appeared in chap. V, 44 of Euler's *Methodus inveniendi,* but not as a minimal surface.

87. From *Mathematical Thought from Ancient to Modern Times* (1972)*

(Taylor Series with Remainder)

MORRIS KLINE

Lagrange . . . showed some awareness of the distinction between convergence and divergence. In his earlier writings he was indeed lax on this matter. In one paper [in 1770] he says that a series will represent a number if it converges to its extremity, that is, if its nth term approaches 0. Later, toward the end of the eighteenth century, when he worked with Taylor's series, he gave what we call Taylor's theorem, namely,

$$f(x + h) = f(x) + f'(x)h + f''(x)\frac{h^2}{2!} + \cdots + f^{(n)}(x)\frac{h^n}{n!} + R_n$$

where

$$R_n = f^{(n+1)}(x + \theta h)\frac{h^{n+1}}{(n + 1)!}$$

and θ is between 0 and 1 in value. This expression for R_n is still known as Lagrange's form of the remainder. Lagrange said that the Taylor (infinite) series should not be used without consideration of the remainder. However, he did not investigate the idea of convergence or the relation of the value of the remainder to the convergence of the infinite series. He thought that one need consider only a finite number of terms of the series, enough to make the remainder small. Convergence was considered later by Cauchy, who stressed Taylor's theorem as primary, as well as the fact that to obtain a convergent series the remainder must approach 0.

*Source: Morris Kline, *Mathematical Thought from Ancient to Modern Times* © 1972, 464-465. Reprinted by permission of Oxford University Press, Inc. Footnotes omitted.

Chapter VII
The Enlightenment (1720–1800)

Section B
Topology, Number Theory, and Probability

88. From the Problem of the Seven Bridges of Königsberg (1736)*

(The Origins of Topology)

LEONHARD EULER

THE SOLUTION OF A PROBLEM BELONGING TO THE *GEOMETRIA SITUS*

1. Besides that part of geometry which treats of quantities and has been studied eagerly at all times, there is another, so far almost unknown, first mentioned by Leibniz, which he named *Geometria situs*. This part concerns itself with that which can be determined by position [*situs*] alone, and with the analysis of the properties of position; here quantities will be ignored and the calculus of quantities not used. But what kind of problems belong to this geometry, and what method has to be utilized for their solution, is not yet certain enough. Thus, when recently a problem was mentioned, seemingly belonging to geometry, but such that it did not call for the determination of a quantity, now admitted of a solution by the calculus of quantities, I did not hesitate to refer it to the Geometria situs, especially since in its solution position only came into consideration, whereas calculus was of no use. Hence I shall set forth the method that I discovered for the solution of such problems, to serve here as a sample of Geometria situs.

2. The problem, supposedly quite well known, was as follows: At Königsberg in Prussia there is an island *A*, called "der Kneiphof," and the river surrounding it is divided into two branches as can be seen in Fig. 1. Over the branches of this river lead seven bridges, *a*, *b*, *c*, *d*, *e*, *f*, and *g*. Now the question is whether one can plan a walk so as to cross each bridge once and not more than once. I was told that some deny this possibility, others are doubtful, but that nobody affirms it. Wherefrom I formulated the following problem, framed in a very general way for myself: Whatever the shape of the river and its division into branches may be, and whatever the number of bridges, to find out whether it is possible or not to cross each bridge exactly once.

3. As concerns the Königsberg problem of the seven bridges, it could be solved by a complete enumeration of all

*Source: This translation of Euler's article in the *Commentarii* of the St. Petersburg Academy, vol. 8 (1736), 128-140, is reprinted by permission of the publishers from *A Source Book in Mathematics, 1200-1800*, edited by D. J. Struik, Cambridge, Mass.: Harvard University Press, Copyright © 1969 by the President and Fellows of Harvard College. Königsberg is the former seaport of East Prussia, on the Pregel, now Kaliningrad, U.S.S.R. The biography of Euler precedes selection 85.

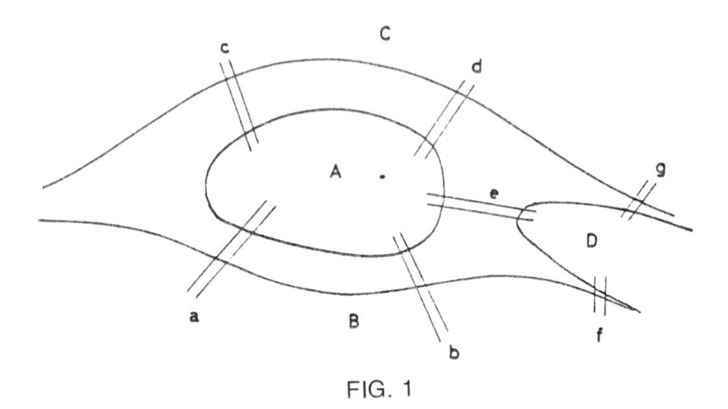

FIG. 1

walks possible; then we would know if one of them fulfills the condition or none. This method, however, is, because of the great number of combinations, too difficult and cumbersome. Moreover, it could not be applied to other questions where still more bridges exist. If the investigation were to be conducted in this way, then there would be found much that was not called for at all; that is the reason, no doubt, why this way would be so arduous. That is why I dropped this method and looked for another, leading only so far that it shows whether such a walk can be found or not; for I suspected that such a method would be much simpler.

4. My whole method is based on the proper designation of the bridges, using the capitals *A, B, C, D* to indicate the single regions separated from each other by the river. If one thus reaches region *B* from region *A*, crossing bridge *a* or *b*, then I denote this transition by the letters AB, where the first gives the region from which the traveler comes, whereas the second gives the region where he arrives after crossing the bridge. If the traveler then goes from the region *B* over the bridge *f* into the region *D*, this transition is denoted by the letters *BD*. These two transitions *AB* and *BD*, carried out in succession, I denote by *ABD* only, because the middle one *B* indicates the region into which the first transition leads, as well as the region out of which the second transition leads.

5. In the same way, if the traveler goes from the region *D* over the bridge *g* to the region *C*, I denote these three successive transitions by the four letters *ABDC*. For these four letters *ABDC* indicate that the traveler, finding himself initially in the region *A*, has passed into the region *B*, whence he proceeded into region *D*, and finally from there arrived at *C*; but as these regions are separated from each other by rivers, the traveler must necessarily cross three bridges. A crossing of four bridges is thus indicated by five letters, and then the traveler crosses an arbitrary number of bridges, then his path will be denoted by a number of letters greater by one than the number of bridges. The crossing of seven bridges requires therefore eight letters for its description.

6. By this method of description I do not pay any attention to what bridges are crossed, that is, when the transition from one region to another can be accomplished on different bridges, then it is irrelevant which one is used, as long as it leads to the region indicated. If therefore the path over the seven bridges could be planned so that it crosses each once and only once, then it could be represented by eight letters, where these letters would have to succeed one another in such a way that the immediate succession of the letters *A*

and *B* appears twice; since there are two bridges *a* and *b* connecting the regions *A* and *B;* for the same reason the succession of letters *A* and *C* should also appear twice in this series of eight letters; moreover, the succession *AD* as well as *BD* and *CD* must each appear once.

7. Our question is now reduced to another, namely, whether from the four letters *A, B, C,* and *D* a series of eight letters can be formed in which all these successions appear as often as is prescribed. However, before one sets out to find such an arrangement one should better attempt to show whether such a one exists or not. For if one could show that no such arrangement is at all possible, then all the effort to find it will be useless. That is why I invented a rule, which permits one to decide in this and all similar questions without difficulty whether such an arrangement of letters is possible.

8. To find such a rule, I observe a single region *A,* to which an arbitrary number of bridges *a, b, c, d,* etc. lead. Of those bridges I first pay attention only to *a.* When the traveler crosses this bridge, he must either have been in *A* before he crossed or arrive at *A* after the crossing; according to our method of notation the letter *A* will appear twice, no matter where the path started, in *A* or not. And if five bridges lead to *A,* then in our notation crossing them all will make the letter *A* appear three times. And when the number of bridges is an arbitrary odd number then, by increasing the number by one and dividing by two, we obtain the number of times the letter *A* must appear.

9. Hence, in the case of the Königsberg bridges (Fig. 1) we have five bridges leading to the island *A,* namely *a, b, d, d, e.* The letter *A* must therefore appear three times in the symbol of the path. As three bridges lead to *B, B* must appear twice, and the same way *D* and *C* will occur twice. In the series of eight letters, which indicates the crossing of the seven bridges. *A* must appear three times but *B, C,* and *D* each twice; and this is in no way possible in a series of eight letters. This shows that the desired crossing of the seven Königsberg bridges cannot be carried out.

10. In a similar way one can always decide whether a path exists that leads over each bridge just once, if only the number of bridges leading to each region is odd. For such a path always exists when the number of bridges, increased by one, equals the sum of all numbers indicating how often each letter must appear. In case this sum, as in our example, is greater than the number of bridges increased by one, then such a path can in no way be laid out. The rule I gave, to determine from the number of bridges leading to *A* how often in the symbol of the path the letter *A* appears, is independent of whether all bridges, as in Fig. 2, come from a single region *B* or from different regions; for I consider only the region *A* and inquire how often the letter *A* must appear.

11. If, however, the number of bridges leading to *A* is even, then one must distinguish whether or not the walk started in *A.* For if two bridges lead to *A,* and the walk starts in *A,* then the letter *A* must appear twice, once to indicate the leaving of *A* over one bridge,

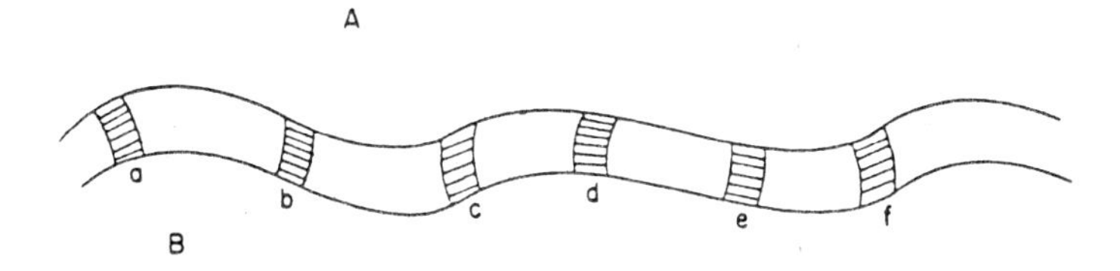

FIG. 2

and a second time to indicate the return to A over the other bridge. But if the traveler starts his path in another region, then the letter A occurs only once, for in my notation the one appearance of A means entrance into A as well as exit out of A.

12. Let now four bridges lead into the region A, and the path begin in A. Then in the symbol of the completed path the letter A must appear three times, if he crosses each bridge only once. But if the path were to start in another region, then A would appear only twice. If six bridges lead into A, then A appears four times, if A is the initial region, otherwise only three times. And in general: if the number of bridges is even, then ½ of it indicates how often A must make its appearance, if A is not the initial region; then one-half increased by one indicates how often A must appear if the walk starts in A.

13. As any walk has to start in some region, I define in the following way from the number of bridges leading to a region the number indicating how often the corresponding letter appears in the path symbol: If the number of bridges is odd, then I increase it by one and take half; if, however, it is even, then I take its half. If the sum of the numbers thus obtained is equal to the number of bridges increased by one, then we shall succeed in finding a path, but one must start in a region to which an odd number of bridges leads. If this sum happens to be smaller by one than the number of bridges increased by one, then the walk succeeds if one starts in a region to which an even number of bridges leads, for in this case our sum must still be increased by one.

[Euler continues with a generalization to more regions $A, B, C, \ldots$ and more bridges, and also deals with the question how, after it has been decided that a solution exists, the actual method of crossing the bridges can be found. Anyone who wishes to study these questions may read chap. IX of W. W. Rouse Ball, *Mathematical recreations and essays,* revised by H. S. M. Coxeter (Macmillan, New York, 1947).]

89. From "Theorems on Residues Obtained by the Division of Powers" (1758/59)*

(Number Theory: Power Residues)

LEONHARD EULER

Theorem 14. *Let p be a prime number and a be prime to p; then the power* a^{p-1} *divided by p has the residue* 1.

Proof. Let a^λ be the lowest power of a giving the residue 1 when divided by p. Then, as we have seen, λ will be $< p$ and we proved above that in this case either $\lambda = p - 1$ or λ is a divisor of the number $p - 1$. In the first case the theorem holds, and a^{p-1} gives, divided by p, the residue 1. In the other case, where λ is a divisor of $p - 1$, we have $p - 1 = n\lambda$; but because the power a^λ gives, divided by p, the residue 1, there-

*Source: The article "Theoremata circa residua ex divisione potestatum relicta" appeared in *Novi Commentarii* of the St. Petersburg Academy, vol. 7, (1758/59, published 1761) 49-82. This translation of an extract is reprinted by permission of the publishers from *A Source Book in Mathematics, 1200-1800,* edited by D. J. Struik, Cambridge, Mass.: Harvard University Press, Footnote renumbered.

fore also all these powers $a^{2\lambda}$, $a^{3\lambda}$, etc. and $a^{n\lambda}$ or a^{p-1} divided by p will give the residue 1. Thus a^{p-1} divided by p will always have the residue 1.

50. *Corollary* 1. Because the power a^{p-1} gives the residue 1 when divided by the prime number p, the formula $a^{p-1} - 1$ is divisible by p, so long as a is a number prime to p, that is, so long as a is not divisible by p.

51. *Corollary* 2. If, therefore, p is a prime, then all powers of exponent $p - 1$, such as n^{p-1}, are divisible by p, or leave 1 as remainder. The latter happens if n is prime to p, the first if this number n is divisible by p.

52. *Corollary* 3. Hence, if p is a prime number, and the numbers a and b are prime to p, then the difference of the powers $a^{p-1} - b^{p-1}$ will be divisible by p. Indeed, since $a^{p-1} - 1$ as well as $b^{p-1} - 1$ are divisible by p, so will also their difference $a^{p-1} - b^{p-1}$ be divisible by p. . . .

53. *Scholium*. This is a new proof of the famous theorem that Fermat once stated, and it is completely different from the one I have given in the *Comment. Acad. Petropol., tome VIII.*[1] There I started out from Newton's series expansion of the binomial $(a + b)^n$, using a reasoning seemingly quite remote from the proposition; here, on the other hand, I prove the theorem starting from properties of the powers alone, which makes the proof seem much more natural. Moreover, other important properties of the residues of the powers when they are divided by a prime number come to light. Indeed, it is shown for a prime number p not only that the expression $a^{p-1} - 1$ is divisible by p, but that, under certain conditions, a simpler expression, $a^{\lambda} - 1$, is divisible by p, and that in that case the exponent λ is always a divisor of $p - 1$.

[In the remaining sections Euler proves several theorems on power residues, of which Theorem 19 states that if $a^m \equiv 1 \pmod{p = mn + 1}$, then there always exist numbers x and y such that $ax^n - y^n = 0$. Here p is a prime. Our notation is Euler's, except that we have written a/b where Euler writes $\frac{a}{b}$.]

NOTE

1. "Theorematum quorundam ad numeros primos spectantium demonstratio," *Commentarii Academiae Scientiarum Petropolitanae 8* (1736, publ. 1741), 141-146, *Opera omnia,* ser. I, vol. 2, 35-37. The proof runs as follows. First it is proved by means of the binomial expansion of $(1 + 1)^{p-1}$ that

$$2^{p-1} - 1 = \frac{p(p-1)}{1 \cdot 2} + \frac{p(p-1)(p-2)(p-3)}{1 \cdot 2 \cdot 3 \cdot 4} + \cdots$$

is divisible by p if p is an odd prime. Then by a similar expansion of $(1 + a)^p$ it is shown that $(1 + a)^p - (1 + a) - (a^p - a) \equiv 0 \pmod{p}$ if a is not a multiple of p. Since $2^p - 2 \equiv 0$, the proof follows by complete induction.

90. From "Demonstrations of Certain Arithmetical Theorems" (1738)*

(A proof of Fermat's great theorem—$x^n + y^n = z^n$ has no positive integral solutions for $n > 2$—for the case $n = 4$.)

LEONHARD EULER

Theorem 1. *The sum of two biquadratic numbers such as* $a^4 + b^4$ cannot be a square number unless one of the two biquadratic numbers vanishes.

Proof. I shall change the theorem to be demonstrated in such a way that I shall show that if in one case $a^4 + b^4$ were a square, no matter how large the numbers a and b, then I can progressively find smaller numbers a and b and at the end can reach the smallest integral numbers. Since there are no such smallest numbers of which the sum of the biquadratic numbers is a square, we must conclude that there are no such among the largest numbers.

Let therefore $a^4 + b^4$ be a square and a and b be relative primes, since if they were not relative primes, then I could reduce them by division to primes. Let a be an odd number; then b must be even, since necessarily one number must be even, the other one odd. Let us therefore write

$$a^2 = p^2 - q^2, \quad b^2 = 2pq;$$

here the numbers p and q must be relative primes, the one even, the other odd. But if $a^2 = p^2 - q^2$, then it is necessary that p be odd, because otherwise $p^2 - q^2$ could not be a square. Hence p is an odd number and q an even one. Since $2pq$ must also be a square it is necessary that both p and $2q$ be a square, because p and $2q$ are relative primes. Since $p^2 - q^2$ is a square, it is necessary that

$$p = m^2 + n^2 \quad \text{and} \quad q = 2mn,$$

where again m and n are relative prime numbers, of which one is even, the other odd. But since $2q$ is a square, $4mn$, or mn is a square, hence m and n are squares. If we therefore put

$$m = x^2, \quad n = y^2$$

then we shall have

$$p = m^2 + n^2 = x^4 + y^4,$$

which must equally be a square. From this it follows that if $a^4 + b^4$ were a square, then also $x^4 + y^4$ would be a square, but it is clear that the numbers x and y would be far smaller than a and b. In the same way we shall from the biquadratic numbers $x^4 + y^4$ again obtain smaller ones, of which the sum is a square, and we progressively reach the smallest biquadratic number among the integers. But since there are no smallest biquadratic numbers of which the sum gives a square, it is clear that there are no very large numbers either. However, if in one pair of the biquadratic numbers one of the terms is zero, then in all re-

Source: The article "Theorematum quorundam arithmeticorum demonstrationes" appeared in the *Commentarii* of the St. Petersburg Academy, vol. 10 (1738, published 1747), 125-146. This translation of an extract is reprinted by permission of the publishers from *A Source Book in Mathematics, 1200-1800,* edited by D. J. Struik, Cambridge, Mass.: Harvard University Press,

maining pairs the one term vanishes, so that here nothing new results.

Corollary 1. Since therefore the sum of two biquadratic numbers cannot be a square, it is a fortiori impossible that the sum of two biquadratic numbers results in a biquadratic number.

Corollary 2. Although this demonstration pertains only to integers, yet it also shows that we cannot find among fractions two biquadratic numbers of which the sum is a square. Indeed, if $(a^4/m^4) + (b^4/n^4)$ were a square, then $a^4n^4 + b^4m^4$, which is a sum of integers, would also be a square, which we have proved to be impossible.

Corollary 3. By means of the same proof we can conclude that no numbers p and q exist such that p, $2q$ and $p^2 - q^2$ are squares; if such numbers existed then there would be values for a and b, which would render $a^4 + b^4$ square; for then $a = \sqrt{p^2 - q^2}$ and $b = \sqrt{2pq}$.

Corollary 4. Suppose therefore $p = x^2$ and $2q = 4y^2$, then $p^2 - q^2 = x^4 - 4y^4$. Then it could not at all happen, that $x^4 - 4y^4$ were a square. Nor could $4x^4 - y^4$ be a square; for then $16x^4 - 4y^4$ would be a square, which reduces it to the former case, because $16x^4$ is a biquadratic number.

Corollary 5. From this it follows that also $ab(a^2 + b^2)$ can never be a square. For the factors a, b, $a^2 + b^2$, all relative primes, would have to be squares, which is impossible.

Corollary 6. In a similar way, there cannot exist relatively prime numbers a and b such as to make $2ab(a^2 - b^2)$ a square. This follows from Corollary 3, where it was proven that no numbers p and q exist such as to make p, $2q$, $p^2 - q^2$ squares. And all this is valid also for numbers that are not relative primes, and the same for fractions according to Corollary 2.

PIERRE-SIMON LAPLACE (1749–1827)

Pierre-Simon Laplace, French astronomer and mathematician, was called "the Newton of France" because of his dynamical investigation that established the long-term stability of the solar system. That is, he was the first to account for the theoretical orbits of the planets and their satellites, including their perturbations, solely in terms of the Newtonian theory of gravitation. Laplace offered a summary of findings in the magisterial *Traité de Mécanique céleste* ("Celestial Mechanics," 5 volumes, 1799–1825), in which he extensively applied the calculus to the problems of celestial mechanics. He also presented a version of the "nebular hypothesis," asserting evolutionary change in the physical universe in his semipopular *Exposition du système du monde* ("System of the World," 1796), and showed the utility of probabilistic interpretations of scientific data.

Laplace was born into a cultivated provincial bourgeois family in Normandy. His father, Pierre, was a prosperous parish syndic, perhaps in the cider business, while his mother, Marie-Anne Sochon, came from a well-to-do farm family. From the age of 7 to 16, Pierre attended the Benedictine *collège* (secondary school) in his hometown of Beaumont-en-Auge. Its graduates usually entered careers in the church or the army, and Laplace's father wanted him to follow an ecclesiastical vocation. Consequently, he went to the University of Caen in 1766 to study the liberal arts as a preparatory step to beginning in theology. These plans soon changed, however, because he displayed unusual mathematical ability. Two of his professors, Christoph Gadbled and Pierre LeCanu, suggested that their gifted pupil pursue mathematics, and he responded to this encouragement by leaving Caen after two years.

Armed with a letter of recommendation from LeCanu to d'Alembert, Laplace departed for Paris at age 19. He impressed d'Alembert, who secured a position for him at the École Militaire. The position, while not intellectually stimulating, allowed him to remain in Paris where he concentrated on earning a reputation in the mathematical sciences.

In 1773, Laplace made a major discovery in dynamical astronomy. He found that the opposite and secular inequalities of Jupiter and Saturn (accelerations and deceleration, respectively) were not cumulative and disruptive but periodic over 929 years. The solution to this complex problem of mutual gravitational interactions within the solar system had eluded Leonhard Euler and Louis Lagrange. He solved it by using astronomical observations dating back to ancient Babylon and improving computational methods, which involved 24 linear equations. His discovery of periodicity, which moved him toward establishing the long-term stability of the solar system, was the principal advance in dynamical astronomy since Newton. It won him election to the Paris Academy of Sciences in 1773, following two unsuccessful attempts that had made him feel slighted. He composed numerous scientific memoirs for the Academy and selected celestial mechanics and probability as the two chief areas for his future research program.

The formative period in his life and studies ended by 1778; between 1778 and 1789, Laplace was at his prime in the sciences. Besides investigating differential operators and definite integrals, he conducted research on the chemical physics of heat with the chemist Antoine Lavoisier. Using an ice calorimeter they invented, the two men demonstrated (1780) that respiration is a form of combustion. In his study of the attraction between spheroids (1784–85), Laplace prepared the mathematical foundations for the scientific study of electricity, heat, and magnetism. In 1785, the Paris Academy promoted him to the senior rank of pensionner. Its members recognized him as one of their members, and their faith was well placed. In 1787, he found that lunar acceleration depends upon the eccentricity of the earth's orbit, thereby removing the last apparent anomaly to the theoretical description of the long-term stability of the solar system based solely on Newton's law of gravitation.

During the 1780s, Laplace also assumed responsibilities in civic life. The government appointed him examiner for the Royal Artillery in 1784, and he examined one of its students, Napoleon Bonaparte, the following year. In the late 1780s, he conducted demographic studies that contributed to the increasing professionalization of 18th-century public administration. In 1788, he married Marie-Charlotte de Courty de Romanges (20 years his junior). They had a son and a daughter.

The French Revolution affected the life of Laplace as well as his research. In 1790, he played a decisive role on the Academic commission that designed the metric system. The commissioners proposed a universal decimal system of standardized weights and measures, not only for the sciences, but also for money, cartography, land registry, and navigation. Laplace also helped to design the Revolutionary calendar, which dropped the Roman names for months. Unfortunately, the new calendar reintroduced a confusing feature of ancient Greek chronology when cities intercalated a day or a month to honor heroes or victories. In 1793, the Revolution became more radical under the Jacobins. Laplace, often abrasive and insensitive, appears to have withdrawn from participation in the Academy when Jean-Paul Marat, whose optical experiments Laplace had spurned in the 1780s, vilified Academy scientists. When the Academy was suppressed in December 1793, Laplace wisely moved to Melun (30 miles southeast of Paris), where he remained until the Jacobins fell from power in July 1794.

Under the Revolutionary Convention and the Directory, Laplace worked enthusiastically on the institutionalization and professionalization of the sciences. In 1795, he helped to establish the first (scientific) class of the French Institut (a reincarnated Academy) and worked with the *ideologues* to make the sciences staple subjects in the schools. From 1795 to 1799, he was an examiner at the École Polytechnique. During these years three of Laplace's books were published, and they were all well received. The *Systéme du monde* was not only a fine scientific popularization but also a model of French prose. With the appearance of the first two volumes of *Mécanique céleste* in 1799, Laplace became a European celebrity.

Under Napoleon and the Restoration, Laplace became wealthy and gained his greatest renown. In 1803, Napoleon appointed him chancellor of the senate, a lucrative position, and he was named to the Legion of Honor in 1805. Although Laplace prepared adulatory dedications to Napoleon in two of his books—the third volume of *Mécanique céleste* (1802) and the

Théorie analytique des probabilités ("Analytical Theory of Probabilities," 1812)—the politically supple astronomer called for his patron's dethronement in 1814. Under Louis XVIII, he was elected to the *Académie française* in 1816 and elevated to the peerage as a marquis the next year.

During his final years, Laplace lived mainly at his country home in Arceuil, next to his friend the chemist C. L. Berthollet. In his study were portraits of Racine, his favorite author, and Newton. The "adopted children of his thoughts"—Arago, Biot, Gay-Lussac, Humboldt, and Poisson—surrounded him. He enjoyed amiable discussions with these men of science and other members of the Société d'Arceuil, which he had helped to form.

Laplace advanced applied mathematics and contributed substantially to the theory of probability. Most notably, this scientist, who viewed mathematics as a tool and disliked the drudgery of calculations, skillfully used ordinary differential equations to solve problems in celestial mechanics. Influenced by Euler and Lagrange, he also worked on an early stage of the "Laplace transform" method of solving differential, difference, and integral equations. Perhaps inspired by Gauss' derivation in 1809 of the least square law from an analysis of normal distribution, he derived the central limit theorem of the least square rule in 1810–11. He drew together his studies on probability dating from the 1780s into the seminal *Théorie analytique des probabilités* and its companion piece *Essais philosophique sur les probabilités* ("A Philosophical Essay on Probabilities," 1814). In these he based the theory of probability on the calculus of generating functions and extended Jakob Bernoulli's work on the law of large numbers. He applied probability to problems involving life expectancy, insurance, moral expectation (prudence), error theory, and decision theory.

91. From *Essais philosophique sur les probabilités* (1814)*

(The Theory of Probability)

PIERRE-SIMON LAPLACE

CHAPTER I INTRODUCTION

This philosophical essay is the development of a lecture on probabilities which I delivered in 1795 to the normal schools whither I had been called, by a decree of the national convention, as professor of mathematics with Lagrange. I have recently published upon the same subject a work entitled *The Analytical Theory of Probabilities.* I present here without the aid of analysis the principles and general results of this theory, applying them to the most important questions of life, which are indeed for the most part only problems of probability. Strictly speaking it may even be said that nearly all our knowledge is problematical; and in the small

**Source:* Pierre-Simon Laplace, *A Philosophical Essay on Probabilities,* trans. by F. W. Truscott and F. L. Emory (1951), 1-19. Reprinted by permission of Dover Publications, Inc.

number of things which we are able to know with certainty, even in the mathematical sciences themselves, the principal means for ascertaining truth—induction and analogy—are based on probabilities; so that the entire system of human knowledge is connected with the theory set forth in this essay. Doubtless it will be seen here with interest that in considering, even in the eternal principles of reason, justice, and humanity, only the favorable chances which are constantly attached to them, there is a great advantage in following these principles and serious inconvenience in departing from them: their chances, like those favorable to lotteries, always end by prevailing in the midst of the vacillations of hazard. I hope that the reflections given in this essay may merit the attention of philosophers and direct it to a subject so worthy of engaging their minds.

CHAPTER II
CONCERNING PROBABILITY

All events, even those which on account of their insignificance do not seem to follow the great laws of nature, are a result of it just as necessarily as the revolutions of the sun. In ignorance of the ties which unite such events to the entire system of the universe, they have been made to depend upon final causes or upon hazard, according as they occur and are repeated with regularity, or appear without regard to order; but these imaginary causes have gradually receded with the widening bounds of knowledge and disappear entirely before sound philosophy, which sees in them only the expression of our ignorance of the true causes.

Present events are connected with preceding ones by a tie based upon the evident principle that a thing cannot occur without a cause which produces it. This axiom, known by the name of *the principle of sufficient reason,* extends even to actions which are considered indifferent; the freest will is unable without a determinative motive to give them birth; if we assume two positions with exactly similar circumstances and find that the will is active in the one and inactive in the other, we say that its choice is an effect without a cause. It is then, says Leibniz, the blind chance of the Epicureans. The contrary opinion is an illusion of the mind, which, losing sight of the evasive reasons of the choice of the will in different things, believes that choice is determined of itself and without motives.

We ought then to regard the present state of the universe as the effect of its anterior state and as the cause of the one which is to follow. Given for one instant an intelligence which could comprehend all the forces by which nature is animated and the respective situation of the beings who compose it—an intelligence sufficiently vast to submit these data to analysis—it would embrace in the same formula the movements of the greatest bodies of the universe and those of the lightest atom; for it, nothing would be uncertain and the future, as the past, would be present to its eyes. The human mind offers, in the perfection which it has been able to give to astronomy, a feeble idea of this intelligence. Its discoveries in mechanics and geometry, added to that of universal gravity, have enabled it to comprehend in the same analytical expressions the past and future states of the system of the world. Applying the same method to some other objects of its knowledge, it has succeeded in referring to general laws observed phenomena and in foreseeing those which given circumstances ought to produce. All these efforts in the search for truth tend to lead it back continually to the vast intelligence which we have just mentioned, but from which it will always remain infinitely removed. This tendency, peculiar to the human race, is that which renders it superior to animals; and their progress in this respect distinguishes nations and ages and constitutes their true glory.

Let us recall that formerly, and at no remote epoch, an unusual rain or an extreme drought, a comet having in train a very long tail, the eclipses, the aurora borealis, and in general all the unusual phenomena were regarded as so many signs of celestial wrath. Heaven was invoked in order to avert their baneful influence. No one prayed to have the planets and the sun arrested in their courses: observation had soon made apparent the futility of such prayers. But as these phenomena, occurring and disappearing at long intervals, seemed to oppose the order of nature, it was supposed that Heaven, irritated by the crimes of the earth, had created them to announce its vengeance. Thus the long tail of the comet of 1456 spread terror through Europe, already thrown into consternation by the rapid successes of the Turks, who had just overthrown the Lower Empire. This star after four revolutions has excited among us a very different interest. The knowledge of the laws of the system of the world acquired in the interval had dissipated the fears begotten by the ignorance of the true relationship of man to the universe; and Halley, having recognized the identity of this comet with those of the years 1531, 1607, and 1682, announced its next return for the end of the year 1758 or the beginning of the year 1759. The learned world awaited with impatience this return which was to confirm one of the greatest discoveries that have been made in the sciences, and fulfill the prediction of Seneca when he said, in speaking of the revolutions of those stars which fall from an enormous height: "The day will come when, by study pursued through several ages, the things now concealed will appear with evidence; and posterity will be astonished that truths so clear had escaped us." Clairaut then undertook to submit to analysis the perturbations which the comet had experienced by the action of the two great planets, Jupiter and Saturn; after immense calculations he fixed its next passage at the perihelion toward the beginning of April 1759, which was actually verified by observation. The regularity which astronomy shows us in the movements of the comets doubtless exists also in all phenomena.

The curve described by a simple molecule of air or vapor is regulated in a manner just as certain as the planetary orbits; the only difference between them is that which comes from our ignorance.

Probability is relative, in part to this ignorance, in part to our knowledge. We know that of three or a greater number of events a single one ought to occur; but nothing induces us to believe that one of them will occur rather than the others. In this state of indecision it is impossible for us to announce their occurrence with certainty. It is, however, probable that one of these events, chosen at will, will not occur because we see several cases equally possible which exclude its occurrence, while only a single one favors it.

The theory of chance consists in reducing all the events of the same kind to a certain number of cases equally possible, that is to say, to such as we may be equally undecided about in regard to their existence, and in determining the number of cases favorable to the event whose probability is sought. The ratio of this number to that of all the cases possible is the measure of this probability, which is thus simply a fraction whose numerator is the number of favorable cases and whose denominator is the number of all the cases possible.

The preceding notion of probability supposes that, in increasing in the same ratio the number of favorable cases and that of all the cases possible, the probability remains the same. In order to convince ourselves let us take two urns, A and B, the first containing four white and two black balls, and the second containing only two white balls and one black one. We may imagine the two black balls of the first urn attached by a thread which breaks at the moment when one of them is seized in order to be drawn out, and the four white balls thus forming two similar systems. All the chances which will favor the seizure of one of the balls of the black system will lead to a black ball. If we conceive now that the threads which unite the balls do not break at all, it is clear that the number of possible chances will not change any more than that of the chances favorable to the extraction of the black balls; but two balls will be drawn from the urn at the same time; the probability of drawing a black ball from the urn A will then be the same as at first. But then we have obviously the case of urn B with the single difference that the three balls of this last urn would be replaced by three systems of two balls invariably connected.

When all the cases are favorable to an event the probability changes to certainty and its expression becomes equal to unity. Upon this condition, certainty and probability are comparable, although there may be an essential difference between the two states of the mind when a truth is rigorously demonstrated to it, or when it still perceives a small source of error.

In things which are only probable the difference of the data, which each man has in regard to them, is one of the principal causes of the diversity of opinions which prevail in regard to the same objects. Let us suppose, for example, that we have three urns, A, B, C, one of which contains only black balls while the two others contain only white balls; a ball is to be drawn from the urn C and the probability is demanded that this ball will be black. If we do not know which of the three urns contains black balls only, so that there is no reason to believe that it is C rather than B or A, these three hypotheses will appear equally possible, and since a black ball can be drawn only in the first hypothesis, the probability of drawing it is equal to one third. If it is known that the urn A contains white balls only, the indecision then extends only to the urns B and C, and the probability that the ball drawn from the urn C will be black is one half. Finally this probability changes to certainty if we are assured that the urns A and B contain white balls only.

It is thus that an incident related to a numerous assembly finds various degrees of credence, according to the extent of knowledge of the auditors. If the man who reports it is fully convinced of it and if, by his position and character, he inspires great confidence, his statement, however extraordinary it may be, will have for the auditors who lack information the same degree of probability as an ordinary statement made by the same man, and they will have entire faith in it. But if some one of them knows that the same incident is rejected by other equally trustworthy men, he will be in doubt and the incident will be discredited by the enlightened auditors, who will reject it whether it be in regard to facts well averred or the immutable laws of nature.

It is to the influence of the opinion of those whom the multitude judges best informed and to whom it has been accustomed to give its confidence in regard to the most important matters of life that the propagation of those errors is due which in times of ignorance have covered the face of the earth. Magic and astrology offer us two great examples. These errors inculcated in infancy, adopted without examination, and having for a basis only universal credence, have maintained themselves during a very long time; but at last the progress of science has destroyed them in the minds of enlightened men, whose opinion consequently has caused them to disappear even among the common people, through the power of imitation and habit which had so generally spread them abroad. This power, the richest resource of the moral world, establishes and conserves in a whole nation ideas entirely contrary to those which it upholds elsewhere with the same authority. What indulgence ought we not then to have for opinions different from ours, when this difference often depends only upon the various points of view where circumstances have placed us! Let us enlighten those whom we judge insufficiently instructed; but first let us examine critically our own opinions and weigh with impartiality their respective probabilities.

The difference of opinions depends, however, upon the manner in which the influence of known data is determined. The theory of probabilities holds to considerations so delicate that it is not surprising that with the same data two persons arrive at different results, especially in very complicated questions. Let us examine now the general principles of this theory.

CHAPTER III
THE GENERAL PRINCIPLES OF THE CALCULUS OF PROBABILITIES

First Principle. The first of these principles is the definition itself of probability, which, as has been seen, is the ratio of the number of favorable cases to that of all the cases possible.

Second Principle. But that supposes the various cases equally possible. If they are not so, we will determine first their respective possibilities, whose exact appreciation is one of the most delicate points of the theory of chance. Then the probability will be the sum of the possibilities of each favorable case. Let us illustrate this principle by an example.

Let us suppose that we throw into the air a large and very thin coin whose two large opposite faces, which we will call heads and tails, are perfectly similar. Let us find the probability of throwing heads at least one time in two throws. It is clear that four equally possible cases may arise, namely, heads at the first and at the second throw; heads at the first throw and tails at the second; tails at the first throw and heads at the second; finally, tails at both throws. The first three cases are favorable to the event whose probability is sought; consequently this probability is equal to $\frac{3}{4}$; so that it is a bet of three to one that heads will be thrown at least once in two throws.

We can count at this game only three different cases, namely, heads at the first throw, which dispenses with throwing a second time; tails at the first throw and heads at the second; finally, tails at the first and at the second throw. This would reduce the probability to $\frac{2}{3}$ if we should consider with d'Alembert these three cases as equally possible. But it is apparent that the probability of throwing heads at the first throw is $\frac{1}{2}$, while that of the other two cases is $\frac{1}{4}$, the first case being a simple event which corresponds to two events combined: heads at the first and at the second throw, and heads at the first throw, tails at the second. If we then, conforming to the second principle, add the possibility $\frac{1}{2}$ of heads at the first throw to the possibility $\frac{1}{4}$ of tails at the first throw and heads at the second, we shall have $\frac{3}{4}$ for the probability sought, which agrees with what is found in the supposition when we play the two throws. This supposition does not change at all the chance of that one who bets on this event; it simply serves to reduce the various cases to the cases equally possible.

Third Principle. One of the most important points of the theory of probabilities and that which lends the most to illusions is the manner in which these probabilities increase or diminish by their mutual combination. If the events are independent of one another, the probability of their combined existence is the product of their respective probabilities. Thus the probability of throwing one ace with a single die is $\frac{1}{6}$; that of throwing two aces in throwing two dice at the same time is $\frac{1}{36}$. Each face of the one being able to combine with the six faces of the other, there are in fact thirty-six equally possible cases, among which one single case gives two aces. Generally the probability that a simple event in the same circumstances will occur consecutively a given number of times is equal to the probability of this simple event raised to the power indicated by this number. Having thus the successive powers of a fraction less than unity diminishing, without ceasing, an event which depends upon a series of very great probabilities may become extremely improbable. Suppose then an incident be transmitted to us by 20 witnesses in such manner that the first has transmitted it to the second, the second to the third, and so on. Suppose again that the probability of each testimony be equal to the fraction $9/10$; that of the incident resulting from the testimonies will be less than $1/8$. We cannot better compare this diminution of the probability than with the extinction of the light of objects by the interposition of several pieces of glass. A relatively small number of pieces suffices to take away the view of an object that a single piece allows us to perceive in a distinct manner. The historians do not appear to have paid sufficient attention to this degradation of the probability of events when seen across a great number of successive generations; many historical events reputed as certain would be at least doubtful if they were submitted to this test.

In the purely mathematical sciences the most distant consequences participate in the certainty of the principle from which they are derived. In the applications of analysis to physics the results have all the certainty of facts or experiences. But in the moral sciences, where each inference is deduced from that which precedes it only in a probable manner, however probable these deductions may be, the chance of error increases with their number and ultimately surpasses the chance of truth in the consequences very remote from the principle.

Fourth Principle. When two events depend upon each other, the probability of the compound event is the product of the probability of the first event and the probability that, this event having occurred, the second will occur. Thus in the preceding case of the three urns A, B, C, of which two contain only white balls and one contains only black balls, the probability of drawing a white ball from the urn C is $2/3$, since of the three urns only two contain balls of that color. But when a white ball has been drawn from the urn C, the indecision relative to that one of the urns which contain only black balls extends only to the urns A and B; the probability of drawing a white ball from the urn B is $1/2$; the product of $2/3$ by $1/2$, or $1/3$, is then the probability of drawing two white balls at one time from the urns B and C.

We see by this example the influence of past events upon the probability of future events. For the probability of drawing a white ball from the urn B, which primarily is $2/3$, becomes $1/2$ when a white ball has been drawn from the urn C; it would change to certainty if a black ball had been drawn from the same urn. We will determine this influence by means of the following principle, which is a corollary of the preceding one.

Fifth Principle. If we calculate *a priori* the probability of the occurred event and the probability of an event composed of that one and a second one which is expected, the second probability divided by the first will be the probability of the event expected, drawn from the observed event.

Here is presented the question raised by some philosophers touching the influence of the past upon the probability of the future. Let us suppose at the play of heads and tails that heads has occurred oftener than tails. By this alone we shall be led to believe that in the constitution of the coin there is a secret cause which favors it. Thus in the conduct of life constant happiness is a proof of competency which would induce us to employ preferably happy persons. But if by the unreliability of circumstances we are constantly brought back to a state of absolute indecision, if, for example, we can change the coin at each throw at the play of heads and tails, the past can shed no light upon the future and it would be absurd to take account of it.

Sixth Principle. Each of the causes to which an observed event may be attributed is indicated with just as much likelihood as there is probability that the event will take place, supposing the event to be constant. The probability of the existence of any one of these causes is then a fraction whose numerator is the probability of the event resulting from this cause and whose denominator is the sum of the similar probabilities relative to all the causes; if these various causes, considered *a priori,* are unequally probable, it is necessary, in place of the probability of the event resulting from each cause, to employ the product of this probability by the possibility of the cause itself. This is the fundamental principle of this branch of the analysis of chances which consists in passing from events to causes.

This principle gives the reason why we attribute regular events to a particular cause. Some philosophers have thought that these events are less possible than others and that at the play of heads and tails, for example, the combination in which heads occurs twenty successive times is less easy in its nature than those where heads and tails are mixed in an irregular manner. But this opinion supposes that past events have an influence on the possibility of future events, which is not at all admissible. The regular combinations occur more rarely only because they are less numerous. If we seek a cause wherever we perceive symmetry, it is not that we regard a symmetrical event as less possible than the others, but, since this event ought to be the effect of a regular cause or that of chance, the first of these suppositions is more probable than the second. On a table we see letters arranged in this order, *C–o–n–s–t–a–n–t–i–n–o–p–l–e,* and we judge that this arrangement is not the result of chance, not because it is less possible than the others, for if this word were not employed in any language we should not suspect it came from any particular cause, but this word being in use among us, it is incomparably more probable that some person has thus arranged the aforesaid letters than that this arrangement is due to chance.

This is the place to define the word *extraordinary.* We arrange in our thought all possible events in various classes; and we regard as *extraordinary* those classes which include a very small number. Thus at the play of heads and tails the occurrence of heads a hundred successive times appears to us extraordinary because of the almost infinite number of combinations which may occur in a hundred throws; and if we divide the combinations into regular series containing an order easy to comprehend, and into irregular series, the latter are incomparably more numerous. The drawing of a white ball from an urn which among a million balls contains only one of this color, the others being black, would appear to us likewise extraordinary, because we form only two classes of events relative to the two colors. But the drawing of the number 475,813, for example, from an urn that contains a million numbers seems to us an ordinary event; because, comparing individually the numbers with one another without dividing them into classes, we have no reason to believe that one of them will appear sooner than the others.

From what precedes, we ought generally to conclude that the more extraordinary the event, the greater the need of its being supported by strong proofs. For, those who attest it being able to deceive or to have been deceived, these two causes are as much more probable as the reality of the event is less. We shall see this particularly when we come to speak of the probability of testimony.

Seventh Principle. The probability of a future event is the sum of the products of the probability of each cause, drawn from the event observed, by the probability that, this cause existing, the future event will occur. The following example will illustrate this principle.

Let us imagine an urn which contains only two balls, each of which may be either white or black. One of these balls is drawn and is put back into the urn before proceeding to a new draw. Suppose that in the first two draws white balls have been drawn; the probability of again drawing a white ball at the third draw is required.

Only two hypotheses can be made here; either one of the balls is white and the other black, or both are white. In the first hypothesis the probability of the event observed is $^1/_4$; it is unity or certainty in the second. Thus in regarding these hypotheses as so many causes, we shall have for the sixth principle $^1/_5$ and $^4/_5$ for their respective probabilities. But if the first hypothesis occurs, the probability of drawing a white ball at the third draw is $^1/_2$; it is equal to certainty in the second hypothesis; multiplying then the last probabilities by those of the corresponding hypotheses, the sum of the products, or $^9/_{10}$, will be the probability of drawing a white ball at the third draw.

When the probability of a single event is unknown we may suppose it equal to any value from zero to unity. The probability of each of these hypotheses, drawn from the event observed, is, by the sixth principle, a fraction whose numerator is the probability of the event in this hypothesis and whose denominator is the sum of the similar probabilities relative to all the hypotheses. Thus the probability that the possibility of the event is comprised within given limits is the sum of the fractions comprised within these limits. Now if we multiply each fraction by the probability of the future event, determined in the corresponding hypothesis, the sum of the products relative to all the hypotheses will be, by the seventh principle, the probability of the future event drawn from the event observed. Thus we find that an event having occurred successively any number of times, the probability that it will happen again the next time is equal to this number increased by unity divided by the same number, increased by two units. Placing the most ancient epoch of history at 5,000 years ago, or at 1,826,213 days, and the sun having risen constantly in the interval at each revolution of 24 hours, it is a bet of 1,826,214 to one that it will rise again tomorrow. But this number is incomparably greater for him who, recognizing in the totality of phenomena the principal regulator of days and seasons, sees that nothing at the present moment can arrest the course of it.

Buffon in his *Political Arithmetic* calculates differently the preceding probability. He supposes that it differs from unity only by a fraction whose numerator is unity and whose denominator is the number 2 raised to a power equal to the number of days which have elapsed since the epoch. But the true manner of relating past events with the probability of causes and of future events was unknown to this illustrious writer.

Chapter VIII

The Nineteenth Century

Introduction by
HELENA M. PYCIOR
University of Wisconsin-Milwaukee

Introduction. The 19th century opened with Western Europe in sharp political transition. The French Revolution that started in 1789 had overthrown the Old Regime and spread the ideas of equality (full legal equality did not apply to women and workers), individual freedom, and basic human rights. The years from 1800 through 1814 witnessed the final act of revolution in the Napoleonic conquest of large portions of Europe. After Napoleon's defeat the major European powers met at the Congress of Vienna (1814–15) and attempted the difficult task of restoring the old social hierarchy and suppressing political liberalism and nationalism—the two chief forces for political change in the first half of the 19th century. Led by Metternich, the statesmen at Vienna succeeded in formulating a peace settlement that helped protect Europe from a major conflagration for 100 years.

Of course, all was not calm in Europe following the Vienna settlement. Revolutions expressing liberal or nationalistic aspirations or a combination of the two were common. After the Greek War for Independence in the 1820s, there were revolutions in France, Belgium, Poland, and the Italian and German states in 1830 and in almost all the nations of Europe, except England and Russia, in 1848. The revolutionary uprisings in 1848, which at first seemed to be succeeding everywhere, were crushed by military force after only a few months. Despite these failures, several leading European nations gained democratic forms of government by the end of the century. In addition, the Italians led by Cavour and Garibaldi and the Germans led by Bismarck achieved national unity by 1871.

The chief forces for change in Europe in the 19th century were not simply political. This century experienced an ongoing Industrial Revolution that was made possible by earlier improvements in agriculture and then in industrial technology. The Industrial Revolution had profound consequences. By the mid-19th century, industrial and population growth had transformed portions of Western Europe into urban societies with larger concentrations of population than ever before. Industrialization gradually raised the standard of living in those nations in which it was most advanced and enhanced the position of the middle class. Its benefits, however, were often purchased at enormous costs in human misery among the laborers—men, women, and children. Consequently, there appeared critics of the middle-class values of the Industrial Revolution, including John Stuart Mill who departed from the doctrine of *laissez-faire* in calling for limited governmental action to assure a more equitable distribution of wealth between employers and employees (the demand for governmental intervention had begun before Mill among disciples of Jeremy Bentham), and Karl Marx who proposed his communist theory of production in 1848.

Changing political, economic, and social conditions in the 19th century were reflected in changing literary and artistic styles. The century began with romanticism in the ascendant, partially in response to the years of the French Revolutionary and Napoleonic Wars. Romanticism, which may be traced back to the ideas of Rousseau and Kant in the late 18th century, rejected the Enlightenment's emphasis on reason as the only way to understand the world. Romantics often stressed the emotional, inner genius, medieval temper, intuition, imagination, and empathy or "modes of feeling." The Romantic movement coincided with expanding literacy in Western Europe. Its greatest achievements were in poetry, with writers like Byron and Goethe, and in music, with composers such as Beethoven. The failures of the revolutions of 1848, in which many romantics had played key roles, and the increasing visibility of the ravages of industrialization on European society led to the emergence of realism as the dominant literary and artistic style in the late 1840s. The novels of Dickens, Dostoyevsky, and Tolstoy are examples of realism at its zenith. Late in the century realism was replaced by naturalism. The materialistic emphasis in industrial society had encouraged some writers to imitate the method and approach of 19th-century science, especially Claude Bernard's mechanical approach to organicism and Charles Darwin's theory of evolution by natural selection. One of the leading naturalistic novelists was Émile Zola.

Amid political revolutions, spreading industrialization, and various literary–artistic styles, the scientists and mathematicians of 19th-century Europe substantially advanced their respective fields. They were not immune from the influences in their society. Of the major mathematicians of the century some were radical liberals like Galois, others conservatives like Gauss. Many, born into the lower and middle classes, benefited from the expansion of educational and professional opportunities for their classes that occurred during the century. This was true of those who studied at the French universities and especially at the German ones, such as Berlin and Göttingen.

The extent to which Kantian philosophy, romanticism, and even realism helped shape the natural science and mathematics of the century still remains to be determined. Some scholars believe that Kant's theory of matter, based on attractive and repulsive forces rather than on material particles, contributed in a fundamental way to the emergence of field theory. In addition, Sir William Rowan Hamilton, the creator of the quaternions, was a romantic who in the early decades of the century hoped for a resurgence of the imagination in science and mathematics. Adapting the Kantian intuition of time to mathematics, Hamilton early in his career defined algebra as the science of pure time.

While recognizing these socio-cultural interactions, one can also study science and mathematics apart from the general culture. Central to the development of 19th-century mathematics were the mathematical problems inherited from 18th-century mathematicians, who clearly formulated but left unresolved such problems as the standing of Euclid's fifth postulate, the lack of adequate definitions of the negative and complex numbers, and the foundations of the calculus. Some of the most important developments of the 19th century originated in the course of work on these problems. Yet, in responding to these and even older problems, 19th-century mathematicians did not merely patch up earlier mathematics. Instead, they opened new, hitherto-unexplored areas of mathematics, radically revising the mathematician's basic understanding of mathematics and establishing unprecedented standards of rigor.

Two of the early mathematical developments of the century were major ones—non-Euclidean geometry and symbolical algebra arose in response to 18th-century problems. Following futile attempts to derive Euclid's fifth postulate from the other four, Gauss, Bolyai, and Lobachevsky decided that the fifth postulate was independent of the others (*i.e.*, could not be proven from them) and that a new kind of geometry (subsequently called non-Euclidean) could be built upon Euclid's first four postulates and a postulate contrary to Euclid's fifth. The new geometry contained strange statements (*e.g.*, the angle-sum of every triangle was strictly less than 180 degrees), yet it appeared to be consistent (*i.e.*, free of contradictions) and was gradually accepted as a geometrical system of equal standing with Euclid's. As a new form of geometry emerged from the problem of the fifth postulate, so a new form of algebra emerged from the problem of the negative and complex numbers. Partially in response to the latter, British algebraists (including George Peacock and Augustus De Morgan), developed symbolical algebra, into which the negative and complex numbers were introduced by assumption and without definition. As a first step toward modern abstract algebra, symbolical algebraists stressed the laws of algebra rather than the meaning of algebraic symbols. Initially, they adopted the laws of arithmetic as the laws of symbolical algebra, because they saw the latter as a generalization or extension of the former. In 1843, however, Hamilton's creation of the quaternions freed algebraists from dependence on arithmetic. The quaternions violated the commutative law of multiplication. While in arithmetic it is always true that $ab = ba$, the basic units i, j, and k of the quaternions (elements of the form $a + bi + cj + dk$, where a, b, c, and d are real numbers) obey the following rules: $ij = -ji = k$, $jk = -kj = i$, and $ki = -ik = j$. Thus, in what are at present thought to have been independent events, 19th-century mathematicians violated two separate laws of traditional mathematics—Euclid's fifth postulate and the commutative property of multiplication.

The development of non-Euclidean geometry and the quaternions forced mathematicians to deal with the fundamental question, What is mathematics? For about two thousand years mathematics had been thought of as a collection of true propositions based on self-evident, absolutely true, first principles. However, it was clear that Euclid's fifth postulate and those used by Gauss, Bolyai, and Lobachevsky (which were contrary to Euclid's) could not *all* be self-evident and true. If the first principles of mathematics were not self-evident and *absolutely* true, where did they come from? Nineteenth-century mathematicians began to argue that mathematicians created mathematics; mathematics was not a collection of absolute truths but rather a system of propositions derived deductively from other propositions or axioms of the mathematician's making. Mathematicians determined the rules of mathematics, subject to certain considerations such as consistency, independence, completeness, and fertility. By the end of the century, this axiomatic development of mathematics was pushed to its extreme in Hilbert's formalist presentation of geometry.

Beginning with the recognition of but one geometry (Euclidean) and one algebra (universal arithmetic), the 19th century witnessed the creation of numerous new geometries and algebras. Riemann, for example, developed the new elliptic geometry, and Boole produced an algebra of logic. Yet, 19th-century mathematicians did not merely revel in the mathematical diversity they created but also searched for similarities among their various creations—developing, for example, the theory of groups and more generally beginning work on the theory of algebraic structures.

Another problem inherited from the 18th century and resolved in the 19th, was providing satisfactory foundations for mathematical analysis. In response to this problem, 19th-century mathematicians freed analysis from dependence on geometry, infinitesimals, and the like, and rigorously grounded it in arithmetic. One of the major achievements of the 19th century—the arithmetization of analysis—was the collective work of some of the century's leading mathematicians, including Abel, Bolzano, Cauchy, Weierstrass, and Dedekind. Through their efforts the concept of the limit was reduced to a straightforward statement about real numbers, and the real numbers, in turn, were presented as human constructs based ultimately on the whole numbers.

By the end of the century, mathematics was the study of not only such traditional entities as points, lines, and numbers—entities which in the course of the century had lost much of their traditional meaning—but also such new creations as groups, n-dimensional linear algebras, Gaussian integers, Kummer's ideals, sets, and even the infinite, which, while previously a somewhat vague philosophical and theological concept, had been successfully mathematized by Cantor. Hilbert's figurative description of Cantor's mathematics of the infinite might appropriately be applied to all the major mathematical work of the 19th century. In reaction to harsh criticism of Cantor's work, Hilbert stated that he refused to leave the paradise created by Cantor. Mathematicians of the 19th century produced a mathematical paradise whose fruits are still enjoyed by 20th-century mathematicians.

SUGGESTIONS FOR FURTHER READING

Roberto Bonola, *Non-Euclidean Geometry,* Trans. by H. S. Carslaw (New York: Dover reprint, 1955).

Michael J. Crowe, *A History of Vector Analysis: The Evolution of the Idea of a Vectorial System* (Notre Dame: University of Notre Dame Press, 1967).

Joseph W. Dauben, *Georg Cantor: His Mathematics and Philosophy of the Infinite* (Cambridge: Harvard University Press, 1979).

J. M. Dubbey, *The Mathematical Work of Charles Babbage* (Cambridge: Cambridge University Press, 1978).

Pierre Dugac, *Histoire du théorème des accroissements finis.* Paris: Université Pierre et Marie Curie, 1979.

G. Waldo Dunnington, *Carl Friedrich Gauss, Titan of Science: A Study of His Life and Work* (New York: Hafner, 1955).

Harold M. Edwards, "The Background of Kummer's Proof of Fermat's Last Theorem for Regular Primes," *Archive for History of Exact Sciences* 14 (1975): 219-236.

I. Grattan-Guinness, *The Development of the Foundations of Mathematical Analysis from Euler to Riemann* (Cambridge: MIT Press, 1970).

Jeremy Gray, *Ideas of Space: Euclidean, Non-Euclidean, and Relativistic* (Oxford: Clarendon Press, 1979).

Thomas L. Hankins, *Sir William Rowan Hamilton* (Baltimore/London: Johns Hopkins University Press, 1980).

Hubert C. Kennedy, *Peano: Life and Works of Giuseppe Peano* (Dordrecht/Boston/London: D. Reidel, 1980).

B. Melvin Kiernan, "The Development of Galois Theory from Lagrange to Artin," *Archive for History of Exact Sciences* 8 (1971): 40-154.

Kenneth R. Manning, "The Emergence of the Weierstrassian Approach to Complex Analysis," *Archive for History of Exact Sciences* 14 (1975): 297-383.

John T. Merz, *A History of European Thought in the Nineteenth Century* (New York: Dover reprint, 4 vols., 1965).

Ernest Nagel, "'Impossible Numbers': A Chapter in the History of Modern Logic," *Studies in the History of Ideas* 3 (1935): 429-474.

Luboŝ Nový, *Origins of Modern Algebra,* Trans. by Jaroslav Tauer (Prague: Academia, 1973).

Chapter VIII
The Nineteenth Century

Section A
Algebra

CARL FRIEDRICH GAUSS (1777–1855)

Carl Gauss, who ranks with Archimedes and Newton as one of the greatest mathematicians in history, came from a family of modest means. The only son of Gebhard Gauss (a gardener and stone mason) and Dorothea Benze (Gebhard's second wife and a woman of natural intellect and strong character), Gauss was a child prodigy who taught himself to read and reckon before entering St. Catherine elementary school. Through the intercession of Johann Martin Bartels, his tutor at Catharineum Gymnasium, and Professor Zimmermann at Collegium Carolinum, the duke of Brunswick became his patron. This allowed young Gauss to attend the Collegium Carolinum (1792–95) in Brunswick and the University of Göttingen (1795–1800). At first he was undecided between being a philologist or a mathematician, but he chose the latter in 1796 after proving the regular 17-gon could be constructed with ruler and compass alone (a problem that had been unsolved from antiquity). Archimedes and Newton were to be the heroes of his research, while Euler and Lagrange were judged highly. The University of Helmstädt granted him the doctorate in absentia in 1801. In his doctoral dissertation, Gauss skillfully proved the fundamental theorem of algebra, which states that every algebraic equation with complex coefficients has at least one root (*i.e.,* at least one complex number satisfies the equality).

During the years from 1801 to 1810, Gauss moved from mathematics to astronomy and secured a good position. Following the discovery of the planetoids Ceres (1801) and Pallus (1802), he applied his superior computational skills to calculate their ephemerides (daily positions) in a way that improved the theory of perturbation in astronomy. Astronomers, who were previously unable to calculate Ceres' orbit, promptly recognized his achievement. Gauss now established contacts with the German astronomers Olbers and Bessel, as well as with Laplace in Paris and with the German geophysicist Alexander von Humboldt in Berlin. In later life he was less communicative in the sciences. The duke of Brunswick generously supported his research through 1803, enabling him to refuse a post in St. Petersburg. In 1804, his alma mater, the University of Göttingen, appointed him director of its observatory and made him professor of astronomy, a post he held from 1807 until his death. In 1805, he married Johanna Osthoff, who died in 1809; the following year he married Minna Waldeck. In 1806, the duke of Brunswick was one of the German commanders at the Battle of Auerstädt. When Napolean ordered the execution of these commanders, Gauss pleaded for clemency for his 71-year-old ducal patron. When Napoleon scorned the suggestion, Gauss became a staunch German nationalist and royalist. In the late Napoleonic and Metternichean era, he sided with his fellow Germans Goethe and Beethoven.

Gauss diligently pursued research throughout his career. In the late 1820s, he turned his attention to geodesy and differential geometry. From 1821 to 1823, he worked on a triangulation of Hanover, which was not completed until 1847. To obtain more accurate measurements, he invented a heliotrope to use in surveying. As a result of the geodesic studies and his work on the theory of curved surfaces in 1827, he may have come to believe that in very large triangles, the sum of the angles is not 180°—an intimation that the local space of Euclidean geometry was distinct from large, curved space. In the 1830s, Gauss turned more toward physics. With Wilhelm Weber he invented the electric telegraph (1833–34) but only carried out experiments on a small scale because of limited finances. His studies of terrestrial magnetism and optical problems brilliantly applied mathematics to physics.

Gauss was less enthusiastic about teaching than research, but he was known as an open-minded and demanding professor, who even agreed to support a woman Ph.D. at a time when this was uncommon. Although he accepted few students, he inspired such scholars as Abel, Dirichlet, and Jacobi. His colleagues knew Gauss for his penetrating blue eyes, wry humor, moral rectitude, and conservative bent. They also knew that he abhorred violence and disliked public polemics. Gauss mastered many foreign languages and enthusiastically read different literatures, with especial fondness for the British authors Scott, Gibbon, and Macaulay. In the 1840s, his health worsened; after 1850, heart disease troubled him. In 1851–52, he supervised his last doctoral students, Riemann and Dedekind.

Gauss believed that mathematics played a crucial role in the sciences. His dictum aptly conveys his position: "Mathematics, the queen of the sciences, and arithmetic, the queen of mathematics." In his classic *Disquisitiones arithmeticae* (1801), he soundly established number theory as a branch of mathematics. In this masterpiece, he proved by induction the law of biquadratic reciprocity and defined numbers as psychical realities. In other research Gauss invented a non-Euclidean geometry (by 1816) and devised Gaussian curvature, which underlies the general theory of relativity. He also introduced the law of normal distribution (*i.e.,* the Gaussian error curve that showed how to represent probability by a bell-shaped curve). In addition, he advanced the theory of elliptic functions, studied complex integrals, and initiated the theory of differentiable manifolds. Primarily, he systematized and solved old problems with such thoroughness that he laid the groundwork for new departures in number theory, differential geometry, and statistics.

This mathematical prodigy was parsimonious about his own publications. The motto on his seal was *Pauca sed matura* ("few but ripe"). This meant that he sought to achieve elegance, conciseness, and the utmost rigor in his writings, and he succeeded. However, he also left much of his work unpublished in his diaries; his writings fill 12 weighty volumes. For his many achievements his colleagues appropriately dubbed Gauss the "Prince of Mathematicians" *(mathematicorum princeps)*.

92. From "New Proof of the Theorem That Every Integral Rational Algebraic Function of One Variable Can Be Decomposed into Real Factors of the First or Second Degree" (1799)*

CARL FRIEDRICH GAUSS

THE FUNDAMENTAL THEOREM OF ALGEBRA

13. *Lemma.* If m is an arbitrary positive integer, then the function $\sin x^m - \sin m\phi \cdot r^{m-1}x + \sin (m-1)\phi \cdot r^m$ is divisible by $x^2 - 2\cos\phi \cdot rx + r^2$.

[The proof is given by direct division.]

14. *Lemma.* If the quantity r and the angle ϕ are so determined that the equations

$$r^m \cos m\phi + Ar^{m-1}\cos (m-1)\phi + Br^{m-2}\cos (m-2)\phi + \text{etc.} + Krr\cos 2\phi + Lr\cos\phi + M = 0, \quad (1)$$

$$r^m \sin m\phi + Ar^{m-1}\sin (m-1)\phi + Br^{m-2}\sin (m-2)\phi + \text{etc.} + Krr\sin 2\phi + Lr\sin\phi = 0 \quad (2)$$

exist, then the function

$$x^m + Ax^{m-1} + Bx^{m-2} + \text{etc.} + Kx^2 + Lx + M = X$$

will be divisible by the quadratic factor $x^2 - 2\cos rx + r^2$, unless $r\sin\phi = 0$. If $r\sin\phi = 0$, then the same function is divisible by $x - r\cos\phi$.

The proof is given by taking the functions

$$\begin{array}{llll}
\sin\phi\cdot rx^m & - \quad \sin m\phi\cdot r^m x & + & \sin(m-1)\phi\cdot r^{m+1}, \\
A\sin\phi\cdot rx^{m-1} & - \quad A\sin(m-1)\phi\cdot r^{m-1}x & + & A\sin(m-2)\phi\cdot r^m, \\
B\sin\phi\cdot rx^{m-3} & - \quad B\sin(m-2)\phi\cdot r^{m-2}x & + & B\sin(m-3)\phi\cdot r^{m-1}, \\
\ldots\text{etc.} & \ldots\ldots\ldots\ldots & & \ldots\text{etc.} \\
K\sin\phi\cdot rx^2 & - \quad K\sin 2\phi\cdot r^2x & + & K\sin\phi\cdot r^3, \\
L\sin\phi\cdot rx & - \quad L\sin\phi\cdot rx & & *\quad, \\
M\sin\phi\cdot r & \quad * & + & M\sin(-\phi)\cdot r,
\end{array}$$

which are each divisible by $x^2 - 2\cos\phi\cdot xr + r^2$ (according to the first lemma) and which, added up, give $\sin\phi\cdot rX + 0 + 0$. When $r = 0$, X is divisible by $x - r\cos\phi$; when $\sin\phi = 0$, then $\cos\phi = \pm 1$, $\cos 2\phi = \pm 1$, $\cos 3\phi = \pm 1$, etc. and X becomes zero for $x = r\cos\phi$.

*Source: From *Demonstratio nova theorematis omnem functionem algebraicam rationalem integram unius variabilis in factores reales primi vel secundi gradus resolvi posse* ("New Proof of the Theorem That Every Integral Rational Algebraic Function of One Variable Can be Decomposed into Real Factors of the First or Second Degree," 1799). This translation is taken from D. J. Struik (ed.), *A Source Book in Mathematics, 1200-1800* (1969), 115-122. It is reprinted by permission of Harvard University Press. This selection was given a new, shorter title.

15. The previous theorem is usually given with the aid of imaginaries, cf. Euler, *Introductio in analysin infinitorum*, I, p. 110;[1] I found it worth while to show that it can be demonstrated in the same easy way without their aid. Hence it is clear that, in order to prove our theorem, we only have to show: *If some function X of the form $x^m + Ax^{m-1} + Bx^{m-2} +$ etc. $+ Lx + M$ is given, then r and ϕ can be determined in such a way that the equations* (1) *and* (2) *are valid.* Indeed, from this it follows that X possesses a real factor of the first or second degree; division by it necessarily gives a real quotient of lower degree . . . We shall now prove this theorem.

16. We consider a fixed infinite plane (the plane of our Fig. 1) and in it a fixed infinite straight line GG passing through the fixed point C. In order to express all

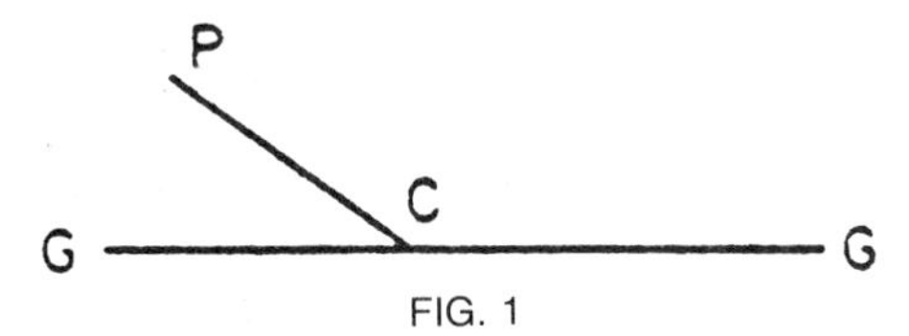

FIG. 1

line segments by numbers we take an arbitrary segment as unit, and erect at an arbitrary point P of the plane, with distance r from center C and with angle $GCP = \mu$, a perpendicular equal to the value of the expression

$$r^m \sin m\phi + Ar^{m-1} \sin (m - 1)\phi + \text{etc.} + Lr \sin \phi.$$

I shall denote this expression by T. I consider the distance r always as positive, and for points on the other side of the axis the angle ϕ must either be taken as larger than two right angles, or (what amounts to the same thing) as negative. The end points of this perpendicular (which have to be taken as above the plane for positive T, as below the plane for negative T, and for vanishing T as in the plane) form a continuous, curved surface, infinite in all directions, which I shall call for the sake of brevity the *first surface*. In exactly the same way we can refer to the same plane, the same center, and the same axis another surface, with altitude above every point of the plane equal to

$$r^m \cos m\phi + Ar^{m-1} \cos (m - 1)\phi + \cdots + Lr \cos \phi + M;$$

this expression I shall always denote by U. This surface, also continuous and infinite in all directions, will be distinguished from the other one by the name of *second surface*. From this it is clear that our entire task is then to prove that there exists at least one point that lies at the same time in the plane, in the first surface, and in the second surface.

17. It can easily be understood that the first surface lies partly above, and partly below the plane, since we can take the distance r from the center so large that the first term $r^m \sin m\phi$ in T surpasses all following terms; if then the angle ϕ is conveniently chosen, this term can become positive as well as negative. The fixed plane must therefore be intersected by the first surface. I shall call this intersection the *first curve*, and it will be determined by $T = 0$. The same reasoning shows that the plane is intersected by the second surface; this intersection will be called the *second curve*, and its equation will be $U = 0$. Both curves will, properly speaking, consist of several branches, which may be entirely separated from each other, but each by itself forms a continuous curve. Indeed, the first curve will always be a so-called reducible curve, since the axis GC must be considered a part of this curve, because $T = 0$ for $\phi = 0$ or $\phi = 180°$ for any value of r. We prefer, however,

to consider the totality of all branches, which pass through all points for which $T = 0$, as one single curve (as is customary in higher geometry). The same happens for all branches passing through the points for which $U = 0$. Now our problem has been reduced to the task of proving that there exists in the plane at least one point at which one of the branches of the first curve is intersected by one of the branches of the second curve. This makes it necessary to study more closely the behavior of these curves.

18. First I observe that each curve is algebraic, and, referred to orthogonal coordinates, of order m. Indeed, if the origin of the abscissae is taken at C and the direction of the abscissa x is measured toward G and that of the ordinate y toward P, then $x = r \cos \phi$, $y = r \sin \phi$, and generally, for arbitrary n:

$$r^n \sin n\phi = nx^{n-1}y - \frac{n(n-1)(n-2)}{1 \cdot 2 \cdot 3} x^{n-3}y^3 + \frac{n \cdots (n-4)}{1 \cdots 5} x^{n-5}y^5 - \text{etc.}$$

$$r^n \cos n\phi = x^n - \frac{n(n-1)}{1 \cdot 2} x^{n-2}y^2 + \frac{n(n-1)(n-2)(n-3)}{1 \cdot 2 \cdot 3 \cdot 4} x^{n-4}y^4 - \text{etc.}$$

T and U consist therefore of several terms of the form $ax^\alpha y^\beta$, where α, β are positive integers, whose sum has m as its maximum value. Moreover, it is easy to see that all terms of T contain the factor y, so that the first curve, to express it exactly, consists of the line with equation $y = 0$ and a curve of order $m - 1$. However, we do not need to take this difference into consideration.

It is of more importance to investigate whether the first and second curves have infinite branches, and what their number and character will be. At an infinite distance from the point C the first curve, with equation

$$\sin m\phi + \frac{A}{r} \sin (m-1)\phi + \frac{B}{rr} \sin (m-2)\phi \text{ etc} = 0$$

coincides with that curve whose equation is $\sin m\phi = 0$. This consists only of m straight lines intersecting at C; the first of these is the axis GCG', the other ones make with this axis the angles $(1/m)180°$, $(2/m)180°$, $(3/m)180°$, ... The first curve therefore has $2m$ infinite branches, which divide the circumference of a circle described with infinite radius into $2m$ equal parts, such that its circumference is intersected by the first branch in the intersection of the circle with the axis, by the second branch at distance $(2/m)180°$, by the third one at distance $(3/m)180°$, etc. It follows similarly that the second curve at infinite distance from the center has the curve represented by the equation $\cos m\phi = 0$ as its asymptote. This curve consists of the totality of m straight lines which also intersect in C at equal angles, but in such a way that the first one forms with the axis CG the angle $(1/m)90°$, the second one the angle $(3/m)90°$, the third one the angle $(5/m)90°$, etc. The second curve therefore also has $2m$ infinite branches, which each form the middle between two neighboring branches of the first curve, so that they intersect the circumference of the circle of infinite radius in points which are $(1/m)90°$, $(3/m)90°$, $(5/m)90°$, . . . away from the axis. It is also clear that the axis itself always forms two infinite branches of the first curve, namely, the first and the $(m + 1)$th. This situation of the branches is well illustrated by Fig. 2, constructed for the case $m = 4$; the branches of the second curve are here dotted to distinguish them from those of the first curve. This also occurs in Fig. 4.[2] Since these results are of the utmost importance, and some readers might be offended by infinitely large quantities, I shall show in the next section how these results can also be obtained without the help of infinite quantities.

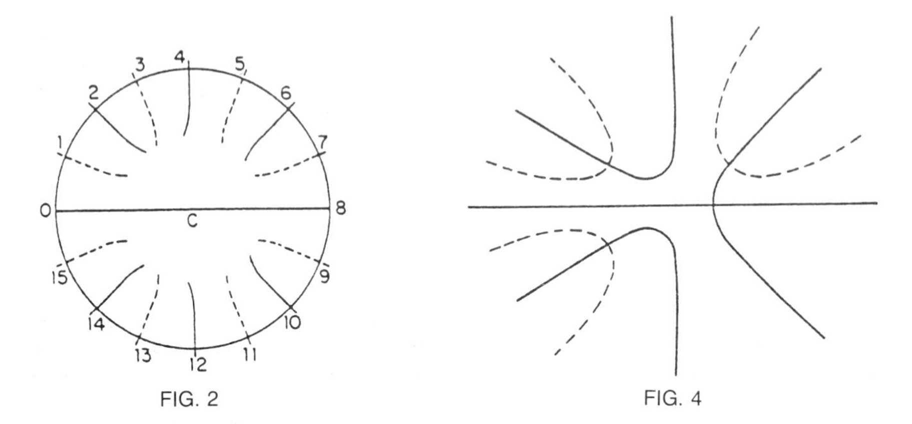

FIG. 2 FIG. 4

19. Theorem. *Under the conditions mentioned before we can construct a circle with center C, on the circumference of which there exist 2m points at which T = 0, and as many points at which U = 0*; they are situated in such a way that each point of the second kind lies between two of the first kind.

Let the sum of all coefficients (taken positive) $A, B, \ldots, K, L, M$ be $= S$; let furthermore R be at the same time $> S\sqrt{2}$ and > 1.[3] I then say that on the circle described with radius R the conditions exist indicated in the theorem. I denote for short by 1 that point of its circumference which is distant $(1/m)45°$ from the point of intersection of the circle with the left-hand side of the axis. Hence for (1) $\phi = (1/m)45°$. Similarly I denote by (3) that point which is distant $(3/m)45°$ from the point of intersection and for which therefore $\phi = (3/m)45°, \ldots$ up to the point $(8m - 1)$ which is distant $[(8m - 1)/m]45°$ from that point of intersection if we always proceed in the same direction, or $(1/m)45°$ if we move in the opposite way. Thus there are in total $4m$ points in the circumference at equal distances from each other. Then there exists between $(8m - 1)$ and (1) a point for which $T = 0$, a similar point lies between (3) and (5), between (7) and (9), . . . ; their number is $2m$. In the same way we see that the single points for which $U = 0$ lie between (1) and (3), between (5) and (7), . . . ; their number is therefore also $2m$. Apart from these $4m$ points there are no other points on the circumference for which T or $U = 0$.

Proof. I. At point (1) $m\phi = 45°$, and therefore

$$T = R^{m-1}\left(R\sqrt{1/2} + A \sin (m - 1)\phi + \frac{B}{R} \sin (m - 2)\phi + \text{etc.} + \frac{L}{R^{m-2}} \sin \phi\right).$$

The sum $A \sin (m - 1)\phi + (B/R) \sin (m - 2)\phi$ etc. certainly cannot be larger than S, and hence must be smaller than $R\sqrt{1/2}$; the value of T at this point is therefore certainly positive. Hence, a fortiori, T is positive when m lies between 45° and 135°, that is, T has always a positive value from point (1) to point (3). The same reasoning shows that T is everywhere positive from point (9) to point (11), and, generally speaking, from some point $(8k + 1)$ to point $(8k + 3)$, where k means any integer. In a similar way we see that T is negative everywhere between (5) and (7), between (13) and (15), etc., and, generally speaking, between $(8k + 5)$ and $(8k + 7)$, so that in all these intervals it can nowhere be $= 0$. But since at (3) the value is positive and at (5) negative, it must be $= 0$ somewhere between (3) and (5)[4] and in the same way between (11) and (13), etc. up to the interval between $(8m - 1)$ and (1) inclusive, so that together $T = 0$ at $2m$ points.

II. That there are, apart from these $2m$ points, no others of the same property can be seen in the following way. Since there are none between (1) and (3), between (5) and (7), etc., other such points would exist only if in one of the intervals from (3) to (5) or from (7) to (9), etc. there were at least two of them. In that case, however, T would have to be in the same interval at some point either a *maximum* or a *minimum*,[5] hence $dT/d\phi = 0$. But

$$\frac{dT}{d\phi} = mR^{m-2}\left(R \cos m\phi + \frac{m-1}{m}A \cos (m - 1)\phi + \text{etc.}\right)$$

and $\cos m\phi$ is always negative between (3) and (5) and [in value] $> \sqrt{1/2}$. From this we can easily see that $dT/d\phi$ is a negative quantity in this whole interval; and in the same way we see that it is everywhere positive between (7) and (9), everywhere negative between (11) and (13), etc. In none of these intervals therefore can it be $=$ 0, so that our assumption was wrong. Hence, etc.

[III. Here Gauss shows in the same way that $dU/d\phi$ cannot be 0 in the intervals (1) and (3), (5) and (7), etc., so that there are on the circumference of the circle no more than $2m$ points where $U = 0$.]

That part of the theorem which teaches that there are no more than $2m$ points at which $T = 0$, and no more than $2m$ points at which $U = 0$, can also be demonstrated by representing $T = 0$, $U = 0$ as curves of order n, which are intersected by a circle, being a curve of the second order, in no more than $2m$ points, as is stated in higher geometry.[6]

20. If another circle with radius larger than R is described around the same center and is divided in the same way, then here also there exists between the points (3) and (5) a single point at which $T = 0$, and similarly between (7) and (9), etc., and it can easily be seen that such points between (3) and (5) on both circumferences are the closer the less the radius of the larger circle differs from the radius R. The same also happens if the circle is described with a radius somewhat smaller than R, but still larger than $S\sqrt{2}$ and 1. From this we see that the circumference of the circle described with radius R is actually *intersected* by a branch of the first curve at that point between (3) and (5) where $T = 0$; the same holds for the other points where $T = 0$. It is also clear that the circumference of this circle is intersected by a branch of the second curve at all $2m$ points for which $U = 0$. These conclusions can also be expressed in the following way: If a circle of sufficient size is described around the center C, then $2m$ branches of the first curve and as many branches of the second curve enter into it, and in such a way that every two neighboring branches of the first curve are separated from each other by a branch of the second curve. See Fig. 2, where the circle is now no longer of infinite, but of finite magnitude; the numbers added to the separate branches should not be confused with the numbers by which I have denoted for short, in the previous and in this paragraph, certain limiting points on the circumference.

21. It is now possible to deduce from the relative position of the branches which enter into the circle that inside the circle there must be an intersection of a branch of the first curve with a branch of the second curve, and this can be done in so many ways that I hardly know which method is to be preferred to another. The following method seems to be the clearest: We indicate by O (Fig. 2) that point of the circumference of the circle in which it is intersected by the left-hand side of the axis (which itself is one of the $2m$ branches of the first curve); the next point, at which a branch of the second curve enters, by 1; the point next to this, at which a branch of the first curve enters, by 2, etc., up to $4m - 1$. At every point indicated by an even number, therefore, a branch of the second curve enters into the circle,

but a branch of the first curve at every point indicated by an odd number. Now it is known from higher geometry that every algebraic curve (or the single parts of an algebraic curve when it happens to consist of several parts) either runs into itself or runs out to infinity in both directions and that therefore, if a branch of an algebraic curve enters into a limited space, it necessarily has to leave it again.[7] From this we can easily conclude that every point indicated by an even number (or, for short, *every even point*) must be connected with another even point by a branch of the first curve inside the circle, and that in a similar way every point indicated by an odd number is connected with another similar point by a branch of the second curve. Although this connection of two points may be quite different because of the nature of the function X, so that it cannot in general be determined, yet it can easily be shown that, *whatever this connection may be, there will always be an intersection of the first with the second curve.*

22. The proof of this necessity can best be given in an indirect way [*apagogice*].[8] We shall assume that the connection of pairs of all even points and of pairs of all odd points can be arranged in such a way that no intersection results of a branch of the first curve with a branch of the second curve. Since the axis is a part of the first curve, point O will clearly be connected with point $2m$. The point 1 therefore cannot be connected with a point situated outside of the axis, that is, with no point indicated by a number larger than $2m$, since otherwise the connecting curve would necessarily intersect the axis. If therefore we suppose that 1 is connected with the point n, then n will be $< 2m$. By a similar reasoning we find that when 2 is connected with n', $n' < n$, since otherwise the branch $2 \cdots n'$ must necessarily intersect the branch $1 \cdots n$. Point 3, for the same reason, must be connected with a point situated between 4 and n', and it is clear that, if we suppose 3, 4, 5, . . . to be connected with n'', n''', n'''', . . . , n''' is situated between 5 and n'', n'''' between 6 and n''', etc. From this it follows that at last we come to a point h which is connected with the point $h + 2$. The branch which at point $h + 1$ enters into the circle must in this case intersect the branch connecting the points h and $h + 2$. But since the one of these two branches belongs to the first, the other to the second curve, it is clear that our assumption is contradictory, and that therefore there exists necessarily somewhere an intersection of the first with the second curve.

If we combine this result with the previous one, then we arrive from all the investigations explained above at the rigorous proof of the theorem that *every integral rational algebraic function of one variable can be decomposed into real factors of the first and second degree.*

[In the last two sections Gauss (1) observes that the same reasoning could have led to the conclusion that there exist at least m intersections of the first and second curve, (2) notes that the proof, here based on geometrical principles, could also have been presented in a purely analytical form, and (3) gives a short sketch of a different proof.

Gauss, during his lifetime, returned to the theorem more than once, and gave three more proofs. The last one, of 1849, took up again the ideas of the first one, but now using imaginaries. Gauss added that he avoided using them in 1799, but in 1849 it seemed to him no longer necessary.]

NOTES

1. See note 8 below.
2. [Footnote by Gauss] Fig. 4 is constructed assuming $X = x^4 - 2xx + 3x + 10$; so that

readers less familiar with general and abstract investigations can study in a concrete example how both curves are situated. The length of line $CG = 10$ ($CN = 1.26255$).

3. [Footnote by Gauss] For $S > \sqrt{1/2}$ the second condition is contained in the first one, for $S < \sqrt{1/2}$ the first condition in the second one.

4. This is one of the places where Gauss accepts on visual evidence a theorem that now requires proof.

5. This theorem is named after Michel Rolle (1652-1719), in whose *Méthode pour résoudre les égalitez* (Paris, 1691) it can be found without proof and without special emphasis. It appeared in other eighteenth-century works, as in Euler's *Institutiones calculi differentialis* (Saint Petersburg, 1755), sec. 298 (*Opera omnia,* ser. I, vol. 18, 503). See F. Cajori, *Bibliotheca Mathematica* (3d ser.) *11* (1910-11), 300-313.

6. Gauss refers to the theorem named after Etienne Bézout (1730-1783), but also announced by other authors of his time. It was only insufficiently proved in Gauss's day.

7. [Footnote by Gauss] It seems to be sufficiently well demonstrated that an algebraic curve can neither be suddenly interrupted (as e.g., occurs with the transcendental curve with equation $y = 1/\log x$), nor lose itself after an infinite number of terms (like the logarithmic spiral), and nobody, to my knowledge, has ever doubted it. But if anybody desires it, then on another occasion I intend to give a demonstration which will leave no doubt. Moreover, it is clear in the present case that if a branch, for instance 2, were nowhere to leave the circle (Fig. 3), one could enter the circle between O and 2, then go around this whole branch (which has to lose

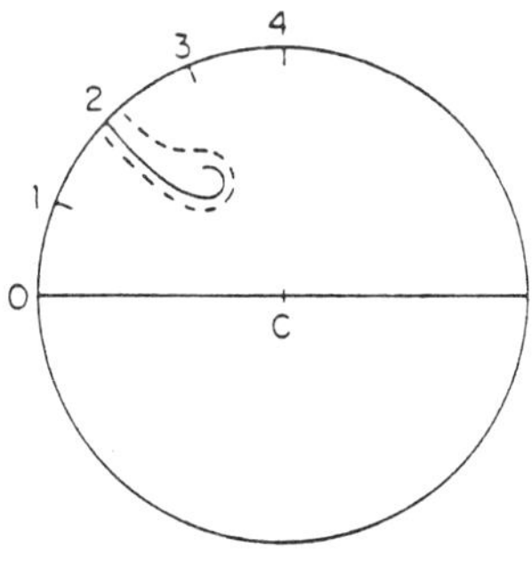

FIG. 3

itself in the space of the circle), and at last leave the circle again between 2 and 4, without meeting the first curve anywhere on the way. But this is patently absurd, since at the point at which you enter the circle you have the first surface above you, but where you leave the circle it is below you. Hence you would necessarily meet the first surface somewhere, and this at a point of the first curve.—From this reasoning, based on the principles of the geometry of position [*geometria situs*], which are no less valid than the principles of the geometry of magnitudes, it also follows that, if you enter the circle on a branch of the first curve, you can leave it at another point by always staying on the first curve, but it does not follow that the path is continuous in the sense accepted in higher geometry [see note 8]. It is here sufficient that the path be a continuous line in a general sense, that is, nowhere interrupted, but everywhere coherent.

8. Gauss may refer here to Euler's definition of a continuous curve in his *Introductio in analysin infinitorum* (Lausanne, 1748), vol. II, cap. 19; *Opera omnia,* ser. I, vol. 9, p. 11: A line is called a continuous curve if its nature is expressed by one definite function, where (vol. I, cap. 14; *Opera omnia,* ser. I, vol. 8, p. 18) a function is a variable quantity z, defined as an analytical expression composed of this variable quantity and constant numbers of quantities, such as $a + 3z$, $az + b\sqrt{(aa - zz)}$, c^z, etc.

On Euler's proof and Gauss's critique, see also A. Speiser's note in Euler, *Opera omnia.* ser. I, vol. 29 (1956), Einleitung, pp. VIII-X.

Chapter VIII
The Nineteenth Century

Section A
Algebra

NIELS (HENRIK) ABEL (1802–29)

Niels Abel, the Norwegian mathematician who worked to establish sound foundations of mathematics with an ingenuity and force of thought rarely equalled, was the son of Soren Georg Abel, a Lutheran pastor. A frail boy, Abel spend his youth in the southeast of Norway (a country that was outside the mainstream of European scientific scholarship), and was educated at home by his father until 1815, when he was sent to study at the Cathedral School in Christiania (Oslo). Under its tyrannical teachers, Abel's initial intellectual enthusiasm diminished until a new mathematics teacher, Bernt Michael Holmboe, discovered his extraordinary ability in 1817 and revived his intellectual interests. Under Holmboe's guidance, he first read the calculus texts of Leonhard Euler and later studied the writings of Joseph-Louis Lagrange, Pierre-Simon Laplace, and Isaac Newton. He was thoroughly challenged by these archetypal mathematicians and rapidly surpassed his teacher in mastering them and detecting errors in their work. In his last year at school, Abel attacked the problem of the solution of the quintic equation (the fifth degree polynomial), which was outstanding since the time of Tartaglia and Cardano in the 16th century.

Abel spent his life in dire financial straits. His father and mother suffered in later life from alcoholism. His father died in 1820, at the age of 48, leaving the family in near poverty. Nevertheless, Abel entered the university (1821) and was granted a free room. The mathematical ability of the impoverished student so impressed his professors that they set aside part of their own salaries to provide him with a cash grant. He received the preliminary degree, *Candidatus Philosophiae,* in 1822. The following year he published his first article (a study of functional equations) in the Norwegian periodical *Magazin fur Naturvidenskaberne*. He also wrote and had published at his own expense a six-page pamphlet dealing with the impossibility of solving the quintic equation. He hoped the pamphlet would gain him recognition. Copies were sent to foreign mathematicians, including Gauss who found it unintelligible. He and its other readers probably failed to recognize that the famous problem had in reality been settled because of its terse reasoning, several errata, and unfamiliar abbreviations.

In 1824, Abel applied to the Norwegian government for a travel grant and received a small stipend to learn the languages needed for study abroad and a modest two-year grant for expenses in Germany and France. Going first to Berlin (September

1825), he met a civil engineer, August-Leopold Crelle, who became a close friend and mentor. At the end of the year, he encouraged Crelle to found the *Journal of Pure and Applied Mathematics* and began publishing in its first volume. Among his notable first papers was one on the generalization of the binomial formula for real or complex exponents. When friends from Norway arrived in Berlin (spring 1826) he turned briefly to social activities, and parties at his room in a boarding house were so noisy they disturbed the philospher Hegel who was a fellow resident.

In the summer of 1826 Abel traveled to Paris, where he met some major mathematicians. His work was still unknown, and he was treated with restrained civility. An important memoir on transcendental functions, which he submitted to the Paris Academy of Sciences in October, was effectively lost for a time by the two referees appointed to report on it. The chairman, Cauchy, laid it aside to do his own research and misplaced it; the other referee, the elderly Legendre, forgot about the long, difficult, and almost illegible paper. Sometime after Christmas Abel returned to Berlin, where he suffered his first attack of tuberculosis.

Heavily in debt, Abel returned to Oslo, where he spent his final two years isolated from the European mathematical community. His prospects were gloomy; there was no academic position for him in Oslo in 1827. The university awarded him a meager stipend that he had to supplement by tutoring schoolboys. In 1828–29 his situation improved. He substituted at the university and the Norwegian Military Academy for the astronomer Hansteen, who had a research grant. In 1828, Abel frantically wrote several papers, principally on equation theory and elliptic functions. He realized that in the field of elliptic functions he had a strong competitor in the person of Karl Jacobi.

Abel traveled through the intense cold for several days to be with his fiancée, Christine Kemp, and his family at Froland estate to celebrate Christmas in 1828. He was feverish when he arrived, and suffered a violent hemorrhage while waiting for the sled to return him to Oslo after the holidays. Abel soon died of tuberculosis, he was only 26 years of age. Two days after his death, in April 1829, Crelle wrote jubilantly that he had secured a position for Abel at the University of Berlin.

As a result of Abel's work, mathematics rests on sounder foundations. In the theory of equations, he first proved the impossibility of solving the general quintic equation by a radical expression (Abel–Ruffini theorem). In extensive studies of elliptic and hyperelliptic integrals, he defined what is today called an Abelian integral. He proposed in 1826 a result known as Abel's theorem that underlay his theory of integrals. It broadly generalized Euler's addition theorem for elliptic integrals. Using inverse functions he was able to transform the theory of elliptic integrals into the theory of elliptic functions, showing them to be a natural generalization of Fourier's trigonometric functions. Abel also worked on the origins of algebraic number theory.

93. From A Memoir on Algebraic Equations, Proving the Impossibility of a Solution of the General Equation of the Fifth Degree (1824)*

(Equations of higher degree than four cannot be solved by root extractions, except for special values of the coefficients)

NIELS ABEL

The mathematicians have been very much absorbed with finding the general solution of algebraic equations, and several of them have tried to prove the impossibility of it. However, if I am not mistaken, they have not as yet succeeded. I therefore dare hope that the mathematicians will receive this memoir with good will, for its purpose is to fill this gap in the theory of algebraic equations.

Let

$$y^5 - ay^4 + by^3 - cy^2 + dy - e = 0$$

be the general equation of fifth degree and suppose that it can be solved algebraically,—*i.e.*, that y can be expressed as a function of the quantities a, b, c, d, and e, composed of radicals. In this case, it is clear that y can be written in the form

$$y = p + p_1R^{\frac{1}{m}} + p_2R^{\frac{2}{m}} + \ldots + p_{m-1}R^{\frac{m-1}{m}},$$

m being a prime number, and R, p, p_1, p_2, etc. being functions of the same form as y. We can continue in this way until we reach rational functions of a, b, c, d, and e. We may also assume that $R^{\frac{1}{m}}$ cannot be expressed as a rational function of a, b, etc., p, p_1, p_2, etc., and substituting $\frac{R}{p_1^m}$ for R, it is obvious that we can make $p_1 = 1$.

Then

$$y = p + R^{\frac{1}{m}} + P_2R^{\frac{2}{m}} + \ldots + p_{m-1}R^{\frac{m-1}{m}}$$

Substituting this value of y in the proposed equation, we obtain, on reducing, a result in the form

$$P = q + q_1R^{\frac{1}{m}} + q_2R^{\frac{2}{m}} + \ldots + q_{m-1}R^{\frac{m-1}{m}} = 0,$$

q, q_1, q_2, etc. being integral rational functions of a, b, c, d, e, p, p_2, etc. and R.

For this equation to be satisfied, it is necessary that $q = 0$, $q_1 = 0$, $q_2 = 0$, . . . $q_{m-1} = 0$. In fact, letting $z = R^{\frac{1}{m}}$, we have the two equations

$$z^m - R = 0, \text{ and } q + q_1z + \ldots + q_{m-1}z^{m-1} = 0.$$

If now the quantities q, q_1, etc. are not equal to zero, these equations must

*Source: This translation of *Mémoire sur les équations algebriques où l'on démontre l'impossibilité de la resolution de l'équation générale du cinquième degré* by W. H. Langdon with notes by Oystein Ore is taken from David Eugene Smith (ed.), *A Source Book in Mathematics* (1929), 261-266. It is reprinted by permission of McGraw-Hill Book Company, Inc. Footnotes renumbered.

necessarily have one or more common roots. If k is the number of these roots, we know that we can find an equation of degree k, whose roots are the k roots mentioned, and whose coefficients are rational functions of R, q, q_1, and q_{m-1}. Let this equation be

$$r + r_1 z + r_2 z^2 + \ldots + r_k z^k = 0.$$

It has all its roots in common with the equation $z^m - R = 0$; now all the roots of this equation are of the form $\alpha_\mu z$, α_μ being one of the roots of the equation $\alpha_\mu^m - 1 = 0$. On substituting, we obtain the following equations

$$r + r_1 z + r_2 z^2 + \ldots + r_k z^k = 0,$$
$$r + \alpha r_1 z + \alpha^2 r_2 z^2 + \ldots + \alpha^k r_k z^k = 0,$$
$$\ldots\ldots\ldots\ldots$$
$$r + \alpha_{k-2} r_1 z + \alpha_{k-2}{}^2 r_2 z^2 + \ldots + \alpha^k{}_{k-2} r_k z^k = 0.$$

From these k equations we can always find the value of z, expressed as a rational function of the quantities $r, r_1, \ldots r_k$; and as these quantities are themselves rational functions of $a, b, c, d, e, R, p, p_2, \ldots$, it follows that z is also a rational function of these latter quantities; but that is contrary to the hypotheses. Thus it is necessary that

$$q = 0, q_1 = 0, \ldots q_{m-1} = 0.$$

If now these equations are satisfied, it is clear that the proposed equation is satisfied by all those values which y assumes when $R^{\frac{1}{m}}$ is assigned the values

$$R^{\frac{1}{m}}, \alpha R^{\frac{1}{m}}, \alpha^2 R^{\frac{1}{m}}, \ldots, \alpha^{m-1} R^{\frac{1}{m}},$$

α being a root of the equation

$$\alpha^{m-1} + \alpha^{m-2} + \ldots + \alpha + 1 = 0.$$

We also note that all the values of y are different; for otherwise we should have an equation of the same form as the equation $P = 0$, and we have just seen that such an equation leads to a contradictory result. The number m cannot exceed 5. Letting y_1, y_2, y_3, y_4, and y_5 be the roots of the proposed equation, we have

$$y_1 = p + R^{\frac{1}{m}} + p_2 R^{\frac{2}{m}} + \ldots + p_{m-1} R^{\frac{m-1}{m}},$$
$$y_2 = p + \alpha R^{\frac{1}{m}} + \alpha^2 p R^{\frac{2}{m}} + \ldots + \alpha^{m-1} p_{m-1} R^{\frac{m-1}{m}},$$
$$\ldots\ldots\ldots\ldots$$
$$y_m = p + \alpha^{m-1} R^{\frac{1}{m}} + \alpha^{m-2} p_2 R^{\frac{2}{m}} + \ldots + \alpha p_{m-1} R^{\frac{m-1}{m}}.$$

Whence it is easily seen that

$$p = \frac{1}{m}(y_1 + y_2 + \ldots + y_m),$$
$$R^{\frac{1}{m}} = \frac{1}{m}(y_1 + \alpha^{m-1} y_2 + \ldots + \alpha y_m),$$
$$p_2 R^{\frac{2}{m}} = \frac{1}{m}(y_1 + \alpha^{m-2} y_2 + \ldots + \alpha^2 y_m),$$
$$\ldots\ldots\ldots\ldots$$
$$p_{m-1} R^{\frac{1}{m}} = \frac{1}{m}(y_1 + \alpha y_2 + \ldots + \alpha^{m-1} y_m).$$

Thus $p, p_2, \ldots p_{m-1}, R$, and $R^{\frac{1}{m}}$ are rational functions of the roots of the proposed equation.

Let us now consider any one of these quantities, say R. Let

$$R = S + v^{\frac{1}{n}} + S_2 v^{\frac{2}{n}} + \ldots + S_{n-1} v^{\frac{n-1}{n}}.$$

Treating this quantity as we have just treated y, we obtain the similar result that the quantities $S, S_2, \ldots, S_{n-1}, v$, and $v^{\frac{1}{n}}$ are rational functions of the different values of R; and since these are rational functions of y_1, y_2, etc., the functions $v^{\frac{1}{n}}, v, S, S_2$ etc. have the same property. Reasoning in this way, we conclude that all the irrational functions contained in the expression for y, are rational functions of the roots of the proposed equation.

This being established, it is not difficult to complete the demonstration. Let us first consider irrational functions of the form $R^{\frac{1}{m}}$, R being a rational function of a, b, c, d, and e. Let $R^{\frac{1}{m}} = r$. Then r is a rational function of y_1, y_2, y_3, y_4, and y_5, and R is a symmetric function of these quantities. Now as we are interested in the solution of the general equation of the fifth degree, it is clear that we can consider y_1, y_2, y_3, y_4, and y_5 as independent variables; thus the equation $R^{\frac{1}{m}} = r$ must be satisfied under this supposition. Consequently we can interchange the quantities y_1, y_2, y_3, y_4, and y_5 in the equation $R^{\frac{1}{m}} = r$; and, remarking that R is a symmetric function, $R^{\frac{1}{m}}$ takes on m different values by this interchange. Thus the function r must have the property of assuming m values, when the five variables which it contains are permuted in all possible ways. Thus either $m = 5$, or $m = 2$, since m is a prime number, (see the memoir by M. Cauchy in the *Journal de l'école polytechnique*, vol. 17).[1] Suppose that $m = 5$. Then the function r has five different values, and hence can be put in the form

$$R^{1/5} = r = p + p_1 y_1 + p_2 y_1^2 + p_3 y_1^3 + p_4 y_1^4,$$

p, p_1, p_2, ... being symmetric functions of y_1, y_2, etc. This equation gives, on interchanging y_1 and y_2,

$$p + p_1 y_1 + p_2 y_1^2 + p_3 y_1^3 + p_4 y_1^4 =$$
$$\alpha p + \alpha p_1 y_2 + \alpha p_2 y_2^2 + \alpha p_3 y_2^3 + \alpha p_4 y_2^4,$$

where

$$\alpha^4 + \alpha^3 + \alpha^2 + \alpha + 1 = 0.$$

But this equation (is impossible);[2] hence m must equal two. Then

$$R^{1/2} = r,$$

and so r must have two different values, of opposite sign. We then have,[3] (see the memoir of M. Cauchy),

$$R^{1/2} = r = v(y_1 - y_2)(y_1 - y_3) \ldots (y_2 - y_3) \ldots (y_4 - y_5) = vS^{1/2}.$$

v being a symmetric function.

Let us now consider irrational functions of the form

$$(p + p_1 R^{\frac{1}{\nu}} + p_2 R_1^{\frac{1}{\mu}} + \ldots)^{\frac{1}{m}},$$

p, p_1, p_2, etc., R, E_1, etc., being rational functions of a, b, c, d, and e, and consequently symmetric functions of y_1, y_2, y_3, y_4, and y_5. We have seen that it is necessary that $\nu = \mu = \ldots = 2$, $R = v^2 S$, $R_1 = v_1^2 S$, etc. The preceding function can thus be written in the form

$$(p + p_1 S^{\frac{1}{2}})^{\frac{1}{m}},$$

Let

$$r = (p + p_1 S^{\frac{1}{2}})^{\frac{1}{m}}$$
$$r_1 = (p^2 - p_1 S^{\frac{1}{2}})^{\frac{1}{m}}.$$

Multiplying, we have

$$rr_1 = (p^2 - p_1^2 S)^{\frac{1}{m}}.$$

If now rr_1 is not a symmetric function, m must equal two; but then r would have four different values, which is impossible; hence rr_1 must be a symmetric function. Let v be this function, then

$$r + r_1 = (p + p_1 S^{\frac{1}{2}})^{\frac{1}{m}} + v(p + p_1 S^{\frac{1}{2}})^{-\frac{1}{m}} = z.$$

This function having m different values, m must equal five, since m is a prime number. We thus have

$$z = q + q_1 y + q_2 y^2 + q_3 y^3 + q_4 y^4$$
$$= (p + p_1 S^{1/2})^{1/5} + v(p + p_1 S^{1/2})^{-1/5},$$

q, q_1, q_2, etc. being symmetric functions of y_1, y_2, y_3, etc., and consequently rational functions of a, b, c, d, and e. Combining this equation with the proposed equation, we can find y expressed as a rational function of z, a, b, c, d, and e. Now such a function can always be reduced to the form

$$y = P + R^{1/5} + P_2 R^{2/5} + P_3 R^{3/5} + P_4 R^{4/5},$$

where P, R, P_2, P_3, and P_4 are functions of the form $p + p_1 S^{1/2}$, where p, p_1, and S are rational functions of a, b, c, d, and e. From this value of y we obtain

$$R^{1/5} = \tfrac{1}{5}(y_1 + \alpha^4 y_2 + \alpha^3 y_3 + \alpha^2 y_4 + \alpha y_5)$$
$$= (p + p_1 S^{1/2})^{1/5},$$

where

$$\alpha^4 + \alpha^3 + \alpha^2 + \alpha + 1 = 0.$$

Now the first member has 120 different values, while the second member has only 10; hence y can not have the form that we have found: but we have proved that y must necessarily have this form, if the proposed equation can be solved: hence we conclude that

It is impossible to solve the general equation of the fifth degree in terms of radicals.

It follows immediately from this theorem, that it is also impossible to solve the general equations of degrees higher than the fifth, in terms of radicals.

NOTES

1. ["Mémoire sur le nombre des valeurs qu'une fonction peut acquérir," etc.

Let p be the greatest prime dividing n. Cauchy then proves (p. 9) that a function of n variables, taking less than p values, either is symmetric or takes only two values. In the latter case the function can be written in the form $A + B\Delta$ where A and B are symmetric, and Δ is the special two-valued function

$$\Delta = (y_1 - y_2)(y_1 - y_3) \ldots (y_{n-1} - y_n).]$$

2. [In a later paper (*Journal für die reine und angewandte Mathematik* Vol. 1, 1826) Abel gives a more detailed proof of the main theorem, based on the same principles. At the corresponding point he gives the following more elaborate proof. By considering y_1 as a common root of the given equation, the relation defining R, y_1 can be expressed in the form

$$y_1 = s_0 + s_1 R^{1/5} + s_2 R^{2/5} + s_3 R^{3/5} + s_4 R^{4/5}.$$

Substituting $\alpha^t R^{1/5}$ for R we obtain the other roots of the equation, and solving the corresponding system of five linear equations gives

$$s_1 R^{1/5} = \tfrac{1}{5}(y_1 + \alpha^4 y_2 + \alpha^3 y_3 + \alpha^2 y_4 + \alpha y_5).$$

This identity is impossible, however, since the right-hand side has 120 values, and the left-hand side has only 5.]

3. [Compare 2.]

Chapter VIII
The Nineteenth Century

Section A
Algebra

ÉVARISTE GALOIS (1811–32)

Evariste Galois, who laid the foundations for modern algebra, had a brief, tormented life. He was the son of Nicolas-Gabriel Galois, an imaginative and liberal thinker, who was elected mayor of Bourg-la-Reine, a suburb of Paris, during Napoleon's "Hundred Days" Regime in 1815. Galois' mother was the headstrong and eccentric Adelaid-Marie Demante, who came from a family of distinguished jurists. She educated her son in classical culture, the principles of austere religion, and Stoic morality. His childhood seems to have been happy and studious.

In 1823, Galois entered Collège Louis-le-Grand in Paris and was troubled by that school's harsh discipline and elementary, uninspiring character of instruction. Finding many textbooks inadequate, he turned to reading original works. In 1827, his mathematical ability suddenly appeared when he read Legendre's *Géométrie* (whose rigor appealed to him) and acquired a firm grounding in the writings of Louis Lagrange on algebra. Guided by one of his more perceptive teachers, Louis Richard, he carefully studied recent works on the theory of equations, the theory of numbers, and elliptic functions in 1828. During that year he wrongly believed, like Niels Abel earlier, that he had solved the general fifth degree (quintic) equation. He had begun his pursuit of "Galois theory" with its deeper understanding of the basic conditions that an equation must satisfy to be solvable by radicals.

Disappointment and tragedy filled Galois' final years. In July 1829, his father, who had been persecuted for liberal opinions, committed suicide. A month later, Galois twice failed the entrance examination for the prestigious École Polytechnique, possibly because he refused to follow the method of exposition required by the examiner. In November, however, he gained admission to the École Normale Supérieure, which trained secondary school teachers. Shortly thereafter he learned of Abel's recent death and became acquainted with his writings. Galois saw that Abel's last publication contained many results that he had found independently and put in a memoir on group theory submitted to the Paris Academy of Sciences. Augustin Cauchy, who was appointed to review the memoir, counseled Galois to revise it in light of Abel's results. Galois thus composed for the Academy a second memoir on algebraic functions, which he submitted in February 1830 for the *grand prix* in mathematics. Unfortunately the second memoir was lost upon the death of its examiner, Joseph Fourier, and Galois was brusquely eliminated from the prize competition.

Galois continued his research during 1830 but turned increasingly to political activism. The reactionary policies of Charles X, especially the four laws attempting to curb freedom of the press and to limit further the suffrage, had brought political agitation in Paris. Galois became a friend of several republican leaders, particularly Louis-Auguste Blanqui. The July Revolution of 1830 toppled Charles X but, to the dismay of many republicans and students, a new citizen-king, Louis Philippe, ascended the throne. Opposed to the strict discipline meted out at his school under the new monarch, Galois wrote an article criticizing its director and was expelled in December 1830.

After 1830, Galois devoted himself chiefly to political propaganda. Royalists detested his republican views, and the police arrested him following a regicide toast given at a republican banquet in May 1831. A court acquitted him in June, however; and throughout the turmoil he was still able to pursue some mathematical research. At the request of Siméon-Denis Poisson, he hastily wrote a third memoir (dealing with the resolution of algebraic equations) for the Paris Academy. In July, however, Poisson hinted that some of the memoir's results were in Abel's posthumous publications and the rest were incomprehensible. Galois' rebellion deepened. Arrested again at the Bastille Day celebration (July 14) in 1831, he was detained at the prison of Sainte Pélagie. It was a painful experience. During the terrible cholera epidemic in March 1832, he was transferred to a nursing home, where he resumed his research and even fell in love with a young woman named Stéphanie.

The romance ended unhappily on May 14, and Galois was provoked into a duel shortly thereafter. The circumstances are unclear. It may have resulted from a quarrel over Stéphanie or been the work of a police agent. Galois, sensing that his death was near, made preparations. Feverishly, he wrote a mathematical last testament and addressed it to his schoolmate Auguste Chevalier. The testamentary letter, which sketched his principal results, was intended for Gauss and Jacobi. The duel took place on May 30; Galois was mortally wounded and died the next day. (In his autobiography, Alexandre Dumas revealed the other duelist's identity.) Galois' funeral on June 2 heralded the bloody Paris riots that were soon to occur.

Galois' name is perpetuated through the terms Galois field, Galois group, and Galois theory—all among the chief topics in modern algebra. He identified three key elements for a group: associativity $[r \times (s \times t) = (r \times s) \times t]$, identity element $[r \times I = I \times r = r]$, and the inverse $[r \times r^{-1} = r^{-1} \times r = I]$. He introduced the notions of a normal (*i.e.,* invariant or subconjugate) subgroup and of simple and composite groups. (By definition a group that has no invariant subgroup is simple.) The work of Lagrange and Gauss on the solvability of certain types of algebraic equations inspired his early research. Like Abel and Jacobi, he moved toward the theory of elliptic functions and selected types of integrals, especially what are today called Abelian integrals. Forty years later, Bernhard Riemann fully developed the theory of algebraic functions.

Galois' writing style was clear and terse, his thought highly original. Joseph Liouville published a collection of his manuscripts with annotations in the *Journal de Mathematiques* in 1846, while Camille Jordan published the first full treatment of Galois' theory in *Traité des substitutions et des équations algébriques* (1870).

94. From the Testamentary Letter Sent to Auguste Chevalier (May 29, 1832)*

(Group Theory and Abelian Integrals)

ÉVARISTE GALOIS

My dear friend,

I have made some new discoveries in analysis.

Some are concerned with the theory of equations; others with integral functions.

In the theory of equations, I have sought to discover the conditions under which equations are solvable by radicals, and this has given me the opportunity to study the theory and to describe all possible transformations on an equation even when it is not solvable by radicals.

It will be possible to make three memoirs of all this.

The first is written, and, despite what Poisson has said of it, I am keeping it, with the corrections I have made.

The second contains some interesting applications of the theory of equations. The following is a summary of the most important of these:

1. From propositions II and III of the first memoir, we perceive a great difference between adjoining to an equation one of the roots of an auxiliary equation and adjoining all of them.

In both cases the group of the equation breaks up by the adjunction in sets such that one passes from one to the other by the same substitution, but the condition that these sets have the same substitutions holds with certainty only in the second case. This is called the *proper decomposition.*[1]

In other words, when a group G contains another H, the group G can be divided into sets, each of which is obtained by multiplying the permutations of H by the same substitution; so that

$$G = H + HS + HS' + \ldots$$

And it can also be divided into sets which contain the same substitutions, so that

$$G = H + TH + T'H + \ldots$$

These two methods of decomposition are usually not identical. When they are identical, the decomposition is *proper.*

It is easy to see that when the group of an equation is not susceptible of any proper decomposition, then, however, the equation be transformed, the groups of the transformed equations will always have the same number of permutations.

On the other hand, when the group of an equation admits a proper decomposition, in which it has been separated into M groups of N permutations, then we can solve the given equation by means of two equations, one having a group of M permutations, the other N.

When therefore we have exhausted in the group of an equation all the possible proper decompositions, we shall arrive at groups which can be transformed, but

Source: This translation by Louis Weiner is taken from David Eugene Smith (ed.), *A Source Book in Mathematics* (1929), 278-285. It is reprinted by permission of McGraw-Hill Book Company, Inc. Footnotes are renumbered.

whose permutations will always be the same in number.

If each of these groups has a prime number of permutations, the equation will be solvable by radicals; otherwise not.

The smallest number of permutations which an indecomposable group can have, when this number is not a prime, is 5 . 4 . 3.

2. The simplest decompositions are those which occur in the method of M. Gauss.

As these decompositions are obvious, even in the actual form of the group of the equation, it is useless to spend time on this matter.

What decompositions are practicable in an equation which is not simplified by the method of M. Gauss?

I have called those equations *primitive* which cannot be simplified by M. Gauss's method; not that the equations are really indecomposable, as they can even be solved by radicals.

As a lemma in the theory of primitive equations solvable by radicals, I made in June 1830, in the *Bulletin de Férussac* an analysis of imaginaries in the theory of numbers.

There will be found herewith[2] the proof of the following theorems:

1. In order that a primitive equation be solvable by radicals its degree must be p^ν, p being a prime.

2. All the permutations of such an equation have the form

$$\chi_{k,l,m,\ldots}$$
$$\Big| \chi_{ak+bl+cm+\cdots+h,\; a'k+b'l+c'm+\cdots+h',\; a''k+\cdots},$$

$k, l, m, \ldots$ being ν indices, which, taking p values each, denote all the roots. The indices are taken with respect to a modulus p; that is to say, the root will be the same if we add a multiple of p to one of the indices.

The group which is obtained on applying all the substitutions of this linear form contains in all

$$p^\nu(p^\nu - 1)(p^\nu - p) \ldots (p^\nu - p^{\nu-1})$$

permutations.

It happens that in general the equations to which they belong are not solvable by radicals.

The condition which I have stated in the *Bulletin de Férussac* for the solvability of the equation by radicals is too restricted; there are few exceptions, but they exist.[3]

The last application of the theory of equations is relative to the modular equations of elliptic functions.

We know that the group of the equation which has for its roots the sines of the amplitude[4] of the $p^3 - 1$ divisions of a period is the following:

$$\chi_{k,l},\ \chi_{ak+bl,\, ck+dl};$$

consequently the corresponding modular equation has for its group

$$\chi_{\frac{k}{l}},\ \chi_{\frac{ak+bl}{ck+dl}},$$

in which $\frac{k}{l}$ may have the $p + 1$ values

$$\infty, 0, 1, 2, \ldots, p - 1.$$

Thus, by agreeing that k may be infinite, we may write simply

$$\chi_k,\ \chi_{\frac{ak+b}{ck+d}}.$$

By giving to a, b, c, d all the values, we obtain

$$(p + 1)p(p - 1)$$

permutations.

Now this group decomposes *properly* in two sets, whose substitutions are

$$\chi_k,\ \chi_{\frac{ak+b}{ck+d}},$$

$ad - bc$ being a quadratic residue of p.

The group thus simplified has $\frac{(p + 1)p(p - 1)}{2}$ permutations.

But it is easy to see that it is not further properly decomposable, unless $p = 2$ or $p = 3$.

Thus, in whatever manner we transform the equation, its group will always have the same number of substitutions.

But it is interesting to know whether the degree can be lowered.

First, it cannot be made less than p, as an equation of degree less than p cannot have p as a factor of the number of permutations of its group.

Let us see then whether the equation of degree $p + 1$, whose roots are denoted by x_k on giving k all its values, including infinity, and has for its group of substitutions

$$x_k, \; x_{\frac{ak+b}{ck+d}}$$

$ad - bc$ being a square, can be lowered to degree p.

Now this can happen only if the group decomposes (improperly, of course) in p sets of $\frac{(p+1)(p-1)}{2}$ permutations each.

Let 0 and ∞ be two conjoint letters of one of these groups. The substitutions which do not change 0 and ∞ are of the form

$$x_k, \; x_{m^2k}.$$

Therefore if M is the letter conjoint to 1, the letter conjoint to m^2 will be m^2M. When M is a square, we shall have $M^2 = 1$. But this simplification can be affected only for $p = 5$.

For $p = 7$ we find a group of $\frac{(p+1)(p-1)}{2}$ permutations, where

$$\infty, 1, 2, 4$$

have respectively the conjoints

$$0, 3, 6, 5.$$

The substitutions of this group are of the form

$$x_k, \; x_{a\frac{(k-b)}{k-c}},$$

b being the letter conjoint to c, and a a letter which is a residue or a non-residue simultaneously with c.

For $p = 11$, the same substitutions will occur with the same notations,

$$\infty, 1, 3, 4, 5, 9,$$

having respectively for conjoints

$$0, 2, 6, 8, 10, 7.$$

Thus for the cases $p = 5, 7, 11$, the modular equation can be reduced to degree p.

In all rigor, this equation is not possible in the higher cases.

The third memoir concerns integrals.

We know that a sum of terms of the same elliptic function[5] always reduces to a single term, plus algebraic or logarithmic quantities.

There are no other functions having this property.

But absolutely analogous properties are furnished by all integrals of algebraic functions.

We treat at one time every integral whose differential is a function of a variable and of the same irrational function of the variable, whether this irrationality is or is not a radical, or whether it is expressible or not expressible by means of radicals.

We find that the number of distinct periods of the most general integral relative to a given irrationality is always an even number.

If $2n$ is this number, we have the following theorem:

Any sum of terms whatever reduces to n terms plus algebraic and logarithmic quantities.

The functions of the first species are those for which the algebraic and logarithmic parts are zero.

There are n distinct functions of the first species.

The functions of the second species are those for which the complementary part is purely algebraic.

There are n distinct functions of the second species.[6]

We may suppose that the differentials of the other functions are never infinite except once for $x = a$, and moreover, that their complementary part reduces to a single logarithm, log P, P being an algebraic quantity. Denoting these functions by $\pi(x, a)$, we have the theorem

$$\pi(x, a) - \pi(a, x) = \Sigma\phi a \cdot \psi x,$$

$\phi(a)$ and $\psi(x)$ being functions of the first and of the second species.

We infer, calling $\pi(a)$ and ψ the periods of $\pi(x, a)$ and ψx relative to the same variation of x,

$$\pi(a) = \Sigma\psi \times \phi a.$$

Thus the periods of the functions of the third species are always expressible in terms of the first and second species.

We can also deduce theorems analogous to the theorem of Legendre.[7]

$$FE' + EF' - FF' = \frac{\pi}{2}.$$

The reduction of functions of the third species to definite integrals, which is the most beautiful discovery of M. Jacobi, is not practicable, except in the case of elliptic functions.

The multiplication of integral functions by a whole number is always possible, as is the addition, by means of an equation of degree n whose roots are the values to substitute in the integral to obtain the reduced terms.[8]

The equation which gives the division of the periods in p equal parts is of degree $p^{2n} - 1$. Its group contains in all $(p^{2n} - 1)(p^{2n} - p) \ldots (p^{2n} - p^{2n-1})$ permutations.

The equation which gives the division of a sum of n terms in p equal parts is of degree p^{2n}. It is solvable by radicals.

Concerning the Transformation.—First, by reasoning analogous to that which Abel has indicated in his last memoir, we can show that if, in a given relation among integrals, we have the two functions

$$\int \Phi(x, X)dx, \int \Psi(y, Y)dy,$$

the last integral having $2n$ periods, it will be permissible to suppose that y and Y can be expressed by means of a single equation of degree n in terms of x and X.

Then we may suppose that the transformations are constantly made for two integrals only, since one has evidently, in taking any rational function whatever of y and Y,

$$\Sigma\int f(y, Y)dy = \int F(x, X)dx$$
$$+ \text{ an algebraic and logarithmic quantity.}$$

There are in this equation obvious reductions in the case where the integrals of the two members do not both have the same number of periods.

Thus we have only to compare those integrals both of which have the same number of periods.

We shall prove that the smallest degree of irrationality of two like integrals cannot be greater for one than for the other.

We shall show subsequently that one may always transform a given integral into another in which one period of the first is divided by the prime number p, and the other $2n - 1$ remain the same.

It will only remain therefore to compare integrals which have the same periods, and such consequently, for which n terms of the one can be expressed without any other equation than a single one of degree n, by means of two of the others, and reciprocally. We know nothing about this.

You know, my dear Auguste, that these subjects are not the only ones I have explored. My reflections, for some time, have been directed principally to the application of the theory of ambiguity to transcendental analysis.[9] It is desired to see *a priori* in a relation among quantities or transcendental functions, what transformations one may make, what quantities one may substitute for the given quantities, without the relation ceasing to be valid. This enables us to recognize at once the impossibility of many expressions which we might seek. But I have no time, and my ideas are not developed in this field, which is immense.

Print this letter in the *Revue encyclopédique*.

I have often in my life ventured to advance propositions of which I was uncertain; but all that I have written here has been in my head nearly a year, and it is too much to my interest not to deceive myself that I have been suspected of announcing theorems of which I had not the complete demonstration.

Ask Jacobi or Gauss publicly to give their opinion, not as to the truth, but as to the importance of the theorems.

Subsequently there will be, I hope, some people who will find it to their profit to decipher all this mess.

Je t'embrasse avec effusion.

NOTES

1. A proper decomposition, in modern parlance, is an arrangement of the permutations of a group into cosets with respect to an invariant subgroup.

2. Liouville remarks: "Galois speaks of manuscripts, hitherto unpublished, which we shall publish."

3. Galois stated in the *Bulletin des sciences mathématiques de M. Férussac* (1830), p. 271, that the elliptic modular equation of degree $p + 1$ could not be reduced to one of degree p when p exceeds 5; but $p = 7$ and $p = 11$ are exceptions to this statement, as Galois shows in the next page of his letter.

4. Meaning the elliptic *sn*-function.

5. Galois presumably means a sum of elliptic integrals of the same species.

6. Picard comments: "We thus acquire the conviction that he (Galois) had in his possession the most essential results concerning Abelian integrals which Riemann was to obtain twenty-five years later."

7. According to Tannery, who collated Galois's manuscripts with Liouville's publication of Galois's Works, Galois wrote Legendre's theorem in the form: $E'F'' - E''F' = \frac{\pi}{2}\sqrt{-1}$. Liouville made other alterations of a minor character.

8. Obscure.

9. Picard comments: "We could almost guess what he means by this, and in this field, which, as he says, is immense, there still to this day remain discoveries to make."

Chapter VIII
The Nineteenth Century

Section A
Algebra

WILLIAM ROWAN HAMILTON (1805–65)

William Rowan Hamilton was the fourth child of a Dublin solicitor named Archibald Rowan Hamilton and his wife Sara Hutton. Like his contemporaries Thomas Macauley and John Stuart Mill, the young Hamilton was a precocious child, who excelled in languages, classics, and mathematics. Before the age of three, when he was sent to live with his paternal uncle James, he easily read English and did arithmetical computations. By the age of five, he was able to translate Latin, Greek, and Hebrew, and recite Homer, Milton, and Dryden. His father boasted that by the end of his ninth year Hamilton had also mastered Arabic, Bengali, Chaldee, German, Malay, Persian, Sanskrit, and Syriac. The choice of languages may in part reflect the father's desire to obtain a clerkship for the son in the East India Company. Hamilton studied languages intensely through age 14.

In 1820, Hamilton became deeply interested in mathematics after a second public competition with Zerah Colburn, the American "calculating boy" who computed mentally with astonishing speed. He then read a Latin copy of Euclid and within two years studied Clairaut's *Elements d'algèbre* and Newton's *Arithmetica Universalis* and *Principia Mathematica*, which aroused his interest in astronomy. In 1822, at the age of 17, he immersed himself in mathematics, writing papers on properties of curves and surfaces and discovering an error in Laplace's *Mécanique céleste*. A friend sent Hamilton's memoir on the discovery to John Brinkley, the royal astronomer of Ireland. After reading it Brinkley is supposed to have said of its author: "This young man, I do not say will be, but is, the first mathematician of his age."

In 1823, as expected, Hamilton ranked first in a field of 100 candidates for entrance into Trinity College, Dublin. He distinguished himself as a student by taking highest honors in both mathematics and English verse—an unprecedented feat. The high quality of mathematical education at Trinity was mainly the result of Professor Bartholomew Lloyd's reform, which introduced French textbooks to keep students abreast of Continental methods. In 1827, while still an undergraduate, Hamilton was appointed Andrew professor of astronomy at Trinity (chosen over some other well-qualified candidates, including George Airy). At the same time, Hamilton received a second position, succeeding Brinkley as royal astronomer at Dunsink Observatory. While he excelled in theory, he failed as an observational astronomer, perhaps because he lacked instrumental and technical training.

Hamilton's colleagues knew him as a genial man of exceptional intellect and broad scholarly interests. His candid disposition, good sense of humor, and eloquence appealed to many. His lectures attracted large audiences because of their literary merit, but his poetry was so poor that Wordsworth, who was a friend, urged him to confine himself to writing mathematics! Hamilton was fond of reading Plato and Kant. Coleridge had introduced him to Kant's writings, which in turn led him toward philosophical idealism. In the theory of matter he accepted the point–atomism of the Yugoslav scientist Rudjer Bŏsović.

Hamilton devoted much time and energy to building scientific organizations. After joining the Royal Irish Academy in 1835, he served as its president from 1837 to 1845. The chief organizer in Dublin of the British Association for the Advancement of Science, he brought its annual meeting there in 1835. He was knighted at that meeting, and the Royal Society twice awarded him the Royal (Gold) Medal for his optical and dynamical research. The St. Petersburg Academy of Science named him a corresponding member, and the newly-founded National Academy of Sciences in the United States placed him at the head of 14 foreign associates in 1863.

In his personal, domestic life Hamilton was not fortunate. Catherine Disney, whom he always loved, refused his proposal of marriage in 1825; so did Ellen DeVere in 1831. In 1833, he married Helen Bayly, who suffered from continued ill health, and they had two sons and a daughter. The sickly Helen ran a poor household; Hamilton rarely had regular meals. In 1853, when Catherine Disney neared death, he grew despondent. In his later years he struggled against alcoholism.

Hamilton contributed significantly to optics, dynamics, and, most of all, the algebra of quaternions. He first developed fundamental ideas in geometrical optics by employing the characteristic function, which involves the action of a system moving from its initial to its final point in space. He relied particularly on Fermat's principle of least time. Seeking the greatest generality, he extended the approach in 1833 and 1834 to provide a general method of dynamics. His general method not only unified optics and dynamics but also reduced dynamics to problems in the calculus of variations. His research included a detailed study of the three-body problem in astronomy. His new method was not fully appreciated for many years largely because of the novel, abstract, and obscure nature of his writings. The Hamiltonian method came into its own only with the rise of quantum mechanics.

Hamilton's landmark discovery of quaternions came in 1843. They are ordered sets of four ordinary numbers satisfying special laws of addition, multiplication, and equality but freeing algebra from the commutative law of multiplication (*i.e.*, $a \times b = b \times a$). The discovery followed a decade of patient and systematic research on algebra. Inspired by Kant's *Critique of Pure Reason,* Hamilton considered geometry to be the "science of pure space" and algebra to be based on the intuition of mathematical time. In a pioneer attempt to develop an axiomatic basis for algebra comparable to that of geometry, he developed a rigorous theory for complex numbers (*i.e.*, numbers of the form $a + bi$ where i is $\sqrt{-1}$). This allowed him to handle the two dimensional plane but not three-dimensional space. After examining triplets, he suddenly realized, while walking by Brougham Bridge in Dublin in 1843, that he needed quadruplets. The discovery so excited Hamilton that he carved the fundamental formulas in the bridge's stonework:

$$i^2 = j^2 = k^2 = i\,j\,k = -1.$$

Here was a three dimensional analogue of complex numbers to represent vectors in space, and he spent the final 22 years of his life developing and applying them. His results appeared in *Lectures on Quaternions* (1853) and the two-volume, posthumously published *Elements of Quaternions* (1866).

95. From *Elements of Quaternions* (1866)*

(On Quaternions, a Generalization of Complex Numbers)

WILLIAM ROWAN HAMILTON

108. Already we may see grounds for the application of the *name* QUATERNION, to such a *Quotient of two Vectors* as has been spoken of in recent articles. In the first place, such a quotient cannot *generally* be what we have called a SCALAR: or in other words, it cannot generally be equal to any of the (so-called) *reals of algebra,* whether of the *positive* or of the *negative* kind. For let x denote any such (actual)[1] scalar, and let α denote any (actual) vector; then we have seen that the product $x\alpha$ denotes *another* (actual) *vector,* say β', which is either *similar* or *opposite* in direction to α, according as the scalar coefficient, or *factor,* x, is positive or negative; in *neither* case, then, can it represent any vector, such as β, which is *inclined* to α, at any actual *angle,* whether acute, or right, or obtuse: or in other words, the *equation* $\beta' = \beta$, or $x\alpha = \beta$, is impossible, under the conditions here supposed. But we have agreed to write, as in algebra, $(x\alpha)/\alpha = x$; we must therefore[2] . . . *abstain* from writing *also* $\beta/\alpha = x$, under the same conditions: x still denoting a *scalar.* Whatever *else* a *quotient of two inclined vectors* may be found to be, it is thus, at least, a NON-SCALAR.

109. Now, in forming the conception of the *scalar itself,* as the *quotient of two parallel*[3] *vectors,* we took into account not only *relative length,* or *ratio* of the usual kind, but also *relative direction,* under the form of *similarity or opposition.* In passing from α to $x\alpha$, we *altered* generally the *length* of the line α, in the ratio of $\pm x$ to 1; and we *preserved or reversed* the *direction* of that line, according as the *scalar coefficient* x was *positive* or *negative,* and, in like manner, in proceeding to form, more definitely than we have yet done, the conception of the *non-scalar quotient,* $q = \beta:\alpha =$ OB:OA, of *two inclined vectors,* which for simplicity may be supposed to be *co-initial,* we have *still* to take account both of the *relative length* and of the *relative direction,* of the two lines compared. But while the *former* element of the *complex relation* here considered, between these two lines or vectors, is *still* represented by a simple RATIO (of the kind commonly considered in geometry), or by a *number*[4] expressing that ratio; the *latter element* of

**Source:* This selection edited by Marguerite D. Darkow is taken from David Eugene Smith (ed.), *A Source Book in Mathematics* (1929), 677-683. It is reprinted by permission of McGraw-Hill Book Company, Inc. Footnotes are renumbered.

the same complex relation is *now* represented by an ANGLE, AOB: and not simply (as it was before) by an *algebraical sign* + or −.

110. Again, in estimating this *angle*, for the purpose of *distinguishing* one quotient of vectors from another, we must consider not only its *magnitude* (or *quantity*), but also its PLANE: since otherwise, in violation of the principle[5] . . . , we should have $OB':OA = OB:OA$, if OB and OB' were *two distinct rays* or sides of a *cone* of revolution, with OA for its *axis;* in which case . . . they would necessarily be *unequal vectors*. For a similar reason, we must attend also to the *contrast* between two *opposite angles,* of equal magnitudes, and in one *common plane*. In short, for the purpose of knowing *fully* the *relative direction* of two co-initial lines OA, OB in *space,* we ought to know not only *how many degrees* . . . the *angle* AOB contains; but also . . . the *direction of the rotation* from OA to OB: including a knowledge of the *plane, in which* the rotation is performed; and of the *band* (as *right* or *left,* when *viewed* from a known *side* of the plane), *towards which* the rotation is *directed*.

111. Or, if we agree to *select* some one *fixed band* (suppose the *right* hand), and to call all *rotations positive* when they are directed towards *this* selected hand, but all rotations *negative* when they are directed towards the *other band,* then, for *any given angle* AOB, supposed for simplicity to be less than two right angles, and considered as representing a *rotation in a given plane* from OA to OB, we may speak of *one perpendicular* OC *to that plane* AOB as being the *positive axis* of that rotation; and of the *opposite perpendicular* OC′ to the same plane as being the *negative axis* thereof: the rotation around the positive axis being *itself* positive, and *vice-versa*. And then the *rotation* AOB may be considered to be entirely *known,* if we know, first, its *quantity,* or the *ratio* which it bears to a *right rotation;* and second, the *direction* of its *positive axis,* OC, but not without knowledge of these *two* things, or of some data equivalent to them. But whether we consider the *direction of an* AXIS, or the *aspect of a* PLANE, we find (as indeed is well known) that the *determination* of such a *direction,* or of such an *aspect,* depend on TWO *polar coordinates,* or other *angular elements*.

112. It appears, then, from the foregoing discussion, that *for the complete determination,* of what we have called the *geometrical* QUOTIENT *of two coinitial Vectors,* a *System of Four Elements,* admitting each separately of numerical expression, *is generally required*. Of these four elements, *one* serves to determine the *relative length* of the two lines compared; and the other *three* are in general necessary, in order to determine *fully* their *relative direction*. Again, of these three latter elements, *one* represents the mutual *inclination,* or *elongation,* of the two lines; or the *magnitude* (or quantity) of the *angle* between them; while the *two others* serve to determine the *direction* of the *axis,* perpendicular to their common *plane,* round which a *rotation* through that angle is to be performed, in a *sense* previously selected as the *positive* one (or towards a fixed and previously selected *band*), for the purpose of *passing* (in the simplest way, and therefore in the plane of the two lines) from the *direction* of the *divisor-line, to* the direction of the *dividend-line*. And *no more than four* numerical *elements* are necessary for our present purpose: because the *relative length* of two lines is not changed when their lengths are altered proportionally, nor is their *relative direction* changed, when the *angle* which they form is merely *turned* about, *in its own plane*. On account, then, of this *essential connexion* of that *complex relation* between two lines, which is *compounded* of a *relation of lengths,* and of a *relation of directions,* and to which we have given (by an *extension* from the theory of *scalars*), the name of a *geometrical quotient,* with a *System of*

FOUR *numerical Elements,* we have already a *motive* for saying that *"The Quotient of two Vectors is generally a Quaternion".*[6]
[*Text omitted.*]

181. Suppose that OI, OJ, OK are any three given and coinitial but rectangular unit lines, the rotation around the first from the second to the third being positive; and let OI′, OJ′, OK′ be the three unit vectors respectively opposite to these, so that

$$OI' = -OI,\ OJ' = -OJ,\ OK' = -OK.$$

Let the three new symbols i, j, k denote a *system of three right versors,*[7] *in three mutually rectangular planes,* . . . ; so that . . . $i = OK:OJ$, $j = OI:OK$, $k = OJ:OI$, as the figure may serve to illustrate. We shall then have these other expressions for the same three versors.

$$i = OJ':OK = OK':OJ' = OJ:OK';$$
$$j = OK':OI = OI':OK' = OK:OI';$$
$$k = OI':OJ = OJ':OI' = OI:OJ';$$

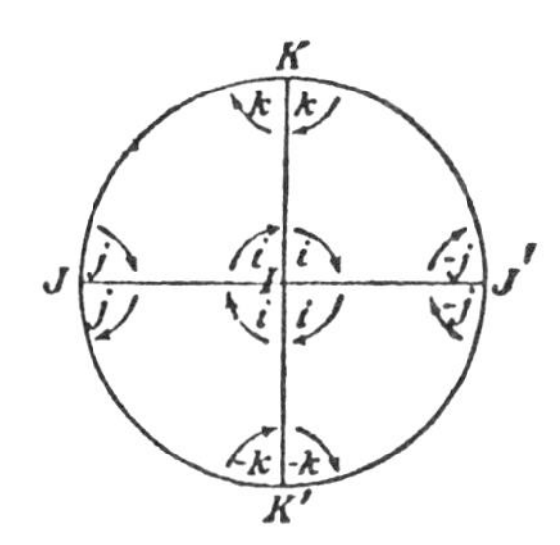

while the three respectively *opposite* versors may be thus expressed:

$$-i = OJ:OK = OK':OJ = OJ':OK' = OK:OJ'$$
$$-j = OK:OI = OI':OK = OK':OI' = OI:OK'$$
$$-k = OI:OJ = OJ':OI = OI':OJ' = OJ:OI'$$

and from the comparison of these different expressions several important symbolical consequences follow . . .

182. In the *first* place, since

$$i^2 = (OJ':OK).(OK:OJ) = OJ':OJ, \text{ etc.,}$$

we deduce the following equal values for the *squares* of the new symbols:

$$\text{I} \qquad i^2 = -1;\ j^2 = -1;\ k^2 = -1,$$

. . . .

In the *second* place, since

$$ij = (OJ:OK').(OK':OI) = OJ:OI, \text{ etc.,}$$

we have the following values for the *products* of the same three symbols, or versors, when taken *two by two,* and in a certain *order* of *succession* . . . :

$$\text{II} \qquad ij = k;\ jk = i;\ ki = j.$$

But in the *third* place . . . , since

$$ji = (OI:OK).(OK:OJ) = OI:OJ, \text{ etc.,}$$

we have these other and *contrasted* formulae, for the *binary products* of the *same* three right versors, when taken as factors with an *opposite order:*

$$\text{III} \qquad ji = -k;\ kj = -i;\ ik = -j.$$

Hence, while the *square of each* of the *three right versors,* denoted by these *three new symbols,* i, j, k, is equal to *negative unity,* the *product of any two* of them is equal either to the *third itself,* or to the *opposite* of that third versor, according as the *multiplier precedes* or *follows* the *multiplicand,* in the *cyclical succession*

$$i, j, k, i, j, \ldots$$

which the annexed figure may give some help towards remembering.

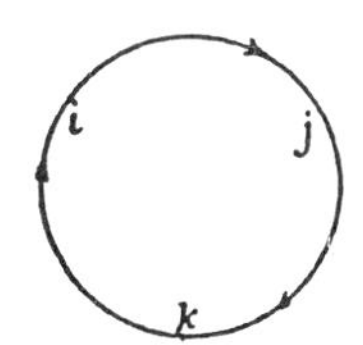

183. Since we have thus $ji = -ij$, . . . we see that the *laws of combination of the new symbols,* i, j, k, are *not in all respects the same* as the corresponding laws in *algebra;* since the *Commutative Property of Multiplication,* or the *convertibility* of the places of the *factors* without change of value of the *product,* does *not here* hold good; which arises from the circumstance that the factors to be combined are here diplanar versors.

It is therefore important to observe that there is a respect in which the *laws of* i, j, k *agree* with usual and *algebraic laws:* namely, in the Associative Property of Multiplication; or in the property that the new symbols always obey the associative formula

$$\iota\kappa.\lambda = \iota.\kappa\lambda,$$

whichever of them may be substituted for ι, for κ, and for λ; in virtue of which equality of values we may *omit the point* in any such symbol of a *ternary product* (whether of equal or unequal factors), and write it simply as $\iota\kappa\lambda$. In particular, we have thus,

$$\text{i.jk} = \text{i.i} = \text{i}^2 = -1;$$
$$\text{ij.k} = \text{k.k} = \text{k}^2 = -1;$$

or briefly

$$\text{ijk} = -1.$$

We may, therefore, . . . establish the following important *Formula:*

$$\text{i}^2 = \text{j}^2 = \text{k}^2 = \text{ijk} = -1;$$

. . . which we shall find to contain (virtually) *all the laws of the symbols* i, j, k, and therefore to be a *sufficient symbolical basis* for the whole *Calculus of Quaternions:* because it will be shown that *every quaternion can be reduced to the Quadrinomial Form,*

$$q = w + ix + jy + kz,$$

where w, x, y, z compose a *system of four scalars,* while i, j, k are the same *three right versors as above.*

If two right versors in two mutually rectangular planes, be multiplied together in two opposite orders, the two resulting products will be two opposite right versors, in a third plane, rectangular to the two former; or in symbols . . .

$$q'q = -qq'$$

. . . In *this* case, therefore, we have what would be in algebra a *paradox* . . . When we come to examine what, in the last analysis, may be said to be the *meaning* of this last equation, we find it to be simply this: that *any two quadrantal* or *right rotations, in planes perpendicular to each other, compound themselves into a third right rotation, as their resultant in a plane perpendicular to each of them:* and that this *third* or *resultant* rotation has one or other of two *opposite directions,* according to the *order* in which the two *component rotations* are taken, so that one shall be *successive* to the other.

NOTES

1. Non zero.
2. By the second assumption.
3. Or collinear.
4. "The tensor of the quotient."
5. Assumption 2.
6. "Quaternion . . . signifies . . . a Set of Four."
7. A right versor is an operator which produces a rotation of a right angle about a given axis in a given direction.

Chapter VIII
The Nineteenth Century

Section A
Algebra

GEORGE BOOLE (1815–64)

George Boole, mathematician and logician, was born at Lincoln, England. His father, John Boole, was a cobbler deeply interested in elementary mathematics and the making of optical instruments. At an early age, the boy learned to assist his father and received his first lessons in mathematics from him. Young Boole attended a local elementary school and, briefly, a small commercial school. His favorite subject was classics. William Brooke, the owner of a scholarly circulating library, taught him Latin; Boole also learned Greek, French, and German on his own by age 14.

At age 15 Boole's future was seriously affected when his father's business declined. He set aside thoughts of taking holy orders and worked to help support his family. He began teaching in the village schools of the West Riding of Yorkshire. At age 20 he opened his own school in Lincoln. It was during these early years of teaching that Boole's talent for mathematics emerged. The Mechanics Institution (founded in Lincoln in 1834) had Royal Society publications in its reading room (Boole's father was curator of that room), and Boole devoted his scant leisure time to studying mathematics there. Almost unaided, he wrestled with Newton's *Principia*, Lagrange's *Mécanique analytique*, and Laplace's *Mécanique céleste*. He quickly earned a local reputation as a learned man. In 1835, at age 19, his first scientific publication, "An Address on the Genius and Discoveries of Sir Isaac Newton," was published.

Boole's work soon became known to a wider audience in the sciences. Beginning in 1839, he published a series of original papers on linear transformations and differential equations in the recently founded *Cambridge Mathematical Journal.* His papers from 1841 through 1843 on linear transformations generalized algebraic studies of Lagrange and Gauss on the relative invariance of discriminants and examined the absolute invariants of transformations. The limited results in these papers, representing a starting point of the theory of algebraic invariants, were extensively developed by one of their readers named Arthur Cayley. Unlike many of his contemporaries, Boole's treatment of differential equations went well beyond successful technical manipulation. He made the foundations of his subject more secure by avoiding the faulty use of analogy and insisting on precise definitions together with a rigorous process of reasoning. Using this approach he increased the power of the operational calculus. Boole also contributed to the *Philosophical Transactions* of the Royal Society, winning its Royal Medal in 1844 for papers on operators in analysis. He was elected a Fellow of the Royal Society in 1857.

In 1849, Boole's friends advised him to apply for the position of professor of mathematics at the newly founded Queen's College in Cork, Ireland. Although he had no university degree, he was appointed on the basis of his publications. The teaching load at Queen's College was heavy, but he now had more time and facilities for research. In 1855, he married Mary Everest (the niece of the surveyor Sir George Everest, after whom Mount Everest was named), and they had five daughters.

At Queen's College, Boole was known as a modest man of independent mind and wide erudition. He avidly read general literature. His favorite poet was Dante; he preferred the *Paradiso* to the *Inferno.* Among the topics he frequently studied were the metaphysics of Aristotle, the writings of Cicero, and the ethics of Spinoza. On occasion he invited people from outside university circles to his home to "talk science." These were people of a similar social background to his own whom he had met at a shop or in a train and whose conversation interested him. Some of his academic colleagues considered these invitations a bit odd. He was also known at the college as a clear and conscientious teacher. His two textbooks, *Treatise on Differential Equations* (1859) and *Treatise on the Calculus of Finite Differences* (1860), which restated and extended results from his research papers, suggest his clarity. In 1864, Boole's health began to fail. He died of pneumonia in December a few days after walking through a rainstorm to a class and lecturing in wet clothes.

In his most important contribution to mathematics, Boole laid the foundation for modern symbolic logic. The tradition of its continuing development began with his creation of the algebra of logic as a new branch of mathematics. The algebra of logic, which reduced Aristotelian logic to an algebraic calculus, is today called Boolean algebra. His major ideas on it appeared in *Mathematical Analysis of Logic* (1847) and *An Investigation of the Laws of Thought* (1854). The idea was not novel with Boole. Leibniz had devised a promising scheme with his universal characteristic, while De Morgan's logic of relations and Hamilton's quaternions, an algebra of quadruplets, proved suggestive. In the *Mathematical Analysis of Logic,* Boole asserted that logic should properly be classified with mathematics rather than metaphysics; he provided a precise symbolism for the Aristotelian syllogism and first achieved a workable logical calculus. His logical calculus emphasized the logic of sets and had as symbols 1 for a universal class, 0 for a null class, and $1 - x$ for the complement of x.

Boole's work has proved a fruitful source in both pure and applied mathematics. The theory of lattices was shown to be a generalization of Boolean algebra, since every Boolean algebra is a lattice and the converse is not true. A lattice is a system with two operations symbolized by $\cup$ and $\cap$ standing for union and intersection having the associative, commutative, and distributive properties. Charles Pierce, Bertrand Russell, and Alfred North Whitehead were among those who refined symbolic logic. Russell, who believed that mathematics could be reduced to logic, once wrote: "Pure mathematics was discovered by Boole, in a work he called *The Laws of Thought.*" This statement, while an exaggeration, points up the unusual merit of Boole's mathematical contribution. In recent years, Boole's two-valued algebra has been successfully applied in designing circuits for high speed computers.

96. From *An Investigation of the Laws of Thought* (1854)*

(The Joining of Algebra and Logic)

GEORGE BOOLE

CHAPTER II
OF SIGNS IN GENERAL, AND OF THE SIGNS APPROPRIATE TO THE SCIENCE OF LOGIC IN PARTICULAR; ALSO OF THE LAWS TO WHICH THAT CLASS OF SIGNS ARE SUBJECT

1. That Language is an instrument of human reason, and not merely a medium for the expression of thought, is a truth generally admitted. It is proposed in this chapter to inquire what it is that renders Language thus subservient to the most important of our intellectual faculties. In the various steps of this inquiry we shall be led to consider the constitution of Language, considered as a system adapted to an end or purpose; to investigate its elements; to seek to determine their mutual relation and dependence; and to inquire in what manner they contribute to the attainment of the end to which, as co-ordinate parts of a system, they have respect.

In proceeding to these inquiries, it will not be necessary to enter into the discussion of that famous question of the schools, whether Language is to be regarded as an *essential* instrument of reasoning, or whether, on the other hand, it is possible for us to reason without its aid. I suppose this question to be beside the design of the present treatise, for the following reason, viz., that it is the business of Science to investigate laws; and that, whether we regard signs as the representatives of things and of their relations, or as the representatives of the conceptions and operations of the human intellect, in studying the laws of signs, we are in effect studying the manifested laws of reasoning. If there exists a difference between the two inquiries, it is one which does not affect the scientific expressions of formal law, which are the object of investigation in the present stage of this work, but relates only to the mode in which those results are presented to the mental regard. For though in investigating the laws of signs, *a posteriori*, the immediate subject of examination is Language, with the rules which govern its use; while in making the internal processes of thought the direct object of inquiry, we appeal in a more immediate way to our personal consciousness,—it will be found that in both cases the results obtained are formally equivalent. Nor could we easily conceive, that the unnumbered tongues and dialects of the earth should have preserved through a long succession of ages so much that is common and universal, were we not assured of the existence of some deep foundation of their agreement in the laws of the mind itself.

2. The elements of which all language consists are signs or symbols. Words are signs. Sometimes they are said to represent things; sometimes the

*Source: *George Boole's Collected Logical Works* (1854), vol. II, *An Investigation of the Laws of Thought,* 26-56.

operations by which the mind combines together the simple notions of things into complex conceptions; sometimes they express the relations of action, passion, or mere quality, which we perceive to exist among the objects of our experience; sometimes the emotions of the perceiving mind. But words, although in this and in other ways they fulfill the office of signs, or representative symbols, are not the only signs which we are capable of employing. Arbitrary marks, which speak only to the eye, and arbitrary sounds or actions, which address themselves to some other sense, are equally of the nature of signs, provided that their representative office is defined and understood. In the mathematical sciences, letters, and the symbols +, −, =, etc., are used as signs, although the term "sign" is applied to the latter class of symbols, which represent the elements of number and quantity. As the real import of a sign does not in any way depend upon its particular form or expression, so neither do the laws which determine its use. In the present treatise, however, it is with written signs that we have to do, and it is with reference to these exclusively that the term "sign" will be employed. The essential properties of signs are enumerated in the following definition.

Definition.—A sign is an arbitrary mark, having a fixed interpretation, and susceptible of combination which other signs in subjection to fixed laws dependent upon their mutual interpretation.

3. Let us consider the particulars involved in the above definition separately.

(1.) In the first place, a sign is an *arbitrary* mark. It is clearly indifferent what particular word or token we associate with a given idea, provided that the association once made is permanent. The Romans expressed by the word "civitas" what we designate by the word "state." But both they and we might equally well have employed any other word to represent the same conception. Nothing, indeed, in the nature of Language would prevent us from using a mere letter in the same sense. Were this done, the laws according to which the letter would require to be used would be essentially the same with the laws which govern the use of "civitas" in the Latin, and of "state" in the English language, so far at least as the use of those words is regulated by any general principles common to all languages alike.

(2.) In the second place, it is necessary that each sign should possess, within the limits of the same discourse or process of reasoning, a fixed interpretation. The necessity of this condition is obvious, and seems to be founded in the very nature of the subject. There exists, however, a dispute as to the precise nature of the representative office of words or symbols used as names in the processes of reasoning. By some it is maintained, that they represent the conceptions of the mind alone; by others, that they represent things. The question is not of great importance here, as its decision cannot affect the laws according to which signs are employed. I apprehend, however, that the general answer to this and such like questions is, that in the processes of reasoning, signs stand in the place and fulfill the office of the conceptions and operations of the mind; but that as those conceptions and operations represent things, and the connexions and relations of things, so signs represent things with their connexions and relations; and lastly, that as signs stand in the place of the conceptions and operations of the mind, they are subject to the laws of those conceptions and operations. This view will be more fully elucidated in the next chapter; but it here serves to explain the third of those particulars involved in the definition of a sign, viz., its subjection to fixed laws of combination depending upon the nature of its interpretation.

4. The analysis and classification of those signs by which the operations of

reasoning are conducted will be considered in the following Proposition:

PROPOSITION I

All the operations of Language, as an instrument of reasoning, may be conducted by a system of signs composed of the following elements, viz.:

1st. *Literal symbols as x, y, etc., representing things as subjects of our conceptions.*

2nd. *Signs of operation, as +, −, ×, standing for those operations of the mind by which the conceptions of things are combined or resolved so as to form new conceptions involving the same elements.*

3rd. *The sign of identity, =.*

And these symbols of Logic are in their use subject to definite laws, partly agreeing with and partly differing from the laws of the corresponding symbols in the science of Algebra.

CHAPTER III
DERIVATION OF THE LAWS OF THE SYMBOLS OF LOGIC FROM THE LAWS OF THE OPERATIONS OF THE HUMAN MIND

1. The object of science, properly so called, is the knowledge of laws and relations. To be able to distinguish what is essential to this end, from what is only accidentally associated with it, is one of the most important conditions of scientific progress. I say, to *distinguish* between these elements, because a consistent devotion to science does not require that the attention should be altogether withdrawn from other speculations, often of a metaphysical nature, with which it is not unfrequently connected. Such questions, for instance, as the existence of a sustaining ground of phaenomena, the reality of cause, the propriety of forms of speech implying that the successive states of things are connected by *operations,* and others of a like nature, may possess a deep interest and significance in relation to science, without being essentially scientific. It is indeed scarcely possible to express the conclusions of natural science without borrowing the language of these conceptions. Nor is there necessarily any practical inconvenience arising from this source. They who believe, and they who refuse to believe, that there is more in the relation of cause and effect than an invariable order of succession, agree in their interpretation of the conclusions of physical astronomy. But they only agree because they recognise a common element of scientific truth, which is independent of their particular views of the nature of causation.

2. If this distinction is important in physical science, much more does it deserve attention in connexion with the science of the intellectual powers. For the questions which this science presents become, in expression at least, almost necessarily mixed up with modes of thought and language, which betray a metaphysical origin. The idealist would give to the laws of reasoning one form of expression; the sceptic, if true to his principles, another. They who regard the phaenomena with which we are concerned in this inquiry as the mere successive *states* of the thinking subject devoid of any causal connexion, and they who refer them to the *operations* of an active intelligence, would, if consistent, equally differ in their modes of statement. Like difference would also result from a difference of classification of the mental faculties. Now the principle which I would here assert, as affording us the only ground of confidence and stability amid so much of seeming and of real diversity, is the following, viz., that if the laws in question are really deduced from observation, they have a real existence as laws of the human mind, independently of any metaphysical theory which may seem to be involved in the mode of their statement. They contain an element of truth which no ulterior criticism upon the nature, or event upon the reality, of the mind's operations, can

essentially affect. Let it even be granted that the mind is but a succession of states of consciousness, a series of fleeting impressions uncaused from without or from within, emerging out of nothing, and returning into nothing again,—the last refinement of the sceptic intellect,—still, as laws of succession, or at least of a past succession, the results to which observation had led would remain true. They would require to be interpreted into a language from whose vocabulary all such terms as cause and effect, operation and subject, substance and attribute, had been banished; but they would still be valid as scientific truths.

Moreover, as any statement of the laws of thought, founded upon actual observation, must thus contain scientific elements which are independent of metaphysical theories of the nature of the mind, the practical application of such elements to the construction of a system or method of reasoning must also be independent of metaphysical distinctions. For it is upon the scientific elements involved in the statement of the laws, that any practical application will rest, just as the practical conclusions of physical astronomy are independent of any theory of the cause of gravitation, but rest only on the knowledge of its phaenomenal effects. And, therefore, as respects both the determination of the laws of thought, and the practical use of them when discovered, we are, for all really scientific ends, unconcerned with the truth or falsehood of any metaphysical speculations whatever.

3. The course which it appears to me to be expedient, under these circumstances, to adopt, is to avail myself as far as possible of the language of common discourse, without regard to any theory of the nature and powers of the mind which it may be thought to embody. For instance, it is agreeable to common usage to say that we converse with each other by the communication of ideas, or conceptions, such communication being the office of words; and that with reference to any particular ideas or conceptions presented to it, the mind possesses certain powers or faculties by which the mental regard may be fixed upon some ideas, to the exclusion of others, or by which the given conceptions or ideas may, in various ways, be combined together. To those faculties or powers different names, as Attention, Simple Apprehension, Conception or Imagination, Abstraction, etc., have been given,—names which have not only furnished the titles of distinct divisions of the philosophy of the human mind, but passed into the common language of men. Whenever, then, occasion shall occur to use these terms, I shall do so without implying thereby that I accept the theory that the mind possesses such and such powers and faculties as distinct elements of its activity. Nor is it indeed necessary to inquire whether such powers or the understanding have a distinct existence or not. We may merge these different titles under the one generic name of *Operations* of the human mind, define these operations so far as is necessary for the purposes of this work, and then seek to express their ultimate laws. Such will be the general order of the course which I shall pursue, though reference will occasionally be made to the names which common agreement has assigned to the particular states or operations of the mind which may fall under our notice.

It will be most convenient to distribute the more definite results of the following investigation into distinct Propositions.

PROPOSITION I

4. *To deduce the laws of the symbols of Logic from a consideration of those operations of the mind which are implied in the strict use of language as an instrument of reasoning.*

In every discourse, whether of the mind conversing with its own thoughts, or of the individual in his intercourse with others, there is an assumed or expressed limit within which the subjects of its operation are confined. The most unfettered discourse is that in which the words we use are understood in the widest possible application, and for them the limits of discourse are co-extensive with those of the universe itself. But more usually we confine ourselves to a less spacious field. Sometimes, in discoursing of men we imply (without expressing the limitation) that it is of men only under certain circumstances and conditions that we speak, as of civilized men, or of men in the vigour of life, or of men under some other condition or relation. Now, whatever may be the extent of the field within which all the objects of our discourse are found, that field may properly be termed the universe of discourse.

5. Furthermore, this universe of discourse is in the strictest sense the ultimate *subject* of the discourse. The office of any name or descriptive term employed under the limitations supposed is not to raise in the mind the conception of all the beings or objects to which that name or description is applicable, but only of those which exist within the supposed universe of discourse. If that universe of discourse is the actual universe of things, which it always is when our words are taken in their real and literal sense, then by men we mean *all men that exist;* but if the universe of discourse is limited by any antecedent implied understanding, then it is of men under the limitation thus introduced that we speak. It is in both cases the business of the word *men* to direct a certain operation of the mind, by which, from the proper universe of discourse, we select or fix upon the individuals signified.

6. Exactly of the same kind is the mental operation implied by the use of an adjective. Let, for instance, the universe of discourse be the actual Universe. Then, as the word *men* directs us to select mentally from that Universe all the beings to which the term "men" is applicable; so the adjective "good," in the combination "good men," directs us still further to select mentally from the class of *men* all those who possess the further quality "good"; and if another adjective were prefixed to the combination "good men," it would direct a further operation of the same nature, having reference to that further quality which it might be chosen to express.

It is important to notice carefully the real nature of the operation here described, for it is conceivable, that it might have been different from what it is. Were the adjective simply *attributive* in its character, it would seem, that when a particular set of beings is designated by *men,* the prefixing of the adjective *good* would direct us to attach mentally to all those beings the quality of goodness. But this is not the real office of the adjective. The operation which we really perform is one of *selection according to a prescribed principle or idea.* To what faculties of mind such an operation would be referred, according to the received classification of its powers, it is not important to inquire, but I suppose that it would be considered as dependent upon the two faculties of Conception or Imagination, and Attention. To the one of these faculties might be referred the formation of the general conception; to the other the fixing of the mental regard upon those individuals within the prescribed universe of discourse which answer to the conception. If, however, as seems not improbable, the power of Attention is nothing more than the power of continuing the exercise of any other faculty of the mind, we might properly regard the whole of the mental process above described as referrible to the mental faculty of Imagination or Conception, the first step of the process being the conception of the Universe itself, and each succeeding step limiting in a definite manner the conception thus formed. Adopting this view, I shall describe each such step, or any definite combination of such steps, as a *definite act of conception.* And the use of this term I shall extend so as to include in its meaning not only the conception of classes of objects represented by particular names or simple attributes of quality, but also the combination of such conceptions in any manner consistent with the powers and limitations of the human mind; indeed, any intellectual operation short of that which is involved in the structure of a sentence or proposition. The general laws to which such operations of the mind are subject are now to be considered.

7. Now it will be shown that the laws which in the preceding chapter have been determined *a posteriori* from the constitution of language, for the use of the literal symbols of Logic, are in reality the laws of that definite mental operation which has just been described. We commence our discourse with a certain understanding as to the limits of its subject, *i.e.,* as to the limits of its Universe. Every name, every term of description that we employ, directs him whom we address to the performance of a certain mental operation upon that subject. And thus is thought communicated. But as each name or descriptive term is in this view but the representative of an intellectual operation, that operation being also prior in the order of nature, it is clear that the laws of the name or symbol must be of a derivative character,—must, in fact, originate in those of the operation which they represent. That the laws of the symbol and of the mental process are identical in expression will now be shown.

8. Let us then suppose that the universe of our discourse is the actual universe, so that words are to be used in the full extent of their meaning, and let us consider the two mental operations implied by the words "white" and "men." The word "men" implies the operation of selecting in thought from its subject, the universe, all men; and the resulting conception, *men,* becomes the subject of the next operation. The operation implied by the word "white" is that of selecting from its subject, "men," all of that class which are white. The final resulting conception is that of "white men." Now it is perfectly apparent that if the operations above described had been performed in a converse order, the result would have been the same. Whether we begin by forming the conception of *"men,"* and then by a second intellectual act limit that conception to "white men," or whether we begin by forming the conception of "white objects," and then limit it to such of that class as are "men," is perfectly indifferent so far as the result is concerned. It is obvious that the order of the mental processes would be equally indifferent if for the words "white" and "men" we substituted any other descriptive or appellative terms whatever, provided only that their meaning was fixed and absolute. And thus the indifference of the order of two successive acts of the faculty of Conception, the one of which furnishes the subject upon which the other is supposed to operate, is a general condition of the exercise of that faculty. It is a law of the mind, and it is the real origin of that law of the literal symbols of Logic which constitutes its formal expression [(1) Chap. II.].

9. It is equally clear that the mental operation above described is of such a nature that its effect is not altered by repetition. Suppose that by a definite act of conception the attention has been fixed upon men, and that by another exercise of the same faculty we limit it to those of the race who are white. Then any further repetition of the latter mental act, by which the attention is limited to white objects, does not in any way modify the conception arrived at, viz., that of white men. This is also an example of a general law of the mind, and it has its formal expression in the law [(2) Chap. II.] of the literal symbols.

10. Again, it is manifest that from the conceptions of two distinct classes of things we can form the conception of that collection of things which the two classes taken together compose; and it

is obviously indifferent in what order of position or of priority those classes are presented to the mental view. This is another general law of the mind, and its expression is found in (3) Chap. II.

11. It is not necessary to pursue this course of inquiry and comparison. Sufficient illustration has been given to render manifest the two following positions, viz.:

First, that the operations of the mind, by which, in the exercise of its power of imagination or conception, it combines and modifies the simple ideas of things or qualities, not less than those operations of the reason which are exercised upon truths and propositions, are subject to general laws.

Secondly, that those laws are mathematical in their form, and that they are actually developed in the essential laws of human language. Wherefore the laws of the symbols of Logic are deducible from a consideration of the operations of the mind in reasoning.

12. The remainder of this chapter will be occupied with questions relating to that law of thought whose expression is $x^2 = x$ (II. 9), a law which, as has been implied (II. 15), forms the characteristic distinction of the operations of the mind in its ordinary discourse and reasoning, as compared with its operations when occupied with the general algebra of quantity. An important part of the following inquiry will consist in proving that the symbols 0 and 1 occupy a place, and are susceptible of an interpretation, among the symbols of Logic; and it may first be necessary to show how particular symbols, such as the above, may with propriety and advantage be employed in the representation of distinct systems of thought.

The ground of this propriety cannot consist in any community of interpretation. For in systems of thought so truly distinct as those of Logic and Arithmetic (I use the latter term in its widest sense as the science of Number), there is, properly speaking, no community of subject. The one of them is conversant with the very conceptions of things, the other takes account solely of their numerical relations. But inasmuch as the forms and methods of any system of reasoning depend immediately upon the laws to which the symbols are subject, and only mediately, through the above link of connexion, upon their interpretation, there may be both propriety and advantage in employing the same symbols in different systems of thought, provided that such interpretations can be assigned to them as shall render their formal laws identical, and their use consistent. The ground of that employment will not then be community of interpretation, but the community of the formal laws, to which in their respective systems they are subject. Nor must that community of formal laws be established upon any other ground than that of a careful observation and comparison of those results which are seen to flow independently from the interpretations of the systems under consideration.

These observations will explain the process of inquiry adopted in the following Proposition. The literal symbols of Logic are universally subject to the law whose expression is $x^2 = x$. Of the symbols of Number there are two only, 0 and 1, which satisfy, this law. But each of these symbols is also subject to a law peculiar to itself in the system of numerical magnitude, and this suggests the inquiry, what interpretations must be given to the literal symbols of Logic, in order that the same peculiar and formal laws may be realized in the logical system also.

PROPOSITION II

13. *To determine the logical value and significance of the symbols* 0 *and* 1.

The symbol 0, as used in Algebra, satisfies the following formal law,

$$0 \times y = 0, \text{ or } 0y = 0, \qquad (1)$$

whatever *number* y may represent. That this formal law may be obeyed in the system of Logic, we must assign to the symbol 0 such an interpretation that the *class* represented by $0y$ may be identical with the class represented by 0, whatever the class y may be. A little consideration will show that this condition is satisfied if the symbol 0 represents Nothing. In accordance with a previous definition, we may term Nothing a class. In fact, Nothing and Universe are the two limits of class extension, for they are the limits of the possible interpretations of general names, none of which can relate to few individuals than are comprised in Nothing, or to more than are comprised in the Universe. Now whatever the class y may be, the individuals which are common to it and to the class "Nothing" are identical with those comprised in the class "Nothing," for they are none. And thus by assigning to 0 the interpretation Nothing, the law (1) is satisfied; and it is not otherwise satisfied consistently with the perfectly general character of the class y.

Secondly, the symbol 1 satisfies in the system of Number the following law, viz.,

$$1 \times y = y, \text{ or } 1y = y,$$

whatever number y may represent. And this formal equation being assumed as equally valid in the system of this work, in which 1 and y represent classes, it appears that the symbol 1 must represent such a class that all the individuals which are found in *any* proposed class y are also all the individuals $1y$ that are common to that class y and the class represented by 1. A little consideration will here show that the class represented by 1 must be "the Universe," since this is the only class in which are found *all* the individuals that exist in *any* class. Hence the respective interpretations of the symbols 0 and 1 in the system of Logic are *Nothing* and *Universe*.

14. As with the idea of any class of objects as "men," there is suggested to the mind the idea of the contrary class of beings which are not men; and as the whole Universe is made up of these two classes together, since of every individual which it comprehends we may affirm either that it is a man, or that it is not a man, it becomes important to inquire how such contrary names are to be expressed. Such is the object of the following Proposition.

PROPOSITION III

If x represent any class of objects, then will $1 - x$ represent the contrary or supplementary class of objects, i.e. the class including all objects which are not comprehended in the class x.

For greater distinctness of conception let x represent the class *men,* and let us express, according to the last Proposition, the Universe by 1; now if from the conception of the Universe, as consisting of "men" and "not-men," we exclude the conception of "men," the resulting conception is that of the contrary class, "not-men." Hence the class "not-men" will be represented by $1 - x$. And, in general, whatever class of objects is represented by the symbol x, the contrary class will be expressed by $1 - x$.

15. Although the following Proposition belongs in strictness to a future chapter of this work, devoted to the subject of *maxims* or *necessary truths,* yet, on account of the great importance of that law of thought to which it relates, it has been thought proper to introduce it here.

PROPOSITION IV

That axiom of metaphysicians which is termed the principle of contradiction, and which affirms that it is impossible for any being to possess a quality, and at the same time not to possess it, is a consequence of the fundamental law of thought, whose expression is $x^2 = x$.

Let us write this equation in the form

$$x - x^2 = 0,$$

whence we have

$$x(1 - x) = 0; \tag{1}$$

both these transformations being justified by the axiomatic laws of combination and transposition (II. 13). Let us, for simplicity of conception, give to the symbol x the particular interpretation of *men*, then $1 - x$ will represent the class of "not-men" (Prop. III.). Now the formal product of the expressions of two classes represents that class of individuals which is common to them both (II. 6). Hence $x(1 - x)$ will represent the class whose members are at once "men," and "not men," and the equation (1) thus express the principle, *that a class whose members are at the same time men and not men does not exist.* In other words, that *it is impossible for the same individual to be at the same time a man and not a man.* Now let the meaning of the symbol x be extended from the representing of "men," to that of any class of beings characterized by the possession of any quality whatever; and the equation (1) will then express that it is impossible for a being to possess a quality and not to possess that quality at the same time. But this is identically that "principle of contradiction" which Aristotle has described as the fundamental axiom of all philosophy. "It is impossible that the same quality should both belong and not belong to the same thing. . . . This is the most certain of all principles. . . . Wherefore they who demonstrate refer to this as an ultimate opinion. For it is by nature the source of all the other axioms."

The above interpretation has been introduced not on account of its immediate value in the present system, but as an illustration of a significant fact in the philosophy of the intellectual powers, viz., that what has been commonly regarded as the fundamental axiom of metaphysics is but the consequence of a law of thought, mathematical in its form. I desire to direct attention also to the circumstances that the equation (1) in which that fundamental law of thought is expressed is an equation of second degree.[1] Without speculating at all in this chapter upon the question, whether that circumstance is necessary in its own nature, we may venture to assert that if it had not existed, the whole procedure of the understanding would have been different from what it is. Thus it is a consequence of the fact that the fundamental equation of thought is of the second degree, that we perform the operation of analysis and classification, by division into pairs of opposites, or, as it is technically said, by *dichotomy.* Now if the equation in question had been of the third degree, still admitting of interpretation as such, the mental division must have been threefold in character, and we must have proceeded by a species of *trichotomy,* the real nature of which it is impossible for us, with our existing faculties, adequately to conceive, but the laws of which we might still investigate as an object of intellectual speculation.

16. The law of thought expressed by the equation (1) will, for reasons which are made apparent by the above discussion, be occasionally referred to as the "law of duality."

NOTE

1. Should it here be said that the existence of the equation $x^2 = x$ necessitates also the existence of the equation $x^3 = x$, which is of the third degree, and then inquired whether that equation does not indicate a process of *trichotomy;* the answer is, that the equation $x^3 = x$ is not interpretable in the system of logic. For writing it in either of the forms

$$x(1 - x)(1 + x) = 0, \quad (2)$$

$$x(1 - x)(-1 - x) = 0, \quad (3)$$

we see that its interpretation, if possible at all, must involve that of the factor $1 + x$, or of the factor $-1 - x$. The former is not interpretable, because we cannot conceive of the addition of any class x to the universe 1; the later is not interpretable, because the symbol -1 is not subject to the law $x(1 - x) = 0$, to which all class symbols are subject. Hence the equation $x^3 = x$ admits of no interpretation analogous to that of the equation $x^2 = x$.

Were the former equation, however, true independently of the latter, *i.e.*, were that act of the mind which is denoted by the symbol x, such that its second repetition should reproduce the result of a single operation, but not its first or mere repetition, it is presumable that we should be able to interpret one of the forms (2), (3), which under the actual conditions of thought we cannot do. There exist operations, known to the mathematician, the law of which may be adequately expressed by the equation $x^3 = x$. But they are of a nature altogether foreign to the province of general reasoning.

In saying that it is conceivable that the law of thought might have been different from what it is, I mean only that we can frame such an hypothesis, and study its consequences. The possibility of doing this involves no such doctrine as that the actual law of human reason is the product either of chance or of arbitrary will.

Chapter VIII
The Nineteenth Century

Section B
Non-Euclidean Geometries

NIKOLAI IVANOVICH LOBACHEVSKY (1792–1856)

Nikolai Ivanovich Lobachevsky, a founder of non-Euclidean geometry, was born in Nizhni Novgorod (now Gorki), Russia. His parents, who were of Polish origin, had migrated to Russia a few years earlier. The father, Ivan Maksimovich Lobachevsky, an impecunious clerk in a government land-survey office, died when Nikolai was a young child. The mother, Praskovia Aleksandrovna, raised her three sons alone, moving to Kazan about 1800. She taught her sons before they entered school; they were all bright and won public scholarships to study at the gymnasium and university in Kazan.

At age 14, Lobachevsky entered the University of Kazan. Among the four distinguished professors the recently-founded university had recruited from Germany was J. Martin Bartels, a tutor and friend of Gauss. Bartels persuaded Lobachevsky to change his major from pre-medicine to mathematics and taught him about the contributions of Euler to the calculus, Lagrange to analytical mechanics, Monge to descriptive geometry, and Gauss to number theory. Although Lobachevsky was an honor student, his penchant for practical jokes caused a temporary withholding of his master's degree in 1811 until he apologized for his behavior.

Almost the entire adult life of Lobachevsky centered around the University of Kazan, as the six years of teaching he pledged in return for a scholarship turned into forty. He became adjunct in the mathematical and physical sciences in 1814, associate professor in 1816, and full professor in 1822. He was an able teacher who independently prepared class notes rather than reading lectures from a text as did many of his colleagues. A warm person, he recognized the emotional as well as intellectual needs of his students. Lobachevsky also displayed skill in administrative assignments. He was twice dean of the department of physics and mathematics (1820–21 and 1823–25), librarian of the university (1825–35), and rector (1827–46). A distrust in Russia of modern science, which many viewed as a menace to Orthodox religion, and of Kantian philosophy, which some associated with what were then considered the evils of the French Revolution, posed troublesome problems for him during the final years of the reign of Tsar Alexander I (d. 1825) and beyond. This distrust produced factionalism and even chaos at Russian universities. As a result, many German-born faculty, including Bartels, left or were dismissed. As dean and rector at Kazan, Lobachevsky opposed factionalism among the faculty, improved academic standards, and strove to retain the best professors.

Lobachevsky worked tirelessly on behalf of the University of Kazan in many ways. As librarian, he personally catalogued books and put them on the shelves. During a cholera epidemic in 1830, he invited faculty, students, and their families within the university grounds. He closed the gates to all except medical doctors and strictly enforced sanitary regulations. This helped to save the lives of members of the university community. After several university buildings, including the observatory, were destroyed by a devastating fire in 1842, Lobachevsky supervised their rebuilding. To carry out his task well, he learned architecture.

In recognition of Lobachevsky's contributions to higher education, Tsar Nicholas I raised him to the hereditary nobility (boyars) in 1837. Lobachevsky, who had married the wealthy Lady Vavara Aleksivna Moisieva in 1832, lived comfortably with her and their seven children on an estate, but the cost of improving the estate took most of their funds. Thus, when the government stripped him of his positions as professor and rector at Kazan without explanation in 1846, he was left with little except a modest pension. The university faculty protested his removal but to no avail.

A worsening sclerotic condition caused Lobachevsky to spend his last years nearly blind. This did not stop his mathematical research nor his raising of prize sheep and fruit on his farm. In 1855 he planted a fruit orchard. He died the following year and was buried there.

Lobachevsky's outstanding contribution to mathematics was the discovery of a non-Euclidean geometry. He was one of three men who made this discovery independently and nearly at the same time. The other two were Gauss, who was the first to make the discovery but did not publish, and the Hungarian Janos Bolyai whose paper "The Science of Absolute Space" (1832) was not appreciated until much later. Lobachevsky deserved priority based on a lost paper of 1826 and two papers in Russian in the *Kazan Messenger* in 1829–30. In these papers and a third in the *Journal für Mathematik* (1835–37) he resolved the two-millennia-old issue of whether Euclid's fifth (parallel) postulate could be proven from Euclid's other nine or whether it was an unprovable assumption independent of them. Lobachevsky showed that it was an unprovable assumption defining one type of space by establishing another consistent geometry in which all of Euclid's axioms and postulates, except for the fifth, were valid. In place of Euclid's fifth postulate, he substituted a new parallel postulate, which holds that through a point C not on a line AB, one can draw more than one coplanar line not intersecting AB. He called the boundary lines for the class of lines that do not meet AB, in

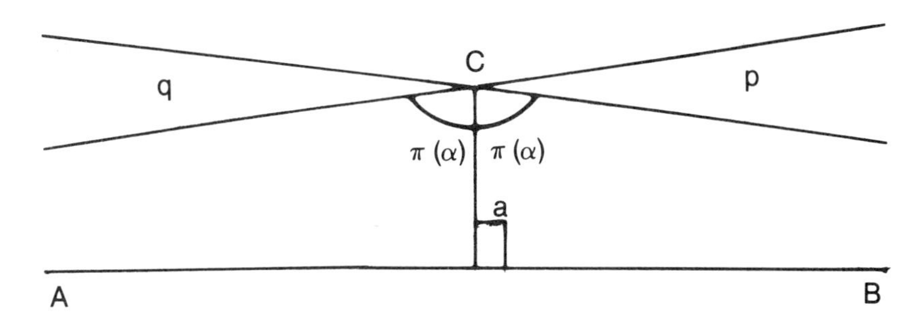

this instance *p* and *q*, parallel lines and the angle π(a) the angle of parallelism. As the length of *a* approaches zero, π(a) increases and approaches 90°. For $\pi(a) = 90°$, one has Euclidean Geometry. As *a* approaches infinity π(a) decreases and approaches zero.

Today, the geometry of Lobachevsky, Gauss, and Bolyai is called hyperbolic geometry. It is one of two non-Euclidean geometries. The other, elliptic geometry, in which no lines are parallel, was developed by Bernhard Riemann in 1854. Lobachevsky did not satisfactorily prove the equal consistency of his geometry with Euclid's. That was done by the Italian mathematician Eugenio Beltrami in the 1860s. He showed that Lobachevsky's geometry applied to geodesics on pseudospherical surfaces (*i.e.*, surfaces of constant negative curvature formed by rotating a tractrix about its asymptote). In it the sum of the angles in a triangle is less than 180°. By mapping one geometry on to the other, Beltrami showed that contradictions in non-Euclidean geometry would also reveal themselves in the Euclidean geometry of surfaces. Lobachevsky believed that Euclidean geometry was a special case of his new system. Consequently, he called his geometric system "imaginary geometry," using an analogy with imaginary (complex) numbers, which are the most general. In the 1870s, the German mathematician Felix Klein proved that in a sufficiently small domain, Euclidean geometry is indeed a limiting case of Lobachevsky's geometry. In his "imaginary geometry" Lobachevsky also differentiated large (cosmic) space from local space and opposed Kant's idealism, which asserted that space was a pure intuition in the mind imposing order on sense experience. He maintained instead that space was an *a posteriori* concept and geometry an empirical science.

Lobachevsky's noteworthy contributions to mathematics were not limited exclusively to geometry. In algebra, he found a method for approximating roots of equations by a repeated squaring process. In the calculus, he presented a more general definition of function based on his study of trigonometric series, a definition similar to Dirichlet's. He was also among the first to define rigorously continuity and differentiability.

The fame of Lobachevsky, like that of Bolyai, was posthumous, even though Gauss praised his work. Few other mathematicians read Lobachevsky's early papers in Russian; nor were his later presentations quickly accepted. These writings, published in French and German, included the article "Géométrie imaginaire" in Crelle's *Journal* (1837), *Geometrische Untersuchungen zur Theorie Parallelinen* ("Geometrical Researches on the Theory of Parallels"), the best exposition of his new geometry (1840), and his last book *Pangéométrie* (1855). His achievement was not widely appreciated among mathematicians until after Hoüel published a French translation of *The Theory of Parallels* in 1865; Beltrami firmly established the consistent foundations of hyperbolic geometry thereby making it "respectable" in the 1860s; and Weierstrass offered a seminar on Lobachevsky's geometry at the University of Berlin in 1870.

97. From *The Theory of Parallels* (1840)*

(Hyperbolic Geometry)

NIKOLAI IVANOVICH LOBACHEVSKY

In geometry I find certain imperfections which I hold to be the reason why this science, apart from transition into analytics, can as yet make no advance from that state in which it has come to us from Euclid.

As belonging to these imperfections, I consider the obscurity in the fundamental concepts of the geometrical magnitudes and in the manner and method of representing the measuring of these magnitudes, and finally the momentous gap in the theory of parallels, to fill which all efforts of mathematicians have been so far in vain.

For this theory Legendre's endeavors have done nothing, since he was forced to leave the only rigid way to turn into a side path and take refuge in auxiliary theorems which he illogically strove to exhibit as necessary axioms. My first essay on the foundations of geometry I published in the Kasan *Messenger* for the year 1829. In the hope of having satisfied all requirements, I undertook hereupon a treatment of the whole of this science, and published my work in separate parts in the *"Gelehrten Schriften der Universitaet Kasan"* for the years 1836, 1837, 1838, under the title "New Elements of Geometry, with a complete Theory of Parallels." The extent of this work perhaps hindered my countrymen from following such a subject, which since Legendre has lost its interest. Yet I am of the opinion that the Theory of Parallels should not lose its claim to the attention of geometers, and therefore I aim to give here the substance of my investigations, remarking beforehand that contrary to the opinion of Legendre, all other imperfections—for example, the definition of a straight line—show themselves foreign here and without any real influence on the Theory of Parallels.

In order not to fatigue my reader with the multitude of those theorems whose proofs present no difficulties, I prefix here only those of which a knowledge is necessary for what follows.

1. A straight line fits upon itself in all its positions. By this I mean that during the revolution of the surface containing it the straight line does not change its place if it goes through two unmoving points in the surface: (*i.e.,* if we turn the surface containing it about two points of the line, the line does not move.)

2. Two straight lines can not intersect in two points.

3. A straight line sufficiently produced both ways and must go out beyond all bounds, and in such way cuts a bounded plain into two parts.

4. Two straight lines perpendicular to a third never intersect, how far soever they be produced.

5. A straight line always cuts another in going from one side of it over to the other side: (*i.e.,* one straight line must cut another if it has points on both sides of it.)

6. Vertical angles, where the sides of one are productions of the sides of the

*Source: *Geometrical Researches on the Theory of Parallels,* trans. by George B. Halsted (1914, copyright © 1942), 11-19. This translation is reprinted by permission of the Open Court Publishing Co., Lasalle, Illinois.

other, are equal. This holds of plane rectilineal angles among themselves, as also of plane surface angles: (*i.e.*, dihedral angles.)

7. Two straight lines can not intersect, if a third cuts them at the same angle.

8. In a rectilineal triangle equal sides lie opposite equal angles, and inversely.

9. In a rectilineal triangle, a greater side lies opposite a greater angle. In a right-angled triangle the hypothenuse is greater than either of the other sides, and the two angles adjacent to it are acute.

10. Rectilineal triangles are congruent if they have a side and two angles equal, or two sides and the included angle equal, or two sides and the angle opposite the greater equal, or three sides equal.

11. A straight line which stands at right angles upon two other straight lines not in one plane with it is perpendicular to all straight lines drawn through the common intersection point in the plane of those two.

12. The intersection of a sphere with a plane is a circle.

13. A straight line at right angles to the intersection of two perpendicular planes, and in one, is perpendicular to the other.

14. In a spherical triangle equal sides lie opposite equal angles, and inversely.

15. Spherical triangles are congruent (or symmetrical) if they have two sides and the included angle equal, or a side and the adjacent angles equal.

From here follow the other theorems with their explanations and proofs.

16. All straight lines which in a plane go out from a point can with reference to a given straight line in the same plane, be divided into two classes—into *cutting* and *not-cutting*.

The *boundary lines* of the one and the other class of those lines will be called *parallel to the given line*.

From the point *A* (Fig. 1) let fall upon the line *BC* the perpendicular *AD*, to which again draw the perpendicular *AE*.

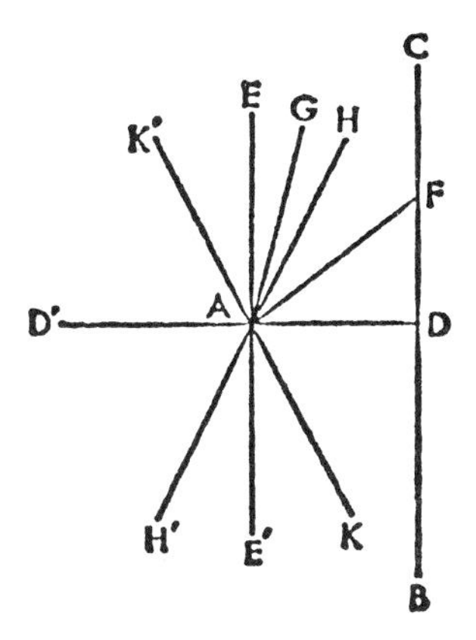

FIG. 1

In the right angle *EAD* either will all straight lines which go out from the point *A* meet the line *DC*, as for example *AF*, or some of them, like the perpendicular *AE*, will not meet the line *DC*. In the uncertainty whether the perpendicular *AE* is the only line which does not meet *DC*, we will assume it may be possible that there are still other lines, for example *AG*, which do not cut *DC*, how far soever they may be prolonged. In passing over from the cutting lines, as *AF*, to the not-cutting lines, as *AG*, we must come upon a line *AH*, parallel to *DC*, a boundary line, upon one side of which all lines *AG* are such as do not meet the line *DC*, while upon the other side every straight line *AF* cuts the line *DC*.

The angle *HAD* between the parallel *HA* and the perpendicular *AD* is called the parallel angle (angle of parallelism), which we will here designate $\Pi(p)$ for $AD = p$.

If $\Pi(p)$ is a right angle, so will the prolongation *AE'* of the perpendicular *AE* likewise be parallel to the prolongation *DB* of the line *DC*, in addition to which we remark that in regard to the four right angles, which are made at the point *A* by the perpendiculars *AE* and *AD*, and their prolongations *AE'* and *AD'*, every straight line which goes out from the point *A*, either itself or at least its prolongation, lies in one of the two right angles, which are turned toward *BC*, so that except the parallel *EE'* all others, if they are sufficiently produced both ways, must intersect the line *BC*.

If $\Pi(p) < \frac{1}{2}\pi$, then upon the other side of *AD*, making the same angle $DAK = \Pi(p)$ will lie also a line *AK*, parallel to the prolongation *DB* of the line *DC*, so that under this assumption we must also make a distinction of *sides in parallelism*.

All remaining lines or their prolongations within the two right angles turned toward *BC* pertain to those that intersect, if they lie within the angle $HAK = 2\,\Pi(p)$ between the parallels; they pertain on the other hand to the non-intersecting *AG*, if they lie upon the other sides of the parallels *AH* and *AK*, in the opening of the two angles $EAH = \frac{1}{2}\pi - \Pi(p)$, $E'AK = \frac{1}{2}\pi - \Pi(p)$, between the parallels and *EE'* the perpendicular to *AD*. Upon the other side of the perpendicular *EE'* will in like manner the prolongations *AH'* and *AK'* of the parallels *AH* and *AK* likewise be parallel to *BC*; the remaining lines pertain, if in the angle *K'AH'*, to the intersecting, but if in the angles *K'AE*, *H'AE'* to the non-intersecting.

In accordance with this, for the assumption $\Pi(p) = \frac{1}{2}\pi$, the lines can be only intersecting or parallel; but if we assume that $\Pi(p) < \frac{1}{2}\pi$, then we must allow two parallels, one on the one and one on the other side; in addition we must distinguish the remaining lines into non-intersecting and intersecting.

For both assumptions it serves as the mark of parallelism that the line becomes intersecting for the smallest deviation toward the side where lies the parallel, so that if *AH* is parallel to *DC*, every line *AF* cuts *DC*, how small soever the angle *HAF* may be.

17. *A straight line maintains the characteristic of parallelism at all its points.*

Given *AB* (Fig. 2) parallel to *CD*, to which latter *AC* is perpendicular. We will consider two points taken at random on the line *AB* and its production beyond the perpendicular.

Let the point *E* lie on that side of the perpendicular on which *AB* is looked upon as parallel to *CD*.

Let fall from the point *E* a perpendicular *EK* on *CD* and so draw *EF* that it falls within the angle *BEK*.

Connect the points *A* and *F* by a straight line, whose production then (by Theorem 16) must cut *CD* somewhere in *G*. Thus we get a triangle *ACG*, into which the line *EF* goes; now since this latter, from the construction, can not cut *AC*, and can not cut *AG* or *EK* a second time (Theorem 2), therefore it must meet *CD* somewhere at *H* (Theorem 3).

Now let *E'* be a point on the production of *AB* and *E'K'* perpendicular to the production of the line *CD*; draw the line *E'F'* making so small an angle *AE'F'* that it cuts *AC* somewhere in *F'*; making the same angle with *AB*, draw also from *A* the line *AF*, whose production will cut *CD* in *G* (Theorem 16).

Thus we get a triangle *AGC*, into which goes the production of the line *E'F'*; since now this line can not cut *AC* a second time, and also can not cut *AG*, since the angle $BAG = BE'G'$ (Theorem

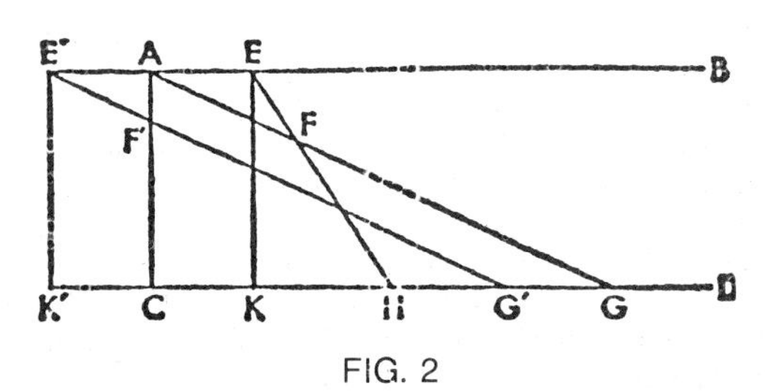

FIG. 2

7), therefore must it meet *CD* somewhere in *G'*.

Therefore from whatever points *E* and *E'* the lines *EF* and *E'F'* go out, and however little they may diverge from the line *AB*, yet will they always cut *CD*, to which *AB* is parallel.

18. *Two lines are always mutually parallel.*

Let *AC* be a perpendicular on *CD*, to which *AB* is parallel; if we draw from *C* the line *CE* making any acute angle *ECD* with *CD*, and let fall from *A* the perpendicular *AF* upon *CE*, we obtain a right-angled triangle *ACF*, in which *AC*, being the hypothenuse, is greater than the side *AF* (Theorem 9).

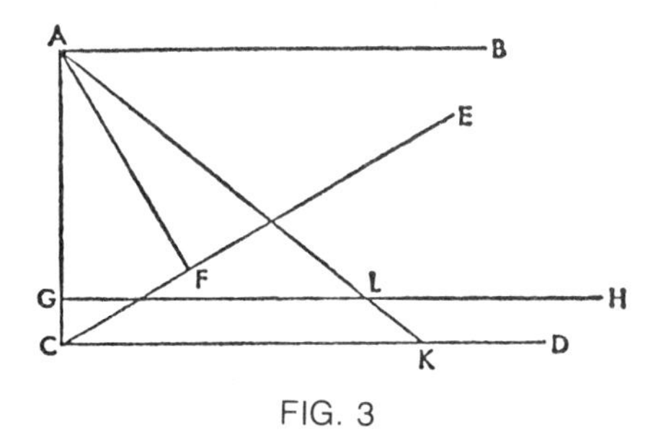

FIG. 3

Make *AG* = *AF*, and slide the figure *EFAB* until *AF* coincides with *AG*, when *AB* and *FE* will take the position *AK* and *GH*, such that the angle *BAK* = *FAC*, consequently *AK* must cut the line *DC* somewhere in *K* (Theorem 16), thus forming a triangle *AKC*, on one side of which the perpendicular *GH* intersects the line *AK* in *L* (Theorem 3), and thus determines the distance *AL* of the intersection point of the lines *AB* and *CE* on the line *AB* from the point *A*.

Hence it follows that *CE* will always intersect, *AB*, how small soever may be the angle *ECD*, consequently *CD* is parallel to *AB* (Theorem 16).

19. *In a rectilineal triangle the sum of the three angles can not be greater than two right ang,es.*

Suppose in the triangle *ABC* (Fig. 4) the sum of the three angles is equal to $\pi + a$; then choose in case of the inequality of the sides the smallest *BC*, halve it in *D*, draw from *A* through *D* the line *AD* and make the prolongation of it, *DE*, equal to *AD*, then join the point *E* to the point *C* by the straight line *EC*. In the congruent triangles *ADB* and *CDE*, the angle *ABD* = *DCE*, and *BAD* = *DEC* (Theorems 6 and 10); whence follows that also in the triangle *ACE* the sum of the three angles must be equal to $\pi + a$; but also the smallest angle *BAC* (Theorem 9) of the triangle *ABC* in passing over into the new triangle *ACE* has been cut up into the two parts *EAC* and *AEC*. Continuing this process, continually halving the side opposite the smallest angle, we must finally attain to a triangle in which the sum of the three angles is $\pi + a$, but wherein are two angles, each of which in absolute magnitude is less than $\frac{1}{2}a$; since now, however, the third angle can not be greater than π, so must *a* be either null or negative.

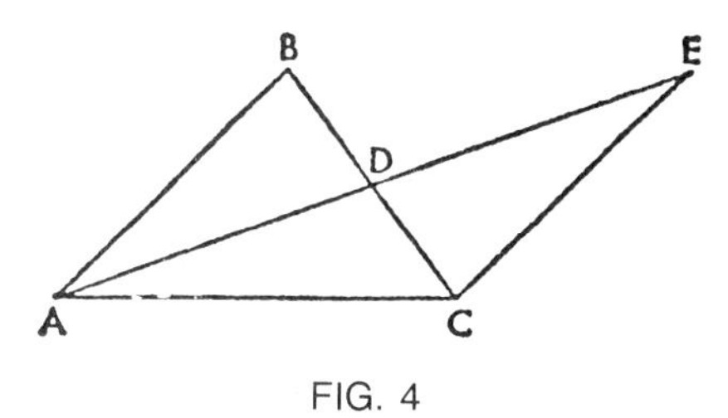

FIG. 4

20. *If in any rectilineal triangle the sum of the three angles is equal to two right angles, so is this also the case for every other triangle.*

If in the rectilineal triangle *ABC* (Fig. 5) the sum of the three angles = π, then must at least two of its angles, *A* and *C*, be acute. Let fall from the vertex of the third angle *B* upon the opposite side *AC*

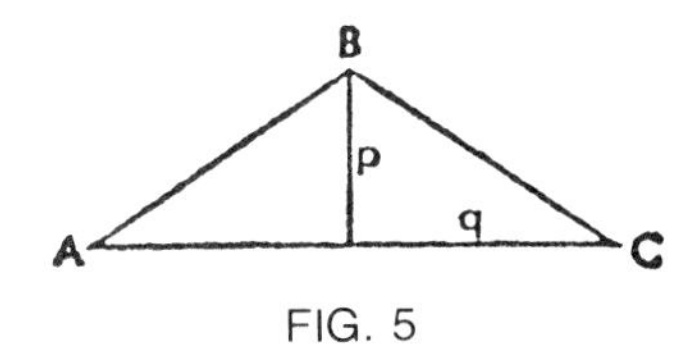

FIG. 5

the perpendicular p. This will cut the triangle into two right-angled triangles, in each of which the sum of the three angles must also be π, since it can not in either be greater than π, and in their combination not less than π.

So we obtain a right-angled triangle with the perpendicular sides p and q, and from this quadrilateral whose opposite sides are equal and whose adjacent sides p and q are at right angles (Fig. 6).

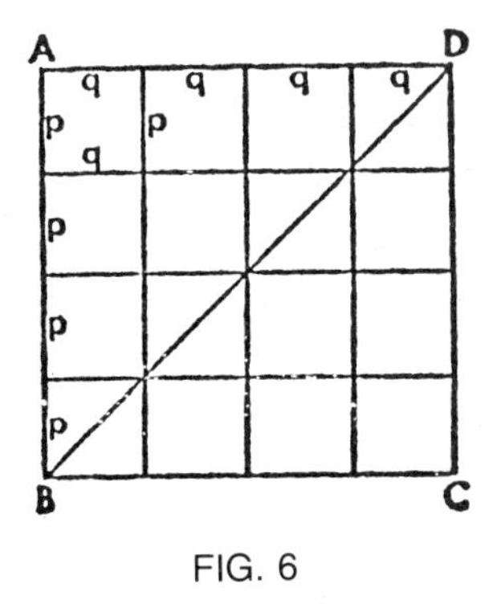

FIG. 6

By repetition of this quadrilateral we can make another with sides np and mq. and finally a quadrilateral $ABCD$ with sides at right angles to each other, such that $AB = np$, $AD = mq$. $DC = np$, $BC = mq$, where m and n are any whole numbers. Such a quadrilateral is divided by the diagonal DB into two congruent right-angled triangles, BAD and BCD, in each of which the sum of the three angles $= \pi$.

The numbers n and m can be taken sufficiently great for the right-angled triangle ABC (Fig. 7) whose perpendicular sides $AB = np$, $BC = mq$, to enclose within itself another given (right-angled) triangle BDE as soon as the right-angles fit each other.

Drawing the line DC, we obtain right-angled triangles of which every successive two have a side in common.

The triangle ABC is formed by the union of the two triangles ACD and DCB, in neither of which can the sum of the angles be greater than π; con-

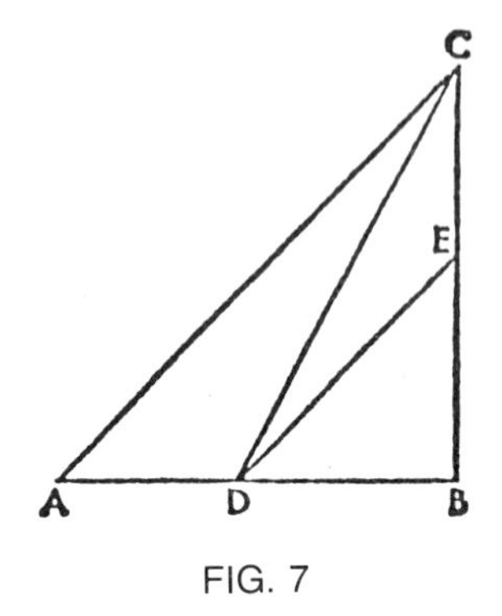

FIG. 7

sequently it must be equal to π, in order that the sum in the compound triangle may be equal to π.

In the same way the triangle BDC consists of the two triangles DEC and DBE, consequently must in DBE the sum of the three angles be equal to π, and in general this must be true for every triangle, since each can be cut into two right-angled triangles.

From this it follows that only two hypotheses are allowable: Either is the sum of the three angles in all rectilineal triangles equal to π, or this sum is in all less than π.

21. *From a given point we can always draw a straight line that shall make with a given straight line an angle as small as we choose.*

Let fall from the given point A (Fig. 8) upon the given line BC the perpendicular AB; take upon BC at random the point D; draw the line AD; make $DE = AD$, and draw AE.

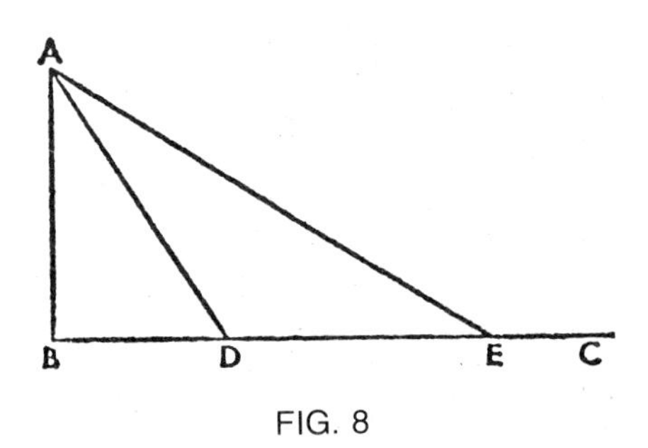

FIG. 8

In the right-angled triangle *ABD* let the angle *ADB* = *a;* then must in the isosceles triangle *ADE* the angle *AED* be either ½*a* or less (Theorems 8 and 20). Continuing thus we finally attain to such an angle, *AEB,* as is less than any given angle.

22. *If two perpendiculars to the same straight line are parallel to each other, then the sum of the three angles in a rectilineal triangle is equal to two right angles.*

Let the lines *AB* and *CD* (Fig. 9) be parallel to each other and perpendicular to *AC*.

Draw from *A* the lines *AE* and *AF* to the points *E* and *F*, which are taken on the line *CD* at any distance *FC* > *EC* from the point *C*.

Suppose in the right-angled triangle *ACE* the sum of the three angles is equal to $\pi - \alpha$, in the triangle *AEF* equal to $\pi - \beta$, then must it in the triangle *ACF* equal $\pi - \alpha - \beta$, where α and β can not be negative.

Further, let the angle *BAF* = *a*, *AFC* = *b*, so is $\alpha + \beta = a - b$; now by revolving the line *AF* away from the perpendicular *AC* we can make the angle *a* between *AF* and the parallel *AB* as small as we choose; so also can we lessen the angle *b*, consequently the two angles α and β can have no other magnitude than $\alpha = 0$ and $\beta = 0$.

It follows that in all rectilineal triangles the sum of the three angles is either π and at the same time also the parallel angle *II (p)* = ½ π for every line *p*, or for all triangles this sum is < π and at the same time also *II (p)* < ½ π.

The first assumption serves as *foundation for the ordinary geometry and plane trigonometry.*

The second assumption can likewise be admitted without leading to any contradiction in the results, and founds a new geometric science, to which I have given the name *Imaginary Geometry,* and which I intend here to expound as far as the development of the equations between the sides and angles of the rectilineal and spherical triangle.

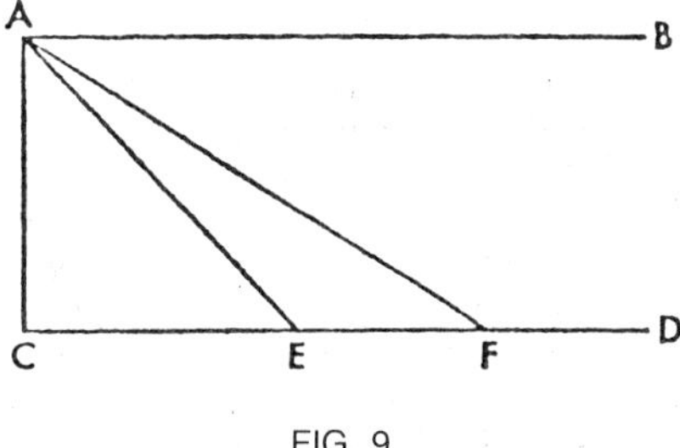

FIG. 9

(GEORG FRIEDRICH) BERNHARD RIEMANN (1826–66)

The son of a Lutheran pastor, Bernhard Riemann grew up in Hanover. He was a shy child whose health was frail. Despite the modest family income, he was able to obtain a fine education. In 1846, he matriculated at the University of Göttingen to study theology and philology, as his father wished. But he soon switched, with his father's consent, to mathematics. With the exception of Gauss, who taught only elementary courses, the general quality of mathematical education at Göttingen was poor, so Riemann went to the University of Berlin and studied under Dirichlet, Jacobi, and Steiner from 1847 to 1849. He then returned to Göttingen to complete his studies with the aged Gauss. In 1851, he submitted to Gauss his doctoral thesis on complex number theory and what are today called Riemann surfaces. The thesis was brilliant. It contained the Cauchy-Riemann differential equations that guarantee the analyticity of a function, improved upon the definition of an integral given by Cauchy in 1823, and, with the Riemann surface, introduced topological considerations into analysis.

It took Riemann over two years to complete his *Habilitationsschrift* to qualify as a *Privatdozent* (an official but unpaid lecturer) at Göttingen. The *Habilitationsschrift* dealt with Fourier series and the necessary conditions of continuity and limits. At the suggestion of Gauss, the author also examined the foundations of geometry. This paper, which is probably the richest for its size ever written in mathematics, gave a broad generalization of space and geometry. In it Riemann presented his non-Euclidean (elliptic) geometry, showing that Euclidean space was but a limited, special case of his more general system. The defense of the *Habilitationsschrift* must have been remarkable as the young and timid Riemann lectured the legendary Gauss on ideas that the latter had long secretly developed. Gauss was impressed with the young man's achievement.

Riemann was to spend his tragically brief teaching career on the faculty of the University of Göttingen; he was *Privatdozent* from 1855 to 1856, associate professor from 1857 to 1859, and full professor from 1859 to 1866. He married Elise Koch in 1862; they had a daughter. In July 1862, he suffered from pleuritis, which, although alleviated by sporadic recoveries, lead to a fatal case of tuberculosis. Tuberculosis had previously struck down his mother, brother, and three sisters. In an attempt to recuperate, he spent the winter of 1862–63 in Sicily. This and subsequent travel through Italy allowed him to indulge in his love of fine art. In June 1863, he returned to Göttingen, but his health deteriorated so rapidly he had to return quickly to Italy. Work drew him back to Göttingen in 1865. In June 1866, he went to Selasca on Lake Maggiore, where he died the following month.

Riemann was a profound and imaginative mathematician whose strength was conceptual rather than algorithmic. This strength is evident in his papers, which were edited for publication by Heinrich Weber and Richard Dedekind. In these his powerful thought is never concealed in a thicket of mere formulas. Among his contributions to mathematics was his invention of the elliptic version of non-Euclidean geometry that included Riemannian space, a multidimensional space which is a source of the tensor calculus. In his geometry he employed as an analytical tool the Dirichlet principle, which states that when a mathematical formalism with an associated function is created to conform to a real physical problem the function must exit. Riemannian geometry was later extended by Klein, Poincaré, and Minkowski and became the mathematical basis for Einstein's general theory of relativity.

There were other areas of mathematics that Riemann advanced. With Dirichlet he was the first to make a deep and deliberate use of analysis to tackle number theory. He investigated the Zeta function, introduced by Euler $\zeta(s) = \sum_{n=0}^{\infty} n^{-s}$, and established that the sum could be found for real s. He attempted to use the Zeta function to prove the prime number theorem: $\lim_{x\to\infty} \dfrac{\pi(x)}{\dfrac{x}{\log x}} = 1$. In mathematical physics he principally did research on sound waves.

98. From "On the Hypotheses Which Lie at the Foundations of Geometry" (1854)*

(Elliptic Geometry and a Distinction between Boundlessness and Infinitude of Straight Lines)

BERNHARD RIEMANN

PLAN OF THE INVESTIGATION

It is well known that geometry presupposes not only the concept of space but also the first fundamental notions for constructions in space as given in advance. It gives only nominal definitions for them, while the essential means of determining them appear in the form of axioms. The relation of these presuppositions is left in the dark; one sees neither whether and in how far their connection is necessary, nor a priori whether it is possible.

From Euclid to Legendre, to name the most renowned of modern writers on geometry, this darkness has been lifted neither by the mathematicians nor by the philosophers who have labored upon it. The reason of this lay perhaps in the fact that the general concept of multiply extended magnitudes, in which spatial magnitudes are comprehended, has not been elaborated at all. Accordingly I have proposed to myself at first the problem of constructing the concept of a multiply extended magnitude out of general notions of quantity. From this it will result that a multiply extended magnitude is susceptible of various metric relations and that space accordingly constitutes only a particular case of a

*Source: This translation of "Über die Hypothesen welche der Geometrie zu Grunde liegen" by Henry S. White is taken from David Eugene Smith (ed.), *A Source Book in Mathematics* (1929), 411-425. It is reprinted by permission of McGraw-Hill Book Company, Inc.

triply extended magnitude. A necessary sequel of this is that the propositions of geometry are not derivable from general concepts of quantity, but that those properties by which space is distinguished from other conceivable triply extended magnitudes can be gathered only from experience. There arises from this the problem of searching out the simplest facts by which the metric relations of space can be determined, a problem which in nature of things is not quite definite; for several systems of simple facts can be stated which would suffice for determining the metric relations of space; the most important for present purposes is that laid down for foundations by Euclid. These facts are, like all facts, not necessary but of a merely empirical certainty; they are hypotheses; one may therefore inquire into their probability, which is truly very great within the bounds of observation, and thereafter decide concerning the admissibility of protracting them outside the limits of observation, not only toward the immeasurably large, but also toward the immeasurably small.

I. THE CONCEPT OF N-FOLD EXTENDED MANIFOLD

While I now attempt in the first place to solve the first of these problems, the development of the concept of manifolds multiply extended, I think myself the more entitled to ask considerate judgment inasmuch as I have had little practise in such matters of a philosophical nature, where the difficulty lies more in the concepts than in the construction, and because I have not been able to make use of any preliminary studies whatever aside from some very brief hints which Privy Councillor Gauss has given on the subject in his second essay on biquadratic residues and in his Jubilee booklet, and some philosophical investigations of Herbart.

1

Notions of quantity are possible only where there exists already a general concept which allows various modes of determination. According as there is or is not found among these modes of determination a continuous transition from one to another, they form a continuous or a discrete manifold; the individual modes are called in the first case points, in the latter case elements of the manifold. Concepts whose modes of determination form a discrete manifold are so numerous, that for things arbitrarily given there can always be found a concept, at least in the more highly developed languages, under which they are comprehended (and mathematicians have been able therefore in the doctrine of discrete quantities to set out without scruple from the postulate that given things are to be considered as all of one kind); on the other hand there are in common life only such infrequent occasions to form concepts whose modes of determination form a continuous manifold, that the positions of objects of sense, and the colors, are probably the only simple notions whose modes of determination form a multiply extended manifold. More frequent occasion for the birth and development of these notions is first found in higher mathematics.

Determinate parts of a manifold, distinguished by a mark or by a boundary, are called quanta. Their comparison as to quantity comes in discrete magnitudes by counting, in continuous magnitude by measurement. Measuring consists in superposition of the magnitudes to be compared; for measurement there is requisite some means of carrying forward one magnitude as a measure for the other. In default of this, one can compare two magnitudes only when the one is a part of the other, and even then one can only decide upon the question of more and less, not upon the question of how many. The investigations which can be set on foot about them in this case form a general part of the doctrine of quantity independent of metric determinations, where magnitudes are thought of not as existing independent of position and not as expressible by a unit, but only as regions in a manifold. Such inquiries have become a necessity for several parts of mathematics, namely for the treatment of many-valued analytic functions, and the lack of them is likely a principal reason why the celebrated theorem of Abel and the contributions of Lagrange, Pfaff, and Jacobi to the theory of differential equations have remained so long unfruitful. For the present purpose it will be sufficient to bring forward conspicuously two points out of this general part of the doctrine of extended magnitudes, wherein nothing further is assumed than what was already contained in the concept of it. The first of these will make plain how the notion of a multiply extended manifold came to exist; the second, the reference of the determination of place in a given manifold to determinations of quantity and the essential mark of an n-fold extension.

2

In a concept whose various modes of determination form a continuous manifold, if one passes in a definite way from one mode of determination to another, the modes of determination which are traversed constitute a simply extended manifold and its essential mark is this, that in it a continuous progress is possible from any point only in two directions, forward or backward. If now one forms the thought of this manifold again passing over into another entirely different, here again in a definite way, that is, in such a way that every point goes over into a definite point of the other, then will all the modes of determination thus obtained form a doubly extended manifold. In similar procedure one obtains a triply extended manifold when one represents to oneself that a double extension passes over in a definite way into one entirely different, and it is easy to see how one can prolong this construction indefinitely. If one considers his object of thought as variable instead of regarding the concept as determinable, then this construction can be characterized as a synthesis of a variability of $n + 1$ dimensions out of a variability of n dimensions and a variability of one dimension.

3

I shall now show how one can conversely split up a variability, whose domain is given, into a variability of one dimension and a variability of fewer dimensions. To this end let one think of a variable portion of a manifold of one dimension,—reckoning from a fixed starting-point or origin, so that its values are comparable one with another—which has for every point of the given manifold a definite value changing continuously with that point; or in other words, let one assume within the given manifold a continuous function of place, and indeed a function such that it is not constant along any portion of this manifold. Every system of points in which the function has a constant value constitutes now a continuous manifold of fewer dimensions than that which was given. By change in the value of the function these manifolds pass over, one into another, continuously; hence one may assume that from one of them all the rest emanate, and this will come about, speaking generally, in such a way that every point of one passes over into a definite point of the other. Exceptional cases, and it is important to investigate them,—can be left out of consideration here. By this means the fixing of position in the given manifold is referred to the determination of one quantity and the fixing of position in a manifold of fewer dimensions. It is easy now to show that this latter has $n - 1$ dimensions if the given manifold was n-fold extended. Hence by repetition of this procedure, to n times, the fixing of position in an n-dimensional manifold is reduced to n determinations of quantities, and therefore the fixing of position in a given manifold is reduced, whenever this is possible, to the determination of a finite number of quantities. There are however manifolds in which the fixing of position requires not a finite number but either an infinite series or a continuous manifold of determinations of quantity. Such manifolds are constituted for example by the possible determinations of a function for a given domain, the possible shapes of a figure in space, et cetera.

II. RELATIONS OF MEASURE, OF WHICH AN N-DIMENSIONAL MANIFOLD IS SUSCEPTIBLE, ON THE ASSUMPTION THAT LINES POSSESS A LENGTH INDEPENDENT OF THEIR POSITION: THAT IS, THAT EVERY LINE CAN BE MEASURED BY EVERY OTHER

Now that the concept of an n-fold extended manifold has been constructed and its essential mark has been found to be this, that the determination of position therein can be referred to n determinations of magnitude, there follows as second of the problems proposed above, an investigation into the relations of measure that such a manifold is susceptible of, also into the conditions which suffice for determining these metric relations. These relations of measure can be investigated only in abstract notions of magnitude and can be exhibited connectedly only in formulae; upon certain assumptions, however, one is able to resolve them into relations which are separately capable of being represented geometrically, and by this means it becomes possible to express geometrically the results of the calculation. Therefore if one is to reach solid ground, an abstract investigation in formulae is indeed unavoidable, but its results will allow an exhibition in the clothing of geometry. For both parts the foundations are contained in the celebrated treatise of Privy Councillor Gauss upon curved surfaces.

1

Determinations of measure require magnitude to be independent of location, a state of things which can occur in more than one way. The assumption that first offers itself, which I intend here to follow out, is perhaps this, that the length of lines be independent of their situation, that therefore every line be measurable by every other. If the fixing of the location is referred to determinations of magnitudes, that is, if the location of a point in the n-dimensional manifold be expressed by n variable quantities x_1, x_2, x_3, and so on to x_n, then the determination of a line will reduce to this, that the quantities x be given as functions of a single variable. The problem is then, to set up a mathematical expression for the length of lines, and for this purpose the quantities x must be thought of as expressible in units. This problem I shall treat only under certain restrictions, and limit myself first to such lines as have the ratios of the quantities dx—the corresponding changes in the quantities x—changing continuously; one can in that case think of the lines as laid off into elements within which the ratios of the quantities dx may be regarded as constant, and the problem reduces then to this: to set up for every point a general expression for a line-element which begins there, an expression which will therefore contain the quantities x and the quantities dx. In the second place I now assume that the length of the line-element, neglecting quantities of the second order, remains unchanged when all its points undergo infinitely small changes of position; in this it is implied that if all the quantities dx increase in the same ratio, the line-element likewise changes in this ratio. Upon these assumptions it will be possible for the line-element to be an arbitrary homogeneous function of the first degree in the quantities dx which remains unchanged when all the dx change sign, and in which the arbitrary constants are continuous functions of the quantities x. To find the simplest cases, I look first for an expression for the $(n - 1)$-fold extended manifolds which are everywhere equally distant from the initial point of the line-element, that is, I look for a continuous function of place, which renders them distinct from one another. This will have to diminish or increase from the initial point out in all directions; I shall assume that it increases in all directions and therefore has a minimum in that point. If then its first and second differential quotients are finite, the differential of the first order must vanish and that of the second order must never be negative; I assume that it is always positive. This differential expression of the second order accordingly remains constant if ds remains constant, and increases in squared ratio when the quantities dx and hence also ds all change in the same ratio. That expression is therefore = const. ds^2, and consequently ds = the square root of an everywhere positive entire homogeneous function of the second degree in quantities dx having as coefficients continuous functions of the quantities x. For space this is, when one expresses the position of a point by rectangular coordinates, $ds = \sqrt{\Sigma(dx)^2}$; space is therefore comprised under this simplest case. The next case in order of simplicity would probably contain the manifolds in which the line-element can be expressed by the fourth root of a differential expression of the fourth degree. Investigation of this more general class indeed would require no essentially different principles, but would consume considerable time and throw relatively little new light upon the theory of space, particularly since the results cannot be expressed geometrically. I limit myself therefore to those manifolds in which the line-element is expressed by the square root of a differential expression of the second degree. Such an expression one can transform into another similar one by substituting for the n independent variables functions of n new independent variables. By this means however one cannot transform every expression into every other; for the expression contains $n \cdot \frac{n+1}{2}$ coefficients which are arbitrary functions of the independent variables; but by introducing new variables one can satisfy only n relations (conditions), and so can make only n of the coefficients equal to given quantities. There remain then $n \cdot \frac{n-1}{2}$ others completely determined by the nature of the manifold that is to be represented, and therefore for determining its metric relations $n \cdot \frac{n-1}{2}$ functions of position are requisite. The manifolds in which, as in the plane and in space, the line-element can be reduced to the form $\sqrt{\Sigma(dx)^2}$ constitute therefore only a particular case of the manifolds under consideration here. They deserve a particular name, and I will therefore term *flat* these manifolds in which the square of the line-element can be reduced to the sum of squares of total differentials. Now in order to obtain a conspectus of the essential differences of the manifolds representable in this prescribed form it is necessary to remove those that spring from the mode of representation, and this is accomplished by choosing the variable quantities according to a definite principle.

2

For this purpose suppose the system of shortest lines emanating from an arbitrary point to have been constructed. The position of an undetermined point will then be determinable by specifying the direction of that shortest line in which it lies and its distance, in that line, from the starting-point; and it can therefore be expressed by the ratios of the quantities dx^0, that is the limiting ratios of the dx at the starting point of this shortest line and by the length s of this line. Introduce now instead of the dx^0 such linear expressions $d\alpha$ formed from them, that the initial value of the square of the line-element equals the sum of the squares of these expressions, so that the independent variables are: the quantity s and the ratios of quantities $d\alpha$. Finally, set in place of the $d\alpha$ such quantities proportional to them, x_1, $x_2, \ldots, x_n$, that the sum of their squares $=s^2$. After introducing these quantities, the square of the line-element for indefinitely small values of x becomes $= \Sigma(dx)^2$, and the term of next order in that $(ds)^2$ will be equal to a homogeneous expression of the second degree in the $n\frac{n-1}{2}$ quantities $(x_1dx_2 - x_2dx_1)$, $(x_1dx_3 - x_3dx_1), \ldots$, that is, an indefinitely small quantity of dimension four; so that one obtains a finite magnitude when one divides it by the square of the indefinitely small triangle-area in whose vertices the values of the variables are $(0, 0, 0, \ldots)$, $(x_1, x_2, x_3, \ldots)$, $(dx_1, dx_2, dx_3, \ldots)$. This quantity retains the same value, so long as the quantities x and dx are contained in the same binary linear forms, or so long as the two shortest lines from the values 0 to the values x and from the values 0 to the values dx stay in the same element of surface, and it depends therefore only upon the place and the direction of that element. Plainly it is $= 0$ if the manifold represented is flat, that is if the square of the line-element is reducible to $\Sigma(dx)^2$, and it can accordingly be regarded as the measure of the divergence of the manifold from flatness in this point and in this direction of surface. Multiplied by $-\frac{3}{4}$ it becomes equal to the quantity which Privy Councillor Gauss has named the measure of curvature of a surface.

For determining the metric relations of an n-fold extended manifold representable in the prescribed form, in the foregoing discussion $n \cdot \frac{n-1}{2}$ functions of position were found needful; hence when the measure of curvature in every point in $n \cdot \frac{n-1}{2}$ surface-directions is given, from them can be determined the metric relations of the manifold, provided no identical relations exist among these values, and indeed in general this does not occur. The metric relations of these manifolds that have the line-element represented by the square root of a differential expression of the second degree can thus be expressed in a manner entirely independent of the choice of the variable quantities. A quite similar path to this goal can be laid out also in case of the manifolds in which the line-element is given in a less simple expression; e.g., as the fourth root of a differential expression of the fourth degree. In that case the line-element, speaking generally, would no longer be reducible to the form of a square root of a sum of squares of differential expressions; and therefore in the expression for the square of the line-element the divergence from flatness would be an indefinitely small quantity of the dimension two, while in the former manifolds it was indefinitely small of the dimension four. This peculiarity of the latter manifolds may therefore well be called flatness in smallest parts. The most important peculiarity of these manifolds, for present purposes, on whose account solely they have been investigated here, is however this, that the relations of those doubly extended can be represented geometrically by surfaces, and those of more dimensions can be referred to those of the surfaces contained in them; and this requires still a brief elucidation.

3

In the conception of surfaces, along with the interior metric relations, in which only the length of the paths lying in them comes into consideration, there is always mixed also their situation with respect to points lying outside them.

One can abstract however from external relations by carrying out such changes in the surfaces as leave unchanged the length of lines in them; *i.e.*, by thinking of them as bent in any arbitrary fashion,—without stretching—and by regarding all surfaces arising in this way one out of another as equivalent. For example, arbitrary cylindrical or conical surfaces are counted as equivalent to a plane, because they can be formed out of it by mere bending, while interior metric relations remain unchanged; and all theorems regarding them—the whole of planimetry—retain their validity; on the other hand they count as essentially distinct from the sphere, which cannot be converted into a plane without stretching. According to the above investigation in every point the interior metric relations of a doubly extended manifold are characterized by the measure of curvature if the line-element can be expressed by the square root of a differential expression of the second degree, as is the case with surfaces. An intuitional significance can be given to this quantity in the case of surfaces, namely that it is the product of the two curvatures of the surface in this point; or also, that its product into an indefinitely small triangle-area formed of shortest lines is equal to half the excess of its angle-sum above two right angles, when measured in radians. The first definition would presuppose the theorem that the product of the two radii of curvature is not changed by merely bending a surface; the second, the theorem that at one and the same point the excess of the angle-sum of an indefinitely small triangle above two right angles is proportional to its area. To give a tangible meaning to measure of curvature of an n-dimensional manifold at a given point and in a surface direction passing through that point, it is necessary to start out from the principle that a shortest line, originating in a point, is fully determined when its initial direction is given. According to this, a determinate surface is obtained when one prolongs into shortest lines all the initial directions going out from a point and lying in the given surface element; and this surface has in the given point a determinate measure of curvature, which is also the measure of curvature of the n-dimensional manifold in the given point and the given direction of surface.

4

Now before applications to space some considerations are needful regarding flat manifolds in general, *i.e.*, regarding those in which the square of the line-element is representable by a sum of squares of total differentials.

In a flat n-dimensional manifold the measure of curvature at every point is in every direction zero; but by the preceding investigation it suffices for determining the metric relations to know that at every point, in $n \cdot \frac{n-1}{2}$ surface directions whose measures of curvature are independent of one another, that measure is zero. Manifolds whose measure of curvature is everywhere zero may be regarded as a particular case of those manifolds whose curvature is everywhere constant. The common character of those manifolds of constant curvature can also be expressed thus: that the figures lying in them can be moved without stretching. For it is evident that the figures in them could not be pushed along and rotated at pleasure unless in every point the measure of curvature were the same in all directions. Upon the other hand, the metric relations of the manifold are completely determined by the measure of curvature. About any point, therefore, the metric relations in all directions are exactly the same as about any other point, and so the same constructions can be carried out from it, and consequently in manifolds with constant curvature every arbitrary position can be given to the figures. The metric relations of these manifolds depend only upon the value of the measure of curvature, and it may be mentioned, with reference to analytical presentation, that if one denotes this value by α, the expression for the line element can be given the form

$$\frac{1}{1 + \frac{\alpha}{4}\sqrt{\Sigma dx^2}}$$

5

Consideration of surfaces with constant measure of curvature can help toward a geometric exposition. It is easy to see that those surfaces whose curvature is positive will always permit themselves to be fitted upon a sphere whose radius is unity divided by the square root of the measure of curvature; but to visualize the complete manifold of these surfaces one should give to one of them the form of a sphere and to the rest the form of surfaces of rotation which touch it along the equator. Such surfaces as have greater curvature than this sphere will then touch the sphere from the inner side and take on a form like that exterior part of the surface of a ring which is turned away from the axis (remote from the axis); they could be shaped upon zones of spheres having a smaller radius, but would reach more than once around. Surfaces with lesser positive measure of curvature will be obtained by cutting out of spherical surfaces of greater radius a portion bounded by two halves of great circles, and making its edges adhere together. The surface with zero curvature will be simply a cylindrical surface standing upon the equator; the surfaces with negative curvature will be tangent to this cylinder externally and will be formed like the inner part of the surface of a ring, the part turned toward the axis.

If one thinks of these surfaces as loci for fragments of surface movable in them, as space is for bodies, then the fragments are movable in all these surfaces without stretching. Surfaces with positive curvature can always be formed in such wise that those fragments can be moved about without even bending, namely as spherical surfaces, not so however those with negative curvature. Beside this independence of position shown by fragments of surface, it is found in the surface with zero curvature that direction is independent of position, as is not true in the rest of the surfaces.

III. Application to Space

1

Following these investigations concerning the mode of fixing metric relations in an n-fold extended magnitude, the conditions can now be stated which are sufficient and necessary for determining metric relations in space, when it is assumed in advance that lines are independent of position and that the linear element is representable by the square root of a differential expression of the second degree; that is if flatness in smallest parts is assumed.

These conditions in the first place can be expressed thus: that the measure of the curvature in every point is equal to zero in three directions of surface; and therefore the metric relations of the space are determined when the sum of the angles in a triangle is everywhere equal to two right angles.

In the second place if one assumes at the start, like Euclid, an existence independent of situation not only for lines but also for bodies, then it follows that the measure of curvature is everywhere constant; and then the sum of the angles in all triangles is determined as soon as it is fixed for one triangle.

In the third place, finally, instead of assuming the length of lines to be independent of place and direction, one might even assume their length and direction to be independent of place. Upon this understanding the changes in place or differences in position are complex quantities expressible in three independent units.

2

In the course of preceding discussions, in the first place relations of extension (or of domain) were distinguished from those of measurement, and it was found that different relations of measure were conceivable along with identical relations of extension. Then were sought systems of simple determinations of measure by means of which the metric relations of space are completely determined and of which all theorems about such relations are a necessary consequence. It remains now to examine the question how, in what degree and to what extent these assumptions are guaranteed by experience. In this connection there subsists an essential difference between mere relations of extension and those of measurement: in the former, where the possible cases form a discrete manifold the declarations of experience are indeed never quite sure, but they are not lacking in exactness; while in the latter, where possible cases form a continuum, every determination based on experience remains always inexact, be the probability that it is nearly correct ever so great. This antithesis becomes important when these empirical determinations are extended beyond the limits of observation into the immeasurably great and the immeasurably small; for the second kind of relations obviously might become ever more inexact, beyond the bounds of observation, but not so the first kind.

When constructions in space are extended into the immeasurably great, unlimitedness must be distinguished from infiniteness; the one belongs to relations of extension, the other to those of measure. That space is an unlimited, triply extended manifold is an assumption applied in every conception of the external world; by it at every moment the domain of real perceptions is supplemented and the possible locations of an object that is sought for are constructed, and in these applications the assumption is continually being verified. The unlimitedness of space has therefore, a greater certainty, empirically, than any experience of the external. From this, however, follows in no wise its infiniteness, but on the contrary space would necessarily be finite, if one assumes that bodies are independent of situation and so ascribes to space a constant measure of curvature, provided this measure of curvature had any positive value however small. If one were to prolong the elements of direction, that lie in any element of surface, into shortest lines (geodetics), one would obtain an unlimited surface with constant positive measure of curvature, consequently a surface which would take on, in a triply extended manifold, the form of a spherical surface, and would therefore be finite.

3

Questions concerning the immeasurably large area, for the explanation of Nature, [are] useless questions. Quite otherwise is it however with questions concerning the immeasurably small. Knowledge of the causal connection of phenomena is based essentially upon the precision with which we follow them down into the infinitely small. The progress of recent centuries in knowledge of the mechanism of Nature has come about almost solely by the exactness of the syntheses rendered possible by the invention of Analysis of the infinite and by the simple fundamental concepts devised by Archimedes, Galileo, and Newton, and effectively employed by modern Physics. In the natural sciences however, where simple fundamental concepts are still lacking for such syntheses, one pursues phenomen into the spatially small, in order to perceive causal connections, just as far as the microscope permits. Questions concerning spatial relations of measure in the indefinitely small are therefore not useless.

If one premise that bodies exist independently of position, then the measure of curvature is everywhere constant; then from astronomical measurements it follows that it cannot differ from zero; at any rate its reciprocal value would have to be a surface in comparison with which the region accessible to our telescopes would vanish. If however bodies have no such non-dependence upon position, then one cannot conclude to relations of measure in the indefinitely small from those in the large. In that case the curvature can have at every point arbitrary values in three directions, provided only the total curvature of every metric portion of space be not appreciably different from zero. Even greater complications may arise in case the line element is not representable, as has been premised, by the square root of a differential expression of the second degree. Now however the empirical notions on which spatial measurements are based appear to lose their validity when applied to the indefinitely small, namely the concept of a fixed body and that of a light-ray; accordingly it is entirely conceivable that in the indefinitely small the spatial relations of size are not in accord with the postulates of geometry, and one would indeed be forced to this assumption as soon as it would permit a simpler explanation of the phenomena.

The question of the validity of the postulates of geometry in the indefinitely small is involved in the question concerning the ultimate basis of relations of size in space. In connection with this question, which may well be assigned to the philosophy of space, the above remark is applicable, namely that while in a discrete manifold the principle of metric relations is implicit in the notion of this manifold, it must come from somewhere else in the case of a continuous manifold. Either then the actual things forming the groundwork of a space must constitute a discrete manifold, or else the basis of metric relations must be sought for outside that actuality, in colligating forces that operate upon it.

A decision upon these questions can be found only by starting from the structure of phenomena that has been approved in experience hitherto, for which Newton laid the foundation, and by modifying this structure gradually under the compulsion of facts which it cannot explain. Such investigations as start out, like this present one, from general notions, can promote only the purpose that this task shall not be hindered by too restricted conceptions, and that progress in perceiving the connection of things shall not be obstructed by the prejudices of tradition.

This path leads out into the domain of another science, into the realm of physics, into which the nature of this present occasion forbids us to penetrate.

Chapter VIII
The Nineteenth Century

Section C
The Arithmetization of Mathematical Analysis

(JEAN-BAPTISTE-) JOSEPH FOURIER (1768–1830)

Joseph Fourier, French mathematician, Egyptologist, and public servant, was the 19th child of Joseph Fourier (a tailor) and his wife Edmée. Both parents died before he was nine. On the recommendation of a family friend, the bishop of Auxerre placed him in the local military school conducted by the Benedictines. At about age 13, his passion for mathematics developed. His emerging mathematical proclivity should have helped him realize his main youthful ambition—to become an army artillery or engineering officer. It did not. Despite a strong recommendation from Adrien-Marie Legendre, he was refused admission to the officer corps because he was "not . . . of noble birth" and it made no difference "even if he is a second Newton." Fourier thereupon entered the Benedictine abbey of Benoît-sur-Loire as a novice in 1787. With the outbreak of the French Revolution two years later, Fourier ended his novitiate and returned to his old school in Auxerre to teach mathematics, rhetoric, philosophy, and history.

The years of the French Revolution and the early Napoleonic period were turbulent for Fourier. Inspired by the ideals of the Revolution, he joined a group associated with the Jacobins and became prominent in local affairs. A courageous defense of victims of the Terror and a criticism of corrupt practices of governmental officials led to his arrest and brief imprisonment by the Jacobin government of Robespierre in 1794. After his release from prison, he attended the short-lived École Normale in Paris. When it closed in May 1795, he was appointed to an assistant teaching post at the École Polytechnique. He was assigned to aid Monge in his course on descriptive geometry and Lagrange in his analysis courses. A new political regime, the Directory, arrested him on the charge that he had supported Robespierre, but his colleagues at the École Polytechnique soon won his release. In 1798, Monge chose him to accompany Napoleon's Egyptian expedition. For the next three years, Fourier organized factories for the army, advised on engineering and diplomatic projects, and was secretary of the Institut d'Egypte, which Napoleon established in Cairo. As secretary of that institute he conducted extensive research on Egyptian antiquities.

After returning to France in 1801, Fourier continued to serve Napoleon's government with distinction. Although he hoped to resume a teaching and research position at the École

Polytechnique, Napoleon, who had noted his administrative genius, appointed him *prefect* (administrator) for the *département* of Isére with its headquarters in Grenoble in southeastern France. From 1802 to 1814, he vigorously directed a policy of public improvement that included the drainage of large marshlands and the planning of a new road from Grenoble to Turin (now route N91—strada 23). His administrative duties did not interrupt his mathematical and Egyptological work. Indeed, Napoleon placed him in charge of publishing the enormous mass of materials gathered on the Egyptian expedition in the *Description de l'Egypte*. His lengthy and, in places, controversial historical preface on ancient Egypt established his literary reputation. For his many services, Fourier was created a Baron of the Empire in 1809. With Napoleon's defeat after the "Hundred Days" in 1815, Fourier's career reached a low point. He found himself in disgrace with a Bonapartist's reputation, without a job, and with only a small pension that was soon annulled.

After 1815, Fourier's personal fortunes and scientific reputation once more rose although not without opposition. The new king, Louis XVIII, refused to approve his election to the Institut de France in 1816 but allowed it to proceed the following year. Through the intercession of a former pupil who was the prefect of Paris, Fourier was also appointed director of the Statistical Bureau of the Seine, a position that allowed him ample time for scholarly studies. With the support of Laplace, his status in the scientific community improved despite continuing criticism of his work by his outclassed rivals Siméon Denis Poisson and Jean Baptiste Biot. In 1822, he was elected to the powerful position of perpetual secretary of the Paris Academy of Sciences, a section of the Institut. In 1827, he was elected to the French Academy for his contributions in Egyptology. The Royal Society of London also elected him a foreign member.

Fourier encouraged and supported younger scientists, such as Oersted, Dirichlet, Sturm, and Liouville. Unfortunately, he also mishandled important papers that two young geniuses submitted to the Paris Academy. As its secretary he had to distribute these to referees. In 1826, he sent Abel's masterpiece on transcendental functions to Cauchy but failed to request its return when Cauchy misplaced it. In 1830, he lost a second paper on the resolution of equations which Galois had submitted to the Academy. Mishandling of these appears to have been the result of disorganization and illness rather than of suppression. During his last five years, Fourier suffered from chronic rheumatism and myxoedema, a disease of the thyroid gland which he may have contacted in Egypt. To keep warm, even in summer he wrapped himself in thick woolen clothes like "an Egyptian mummy" and kept the temperature in his apartment very high. After two weeks of suffering from nervous angina complicated by a heart problem, he died in May 1830.

The most famous scientific achievements of Fourier were his studies of heat diffusion and the introduction of a new range of powerful mathematical techniques for solving related partial differential equations. Seeing heat as a primary physical agent in the universe, he began the investigation of heat diffusion in 1807. This led to a prize winning paper submitted to the Institut de France in 1811. Although the referees awarded his paper the Institut prize, they criticized it for lack of mathematical "rigor and generality." Fourier considered the reproach unjustified and with reason. Among his contemporaries only Abel and Cauchy surpassed him in their adherence to standards of rigor. He set out to expand the mathematical portions of the paper. The result was the classic book, *Théorie analytique de la chaleur* ("The Analytical Theory of Heat," 1822). In the tradition of the Bernoullis and Euler, he stressed simplicity as a criterion for physical explanation. His analysis of heat diffusion went beyond the scope of the rational and celestial mechanics dominant since Newton's *Principia* and thus provided a major extension of mathematical physics.

Fourier's efforts to further the study of the rational mechanics of heat culminated in his chief contributions to mathematics. For example, he provided the Fourier series as a series solution to the equation of heat flow. Although Euler and other 18th-century geometers had occasionally employed these trigonometric expansions of functions, he first established their significance in modern mathematics. (Later studies of their validity by Dirichlet, Riemann, Lebesgue, and others fundamentally renewed the concept of real function.) He derived the Fourier integral

$$[\pi f(x) = \int_{-\infty}^{\infty} f(t)dt \int_{-\infty}^{\infty} \cos q\,(x - t)\, dq]$$

in 1816 and his operator calculus in the early 1820s. With these results he created a general method for solving linear differential equations. In other research, he developed the basic theory of the transcendental function known today by the misnomer "Bessel function." In the theory of equations, he discovered an inductive proof of Descartes rule of signs, generalized the estimate of the number of real roots of a polynomial equation $f(x) = 0$ for $a \leqslant x \leqslant b$, and introduced linear programming. In probability theory, he estimated the errors of measurement from a large number of observations.

Fourier's superb mastery of analytical technique, his refined notation (including his invention of $\int_a^b$), and his powerful physical intuition underlay his success in mathematics. He believed that every mathematical statement had a possible physical meaning and conversely that "the profound study of nature is the most fertile source of mathematical discoveries." Because of his mathematization of physical theory, Auguste Comte, the founder of positivism who lectured at the École Polytechnique beginning in 1829, adopted Fourier as a philosophical patron.

99. From *Joseph Fourier 1768-1830**

I. GRATTAN-GUINNESS

11. Solution for the Annulus: The Full Fourier Series for an Arbitrary Function

Fourier now . . . turned to his next body, the annulus of radius R, for which he had found in article 23 . . . the equation

$$(11.1)\qquad \frac{\partial z}{\partial t} = \frac{K}{CD}\frac{\partial^2 z}{\partial x^2} - \frac{hl}{CDS}z,$$

where x is the angular variable on the annulus. He began his solution with the transformation

$$(11.2)\qquad z = e^{-ht}v,$$

which converted (11.1) into the orthodox diffusion equation

$$(11.3)\qquad \frac{\partial v}{\partial t} = K\frac{\partial^2 v}{\partial x^2}$$

(where K represents the previous K/CD), and effectively eliminated external diffusion for the model. This was the first place in his paper that he tackled the diffusion equation: the (function of x) $\times$ (function of t) solution form which applied to it was either $e^{-kn^2t}\sin nx$ or $e^{-kn^2t}\cos nx$, and therefore the general solution would be made up of a linear combination of such terms. The arbitrary initial temperature distribution would lead, therefore, to the full "Fourier series":

$$(11.4)\qquad \phi(x) = b_o + \sum_{r=1}^{\infty}(a_r \sin rx + b_r \cos rx),$$

in which the integration term-by-term method over $[0, 2\pi]$ would quickly find the values of the coefficients, and thus give the solution to (11.3):

$$(11.5)\qquad \pi Rz = e^{-ht}\left\{\frac{1}{2}\int_0^{2\pi}\phi\left(\frac{q}{R}\right)dq + \sum_{r=1}^{\infty}\left[\sin rx\int_0^{2\pi}\phi\left(\frac{q}{R}\right)\sin rq\,dq + \cos rx\int_0^{2\pi}\phi\left(\frac{q}{R}\right)\cos rq\,dq\right]e^{-\frac{r^2Kt}{R^2}}\right\}.$$

The first application that Fourier made of (11.5) was to the steady-state version of (11.1):

$$(11.6)\qquad \frac{d^2z}{dx^2} = \frac{hl}{KS}z,$$

and to the solution

$$(11.7)\qquad z = a\alpha^x + b\alpha^{-x},\ \alpha = e^{-\sqrt{hl/KS}},$$

. . . He put (11.7) into his general result (11.5) with $t = 0$ and thus obtained the cosine series for $(e^u + e^{-u})$ similar to the sine series for $(e^x - e^{-x})$ . . . From (11.5) he was also able to find the general time-dependent solution for this case.

The next application was to a solution where half the annulus was at 1 unit of temperature and the other half at 0 units, that is,

$$(11.8)\qquad \phi(x) = \begin{cases}1 \text{ when } & 0 \leq x \leq \pi\\ 0 \text{ when } & \pi < x \leq 2\pi.\end{cases}$$

Then he turned his attention to the mean temperature, showing that it would be shared by all points of the annulus if the interior conductivity was infinitely large or the radius infinitely small, and that in general it followed an exponential law of decay with time. Next he formulated the quantity of heat flow through a section of the annulus

Source: I. Grattan-Guinness, *Joseph Fourier 1768-1830* (1972), 254-255. This selection is reprinted by permission of the MIT Press. Comparative references omitted.

over a given period of time as a double integral, and then showed that the final distribution of temperature was such that the average for diametrically opposed points equaled the mean. In fact one pair of points would maintain the mean value constantly and divide the annulus into two halves which would mirror the variation of temperature along their respective lengths; each term of the general solution (11.5) could be interpreted this way and so the general solution could be seen as a combination of an infinity of examples of this kind of behavior.

The purpose of these articles was partly to prepare for later experiments, partly to interpret the series solution as heat flow, and always to show the further power of his method in its handling (for the first time in the paper) of a time-dependent, as opposed to a steady-state, diffusion problem.

100. From *Théorie analytique de la chaleur* (1822)*

JOSEPH FOURIER

Primary causes are unknown to us; but are subject to simple and constant laws, which may be discovered by observation, the study of them being the object of natural philosophy.

Heat, like gravity, penetrates every substance of the universe, its rays occupy all parts of space. The object of our work is to set forth the mathematical laws which this element obeys. The theory of heat will hereafter form one of the most important branches of general physics.

The knowledge of rational mechanics, which the most ancient nations had been able to acquire, has not come down to us, and the history of this science, if we except the first theorems in harmony, is not traced up beyond the discoveries of Archimedes. This great geometer explained the mathematical principles of the equilibrium of solids and fluids. About eighteen centuries elapsed before Galileo, the originator of dynamical theories, discovered the laws of motion of heavy bodies. Within this new science Newton comprised the whole system of the universe. The successors of these philosophers have extended these theories, and given them an admirable perfection: they have taught us that the most diverse phenomena are subject to a small number of fundamental laws which are reproduced in all the acts of nature. It is recognised that the same principles regulate all the movements of the stars, their form, the inequalities of their courses, the equilibrium and the oscillations of the seas, the harmonic vibrations of air and sonorous bodies, the transmission of light, capillary actions, the undulations of fluids, in fine the most complex effects of all the natural forces, and thus has the thought of Newton been confirmed: *quod tam paucis tam multa prœstet geometria gloriatur.*

But whatever may be the range of mechanical theories, they do not apply to the effects of heat. These make up a

*Source: This translation by Alexander Freeman is taken from Robert M. Hutchins (ed.), *Great Books of the Western World* (1955), vol. 45, 169-173.

special order of phenomena, which cannot be explained by the principles of motion and equilibrium. We have for a long time been in possession of ingenious instruments adapted to measure many of these effects; valuable observations have been collected; but in this manner partial results only have become known, and not the mathematical demonstration of the laws which include them all.

I have deduced these laws from prolonged study and attentive comparison of the facts known up to this time: all these facts I have observed afresh in the course of several years with the most exact instruments that have hitherto been used.

To found the theory, it was in the first place necessary to distinguish and define with precision the elementary properties which determine the action of heat. I then perceived that all the phenomena which depend on this action resolve themselves into a very small number of general and simple facts; whereby every physical problem of this kind is brought back to an investigation of mathematical analysis. From these general facts I have concluded that to determine numerically the most varied movements of heat, it is sufficient to submit each substance to three fundamental observations. Different bodies in fact do not possess in the same degree the power to *contain* heat, *to receive or transmit it across their surfaces,* nor to *conduct* it through the interior of their masses. These are the three specific qualities which our theory clearly distinguishes and shows how to measure.

It is easy to judge how much these researchers concern the physical sciences and civil economy, and what may be their influence on the progress of the arts which require the employment and distribution of heat. They have also a necessary connection with the system of the world, and their relations become known when we consider the grand phenomena which take place near the surface of the terrestrial globe.

In fact, the radiation of the sun in which this planet is incessantly plunged, penetrates the air, the earth, and the waters; its elements are divided, change in direction every way, and, penetrating the mass of the globe, would raise its mean temperature more and more, if the heat acquired were not exactly balanced by that which escapes in rays from all points of the surface and expands through the sky.

Different climates, unequally exposed to the action of solar heat, have, after an immense time, acquired the temperatures proper to their situation. This effect is modified by several accessory causes, such as elevation, the form of the ground, the neighbourhood and extent of continents and seas, the state of the surface, the direction of the winds.

The succession of day and night, the alternations of the seasons occasion in the solid earth periodic variations, which are repeated every day or every year: but these changes become less and less sensible as the point at which they are measured recedes from the surface. No diurnal variation can be detected at the depth of about three metres [ten feet]; and the annual variations cease to be appreciable at a depth much less than sixty metres. The temperature at great depths is then sensibly fixed at a given place: but it is not the same at all points of the same meridian; in general it rises as the equator is approached.

The heat which the sun has communicated to the terrestrial globe, and which has produced the diversity of climates, is now subject to a movement which has become uniform. It advances within the interior of the mass which it penetrates throughout, and at the same time recedes from the plane of the equator, and proceeds to lose itself across the polar regions.

In the higher regions of the atmosphere the air is very rare and transparent, and retains but a minute part of the heat of the solar rays; this is the cause of the excessive cold of elevated places.

The lower layers, denser and more heated by the land and water, expand and rise up: they are cooled by the very fact of expansion. The great movements of the air, such as the trade winds which blow between the tropics, are not determined by the attractive forces of the moon and sun. The action of these celestial bodies produces scarcely perceptible oscillations in a fluid so rare and at so great a distance. It is the changes of temperature which periodically displace every part of the atmosphere.

The waters of the ocean are differently exposed at their surface to the rays of the sun, and the bottom of the basin which contains them is heated very unequally from the poles to the equator. These two causes, ever present, and combined with gravity and the centrifugal force, keep up vast movements in the interior of the seas. They displace and mingle all the parts, and produce those general and regular currents which navigators have noticed.

Radiant heat which escapes from the surface of all bodies, and traverses elastic media, or spaces void of air, has special laws, and occurs with widely varied phenomena. The physical explanation of many of these facts is already known; the mathematical theory which I have formed gives an exact measure of them. It consists, in a manner, in a new catoptrics which has its own theorems, and serves to determine by analysis all the effects of heat direct or reflected.

The enumeration of the chief objects of the theory sufficiently shows the nature of the questions which I have proposed to myself. What are the elementary properties which it is requisite to observe in each substance, and what are the experiments most suitable to determine them exactly? If the distribution of heat in solid matter is regulated by constant laws, what is the mathematical expression of those laws, and by what analysis may we derive from this expression the complete solution of the principal problems? Why do terrestrial temperatures cease to be variable at a depth so small with respect to the radius of the earth? Every inequality in the movement of this planet necessarily occasioning an oscillation of the solar heat beneath the surface, what relation is there between the duration of its period, and the depth at which the temperatures become constant?

What time must have elapsed before the climates could acquire the different temperatures which they now maintain; and what are the different causes which can now vary their mean heat? Why do not the annual changes alone in the distance of the sun from the earth, produce at the surface of the earth very considerable changes in the temperatures?

From what characteristic can we ascertain that the earth has not entirely lost its original heat; and what are the exact laws of the loss?

If, as several observations indicate, this fundamental heat is not wholly dissipated, it must be immense at great depths, and nevertheless it has no sensible influence at the present time on the mean temperature of the climates. The effects which are observed in them are due to the action of the solar rays. But independently of these two sources of heat, the one fundamental and primitive proper to the terrestrial globe, the other due to the presence of the sun, is there not a more universal cause, which determines *the temperature of the heavens,* in that part of space which the solar system now occupies? Since the observed facts necessitate this cause, what are the consequences of an exact theory in this entirely new question; how shall we be able to determine that constant value of *the temperature of space*, and deduce from it the temperature which belongs to each planet?

To these questions must be added others which depend on the properties of radiant heat. The physical cause of the reflection of cold, that is to say the reflection of a lesser degree of heat, is very distinctly known; but what is the mathematical expression of this effect?

On what general principles do the

atmospheric temperatures depend, whether the thermometer which measures them receives the solar rays directly, on a surface metallic or unpolished, or whether this instrument remains exposed, during the night, under a sky free from clouds, to contact with the air, to radiation from terrestrial bodies, and to that from the most distant and coldest parts of the atmosphere?

The intensity of the rays which escape from a point on the surface of any heated body varying with their inclination according to a law which experiments have indicated, is there not a necessary mathematical relation between this law and the general fact of the equilibrium of heat; and what is the physical cause of this inequality in intensity?

Lastly, when heat penetrates fluid masses, and determines in them internal movements by continual changes of the temperature and density of each molecule, can we still express, by differential equations, the laws of such a compound effect; and what is the resulting change in the general equations of hydrodynamics?

Such are the chief problems which I have solved, and which have never yet been submitted to calculation. If we consider further the manifold relations of this mathematical theory to civil uses and the technical arts, we shall recognize completely the extent of its applications. It is evident that it includes an entire series of distinct phenomena, and that the study of it cannot be omitted without losing a notable part of the science of nature.

The principles of the theory are derived, as are those of rational mechanics, from a very small number of primary facts, the causes of which are not considered by geometers, but which they admit as the results of common observations confirmed by all experiment.

The differential equations of the propagation of heat express the most general conditions, and reduce the physical questions to problems of pure analysis and this is the proper object of theory. They are not less rigorously established than the general equations of equilibrium and motion. In order to make this comparison more perceptible, we have always preferred demonstrations analogous to those of the theorems which serve as the foundation of statics and dynamics. These equations still exist, but receive a different form, when they express the distribution of luminous heat in transparent bodies, or the movements which the changes of temperature and density occasion in the interior of fluids. The coefficients which they contain are subject to variations whose exact measure is not yet known, but in all the natural problems which it most concerns us to consider, the limits of temperature differ so little that we may omit the variations of these coefficients.

The equations of the movement of heat, like those which express the vibrations of sonorous bodies, or the ultimate oscillations of liquids, belong to one of the most recently discovered branches of analysis, which it is very important to perfect. After having established these differential equations their integrals must be obtained; this process consists in passing from a common expression to a particular solution subject to all the given conditions. This difficult investigation requires a special analysis founded on new theorems, whose object we could not in this place make known. The method which is derived from them leaves nothing vague and indeterminate in the solutions, it leads them up to the final numerical applications, a necessary condition of every investigation, without which we should only arrive at useless transformations.

The same theorems which have made known to us the equations of the movement of heat, apply directly to certain problems of general analysis and dynamics whose solution has for a long time been desired.

Profound study of nature is the most fertile source of mathematical discoveries. Not only has this study, in offering a determinate object to investigation, the advantage of excluding vague questions and calculations without issue; it is besides a sure method of forming analysis itself, and of discovering the elements which it concerns us to know, and which natural science ought always to preserve: these are the fundamental elements which are reproduced in all natural effects.

We see, for example, that the same expression whose abstract properties geometers had considered, and which in this respect belongs to general analysis, represents as well the motion of light in the atmosphere, as it determines the laws of diffusion of heat in solid matter, and enters into all the chief problems of the theory of probability.

The analytical equations, unknown to the ancient geometers, which Descartes was the first to introduce into the study of curves and surfaces, are not restricted to the properties of figures, and to those properties which are the object of rational mechanics; they extend to all general phenomena. There cannot be a language more universal and more simple, more free from errors and from obscurities, that is to say more worthy to express the invariable relations of natural things.

Considered from this point of view, mathematical analysis is as extensive as nature itself; it defines all perceptible relations, measures times, spaces, forces, temperatures; this difficult science is formed slowly, but it preserves every principle which it has once acquired; it grows and strengthens itself incessantly in the midst of the many variations and errors of the human mind.

Its chief attribute is clearness; it has no marks to express confused notions. It brings together phenomena the most diverse, and discovers the hidden analogies which unite them. If matter escapes us, as that of air and light, by its extreme tenuity, if bodies are placed far from us in the immensity of space, if man wishes to know the aspect of the heavens at successive epochs separated by a great number of centuries, if the actions of gravity and of heat are exerted in the interior of the earth at depths which will be always inaccessible, mathematical analysis can yet lay hold of the laws of these phenomena. It makes them present and measurable, and seems to be a faculty of the human mind destined to supplement the shortness of life and the imperfection of the senses; and what is still more remarkable, it follows the same course in the study of all phenomena; it interprets them by the same language, as if to attest the unity and simplicity of the plan of the universe, and to make still more evident that unchangeable order which presides over all natural causes.

Chapter VIII
The Nineteenth Century

Section C
The Arithmetization of Mathematical Analysis

101. From "On the Continuity of Functions Defined by Power Series" (1826)[1]*

(The Binomial Series; Convergence of Power Series)

NIELS ABEL

[After some critical paragraphs concerning the contemporary state of knowledge and rigor in using infinite series, Abel continues as follows.]

One of the most remarkable series of algebraic analysis is the following [binomial series[2]]:

$$1+\frac{m}{1}x+\frac{m(m-1)}{1\cdot 2}x^2+\frac{m(m-1)(m-2)}{1\cdot 2\cdot 3}x^3+\cdots+\frac{m(m-1)\cdots[m-(n-1)]}{1\cdot 2\cdots n}x^n+\cdots.$$

When m is a positive integer, the sum of the series, which is then finite, represents $(1+x)^m$, as is known. When m is not an integer, the series is infinite, and it converges or diverges depending on the values of m and x. In this case, one can write the same equality

$$(1+x)^m=1+\frac{m}{1}x+\frac{m(m-1)}{1\cdot 2}x^2+\cdots,$$

but then the equation shows only that the two expressions $(1+x)^m$ and $1+(m/1)x+[m(m-1)/1\cdot 2]x^2+\cdots$ have certain common properties, from which, for certain values of m and x, follows the *numerical* equality of the expressions. It is assumed that numerical equality will occur whenever the series is convergent, but this has never yet been proved. Hitherto, not even all the cases in which the series converges have been examined. Even if the existence of the above equality is assumed, one must still find the *value* of $(1+x)^m$, for in general this expression has infinitely many different values, whereas the series $1+m+\cdots$ has only one.

The aim of this memoir is to . . . solve completely the problem of summing the [binomial] series for all real and complex values of x and m from which it is convergent.

We will first establish some necessary theorems on series. The excellent work of

*Source: Reprinted by permission of the publishers from *A Source Book in Classical Analysis*, edited by Garrett Birkhoff, Cambridge, Mass.: Harvard University Press, Copyright © 1973 by the President and Fellows of Harvard College. The biography of Abel appears before selection 93.

In parts III and IV of this paper Abel stringently proves the binomial formula for arbitrary real and complex exponents.

Cauchy, *Cours d'analyse de l'École Polytechnique,*[3] which should be read by every analyst who likes rigor in mathematical investigations, will serve to guide us.

Definition. An arbitrary series

$$v_0 + v_1 + v_2 + \cdots + v_m + \cdots$$

will be called *convergent* if the [partial] sum $v_0 + v_1 + \cdots + v_m$ tends to some limit as m increases [without limit]. This limit will be called the *sum of the series.* In the contrary case the series will be called *divergent,* and then it has no sum. From this definition it follows that, for a series to be convergent, it is necessary and sufficient that the sum $v_m + v_{m+1} + \cdots v_{m+n}$ shall tend to zero as m increases [without limit], regardless of the value of n.

In any convergent series, therefore, the general term v_m tends to zero.[4]

Theorem I. If a series of positive quantities is denoted by $r_0, r_1, r_2, \ldots$, and if, for monotonically increasing values of m, the ratio r_{m+1}/r_m approaches a limit α which is greater than 1, then the series

$$c_0r_0 + c_1r_1 + c_2r_2 + \cdots + c_mr_m + \cdots$$

will necessarily *diverge* as m increases without limit, unless the sequence $\{c_m\}$ tends to zero.

Theorem II. Given the series of positive quantities $r_0 + r_1 + r_2 + \cdots + r_m + \cdots$, if the quotient r_{m+1}/r_m tends to a limit β which is *less than* 1 as m increases without limit, then the series

$$c_0r_0 + c_1r_1 + c_2r_2 + \cdots + c_mr_m + \cdots,$$

where $c_0, c_1, c_2, \ldots$ are quantities which *do not exceed* 1, will necessarily converge.

Indeed, by hypothesis m can always be taken large enough that one will have $r_{m+1} < \alpha r_{m+2} < r_{m+2} < \cdots < r_{m+n} < \alpha r_{m+n-1}$ [where $\beta < \alpha < 1$].[5] Thence it follows that $r_{m+k} < \alpha^k r_m$, and hence

$$r_m + r_{m+1} + \cdots + r_{m+n} < r_m(1 + \alpha + \cdots + \alpha^n) < \frac{r_m}{(1-\alpha)},$$

so that *a fortiori*

$$c_mr_m + c_{m+1}r_{m+1} + \cdots + c_{m+n}r_{m+n} < \frac{r_m}{(1-\alpha)}.$$

Since, however, $r_{m+k} < \alpha^k r_m$ and $\alpha < 1$, it is clear that r_m, and hence also the sum

$$c_mr_m + c_{m+1}r_{m+1} + \cdots + c_{m+n}r_{m+n},$$

will have zero as limit. Hence the series given above is convergent.

Theorem III. If $u_0, u_1, u_2, \ldots, u_m, \ldots$ denote a sequence of arbitrary quantities, and if

$$s_m = u_0 + u_1 + u_2 + \cdots + u_m$$

is always less than some fixed quantity δ, then one has

$$R = c_0u_0 + c_1u_1 + c_2u_2 + \cdots + c_mu_m < \delta c_0,$$

whenever $c_0, c_1, c_2, \ldots$ are positive decreasing quantities.

Indeed, one has

$$u_0 = s_0, \quad u_1 = s_1 - s_0, \quad u_2 = s_2 - s_1, \ldots,$$

hence

$$R = c_0 s_0 + c_1(s_1 - s_0) + c_2(s_2 - s_1) + \cdots + c_m(s_m - s_{m-1}),$$

or else

$$R = s_0(c_0 - c_1) + s_1(c_1 - c_2) + \cdots + s_{m-1}(c_{m-1} - c_m) + s_m c_m.$$

Now since the differences $c_0 - c_1$, $c_1 - c_2$, ... are positive, the quantity r is obviously smaller than δc_0.

Definition. A function $f(x)$ is called a continuous function of x between the limits $x = a$ and $x = b$ if, for any value of x between these limits, the quantity $f(x - \beta)$ tends to the limit $f(x)$ as β tends to zero ["decreases continually"].

Theorem IV. If the series

$$f(\alpha) = v_0 + v_1\alpha + v_2\alpha^2 + \cdots + v_m\alpha^m + \cdots$$

converges for a fixed value δ of α, it will also converge for every smaller value of α, in such a way that as β tends to zero $f(\alpha - \beta)$ tends to the limit $f(\alpha)$, assuming that $\alpha \leqslant \delta$.[6]

Let

$$v_0 + v_1\alpha + \cdots + v_{m-1}\alpha^{m-1} = \phi(\alpha),$$

$$v_m\alpha^m + v_{m+1}\alpha^{m+1} + \cdots = \psi(\alpha);$$

then one has

$$\psi(\alpha) = \left(\frac{\alpha}{\delta}\right)^m v_m\delta^m + \left(\frac{\alpha}{\delta}\right)^{m+1} v_{m+1}\,\delta^{m+1} + \cdots;$$

hence, by Theorem III, $\psi(\alpha) < (\alpha/\delta)^m s$, where s denotes the largest of the quantities $v_m\delta^m$, $v_m\delta^m + v_{m+1}\,\delta^{m+1}$, $v_m\delta^m + v_{m+1}\delta^{m+1} + v_{m+2}\,\delta^{m+2}$, ... *Then for each value of* α such that $\alpha \leqslant \delta$, one can take m sufficiently large that $\psi(\alpha) = \omega$. Now $f(\alpha) = \phi(a) + \psi(\alpha)$, and therefore

$$f(\alpha) - f(\alpha - \beta) = \phi(\alpha) - \phi(\alpha - \beta) + \omega.$$

Since, moreover, $\phi(\alpha)$ is a polynomial ["entire"] function of α, one can take β so small that $\phi(\alpha) - \phi(\alpha - \beta) = \omega$, and therefore also $f(\alpha) - f(\alpha - \beta) = \omega$; Q.E.D.

[The paper continues, giving an imperfect discussion of power series with variable coefficients and Theorem V, and in Theorem VI disposes of the product of two convergent series. Parts III and IV, which form the main substance of the paper, deal strictly with the binomial series. After "proving" his Theorem V, Abel makes the following interesting historical comment:]

In the [*Cours d'analyse*, p. 131][7] of Mr. Cauchy one finds the following theorem:

"If the different terms of the series

$$u_0(x) + u_1(x) + u_2(x) + u_3(x) + \cdots$$

are functions of one and the same variable x, and indeed continuous functions of this variable in the neighborhood of a particular value for which the series converges, then the sum s of the series is also a continuous function of x in the neighborhood of this particular value."

It appears to me that this theorem suffers exceptions. Thus, for example, the series

$$\sin\phi - \tfrac{1}{2}\sin 2\phi + \tfrac{1}{3}\sin 3\phi - \cdots, \text{ etc.}$$

is discontinuous for each value $(2m + 1)\pi$ of ϕ, where m is an integer. It is well known that there are many series with similar properties.[8]

NOTES

1. The translation . . . is adapted and extended from that of A. A. Bennett in Smith, *Source Book,* pp. 286-291. For the original, see Abel's *Oeuvres,* I, pp. 219-50.
2. See Struik, *Source Book,* §V. 4, for early work on this series.
3. [Birkhoff's] Selections 1a-1c were taken from this work.
4. For brevity, in this article, by ϵ will be meant a quantity which can be smaller than any given quantity no matter how small. (N. H. A.) [Abel's ω and ϵ have been changed to ϵ and c_i, *respectively.*]
5. The context shows that this somewhat ambiguous statement is to be understood in the required sense of $\lim_{m\to\infty} [\lim_{n\to\infty}(c_m r_m + c_{m+k} r_{m+1} + \cdots + c_{m+n} r_{m+n})] = 0$. (A. A. B.)
6. Presumably, Abel means $|\alpha| \leqslant |\delta|$.
7. In the original u_k is written instead of $u_k(x)$.
8. In similar vein, Dirichlet (*Werke,* p. 132) noted a few years later that the limit $f(x) = \lim_{n\to\infty} [\lim_{m\to\infty} \cos(m!\pi x^n)]$ was 1 for all rational x and 0 for irrational x.

Chapter VIII
The Nineteenth Century

Section C
The Arithmetization of Mathematical Analysis

AUGUSTIN-LOUIS CAUCHY (1789–1857)

Augustin-Louis Cauchy, the eldest son of Louis-François and Marie Madelene Cauchy, was born in Paris in the year the French Revolution began. His father was a barrister and police Lieutenant. To escape the Reign of Terror (1793-94), his parents moved from Paris to the nearby village of Arceuil, where they were neighbors of Laplace and Berthollet. Lagrange reportedly forecast the scientific genius of the young boy. Augustin-Louis received an excellent education. At age 15 he completed classical studies—with distinction—at the École Centrale du Panthéon and then studied engineering at the École Polytechnique (1805-1807) and the École des Pontes et Chaussées (1807-1809). He worked briefly as a military engineer—first at the Ourcq Canal, then at the St. Cloud bridge (1810), and finally at the harbor of Cherbourg (1810–13) where Napoleon was building a fleet for his intended invasion of Britain. Young Cauchy's work as a mathematician began in 1811 when he solved a problem set by Lagrange of whether the faces of a convex polyhedron determine its angles. The next year he solved Fermat's problem on polygonal numbers, that is, if any number is a sum of n $n$gonal numbers. In these and subsequent contributions to mathematics, he proved clever, exact, and prolific. His reputation quickly grew.

In 1813, when Cauchy returned to Paris, perhaps for reasons of health, Lagrange and Laplace persuaded him to leave engineering and turn exclusively to mathematics. With their help he became an instructor at the École Polytechnique. He became a full professor there in 1816—a year that had other good news for Cauchy. He won the grand prize of the Institut de France for a paper on wave propagation at the surface of a liquid, and he was appointed a member of the Paris Academy of Sciences after the Bonapartist Gaspard Monge and the regicide Lazare Carnot were expelled for political reasons. He also appears to have been appointed to teach at the Faculté des Sciences and the Collége de France, both of which were in Paris. In 1818 he married Aloïse de Bure, the daughter of a publisher who published most of Cauchy's work. The couple had two daughters.

Until the July Revolution of 1830, Cauchy led a quiet life. An ardent royalist who strictly supported the Bourbons, he refused in 1830 to take the oath of allegiance to the new Orléans king, Louis Philippe, who replaced the exiled Bourbon Charles X. Cauchy instead went into a self-imposed exile. He lived briefly with the Jesuits at Fribourg in Switzerland and then taught at the University of Turin (1831–33). Seeking to emulate Bossuet and Fenelon as princely educators, he went to Prague to tutor the grandson of the deposed monarch (1833–38). However, he returned to Paris in 1838 to resume his work at the Academy. Immediately after the Second Republic was proclaimed following the Revolution of 1848, an oath of allegiance was no longer required, and Cauchy resumed the chair of celestial mechanics at the Sorbonne, where he remained for the rest of his career. When Napoleon III reinstituted an oath of allegiance in 1852, Cauchy was exempted from it. A gifted teacher, Cauchy at times presented his material with brilliance, despite an occasional tendency to wander from the subject.

Devoutly Catholic and at times naïve, Cauchy took the lead in projects to aid unwed mothers, feed Irish paupers, and rehabilitate French criminals. In his later years he was a social worker in the town of Sceaux. Quixotic and often melodramatic, he occasionally criticized scientists for research which he considered dangerous to religion.

Fascinated and essentially mastered by mathematics, Cauchy published 789 papers and seven books in the field—ranking second only to Euler in prolificity. Unlike Gauss, he published rapidly, but never carelessly, over a wide range of pure and applied mathematics. There are more concepts and theorems named for him (16 in elasticity alone) than for any other mathematician.

In his greatest contribution to mathematics, Cauchy provided a first phase of satisfactory foundations for mathematical analysis. He presented some of these foundations in his classic *Cours d'analyse de l'École Royale Polytechnique* ("Courses on Analysis for the École Royal Polytechnic," 1821) by refining the notions of limit, continuity, function, and convergence. In this and later works, such as *Résume des leçons . . . sur le calcul infinitésimal* (1823), he established the limit concept as the cornerstone for the calculus.

Cauchy's writings were characterized by rigorous methods significantly beyond those of his predecessors. Thus, while in the *Résumé* he retained the mean-value theorem much as Lagrange had derived it, he made departures elsewhere. Unlike Lagrange, Cauchy used tools from the integral calculus to get Taylor's theorem. Cauchy also did more with convergence than using Lagrange's remainder as a tool for investigating the convergence properties of Taylor's series. In the *Cours d'analyse,* he had formulated the Cauchy criterion independently of Bolzano, who had it as early as 1817. This criterion states that a sequence $\{a_n\}$ converges if and only if for each $\epsilon > 0$ there exists a number N, such that $|a_p - a_q| < \epsilon$ for all $p, q > N$. In his writings Cauchy proved and often used the ratio, root, logarithm, and integral tests. As a result it may be said that he established the first general theory of convergence.[1]

Taken together, the *Cours d'analyse,* the *Résumé,* and a series of papers, also make Cauchy a principal founder of complex function theory built upon an appeal to residues, a subject pioneered by Euler, Laplace, and Gauss. In this work, he devised the first systematic theory of complex numbers. In the *Résumé,* he built a theory of derivatives on his definition of the derivative as a limit and defined the definite integral as the limit of sums. Subsequently, Weierstrass and Riemann extended his work on complex function theory, while Riemann and Lebesgue improved upon his definition of the integral.

Cauchy made other contributions to mathematics. He skillfully developed the Fourier transform in differential equations, elaborated the error theory of Laplace, wrote the first comprehensive treatise on determinants, and attempted to prove Fermat's last theorem in number theory. In applied mathematics, he founded elasticity theory and made minor contributions to the mechanics of rigid bodies.

NOTE

1. Cauchy achieved a partial rigorization of real analysis. The distinction between continuity and uniform continuity and convergence and uniform convergence appeared later. By the 1880s Weierstrass, Cantor, and Dedekind attained a higher standard of rigor in analysis and completed its arithmetization.

102. From *Cours d'analyse de l'École Royale Polytechnique* (1821)*

AUGUSTIN-LOUIS CAUCHY

1A. ON LIMITS AND CONTINUITY[1]

When the values successively attributed to the same variable approach indefinitely a fixed value, eventually differing from it by as little as one could wish, that fixed value is called the *limit* of all the others.

When the successive absolute values[2] of a variable decrease indefinitely, in such a way as to become less than any given quantity, that variable becomes what is called an *infinitesimal*.[3] Such a variable has zero for its limit. . . .

Let $f(x)$ be a function of [the real] variable x and suppose that this function . . . has a unique and finite value for each value of x in a given interval.[4] If, to a value of x in this interval, one adds an infinitesimal increment h, the function itself increases by the difference $f(x + h) - f(x)$; this depends on both the new variable h and the value of x. Given this, the function $f(x)$ will be a *continuous* function of the variable x in the interval when, for each value of x in the interval, the magnitude of the difference $f(x + h) - f(x)$ decreases indefinitely with that of h.

In other words, the function $f(x)$ will remain continuous relative to x in a given interval if [in this interval] an infinitesimal increment in the variable always produces an infinitesimal increment in the function itself.

1B. ON CONVERGENCE

A *sequence* is an infinite succession of quantities $u_0, u_1, u_2, u_3, \ldots$ which succeed each other according to some fixed law. These quantities themselves are the different terms of the sequence considered.[5] Let

$$s_n = u_0 + u_1 + u_2 + \cdots + u_{n-1}$$

be the sum of the first n terms, where n is some integer. If the sum s_n tends to a certain limit s for increasing values of n, then the series is said to be *convergent*,

*Source: This translation, taken from Garrett Birkhoff (ed.), *A Source Book in Classical Analysis* (1973), 2-6, is reprinted by permission of Harvard University Press.

and the limit in question is called the *sum* of the series. On the contrary, if the [partial] sum s_n approaches no fixed limit as n increases indefinitely, the series is *divergent* and has no sum. In either case, the term corresponding to the index n, namely u_n, is called the *general term*. It suffices to give this general term as a function of the index n in order for the sequence to be completely determined.

One of the simplest sequences is the geometric progression $1, x, x^2, x^3, \ldots,$ whose general term is x^n, that is, the nth power of x. If one sums the first n terms of this sequence, one finds

$$1 + x + x^2 + \cdots + x^{n-1} = 1/(1 - x) - x^n/(1 - x)$$

and, since the magnitude of the fraction $x^n/(1 - x)$ either converges to zero for increasing values of n or increases beyond all limits, depending on whether one supposes the magnitude of x to be less than or greater than unity; one must conclude that under the first hypothesis the progression $1, x, x^2, x^3, \ldots$ defines ["is"] a convergent series whose sum is $1/(1 - x)$, while under the second hypothesis the same progression defines a divergent series which has no sum.

By the principles established above, for the series

$$u_0 + u_1 + u_2 + \cdots + u_n + u_{n+1} + \cdots \tag{1}$$

to converge it is necessary and sufficient that the sum[s] $s_n + u_0 + u_1 + u_2 + \cdots + u_{n-1}$ converge to a fixed limit s as n increases; in other words, it is necessary and sufficient that for infinitely large values of n the sums $s_n, s_{n+1}, s_{n+2}, \ldots$ differ from the limit s, and hence from each other, by infinitesimal quantities. Besides, the successive differences between the first sum s_n and each of the following are respectively determined by the following equations.:

$$s_{n+1} - s_n = u_n, \; s_{n+2} - s_n = u_n + u_{n+1}, \; s_{n+3} - s_n = u_n + u_{n+1} + u_{n+2} \ldots .$$

Hence, for the series (1) to converge, it is necessary that the general term u_n decrease indefinitely as n increases; but this condition is not sufficient, and it must also be true that, for increasing values of n, the different sums $u_n + u_{n+1}$, $u_n + u_{n+1} + u_{n+2}, \ldots,$ that is, the sums of quantities $u_n, u_{n+1}, u_{n+2}, \ldots$ taken in arbitrary number from the first will always have a magnitude which is less than any assignable limit. Conversely, when these various conditions are satisfied, the convergence of the series is assured.

1C. ON THE RADIUS OF CONVERGENCE

Let z be a complex ["imaginary"] variable.[6] Every complex power series in z will be of the form

$$a_0 + ib_0 + (a_1 + ib_1)z + (a_2 + ib_2)z^2 + \cdots + (a_n + ib_n)z^n + \cdots,$$

where $a_0, a_1, a_2, \ldots, a_n, \ldots, b_0, b_1, b_2, \ldots, b_n, \ldots$ denote two sequences of [real] constants. In the case where the constants of the second sequence vanish, the preceding series is reduced to

$$a_0 + a_1 z + a_2 z^2 + \cdots + a_n z^n + a_n z^n \cdots . \tag{1}$$

In this section in particular, we will consider series of this last kind. If, for greater convenience, we set

(2) $$z = r(\cos \theta + i \sin \theta),$$

where r denotes a real variable and θ a real arc, series (1) becomes

(3)
$$a_0 + a_1 r(\cos \theta + i \sin \theta) + a_2 r^2(\cos 2\theta + i \sin 2\theta) + \cdots + a_n r^n(\cos n\theta + i \sin n\theta) + \cdots.$$

Now . . . let A be the largest of the limits to which the nth root of the absolute value of a_n tends as n increases indefinitely.[7] The largest of the limits to which the nth root of the absolute value of the complex expression $a_n z^n = a_n r^n(\cos n\theta + i \sin n\theta)$ converges under the same assumption will be equivalent to the magnitude of the product Az; hence . . . the series (3) will be convergent or divergent depending on whether the product Az has a magnitude less than or greater than 1. From this remark, one immediately deduces the following proposition:

THEOREM. The series (3) is convergent for all values of z contained between the limits $z = -1/A$ and $z = +1/A$, and divergent for all values of z lying outside the same limits. In other words, the series (1) is convergent or divergent [according as] the absolute value of the complex expression z is less than or greater than $1/A$.

NOTES

1. A. L. Cauchy, *Oeuvres* (2), III, which reproduces his *Cours d'analyse de l'École Polytechnique. I. Analyse algebrique.* Selection 1a is from pp. 19, 43; 1*b* is from pp. 114-116; 1c is from pp. 239-240.
2. Here and elsewhere, Cauchy's "valeur numérique" has been translated as "absolute value" or "magnitude."
3. Cauchy writes "quantité infiniment petit." Note his clear specification of an infinitesimal as a *variable* quantity tending to zero, not a constant set equal to zero.
4. Here and elsewhere, Cauchy uses the phrase "entre des limites données" for "in a given interval."
5. Cauchy uses the word "série" for "sequence" and "series" alike; moreover, he puts a comma in (1) where today one writes a plus sign.
6. "Imaginaire" has been translated "complex" throughout this book. Cauchy uses x for z, z for r, and $\sqrt{-1}$ for i.
7. In modern notation, let $A = \limsup_{n\to\infty} |a_n|^{1/n}$.

103. From *Résumé des leçons . . . sur le calcul infinitésimal* (1823)*

(On the Derivative as a Limit[1])

AUGUSTIN-LOUIS CAUCHY

THIRD LESSON
DERIVATIVES OF FUNCTIONS OF ONE VARIABLE

When the function $y = f(x)$ is continuous between two given limits of the variable x, and one assigns a value between these limits to the variable, an infinitesimal increment Δx of the variable produces an infinitesimal increment in the function itself. Consequently, if we then set $\Delta x = h$,[2] the two terms of the *difference quotient* ["rapport aux différences"]

$$\frac{\Delta y}{\Delta x} = \frac{f(x + h) - f(x)}{h} \tag{1}$$

will be infinitesimals. But whereas these terms tend to zero simultaneously, the ratio itself may converge to another limit, either positive or negative. This limit, when it exists, has a definite value for each particular value of x; but it varies with x. Thus, for example, if we take $f(x) = x^m$, m being a [positive] integer, the ratio of the infinitesimal differences will be

$$\frac{(x + h)^m - x^m}{h} = mx^{m-1} + \frac{m(m - 1)}{1 \cdot 2} x^{m-2}h + \cdots + h^{m-1},$$

and it will have for [its] limit the quantity mx^{m-1}, that is to say, a new function of the variable x. The same will hold generally; only the form of the new function which serves as the limit of the ratio $[f(x + h) - f(x)]/h$ will depend upon the form of the given function $y = f(x)$. In order to indicate this dependence, we give to the new function the name derivative ["fonction dérivée"] and we designate it, using a prime, by the notation y' or $f'(x)$.[3]

FOURTH LESSON
DIFFERENTIALS OF FUNCTIONS OF A SINGLE VARIABLE

Let $y' = f(x)$ remain a function of the independent variable x; let h be an infinitesimal and k a finite quantity. If we set $h = \alpha k$, α will also be an infinitesimal quantity, and we will have identically

$$\frac{f(x + h) - f(x)}{h} = \frac{f(x + \alpha k) - f(x)}{\alpha k},$$

*Source: This translation, taken from Garrett Birkhoff (ed.), *A Source Book in Classical Analysis* (1973), 2-6, is reprinted by permission of Harvard University Press.

whence one concludes that

$$\frac{f(x+\alpha k)-f(x)}{\alpha}=\frac{f(x+h)-f(x)}{h}k. \tag{1}$$

The limit toward which the left side of equation (1) converges as the variable α tends to zero, the quantity k remaining constant, is called the *differential* of the function $y = f(x)$. We indicate this differential by the symbol d, as follows:

$$dy \quad \text{or} \quad df(x).$$

It is easy to obtain its value when we know that of the derivative y' or $f'(x)$. Indeed, taking the limits of the two sides of equation (1), we shall find generally

$$df(x) = kf'(x). \tag{2}$$

In the special case where $f(x) = x$, equation (2) reduces to

$$dx = k. \tag{3}$$

Thus the differential of the independent variable x is just the finite constant k. Granting this, equation (2) becomes

$$df(x) = f'(x)dx \tag{4}$$

or, what amounts to the same thing,

$$dy = y'\,dx. \tag{5}$$

It follows from these last [equations] that the derivative $y' = f'(x)$ of any function $y = f(x)$ is precisely equal to dy/dx, that is, to the ratio of the differential of the function to that of the variable, or, if one wishes, to the coefficient by which the second differential must be multiplied in order to obtain the first. It is for this reason that we sometimes give to the derivative the name of *differential coefficient.*[4]

NOTES

1. A. L. Cauchy, *Résumé des leçons . . . sur le calcul infinitésimal* (Paris, 1823); *Oeuvres* (2), IV, 22ff, 27ff. Our translation has been adapted from the translation by Evelyn Walker (E. W.) in Smith, *Source Book*.
2. Cauchy uses i for h and h for k.
3. The phrase "fonction dérivée" and the notation $f'(x)$ were due to Lagrange.
4. After this Cauchy gives the rules for differentiating various elementary functions: algebraic, exponential, trigonometric, and inverse trigonometric. (E. W.)

104. From *Résumé des leçons . . . sur le calcul infinitésimal* (1823)*

AUGUSTIN-LOUIS CAUCHY

SEVENTH LESSON
THE FIRST RIGOROUS PROOF ABOUT DERIVATIVES

THEOREM. If the function $f(x)$ is continuous between the limits[1] $x=x_0$, $x=x$, and if we let A be the smallest, B the largest, value of the derivative $f'(x)$ in that interval, the ratio of the finite differences

$$[f(X) - f(x_0)]/X-x_0 \tag{4}$$

must be included[2] between A and B.

PROOF. Let δ, ϵ be two very small numbers; the first is chosen so that, for all numerical [i.e., absolute] values of i less than δ, and for any value of x included between the limits x_0, X, the ratio

$$f(x+i) - f(x)/i$$

will always be greater than $f'(x) - \epsilon$, [3]and less than $f'(x) + \epsilon$. If we interpose $n-1$ new values of the variable x between the limits x_0, X, that is

$$x_1, x_2, \ldots, x_{n-1},$$

so that the difference $X-x_0$ is divided into elements

$$x_1-x_0, x_2-x_1, \ldots, X-x_{n-1},$$

which all have the same sign and which have numerical values less than δ; then, since of the fractions

$$f(x_1)-f(x_0)/x_1-x_0, f(x_2)-f(x_1)/x_2-x_1, \ldots, f(X)-f(x_{n-1})/X-x_{n-1}. \tag{5}$$

the first will be included between the limits $f'(x_0) - \epsilon$, $f'(x_0) + \epsilon$, the second between the limits $f'(x_1) - \epsilon$, $f'(x_1) + \epsilon$, . . . , etc., each of the fractions will be greater than $A-\epsilon$, and less than $B+\epsilon$. Moreover, since the fractions (5) have denominators of the same sign, if we divide the sum of their numerators by the sum of their denominators, we obtain a mean fraction, that is, one included between the small-

*Source: This translation of the Seventh Lesson by Judith V. Grabiner is taken from *Historia Mathematia* (1978), 406-407. It is reprinted by permission of Academic Press.

est and the largest of those under consideration [see *Analyse algébrique,* Note II, Theorem XII].

[Note: Here Cauchy was referring to this theorem: "If b, b', b'' . . . are n quantities with the same sign, and if a, a', a'' . . . are any n quantities, we have

$$\frac{a+a'+a''+\ldots}{b+b'+b''+\ldots} = M(a/b,\, a'/b',\, a''/b'',\, \ldots)."$$

He had defined a mean of $c, c', c'', \ldots$. $M(c, c', c'', \ldots)$, as "a new quantity included between the smallest and the largest of those under consideration." *Cours d'analyse,* in *Oeuvres* (2) III, p. 27.]

The expression (4), with which that mean coincides, will thus itself lie between the limits $A-\epsilon$, $B+\epsilon$, and since this conclusion holds no matter how small ϵ may be, we can conclude that the expression (4) will be included between A and B.

NOTES

1. Cauchy uses "limites" for "bounds" or "endpoints."
2. I have translated Cauchy's "comprise" as "included," and his "renfermé" as "lying between." The context of the proof makes clear that c "included" between a and b means $a \leq c \leq b$, and that c "lies between" a and b means $a < c < b$.
3. Cauchy's *Oeuvres* has $f(x) - \epsilon$, a misprint.

Chapter VIII
The Nineteenth Century

Section C
The Arithmetization of Mathematical Analysis

KARL (THEODOR WILHELM) WEIERSTRASS (1815–97)

The German mathematician Karl Weierstrass was the eldest son of Wilhelm and Theodora Vonderforst Weierstrass. His parents were both liberal Catholics. The father served in the Prussian taxation service and later at Paderborn (Westphalia) as the main customs house treasurer. After studies at the Catholic Gymnasium (Latin high school) in Paderborn, where he made a brilliant record in German, Latin, Greek, and mathematics, Weierstrass attended the University of Bonn from 1834 to 1838. At his father's bidding he majored in public finance and administration in order to prepare for a position in the Prussian civil service. However, he preferred mathematics and read Laplace, Legendre, Jacobi, and Abel; Laplace's *Celestial Mechanics* aroused his interest in dynamics and differential equations. He studied little in his major, preferring the camaraderie of taverns with their opportunities for social intercourse. With his great physical strength and quick reflexes, he also became an expert in fencing. To the chagrin of his family Weierstrass left Bonn with poor grades and no degree in 1838. To earn a living he decided to study for a teaching certificate at Theological and Philosophical Academy at Münster from 1838 to 1841. One of its teachers, Christof Gudermann, profoundly influenced him. The two men examined the theory of elliptic functions and the expansion of functions by power series.

After receiving a certificate to teach, Weierstrass began a 14-year career in secondary education at age 26. He taught first at the Münster Gymnasium (1841-42), at the Royal Catholic Gymnasium in Deutsch-Krone (1842–48), and at Braunsberg (1848–55). During this period he was isolated from contact with other mathematicians and had insufficient funds even for correspondence. Nevertheless, he worked steadily on mathematical analysis and largely worked out his program for its arithmetization. In addition to elementary mathematics, he taught subjects ranging from botany to calligraphy to gymnastics. Although not directly involved in the Revolution of 1848 he, like Dirichlet, expressed democratic leanings. As a censor at Braunsberg he permitted the publication of freedom songs and poetry in the Deutsch-Krone newspaper.

After a series of brilliant papers on Abelian functions in Crelle's *Journal*, the obscure schoolmaster was at last recognized in 1856 by the European mathematical community. In that year Weierstrass was appointed professor of mathematics at the Industry Institute in Berlin–a post he held until 1864–and also became an associate professor at the University of Berlin and a member of the Berlin Academy. As a result of overwork he suffered a nervous breakdown in 1859, and he would continue to experience dizzy spells intermittently for the rest of his life. In 1861 he and Ernst Kummer founded the first seminar in Germany devoted exclusively to pure mathematics. After 1862 he could lecture only while seated in a chair because of "brain spasms" and the onset of recurrent attacks of bronchitis and phlebitis. During his classes an advanced student assisted him by writing on the chalkboard. His health problems did not lessen the quality of his teaching, and he gradually gained a reputation as a master teacher through a continuous revision and expansion of his lectures. By the 1870s as many as 250 students attended his classes each year to listen to the presentation of his new ideas. This enrollment was exceptionally high for advanced mathematics courses in his time. At the University of Berlin he also improved the quality of doctoral examinations by insisting that careful records be kept and that the reasons for approval or failure of a student's work be duly recorded. In addition, he removed the requirement that doctoral dissertations be in Latin.

By the late 1870s Weierstrass had become involved in a famous quarrel with another Berlin professor, Leopold Kronecker, who vigorously opposed his version of the "arithmetization of analysis." Kronecker wanted all analysis based upon the whole numbers only, while Weierstrass granted equal validity to whole numbers and irrational numbers. In addition, Kronecker absolutely opposed Georg Cantor's transfinite numbers, while Weierstrass defended them. By 1888 the break between the two men was complete. With Kummer's retirement in 1883 and Kronecker's promotion to full professor and co-director of the mathematics seminar, Kronecker held the more powerful faculty position. Still, Weierstrass's 70th and 80th birthdays were times of public honors and gatherings by former students. After a long bout with influenza, he died in 1897.

Weierstrass played a crucial role in the development of modern mathematical analysis. Perhaps he is most famous for his painstaking, constructive criticism of the foundations of mathematical analysis, which he soundly based on a carefully developed real number system. In his study of functions he provided the delta–epsilon definition of limits that surpassed those of Bolzano and Cauchy, devised new convergency tests, and with the assistance of his students refined the notion of uniform convergence. Following upon similar studies by Riemann in the theory of elliptic functions, he also developed in the 1870s a continuous "nondifferentiable" function:

$$f(x) = \sum_{n=1}^{\infty} b^n \cos(a^n \pi x),$$

where $a > 1$ and $\frac{1}{a} < b < 1$. That is to say, he discovered a continuous function that has no derivatives at any point. This idiosyncracy of an apparently derivable function troubled those analysts who relied greatly on intuition. Desiring to complete studies begun by Niels Abel and Karl Jacobi, Weierstrass worked to construct a theory of functions based on Dirichlet's principle. His theory of functions, which began with power series, was accepted over Riemann's, which began with complex differentiation. In these and his other mathematical studies he concentrated on achieving "Weierstrassian rigor" with its emphasis on clarity and logical stringency.

Weierstrass believed like Gauss that mathematics must be an active friend of science and engineering. Thus he did not limit his investigations to "pure mathematics." He also applied Fourier series and integrals to problems in mathematical physics, added to the calculus of variations, and contributed to the n-body problem in astronomy as well as to the theory of light in optics. Driven by an intensely critical sense and a continual striving for an improved rigor that permitted a higher degree of maturity of his ideas, Weierstrass published comparatively little. As a result his greatest immediate influence came through his students, who included Georg Frobenius, Sonya Kovalevsky, and Edmund Husserl. His students often were the first to publish his major new ideas, methodology, and his lectures. Seven volumes of his *Gesammelte Abhandlungen* (collected works) subsequently appeared in print between 1894 and 1927.

105. From Lectures on the Differential Calculus (1861)*

KARL WEIERSTRASS

In contrast to a non-variable magnitude or constant, which can assume only one value, a variable *magnitude* is defined to be one which cannot only assume several particular values, but infinitely many ones. It may happen that a variable magnitude can assume every possible positive and negative value; then it is called an *unrestrictedly* variable magnitude. A variable magnitude may also be restrictedly variable and have a lower or upper bound, or both. The values which a variable magnitude can assume may belong to one or several *continuous* sequences if the variable magnitude can assume all possible values between two bounds. The differential calculus deals only with such continuously variable magnitudes.[1]

Two variable magnitudes may be related in such a way that to every definite value of one there corresponds a definite value of the other; then the latter is called a *function* of the former. This relationship may extend to several variable magnitudes; accordingly one distinguishes functions with one and with several variable magnitudes. If to one value of the one variable magnitude there always corresponds only one value of another, then the latter is called an unambiguous function and single-valued function of the former. If to one value of the one magnitude there correspond several values of another, then the latter is called a multi-valued function of the former. The criterion of a function is that the one variable magnitude changes in general by a definite amount as soon as a definite change of the other one is assumed.

(2)

. . .

If $f(x)$ is a function of x and if x is a definite value, then the function will change to $f(x+h)$ if x changes to $x+h$;

*Source: This account of lectures given by Weierstrass at the Berlin Industrial Institute in 1861 is based on class notes taken by Hermann Amandus Schwarz. Dr. Kurt Bing, Professor Emeritus of Mathematical Sciences at Rensselaer Polytechnic Institute, translated this selection from Weierstrass and provided explanatory notes. To more clearly separate parenthetical examples and bibliographic citations, some punctuation has been changed. The German original appears in *Elements d'Analyse de Karl Weierstrass,* edited and biography written by Pierre Dugac (1972), 101-104. This selection is reprinted by permission of Université de Paris VI.

one calls the difference $f(x+h) - f(x)$ the change which the function undergoes by the change of the argument from x to $x+h$. If it is now possible to determine for h a bound δ such that for *all* values of h which in their absolute value are smaller than δ, $f(x+h) - f(x)$ becomes smaller than any magnitude ϵ, however small, then one says that infinitely small changes of the argument correspond to infinitely small changes of the function. For one says, if the absolute value of a magnitude can become smaller than any arbitrarily chosen magnitude, however, small, then it can become infinitely small.[2] If now

(3)

a function is such that to infinitely small changes of the argument there correspond infinitely small changes of the function, one then says that it is a *continuous function* of the argument, or that it changes continuously with this argument.

. . .

Theorem. If a continuous function of x has, for a definite value x_1 of the argument, a definite value of the function y_1, and has for another definite value x_2 a definite value of the function y_2, and if y_3 is an arbitrary value between y_1 and y_2, then there must be between x_1 and x_2 at least one value x_3 for which the function assumes the value y_3.

The following auxiliary theorems serve as proof.

If $y = f(x)$ is a continuous function of x and $y_0 = f(x_0)$ is not zero, then the values $f(x)$ of the function will, for all values of x which lie in the neighborhood of x_0, *i.e.*, for which the difference $x - x_0$ in its absolute value does not exceed a definite bound, have the same sign as $f(x_0)$.

. . .

(4)

. . .

The values which a continuous function takes on or can take on also belong to a continuous sequence, therefore its name is justified.

(5)

FUNDAMENTAL CONCEPTS OF THE DIFFERENTIAL CALCULUS

The complete change $f(x+h) - f(x)$ which a function $f(x)$ undergoes if x changes to $x+h$ can in general be decomposed into two parts, the first of which is proportional to the change h of the argument, thus consists of h and a factor independent of h (*i.e.*, constant relative to h), thus becomes infinitely small when h becomes infinitely small, or becomes infinitely small simultaneously with h. The other, however, not only becomes infinitely small by itself when h becomes infinitely small (*i.e.*, it still becomes infinitely small when one divides it by h).

If h denotes a magnitude which can assume infinitely small values, and if $\phi(h)$ is an arbitrary function of h with the property that for infinitely small values of h it also becomes infinitely small (*i.e.*, that always, as soon as a definite arbitrarily small magnitude ϵ is chosen, a magnitude δ can be determined such that for all values of h whose absolute value is smaller than δ, $\phi(h)$ becomes smaller than ϵ), then it may happen that $\frac{\phi(h)}{h}$ is also still a function of h which for infinitely small values of h becomes itself infinitely small; in this case one says $\phi(h)$ becomes, for infinitely small values of h, infinitely small relative to h.[3]

The first part of the whole change of the function, which is proportional to the change of the argument, is called *differential change* or *differential* and is denoted by a characteristic d prefixed to the function, while a prefixed Δ signifies the whole change. Analogously to this one also writes for h, since the simplest function of x is x itself, dx, a magnitude completely independent of x, which can become infinitely small. The smaller dx or h is taken, the less the differential change will differ from the total change; by decreasing dx, one can make the difference smaller than every

(6)

magnitude, however small; hence one has defined the differential as the change which a function undergoes when its argument changes by an infinitely small magnitude.

The differential of a function has thus in general the form $df(x) = p \cdot dx$; the factor p, by which one must multiply the differential of the argument in order to obtain the differential of the function is called *differential coefficient* or differential quotient. It is, in general, again a function of x and since it is derived from $f(x)$ in a definite manner, it is called the *derivative* of the function and is written $f'(x)$. This function is therefore complete and independent of the change of the argument.

Theorem. If, for a definite value x of the argument, a function has a differential quotient, then there can exist no second one different from it for the function and the same value of x.

Suppose one had succeeded to decompose $f(x+h)-f(x)$ in the manner described $= ph + h(h)$, where (h) denotes a magnitude which becomes infinitely small with h; it is to be shown that this decomposition is the only possible one.

. . .

If one divides the differential of a function by the differential of the argument, then one obtains the differential coefficient; for this reason it is also called differential quotient:

$$\frac{df(x)}{dx} = p = f'(x).$$

. . .

(7)

. . .

AUXILIARY THEOREMS FOR THE DETERMINATION OF THE DIFFERENTIAL

1. If $f_1(h)$ and $f_2(h)$ are two functions which become infinitely small simultaneously with their argument h, then their sum $f_1(h)+f_2(h)$ is also such a function.

Suppose ϵ is an arbitrary magnitude, however small; let one decompose it into two parts $\epsilon_1+\epsilon_2$ and determine δ such that for all values of h which in their absolute value do not exceed δ, both $f_1(h)<\epsilon_1$ and $f_2(h)<\epsilon_2$ hold, then $f_1(h) + f_2(h) < \epsilon_1 + \epsilon_2$ holds. The same theorem is true also for the addition of several such functions which become infinitely small simultaneously with h.

2. If $f(h)$ is a function which becomes infinitely small simultaneously with h and if A signifies a magnitude which is constant relative to h, then $A \cdot f(h)$ likewise becomes infinitely small with h.

3. The product $F(h) \cdot f(h)$, in which $f(h)$ is a function which becomes infinitely small with h and in which $F(h)$ is any function of

(8)

h which for infinitely small values of h does not become infinitely large, is likewise such a function which becomes infinitely small simultaneously with h.

. . .

4. If $f_1(h)$ and $f_2(h)$ are functions of h which become infinitely small for infinitely small values of h and if A is a magnitude independent of h which is not zero, then the quotient $\dfrac{f_1(h)}{A+f_2(h)}$ is also a function which becomes infinitely small for infinitely small values of h.

. . .

(9)

. . .

DEVELOPMENT OF SOME RULES FOR THE DETERMINATION OF THE DIFFERENTIALS OF GIVEN FUNCTIONS

Let $f_1(x)=p$, $f_2(x)=q$, $f_3(x)=r$, . . . be

functions of x which are continuous between given bounds and have a differential which one knows; it is to be shown that in general all functions compounded from these functions by the fundamental operations of adding, subtracting, multiplying, dividing, raising to a power and extracting a root have a differential, and it is to be proved how one can derive it from the given differentials.

. . .

(15)

. . .

DIFFERENTIALS OF HIGHER ORDER

If one has found for a function the derivative $f'(x)$, then $f'(x) \cdot dx$ is its differential. The magnitudes x and dx are completely independent of each other, so that thus dx can be considered as a constant relative to x. One can therefore differentiate the product $f(x) \cdot dx$ again with respect to x and one has $d(f'(x) \cdot dx) = dx \cdot df'(x)$. One calls the result the second differential of the given function or the differential of the second order and writes it $d^2f(x)$. If the derivative of $f'(x) = f''(x)$, then one has $d^2f(x) = d(df(x)) = f''(x) \cdot dx^2$ . . .

NOTES BY TRANSLATOR

1. The introductory paragraph and the following one give the impression that when delivering these lectures, Weierstrass was aware of the need for giving a precise definition of the notion of a real number, but had not yet succeeded in doing so. What is meant by a continuous sequence (above or at the end of the present section) is not explained, and the values which a variable magnitude can assume are left undefined.

The notion of a variable magnitude is in itself not clear. Morris Kline, in his discussion of Weierstrass' work on the rigorization of analysis, in his work *Mathematical Thought from Ancient to Modern Times* (Oxford University Press, New York, 1972), p. 952, states that Weierstrass "interprets a variable simply as a letter standing for any one of a set of values which the letter may be given," and that for him a continuous variable is one such that if x_0 is any value of the set of values of the variable and δ any positive number, there are other values of the variable in the interval between $x_0-\delta$ and $x_0+\delta$, the endpoints excluded. These definitions, for which Kline does not give the source, may indicate what Weierstrass may have meant in the present passage by a continuous sequence and by a variable magnitude. Yet the notion of a variable magnitude, rather than a variable, and its use in the definition of a function, suggests a mathematical object named by a symbol rather than a symbol.

Confusion between an object and its name is still found in mathematical writing, and it occurs sometimes when the subject of functions and the notation for functions is discussed. It may be harmless, but is likely to be damaging in foundational research. Gottlob Frege, the founder of modern mathematical logic, used the notion of function and a notation for functions in his work on mathematical logic and the foundations of arithmetic in the three decades preceding the first years of this century; he emphasized the importance of distinguishing between an object and its name. A modified version of Frege's treatment of the concept of function and of the functional notation is found in the introduction to Alonzo Church's work *Introduction to Mathematical Logic*, vol. I (Princeton University Press, Princeton, New Jersey, 1956), pp. 3-23. A thorough discussion of Frege's ideas on this matter, with a bibliographical note, is presented in the Editor's Introduction to: *Gottlob Frege, The Basic Laws of Arithmetic, Exposition of the System, Translated and Edited, with an Introduction, by Montgomery Furth* (University of California Press, Berkeley, California, 1964, 1967).

Today it has become more usual to treat variables as symbols and to make careful distinctions between a function, the value of a function for a given argument, and the notation for these objects. Also, what Weierstrass called a single-valued function is now usually called a function, and Weierstrass' multi-valued function is not called a function, but is taken as a special kind of relation (as is Weierstrass' single-valued function), without a special name.

2. In the preceding passage, Weierstrass gives what is essentially the now accepted definition of the limit of a function: One says that $\lim_{h \to a} g(h) = L$ if for every positive ϵ there

exists a positive δ such that $|g(h) - L| < \epsilon$ whenever $0 < |h - a| < \delta$.

This is done for the special case in which $g(h)$ is $f(x+h) - f(x)$ and a and L are zero, *i.e.*, $\lim_{h\to 0} [f(x + h) - f(x)] = 0$; the statement is made in words rather than symbols, and it appears as a definition of the phrase: "infinitely small changes of the argument correspond to infinitely small changes of the function," which, in turn, is used in what follows to define the notion of a continuous function of one argument.

The last sentence of the passage makes it clear that in the defining sentence the absolute value of $f(x+h) - f(x)$ is intended, and it implies that ϵ is thought of as positive. Also, the wording implies that a δ as described must exist for every ϵ chosen; this is made explicit in the subsequent discussion of the function $\phi(h)$.

Weierstrass' definition of a continuous function takes it for granted that the function is defined at x, and indeed does not mention continuity *at* x, possibly because it may be intended as a definition of continuity (for all values of x) in an interval. If the function is assumed to be defined at x, the stipulation that $0 < |h-a|$ is not necessary for the special case in question (since then $g(a) = g(0) = 0$ is defined and equals L) and, indeed, the stipulation is not made in the text.

H. E. Heine, a student of Weierstrass, gave the now accepted definition of the limit of a function in the article "Die Elemente der Funktionenlehre" in Crelle's *Journal* for 1872—see Carl B. Boyer, *A History of Mathematics* (John Wiley & Sons, Inc., New York, 1968), pp. 608-609 (read "η_0" for "η"), 604, 607.

As mentioned above, the actual idiom defined by Weierstrass is "infinitely small changes of the argument correspond to infinitely small changes of the function." In justifying this idiom, he tries to define the notion of a magnitude becoming infinitely small. (This notion is not actually needed for the sequel; see also the first two paragraphs of the preceding note.) It would seem that, while giving a precise definition of the above imprecise idiom, Weierstrass, as a teacher, was also aiming at bridging the gap between the technical definition and the intuitive content expressed in the way usual at the time; a comment on this usage is still found in R. Courant's classical work *Differential and Integral Calculus* (English version by J. E. McShane, vol. 1, second edition, 1937, vol. 2, 1936, Blackie & Son, London), reprinted many times and distributed in the United States by Interscience-Wiley, New York (see vol. 2. p. 48, first footnote). It may be that the use of words instead of symbols throughout this and subsequent passages (often in connection with the notion of infinite smallness), which must be striking to the modern reader, was motivated by the same intention.

It may be mentioned also that the (direct) use of infinitely small or infinitesimal magnitudes as such (as envisaged by Leibniz and suggested by his notation for the calculus about 300 years ago) was justified about 20 years ago by Abraham Robinson's theory of non-standard analysis—see the textbook, based on Robinson's discovery, H. Jerome Keisler, *Elementary Calculus* (Prindle, Weber, and Schmidt, Boston, 1976), pp. VII-VIII, 873-880, for a survey, remarks on history and background, and suggestions for further reading.

3. In the present section Weierstrass explains that the increase $f(x+h) - f(x)$ of the value of a function f corresponding to a change h of the argument can be approximated near x by a linear function and hence defines the derivative f' of f at x. In more detail, and, partly in the notation and terminology of the next few paragraphs (some of which is still used today), $\Delta f(x) = f(x+h) - f(x) = p\cdot h + \phi(h)$ [or $p\cdot h + h\cdot(h)$]; here p is to be independent of h for given x and $\lim_{h\to 0} \frac{\phi(h)}{h} = \lim_{h\to 0} (h) = 0$, and the derivative $f'(x)$ of f at x is defined to be the number p, unique if it exists for a given x by the theorem below. Again, essentially in the author's notation and terminology, $\Delta f(x) = df(x) + \phi(h)$, $h = dx$, $df(x) = p\cdot dx$ (a linear function of dx), $dx(x)$ is the differential, and $f'(x) = p = \frac{df(x)}{dx}$ is the differential coefficient or differential quotient or derivative of the function.

The linear function $df(x)$ may be said to approximate $\Delta f(x)$ [and hence the linear function $f(x) + df(x)$ whose graph is the tangent at the point $(x, f(x))$ to the graph of f may be said to approximate the function $f(x+h) = f(x) + \Delta f(x)$] near x in the precise sense that as h approaches 0, the difference $\phi(h)$ between Δf and df approaches 0 even when divided by h, a situation which one expresses by saying that $\phi(h)$ vanishes to a higher order than h (see Courant, loc. cit., vol. 1, pp. 106 and 195, vol. 2, pp. 59-60 and 48).

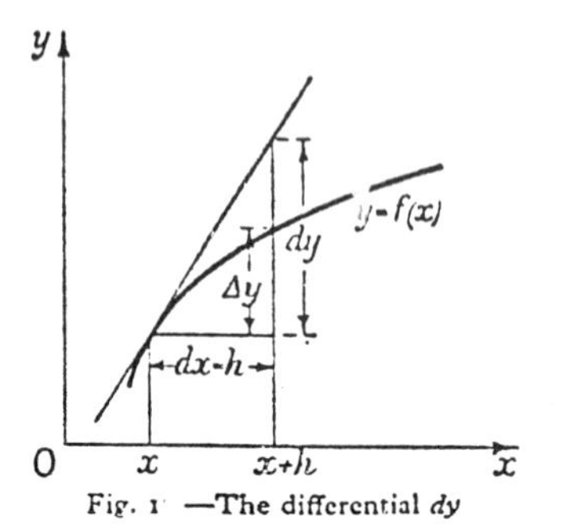

Fig. 1 —The differential dy

[Weierstrass uses "$df(x)$" instead of "dy," and his notation would call for "$\Delta f(x)$" instead of "Δy."]

Weierstrass' definition of the derivative of a function f at x as $\frac{df(x)}{dx}$, where the differential $df(x)$ is first described as approximating $\Delta f(x)$ linearly at x in the precise sense mentioned, differs from the now usual definition as $\lim_{h\to 0} \frac{f(x+h) - f(x)}{h}$ and requires the uniqueness theorem stated further on. The two definitions are logically equivalent, as follows from Courant's remarks in two of the passages just cited. For functions of more than one argument, today's definition of differentiability is essentially that of Weierstrass as the existence of a linear approximation, see, for example, Courant, loc. cit., vol. 2, pp. 60-61, 50-51, 64-65. But for functions of one argument, the relation of differentiability to the existence of a linear approximation in the precise sense is rarely mentioned in today's elementary texts.

106. From a letter to Hermann Amandus Schwarz (May 5, 1875)*

KARL WEIERSTRASS

. . .

Moreover, I can tell you how happy I am that my long-held desire to see you active again at a distinguished German university is now to be fulfilled. *Göttingen* is exactly the right place for you now. The mathematical sciences are respected there as they are hardly anywhere else, and there will never be a lack of capable students if only there are qualified teachers. For the faculty the connection of the University with the Society of Sciences (Societät der Wissenschaften) is of the very greatest significance, and at the same time being reminded of the great predecessors is the most powerful motive for scientific activity. You have often told me that you wished to remain my student even after graduating from the university. I accept this in the sense that you are one of the few who in their later development have never reneged on the principles which I consider my main task to establish firmly in my students and the essence of which is expressed in the demand of science *(Wissenschaft)* that clarity and truth are what are most necessary; to eschew and to hate nothing more than empty talk about what is only half understood, not to mention the fraudulent activity which—regrettably—seeks to assert itself also in the

Source: Drs. Kurt Bing, Professor Emeritus of Mathematical Sciences at Rensselaer Polytechnic Institute, and Ronald Calinger, editor of this volume, translated this letter from Weierstrass to Schwarz. The German original appears in *Elements d'Analyse de Karl Weierstrass*, edited and biography written by Pierre Dugac (1972), 157. This selection is reprinted by permission of Université de Paris.

sternest and purest of the sciences; but instead in one's own work to let oneself be guided by the reflection that obtaining general results is the highest goal which, however, can be reached only by way of thorough investigation.

107. From *Encounters with Mathematics* (1977)*

(Riemann on physics and partial differential equations)

LARS GARDING

Riemann took a lively interest in physics. His lectures on partial differential equations given around 1860 and published in 1882 are still a model of clarity and simplicity. Here are some passages from his introduction.

"As is well known, physics became a science only after the invention of differential calculus. It was only after realizing that natural phenomena are continuous that attempts to construct abstract models were successful. The task is two-fold: to devise simple basic concepts referring to time and space and to find a method to deduce from the processes which can be checked against experiments.

"The first such basic concept, that of an accelerating force, is due to Galilei. He found it to be the simple time independent cause of motion in a free fall. Newton took the second step: he found the concept of an attracting center, a simple cause of force. Contemporary physics still works with these two concepts, an accelerating force and an attracting or repellent center . . . all attempts to bypass these concepts have failed.

"But the methods—differential calculus—through which one passes from the concepts to the processes have been improved in an essential way. In the first period after the invention of differential calculus, only certain abstract cases were treated: in a free fall, the mass of the body was considered to be concentrated at its center of gravity, the planets were mathematical points, . . . so that the passage from the infinitely near to the finite was made only in one variable, the time. In general, however, this passage has to be done in several variables. For the basic concepts deal only with points in time and space and the processes describe actions over finite times and distances. Such passages lead to partial differential equations. [Riemann then reviews the work of d'Alembert and Fourier, both having dealt with such equations.]

"Since then, partial differential equations are the basis of all physical theorems. In the theory of sound in gases, liquids and solids, in the investigations of elasticity, in optics, everywhere partial differential equations formulate basic laws of nature which can be checked against experiments. It is true that in most cases, these theories start from the assumption of molecules subject to certain forces. The constants of the partial differential equations then depend on the distribution of molecules and how they influence each other at a distance. But one is far from being able

*Source: From Lars Garding, *Encounters with Mathematics* (1977), 166-167. This extract is reprinted by permission of Springer-Verlag. The biography of Riemann is given before selection 98.

to draw definite conclusions from these distributions. . . . In all physical theories and by all phenomena explained by molecular forces, partial differential equations constitute the only verifiable basis.

These facts, established by induction, must also hold *a priori.* True basic laws can only hold in the small and must be formulated as partial differential equations. Their integration provides the laws for extended parts of time and space."

Note that this was written before the work of Maxwell and before relativity theory. Riemann's vision is only partly true in quantum mechanics but on the whole it has held up remarkably well. What he says about classical physics and partial differential equations can be said today.

Chapter VIII
The Nineteenth Century

Section D
Number Theory, Set Theory, and Symbolic Logic

108. From *Carl Friedrich Gauss: A Biography* (1970)*

(A Discussion of the *Disquisitiones arithmeticae*, including Congruences and the Fundamental Theorem of Arithmetic)

TORD HALL

ARITHMETICAL INVESTIGATIONS

The number-theoretical ideas which streamed from Gauss during the fruitful years 1795-1801 were for the most part collected together in the work that he published in Leipzig in 1801. *Disquisitiones arithmeticae*—Arithmetical investigations. It is Gauss's *magnum opus*, which in one stroke made number theory a firmly grounded and coherent part of mathematics.

The cost of printing, as was the case of his doctor's thesis, was paid by Duke Ferdinand. The work opens with a dedication to "His most Graceous Highness, Prince and Lord Carl Wilhelm Ferdinand, Duke of Braunschweig and Lüneburg." In it Gauss says, among other things, that without the Duke's goodness "I should never have been able completely to devote myself to mathematics, to which I have always been drawn with passionate love." The dedication is supplied in the rococo style demanded by the customs of the period, but in this case the homage was not empty flattery. For these words reflected Gauss's real feelings. . . .

Before proceeding, let us say a few words about the language of the work. Like Gauss's work of earlier years, it is written in Latin, which was then still the international language of science. Under the influence of nationalist feelings at the beginning of the nineteenth century, however, Gauss later changed over and began to write in German. If he and other researchers had stood their ground with Latin, perhaps we could have avoided the present-day confusion of languages. Perhaps we would still have a reasonably easy to understand "scientific Latin" which could be read by all scientists on earth. . . .

Arithmetical investigations is divided into seven parts:

- Congruences in general
- Congruences of the first degree
- Residues of powers
- Congruences of the second degree
- Quadratic forms
- Applications
- Division of the circle

Source: From Tord Hall, *Carl Friedrich Gauss: A Biography* (1970), 48-50, 53-57, and 59. This selection is reprinted by permission of the MIT Press. The biography of Gauss appears before selection 92.

The work has generally been judged extremely hard to read. The distinguished German mathematician Peter Gustav Dirichlet (1805-1859) who later became Gauss's successor at Göttingen, was the first to disseminate their contents in his lectures. We shall here set down only a few samples of Gauss's exploits in number theory, illustrating these results with some examples.

CONGRUENCES IN GENERAL AND CONGRUENCES OF FIRST DEGREE

On the first page Gauss introduced a new mathematical symbol, one of his revolutionary improvements in the nomenclature of number theory, which simplifies immensely the handling of the notion of arithmetic divisibility. (This chapter is concerned solely with whole numbers, that is to say, the number 0, $\pm 1, \pm 2, \pm 3, \ldots$.)

If the number m divides the difference $a - b$ (or $b - a$) of two numbers a and b without remainder, then a and b are said to be *congruent modulo m*, and Gauss wrote $a \equiv b \pmod{m}$. This expression is read: *a is congruent to b modulo m*. The relation is called a *congruence;* and m is called the *modulus* of the congruence. The number b is called the *residue* of a modulo m, and conversely a is called a residue of b modulo m. If the difference $a - b$ is not divisible by m, then a and b are said to be *incongruent modulo m*, and a and b are said to be *nonresidues* of each other, modulo m. We have for example $25 \equiv 10 \pmod 3$ and $-7 \equiv 15 \pmod{11}$, since $25 - 10 = 15$ is divisible by 3, and $-7 - 15 = -22$ is divisible by 11. The number 10 is a residue of 25 with respect to the modulus 3, while 10 is a nonresidue of 7 with respect to the modulus 11.

According to the definition, $a \equiv b \pmod{m}$ means the same as $a - b = m \times y$, where y, is some whole number. If for the given numbers a, b, and m there is no number y that fulfills this condition, then b is a nonresidue of a modulo m. The congruence $a \equiv 0 \pmod{m}$ means that a is divisible by m, for example $35 \equiv 0 \pmod 7$.

Gauss chose the symbol $\equiv$ (which reminds us of the symbol $=$) with great foresight, since there is close analogy between congruences and equalities. The notion of congruence is however more inclusive, since one may interpret an equality as a congruence with modulus 0.

In questions about addition, subtraction, and multiplication of congruences with the same modulus, there are arithmetical laws analogous to those for equalities. For example, we shall show that the Fermat number $2^{2^5} + 1 = 2^{32} + 1 \ldots$ has the factor 641.

First we calculate with equalities, then with congruences. We start with the equalities.

$$5 \times 2^7 = 641 - 1, \tag{1}$$

$$5^4 = 641 - 2^4. \tag{2}$$

If we raise (1) to the fourth power we have $5^4 \times 2^{28} = (641 - 1)^4 = 641^4 - 4 \times 641^3 + 6 \times 641^2 - 4 \times 641 + 1$, so that $5^4 \times 2^{28} - 1 = 641\,(641^3 - 4 \times 641^2 + 6 \times 641 - 4)$, or, if we denote the number inside the parentheses by k,

$$5^4 \times 2^{28} - 1 = 641 \times k, \tag{3}$$

where k is a natural number. If we combine (2) and (3) we obtain

$$(641 - 2^4) \times 2^{28} - 1 = 641 \times k,$$

$$641 \times 2^{28} - 2^{32} - 1 = 641 \times k,$$

and

$$2^{32} + 1 = 641(2^{28} - k) = 641m, \tag{4}$$

where m is a natural number. From (4) it follows directly that 641 is a factor of $2^{32} + 1$.

In congruence notation, (1) is written $5 \times 2^7 \equiv -1 \pmod{641}$; any by raising this to the fourth power we obtain (3) as a congruence. That is to say,

$$5^4 \times 2^{28} \equiv 1 \pmod{641}. \tag{5}$$

Expression (2) becomes the congruence $5^4 \equiv -2^4 \pmod{641}$, and in combination with (5) it becomes $-2^{32} \equiv 1$ (mod 641) or $2^{32} + 1 \equiv 0$ (mod 641); in other words, $2^{32} + 1$ is divisible by 641.

In the second section of his Arithmetical Investigations Gauss first proves several theorems from which emerge what is usually called *the fundamental theorem of arithmetic: Every natural number larger than 1 can, except for the order of the factors, be written in only one way as the product of prime numbers.* Examples: $130 = 2 \times 5 \times 13$; $250 = 2 \times 5 \times 5 \times 5$.

This theorem, like the fundamental theorem of algebra, appears more or less obvious, but it is just such obvious theorems that are often difficult to prove, because circular proofs lie near at hand: more or less unconsciously we assume the truth of what we wish to prove.

By using the fundamental theorem Gauss then determined the *greatest common divisor* (a,b) and the *least common multiple* $\{a,b\}$ of two numbers a and b. (Naturally one can also consider more than two numbers.) For example, if $a = 24 = 2^3 \times 3$ and $b = 90 = 2 \times 3^2 \times 5$, we have $(24,90) = 2 \times 3 = 6$ and $\{24,90\} = 2^3 \times 3^2 \times 5 = 360$.

If a and b are natural numbers, then $a \times b = (a,b) \times \{a,b\}$. We can verify this in our example, where we find that $24 \times 90 = 6 \times 360 = 2{,}160$. If the greatest common divisor (a,b) of two whole numbers is 1, then a and b are said to be *relatively prime.* We see that $(6,35) = 1$, for example.

In analogy with algebraic equations of the first degree, Gauss then went on to congruences of the first degree $ax \equiv b \pmod{m}$. This expression is equivalent to the equality $ax - my = b$, where a, b and m are given whole numbers and x and y are unknown whole numbers. Such equations are called Diophantine equations after the Alexandrian Greek Diophantos (c. A.D. 250-330). *If $(a,m) = d$, then a necessary and sufficient condition for the congruence $ax \equiv b \pmod{m}$ to be solvable is that d be a divisor of b. There then exist d different sequences of solutions, or, more succinctly, d solutions.* We shall give several examples.

Example 1. The congruence $x \equiv 4 \pmod{7}$ needs little analysis, since we see immediately that the only solutions are the numbers $x = 4 + 7k$, where k is a whole number.

Example 2. The congruence $6x \equiv 5 \pmod{8}$ is not solvable, because $d = (6,8) = 2$ and 2 is not a divisor of 5.

Example 3. The congruence $2x \equiv 3 \pmod{5}$ is solvable, since $d = (2,5) = 1$; that is to say, the numbers 2 and 5 are relatively prime, and we have one solution. Through substitution we find that $x = 4$ answers the requirements. The Diophantine equation $2x - 5y = 3$ gives $x = 4 + 5k$, where k is an arbitrary whole number. We write this in convenient notation as $x \equiv 4 \pmod{5}$.

Example 4. The congruence $34x \equiv 60 \pmod{98}$ is solvable, since $d = (34,98) = 2$ and 2 is a divisor of 60. Thus there exist *two* solutions: $x \equiv -4 \pmod{98}$ and $x \equiv 45 \pmod{98}$. We may write $x \equiv -4, 45 \pmod{98}$.

CONGRUENCES OF SECOND DEGREE

In the third and fourth sections Gauss continued on to congruences of higher degree. Especially important is the binomial congruence $x^n \equiv 1 \pmod{m}$, of which a simple example is $x^2 \equiv 1 \pmod{8}$.

It has four solutions: $x \equiv 1, 3, 5, 7 \pmod{8}$.

Another example is the result known as the *little theorem of Fermat,* which was stated by Fermat, but probably known much earlier by the Hindus and Chinese. Gauss formulated and proved it with congruences. It runs as follows:

If p is a prime number and a is a whole number that is not divisible by p, then $a^{p-1} \equiv 1 \pmod{p}$. Consider two examples. For $p = 7$ and $a = 2$ we have $2^6 \equiv 1 \pmod{7}$; that is to say, $2^6 - 1 = 63$ is divisible by 7, an obvious fact. For $p = 37$ and $a = 10$ we find that the number $10^{36} - 1$ is divisible by 37; this we can check by calculation.

The fourth section deals with one of

the most interesting parts of number theory, the *theory of quadratic residues*. A number a is called a quadratic residue of the number m if the congruence $x^2 \equiv a \pmod{m}$ has a solution. If the congruence has no solution, then a is called a quadratic nonresidue of m.

These names were not introduced by Gauss but by Euler, who had worked earlier with the question of determining whether a number is a quadratic residue or nonresidue of another number.

For example, is 3 a quadratic residue of 11? If so, then we should be able to solve the congruence $x^2 \equiv 3 \pmod{11}$. If we experiment with $x = 1, 2, 3$ and so forth, we find that $x = 5$ is a solution. Thus 3 is a quadratic residue of 11. On the other hand, the congruence $x^2 \equiv 6 \pmod{11}$ is not solvable: 6 is a quadratic nonresidue of 11.

In the general investigation one may restrict one's attention to the case in which a and m are (different) prime numbers. Then the question becomes: *If p is a given prime number, for which prime numbers q can we solve $x^2 \equiv q$* $\pmod{p}$?

In this general form Gauss could not solve the question—nor has anyone else been able to. But there is a curious relationship between this congruence and the congruence $x^2 \equiv p \pmod{q}$. This relation had been set forth in turn by Euler and Legendre (the latter called it the *law of quadratic reciprocity*), but neither had been able to give a rigorous proof for what they had discovered by numerical calculations. As mentioned before, Gauss discovered this theorem on his own in 1795 and published the first correct proof. He called it the *fundamental theorem in the theory of quadratic residues*. It runs as follows:

For the pair of congruences $x^2 \equiv q$ $\pmod{p}$ and $x^2 \equiv p \pmod{q}$, *where p and q are different primes, the following holds: either both congruences are solvable or both are unsolvable, except in the case in which both p and q have remainder 3 when divided by 4, in which case one of the congruences is solvable while the other is unsolvable.*

The foundation of Gauss's proof of this theorem was his experimentation with numbers, which certainly is unique in the history of mathematics. Of course we are speaking here of people and not of machines. During the year 1795 and earlier, while Gauss still did not know that he had predecessors in this area, he made up a huge table of prime numbers; of quadratic residues and nonresidues; and of fractions $1/n$ from $n = 1$ to $n = 1{,}000$, expressed as periodic decimals with the entire period given.

The longest period that the decimal expansion of the fraction $1/n$ can have is $(n - 1)$ digits. For example,

$$\frac{1}{7} = 0.142857\ 142857\ \ldots,$$

has the maximal length of period, 6 digits, while

$$\frac{1}{9} = 0.11111\ \ldots$$

has only one digit in its period.

In computing the entire period of the decimal expansions of $1/n$ for $n = 1$ up to $n = 1{,}000$, in a good many cases Gauss had to calculate several hundred decimals. For example, he determined 1/811 to 822 decimals, the last few being put in as a check of the calculations. Gauss wrote up this table to find the connection between the period of the decimal expression and the denominator n. It was a frightfully laborious path that he set out upon, but it finally led him to his goal.

On April 8, 1796, a short notice in the journal states that he had found the first exact proof of the fundamental theorem of quadratic residues. It was a very long proof, which contained eight different cases and was carried out with obstinate logic. The great German mathematician Leopold Kronecker (1823-1891) later characterized it as the "test of strength of Gauss's genius."

True to his principle of the finished

work of art, Gauss worked on and presented a total of eight proofs of "the gem of arithmetic," which really deserves its name; it plays a fundamental role in higher number theory and in several areas of algebra.

The law of quadratic reciprocity can be formulated in various ways. The shortest is probably the following: *A prime number p is a quadratic residue or nonresidue of another prime number q according to whether* $q \times (-1)^{(q-1)/2}$ *is a residue or nonresidue of p.*

We shall illustrate the different cases that can occur (in the first formulation) with a few examples.

Example 1. $p = 23$, $q = 13$. Note that 23 yields the remainder 3 while 13 yields the remainder 1 when divided by 4. Thus the two congruences $x^2 \equiv 13 \pmod{23}$ and $x^2 \equiv 23 \pmod{13}$ are either both solvable or both unsolvable. They are solvable. For example, the first one has the solution $x = 6$, and the second one also has the solution $x = 6$.

Example 2. $p = 5$, $q = 17$. Since both of these numbers have remainder 1 when divided by 4, it must be true that either both of the congruences $x^2 \equiv 17 \pmod 5$ and $x^2 \equiv 5 \pmod{17}$ are solvable or both are unsolvable. They are unsolvable.

Example 3. $p = 23$, $q = 11$. In this case both numbers have remainder 3 when divided by 4. Of the two congruences $x^2 \equiv 11 \pmod{23}$ and $x^2 \equiv 23 \pmod{11}$, one must be solvable while the other is unsolvable. The first is unsolvable. The second has the solutions 10, 21, 32, and 43, for example.

DIVIDING THE CIRCLE

In the seventh and final section Gauss applies his earlier results to the binomial congruence $x^n \equiv 1 \pmod p$, where p is a prime number and n a natural number. The relation between these arithmetic congruences and the binomial equation $x^n = 1$ gives the solution to the problem of dividing the circle and constructing the regular 17-gon. . . . The binomial congruence $x^n \equiv 1 \pmod p$ unites arithmetic, algebra, and geometry in one of the great syntheses which Gauss pursued and which he achieved here in a fashion that has few counterparts in the history of mathematics.

ERNST EDUARD KUMMER (1810–93)

"One of the creative pioneers of 19th-century mathematics," Kummer was the son of physician Carl Gotthelf Kummer and Friedrike Sophie (née Rothe). He enrolled in the Gymnasium in Sorau in 1819 and the University of Halle in 1823. Under the influence of the mathematics professor Heinrich F. Scherk, he quickly turned from his first major, Protestant theology, to mathematics. In 1831 he passed the examination for Gymnasium teaching and was awarded a doctorate.

Kummer had a distinguished academic career. After spending a year's probationary period in Sorau (1831–32), he taught at the Gymnasium in Liegnitz (now Legnica, Poland) from 1832 to 1842. Two of his Liegnitz students, who became interested in mathematics because of him, were Friedrich Joachimsthal and Leopold Kronecker. He inspired his students to pursue independent scientific work. While at Liegnitz he concentrated on examining function theory. A little later in military service, Kummer sent a paper on hypergeometric series to Jacobi, who introduced his ideas to Dirichlet at the University of Berlin. Dirichlet recommended him for a corresponding membership in the Berlin Academy of Sciences in 1839. The next year Kummer married Ottilie Mendelssohn who died eight years later. Shortly afterward he married Bertha Cauer. He was the father of nine children.

From 1842 to 1855 Kummer served as full professor at the University of Breslau (now Wroclaw, Poland). Dirichlet and Jacobi had recommended him for the position. For 20 years after 1842 he concentrated on number theory. He was an honest conservative who during the Revolution of 1848 advocated a constitutional monarchy for Prussia.

When, in 1855, Dirichlet left the University of Berlin to succeed Gauss in Göttingen, he proposed that Kummer succeed him in Berlin. Kummer was appointed. In turn Kummer recommended his student Joachimsthal for the Breslau position he vacated, which blocked Karl Weierstrass's application there. He worked instead to draw Weierstrass to Berlin in 1856. His student Leopold Kronecker had moved to Berlin in 1855. The University of Berlin was about to experience a full flowering in mathematics.

At Berlin, Kummer continued to be an exceptional teacher as well as an able administrator. The clarity and vividness of his carefully prepared lectures drew as many as 250 students to hear them. On his and Weierstrass's recommendation the University of Berlin established Germany's first seminar in pure mathematics in 1861. Kummer guided 39 doctoral dissertations, including those of Paul Gordan and Georg Cantor. His students knew him for his penetrating thought, charm, sense of humor, and willingness to aid them, even in matters of material difficulty. Kummer reveled in teaching as well as serving on the faculty of the Kriegschule until 1874. Additional work did not limit his creative achievements. He was perpetual secretary of the physics-mathematics section of the Berlin Academy from 1863 to 1878 and was dean for two years (1857–58 and 1865–66) and rector for one (1868–69) at the University of Berlin. On the faculty his continuing friendship with Kronecker led to some tensions with Weierstrass in the

1870s. In 1882 he detected a weakening of memory and of his ability to develop freely logical, coherent, and abstract arguments. Though his colleagues failed to see these, he compelled them to accept his retirement in 1883.

During the three creative mathematical periods in his adult life, Kummer examined function theory, arithmetic, and geometry. Inspired most of all by Gauss and Dirichlet, he wrote a number of articles but never published a textbook. In his arithmetical period, he contributed to the theory of algebraic numbers, appropriately defined integer and divisibility, and formulated the theory of ideal prime factors. With the help of this theory Kummer was able to demonstrate in a number of cases the so-called great theorem of Fermat which states that for $n > 2$, there are no integral solutions of $x^n + y^n = z^n$, except when x, y, or $z = 0$, and to prove the law of biquadratic reciprocity. Through Kronecker and Richard Dedekind, who generalized Gauss's complex integers and Kummer's algebraic numbers into a new theory of algebraic numbers, these ideas were developed further. During his geometric period, Kummer followed William Rowan Hamilton in optics by treating the theory of general ray systems purely algebraically. He also discovered the fourth-order surface.

109. From "On the Theory of Complex Numbers" (1847)*

(Theory of Ideal Prime Factors)

ERNST EDUARD KUMMER

I have succeeded in completing and in simplifying the theory of those complex numbers which are formed from the higher roots of unity and which, as is well known, play an important rôle in cyclotomy and in the study of power residues and of forms of higher degree; this I have done through the introduction of a peculiar kind of imaginary divisors which I call *ideal complex numbers* and concerning which I take the liberty of making a few remarks.

If α is an imaginary root of the equation $\alpha^\lambda = 1$, λ a prime number, and a, a_1, a_2, etc. whole numbers, then $f(\alpha) = a + a_1\alpha + a_2\alpha^2 + \ldots + a_{\lambda-1}\alpha^{\lambda-1}$ is a complex whole number. Such a complex number can either be broken up into factors of the same kind or such a decomposition is not possible. In the first case, the number is a composite number; in the second case, it has hitherto been called a complex prime number. I have observed, however, that, even though $f(\alpha)$ cannot in any way be broken up into complex factors, it still does not possess the true nature of a complex prime number, for, quite commonly, it lacks the first and most important property of prime numbers;

*Source: This translation by Thomas Freeman Cope is taken from David Eugene Smith (ed.), *A Source Book in Mathematics* (1929), 119-126. It is reprinted by permission of McGraw-Hill Book Company, Inc. The German original of this article appeared in Crelle's *Journal*, 35 (1847), 319-326.

namely, that the product of two prime numbers is divisible by no other prime numbers. Rather, such numbers $f(\alpha)$, even if they are not capable of decomposition into complex factors, have nevertheless the nature of composite numbers; the factors in this case are, however, not actual but ideal complex numbers. For the introduction of such ideal complex numbers, there is the same, simple, basal motive as for the introduction of imaginary formulas into algebra and analysis; namely, the decomposition of integral rational functions into their simplest factors, the linear. It was, moreover, such a desideratum which prompted Gauss, in his researches on biquadratic residues (for all such prime factors of the form $4m + 1$ exhibit the nature of composite numbers), to introduce for the first time complex numbers of the form $a + b\sqrt{-1}$.

In order to secure a sound definition of the true (usually ideal) prime factors of complex numbers, it was necessary to use the properties of prime factors of complex numbers which hold in every case and which are entirely independent of the contingency of whether or not actual decomposition takes place: just as in geometry, if it is a question of the common chords of two circles even though the circles do not intersect, one seeks an actual definition of these ideal common chords which shall hold for all positions of the circles. There are several such permanent properties of complex numbers which could be used as definitions of ideal prime factors and which would always lead to essentially the same result; of these, I have chosen *one* as the simplest and the most general.

If p is a prime number of the form $m\lambda + 1$, then it can be represented, in many cases, as the product of the following $\lambda - 1$ complex factors: $p = f(\alpha) \cdot f(\alpha^2) \cdot f(\alpha^3) \ldots f(\alpha^{\lambda-1})$; when, however, a decomposition into actual complex prime factors is not possible, let ideals make their appearance in order to bring this about. If $f(\alpha)$ is an actual complex number and a prime factor of p, it has the property that, if instead of the root of the equation $\alpha^\lambda = 1$ a definite root of the congruence $\xi^\lambda \equiv 1$, mod. p, is substituted, then $f(\xi) \equiv 0$, mod. p. Hence too if the prime factor $f(\alpha)$ is contained in a complex number $\Phi(\alpha)$, it is true that $\Phi(\xi) \equiv 0$, mod. p; and conversely, if $\Phi(\xi) \equiv 0$, mod. p, and p is factorable into $\lambda - 1$ complex prime factors, then $\Phi(\alpha)$ contains that prime factor $f(\alpha)$. Now the property $\Phi(\xi) \equiv 0$, mod. p, is such that it does not depend in any way on the factorability of the number p into prime factors; it can accordingly be used as a definition, since it is agreed that the complex number $\Phi(\alpha)$ shall contain the ideal prime factor of p which belongs to $\alpha = \xi$, if $\Phi(\xi) \equiv 0$, mod. p. Each of the $\lambda - 1$ complex prime factors of p is thus replaced by a congruence relation. This suffices to show that complex prime factors, whether they be actual or merely ideal, give to complex numbers the same definite character. In the process given here, however, we do not use the congruence relations as the definitions of ideal prime factors because they would not be sufficient to represent several equal ideal prime factors of a complex number, and because, being too restrictive, they would yield only ideal prime factors of the real prime numbers of the form $m\lambda - 1$.

Every prime factor of a complex number is also a prime factor of every real prime number q, and the nature of the ideal prime factors is, in particular, dependent on the exponent to which q belongs for the modulus λ. Let this exponent be f, so that $q^f \equiv 1$, mod. λ, and $\lambda - 1 = e \cdot f$. Such a prime number q can never be broken up into more than e complex prime factors which, if this decomposition can actually be carried out, are represented as linear functions of the e periods of each set of f terms. These periods of the roots of the equation $\alpha^\lambda = 1$, I denote by $\eta, \eta_1, \eta_2, \ldots$

η_{e-1}; and indeed in such an order that each goes over into the following one whenever α is transformed into α^γ, where γ is a primitive root of λ. As is well known, the periods are the e roots of an equation of the eth degree; and this equation, considered as a congruence for the modulus q, has always e real congruential roots which I denote by $u, u_1, u_2, \ldots u_{e-1}$ and takes in an order corresponding to that of the periods, for which, besides the congruence of the eth degree, still other easily found congruences may be used. If now the complex number $c'\eta + c_1'\eta_1 + c_2'\eta_2 + \ldots + c'_{e-1}\eta_{e-1}$, constructed out of periods, is denoted shortly by $\Phi(\eta)$, then among the prime numbers q which belong to the exponent f, there are always such that can be brought into the form

$$q = \Phi(\eta)\Phi(\eta_1)\Phi(\eta_2) \ldots \Phi(\eta_{e-1}),$$

in which, moreover, the e factors never admit a further decomposition. If one replaces the periods by the congruential roots corresponding to them, where a period can arbitrarily be designated to correspond to a definite congruential root, then one of the e prime factors always becomes congruent to zero for the modulus q. Now if any complex number $f(\alpha)$ contains the prime factor $\Phi(\eta)$, it will always have the property, for $\eta = u_k$, $\eta_1 = u_{k+1}$, $\eta_2 = u_{k+2}$, etc., of becoming congruent to zero for the modulus q. This property (which implies precisely f distinct congruence relations, the development of which would lead too far) is a permanent one even for those prime numbers q which do not admit an actual decomposition into e complex prime factors. It could therefore be used as a definition of complex prime factors; it would, however, have the defect of not being able to express the equal ideal prime factors of a complex number.

The definition of ideal complex prime factors which I have chosen and which is essentially the same as the one described but is simpler and more general, rests on the fact that, as I prove separately, one can always find a complex number $\psi(\eta)$, constructed out of periods, which is of such a nature that $\psi(\eta)\, \psi(\eta_1)\, \psi(\eta_2) \ldots \psi(\eta_{e-1})$ (this product being a whole number) is divisible by q but not by q^2. This complex number $\psi(\eta)$ has always the abovementioned property, namely, that it is congruent to zero, modulo q, if for the periods are substituted the corresponding congruential roots, and therefore $\psi(\eta) \equiv 0$, mod. q, for $\eta = u$, $\eta_1 = u_1$, $\eta_2 = u_2$, etc. I now set $\psi(\eta_1)\psi(\eta_2) \ldots \psi(\eta_{e-1}) = \Psi(\eta)$ and define ideal prime numbers in the following manner:—

If $f(\alpha)$ has the property that the product $f(\alpha) \cdot \Psi(\eta_r)$ is divisible by q, this shall be expressed as follows: $f(\alpha)$ contains the ideal prime factor of q which belongs to $u = \eta_r$. Furthermore, if $f(\alpha)$ has the property that $f(\alpha) \cdot (\Psi(\eta_r))^\mu$ is divisible by q^μ but $f(\alpha)(\Psi(\eta_r))^{\mu+1}$ is not divisible by $q^{\mu+1}$, this shall be described thus: $f(\alpha)$ contains the ideal prime factor of q which belongs to $u = \eta_r$, exactly μ times.

It would lead too far if I should develop here the connection and the agreement of this definition with those given by congruence relations as described above; I simply remark that the relation: $f(\alpha)\Psi(\eta_r)$ divisible by q, is completely equivalent to f distinct congruence relations, and that the relation: $f(\alpha)\,(\Psi(\eta_r))^\mu$ divisible by q^μ, can always be entirely replaced by $u \cdot f$ congruence relations. The whole theory of ideal complex numbers which I have already perfected and of which I here announce the principal theorems, is a justification of the definition given as well as of the nomenclature adopted. The principal theorems are the following:

The product of two or more complex numbers has exactly the same ideal prime factors as the factors taken together.

If a complex number (which is a product of factors) contains all the e prime factors of q, it is also divisible by

q itself; if, however, it does not contain some one of these e ideal prime factors, it is not divisible by q.

If a complex number (in the form of a product) contains all the e ideal prime factors of q and, indeed, each at least μ times, it is divisible by q^{μ}.

If $f(\alpha)$ contains exactly m ideal prime factors of q, which may all be different, or partly or wholly alike, then the norm $Nf(\alpha) = f(\alpha)f(\alpha^2) \ldots f(\alpha^{\lambda-1})$ contains exactly the factor q^{mf}.

Every complex number contains only a finite, determinate number of ideal prime factors.

Two complex numbers which have exactly the same ideal prime factors differ only by a complex unit which may enter as a factor.

A complex number is divisible by another if all the ideal prime factors of the divisor are contained in the dividend; and the quotient contains precisely the excess of the ideal prime factors of the dividend over those of the divisor.

From these theorems it follows that computation with complex numbers becomes, by the introduction of ideal prime factors, entirely the same as computation with integers and their real integral prime factors. Consequently, the grounds for the complaint which I voiced in the *Breslauer Programm zur Jubelfeier der Universität Königsberg* S. 18, are removed:—

It seems a great pity that this quality of real numbers, namely, that they can be resolved into prime factors which for the same number are always the same, is not shared by complex numbers; if now this desirable property were part of a complete doctrine, the effecting of which is as yet beset with great difficulties, the matter could easily be resolved and brought to a successful conclusion. Etc. One sees therefore that ideal prime factors disclose the inner nature of complex numbers, make them transparent, as it were, and show their inner crystalline structure. If, in particular, a complex number is given merely in the form $a + a_1\alpha + a_2\alpha^2 + \ldots + a_{\lambda-1}\alpha^{\lambda-1}$, little can be asserted about it until one has determined, by means of its ideal prime factors (which in such a case can always be found by direct methods), its simplest qualitative properties to serve as the basis of all further arithmetical investigations.

Ideal factors of complex numbers arise, as has been shown, as factors of actual complex numbers: hence ideal prime factors multiplied with others suitably chosen must always give actual complex numbers for products. This question of the combination of ideal factors to obtain actual complex numbers is, as I shall show as a consequence of the results which I have already found, of the greatest interest, because it stands in an intimate relationship to the most important sections of number theory. The two most important results relative to this question are the following:

There always exists a finite, determinate number of ideal complex multipliers which are necessary and sufficient to reduce all possible ideal complex numbers to actual complex numbers.[1]

Every ideal complex number has the property that a definite integral power of it will give an actual complex number.

I consider now some more detailed developments from these two theorems. Two ideal complex numbers which, when multiplied by one and the same ideal number, form actual complex numbers, I shall call *equivalent* or of the same class, because this investigation of actual and ideal complex numbers is identical with the classification of a certain set of forms of the $\lambda - 1$st degree and in $\lambda - 1$ variables; the principal results relative to this classification have been found by Dirichlet but not yet published so that I do not know precisely whether or not his principle of classification coincides with that resulting from the theory of complex numbers. For example, the theory of a form of the second degree in two variables with determinant, however, a prime number λ, is closely interwoven with these investigations, and our classification in this case coincides with that of Gauss but not with that of Legendre. The same considerations also throw great light upon Gauss's classification of forms of the second degree and upon the true basis for the differentiation between *Aequivalentia propria et impropria*,[2] which, undeniably, has always an appearance of impropriety when it presents itself in the *Disquisitiones arithmeticae*. If, for example, two forms such as $ax^2 + 2bxy + cy^2$ and $ax^2 - 2bxy + cy^2$, or $ax^2 + 2bxy + cy^2$ and $cx^2 + 2bxy + ay^2$, are considered as belonging to different classes, as is done in the above-mentioned work, while in fact no essential difference between them is to be found; and if on the other hand Gauss's classification must notwithstanding be admitted to be one arising for the most part out of the very nature of the question: then one is forced to consider forms such as $ax^2 + 2bxy + cy^2$ and $ax^2 - 2bxy + cy^2$ which differ from each other in outward appearance only, as merely representative of two new but essentially different concepts of number theory. These however, are in reality nothing more than two different ideal prime factors which belong to one and the same number. The entire theory of forms of the second degree in two variables can be thought of as the theory of complex numbers of the form $x + y\sqrt{D}$ and then leads necessarily to ideal complex numbers of the same sort. The latter, however, classify themselves according to the ideal multipliers which are necessary and sufficient to reduce them to actual complex numbers of the form $x + y\sqrt{D}$. Because of this agreement with the classification of Gauss, ideal complex numbers thus constitute the true basis for it.

The general investigation of ideal complex numbers presents the greatest analogy with the very difficult section by Gauss: *De compositione formarum*, and the principal results which Gauss proved for quadratic forms, pp. 337 and following, hold true also for the combination of general ideal complex numbers. Thus there belongs to every class of ideal numbers another class which, when multiplied by the first class, gives rise to actual complex numbers (here the actual complex numbers are the analogue of the *Classis principalis*).[3] Likewise, there are classes which, when multiplied by themselves, give for the result actual complex numbers (the *Classis principalis*), and these classes are therefore *ancipites*;[4] in particular, the *Classis principalis* itself is always a *Classis anceps*. If one takes an ideal complex number and raises it to powers, then in accordance with the second of the foregoing theorems, one will arrive at a power which is an actual complex number; if h is the smallest number for which $(f(\alpha))^h$ is an actual complex number, then $f(\alpha)$, $(f(\alpha))^2$, $(f(\alpha))^3$, . . . $(f(\alpha))^h$ all belong to different classes. It now may happen that, by a suitable choice of $f(\alpha)$, these exhaust all existing classes: if such is not the case, it is easy to prove that the number of classes is at least always a multiple of h. I have not gone deeper yet into this domain of complex numbers; in particular, I have not undertaken an investigation of the exact number of classes because I have heard that Dirichlet, using principles similar to those employed in his famous treatise on quadratic forms, has already found this number. I shall make only one additional remark about the character of ideal complex numbers, namely, that by the second of the foregoing theorems they can always be considered and represented as definite roots of actual complex numbers that is, they always take the form $\sqrt[h]{\Phi(\alpha)}$ where $\Phi(\alpha)$ is an actual complex number and h an integer.

Of the different applications which I have already made of this theory of complex number, I shall refer only to the application to cyclotomy to complete the results which I have already announced in the above-mentioned *Programm*. If one sets

$$(\alpha, x) = x + \alpha x^{g} + \alpha^2 x^{g^2} + \ldots + \alpha^{p-2} x^{g^{p-2}},$$

where $\alpha^\lambda = 1$, $x^p = 1$, $p = m\lambda + 1$, and g is a primitive root of the prime number p, then it is well known that $(\alpha, x)^\lambda$ is a complex number independent of x and formed from the roots of the equation $\alpha^\lambda = 1$. In the *Programm* cited, I have found the following expression for this number, under the assumption that p can be resolved into $\lambda - 1$ actual complex prime factors, one of which is $f(\alpha)$:

$$(\alpha, x)^\lambda = \pm\alpha^h f^{m_1}(\alpha) \cdot f^{m_2}(\alpha^2) \cdot f^{m_3}(\alpha^3) \ldots f^{m_{\lambda-1}}(\alpha^{\lambda-1}),$$

where the power-exponents m_1, m_2, m_3, etc. are so determined that the general m_κ, positive, is less than λ and $k \cdot m_k \equiv 1$, mod. λ. Exactly the same simple expression holds in complete generality, as can easily be proved, even when $f(\alpha)$ is not the actual but only the ideal prime factor of p. In order, however, in the latter case, to maintain the expression for $(\alpha, x)^\lambda$ in the form for an actual complex number, one need only represent the ideal $f(\alpha)$ as a root of an actual complex number, or apply one of the methods (although indirect) which serve to represent an actual complex number whose ideal prime factors are given.

NOTES

1. A proof of this important theorem, although in far less generality and in an entirely different form, is found in the dissertation: L. Kronecker, *De unitatibus complexis*, Berlin, 1845.
2. *i. e.*, proper and improper equivalence.
3. Principal class.
4. Dual, or of a double nature.

Chapter VIII
The Nineteenth Century

Section D
Number Theory, Set Theory, and Symbolic Logic

(JULIUS WILHELM) RICHARD DEDEKIND (1831–1916)

Richard Dedekind, who is sometimes called "a modern Eudoxus," was the son of the jurist Julius Dedekind and Caroline Henriette (*née* Emperius). He grew up in Brunswick, Germany. In his last years as a pupil at the Gymnasium Martino-Catherineum, his attention turned increasingly from chemistry and physics toward mathematics. In 1848 he entered the Collegium Carolinum, an institute offering courses between the high school and university levels. To qualify to study advanced mathematics at the University of Göttingen, he took courses in analytic geometry, algebra, the calculus, and mechanics for two years. In 1850 he entered Göttingen, where he met Bernhard Riemann and, after only four semesters, earned his doctorate under Carl Gauss with a thesis on the theory of Eulerian integrals. Gauss wrote prophetically of having "favorable expectations of his future performance."

Dedekind pursued an academic career. In 1854 he qualified as a *Privatdozent* at Göttingen. He first lectured on probability and then introduced students, perhaps for the first time, to Galois' theory. He was a pallbearer at Gauss's funeral in 1855. To continue his mathematical education Dedekind attended the lectures of Riemann on Abelian and elliptic functions and those of Gauss's successor, P. G. L. Dirichlet, on the theory of numbers. The latter showed him the need to redefine irrational numbers in terms of arithmetical properties rather than in terms of the Eudoxian geometric approach. Dedekind credited Dirichlet with enlarging his scholarly horizons and making "a new man" of him. In 1858 Dedekind went to teach at Zürich Polytechnic, staying for four years. In 1859 he accompanied Riemann on a trip to Berlin to meet Weierstrass, Kummer, and Kronecker.

In 1862 Dedekind accepted a professorship at Brunswick Polytechnic which had been created out of the Collegium Carolinum. Despite major offers from such universities as Göttingen and Halle, he remained at the Polytechnic for the rest of his career, living in comparative isolation and without full recognition of his mathematical achievements. He chose to stay close to his brother and sister in familiar surroundings, where he found sufficient freedom and leisure for basic mathematical research. He never married. Dedekind was director of Brunswick Polytechnic from 1872 to 1875 and for a time chaired its building commission. He became professor emeritus in 1894 but continued to give occasional lectures.

Besides mathematics Dedekind enjoyed music, reading, and travel. An accomplished pianist and cellist, he composed a chamber opera. He took recreational trips to the Austrian Tyrol, to Switzerland, and to the Black Forest. In 1878 he visited the Paris Exposition.

In character, views, and way of life Dedekind had much in common with Gauss who also came from Brunswick. Both were strong-willed conservatives who opposed rapid innovation, possessed unshakable principles, and led simple lives. Combined with a sense of duty each had a distinct sense of humor. Both listed Walter Scott among their favorite authors. Not surprisingly the similarities extended to mathematics, where both preferred to study the theory of numbers, had reservations about the algorithm, and valued concrete "notions" above "notations."

Throughout his career Dedekind concentrated on providing satisfactory foundations for mathematics. In number theory, for example, he made the first major redefinition of irrational numbers since Eudoxus (Euclid V.5) by supplying the missing theory of continuity. He proposed that rational and irrational numbers form an ordered continuum of real numbers wherein the irrationals are boundaries between the rationals; then he divided the continuum into two classes or sets. By the theory of order each real number in one class has to be less than each real number in the higher class. The theory of order and a method now known as the "Dedekind cut" allowed Dedekind to trace the real numbers to the rational numbers. Any rational number a divides the system of rational numbers R into two classes, A_1 and A_2, such that each a_1 in class A_1 is indeed less than each a_2 in class A_2 and a is either the largest number in A_1 or the smallest number in A_2. Any number which similarly divides R into classes A_1 and A_2 but is neither the largest number in A_1 nor the smallest in A_2 is a "cut (A_1, A_2) which . . . create(s) a new irrational number α, which we regard as completely defined by this cut." (*Stetigkeit*, Prop. 4). Dedekind clearly and rigorously presented his novel method in his monograph on the foundations of analysis, *Stetigkeit und irrationale Zahlen* (Continuity and Irrational Numbers, 1872).

At first Dedekind's highly original concepts in the theory of numbers and mathematical analysis were not received enthusiastically by leading mathematicians. Georg Cantor and Camille Jordan were two notable exceptions. Dedekind met Cantor in 1872, and they both explored the concept of the infinite. But Kronecker challenged Dedekind's militant defense of Gauss's view that numbers are free creations of the human intellect, while Weierstrass believed that his work on complex numbers was not fully understood by Dedekind. In his monograph on the foundations of arithmetic, *Was sind und was sollen die Zahlen?* (What are and what should numbers be? 1888), Dedekind offered a logical theory of number and maintained that the ordinal, not cardinal, number was the original numerical concept. Later Hilbert criticized his solely logical approach, while Frege and Russell adhered to it. Among his other achievements, Dedekind proposed a celebrated theory of ideals that went beyond Kummer's theory of "ideal numbers." The ideal classes were collections of algebraic integer multiples. Dedekind showed that each ideal aside from the unit ideal R can be represented as the product of prime numbers. This was the first application of unique factorization to algebraic structures. He also developed with Heinrich Weber an arithmetic approach to algebraic geometry and introduced the abstract structure of a lattice, which provided a single framework for working with the notions of cut, chain, and ideal. He further set forth the fundamental concepts of *ring* and *unit*.

Recognizing the thinkers who most influenced him, Dedekind edited the posthumous manuscripts of Gauss, Dirichlet, and Riemann. In assessing his own achievement he shared with them a modesty bordering on shyness; he was embarrassed when people cited his brilliance. Dedekind wrote, "For what I have accomplished and what I have become, I have to thank my industry much more, my indefatigable working rather than any outstanding talent."

110. From *Stetigkeit und irrationale Zahlen* (1872)*

(Continuity, Irrational Numbers, and Dedekind Cuts)

RICHARD DEDEKIND

My attention was first directed toward the considerations which form the subject of this pamphlet in the autumn of 1858. As professor in the Polytechnic School in Zürich I found myself for the first time obliged to lecture upon the elements of the differential calculus and felt more keenly than ever before the lack of a really scientific foundation for arithmetic. In discussing the notion of the approach of a variable magnitude to a fixed limiting value, and especially in proving the theorem that every magnitude which grows continually but not beyond all limits, must certainly approach a limiting value, I had recourse to geometric evidences. Even now such resort to geometric intuition in a first presentation of the differential calculus, I regard as exceedingly useful, from the didactic standpoint, and indeed indispensable, if one does not wish to lose too much time. But that this form of introduction into the differential calculus can make no claim to being scientific, no one will deny. For myself this feeling of dissatisfaction was so overpowering that I made the fixed resolve to keep meditating on the question till I should find a purely arithmetic and perfectly rigorous foundation for the principles of infinitesimal analysis. The statement is so frequently made that the differential calculus deals with continuous magnitude, and yet an explanation of this continuity is nowhere given; even the most rigorous expositions of the differential calculus do not base their proofs upon continuity but, with more or less consciousness of the fact, they either appeal to geometric notions or those suggested by geometry, or depend upon theorems which are never established in a purely arithmetic manner. Among these, for example, belongs the above-mentioned theorem, and a more careful investigation convinced me that this theorem, or any one equivalent to

*Source: From Richard Dedekind, *Essays on the Theory of Numbers*, trans. by Wooster Woodruff Berman (1924), 1-19. This translation is reprinted by permission of the Open Court Publishing Company.

it, can be regarded in some way as sufficient basis for infinitesimal analysis. It then only remained to discover its true origin in the elements of arithmetic and thus at the same time to secure a real definition of the essence of continuity. I succeeded November 24, 1858, and a few days afterward I communicated the results of my meditations to my dear friend Durege with whom I had a long and lively discussion. Later I explained these views of a scientific basis of arithmetic to a few of my pupils, and here in Braunschweig read a paper upon the subject before the scientific club of professors, but I could not make up my mind to its publication, because, in the first place, the presentation did not seem altogether simple, and further, the theory itself had little promise. Nevertheless I had already half determined to select this theme as subject for this occasion, when a few days ago, March 14, by the kindness of the author, the paper *Die Elemente der Funktionenlehre* by E. Heine (*Crelle's Journal*, Vol. 74) came into my hands and confirmed me in my decision. In the main I fully agree with the substance of this memoir, and indeed I could hardly do otherwise, but I will frankly acknowledge that my own presentation seems to me to be simpler in form and to bring out the vital point more clearly. While writing this preface (March 20, 1872), I am just in receipt of the interesting paper *Ueber die Ausdehnung eines Satzes aus der Theorie der trigonometrischen Reihen*, by G. Cantor (*Math. Annalen*, Vol. 5), for which I owe the ingenious author my hearty thanks. As I find on a hasty perusal, the axiom given in Section II of that paper, aside from the form of presentation, agrees with what I designate in Section III as the essence of continuity. But what advantage will be gained by even a purely abstract definition of real numbers of a higher type, I am as yet unable to see, conceiving as I do of the domain of real numbers as complete in itself.

I
PROPERTIES OF RATIONAL NUMBERS

The development of the arithmetic of rational numbers is here presupposed, but still I think it worth while to call attention to certain important matters without discussion, so as to show at the outset the standpoint assumed in what follows. I regard the whole of arithmetic as a necessary, or at least natural, consequence of the simplest arithmetic act, that of counting, and counting itself as nothing else than the successive creation of the infinite series of positive integers in which each individual is defined by the one immediately preceding; the simplest act is the passing from an already-formed individual to the consecutive new one to be formed. The chain of these numbers forms in itself an exceedingly useful instrument for the human mind; it presents an inexhaustible wealth of remarkable laws obtained by the introduction of the four fundamental operations of arithmetic. Addition is the combination of any arbitrary repetitions of the above-mentioned simplest act into a single act; from it in a similar way arises multiplication. While the performance of these two operations is always possible, that of the inverse operations, subtraction and division, proves to be limited. Whatever the immediate occasion may have been, whatever comparisons or analogies with experience, or intuition, may have led thereto; it is certainly true that just this limitation in performing the indirect operations has in each case been the real motive for a new creative act; thus negative and fractional numbers have been created by the human mind; and in the system of all rational numbers there has been gained an instrument of infinitely greater perfection. This system, which I shall denote by R, possesses first of all a completeness and self-containedness which I have designated in another place[1] as characteristic of a *body of numbers* [Zahlkörper] and which consists in this that the four fundamental operations are always performable with any two individuals in R, *i.e.*, the result is always an individual of R, the single case of division by the number zero being excepted.

For our immediate purpose, however, another property of the system R is still more important; it may be expressed by saying that the system R forms a well-arranged domain of one dimension extending to infinity on two opposite sides. What is meant by this is sufficiently indicated by my use of expressions borrowed from geometric ideas; but just for this reason it will be necessary to bring out clearly the corresponding purely arithmetic properties in order to avoid even the appearance as if arithmetic were in need of ideas foreign to it.

To express that the symbols a and b represent one and the same rational number we put $a = b$ as well as $b = a$. The fact that two rational numbers a, b are different appears in this that the difference $a - b$ has either a positive or negative value. In the former case a is said to be *greater* than b, b *less* than a; this is also indicated by the symbols $a > b$, $b < a$.[2] As in the latter case $b - a$ has a positive value it follows that $b > a$, $a < b$. In regard to these two ways in which two numbers may differ the following laws will hold:

I. If $a > b$, and $b > c$, then $a > c$. Whenever a, c are two different (or unequal) numbers and b is greater than the one and less than the other, we shall, without hesitation because of the suggestion of geometric ideas, express this briefly by saying: b lies between the two numbers a, c.

II. If a, c are two different numbers, there are infinitely many different numbers lying between a, c.

III. If a is any definite number, then all numbers of the system R fall into two classes, A_1 and A_2, each of which contains infinitely many individuals; the first class A_1 comprises all numbers a_1 that are $< a$, the second class A_2 comprises all numbers a_2 that are $> a$; the number a by itself may be assigned at pleasure to the first or second class, being respectively the greatest number of the first class or the least of the second. In every case the separation of the system R into two classes, A_1, A_2 is such that every number of the first class A_1 is less than every number of the second class A_2.

II
COMPARISON OF THE RATIONAL NUMBERS WITH THE POINTS OF A STRAIGHT LINE

The above-mentioned properties of rational numbers recall the corresponding relations of position of the points of a straight line L. If the two opposite directions existing upon it are distinguished by "right" and "left," and p, q are two different points, then either p lies to the right of q, and at the same time q to the left of p, or conversely q lies to the right of p and at the same time p to the left of q. A third case is impossible, if p, q are actually different points. In regard to this difference in position the following laws hold:

I. If p lies to the right of q, and q to the right of r, then p lies to the right of r; and we say that q lies between the points p and r.

II. If p, r are two different points then there always exist infinitely many points that lie between p and r.

III. If p is a definite point in L, then all points in L fall into two classes, P_1, P_2, each of which contains infinitely many individuals; the first class P_1 contains all the points p_1, that lie to the left of p, and the second class P_2 contains all the points p_2 that lie to the right of p; the point p itself may be assigned at pleasure to the first or second class. In every case the separation of the straight line L into the two classes or portions P_1, P_2, is of such a character that every point of the first class P_1 lies to the left of every point of the second class P_2.

This analogy between rational numbers and the points of a straight line, as is well known, becomes a real correspondence when we select upon the straight line a definite origin or zero-point *0* and a definite unit of length for the measurement of segments. With the aid of the latter to every rational number *a* a corresponding length can be constructed and if we lay this off upon the straight line to the right or left of *0* according as *a* is positive or negative, we obtain a definite end-point *p*, which may be regarded as the point corresponding to the number *a*; to the rational number zero corresponds the point *0*. In this way to every rational number *a*, *i.e.*, to every individual in *R*, corresponds one and only one point *p*, *i.e.*, an individual in *L*. To the two numbers *a*, *b* respectively correspond the two points *p*, *q*, and if $a > b$, then *p* lies to the right of *q*. To the laws I, II, III of the previous Section correspond completely the laws I, II, III of the present.

III CONTINUITY OF THE STRAIGHT LINE

Of the greatest importance, however, is the fact that in the straight line *L* there are infinitely many points which correspond to no rational number. If the point *p* corresponds to the rational number *a*, then, as is well known, the length *0p* is commensurable with the invariable unit of measure used in the construction, *i.e.*, there exists a third length, a so-called common measure, of which these two lengths are integral multiples. But the ancient Greeks already knew and had demonstrated that there are lengths incommensurable with a given unit of length, *e.g.*, the diagonal of the square whose side is the unit of length. If we lay off such a length from point *0* upon the line we obtain an end-point which corresponds to no rational number. Since further it can be easily shown that there are infinitely many lengths which are incommensurable with the unit of length, we may affirm: the straight line *L* is infinitely richer in point-individuals than the domain *R* of rational numbers in number-individuals.

If now, as is our desire, we try to follow up arithmetically all phenomena in the straight line, the domain of rational numbers is insufficient and it becomes absolutely necessary that the instrument *R* constructed by the creation of the rational numbers be essentially improved by the creation of new numbers such that the domain of numbers shall gain the same completeness, or as we may say at once, the same *continuity*, as the straight line.

The previous considerations are so familiar and well known to all that many will regard their repetition quite superfluous. Still I regarded this recapitulation as necessary to prepare properly for the main question. For, the way in which the irrational numbers are usually introduced is based directly upon the conception of extensive magnitudes—which itself is nowhere carefully defined—and explains numbers as the result of measuring such a magnitude by another of the same kind.[3] Instead of this I demand that arithmetic shall be developed out of itself.

That such comparison with non-arithmetic notions have furnished the immediate occasion for the extension of the number-concept may, in a general way, be granted (though this was certainly not the case in the introduction of complex numbers); but this surely is no sufficient ground for introducing these foreign notions into arithmetic, the science of numbers. Just as negative and fractional rational numbers are formed by a new creation, and as the laws of operating with these numbers must and can be reduced to the laws of operating with positive integers, so we must endeavor completely to define irrational numbers by means of the rational numbers alone. The question only remains how to do this.

The above comparison of the domain *R* of rational numbers with a straight line has led to the recognition of the existence of gaps, of a certain incompleteness or discontinuity of the former, while we ascribe to the straight line completeness, absence of gaps, or continuity. In what then does this continuity consist? Everything must depend on the answer to this question, and only through it shall we obtain a scientific basis for the investigation of *all* continuous domains. By vague remarks upon the unbroken connection in the smallest parts obviously nothing is gained; the problem is to indicate a precise characteristic of continuity that can serve as the basis for valid deductions. For a long time I pondered over this in vain, but finally I found what I was seeking. This discovery will, perhaps, be differently estimated by different people; the majority may find its substance very commonplace. It consists of the following. In the preceding section attention was called to the fact that every point p of the straight line produces a separation of the same into two portions such that every point of one portion lies to the left of every point of the other. I find the essence of continuity in the converse, *i.e.*, in the following principle:

"If all points of the straight line fall into two classes such that every point of the first class lies to the left of every point of the second class, then there exists one and only one point which produces this division of all points into two classes, this severing of the straight line into two portions."

As already said I think I shall not err in assuming that every one will at once grant the truth of this statement; the majority of my readers will be very much disappointed in learning that by this commonplace remark the secret of continuity is to be revealed. To this I may say that I am glad if every one finds the above principle so obvious and so in harmony with his own ideas of a line; for I am utterly unable to adduce any proof of its correctness, nor has any one the power. The assumption of this property of the line is nothing else than an axiom by which we attribute to the line its continuity, by which we find continuity in the line. If space has at all a real existence it is *not* necessary for it to be continuous; many of its properties would remain the same even were it discontinuous. And if we knew for certain that space was discontinuous there would be nothing to prevent us, in case we so desired, from filling up its gaps, in thought, and thus making it continuous; this filling up would consist in a creation of new point-individuals and would have to be effected in accordance with the above principle.

IV
CREATION OF IRRATIONAL NUMBERS

From the last remarks it is sufficiently obvious how the discontinuous domain R of rational numbers may be rendered complete so as to form a continuous domain. In Section I it was pointed out that every rational number a effects a separation of the system R into two classes such that every number a_1 of the first class A_1 is less than every number a_2 of the second class A_2; the number a is either the greatest number of the class A_1 or the least number of the class A_2. If now any separation of the system R into two classes A_1, A_2, is given which possesses only *this* characteristic property that every number a_1 in A_1 is less than every number a_2 in A_2, then for brevity we shall call such a separation a *cut* [Schnitt] and designate it by (A_1, A_2). We can then say that every rational number a produces one cut or, strictly speaking, two cuts, which, however, we shall not look upon as essentially different; this cut possesses, *besides*, the property that either among the numbers of the first class there exists a greatest or among the numbers of the second class a least number. And conversely, if a cut possesses this property, then it is produced by this greatest or least rational number.

But it is easy to show that there exist

infinitely many cuts not produced by rational numbers. The following example suggests itself most readily.

Let D be a positive integer but not the square of an integer, then there exists a positive integer λ such that

$$\lambda^2 < D < (\lambda + 1)^2.$$

If we assign to the second class A_2, every positive rational number a_2 whose square is $> D$, to the first class A_1 all other rational numbers a_1, this separation forms a cut (A_1, A_2), *i.e.*, every number a_1 is less than every number a_2. For if $a_1 = 0$, or is negative, then on that ground a_1 is less than any number a_2, because, by definition, this last is positive; if a_1 is positive, then is its square $\leqq D$, and hence a_1 is less than any positive number a_2 whose square is $> D$.

But this cut is produced by no rational number. To demonstrate this it must be shown first of all that there exists no rational number whose square $= D$. Although this is known from the first elements of the theory of numbers, still the following indirect proof may find place here. If there exists a rational number whose square $= D$, then there exist two positive integers, t, u, that satisfy the equation

$$t^2 - Du^2 = 0,$$

and we may assume that u is the *least* positive integer possessing the property that its square, by multiplication by D, may be converted into the square of an integer t. Since evidently

$$\lambda u < t < (\lambda + 1)u,$$

the number $u' = t - \lambda u$ is a positive integer certainly *less* than u. If further we put

$$t' = Du - \lambda t,$$

t' is likewise a positive integer, and we have

$$t^2 - Du'^2 = (\lambda^2 - D)(t^2 - Du^2) = 0,$$

which is contrary to the assumption respecting u.

Hence the square of every rational number x is either $< D$ or $> D$. From this it easily follows that there is neither in the class A_1 a greatest, nor in the class A_2 a least number. For if we put

$$y = \frac{x(x^2 + 3D)}{3x^2 + D},$$

we have

$$y - x = \frac{2x(D - x^2)}{3x^2 + D}$$

and

$$y^2 - D = \frac{(x^2 - D)^3}{(3x^2 + D)^2}$$

If in this we assume x to be a positive number from the class A_1, then $x^2 < D$, and hence $y > x$ and $y^2 < D$. Therefore y likewise belongs to the class A_1. But if we assume x to be a number from the class A_2, then $x^2 > D$, and hence $y < x$, $y > 0$, and $y^2 > D$. Therefore y likewise belongs to the class A^2. This cut is therefore produced by no rational number.

In this property that not all cuts are produced by rational numbers consists the incompleteness or discontinuity of the domain R of all rational numbers.

Whenever, then, we have to do with a cut (A_1, A_2) produced by no rational number, we create a new, an *irrational* number α, which we regard as completely defined by this cut (A_1, A_2); we shall say that the number α corresponds to this cut, or that it produces this cut. From now on, therefore, to every definite cut there corresponds a definite rational or irrational number, and we regard two numbers as *different* or *unequal* always and only when they correspond to essentially different cuts.

In order to obtain a basis for the orderly arrangement of all *real, i.e.,* of all rational and irrational numbers we must investigate the relation between any two cuts (A_1, A_2) and (B_1, B_2) produced by any two numbers α and β. Obviously a cut (A_1, A_2) is given completely when one of the two classes, e.g., the first, A_1, is known, because the second, A_2, consists of all rational numbers not contained in A_1, and the characteristic property of such a first class lies in this: that if the number a_1 is contained in it, it also contains all numbers less than a_1. If now we compare two such first classes A_1, B_1 with each other, it may happen

1. That they are perfectly identical, *i.e.*, that every number contained in A_1 is also contained in B_1, and that every number contained in B_1 is also contained in A_1. In this case A_2 is necessarily identical with B_2, and the two cuts are perfectly identical, which we denote in symbols by $\alpha = \beta$ or $\beta = \alpha$.

But if the two classes, A_1, B_1 are not identical, then there exists in the one, e.g., in A_1, a number $a'_1 = b'_2$ not contained in the other B_1 and consequently found in B_2; hence all numbers b_1 contained in B_1 are certainly less than this number $a'_1 = b'_2$ and therefore all numbers b_1 are contained in A_1.

2. If now this number a'_1 is the only one in A_1 that is not contained in B_1, then is every other number a_1 contained in A_1 also contained in B_1 and is consequently $< a'_1$, *i.e.*, a'_1 is the greatest among all the numbers a_1, hence the cut (A_1, A_2) is produced by the rational number $\alpha = a'_1 = b'_2$. Concerning the other cut (B_1, B_2) we know already that all numbers b_1 in B_1 are also contained in A_1 and are less than the number $a'_1 = b'_2$ which is contained in B_2; every other number b_2 contained in B_2 must, however, be greater than b'_2, for otherwise it would be less than a'_1, therefore contained in A_1 and hence in B_1; hence b'_2 is the least among all numbers contained in B_2, and consequently the cut (B_1, B_2) is produced by the same rational number $\beta = b'_2 = a'_1 = \alpha$. The two cuts are then only unessentially different.

3. If, however, there exist in A_1 at least two different numbers $a'_1 = b'_2$ and $a''_1 = b''_2$, which are not contained in B_1, then there exist infinitely many of them, because all the infinitely many numbers lying between a'_1 and a''_1 are obviously contained in A_1 (Section I, II) but not in B_1. In this case we say that the numbers α and β corresponding to these two essentially different cuts (A_1, A_2) and (B_1, B_2) are *different*, and further that α is *greater* than β, that β is *less* than α, which we express in symbols by $\alpha > \beta$ as well as $\beta < \alpha$. It is to be noticed that this definition coincides completely with the one given earlier, when α, β are rational.

The remaining possible cases are these:

4. If there exists in B_1 one and only one number $b' = a'_2$, that is not contained in A_1 then the two cuts (A_1, A_2) and (B_1, B_2) are only unessentially different and they are produced by one and the same rational number $\alpha = a'_2 = b'_1 = \beta$.

5. But if there are in B_1 at least two numbers which are not contained in A_1, then $\beta > \alpha$, $\alpha < \beta$.

As this exhausts the possible cases, it follows that of two different numbers one is necessarily the greater, the other the less, which gives two possibilities. A third case is impossible. This was indeed involved in the use of the *comparative* (greater, less) to designate the relation between α, β; but this use has only now been justified. In just such investigations one needs to exercise the greatest care so that even with the best intention to be honest he shall not, through a hasty choice of expressions borrowed from other notions already developed, allow himself to be led into the use of inadmissible transfers from one domain to the other.

If now we consider again somewhat carefully the case $\alpha > \beta$ it is obvious that the less number β, if rational, certainly belongs to the class A_1; for since there is in A_1 number $a'_1 = b'_2$ which belongs to the class B_2, it follows that the number β, whether the greatest number in B_1 or the least in B_2 is certainly $\leqq a'_1$ and hence contained in A_1. Likewise it is obvious from $\alpha > \beta$ that the greater number α, if rational, certainly belongs to the class B_2, because $\alpha \geqq a'_1$. Combining these two considerations we get the following result: If a cut is produced by the number α then any rational number belongs to the class A_1 or to the class A_2 according as it is less or greater than α; if the number α is itself rational it may belong to either class.

From this we obtain finally the following: If $\alpha > \beta$, *i.e.*, if there are infinitely many numbers in A_1 not contained in B_1 then there are infinitely many such numbers that at the same time are different from α and from β; every such rational number c *is* $< \alpha$, because it is contained in A_1 and at the same time it is $> \beta$ because contained in B_2.

NOTES

1. *Vorlesungen uber Zahlentheorie*, by P. G. Lejeune Dirichlet, 2d ed. § 159.
2. Hence in what follows the so-called "algebraic" greater and less are understood unless the word "absolute" is added.
3. The apparent advantage of the generality of this definition of number disappears as soon as we consider complex numbers. According to my view, on the other hand, the notion of the ratio between two numbers of the same kind can be clearly developed only after the introduction of irrational numbers.

111. From *Was sind und was sollen die Zahlen?* (1888)*

(Simply Infinite Systems)

RICHARD DEDEKIND

CHAPTER VI

71. Definition. A system N is said to be *simply infinite* when there exists a similar transformation ϕ of N in itself such that N appears as chain (44) of an element not contained in $\phi(N)$. We call this element, which we shall denote in what follows by the symbol 1, the *base-element* of N and say the simply infinite system N is *set in order* [*geordnet*] by this transformation ϕ. If we retain the earlier convenient symbols for transforms and chains (IV) then the essence of a simply infinite system N consists in the existence of a transformation ϕ of N and an element 1 which satisfy the following conditions $\alpha, \beta, \gamma, \delta$:

α. $N' \ni N$.

β. $N = 1_0$.

γ. The element 1 is not contained in N'.

δ. The transformation ϕ is similar.

Obviously it follows from α, γ, δ that every simply infinite system N is actually an infinite system (64) because it is similar to a proper part N' of itself. . . .

(Definition of a Transformation of the Number-Series by Induction)

CHAPTER IX

126. Theorem of the definition by induction. If there is given an arbitrary (similar or dissimilar) transformation θ of a system Ω in itself, and besides a determinate element ω in Ω, then there exists one and only one transformation

*Source: From Richard Dedekind, *Essays in the Theory of Numbers*, trans. by Wooster Woodruff Berman (1924), 67 and 85-86. This translation is reprinted by permission of the Open Court Publishing Company.

ψ of the number-series N. which satisfies the conditions

I. $\psi(N) \ni \Omega$

II. $\psi(1) = \omega$

III. $\psi(n') = \theta\psi(n)$, where n represents every number.

Proof. Since, if there actually exists such a transformation ψ, there is contained in it by (21) a transformation ψ_n of the system Z_n, which satisfies the conditions I, II, III stated in (125), then because there exists one and only one such transformation ψ_n must necessarily

$$\psi(n) = \psi_n(n). \qquad (n)$$

Since thus ψ is completely determined it follows also that there can exist only one such transformation ψ (see the closing remark in (130)). That conversely the transformation ψ determined by (n) also satisfies our conditions I, II, III, follows easily from (n) with reference to the properties I, II and (p) shown in (125), which was to be proved.

Chapter VIII
The Nineteenth Century

Section D
Number Theory, Set Theory, and Symbolic Logic

GEORG (FERDINAND) CANTOR (1845–1918)

The founder of set theory was born into a cosmopolitan merchant family in St. Petersburg (now Leningrad), Russia. His parents were Danish. His artistic mother, Marie Boehm, a Roman Catholic, came from a family of musicians; his father, Georg Waldemar Cantor, was a Protestant. The young Georg was an artistically inclined child, deeply interested in the violin. Following the father's illness, the family moved to Frankfurt, Germany, in 1856. While studying at the Wiesbaden Gymnasium, the son's mathematical talents emerged before his fifteenth birthday. On his father's recommendation he began his university studies in engineering at Zürich Polytechnic in 1862. Drawn by the reputation of Weierstrass, he transferred the next year to the University of Berlin, where he specialized in mathematics and physics. He became a member of the "Berlin school" of mathematics.

Cantor received a fine education at the University of Berlin. Karl Weierstrass taught him analysis and real numbers. From Ernst Kummer he learned higher arithmetic and from Leopold Kronecker the theory of numbers. In 1867 Cantor completed his doctoral dissertation for Kummer on a problem that Gauss had left unsettled in the *Disquisitiones Arithmeticae* (1801). He showed a unique ability in "the art of asking questions" that opened vast new areas of mathematical inquiry, an ability that he considered "more valuable than solving questions." In Berlin he was a close friend of H. A. Schwarz and was from 1864 to 1865 president of the Mathematical Society.

After briefly teaching at a Berlin girls' school, Cantor became a member of the faculty of the University of Halle in 1869. He remained there for the rest of his life. He became a *Privatdozent* in 1869, an associate professor in 1872, and a full professor in 1879, a position he held until 1913. In 1872 he married Vally Guttmann, whose cheerfulness happily offset his own melancholy temperament. On his honeymoon in Switzerland, he met Richard Dedekind and discussed his new set theory with him. Although salaries at Halle were low, Cantor was eventually able to build a house for his wife and five children from the estate left to him by his father. He hoped to obtain a better endowed, more prestigious professorial chair in Berlin but Kronecker, who heatedly opposed his views on transfinite numbers, effectively thwarted his efforts.

Mental illness afflicted Cantor in the final decades of his life. Beginning in 1884 he suffered sporadically from depression. The precise causes are not known. His exhausting efforts to solve mathematical problems and the rejection of his pioneering work on the linear continuum by eminent mathematicians like Kronecker must have aggravated his illness. There were family concerns as well. Trying to balance his consuming studies in mathematics he turned, after 1884, to the study of other subjects. He examined history and Patristic literature, tried to prove that Francis Bacon had written Shakespeare's plays, and devoted himself to Freemasonry and the teachings of the Rosicrucians.

Despite these interests and his recurring bouts of mental illness, Cantor continued to work actively in mathematics for the remainder of his life. In 1890 he overcame the resistance of critics and founded the Association of German Mathematicians. He was its president until 1893. He advocated international congresses of mathematicians and made arrangements for the first of these, which was held in Zürich in 1897. His death came in January 1918 at the University of Halle *Nervenklinik* (neuropathic hospital).

Cantor's chief contributions to mathematics lie in classical analysis and his founding of set theory. His ten papers written from 1869 to 1873 examined the theory of numbers and its relation to analysis. Most of his early papers were published in Sweden in *Acta Mathematica* by Mittag Leffler who was among the first to recognize his ability. In 1872 in a paper on trigonometric functions, he defined irrational numbers with the aid of convergent sequences of rational numbers (quotients of integers), now known as Cauchy sequences. He improved the definition of limit and represented any positive real number r as $r = c_1 + \frac{c_2}{2!} + \frac{c_3}{3!} + \frac{c_4}{4!} + \ldots$, where coefficients c_k satisfy the inequality $0 \leq c_k \leq 1$. His early research expanded upon and extended that of Weierstrass in the arithmetization of analysis.

The major fruit of Cantor's life work was set theory, wherein he propounded a refined view of continuity and the infinite. In a paper published in Crelle's *Journal* in 1874 through the intercession of Dedekind, he distinguished between differences in infinite sets with the aid of the device of one-to-one correspondence. By definition a set is infinite if it can be put into a one-to-one correspondence with one of its subsets. In his 1874 paper Cantor demonstrated that the rational numbers, although infinite, are countable (denumerable)—*i.e.*, they can be put into one-to-one correspondence with the positive integers; while the set of real numbers is uncountable (non-denumerable)—*i.e.*, it possesses the power of a continuum. In set theory Cantor not only supplied proofs he also refined definitions of the concepts of "dense" and "closure" and proposed crude definitions of cardinal- and ordinal- numbers. The smallest cardinal number was denoted by $\aleph_0$ (aleph-null). His two major books on sets were *Grundlagen* . . . ("Foundations of a General Theory of Aggregates," 1883) and *Beitrage* . . . ("Contributions to the Founding of the Theory of Transfinite Numbers," 1895-97).

Cantor's set theory was controversial and took time to gain acceptance. Through his study of Plato, Aquinas, Spinoza, and Leibniz he had come to accept different orders of actual infinities. This put him at odds with a tradition stretching from Aristotle to Gauss that there was only potential infinity. Despite the existence of some antinomies in set theory, Cantor's work was recognized as fundamental to analysis, function theory, and topology by the turn of the twentieth century. After Cantor's death, Russell lauded his mathematical achievement as ". . . probably the greatest of which the age can boast." Speaking of Cantor's ideas David Hilbert stated, "No one shall expel us from the paradise which Cantor created for us."

112. From *Grundlagen einer Allgemeinen Mannigfaltigkeitslehre* (1883)*

(Fundamental Series)

GEORG CANTOR

Mathematics is entirely free in its development, bound only by the self-evident concern that its concepts be both internally without contradiction and stand in definite relations, organized by means of definitions, to previously formed, already existing and proven concepts. (7) In particular, in introducing new numbers mathematics is obliged only to give such definitions of them as will lend them the kind of determiniteness and, under certain circumstances, their kind of relationship to the older numbers, which in a given case will definitely permit them to be distinguished from one another. As soon as a number satisfies all these conditions, mathematics can and must regard it as existent and real. Here I see the reason, suggested in Section 4 why the rational, irrational, and complex numbers should be regarded just as much as existent as the finite positive whole numbers.

I believe that it is not necessary to fear, as many do that these principles contain any danger to science. On one hand the designated conditions under which the freedom of the formation of numbers can alone be exercised, are such that they leave extremely little room for arbitrariness. And then every mathematical concept also carries within itself the necessary corrective; if it is unfruitful and inapt this is soon demonstrated by its uselessness, and it will then be dropped because of its lack of success. Any superfluous confinement of mathematical research work, on the other hand, seems to me to carry with it a much greater danger, a danger that is so much the greater as there is really no justification for it that could be deduced from the essence of the science, for the *essence of mathematics* lies precisely in its *freedom*.

If this quality of mathematics had not presented itself to me for the reasons mentioned, still the whole development of the science itself, as we perceive it in our century, would necessarily lead me to exactly the same views.

If Gauss, Cauchy, Abel, Jacobi, Dirichlet, Weierstrass, Hermite, and Riemann had been bound to constantly subject their new ideas to metaphysical control, then we would not be able to rejoice in the magnificent structure of modern function theory, which, while designed and erected entirely freely and without transient purposes, nonetheless has already revealed its transient significance in applications to mechanics, astronomy, and mathematical physics, as was to be expected. We would not have seen Fuchs, Poincaré and many others

**Source:* This translation by Uwe Parpart is taken from the *Campaigner,* vol. 9 (1976), 79-84. It is reprinted from "The Concept of the Transfinite" by Uwe Parpart, by permission of Campaigner Publications. Mr. Parpart's translation, the first into English, follows the *Grundlagen* in Georg Cantor's *Gesammelte Abhandlungen* (Berlin, 1932) edited by Ernst Zermelo.

bring about the great forward thrust in the theory of differential equations if these excellent intellects had been hemmed in and constricted by extraneous influences; and if Kummer had not taken the liberty, rich in consequences, of introducing the so-called "ideal" numbers into number theory, we would not today be in the position to admire the very important and excellent algebraic and arithmetical works of Kronecker and Dedekind.

As justified, therefore, as mathematics is to move entirely free from all metaphysical fetters, I do not find it possible on the other hand to grant the same right to "applied" mathematics, for example analytical mechanics or mathematical physics. These disciplines, in my opinion, are *metaphysical* both in their foundations and in their goals; if they try to free themselves from this, as has been proposed of late by a famous physicist, they degenerate into a "description of nature" which must lack both the fresh breeze of free mathematical thought and the power of the *explanation* and *exploration (Erklärung und Ergründung)* of natural phenomena.

SECTION 9

Given the great significance which attaches to the so-called real, rational, and irrational numbers in the theory of manifolds, I would not want to neglect to say here what is most important concerning their definition. I will not go into the introduction of the rational numbers more closely, since rigorously arithmetical presentations of this have frequently been formulated. Among the ones close to my own view I call special attention to those of H. Grassmann, *Lehrbuch der Arithmetik* (Berlin 1861) and J. H. T. Mueller, *Lehrbuch der Allgemeinen Arithmetik* (Halle 1855). Yet I want to briefly discuss in more detail the three forms known to me, probably essentially the only major forms, of the rigorously arithmetical introduction of the general real numbers. These are *first*, the mode of introduction which has been followed for many years by Prof. Weierstrass in his lectures on analytical functions, of which a few hints can be found in Herr E. Kossak's programmatic treatise *Die Elemente der Arithmetik* (Berlin 1872). *Second,* Herr. R. Dedekind, in his essay *Stetigkeit und Irrationale Zahlen* (Braunschweig 1872), has published a peculiar form of definition, and *third*, I put forth a form of definition in the year 1871 (*Mathematische Annalen*, Vol. 5, p. 123) which externally bears a certain resemblance to the Weierstrass definition, so that it was possible for Herr H. Weber (*Zeitschrift für Mathematik und Physik*, 27th year, Historical Literature Division, p. 163) to confuse it with the latter. In my opinion, however, this *third* form of definition, which later was also developed by Herr Lipschitz, (*Grundlagen der Analysis*, Bonn 1877), is the simplest and most natural of all, and has the advantage that it is most immediately adapted to the analytic calculus.

Part of the definition of an irrational real number is always a well-defined infinite aggregate of the first power of rational numbers; this is the common characteristic of all forms of definition. Their difference lies in the generative moment *(Erzeugungs-moment)* through which the aggregate is tied to the number it defines, and in the conditions which the aggregate must satisfy in order to be a suitable basis for the number definition in question.

In the case of the *first* definition, an aggregate of positive rational numbers a_ν, denoted by (a_ν), is taken as a basis, which satisfies the condition that no matter how many or which of a finite number of a_ν are summed up, this sum always remains below a specifiable bound. If now we have two such aggregates (a_ν) and (a'_ν), then it is rigorously shown that they can present three cases: either every part $\frac{1}{n}$ of unity is always contained equally often in both aggregates so long as a sufficient, augmentable, finite number of their elements are summed up; or, from a certain n on, $\frac{1}{n}$ is always contained more frequently in the first aggregate than in the second; or thirdly, from a certain n on, $\frac{1}{n}$ is always contained more frequently in the second than in the first. In accordance with these occurrences. if b and b' are the numbers to be defined by means of the two aggregates (a_ν) and (a'_ν), then in the first case we set

$$b = b',$$

in the second

$$b > b',$$

in the third

$$b < b'.$$

If the two aggregates are joined together in a new one

$$(a_\nu, a'_\nu),$$

then this provides the basis for the definition of

$$b + b';$$

if, however, we form of the two aggregates (a_ν) and (a'_ν) the new one

$$(a_\nu \cdot a'_\nu),$$

the elements of which are the products of all the a_ν, and all the a'_ν, then this new aggregate is taken as a basis for the definition of the product bb'.

We see that here the generative moment which ties the aggregate to the number to be defined by it, lies in the *formation of sums;* however, it has to be stressed as *essential* that only the summation of an always finite number of rational elements is utilized so that the number b to be defined is *not* already posited from the outset as the sum

$$\Sigma a_\nu$$

of the infinite series (a_ν); this would embody a *logical mistake* since, on the contrary, the definition of the sum

$$\Sigma a_\nu$$

is attained only by setting it equal to the *completed* number b which must of necessity already have been defined in advance. I believe that this logical mistake, which was first avoided by Herr Weierstrass, was committed almost universally in previous times, and not noticed because it belongs among those rare cases in which actual mistakes cannot cause any significant damage to the calculus.

I am nonetheless convinced that all the difficulties which have been found in the concept of the irrational are linked to this mistake, whereas when this mistake is avoided, the irrational number will implant itself in our mind with the same determinateness, distinctness, and clarity as the rational number.

Herr Dedekind's form of definition takes as its basis the *entirety of all* rational numbers, partitioned into two groups in such a way that, if the numbers of the first group are denoted by A_ν, those of the second group by B_μ, then always

$$A_\nu < B_\mu .$$

Herr Dedekind calls such a partition of the rational number aggregate a "cut" of the latter, denotes it by

$$(A_\nu \mid B_\nu) ,$$

and associates a number b with it. If we compare two such cuts

$$(A_\nu \mid B_\mu)$$

and

$$(A'_\nu \mid B'_\mu) ,$$

we find that just as in the *first* form of definition there exist altogether *three* possibilities in accordance with which the numbers b and b' represented by the two cuts are posited as equal to each other, or as

$$b > b' ,$$

or as

$$b < b'$$

The first case, apart from certain easily adjustable exceptions which occur in the case of the being-rational *(Rationalsein)* of the numbers to be defined, takes place only in the case of the total identity of the two cuts, and in this the undeniable, decisive preferability of this form of definition over the two others comes to the fore, that to each number b there corresponds a unique cut. This, however, is counterbalanced by the great disadvantage, that in analysis numbers *never* present themselves in the form of "cuts," into which form they must first be brought with great skill and trouble.

Here, as well, the definitions for the sum

$$b + b'$$

and the product

$$bb'$$

follow on the basis of new cuts arising from the two given ones.

The disadvantage attaching to the *first* and the *third* form of definition—that here the same, i.e. equal, numbers present themselves infinitely often so that an unambiguous overview over the entirety of the real numbers is not immediately obtained—can be removed with the greatest of ease through specification of the base aggregates (a_ν), by drawing on any one of the well-known unique system formations such as the decimal system or simple continued-fraction expansions *(Kettenbruchentwickelung)*.

I come now to the *third* form of the definition of real numbers. Here again an infinite aggregate of rational numbers (a_ν) of the first power is taken as a basis; however, a different character is demanded of it than in the Weierstrass form of definition. I postulate that after the choice of an arbitrarily small rational number ϵ a finite number of members of the aggregate can be separated off, so that those remaining have pairwise a difference which in absolute terms is smaller than ϵ. Every such aggregate (a_ν) which can also be characterized by the postulate

$$\lim_{\nu=\infty} (a_{\nu+\mu} - a_\nu) = 0$$

(for arbitrary μ)

I call a *fundamental series (Fundamentalreihe)*, and associate with it a number b to be defined by it and for which the sign (a_ν) itself could even be fittingly used, as was proposed by Herr Heine, who after numerous discussions had come to agree with me on these questions. (Cf. *Crelle's Journal*, vol 74, p. 172). Such a *fundamental series*, as can rigorously be deduced from the concept, presents three cases: either its members a_ν are, for sufficiently large values of ν, smaller in absolute terms than an arbitrarily preassigned number; or from a certain ν on, the latter are larger than a definitely determinable positive rational number ρ; or from a certain ν on, they are smaller than a definitely determinable negative rational magnitude $-\rho$. In the first case I say that b is equal to zero, in the second that b is greater than zero or positive, in the third that b is smaller than zero or negative.

Now come the elementary operations. If (a_ν) and (a'_ν) are two *fundamental series* by means of which the numbers b and b' are determined, it then turns out that

$$(a_\nu \pm a'_\nu)$$

and

$$(a_\nu \cdot a'_\nu)$$

are also *fundamental series*, which thus determine three new numbers. These serve as definitions for the sum and difference

$$b \pm b'$$

and the product

$$b \cdot b' .$$

If in addition b is different from zero, the definition of which has been given above, then it can be proved that

$$\left(\frac{a'_\nu}{a_\nu}\right)$$

is also a *fundamental series* whose associated number provides the definition for the quotient

$$\frac{b'}{b} .$$

The elementary operations between a number b given by a fundamental series (a_ν) and a directly given rational number a are included in the operations just established, by letting

$$a'_\nu = a, \; b' = a.$$

Only now come the definitions of the equality, the being-smaller, and the being-greater of two numbers b and b' (of which b' can also equal a). In particular we say that

$$b = b'$$

or

$$b > b'$$

or

$$b < b',$$

according to whether

$$b - b'$$

is equal to zero or greater than or smaller than zero.

After all these preparations we get as the first *rigorously provable* theorem that if b is the number determined by a fundamental series (a_ν), then, with increasing ν

$$b - a_\nu$$

will become smaller in absolute terms than any conceivable rational number, or, what is the same, that

$$\operatorname*{Lim}_{\nu=\infty} a_\nu = b.$$

It would be well to observe this cardinal point, whose significance could easily be overlooked: in the case of the *third* form of definition it is not at all true that the number b is defined as the "limit" of the members a_ν of a fundamental series (a_ν). This would be a logical mistake similar to that pointed out in the discussion of the *first* form of definition, for the reason that then the *existence* of the limit

$$\operatorname*{Lim}_{\nu=\infty} a_\nu$$

would be presumed. Rather, the opposite is the case, so that by means of our preceding definitions the concept b has been furnished with properties and with relations to the rational numbers such that it can be concluded with logical certainty:

$$\operatorname*{Lim}_{\nu=\infty} a_\nu$$

exists and is equal to b. May I be forgiven my thoroughness, which I motivate with the perception that most people pass by this unpretentious detail and then easily get entangled in doubts and contradictions with respect to the irrational which, by observing the particulars emphasized here, they could have been spared entirely, for they would then recognize clearly that the irrational number, by virtue of the *characteristics given to it by the definitions,* is just as definite a reality in our mind as the rational number, even as the whole rational number, and that one need not first *obtain* it by a limiting process but on the contrary—through its *possession* one is convinced of the feasibility and evident admissibility of the limiting processes. (8) For now the just-adduced theorem is easily extended to yield the following: If (b_ν) is any aggregate of rational or irrational numbers such that

$$\operatorname*{Lim}_{\nu=\infty} (b_{\nu+\mu} - b_\nu) = 0,$$

(whatever μ may be),

then there is a number b determined by a fundamental series (a_ν) such that

$$\lim_{\nu=\infty} b_\nu = b .$$

Thus it turns out that the same numbers b, which on the basis of fundamental series (a_ν) (I call these fundamental series of the *first* order) are defined in such a way that they prove to be limits of a_ν, are also in manifold ways representable as limits of series (b_ν), where each b_ν is defined by a fundamental series of the first order

$$(a_\mu^{(\nu)}) \text{ (with fixed } \nu).$$

I therefore call such an aggregate (b_ν), if it has the property that

$$\lim_{\nu=\infty} (b_{\nu+\mu} - b_\nu) = 0 \text{ (for arbitrary } \mu)$$

a fundamental series of the *second* order.

Similarly, fundamental series of the *third, fourth, . . . nth order, and fundamental series of the* α *th* order can also be formed, where α is an arbitrary number of the second number-class.

All these fundamental series accomplish exactly the same thing for the determination of a real number b as the fundamental series of the *first* order, the only difference consisting of the more complicated and broader form in which they are given. It nonetheless seems to me highly appropriate, provided that one wants to assume the standpoint of the third form of definition at all, to fix this difference in the form noted, as I have done in similar fashion in the cited works (*Mathematische Annalen*, Vol. V, p. 123). Therefore, I now use the following mode of expression: the numerical magnitude b is given by a fundamental series of the nth or, respectively, the α th order. If we decide upon this, we achieve an extraordinarily free-flowing and simultaneously comprehensible language, enabling us to describe the richness of the multiform and often so complicated webs of analysis in the most simple and distinctive manner, through which, in my opinion, a gain in clarity and transparency is attained which should not be underestimated. In this I oppose the misgivings which Herr Dedekind voiced in the preface to his essay *Continuity and Irrational Numbers (Stetigkeit und Irrationale Zahlen)* concerning these distinctions. It was the farthest thing from my mind to introduce through the fundamental series of the second, the third order, etc., *new* numbers which are not already determinable through fundamental series of the first order; rather, I was merely focusing on the conceptually distinct forms of the being-given *(des Gegebenseins)* of the numbers. This clearly flows from particular parts of my paper itself.

In regard to this I would like to call attention to a remarkable circumstance. These orders of fundamental series, distinguished by me through numbers of the first and second number-classes, exhaust any and all conceivable forms of the usual series-character—whether analysis has already discovered them or not—in the sense that fundamental series, the number of whose order might be denoted by a number of the third number-class, actually do not exist, as I shall rigorously prove on a different occasion.

Now I will attempt to explain in brief the appropriateness of the *third* form of definition.

To denote the fact that a number b is given on the basis of a fundamental series (e_ν) of any order n or α, I will use the formulas

$$b \sim (e_\nu)$$

or

$$(e_\nu) \sim b.$$

If, for example, a convergent series with the general member c_ν is given, then the necessary and sufficient condition for convergence (as is well-known) is:

$$\lim_{\nu=\infty} (c_{\nu+1} + \cdots + c_{\nu+\mu}) = 0$$

$$\text{(for arbitrary } \mu).$$

113. From a letter to Richard Dedekind (1899)*

(Transfinite Cardinal Numbers and Set Theory)

GEORG CANTOR

. . . As you know, many years ago I had already arrived at a well-ordered sequence of cardinalities [*Mächtigkeiten*], or transfinite cardinal numbers, which I call "alephs":

$$\aleph_0, \aleph_1, \aleph_2, \ldots \aleph_{\omega_0}, \ldots$$

$\aleph_0$ means the cardinality of the sets "denumerable" in the usual sense, $\aleph_1$ is the next greater cardinal number, $\aleph_2$ is the next greater still, and so on; $\aleph_{\omega_0}$ is the one next following (that is, next greater than) all the $\aleph_\nu$ and equals

$$\lim_{\nu \to \omega_0} \aleph_\nu,$$

and so on.

The big question was whether, besides the alephs, there were still other cardinalities of sets; for two years now I have been in possession of a proof that there are no others, so that, for example, the arithmetic linear continuum (the totality of all real numbers) has a definite aleph as its cardinal number.

If we start from the notion of a definite multiplicity [*Vielheit*] (a system, a totality) of things, it is necessary, as I discovered, to distinguish two kinds of multiplicities (by this I always mean *definite* multiplicities).

For a multiplicity can be such that the assumption that *all* of its elements "are together" leads to a contradiction, so that it is impossible to conceive of the multiplicity as a unity, as "one finished thing." Such multiplicities I call *absolutely infinite* or *inconsistent multiplicities*.

As we can readily see, the "totality of everything thinkable," for example, is such a multiplicity; later still other examples will turn up.

If on the other hand the totality of the elements of a multiplicity can be thought of without contradiction as "being together," so that they can be gathered together into "*one* thing," I call it a *consistent multiplicity* or a "set." (In French and in Italian this notion is aptly expressed by the words "ensemble" and "insieme.")

Two equivalent multiplicities either are both "sets" or are both inconsistent.

Every submultiplicity [Teilvielheit] of a set is a set.

Whenever we have a set of sets, the elements of these sets again form a set.

If a set M is given, I call the general notion that applies to it and all sets equivalent to it, and to these alone, its *cardinal number* or also its *cardinality*, and I denote it by m. I then arrive at the system of all cardinalities—which will later turn out to be an *inconsistent* multiplicity—in the following way.

A multiplicity is said to be "simply ordered" if between its elements there exists a rank order such that, for any two of its elements, one is the earlier and the other the later, and that, for any three of its elements, one is the earliest,

*Source: This translation by Stefan Bauer-Mengelberg and Jean van Heijenoort is reprinted by permission of the publisher from *From Frege to Gödel: A Source Book in Mathematical Logic, 1879–1931*, edited by Jean van Heijenoort, Cambridge, Mass.: Harvard University Press, Copyright © 1967 by the President and Fellows of Harvard College.

another is the middle one, and the remaining one is the last by rank among them.

If a simply ordered multiplicity is a *set*, then by its *type* μ I understand the general notion that applies to it and to all ordered sets *similar* to it, and to these alone. (I use the notion of *similarity* in a more restricted sense than you do;[1] I say that two simply ordered multiplicities are *similar* if they can be brought into a one-to-one relation such that the rank order of corresponding elements is the same in both.)

A multiplicity is said to be *well-ordered* if it satisfies the condition that every *sub-multiplicity* of it has a *first* element; I call such a multiplicity a "sequence" for short.

Every part [*Teil*] of a sequence is a sequence.

If now a sequence F is a set, I call the type of F its *ordinal number* or, more briefly, its *number*; thus, when in what follows I speak simply of numbers, I shall have in mind only ordinal numbers, that is, types of well-ordered sets.

I now consider the system of *all numbers* and denote it by Ω.

I proved (*1897*, p. 216) that, of two distinct numbers α and β, one is always the smaller, the other the greater, and that, if for three numbers we have $\alpha < \beta$ and $\beta < \gamma$, we also have $\alpha < \gamma$.

Ω is therefore a simply ordered system.

But it also follows easily from the theorems on well-ordered sets proved in § 13 that every multiplicity of numbers, that is, every part of Ω, contains a *least* number.

Hence the system Ω, when naturally ordered according to magnitude, forms a sequence.

If we then add 0 to this sequence as an element—putting it first, of course—we obtain a sequence Ω',

$$0, 1, 2, 3, \ldots, \omega_0, \omega_0 + 1, \ldots, \gamma, \ldots,$$

in which, as we can readily see, *every* number is the *type* of the *sequence of all elements preceding it* (including 0). (The sequence Ω has this property only from $\omega_0 + 1$ on [in fact, from ω_0 on].)

Ω' *cannot* be a *consistent* multiplicity (and therefore neither can Ω); if Ω' were consistent, then, since it is a well-ordered set, there would correspond to it a number δ greater than all numbers of the system Ω; but the number δ also occurs in the system Ω, because this system contains *all* numbers; δ would thus be greater than δ, which is a contradiction. Therefore

A. *The system Ω of all numbers is an inconsistent, absolutely infinite multiplicity.*

Since the *similarity* of well-ordered sets establishes at the same time their *equivalence*, to every number γ there corresponds a definite cardinal number $\aleph(\gamma) = \bar{\gamma}$, namely, the cardinal number of any well-ordered set whose type is γ.

The cardinal numbers that correspond in this sense to the *transfinite* numbers of the system Ω I call "alephs," and the *system of all alephs* is denoted by ת (*tav*, the last letter of the Hebrew alphabet).

I call the system of all numbers γ corresponding to one and the same cardinal number c a "number class," and, more specifically, the number class Z(c). We readily see that in every number class there occurs a least number γ_0 and that there is a number γ_1 falling outside of Z(c) such that the condition

$$\gamma_0 \leqq \gamma < \gamma_1$$

is equivalent to the fact that the number γ belongs to the number class Z(c). Every number class is therefore a definite "segment" of the sequence Ω.[2]

Certain numbers of the system Ω form, each one by itself, a number class; they are the *finite* numbers, 1, 2, 3, . . . , ν, . . . , to which correspond the various "finite" cardinal numbers, $\bar{1}, \bar{2}, \bar{3}, \ldots, \bar{\nu}, \ldots$.

Let ω_0 be the least transfinite number; I call the aleph corresponding to it $\aleph_0$, so that

$$\aleph_0 = \bar{\omega}_0;$$

$\aleph_0$ is the *least* aleph and determines the number class

$$Z(\aleph_0) = \Omega_0.$$

The numbers α of $Z(\aleph_0)$ satisfy the condition

$$\omega_0 \leqq \alpha < \omega_1$$

and are characterized by it; here ω_1 is the least transfinite number whose cardinal number is not equal to $\aleph_0$. If we put

$$\bar{\omega}_1 = \aleph_1,$$

then $\aleph_1$ is not only distinct from $\aleph_0$, but it is also the next greater aleph, for we can prove that there is no cardinal number at all that would lie between $\aleph_0$ and $\aleph_1$. We thus obtain the number class $\Omega_1 = Z(\aleph_1)$, which immediately follows Ω_0. It contains all numbers β' that satisfy the condition

$$\omega_1 \leqq \beta < \omega_2;$$

here ω_2 is the least transfinite number whose cardinal number differs from $\aleph_0$ and $\aleph_1$.

$\aleph_2$ is the next greater aleph after $\aleph_1$; it determines the number class $\Omega_2 = Z(\aleph_2)$ that immediately follows Ω_1 and consists of all numbers γ that are $\geqq \omega_2$ and $< \omega_3$, where ω_3 is the least transfinite number whose cardinal number differs from $\aleph_0$, $\aleph_1$, and $\aleph_2$; and so on.

I would still like to stress the following:

$$\bar{\bar{\Omega}}_0 = \aleph_1, \bar{\bar{\Omega}}_1 = \aleph_2, \ldots, \bar{\bar{\Omega}}_\nu = \aleph_{\nu+1},$$

$$\sum_{\nu' = 0,1,2,\ldots,\nu} \aleph_{\nu'} = \aleph_\nu;$$

all this is easy to prove.

Among the transfinite numbers of the system Ω to which no $\aleph_\nu$ [with finite ν] corresponds as a cardinal number, there is again a least, which we call ω_{ω_0}, and with it we obtain a new aleph,

$$\aleph_{\omega_0} = \bar{\omega}_{\omega_0},$$

which is also definable by means of the equation

$$\aleph_{\omega_0} = \sum_{\nu = 1,2,3,\ldots} \aleph_\nu$$

and which we recognize as the next greater cardinal number after all the $\aleph_\nu$.

We see that this process of formation of the alephs and of the number classes of the system Ω that correspond to them is *absolutely* limitless.

B. *The system* ת *of all alephs, when ordered according to magnitude,*

$$\aleph_0, \aleph_1, \ldots, \aleph_{\omega_0}, \aleph_{\omega_0+1}, \ldots, \aleph_{\omega_1}, \ldots,$$

forms a sequence that is similar to the system Ω and therefore likewise inconsistent, or absolutely infinite.

The question now arises whether *all transfinite cardinal numbers* are contained in the system ת. In other words, is there a *set* whose cardinality is *not an aleph?*

This question is to be answered *negatively*, and the reason for this lies in the *inconsistency* that we discerned in the systems Ω and ת.

Proof. If we take a definite multiplicity V and assume that *no aleph* corresponds to it *as its cardinal number*, we conclude that V must be *inconsistent*.

For we readily see that, on the assumption made, the whole system Ω is projectible into the multiplicity V, that is, there must exist a submultiplicity V' of V that is equivalent to the system Ω.[3]

V' is *inconsistent* because Ω is, and the same must therefore be asserted of V. . . .

Accordingly, every transfinite *consistent multiplicity*, that is, every transfinite set, must have a *definite aleph* as its cardinal number. Hence

C. *The system* ת *of all alephs is nothing but the system of all transfinite cardinal numbers.*

All sets, and in particular all *"continua,"* are therefore *"denumerable"* in *an extended sense*.

Furthermore C makes it clear that I was right when I stated (*1895*, p. 484) the theorem:

> "If a and b are arbitrary cardinal numbers, then $a = b$ or $a < b$ or $a > b$."

For, as we have seen, these relations of magnitude obtain between the alephs.

NOTES

1. Dedekind uses "ähnlich" in the sense of "equivalent."

2. Here we constantly use the theorem mentioned a few paragraphs above according to which *every* totality of numbers, hence *every* submultiplicity of Ω, has a *minimum, a least number*.

3. It is precisely at this point that the weakness of the proof sketched here lies. It has *not* been proved that the entire number sequence Ω would necessarily be "projectible" into every multiplicity V that has no aleph as its cardinal number. Cantor apparently thinks that successive and arbitrary elements of V are assigned to the numbers of Ω in such a way that every element of V is used only *once*. *Either* this procedure would of necessity come to an end once all elements of V had been exhausted, and then V would be mapped onto a *segment* of the number sequence and its cardinality would be an aleph, contrary to the assumption, *or* V would remain inexhaustible, hence contain a constituent part that is equivalent to all of Ω and therefore inconsistent. Thus the intuition of time is applied here to a process that goes beyond all intuition, and a fictitious entity is posited of which it is assumed that it could make *successive* arbitrary choices and thereby define a subset V' of V that, by the conditions imposed, is precisely *not* definable. Only through the use of the "axiom of choice," which postulates the possibility of a *simultaneous* choice and which Cantor uses unconsciously and instinctively everywhere but does not formulate explicitly anywhere, could V' be defined as a subset of V. But even then there would still remain a doubt: perhaps the proof involves "inconsistent" multiplicities, indeed possibly contradictory notions, and is logically inadmissible already because of that. It is precisely doubts of this kind that impelled the editor a few years later to base his own proof of the well-ordering theorem *(1904)* purely upon the axiom of choice without using inconsistent multiplicities.

Chapter VIII
The Nineteenth Century

Section D
Number Theory, Set Theory, and Symbolic Logic

(FRIEDRICH LUDWIG) GOTTLOB FREGE (1848–1925)

The founder of modern mathematical logic was the son of Karl Alexander Frege, a principal at a private, girls high school, and Auguste Bialloblotzke. After graduating from the Gymnasium in his hometown of Wismar, Germany, young Gottlob attended the University of Jena from 1869 to 1871. During the next two years he studied mathematics, physics, chemistry, and philosophy at the University of Göttingen, which awarded him the doctorate in 1873. The following year he qualified for teaching at the University of Jena with a paper on one-parameter groups of functions.

Frege spent his entire working life as a teacher of mathematics at the University of Jena. He began as a *Privatdozent* and became an associate professor in 1879, professor *(ordentlichen Honorarprofessor)* in 1896, and professor emeritus in 1917. He taught all branches of mathematics as well as his own logical system. He once stated that "Every good mathematician is at least half a philosopher and every good philosopher at least half a mathematician." As a teacher and researcher he remained aloof from students and even more from colleagues. His work was original and highly independent of other influences.

Frege was married to Margarete Lieseberg; they had one adopted son named Alfred. In religion he was a liberal Lutheran; in politics a reactionary who loved the monarchy and the grand dukes of Mecklenburg. During and after World War I he developed an intense hatred of socialism and democracy, blaming them for the German defeat and the harsh Treaty of Versailles. The diary that he kept in later life shows that he loathed the French, Catholics, and Jews—a not uncommon view among right-wing Germans during the Weimar Republic.

In his research Frege delved into logic as well as the borderline and interrelationships between the sciences of logic and mathematics. Boole, De Morgan, and others had provided some pioneering studies in modern mathematical logic in the mid-nineteenth century, but it was Frege's 88-page booklet *Begriffsschrift* ("Conceptual Script," 1879) with its well-developed propositional calculus and quantification theory that marked the beginning of a new epoch in the history of logic. It built logic on explicit axioms and presented a new ideography or "formula language" that included the symbol ⊢ before a proposition to indicate that it is an assertion. With this ideography Frege sought to overcome the imprecision of ordinary language as well as the sometimes ambiguous and confusing signs of Boole. He wanted to create a *lingua characteristica* in Leibniz's sense—that is, a symbolic language for pure thought that did not have to be supplemented by intuitive reasoning.

After a period of intense studies in the philosophy of logic and of mathematics, Frege wrote his masterpiece *Die Grundlagen der Arithmetik* ("The Foundations of Arithmetic," 1884). It was an ambitious program for the algebraic construction of arithmetic, appealing at times to the ideas of Dedekind. Frege strongly criticized other theories of number and offered his celebrated definition of cardinal number: "The number which belongs to the concept *F* is the extension of the concept of being equal to the concept *F*." In the only review that the *Grundlagen* received, Georg Cantor (whose ideas were close to Frege's) wrote a devastating critique without endeavoring to understand the material in the book. Otherwise it was neglected. (Dedekind and Peano independently contributed axioms for arithmetic.) The lack of response to the *Grundlagen* stung Frege, who realized its importance.

The last major work of Frege was the two-volume *Grundgesetze der Arithmetik* ("The Fundamental Laws of Arithmetic," 1893 and 1903). The first volume improved upon his symbolism from the *Begriffsschrift* and rigorously developed ideas from the *Grundlagen*. The goal of the *Begriffsschrift* was met; Frege's symbolism together with the construction of cardinals introduced in to logic the direct method of formal deduction that was so important in the calculus. Nevertheless, the first volume of the *Grundgesetze* also received only a single review (written by Peano). The years of neglect for his ideas began to embitter Frege. In volume two he abusively attacked other writers.

Despite his books and extensive correspondence with other mathematicians, Frege's abstruse doctrines together with his complex and unfamiliar symbolism were slow in winning recognition. They were mostly transmitted by others, such as Giuseppe Peano and Bertrand Russell. Among Frege's contemporaries, Russell was one of the very few to support his ideas, which were discussed in Appendix B of the *Principles of Mathematics* (1903). Ironically, Russell discovered a flaw in the Fregean theory of arithmetic. He arrived at a paradox in using the terms class and element. He communicated this paradox of set theory to Frege as the second volume of the *Grundgesetze* was at the printer's. Frege found no resolution and afterward his creative scientific activity largely ended. This remained true even after Russell resolved the paradox with his theory of types (1908).

In his last years Frege confronted new developments in the foundations of mathematics, but failed to grasp Hilbert's decisive advance in axiomatics and his reduction of geometry to the concepts of point and betweenness. After retiring in 1917, Frege worked on an extension of his earlier work entitled *Logische Untersuchungen* ("Logical Investigations," 1918–1925). He did not live to appreciate his influence upon Ludwig Wittgenstein, one of the most original and imaginative philosophical thinkers of the 20th century.

114. From *Begriffsschrift* (1879)*

(Symbolic Logic)

GOTTLOB FREGE

PREFACE

In apprehending a scientific truth we pass, as a rule, through various degrees of certitude. Perhaps first conjectured on the basis of an insufficient number of particular cases, a general proposition comes to be more and more securely established by being connected with other truths through chains of inferences, whether consequences are derived from it that are confirmed in some other way or whether, conversely, it is seen to be a consequence of propositions already established. Hence we can inquire, on the one hand, how we have gradually arrived at a given proposition and, on the other, how we can finally provide it with the most secure foundation. The first question may have to be answered differently for different persons; the second is more definite, and the answer to it is connected with the inner nature of the proposition considered. The most reliable way of carrying out a proof, obviously, is to follow pure logic, a way that, disregarding the particular characteristics of objects, depends solely on those laws upon which all knowledge rests. Accordingly, we divide all truths that require justification into two kinds, those for which the proof can be carried out purely by means of logic and those for which it must be supported by facts of experience. But that a proposition is of the first kind is surely compatible with the fact that it could nevertheless not have come to consciousness in a human mind without any activity of the senses.[1] Hence it is not the psychological genesis but the best method of proof that is at the basis of the classification. Now, when I came to consider the question to which of these two kinds the judgments of arithmetic belong, I first had to ascertain how far one could proceed in arithmetic by means of inferences alone, with the sole support of those laws of thought that transcend all particulars. My initial step was to attempt to reduce the concept of ordering in a sequence to that of *logical* consequence, so as to proceed from there to the concept of number. To prevent anything intuitive [*Anschauliches*] from penetrating here unnoticed, I had to bend every effort to keep the chain of inferences free of gaps. In attempting to comply with this requirement in the strictest possible way I found the inadequacy of language to be an obstacle; no matter how unwieldy the expressions I was ready to accept, I was less and less able, as the relations became more and more complex, to attain the precision that my purpose required. This deficiency led me to the idea of the present ideography. Its first purpose, therefore, is to provide us with the most reliable test of the validity of a chain of inferences and to point out every presupposition that tries to sneak in unnoticed, so that its origin can be investigated. That is why I decided to forego expressing anything that is without significance for the *inferential sequence*. In § 3 I called what alone mattered to me the *conceptual content* [*begrifflichen Inhalt*]. Hence this definition must always be kept in mind if one wishes to gain a proper understanding of what my formula language is. That, too, is what led me to the name "Begriffsschrift." Since I confined myself for the time being to expressing relations that are independent of the particular characteristics of objects, I was also able to use the expression "formula language for pure thought." That it is modeled upon the formula language of arithmetic, as I indicated in the title, has to do with fundamental ideas rather than with details of execution. Any effort to create an artificial similarity by regarding a concept as the sum of its marks [*Merkmale*] was entirely alien to my thought. The most immediate point of contact between my formula language and that of arithmetic is the way in which letters are employed.

I believe that I can best make the relation of my ideography to ordinary language [*Sprache des Lebens*] clear if I compare it to that which the microscope has to the eye. Because of the range of its possible uses and the versatility with which it can adapt to the most diverse circumstances, the eye is far superior to the microscope. Considered as an optical instrument, to be sure, it exhibits many imperfections, which ordinarily remain unnoticed only on account of its intimate connection with our mental life. But, as soon as scientific goals demand great sharpness of resolution, the eye proves to be insufficient. The microscope, on the other hand, is perfectly suited to precisely such goals, but that is just why it is useless for all others.

This ideography, likewise, is a device invented for certain scientific purposes, and one must not condemn it because it is not suited to others. If it answers to these purposes in some degree, one should not mind the fact that there are no new truths in my work. I would console myself on this point with the realization that a development of method, too, furthers science. Bacon, after all, thought it better to invent a means by which everything could easily be discovered than to discover particular truths, and all great steps of scientific progress in recent times have had their origin in an improvement of method.

Leibniz, too, recognized—and perhaps overrated—the advantages of an adequate system of notation. His idea of a universal characteristic, of a *calculus philosophicus* or *ratiocinator*,[2] was so gigantic that the attempt to realize it could not go beyond the bare preliminaries. The enthusiasm that seized its originator when he contemplated the immense increase in the intellectual power of mankind that a system of notation directly appropriate to objects themselves would bring about led him to underestimate the difficulties that stand in the way of such an enterprise. But, even if this worthy goal cannot be reached in one leap, we need not despair of a slow, step-by-step approximation. When a problem appears to be unsolvable in its full generality, one should temporarily restrict it; perhaps it can then be conquered by a gradual advance. It is possible to view the signs of arithmetic, geometry, and chemistry as realizations, for specific fields, of Leibniz's idea. The ideography proposed here adds a new one to these fields, indeed the central one, which borders on all the others. If we take our departure from there, we can with the greatest expectation of success proceed to fill the gaps in the existing formula languages, connect their hitherto separated fields into a single domain, and extend this domain to include fields that up to now have lacked such a language.[3]

I am confident that my ideography can be successfully used wherever special value must be placed on the validity of proofs, as for example when the foundations of the differential and integral calculus are established.

It seems to me to be easier to extend the domain of this formula language to include geometry. We would only have to add a few signs for the intuitive relations that occur there. In this way we would obtain a kind of *analysis situs*.

The transition to the pure theory of motion and then to mechanics and physics could follow at this point. The latter two fields, in which besides rational necessity [*Denknothwendigkeit*] empirical necessity [*Naturnothwendigkeit*] asserts itself, are the first for which we can predict a further development of the notation as knowledge progresses. That is no reason, however, for waiting until such progress appears to have become impossible.

If it is one of the tasks of philosophy to break the domination of the word over the human spirit by laying bare the misconceptions that through the use of language often almost unavoidably arise concerning the relations between concepts and by freeing thought from that with which only the means of expression of ordinary language, constituted as they are, saddle it, then my ideography, further developed for these purposes, can become a useful tool for the philosopher. To be sure, it too will fail to reproduce ideas in a pure form, and this is probably inevitable when ideas are represented by concrete means; but, on the one hand, we can restrict the discrepancies to those that are unavoidable and harmless, and, on the other, the fact that they are of a completely different kind from those peculiar to ordinary language already affords protection against the specific influence that a particular means of expression might exercise.

The mere invention of this ideography has, it seems to me, advanced logic. I hope that logicians, if they do not allow themselves to be frightened off by an initial impression of strangeness, will not withhold their assent from the innovations that, by a necessity inherent in the subject matter itself, I was driven to make. These deviations from what is traditional find their justification in the fact that logic has hitherto always followed ordinary language and grammar too closely. In particular, I believe that the replacement of the concepts *subject* and *predicate* by *argument* and *function*, respectively, will stand the test of time. It is easy to see how regarding a content as a function of an argument leads to the formation of concepts. Furthermore, the demonstration of the connection between the meanings of the words *if, and, not, or, there is, some, all,* and so forth, deserves attention.

Only the following point still requires special mention. The restriction, in § 6, to a single mode of inference is justified by the fact that, when the foundations for such an ideography are laid, the primitive components must be taken as simple as possible, if perspicuity and order are to be created. This does not preclude the possibility that *later* certain transitions from several judgments to a new one, transitions that this one mode of inference would not allow us to carry out except mediately, will be abbreviated into immediate ones. In fact this would be advisable in case of eventual application. In this way, then, further modes of inference would be created.

I noticed afterward that formulas (31) and (41) can be combined into a single one,

$$\vdash (\Pi a \equiv a),$$

which makes some further simplifications possible.

As I remarked at the beginning, arithmetic was the point of departure for the train of thought that led me to my ideography. And that is why I intend to apply it first of all to that science, attempting to provide a more detailed analysis of the concepts of arithmetic and a deeper foundation for its theorems. For the present I have reported in the third chapter some of the developments in this direction. To proceed farther along the path indicated, to elucidate the concepts of number, magnitude, and so forth—all this will be the object of further investigations, which I shall publish immediately after this booklet.

NOTES

1. Since without sensory experience no mental development is possible in the beings known to us, that holds of all judgments.

2. On that point see *Trendelenburg 1867* [pp. 1-47, *Ueber Leibnizens Entwurf einer allgemeinen Charakteristik*].

3. On that point see *Frege 1879a*.

115. From *Die Grundlagen der Arithmetik* (1884)*

(Definition of Number in Logical Terms)

GOTTLOB FREGE

TO OBTAIN THE CONCEPT OF NUMBER, WE MUST FIX THE SENSE OF A NUMERICAL IDENTITY

§ 62. How, then, are numbers to be given to us, if we cannot have any ideas or intuitions of them? Since it is only in the context of a proposition that words have any meaning, our problem becomes this: To define the sense of a proposition in which a number word occurs. That, obviously, leaves us still a very wide choice. But we have already settled that number words are to be understood as standing for self-subsistent objects. And that is enough to give us a class of propositions which must have a sense, namely those which express our recognition of a number as the same again. If we are to use the symbol *a* to signify an object, we must have a criterion for deciding in all cases whether *b* is the same as *a*, even if it is not always in our power to apply this criterion. In our present case, we have to define the sense of the proposition "the number which belongs to the concept *F* is the same as that which belongs to the concept *G*;" that is to say, we must reproduce the content of this proposition in other terms, avoiding the use of the expression "the Number which belongs to the concept *F*." In doing this, we shall be giving a general criterion for the identity of numbers. When we have thus acquired a means of arriving at a determinate number and of recognizing it again as the same, we can assign it a number word as its proper name.

§ 63. Hume[1] long ago mentioned such a means: "When two numbers are so combined as that the one has always an unit answering to every unit of the other, we pronounce them equal." This opinion, that numerical equality or identity must be defined in terms of one-one correlation, seems in recent years to have gained widespread acceptance among mathematicians.[2] But it raises at once certain logical doubts and difficulties, which ought not to be passed over without examination.

It is not only among numbers that the relationship of identity is found. From which it seems to follow that we ought not to define it specially for the case of numbers. We should expect the concept of identity to have been fixed first, and that then, from it together with the concept of Number, it must be possible to deduce when Numbers are identical

*Source: Gottlob Frege, *The Foundations of Arithmetic* trans. by J. L. Austin (1953), 73e - 92e. This section is reprinted by permission of the Philosophical Library.

with one another, without there being need for this purpose of a special definition of numerical identity as well.

As against this, it must be noted that for us the concept of Number has not yet been fixed, but is only due to be determined in the light of our definition of numerical identity. Our aim is to construct the content of a judgement which can be taken as an identity such that each side of it is a number. We are therefore proposing not to define identity specially for this case, but to use the concept of identity, taken as already known, as a means for arriving at that which is to be regarded as being identical. Admittedly, this seems to be a very odd kind of definition, to which logicians have not yet paid enough attention; but that it is not altogether unheard of, may be shown by a few examples.

§ 64. The judgement "line *a* is parallel to line *b*," or, using symbols,

$$a \,//\, b,$$

can be taken as an identity. If we do this, we obtain the concept of direction, and say: "the direction of line *a* is identical with the direction of line *b*." Thus we replace the symbol // by the more generic symbol =, through removing what is specific in the content of the former and dividing it between *a* and *b*. We carve up the content in a way different from the original way, and this yields us a new concept. Often, of course, we conceive of the matter the other way round, and many authorities define parallel lines as lines whose directions are identical. The proposition that "straight lines parallel to the same straight line are parallel to one another" can then be very conveniently proved by invoking the analogous proposition about things identical with the same thing. Only the trouble is, that this is to reverse the true order of things. For surely everything geometrical must be given originally in intuition. But now I ask whether anyone has an intuition of the direction of a straight line. Of a straight line, certainly; but do we distinguish in our intuition between this straight line and something else, its direction? That is hardly plausible. The concept of direction is only discovered at all as a result of a process of intellectual activity which takes its start from the intuition. On the other hand, we do have an idea of parallel straight lines. Our convenient proof is only made possible by surreptitiously assuming, in our use of the word "direction," what was to be proved; for if it were false that "straight lines parallel to the same straight line are parallel to one another," then we could not transform $a \,//\, b$ into an identity.

We can obtain in a similar way from the parallelism of planes a concept corresponding to that of direction in the case of straight lines; I have seen the name "orientation" [*Stellung*] used for this. From geometrical similarity is derived the concept of shape, so that instead of "the two triangles are similar" we say "the two triangles are of identical shape" or "the shape of the one is identical with that of the other." It is possible to derive yet another concept in this way, to which no name has yet been given, from the collineation of geometrical forms.

§ 65. Now in order to get, for example, from parallelism[3] to the concept of direction, let us try the following definition:

> The proposition "line *a* is parallel to line *b*" is to mean the same as "the direction of line *a* is identical with the direction of line *b*."

This definition departs to some extent from normal practice, in that it serves ostensibly to adapt the relation of identity, taken as already known, to a special case, whereas in reality it is designed to introduce the expression "the direction of line *a*," which only comes into it incidentally. It is this that gives rise to a second doubt—are we not liable, through using such methods, to become involved in conflict with the well-known laws of identity? Let us see what these are. As analytic truths they should be capable of being derived from the concept itself alone. Now Leibniz's[4] definition is as follows:

> Things are the same as each other, of which one can be substituted for the other without loss of truth.[5]

This I propose to adopt as my own definition of identity. Whether we use "the same," as Leibniz does, or "identical," is not of any importance. "The same" may indeed be thought to refer to complete agreement in all respects, "identical"[6] only to agreement in this respect or that; but we can adopt a form of expression such that this distinction vanishes. For example, instead of "the segments are identical in length," we can say "the length of the segments is identical" or "the same," and instead of "the surfaces are identical in colour," "the colour of the surfaces is identical." And this is the way in which the word has been used in the examples above. Now, it is actually the case that in universal substitutability all the laws of identity are contained.

In order, therefore, to justify our proposed definition of the direction of a line, we should have to show that it is possible, if line *a* is parallel to line *b*, to substitute "the direction of *b*" everywhere for "the direction of *a*." This task is made simpler by the fact that we are being taken initially to know of nothing that can be asserted about the direction of a line except the one thing, that it coincides with the direction of some other line. We should thus have to show only that substitution was possible in an identity of this one type, or in judgement-contents containing such identities as constituent elements.[7] The meaning of any other type of assertion about directions would have first of all to be defined, and in defining it we can make it a rule always to see that it must remain possible to substitute for the direction of any line the direction of any line parallel to it.

§ 66. But there is still a third doubt which may make us suspicious of our proposed definition. In the proposition "the direction of *a* is identical with the direction of *b*" the direction of *a* plays the part of an object,[8] and our definition affords us a means of recognizing this object as the same again, in case it should happen to crop up in some other guise, say as the direction of *b*. But this means does not provide for all cases. It will not, for instance, decide for us whether England is the same as the direction of the Earth's axis—if I may be forgiven an example which looks nonsensical. Naturally no one is going to confuse England with the direction of the Earth's axis; but that is no thanks to our definition of direction. That says nothing as to whether the proposition "the direction of *a* is identical with *q*" should be affirmed or denied, except for the one case where *q* is given in the form of "the direction of *b*." What we lack is the concept of direction; for if we had that, then we could lay it down that, if *q* is not a direction, our proposition is to be denied, while if it is a direction, our original definition will decide whether it is to be denied or affirmed. So the temptation is to give as our definition: *q* is a direction, if there is a line *b* whose direction is *q*. But then we have obviously come round in a circle. For in order to make use of this definition, we should have to know already in every case whether the proposition "*q* is identical with the direction of *b*" was to be affirmed or denied.

§ 67. If we were to try saying: *q* is a direction if it is introduced by means of the definition set out above, then we should be treating the way in which the object *q* is introduced as a property of *q*, which it is not. The definition of an object does not, as such, really assert anything about the object, but only lays down the meaning of a symbol. After this has been done, the definition transforms itself into a judgement, which does assert about the object; but now it no longer introduces the object, it is exactly on a level with other assertions made about it. If, moreover, we were to adopt this way out, we should have to be presupposing that an object can only be given in one single way; for otherwise it would not follow, from the fact that q *was* not introduced by means of our definition, that it *could* not have been introduced by means of it. All identities would then amount simply to this, that whatever is given to us in the same way is to be reckoned as the same. This, however, is a principle so obvious and so sterile as not to be worth stating. We could not, in fact, draw from it any conclusion which was not the same as one of our premises. Why is it, after all, that we are able to make use of identities with such significant results in such diverse fields? Surely it is rather because we are able to recognize something as the same again even although it is given in a different way.

§ 68. Seeing that we cannot by these methods obtain any concept of direction with sharp limits to its application, nor therefore, for the same reasons, any satisfactory concept of Number either, let us try another way. If line a is parallel to line b, then the extension of the concept "line parallel to line a" is identical with the extension of the concept "line parallel to line b"; and conversely, if the extensions of the two concepts just named are identical, then a is parallel to b. Let us try, therefore, the following type of definition: the direction of line a is the extension of the concept "parallel to line a"; the shape of triangle t is the extension of the concept "similar to triangle t."

To apply this to our own case of Number, we must substitute for lines or triangles concepts, and for parallelism or similarity the possibility of correlating one to one the objects which fall under the one concept with those which fall under the other. For brevity, I shall, when this condition is satisfied, speak of the concept F being *equal*[9] to the concept G; but I must ask that this word be treated as an arbitrarily selected symbol, whose meaning is to be gathered, not from its etymology, but from what is here laid down.

My definition is therefore as follows: the Number which belongs to the concept F is the extension[10] of the concept "equal to the concept F."

§ 69. That this definition is correct will perhaps be hardly evident at first. For do we not think of the extensions of concepts as something quite different from numbers? How we do think of them emerges clearly from the basic assertions we make about them. These are as follows:

1. that they are identical,
2. that one is wider than the other.

But now the proposition:

> the extension of the concept "equal to the concept F" is identical with the extension of the concept "equal to the concept G"

is true if and only if the proposition

> "the same number belongs to the concept F as to the concept G"

is also true. So that here there is complete agreement.

Certainly we do not say that one number is wider than another, in the sense in which the extension of one concept is wider than that of another; but then it is also quite impossible for a case to occur where

> the extension of the concept "equal to the concept F" would be wider than
> the extension of the concept "equal to the concept G."

For on the contrary, when all concepts equal to G are also equal to F, then conversely all concepts equal to F are equal to G. "Wider" as used here must not, of course, be confused with "greater" as used of numbers.

Another type of case is, I admit, conceivable, where the extension of the concept "equal to the concept F" might be wider or less wide than the extension of some other concept, which then could not, on our definition, be a Number; and it is not usual to speak of a Number as wider or less wide than the extension of a concept; but neither is there anything to prevent us speaking in this way, if such a case should ever occur.

OUR DEFINITION COMPLETED AND ITS WORTH PROVED

§ 70. Definitions show their worth by proving fruitful. Those that could just as well be omitted and leave no link missing in the chain of our proofs should be rejected as completely worthless.

Let us try, therefore, whether we can derive from our definition of the Number which belongs to the concept *F* any of the well-known properties of numbers. We shall confine ourselves here to the simplest.

For this it is necessary to give a rather more precise account still of the term "equality." "Equal" we defined in terms of one-one correlation, and what must now be laid down is how this latter expression is to be understood, since it might easily be supposed that it had something to do with intuition.

We will consider the following example. If a waiter wishes to be certain of laying exactly as many knives on a table as plates, he has no need to count either of them; all he has to do is to lay immediately to the right of every plate a knife, taking care that every knife on the table lies immediately to the right of a plate. Plates and knives are thus correlated one to one, and that by the identical spatial relationship. Now if in the proposition "*a* lies immediately to the right of *A*" we conceive first one and then another object inserted in place of *a* and again of *A*, then that part of the content which remains unaltered throughout this process constitutes the essence of the relation. What we need is a generalization of this.

If from a judgement-content which deals with an object *a* and an object *b* we subtract *a* and *b*, we obtain as remainder a relation-concept which is, accordingly, incomplete at two points. If from the proposition "the Earth is more massive than the Moon" we subtract "the Earth," we obtain the concept "more massive than the Moon." If, alternatively, we subtract the object, "the Moon," we get the concept "less massive than the Earth." But if we subtract them both at once, then we are left with a relation-concept, which taken by itself has no [assertible] sense any more than a simple concept has: it has always to be completed in order to make up a judgement-content. It can however be completed in different ways: instead of Earth and Moon I can put, for example, Sun and Earth, and this *eo ipso* effects the subtraction.

Each individual pair of correlated objects stands to the relation-concept much as an individual object stands to the concept under which it falls—we might call them the subject of the relation-concept. Only here the subject is a composite one. Occasionally, where the relation in question is convertible, this fact achieves verbal recognition, as in the proposition "Peleus and Thetis were the parents of Achilles."[11] But not always. For example, it would scarcely be possible to put the proposition "the Earth is bigger than the Moon" into other words so as to make "the Earth and the Moon" appear as a composite subject; the "and" must always indicate that the two things are being put in some way on a level. However, this does not affect the issue.

The doctrine of relation-concepts is thus, like that of simple concepts, a part of pure logic. What is of concern to logic is not the special content of any particular relation, but only the logical form. And whatever can be asserted of this, is true analytically and known a priori. This is as true of relation-concepts as of other concepts.

Just as "*a* falls under the concept *F*" is the general form of a judgement-content which deals with an object *a*, so we can take "*a* stands in the relations ϕ to *b*" as the general form of a judgement-content which deals with an object *a* and an object *b*.

§ 71. If now every object which falls under the concept *F* stands in the relation ϕ to an object falling under the concept *G*, and if to every object which falls under *G* there stands in the relation

ϕ an object falling under F, then the objects falling under F and under G are correlated with each other by the relation ϕ.

It may still be asked, what is the meaning of the expression "every object which falls under F stands in the relations ϕ to an object falling under G" in the case where no object at all falls under F. I understand this expression as follows:

> the two propositions "a falls under F" and "a does not stand in the relation ϕ to any object falling under G" cannot, whatever be signified by a, both be true together; so that either the first proposition is false, or the second is, or both are. From this it can be seen that the proposition "every object which falls under F stands in the relation ϕ to an object falling under G" is, in the case where there is no object falling under F, true; for in that case the first proposition "a falls under F" is always false, whatever a may be.

In the same way the proposition "to every object which falls under G there stands in the relation ϕ an object falling under F" means that the two propositions "a falls under G" and "no object falling under F stands to a in the relation ϕ" cannot, whatever a may be, both be true together.

§ 72. We have thus seen when the objects falling under the concepts F and G are correlated with each other by the relation ϕ. But now in our case, this correlation has to be one-one. By this I understand that the two following propositions both hold good:

1. If d stands in the relation ϕ to a, and if d stands in the relation ϕ to e, then generally, whatever d, a and e may be, a is the same as e.
2. If d stands in the relation ϕ to a, and if b stands in the relation ϕ to a, then generally, whatever d, b and a may be, d is the same as b.

This reduces one-one correlation to purely logical relationships, and enables us to give the following definition:

> the expression
>
> "the concept F is equal to the concept G"
>
> is to mean the same as the expression
>
> "there exists a relation ϕ which correlates one to one the objects falling under the concept F with the objects falling under the concept G."

We now repeat our original definition: the Number which belongs to the concept F is the extension of the concept "equal to the concept F" and add further: the expression "n is a Number" is to mean the same as the expression "there exists a concept such that n is the Number which belongs to it."

Thus the concept of Number receives its definition, apparently, indeed, in terms of itself, but actually without any fallacy, since "the Number which belongs to the concept F" has already been defined.

§ 73. Our next aim must be to show that the Number which belongs to the concept F is identical with the Number which belongs to the concept G if the concept F is equal to the concept G. This sounds, of course, like a tautology. But it is not; the meaning of the word "equal" is not to be inferred from its etymology, but taken to be as I defined it above.

On our definition [of "the Number which belongs to the concept F"], what has to be shown is that the extension of the concept "equal to the concept F" is the same as the extension of the concept "equal to the concept G," if the concept F is equal to the concept G. In other words: it is to be proved that, for F equal to G, the following two propositions hold good universally:

> if the concept H is equal to the concept F, then it is also equal to the concept G;
>
> and
>
> if the concept H is equal to the concept G, then it is also equal to the concept F.

The first proposition amounts to this, that there exists a relation which correlates one to one the objects falling under the concept H with those falling under the concept G, if there exists a relation ϕ which correlates one to one the objects falling under the concept F with those falling under the concept G and if there exists also a relation Ψ which correlates one to one the objects falling under the concept H with those falling under the concept F. The following arrangement of letters will make this easier to grasp:

$$H \; \Psi \; F \; \phi \; G.$$

Such a relation can in fact be given: it is to be found in the judgement-content

> there exists an object to which c stands in the relation Ψ and which stands to b in the relation ϕ,

if we subtract from it c and b—take them, that is, as the terms of the relation. It can be shown that this relation is one-one, and that it correlates the objects falling under the concept H with those falling under the concept G.

A similar proof can be given of the second proposition also.[12] And with that, I hope, enough has been indicated of my methods to show that our proofs are not dependent at any point on borrowings from intuition, and that our definitions can be used to some purpose.

§ 74. We can now pass on to the definitions of the individual numbers.

Since nothing falls under the concept "not identical with itself," I define nought as follows: 0 is the Number which belongs to the concept "not identical with itself."

Some may find it shocking that I should speak of a concept in this connection. They will object, very likely, that it contains a contradiction and is reminiscent of our old friends the square circle and wooden iron. Now I believe that these old friends are not so black as they are painted. To be of any use is, I admit, the last thing we should expect of them; but at the same time, they cannot do any harm, if only we do not assume that there is anything which falls under them—and to that we are not committed by merely using them. That a concept contains a contradiction is not always obvious without investigation; but to investigate it we must first possess it and, in logic, treat it just like any other. All that can be demanded of a concept from the point of view of logic and with an eye to rigour of proof is only that the limits to its application should be sharp, that we should be able to decide definitely about every object whether it falls under that concept or not. But this demand is completely satisfied by concepts which, like "not identical with itself," contain a contradiction; for of every object we know that it does not fall under any such concept.[13]

On my use of the word "concept," "*a* falls under the concept *F*" is the general form of a judgement-content which deals with an object *a* and permits of the insertion for *a* of anything whatever. And in this sense "*a* falls under the concept 'not identical with itself'" has the same meaning as "*a* is not identical with itself" or "*a* is not identical with *a*."

I could have used for the definition of nought any other concept under which no object falls. But I have made a point of choosing one which can be proved to be such on purely logical grounds; and for this purpose "not identical with itself" is the most convenient that offers, taking for the definition of "identical" the one from Leibniz given above [(§ 65)], which is in purely logical terms.

§ 75. Now it must be possible to prove, by means of what has already been laid down, that every concept under which no object falls, is equal to every other concept under which no object falls, and to them alone; from which it follows that 0 is the Number which belongs to any such concept, and that no object falls under any concept if the number which belongs to that concept is 0.

If we assume that no object falls under either the concept F or the concept G, then in order to prove them equal we have to find a relation ϕ which satisfies the following conditions:

> every object which falls under F stands in the relation ϕ to an object which falls under G; and to every object which falls

> under G there stands in the relation ϕ an object falling under F.

In view of what has been said above [(§ 71)] on the meaning of these expressions, it follows, on our assumption [that no object falls under either concept], that these conditions are satisfied by every relation whatsoever, and therefore among others by identity, which is moreover a one-one relation; for it meets both the requirements laid down [in § 72] above.

If, to take the other case, some object, say a, does fall under G, but still none falls under F, then the two propositions "a falls under G" and "no object falling under F stands to a in the relation ϕ" are both true together for every relation ϕ; for the first is made true by our first assumption and the second by our second assumption. If, that is, there exists no object falling under F, then a fortiori there exists no object falling under F which stands to a in any relation whatsoever. There exists, therefore, no relation by which the objects falling under F can be correlated with those falling under G so as to satisfy our definition [of equality], and accordingly the concepts F and G are unequal.

§ 76. I now propose to define the relation in which every two adjacent members of the series of natural numbers stand to each other. The proposition:

> there exists a concept F, and an object falling under it x, such that the Number which belongs to the concept F is n and the Number which belongs to the concept 'falling under F but not identical with x' is m

is to mean the same as

> n follows in the series of natural numbers directly after m.

I avoid the expression "n is *the* Number following next after m," because the use of the definite article cannot be justified until we have first proved two propositions. For the same reason I do not yet say at this point "$n = m + 1$," for to use the symbol $=$ is likewise to designate $(m + 1)$ an object.

§ 77. Now in order to arrive at the number 1, we have first of all to show that there is something which follows in the series of natural numbers directly after 0.

Let us consider the concept—or, if you prefer it, the predicate—"identical with 0." Under this falls the number 0. But under the concept "identical with 0 but not identical with 0," on the other hand, no object falls, so that 0 is the Number which belongs to this concept. We have, therefore, a concept "identical with 0" and an object falling under it 0, of which the following propositions hold true:

> the Number which belongs to the concept "identical with 0" is identical with the Number which belongs to the concept "identical with 0";
> the Number which belongs to the concept "identical with 0 but not identical with 0" is 0.

Therefore, on our definition [§ 76], the Number which belongs to the concept "identical with 0" follows in the series of natural numbers directly after 0.

Now if we give the following definition: 1 is the Number which belongs to the concept "identical with 0," we can then put the preceding conclusion thus: 1 follows in the series of natural numbers directly after 0.

It is perhaps worth pointing out that our definition of the number 1 does not presuppose, for its objective legitimacy, any matter of observed fact.[14] It is easy to get confused over this, seeing that certain subjective conditions must be satisfied if we are to be able to arrive at the definition, and that sense experiences are what prompt us to frame it.[15] All this, however, may be perfectly correct, without the propositions so arrived at ceasing to be a priori. One such condition is, for example, that blood of the right quality must circulate in the brain in sufficient volume—at least so far as we know; but the truth of our last proposition does not depend on this; it still holds, even if the circulation stops; and even if all rational beings were to take to hibernating and fall asleep simultaneously, our proposition would not be, say, cancelled for the duration, but would remain quite unaffected. For a proposition to be true is just not the same thing as for it to be thought.

§ 78. I proceed to give here a list of several propositions to be proved by means of our definitions. The reader will easily see for himself in outline how this can be done.

1. If a follows in the series of natural numbers directly after 0, then a is $= 1$.
2. If 1 is the Number which belongs to a concept, then there exists an object which falls under that concept.
3. If 1 is the Number which belongs to a concept F; then, if the object x falls under the concept F and if y falls under the concept F, x is $= y$; that is, x is the same as y.
4. If an object falls under the concept F, and if it can be inferred generally from the propositions that x falls under the concept F and that y falls under the concept F that x is $= y$, that 1 is the Number which belongs to the concept F.
5. The relation of m to n which is established by the proposition:
   "n follows in the series of natural numbers directly after m"
   is a one-one relation.
6. Every Number except 0 follows in the series of natural numbers directly after a Number.

NOTES

1. . . . *Treatise*, Bk. I, Part iii, Sect. 1.

2. Cf. . . . E. Kossak, *Die Elemente der Arithmetik, Programm des Friedrichs-Werder'schen Gymnasiums*, Berlin 1872, p. 16; G. Cantor, *Grundlagen einer allgemeinen Mannichfaltigkeitslebre*, Leipzig 1883.

3. I have chosen to discuss here the case of parallelism, because I can express myself less clumsily and make myself more easily understood. The argument can readily be transferred in essentials to apply to the case of numerical identity.

4. *Non inelegans specimen demonstrandi in abstractis* (Erdmann edn., p. 94).

5. [*Eadem sunt, quorum unum potest substitui alteri salva veritate.*]

6. [Still more "equal" or "similar," which the German *gleich* can also mean.]

7. In a hypothetical judgement, for example, an identity of directions might occur as antecedent or consequent.

8. This is shown by the definite article. A concept is for me that which can be predicate of a singular judgement-content, an object that which can be subject of the same. If in the proposition "the direction of the axis of the telescope is identical with the direction of the Earth's axis" we take the direction of the axis of the telescope as subject, then the predicate is "identical with the direction of the Earth's axis." This is a concept. But the direction of the Earth's axis is only an element in the predicate; it, since it can also be made the subject, is an object.

9. [*Gleichzablig*—an invented word, literally "identinumerate" or "tautarithmic"; but these are too clumsy for constant use. Other translators have used "equinumerous"; "equinumerate" would be better. Later writers have used "similar" in this connection (but as a predicate of "class" not of "concept").]

10. I believe that for "extension of the concept" we could write simply "concept." But this would be open to the two objections:

   1. that this contradicts my earlier statement that the individual numbers are objects, as is indicated by the use of the definite article in expressions like "the number two" and by the impossibility of speaking of ones, twos, etc. in the plural, as also by the fact that the number constitutes only an element in the predicate of a statement of number;

   2. that concepts can have identical extensions without themselves coinciding.

I am, as it happens, convinced that both these objections can be met; but to do this would take us too far afield for present purposes. I assume that it is known what the extension of a concept is.

11. This type of case should not be confused with another, in which the "and" joins the subjects in appearance only, but in reality joins two propositions.

12. And likewise of the converse: If the number which belongs to the concept F is the same as that which belongs to the con-

cept G, then the concept F is equal to the concept G.

13. The definition of an object in terms of a concept under which it falls is a very different matter. For example, the expression "the largest proper fraction" has no content, since the definite article claims to refer to a definite object. On the other hand, the concept "fraction smaller than 1 and such that no fraction smaller than one exceeds it in magnitude" is quite unexceptionable: in order, indeed, to prove that there exists no such fraction, we must make use of just this concept, despite its containing a contradiction. If, however, we wished to use this concept for defining an object falling under it, it would, of course, be necessary first to show two distinct things:

1. that some object falls under this concept;
2. that only one object falls under it.

Now since the first of these propositions, not to mention the second, is false, it follows that the expression "the largest proper fraction" is senseless.

14. Non-general proposition.

15. Cf. B. Erdmann, *Die Axiome der Geometrie*, p. 164.

Chapter VIII
The Nineteenth Century

Section D
Number Theory, Set Theory, and Symbolic Logic

GIUSEPPE PEANO (1858–1932)

The Italian mathematician Giuseppe Peano was the second child of Bartolomeo and Rosa Cavallo Peano. He was a frail youth who grew up in a Catholic family that emphasized schooling. When he was 12 or 13 his maternal uncle, Michele Cavallo, a priest and lawyer, invited him to move to Turin to continue his education. After his uncle tutored him, he attended the Cavour (Secondary) School (1873–76) and won a scholarship to the Collegio delle Provincie, which prepared him for studies at the University of Turin. After graduating from the university with high honors in 1880, he spent the ensuing decade initially as an assistant to two of his professors and subsequently as a lecturer of infinitesimal calculus. In 1887 he married Carola Crosio; they had no children.

Peano spent his entire career at the University of Turin. Following a regular competition, he was named an associate professor of infinitesimal calculus in 1890. He was promoted to full professor in 1895, a position that he retained until his death. From 1886 to 1901 he also taught at the nearby military academy. His gentle personality and keen intellect won the respect of colleagues and students. Despite having a gruff voice, which on occasion made his lectures difficult to follow, he was known at first as an inspiring teacher, who summoned students logically to rigorous thought and a clarity of expression in mathematics. These two traits characterized the calculus textbooks he wrote—*Angelo Genocchi, Calcolo differenziale e principii di calcolo integrale* and the two-volume *Lezioni di analysi infinitesimale* (1893). As he turned his attention more to symbolic logic and the foundations of mathematics, especially in his *Formulario mathematico* (5 editions; 1895-1908), his reputation as a teacher changed for the worse. When he tried to introduce his new symbolism and what seemed an excessive formalism into classes at the military academy, the students rebelled and he was obliged to resign in 1901. Within a few years he stopped most of his lecturing at the university.

During the 1890s Peano concentrated more on research than teaching. A member of the Academy of Sciences of Turin, he founded the journal *Rivista di matematica* in 1891 to publish the results of his research and that of his disciples. The next year he announced the *Formulario* project in the *Rivista*. In this project, which took 16 years to complete, he collected and gave proofs for some 4,200 theorems. In 1897 he presented a paper at the First International Congress of Mathematicians in Zurich on his new ideas in symbolic logic and arithmetic. Three years later he and his disciples, including C. Burali-Forti, dominated the Paris Philosophical Congress. Of this meeting Bertrand Russell wrote, "The Congress was a turning point in my intellectual life, because I there met Peano."

After 1900, Peano's search for an international auxiliary language for the scholarly community eclipsed his mathematical research. This language, which he called "interlingua," was to play the same role that Latin had during the Middle Ages and early modern times in the West. In 1903 Peano proposed a *latino sine flexione* ("Latin without grammar") as the interlingua. In 1908 he was elected president of the Akademi Internasional de Lingua Universal. Two years later he changed its name to Academia pro Interlingua. He continued to serve as its president until his death. His linguistic studies led him in part to examine philology. These studies resulted notably in the *Vocabulario commune ad latino-italiano-français-english-deutsch* (1909), whose expanded second edition in 1915 contained over 14,000 entries. During World War I he also exhibited a strong, continuing interest in mathematical pedagogy. Besides actively participating in the Mathesis Society of school teachers of mathematics, he organized several conferences in Turin to improve the teaching of mathematics in the secondary schools.

Although Peano considered mathematical analysis his most important work, today he is best known for pioneering efforts in symbolic logic and the modern axiomatization of mathematics. He contributed to both of these in *Arithmetices principia, nova methodo exposita* ("The Principles of Arithmetic, Presented by a New Method, 1889"), in which he attempted to write a theory of natural numbers in symbolic notation. In this short booklet he built upon ideas of Dedekind. (He had not yet read Frege.) His notation, which is superior to Boole's and Schröder's, includes ϵ for "belongs to," the inverted C which becomes Ɔ for "implies," N_0 for "the class of natural numbers," a^+ for "the next natural number after a," O for a null set, and ● for the class of everything (universal quantification). With this new ideography, he hoped to sharpen mathematical reasoning. It has now largely been accepted in the original or a modified form.

Proceeding from the undefined arithmetic notions of "class," "number," "one," "successor," and "is equal to," Peano first proposed in the *Arithmetices principia* the five axioms for natural numbers now known as "the Peano axioms." They are given below in his revised order of 1898 with the *Arithmetices* arrangement listed in parentheses.

1. One is a natural number. (Axiom 1)
2. One is not the successor of any other natural number. (Axiom 8)
3. Each natural number a has a successor. (Axiom 6)
4. If the successors of a and b are equal, then so are a and b. (Axiom 7)
5. If a set K of natural numbers contains 1, and if K contains any number x it also contains the successor of x, then K contains all the natural numbers. (Axiom 9)

The last axiom is a translation of the principle of mathematical induction.

Peano played a distinctive role in the development of symbolic logic and the foundations of mathematics. He saw his contributions to symbolic logic, including a new ideography, as responding to Leibniz's call for a universal scientific language *(characteristica universalis)* and a calculus of reasoning *(calculus ratiocinator)*. In this, as in most of his research, Peano considered himself an apostle. Although his ideas strongly influenced the logicists and formalists at the turn of the 20th century, he belonged to neither school. He did not subscribe to the view of Russell and Whitehead that mathematics can be reduced to logic. Instead, he concentrated on the logic of mathematical reasoning, which he called "mathematical logic." Nor did his proofs in *Arithmetices principia* and *Formulario* constitute a formal procedure because they lacked rules of inference and required some intuitive logical arguments. After World War I, his logical ideas influenced other attempts to restructure mathematics, notably the program of those French mathematicians who published under the pseudonym Nicolas Bourbaki.

Peano made other significant contributions to mathematics. His clear presentation of H. Grassmann's abstruse vectorial methods in the 1880s stimulated the Italian school of vector analysis. In 1890 he published the first explicit statement of the axiom of choice: given any collection of nonempty sets, a choice set exists that contains at least one member from each set in the collection. During the same year he presented a famous counterexample to accepted mathematical notions with his discovery of a space-filling curve. To derive this curve he divided a space into an increasing number of squares and connected the center of the squares in a determined order.

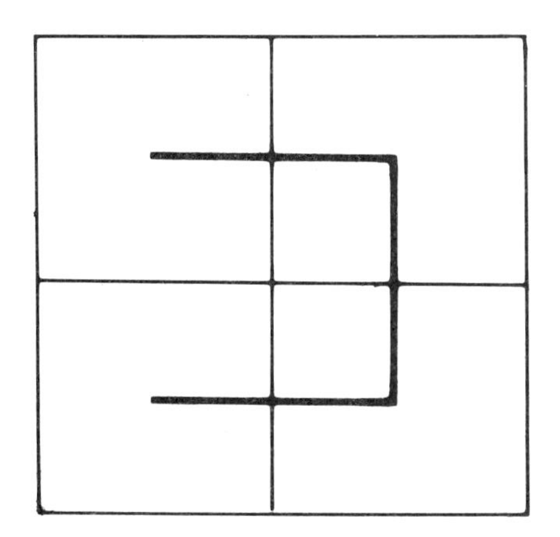

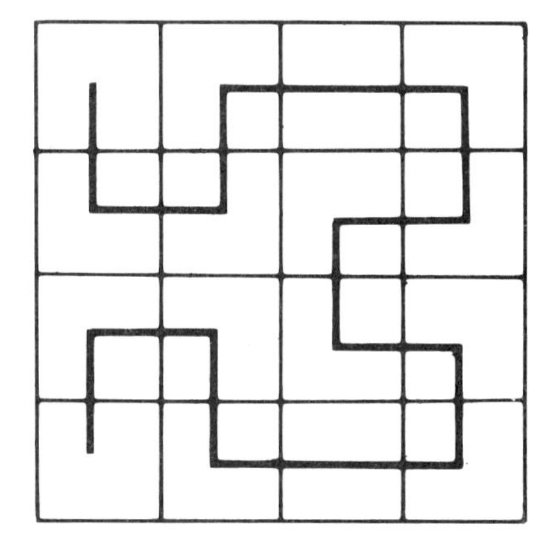

116. From *Arithmeticas principia* (1889)*

(Set of Axioms for Integers)

GIUSEPPE PEANO

PREFACE

Questions that pertain to the foundations of mathematics, although treated by many in recent times, still lack a satisfactory solution. The difficulty has its main source in the ambiguity of language.

That is why it is of the utmost importance to examine attentively the very words we use. My goal has been to undertake this examination, and in this paper I am presenting the results of my study, as well as some applications to arithmetic.

I have denoted by signs all ideas that occur in the principles of arithmetic, so that every proposition is stated only by means of these signs.

The signs belong either to logic or to arithmetic proper. The signs of logic that occur here are ten in number, although not all are necessary. In the first part of the present paper [Logical notations] the use of these signs, as well as some of their properties, is explained in ordinary language. It was not my intention to present their theory more fully there. The signs of arithmetic are explained wherever they occur.

With these notations, every proposition assumes the form and the precision that equations have in algebra; from the propositions thus written other propositions are deduced, and in fact by procedures that are similar to those used in solving equations. This is the main point of the whole paper.

Thus, having introduced the signs with which I can write the propositions of arithmetic, I have, in dealing with these propositions, used a method that, because it will have to be followed in other studies too, I shall present briefly here.

Among the signs of arithmetic, those that can be expressed by other signs of arithmetic together with the signs of logic represent the ideas that we can define. Thus, I have defined all signs except the four that are contained in the explanations of § 1. If, as I think, these cannot be reduced any further, it is not possible to define the ideas expressed by them through ideas assumed to be known previously.

Propositions that are deduced from others by the operations of logic are *theorems;* propositions that are not thus deduced I have called *axioms*. There are nine of these axioms (§ 1), and they express the fundamental properties of the signs that lack definition.

In §§ 1-6 I have proved the ordinary properties of numbers. For the sake of brevity I have omitted proofs that are similar to other proofs given before. In order to express proofs with the signs of logic, the ordinary form of these proofs has to be changed; this transformation is sometimes rather difficult, but it is by means of it that the nature of the proof reveals itself most clearly.

*Source: This translation by Jean van Heijenoort is reprinted by permission of the publisher from *From Frege to Gödel: A Source Book in Mathematical Logic, 1879–1931*, edited by Jean van Heijenoort, Cambridge, Mass.: Harvard University Press, Copyright © 1967 by the President and Fellows of Harvard College and Unione Matematica Italiana.

In subsequent sections I deal with various subjects, so that the power of the method will be more apparent.

§ 7 contains a few theorems that pertain to number theory. In §§ 8-9 the definitions of rational and irrational numbers are found.

Finally, in § 10 I present a few theorems, which, I think, are new, pertaining to the theory of those objects that Cantor has called *Punktmengen (ensembles de points).*

In the present paper I have made use of the studies of other writers. The logical symbols and propositions contained in parts II, III, and IV, except for a few, are to be traced to the works of many writers, especially Boole.[1]

I introduced the sign ϵ, which should not be confused with the sign Ɔ; I also introduced applications of inversion in logic, as well as a few other conventions, in order to be able to express any proposition whatsoever.

For proofs in arithmetic, I used *Grassmann 1861.*

The recent work of Dedekind *(1888)* was also most useful to me; in it, questions pertaining to the foundations of numbers are acutely examined.

This little book of mine is intended to give an example of the new method. With these notations we can state and prove innumerable other propositions, whether they pertain to rational or to irrational numbers. But to deal with other theories new signs denoting new objects must be introduced. However, I think that the propositions of any science can be expressed by these signs of logic alone, provided we add signs representing the objects of that science.

LOGICAL NOTATIONS

I. PUNCTUATION

The letters *a, b, . . . , x, y, . . . , x′, y′, . . .* denote indeterminate objects. We denote well-determined objects by signs or by the letters P, K, N, ...

We shall generally write signs on a single line. To show the order in which they should be taken, we use *parentheses,* as in algebra, or *dots,* ., :, .·., ::, and so on.

To understand a formula divided by dots we first take together the signs that are not separated by any dot, next those separated by one dot, then those separated by two dots, and so on.

For example, let *a, b, c, . . .* be any signs. Then *ab.cd* means *(ab)(cd);* and *ab . cd : ef . gh .·. k* means *(((ab)(cd))((ef)(gh)))k.*

Punctuation signs may be omitted if formulas differing in punctuation have the same meaning or if only one formula, which is just the one we want to write, has meaning.

To avoid the danger of ambiguity we never use . or : as signs for arithmetic operations.

The only kind of parenthesis is ( ); if parentheses and dots occur in the same formula, what is contained within parentheses is taken together first.

II. PROPOSITIONS

The sign P means *proposition.*

The sign ∩ is read *and.* Let *a* and *b* be propositions; then *a* ∩ *b* is the simultaneous affirmation of the propositions *a* and *b.* For the sake of brevity, we ordinarily write *ab* instead of *a* ∩ *b.*

The sign − is read *not.* Let *a* be a P; then −*a* is the negation of the proposition *a.*

The sign ∪ is read *or [vel].* Let *a* and *b* be propositions; then *a* ∪ *b* is the same as − : −*a*. −*b*.

[The sign V means *the true,* or *identity;* but we never use this sign.]

The sign Λ means *the false,* or *the absurd.*

[The sign C means *is a consequence of;* thus b C a is read *b is a consequence of the proposition a.* But we never use this sign.]

The sign $\supset$ means *one deduces* [*deducitur*];[2] thus $a \supset b$ means the same as b C a. If propositions a and b contain the indeterminate objects $x, y, \ldots$, that is, are conditions between these objects, then $a \supset_{x,y,\ldots} b$ means: whatever $x, y, \ldots$ may be, from the proposition a one deduces b. If there is no danger of any ambiguity, we write only $\supset$ instead of $\supset_{x,y,\ldots}$.

The sign $=$ means *is equal to* [*est aequalis*]. Let a and b be propositions; then $a = b$ means the same as $a \supset b \,.\, b \supset a$; the proposition $a =_{x,y,\ldots} b$ means the same as $a \supset_{x,y,\ldots} b \,.\, b \supset_{x,y,\ldots} a$.

III. PROPOSITIONS OF LOGIC

Let $a, b, c, \ldots$ be propositions. Then we have:

1. $a \supset a$.
2. $a \supset b \,.\, b \supset c : \supset .\, a \supset c$.
3. $a = b \,.\, = : a \supset b \,.\, b \supset a$.
4. $a = a$.
5. $a = b \,.\, = .\, b = a$.
6. $a = b \,.\, b \supset c : \supset .\, a \supset c$.
7. $a \supset b \,.\, b = c : \supset .\, a \supset c$.
8. $a = b \,.\, b = c : \supset .\, a = c$.
9. $a = b \,.\, \supset .\, a \supset b$.
10. $a = b \,.\, \supset .\, b \supset a$.
11. $ab \supset a$.
12. $ab = ba$.
13. $a(bc) = (ab)c = abc$.
14. $aa = a$.
15. $a = b \,.\, \supset . ac = bc$.
16. $a \supset b \,.\, \supset .\, ac \supset bc$.
17. $a \supset b \,.\, c \supset d : \supset .\, ac \supset bd$.
18. $a \supset b \,.\, a \supset c : = .\, a \supset bc$.
19. $a = b \,.\, c = d : \supset .\, ac = bd$.

. . .

Let α be a relation sign (for example, $=$, $\supset$), so that $a\, \alpha\, b$ is a proposition. Then, instead of $- .a\, \alpha\, b$, we write $a -\alpha\, b$; that is,

$$\alpha -= b \,.\, = : - \,.\, a = b.$$

$$a - \supset b \,.\, = : - \,.\, a \supset b.$$

Thus the sign $-=$ means *is not equal to.* If proposition a contains the indeterminate x, then $a -=_x \Lambda$ means: there are x that satisfy condition a. The sign $-\supset$ means *one does not deduce.*

Similarly, if α and β are relation signs, instead of $a\,\alpha\, b \,.\, a\,\beta\, b$ and $a\,\alpha\, b \,.\, \cup \,.\, a\,\beta\, b$, we can write $a \,.\, \alpha\beta \,.\, b$ and $a \,.\, \alpha \cup \beta \,.\, b$, respectively. Thus, if a *and* b are propositions, the formula $a \,.\, \supset -= .\, b$ says: from a one deduces b, but not conversely.

$$a \,.\, \supset -= .\, b : = : a \supset b \,.\, b - \supset a.$$

IV. CLASSES

The sign K means *class,* or aggregate of objects.

The sign ϵ means *is.* Thus $a \,\epsilon\, b$ is read *a is a b;* $a \,\epsilon$ K means *a is a class;* $a \,\epsilon$ P means *a is a proposition.*

Instead of $-(a \in b)$ we write $a -\!\in b$; the sign $-\!\in$ means *is not;* that is,

44. $a -\!\in b \,.\, = : -. \, a \in b.$

The sign $a, b, c \in m$ means: a, b, and c are m; that is,

45. $a, b, c \in m \,.\, = : a \in m \,.\, b \in m \,.\, c \in m.$

Let a be a class; then $-a$ means the class composed of the individuals that are not a.

46. $a \in \mathrm{K} \,.\, \supset : x \in -a \,.\, = .\, x -\!\in a.$

Let a and b be classes; $a \cap b$, or ab, is the class composed of the individuals that are at the same time a and b; $a \cup b$ is the class composed of the individuals that are a or b.

47. $a, b \in \mathrm{K} \,.\, \supset .\!:\, x \in .\, ab : = : x \in a \,.\, x \in b.$

48. $a, b \in \mathrm{K} \,.\, \supset .\!:\, x \in .\, a \cup b : = : x \in a \,.\!\cup\!.\, x \in b.$

The sign Λ denotes the class that contains no individuals. Thus,

49. $a \in \mathrm{K} \,.\, \supset .\!:\, a = \Lambda : = : x \in a \,.\, =_x \Lambda.$

[We shall not use the sign V, which denotes the class composed of all individuals under consideration.]

The sign $\supset$ *means is contained in*. Thus $a \supset b$ means *class a is contained in class b.*

50. $a, b \in \mathrm{K} \,.\, \supset .\!:\, a \supset b : = : x \in a \,.\, \supset_x .\, x \in b.$

[The formula $b \subset a$ could mean *class b contains class a;* but we shall not use the sign C.]

The signs Λ and $\supset$ here have a meaning that differs somewhat from the meaning given above; but no ambiguity will arise. If we are dealing with propositions, these signs are read *the absurd* and *one deduces;* but, if we are dealing with classes, they are read *nothing* and *is contained in.*

If a and b are classes, the formula $a = b$ means $a \supset b \,.\, b \supset a$. Thus,

51. $a, b \in \mathrm{K} \,.\, \supset .\!:\, a = b : = : x \in a \,.\, =_x .\, x \in b.$

VI. FUNCTIONS

The symbols of logic introduced above suffice to express any proposition of arithmetic, and we shall use only these. We explain here briefly some other symbols that may be useful.

Let s be a class; we assume that equality is defined between the objects of the system s so as to satisfy the conditions:

$$a = a.$$
$$a = b \,.\, = .\, b = a.$$
$$a = b \,.\, b = c : \supset .\, a = c.$$

Let ϕ be a sign or an aggregate of signs such that, if x is an object of the class s, the expression ϕx denotes a new object; we assume also that equality is defined between the objects ϕx; further, if x and y are objects of the class s and if $x = y$, we assume it is possible to deduce $\phi x = \phi y$. Then the sign ϕ is said to be a *function presign [praesignum] in the class s,* and we write $\phi \in \mathrm{F}'s$:

$$s \in \mathrm{K} \,.\, \supset :: \phi \in \mathrm{F}'s \,.\, = .\!:\, x, y \in s \,.\, x = y : \supset_{x,y} .\, \phi x = \phi y.$$

If, x being any object of the class s, the expression $x\phi$ denotes a new object and $x\phi = y\phi$ follows from $x = y$, then we say that ϕ is a *function postsign [postsignum] in the class s*, and we write $\phi \epsilon s'F$:

$$s \epsilon K . \supset :: \phi \epsilon s'F . = .\because x, y \epsilon s . x = y : \supset_{x . y} . x\phi = y\phi.$$

REMARKS

A *definition*, or *Def.* for short, is a proposition of the form $x = a$ or $\alpha \supset . x = a$, where a is an aggregate of signs having a known meaning, x is a sign or an aggregate of signs, hitherto without meaning, and α is the condition under which the definition is given.

A *theorem* (Theor. or Th.) is a proposition that is *proved*. If a theorem has the form $a \supset \beta$, where α and β are propositions, then α is called the *hypothesis* (Hyp. or, even shorter, Hp.) and β the *thesis* (Thes. or Ts.). Hyp. and Ts. depend on the form of the theorem; in fact, if we write $-\beta \supset -\alpha$ instead of $\alpha \supset \beta$, then $-\beta$ is the Hp. and $-\alpha$ the Ts.; if we write $\alpha - \beta = \Lambda$, Hp. and Ts. do not exist.

In any section below, the sign P followed by a number denotes the proposition indicated by that number in the same section. Propositions of logic are indicated by the sign L and the number of the proposition.

Formulas that do not fit on one line are continued on the next line without any intervening sign.

§ 1. NUMBERS AND ADDITION

EXPLANATIONS

The sign N means *number (positive integer)*.

The sign 1 means *unity*.

The sign $a + 1$ means *the successor of a*, or *a plus* 1.

The sign = means *is equal to*. We consider this sign as new, although it has the form of a sign of logic.

AXIOMS

1. $1 \epsilon N$.
2. $a \epsilon N . \supset . a = a$.
3. $a, b \epsilon N . \supset : a = b . = . b = a$.
4. $a, b, c \epsilon N . \supset .\because a = b . b = c : \supset . a = c$.
5. $a = b . b \epsilon N : \supset . a \epsilon N$.
6. $a \epsilon N . \supset . a + 1 \epsilon N$.
7. $a, b \epsilon N . \supset : a = b . = . a + 1 = b + 1$.
8. $a \epsilon N . \supset . a + 1 -= 1$.
9. $k \epsilon K \therefore 1 \epsilon k \therefore x \epsilon N . x \epsilon k : \supset_x . x + 1 \epsilon k :: \supset . N \supset k$.

DEFINITIONS

10. $2 = 1 + 1$; $3 = 2 + 1$; $4 = 3 + 1$; and so forth.

THEOREMS

11. $2 \epsilon N$.

Proof:

P 1 . $\supset$:	$1 \epsilon N$	(1)
1 [a] (P 6) . $\supset$:	$1 \epsilon N . \supset . 1 + 1 \epsilon N$	(2)
(1) (2) . $\supset$:	$1 + 1 \epsilon N$	(3)
P 10 . $\supset$:	$2 = 1 + 1$	(4)
(4) . (3) . (2, 1 + 1) [a, b] (P 5) : $\supset$:	$2 \epsilon N$	(Theorem).

Note. We have written explicitly all the steps of this very easy proof. For the sake of brevity, we now write it as follows:

$$\text{P } 1.1\ [a]\ (\text{P } 6) : \supset : 1 + 1 \in \text{N}.\text{P } 10.\ (2, 1+1)\ [a, b]\ (\text{P } 5) : \supset : \text{Th}.$$

or

$$\text{P } 1.\text{P } 6 : \supset : 1 + 1 \in \text{N} .\text{P } 10.\text{P } 5 : \supset : \text{Th}.$$

NOTES

1. See *Boole 1847, 1848, 1854,* and *Schröder 1877*. Schröder had already dealt with some questions relevant to logic in an earlier work *(1873)*. I presented the theories of Boole and Schröder very briefly in a book of mine *(1888)*. See also *Peirce 1880, 1885, Jevons 1883, MacColl 1877, 1878, 1878a,* and *1880*.

2. [Peano reads $a \supset b$ "ab *a* deducitur *b*." Translated word for word, this either would be awkward ("from *a* is deduced *b*") or would reverse the relative positions of *a* and *b* ("*b* is deduced from *a*"), which would lead to misinterpretations when the sign is read alone. Peano himself uses "on déduit" for "deducitur" when writing in French (for instance, *1890*, p. 184), and this led to the translation adopted here.]

Chapter VIII
The Nineteenth Century

Section D
Number Theory, Set Theory, and Symbolic Logic

BERTRAND (ARTHUR WILLIAM) RUSSELL (1872–1970)

One of the most profound and widely influential intellects of the 20th century, Bertrand Russell was born into a prominent British family whose liberal, Whig conviction firmly supported civil and religious liberty. His parents were Viscount Amberley and his wife Katherine Stanley, but Bertrand was orphaned by age four. Following a brief court battle, he and his older brother Frank were raised by their strict and exacting Russell grandmother. Like other upper-class Victorian children they were educated privately at home by tutors. Lacking much contact with other children, Bertrand fashioned an intense inner life and searched for certain knowledge. By age 11 the skeptical cast of his intelligence began to appear. Religious doubts emerged. Informed by his brother that the axioms of geometry were not provable but had to be taken on trust, the 11-year-old Bertrand was puzzled and disappointed.

In 1890 Russell entered Trinity College, Cambridge. In general he did not enjoy class lectures. In his *Autobiography* (2 vols., 1967–68) he noted, "I derived no benefits from lectures." His election in 1892 to an exclusive, informal society of intellectuals known to outsiders as "The Apostles" proved a greater stimulus to his thought and showed that his youthful brilliance had been recognized. The Apostles included such talented thinkers as the philosophers G. E. Moore and A. N. Whitehead, the historian G. M. Trevelyan, and the economist J. M. Keynes. Although the formality of examinations did not appeal to him, Russell earned first-class honors in the Mathematical Tripos (honor examination) in 1893 and a first-class degree in moral sciences in 1894. Under the influence of the Cambridge metaphysicist J. M. E. McTaggart, who guided his degree work, he became an Idealist for a time.

In the prologue to his *Autobiography,* Russell asserted that three simple but irresistible passions guided his life: the longing for love, the search for knowledge, and a humanitarian zeal that found human suffering intolerable. In the first—the longing for love and happiness—he was not too successful. Three of his four marriages ended in divorce. In 1894 he married Alys Smith against the wishes of his family. He married Dora Black in 1921; Patricia Spence, a research assistant, in 1936; and Edith Finch, an American, in 1952.

Russell was more successful in pursuing knowledge and in making people more aware of the need for a better condition for mankind. In 1895 he won a six-year fellowship at Trinity for a dissertation on the foundations of mathematics. During that and the next year he traveled to the U.S. to lecture on non-Euclidean geometry and to Germany, where Social Democrats introduced him to Marxist thought (which he rejected). In 1898 at Trinity he rejected Idealism and became a physical Realist, who accepted the current scientific view of the physical world as largely the correct one.

The decade of work beginning in 1900 proved the hardest of Russell's life. He embarked upon a study of logic and mathematics. After writing *A Critical Exposition of the Philosophy of Leibniz* (1900), corresponding with Gottlob Frege and meeting with Giuseppe Peano, he completed *The Principles of Mathematics* in 1903. In collaboration with his former tutor, Alfred North Whitehead, he conducted studies that culminated in the three volumes of *Principia Mathematica* (1910, 1912, and 1913), an immensely influential work among logicians. He set forth the logicist approach to the foundations of mathematics, which maintained that mathematics could be derived from a small number of self-evident logical principles. The decade after 1900 also saw a "mystic illumination" in Russell's life that transformed him into a pacifist and imbued him with "semimystical feelings about beauty." In 1907 he unsuccessfully campaigned for a seat in Parliament as an advocate of women's suffrage and free trade. The next year he was elected a Fellow of the Royal Society.

In 1910 Trinity appointed Russell a special lecturer of logic and the philosophy of mathematics. Two years later the young philosopher Ludwig Wittgenstein attended his classes and was strongly influenced by him. Russell now felt compelled to devote more of his time to human concerns and less to mathematics. During World War I he was a staunch and vocal pacifist, which led to a £100 fine and dismissal from Trinity in 1916. It also brought him a six-month sentence in Brixton Prison in 1918, where he wrote *Introduction to Mathematical Philosophy* (1919). The breach with Trinity was only healed in 1925. Meanwhile, Wittgenstein had also been a prisoner during the war. Serving in the Austrian army, he was captured by the Italians and spent his captivity composing the *Tractatus*. Russell wrote a glowing introduction for its English edition in 1922.

During the 1920s and early 1930s, Russell continued fundamental studies in logic and worked on scientific popularization. In philosophy he contributed to the Analytic Movement, with which he eventually lost sympathy. He wrote *The Analysis of Mind* (1921) and *The Analysis of Matter* (1927), describing a logical atomism that begins with the atomic data that are minimally required for a language. For the general reading public, he wrote with his usual clarity and precision *The ABC of Atoms* (1923), *The ABC of Relativity* (1925), and *The Scientific Outlook* (1931). When his brother died in 1931 he entered the House of Lords.

Russell spent the years 1938 to 1944 in the United States. In 1938 he lectured at the University of Chicago, where Rudolf Carnap and Charles Morris assisted him. In 1940 the City College of New York appointed him a professor, but the courts annulled the appointment because of his advocacy of trial marriages. This led Russell to refer to City College as a "satellite of the Vatican." The court decision reduced him to near poverty. In his writing he completed *An Inquiry into Meaning and Truth* (1940) and did preparatory work on his last major book, *Human Knowledge: Its Scope and Limits* (1948).

After 1944 Russell gained widespread fame and renewed his political activism. When he returned to England, Trinity College reappointed him. His book *A History of Western Philosophy* (1950) became a best seller. His appearances on the British Broadcasting Corporation (BBC) and the Nobel Prize for Literature in 1950 brought respectability. The political activism of his last years was directed against racial segregation, against nuclear warfare, and for peace. He and Albert Einstein warned the Pugwash Conference of Scientists of the dangers of nuclear warfare. In the 1960s he, French existentialist Jean-Paul Sartre, and Yugoslav historian Vladimir Dedijer protested the U.S. war effort in Vietnam, going so far as to establish an international war crimes tribunal in Stockholm.

In mathematics, Russell responded to the crisis in the foundations of mathematics at the turn of the 20th century. Building upon the work of Leibniz, Boole, Frege, and Peano, he and Whitehead set forth the logicist school which, as indicated earlier, held that mathematics can be reduced to logic. It was an attempt to construct classical mathematics on the basis of logic and set theory. Their logicist program rested on propositional functions and the calculus of relations suggested largely by DeMorgan and Peirce. The propositional functions dealt with attributes or, when they covered two or more arguments, relations.

Throughout, Russell sought to avoid and eliminate ambiguities. He improved upon the definition of real number given by Dedekind and in his theory of description sharpened the formulation of contextual definition, an idea that had originated with Jeremy Bentham. Russell also accepted and labored to remove paradoxes, or contradictions, from Cantor's set theory. Indeed, using the bare notions of set and element he discovered the Russell paradox, which he communicated to Frege in 1902. This paradox of the class or totality of all classes that are not members of themselves shook the world of logicians. After examining different theories, Russell believed that he had solved his paradox and that of Burali-Forti concerning the greatest ordinal with his theory of types, or levels (1908). In this theory members of a class had to be of the same type of thing. Russell postulated a formal system of axioms and principles of inferences for the theory of types of propositional functions to be the basis of mathematical logic. There were problems, however, with a proliferation of notation and with the axioms. In 1925 Paul Bernays proved that the fifth axiom could be derived from the others. In 1931 Frank Ramsey wisely urged dropping the axiom of reductibility, wherein every propositional function has a coextensive predicative one, because it introduced a needless complication.

117. From *Introduction to Mathematical Philosophy* (1919)*

(The Definition of Number)

BERTRAND RUSSELL

CHAPTER I
THE SERIES OF NATURAL NUMBERS

Mathematics is a study which, when we start from its most familiar portions, may be pursued in either of two opposite directions. The more familiar direction is constructive, towards gradually increasing complexity: from integers to fractions, real numbers, complex numbers; from addition and multiplication to differentiation and integration, and on to higher mathematics. The other direction, which is less familiar, proceeds, by analysing, to greater and greater abstractness and logical simplicity; instead of asking what can be defined and deduced from what is assumed to begin with, we ask instead what more general ideas and principles can be found, in terms of which what was our starting-point can be defined or deduced. It is the fact of pursuing this opposite direction that characterizes mathematical philosophy as opposed to ordinary mathematics. But it should be understood that the distinction is one, not in the subject matter, but in the state of mind of the investigator. Early Greek geometers, passing from the empirical rules of Egyptian land-surveying to the general propositions by which those rules were found to be justifiable, and thence to Euclid's axioms and postulates, were engaged in mathematical philosophy, according to the above definition; but when once the axioms and postulates had been reached, their deductive employment, as we find it in Euclid, belonged to mathematics in the ordinary sense. The distinction between mathematics and mathematical philosophy is one which depends upon the interest inspiring the research, and upon the stage which the research has reached; not upon the propositions with which the research is concerned.

We may state the same distinction in another way. The most obvious and easy things in mathematics are not those that come logically at the beginning; they are things that, from the point of view of logical deduction, come somewhere in the middle. Just as the easiest bodies to see are those that are neither very near nor very far, neither very small nor very great, so the easiest conceptions to grasp are those that are neither very complex nor very simple (using "simple" in a *logical* sense). And as we need two sorts of instruments, the telescope and the microscope, for the enlargement of our visual powers, so we need two sorts of instruments for the enlargement of our logical powers, one to take us forward to the higher mathematics, the other to take us backward to the logical foundations of the things that we are inclined to take for granted in math-

*Source: Bertrand Russell, *Introduction to Mathematical Philosophy* (1919), 1-28. This selection is reprinted by permission of George Allen & Unwin, Ltd., and the Macmillan Company.

ematics. We shall find that by analysing our ordinary mathematical notions we acquire fresh insight, new powers, and the means of reaching whole new mathematical subjects by adopting fresh lines of advance after our backward journey. It is the purpose of this book to explain mathematical philosophy simply and untechnically, without enlarging upon those portions which are so doubtful or difficult that an elementary treatment is scarcely possible. A full treatment will be found in *Principia Mathematica* [1910–13]; the treatment in the present volume is intended merely as an introduction.

To the average educated person of the present day, the obvious starting-point of mathematics would be the series of whole numbers, 1, 2, 3, 4, . . . etc. Probably only a person with some mathematical knowledge would think of beginning with 0 instead of 1, but we will presume this degree of knowledge; we will take as our starting-point the series: 0, 1, 2, 3, . . . n, $n + 1$, . . . and it is this series that we shall mean when we speak of the "series of natural numbers."

It is only at a high stage of civilisation that we could take this series as our starting-point. It must have required many ages to discover that a brace of pheasants and a couple of days were both instances of the number 2: the degree of abstraction involved is far from easy. And the discovery that 1 is a number must have been difficult. As for 0, it is a very recent addition; the Greeks and Romans had no such digit. If we had been embarking upon mathematical philosophy in earlier days, we should have had to start with something less abstract than the series of natural numbers, which we should reach as a stage on our backward journey. When the logical foundations of mathematics have grown more familiar, we shall be able to start further back, at what is now a late stage in our analysis. But for the moment the natural numbers seem to represent what is easiest and most familiar in mathematics.

But though familiar, they are not understood. Very few people are prepared with a definition of what is meant by "number," or "0" or "1." It is not very difficult to see that, starting from 0, any other of the natural numbers can be reached by repeated additions of 1, but we shall have to define what we mean by "adding 1," and what we mean by "repeated." These questions are by no means easy. It was believed until recently that some, at least, of these first notions of arithmetic must be accepted as too simple and primitive to be defined. Since all terms that are defined are defined by means of other terms, it is clear that human knowledge must always be content to accept some terms as intelligible without definition, in order to have a starting-point for its definitions. It is not clear that there must be terms which are *incapable* of definition: it is possible that, however far back we go in defining, we always *might* go further still. On the other hand, it is also possible that, when analysis has been pushed far enough, we can reach terms that really are simple, and therefore logically incapable of the sort of definition that consists in analysing. This is a question which it is not necessary for us to decide; for our purposes it is sufficient to observe that, since human powers are finite, the definitions known to us must always begin somewhere, with terms undefined for the moment, though perhaps not permanently.

All traditional pure mathematics, including analytical geometry, may be regarded as consisting wholly of propositions about the natural numbers. That is to say, the terms which occur can be defined by means of the natural numbers, and the propositions can be deduced from the properties of the natural numbers—with the addition, in each case, of the ideas and propositions of pure logic.

That all traditional pure mathematics

can be derived from the natural numbers is a fairly recent discovery, though it had long been suspected. Pythagoras, who believed that not only mathematics, but everything else could be deduced from numbers, was the discoverer of the most serious obstacle in the way of what is called the "arithmetising" of mathematics. It was Pythagoras who discovered the existence of incommensurables, and, in particular, the incommensurability of the side of a square and the diagonal. If the length of the side is 1 inch, the number of inches in the diagonal is the square root of 2, which appeared not to be a number at all. The problem thus raised was solved only in our own day, and was only solved *completely* by the help of the reduction of arithmetic to logic, which will be explained in following chapters. For the present, we shall take for granted the arithmetisation of mathematics, though this was a feat of the very greatest importance.

Having reduced all traditional pure mathematics to the theory of the natural numbers, the next step in logical analysis was to reduce this theory itself to the smallest set of premises and undefined terms from which it could be derived. This work was accomplished by Peano. He showed that the entire theory of the natural numbers could be derived from three primitive ideas and five primitive propositions in addition to those of pure logic. These three ideas and five propositions thus became, as it were, hostages for the whole of traditional pure mathematics. If they could be defined and proved in terms of others, so could all pure mathematics. Their logical "weight," if one may use such an expression, is equal to that of the whole series of sciences that have been deduced from the theory of the natural numbers; the truth of this whole series is assured if the truth of the five primitive propositions is guaranteed, provided, of course, that there is nothing erroneous in the purely logical apparatus which is also involved. The work of analysing mathematics is extraordinarily facilitated by this work of Peano's.

The three primitive ideas in Peano's arithmetic are: 0, number, successor.

By "successor" he means the next number in the natural order. That is to say, the successor of 0 is 1, the successor of 1 is 2, and so on. By "number" he means, in this connection, the class of the natural numbers.[1] He is not assuming that we know all the members of this class, but only that we know what we mean when we say that this or that is a number, just as we know what we mean when we say "Jones is a man," though we do not know all men individually.

The five primitive propositions which Peano assumes are:

(1) 0 is a number.
(2) The successor of any number is a number.
(3) No two numbers have the same successor.
(4) 0 is not the successor of any number.
(5) Any property which belongs to 0, and also to the successor of every number which has the property, belongs to all numbers.

The last of these is the principle of mathematical induction. We shall have much to say concerning mathematical induction in the sequel; for the present, we are concerned with it only as it occurs in Peano's analysis of arithmetic.

Let us consider briefly the kind of way in which the theory of the natural numbers results from these three ideas and five propositions. To begin with, we define 1 as "the successor of 0," 2 as "the successor of 1," and so on. We can obviously go on as long as we like with these definitions, since, in virtue of (2), every number that we reach will have a successor, and, in virtue of (3), this cannot be any of the numbers already defined, because, if it were, two different numbers would have the same successor; and in virtue of (4) none of the numbers we reach in the series of successors can be 0. Thus the series of successors gives us an endless series of continually new numbers. In virtue of (5) all numbers come in this series, which begins with 0 and travels on through successive successors: for *(a)* 0 belongs to this series, and *(b)* if a number n belongs to it, so does its successor, whence, by mathematical induction, every member belongs to the series.

Suppose we wish to define the sum of two numbers. Taking any number m, we define $m + 0$ as m, and $m + (n + 1)$ as the successor of $m + n$. In virtue of (5) this gives a definition of the sum of m and n, whatever number n may be. Similarly we can define the product of any two numbers. The reader can easily convince himself that any ordinary elementary proposition of arithmetic can be proved by means of our five premisses, and if he has any difficulty he can find the proof in Peano.

It is time now to turn to the considerations which make it necessary to advance beyond the standpoint of Peano, who represents the last perfection of the "arithmetisation" of mathematics, to that of Frege, who first succeeded in "logicising" mathematics, *i.e.* in reducing to logic the arithmetical notions which his predecessors had shown to be sufficient for mathematics. We shall not, in this chapter, actually give Frege's definition of number and of particular numbers, but we shall give some of the reasons why Peano's treatment is less final than it appears to be.

In the first place, Peano's three primitive ideas—namely, "0," "number," and "successor"—are capable of an infinite number of different interpretations, all of which will satisfy the five primitive propositions. We will give some examples.

(1) Let "0" be taken to mean 100, and let "number" be taken to mean the numbers from 100 onward in the series of natural numbers. Then all our primitive propositions are satisfied, even the fourth, for, though 100 is the successor of 99, 99 is not a "number" in the sense which we are now giving the word "number." It is obvious that any number may be substituted for 100 in this example.

(2) Let "0" have its usual meaning, but let "number" mean what we usually call "even numbers," and let the "successor" of a number be what results from adding two to it. The "1" will stand for the number two, "2" will stand for the number four, and so on; the series of "numbers" now will be. 0, two, four, six, eight . . . All Peano's five premises are satisfied still.

(3) Let "0" mean the number one, let "number" mean the set 1, $1/2$, $1/4$, $1/8$, $1/16$, . . . and let "successor" mean "half." Then all Peano's five axioms will be true of this set.

It is clear that such examples might be multiplied indefinitely. In fact, given any series $x_0, x_1, x_2, x_3, \ldots x_n, \ldots$ which is endless, contains no repetitions, has a beginning, and has no terms that cannot be reached from the beginning in a finite number of steps, we have a set of terms verifying Peano's axioms. This is easily seen, though the formal proof is somewhat long. Let "0" mean x_0, let "number" mean the whole set of terms, and let the "successor" of x_n mean x_{n+1}. Then

(1) "0 is a number," *i.e.* x_0 is a member of the set.

(2) "The successor of any number is a number," *i.e.* taking any term x_n in the set, x_{n+1} is also in the set.

(3) "No two numbers have the same successor," *i.e.* if x_m and x_n are two different members of the set, x_{m+1} and x_{n+1} are different; this results from the fact that (by hypothesis) there are no repetitions in the set.

(4) "0 is not the successor of any number," *i.e.* no term in the set comes before x_0.

(5) This becomes: Any property which belongs to x_0, and belongs to x_{n+1} provided it belongs to x_n, belongs to all the x's.

This follows from the corresponding property for numbers.

A series of the form $x_0, x_1, x_2, \ldots x_n, \ldots$ in which there is a first term, a successor to each term (so that there is no last term), no repetitions, and every term can be reached from the start in a finite number of steps, is called a *progression*. Progressions are of great importance in the principles of mathematics. As we have just seen, every progression verifies Peano's five axioms. It can be proved, conversely, that every series which verifies Peano's five axioms is a progression. Hence these five axioms may be used to define the class of progressions: "progressions" are "those series which verify these five axioms." Any progression may be taken as the basis of pure mathematics: we may give the name "0" to its first term, the name "number" to the whole set of its terms, and the name "successor" to the next in the progression. The progression need not be composed of numbers: it may be composed of points in space, or moments of time, or any other terms of which there is an infinite supply. Each different progression will give rise to a different interpretation of all the propositions of traditional pure mathematics; all these possible interpretations will be equally true.

In Peano's system there is nothing to enable us to distinguish between these different interpretations of his primitive ideas. It is assumed that we know what is meant by "0," and that we shall not suppose that this symbol means 100 or Cleopatra's Needle or any of the other things that it might mean.

This point, that "0" and "number" and "successor" cannot be defined by means of Peano's five axioms, but must be independently understood, is important. We want our numbers not merely to verify mathematical formulae, but to apply in the right way to common objects. We want to have ten fingers and two eyes and one nose. A system in which "1" meant 100, and "2" meant 101, and so on, might be all right for pure mathematics, but would not suit daily life. We want "0" and "number" and "successor" to have meanings which will give us the right allowance of fingers and eyes and noses. We have already some knowledge (though not sufficiently articulate or analytic) of what we mean by "1" and "2" and so on, and our use of numbers in arithmetic must conform to this knowledge. We cannot secure that this shall be the case by Peano's method; all that we can do if we adopt his method, is to say "we know what we mean by '0' and 'number' and 'successor,' though we cannot explain what we mean in terms of other simpler concepts." It is quite legitimate to say this when we must, and at *some* point we all must; but it is the object of mathematical philosophy to put off saying it as long as possible. By the logical theory of arithmetic we are able to put it off for a very long time.

It might be suggested that, instead of setting up "0" and "number" and "successor" as terms of which we know the meaning although we cannot define them, we might let them stand for *any* three terms that verify Peano's five axioms. They will then no longer be terms which have a meaning that is definite though undefined: they will be "variables," terms concerning which we make certain hypotheses, namely, those stated in the five axioms, but which are otherwise undetermined. If we adopt this plan, our theorems will not be proved concerning an ascertained set of terms called "the natural numbers," but concerning all sets of terms having certain properties. Such a procedure is not fallacious; indeed for certain purposes it represents a valuable generalisation. But from two points of view it fails to give an adequate basis for arithmetic. In the first place, it does not enable us to know whether there are any sets of terms verifying Peano's axioms; it does not even give the faintest suggestion of any way of discovering whether there are such sets. In the second place, as already observed, we want our numbers to be such as can be used for counting common objects, and this requires that our numbers should have a *definite* meaning, not merely that they should have certain formal properties. This definite meaning is defined by the logical theory of arithmetic.

CHAPTER II
DEFINITION OF NUMBER

The question "What is a number?" is one which has been often asked, but has only been correctly answered in our own time. The answer was given by Frege in 1884, in his *Grundlagen der Arithmetik*.[2] Although this book is quite short, not difficult, and of the very highest importance, it attracted almost no attention, and the definition of number which it contains remained practically unknown until it was rediscovered by the present author in 1901.

In seeking a definition of number, the first thing to be clear about is what we may call the grammar of our inquiry. Many philosophers, when attempting to define number, are really setting to work to define plurality, which is quite a different thing. *Number* is what is characteristic of numbers, as *man* is what is characteristic of men. A plurality is not an instance of number, but of some particular number. A trio of men, for example, is an instance of the number 3, and the number 3 is an instance of number; but the trio is not an instance of number. This point may seem elementary and scarcely worth mentioning; yet it has proved too subtle for the philosophers, with few exceptions.

A particular number is not identical with any collection of terms having that number: the number 3 is not identical with the trio consisting of Brown, Jones, and Robinson. The number 3 is something which all trios have in common, and which distinguishes them from other collections. A number is something that characterises certain collections, namely, those that have that number.

Instead of speaking of a "collection," we shall as a rule speak of a "class," or sometimes a "set." Other words used in mathematics for the same thing are "aggregate" and "manifold." We shall have much to say later on about classes. For the present, we will say as little as possible. But there are some remarks that must be made immediately.

A class or collection may be defined in two ways that at first sight seem quite distinct. We may enumerate its members, as when we say, "The collection I mean is Brown, Jones, and Robinson." Or we may mention a defining property, as when we speak of "mankind" or "the inhabitants of London." The definition which enumerates is called a definition by "extension," and the one which mentions a defining property is called a definition by "intension." Of these two kinds of definition, the one by intension is logically more fundamental. This is shown by two considerations: (1) that the extensional definition can always be reduced to an intensional one; (2) that the intensional one often cannot even theoretically be reduced to the extensional one. Each of these points needs a word of explanation.

(1) Brown, Jones and Robinson all of them possess a certain property which is possessed by nothing else in the whole universe, namely, the property of being either Brown or Jones or Robinson. This property can be used to give a definition by intension of the class consisting of Brown and Jones and Robinson. Consider such a formula as "x is Brown or x is Jones or x is Robinson." This formula will be true for just three x's, namely, Brown and Jones and Robinson. In this respect it resembles a cubic equation with its three roots. It may be taken as assigning a property common to the members of the class consisting of these three men, and peculiar to them. A similar treatment can obviously be applied to any other class given in extension.

(2) It is obvious that in practice we can often know a great deal about a class without being able to enumerate

its members. No one man could actually enumerate all men, or even all the inhabitants of London, yet a great deal is known about each of these classes. This is enough to show that definition by extension is not *necessary* to knowledge about a class. But when we come to consider infinite classes, we find that enumeration is not even theoretically possible for beings who only live for a finite time. We cannot enumerate all the natural numbers: they are 0, 1, 2, 3, *and so on*. At some point we must content ourselves with "and so on." We cannot enumerate all fractions or all irrational numbers, or all of any other infinite collection. Thus our knowledge in regard to all such collections can only be derived from a definition by intension.

These remarks are relevant, when we are seeking the definition of number, in three different ways. In the first place, numbers themselves form an infinite collection, and cannot therefore be defined by enumeration. In the second place, the collections having a given number of terms themselves presumably form an infinite collection: it is to be presumed, for example, that there are an infinite collection of trios in the world, for if this were not the case the total number of things in the world would be finite, which, though possible, seems unlikely. In the third place, we wish to define "number" in such a way that infinite numbers may be possible; thus we must be able to speak of the number of terms in an infinite collection, and such a collection must be defined by intension, *i.e.* by a property common to all its members and peculiar to them.

For many purposes, a class and a defining characteristic of it are practically interchangeable. The vital difference between the two consists in the fact that there is only one class having a given set of members, whereas there are always many different characteristics by which a given class may be defined. Men may be defined as featherless bipeds, or as rational animals, or (more correctly) by the traits by which Swift delineates the Yahoos. It is this fact that a defining characteristic is never unique which makes classes useful; otherwise we could be content with the properties common and peculiar to their members.[3] Any one of these properties can be used in place of the class whenever uniqueness is not important.

Returning now to the definition of number, it is clear that number is a way of bringing together certain collections, namely, those that have a given number of terms. We can suppose all couples in one bundle, all trios in another, and so on. In this way we obtain various bundles of collections, each bundle consisting of all the collections that have a certain number of terms. Each bundle is a class whose members are collections, *i.e.* classes; thus each is a class of classes. The bundle consisting of all couples, for example, is a class of classes: each couple is a class with two members, and the whole bundle of couples is a class with an infinite number of members, each of which is a class of two members.

How shall we decide whether two collections are to belong to the same bundle? The answer that suggests itself is: "Find out how many members each has, and put them in the same bundle if they have the same number of members." But this presupposes that we have defined numbers, and that we know how to discover how many terms a collection has. We are so used to the operation of counting that such a presupposition might easily pass unnoticed. In fact, however, counting, though familiar, is logically a very complex operation; moreover it is only available, as a means of discovering how many terms a collection has, when the collection is finite. Our definition of number must not assume in advance that all numbers are finite; and we cannot in any case, without a vicious circle, use counting to define numbers, because numbers are used in counting. We need, therefore, some other method of deciding when two collections have the same number of terms.

In actual fact, it is simpler logically to find out whether two collections have the same number of terms than it is to define what that number is. An illustration will make this clear. If there were no polygamy or polyandry anywhere in the world, it is clear that the number of husbands living at any moment would be exactly the same as the number of wives. We do not need a census to assure us of this, nor do we need to know what is the actual number of husbands and of wives. We know the number must be the same in both collections, because each husband has one wife and each wife has one husband. The relation of husband and wife is what is called "one-one."

A relation is said to be "one-one" when, if x has the relation in question to y, no other term x' has the same relation to y, and x does not have the same relation to any term y' other than y. When only the first of these two conditions is fulfilled, the relation is called "one-many"; when only the second is fulfilled, it is called "many-one." It should be observed that the number 1 is not used in these definitions.

In Christian countries, the relation of husband to wife is one-one; in Mohammedan countries it is one-many; in Tibet it is many-one. The relation of father to son is one-many; that of son to father is many-one, but that of eldest son to father is one-one. If n is any number, the relation of n to $n + 1$ is one-one; so is the relation of n to $2n$ or to $3n$. When we are considering only positive numbers, the relation of n to n^2 is one-one; but when negative numbers are admitted, it becomes two-one, since n *and* $-n$ have the same square. These instances should suffice to make clear the notions of one-one, one-many, and many-one relations, which play a great part in the principles of mathematics, not only in relation to the definition of numbers, but in many other connections.

Two classes are said to be "similar" when there is a one-one relation which correlates the terms of the one class each with one term of the other class, in the same manner in which the relation of marriage correlates husbands with wives. A few preliminary definitions will help us to state this definition more precisely. The class of those terms that have a given relation to something or other is called the *domain* of that relation: thus fathers are the domain of the relation of father to child, husbands are the domain of the relation of husband to wife, wives are the domain of the relation of wife to husband, and husbands and wives together are the domain of the relation of marriage. The relation of wife to husband is called the *converse* of the relation of husband to wife. Similarly *less* is the converse of *greater, later* is the converse of *earlier,* and so on. Generally, the converse of a given relation is that relation which holds between y and x whenever the given relation holds between x and y. The *converse domain* of a relation is the domain of its converse: thus the class of wives is the converse domain of the relation of husband to wife. We may now state our definition of similarity as follows:—

> *One class is said to be "similar" to another when there is a one-one relation of which the one class is the domain, while the other is the converse domain.*

It is easy to prove (1) that every class is similar to itself, (2) that if a class α is similar to a class β, then β is similar to α, (3) that if α is similar to β and β to γ, then α is similar to γ. A relation is said to be *reflexive* when it possesses the first of these properties, *symmetrical* when it possesses the second, and *transitive* when it possesses the third. It is obvious that a relation which is symmetrical and transitive must be reflexive throughout its domain. Relations which possess these properties are an important kind, and it is worth while to note that similarity is one of this kind of relations.

It is obvious to common sense that two finite classes have the same number of terms if they are similar, but not otherwise. The act of counting consists in establishing a one-one correlation between the set of objects counted and the natural numbers (excluding 0) that are used up in the process. Accordingly common sense concludes that there are as many objects in the set to be counted as there are numbers up to the last number used in the counting. And we also know that, so long as we confine ourselves to finite numbers, there are just n numbers from 1 up to n. Hence it follows that the last number used in counting a collection is the number of terms in the collection, provided the collection is finite. But this result, besides being only applicable to finite collections, depends upon and assumes the fact that two classes which are similar have the same number of terms; for what we do when we count (say) 10 objects is to show that the set of these objects is similar to the set of numbers 1 to 10. The notion of similarity is logically presupposed in the operation of counting, and is logically simpler though less familiar. In counting, it is necessary to take the objects counted in a certain order, at first, second, third, etc., but order is not of the essence of number: it is an irrelevant addition, an unnecessary complication from the logical point of view. The notion of similarity does not demand an order: for example, we saw that the number of husbands is the same as the number of wives, without having to establish an order of precedence among them. The notion of similarity also does not require that the classes which are similar should be finite. Take, for example, the natural numbers (excluding 0) on the one hand, and the fractions which have 1 for their numerator on the other hand: it is obvious that we can correlate 2 with ½, 3 with ⅓, and so on, thus proving that the two classes are similar.

We may thus use the notion of "similarity" to decide when two collections are to belong to the same bundle, in the sense in which we were asking this question earlier in this chapter. We want to make one bundle containing the class that has no members: this will be for the number 0. Then we want a bundle of all the classes that have one member: this will be for the number 1. Then, for the number 2, we want a bundle consisting of all couples; then one of all trios; and so on. Given any collection, we can define the bundle it is to belong to as being the class of all those collections that are "similar" to it. It is very easy to see that if (for example) a collection has three members, the class of all those collections that are similar to it will be the class of trios. And whatever number of terms a collection may have, those collections that are "similar" to it will have the same number of terms. We may take this as a *definition* of "having the same number of terms." It is obvious that it gives results conformable to usage so long as we confine ourselves to finite collections.

So far we have not suggested anything in the slightest degree paradoxical. But when we come to the actual definition of numbers we cannot avoid what must at first sight seem a paradox, though this impression will soon wear off. We naturally think that the class of couples (for example) is something different from the number 2. But there is no doubt about the class of couples: it is indubitable and not difficult to define, whereas the number 2, in any other sense, is a metaphysical entity about which we can never feel that it exists or that we have tracked it down. It is therefore more prudent to content ourselves with the class of couples, which we are sure of, than to hunt for a problematical number 2 which must always remain elusive. Accordingly we set up the following definition:—

> *The number of a class is the class of all those classes that are similar to it.*

Thus the number of a couple will be

the class of all couples. In fact, the class of all couples will *be* the number 2, according to our definition. At the expense of a little oddity, this definition secures definiteness and indubitableness; and it is not difficult to prove that numbers so defined have all the properties that we expect numbers to have.

We may now go on to define numbers in general as any one of the bundles into which similarity collects classes. A number will be a set of classes such as that any two are similar to each other, and none outside the set are similar to any inside the set. In other words, a number (in general) is any collection which is the number of one of its members; or, more simply still:

A number is anything which is the number of some class.

Such a definition has a verbal appearance of being circular, but in fact it is not. We define "the number of a given class" without using the notion of number in general; therefore we may define number in general in terms of "the number of a given class" without committing any logical error.

Definitions of this sort are in fact very common. The class of fathers, for example, would have to be defined by first defining what it is to be the father of somebody; then the class of fathers will be all those who are somebody's father. Similarly if we want to define square numbers (say), we must first define what we mean by saying that one number is the square of another, and then define square numbers as those that are the squares of other numbers. This kind of procedure is very common, and it is important to realise that it is legitimate and even often necessary.

NOTE

1. We shall use "number" in this sense in the present chapter. Afterwards the word will be used in a more general sense.

2. The same answer is given more fully and with more development in his *Grundgesetze der Arithmetik,* vol. i., 1893.

3. As will be explained later, classes may be regarded as logical fictions, manufactured out of defining characteristics. But for the present it will simplify our exposition to treat classes as if they were real.

Chapter IX

The Early 20th Century to 1932

Introduction by
HELENA M. PYCIOR
University of Wisconsin-Milwaukee

Introduction. Twentieth-century mathematics began optimistically. In 1900 David Hilbert proposed 23 problems for solution. His summons to mathematicians was based on a recognition of the importance of problems in the development of mathematics as well as a belief in the completeness of mathematical systems—that is, that any proposition that could be formulated in a particular system was decidable, or could be proven true or false within that system. "This conviction of the solvability of every mathematical problem," Hilbert declared, "is a powerful incentive to the worker. We hear within us the perpetual call: There is the problem. Seek its solution. You can find it by pure reason, for in mathematics there is no *ignorabimus.*"

Hilbert's second problem concerned the consistency of arithmetic: given the axioms of arithmetic, mathematicians should prove "that a finite number of logical steps based upon them can never lead to contradictory results." As he saw it, the proof of the internal consistency of arithmetic was of utmost significance. According to the formalist approach to mathematics, which Hilbert espoused, mathematics was built on uninterpreted laws and meaningless symbols, and offered, in particular, no guarantee that mathematical concepts even existed in a traditional sense. Rather boldly he therefore tied mathematical existence to consistency, maintaining that a proof of the internal consistency of arithmetic was a proof of the mathematical existence of the system of real numbers.

In 1931 Hilbert's formalism received a serious setback, from which it has never completely recovered. In that year Kurt Gödel disposed of Hilbert's conjectures of both the completeness and consistency of formal mathematical systems. Put simply, in a proof once described as an "amazing intellectual symphony," Gödel proved that arithmetic contains formally undecidable propositions and that there cannot be formulated within the arithmetical calculus a proof of the consistency of arithmetic.

Although revealing some disturbing limits to the formalist approach, Gödel's proof did not deal a deathblow to the foundations of mathematics partially because formalism was but one of three approaches to the foundations of mathematics explored during the first third of the 20th century. While Hilbert pursued the formalist approach, Bertrand Russell and Alfred North Whitehead adopted the logicist approach and L. E. J. Brouwer, the intuitionist one. The logicists viewed mathematics as a branch of logic and tried therefore to derive mathematics from a few fundamental logical concepts and principles. Like formalism logicism was beset by difficulties which prevented its universal acceptance by the mathematical community. In 1901, for example, Russell constructed an antinomy, an apparent contradiction, within basic set theory upon which logic was based. Communicating this paradox to Frege in 1902, he raised the problem presented by the set of all sets which are not members of themselves. Was this set a member of itself? As Russell pointed out, either answer to the preceding question led to a contradiction. To avoid such paradoxes Russell and Whitehead incorporated a theory of types—too sophisticated for explanation here—into *Principia Mathematica* (1910–1913), their three-volume exposition of the logicist approach. Some of their contemporaries criticized not only this theory but also such other fundamental assumptions of the authors as the axiom of infinity and the axiom of choice. (Used but seldom explicitly stated in earlier mathematics, the axiom of choice had in 1904 formed the basis of Ernst Zermelo's controversial proof of the well-ordering theorem, another of Hilbert's problems.)

Brouwer, the founder of the third school of mathematics, intuitionism, was a Dutch mathematician whose rebellion against formalism and logicism went hand in hand with a romantic rejection of industrialization and intellectualism. Dismissing formalist mathematics as a meaningless game, Brouwer maintained that mathematics was the development of the intuition of time and thus inherently meaningful. From the notion of time, he argued, the natural numbers were derived; and from the natural numbers the rest of mathematics was to be derived in a finite number of steps and operations. While Hilbert vainly related valid existence to consistency, Brouwer made this existence synonymous with constructibility. Brouwer, in fact, adhered strictly to the principle of finite constructibility, rejecting the law of the excluded middle for infinite sets. Repudiating this law and hence the indirect method of proof, Brouwer and other intuitionists attempted—with but partial success—the reconstruction of classical mathematics.

While dissatisfaction with formalism and logicism stemmed from problems internal to each of these approaches, many early 20th-century mathematicians spurned the intuitionist approach because of its relative barrenness. Uncomfortable with all three many working mathematicians continued to devote themselves to doing mathematics and left the foundational problems to others.

While some mathematicians struggled with the question of the very foundations of mathematics, physicists were radically revising the foundations of physics during the first third of the 20th century. In 1905 Albert Einstein proposed the special theory of relativity which assumed, despite that designation, the constancy of the speed of light and the restricted principle of relativity. Acceptance of these assumptions, he explained, involved replacing Newton's (or the common sense) notions of absolute time, space, and mass with notions of time, space, and mass relative to coordinate systems. Einstein demonstrated that a rod with a certain length as measured in a coordinate system in which it was at rest would have a smaller length in a coordinate system in which it was moving. That is, rods do not have absolute lengths, only lengths relative to specific coordinate systems or reference frames. Also in 1905, Einstein contributed to the formulation of the quantum theory, a revolutionary theory of submicroscopic particles and their associated waves, which was elaborated during the first third of the century by Max Planck, Werner Heisenberg, Erwin Schrödinger, and others.

At roughly the same time as this re-examination of the foundations of mathematics and physics, there were also criticisms in the public arena of many of the foundations of Western civilization itself. This criticism was largely prompted by the devastation of World War I, in which over eight million men were killed and twice that number wounded. Still, in 1914, most people, including intellectuals, had welcomed the war. This initial enthusiasm soon yielded to a more somber mood as the war dragged on, casualties mounted, and food and fuel grew scarce. Wartime pressures as well as long-term, internal historical factors led to a momentous revolution in Russia in 1917—which overthrew the Romanoff autocracy and established in power a Communist dictatorship under Lenin. When the war finally ended in 1918 the Paris peace settlement, notably the Treaty of Versailles, was harsh. Even so, not all the victors were satisfied; Italian leaders, for example, wanted greater compensation for Italy's wartime efforts. In the Versailles Treaty Germany was charged with guilt for the war, was stripped of its colonies, and had to pay billions of marks in reparations. These terms wreaked havoc on the German economy. The terms of the Paris settlement and the "red scare" of the 1920s, the latter symbolized in a Communist Russia, helped move Europe toward authoritarian governments. Italy in 1922 became a Fascist dictatorship under Benito Mussolini. Coupled with the Great Depression of 1929, the same conditions fostered the rise in the early 1930s of the Nazi dictatorship under Adolf Hitler.

At the turn into the 20th-century avant-garde writers, artists, and musicians had questioned the culture and directions of the West. This group grew into the "Lost Generation" which after giving of itself spiritually and physically had come away from World War I bitterly disillusioned with the Western liberal tradition which had permitted (perhaps spawned) such a war. World War I, the generation's prominent intellectuals claimed, demonstrated that Western civilization—its values and traditions—was dead. Ignoring the constructive aspects of early 20th-century physics, a few even cited the departure from classical physics as further proof of the demise of Western civilization. Some of the disenchanted embraced communism. Others turned to mysticism and traditional religions to fill the void left by the displaced secular liberal creed. Rebelling against the old moral standards, still others adopted, and oftentimes flaunted, a less rigid sexual morality. Many of those who advocated changing norms of behavior appealed to the ideas of the Viennese physician Sigmund Freud, the father of psychoanalysis, who grew famous during the post-war period.

On a different note, the immediate post-war period witnessed the extension of suffrage to American and some European women. In addition, in the first third of the 20th century women were admitted to institutions of higher education in unprecedented numbers. As the condition of women in general improved, so did that of female mathematicians and scientists—albeit slowly. Having obtained the higher education denied most of their female predecessors, the Polish-born chemist and physicist Marie Sklodowska Curie and the German mathematician Emmy Noether joined the ranks of the leading scientists of the early 20th century.

While some of their contemporaries wrestled with the problem of the foundations of mathematics, Noether and other early 20th-century mathematicians continued the 19th-century drive toward ever greater generality and abstraction. At the turn of the century Henri Lebesgue introduced a generalization of the Riemann integral, based on measure-theoretic notions. Noether, building upon the work of Dedekind, offered an axiomatic development of a general theory of ideals and, later in her career, building upon the work of Hamilton and others, probed the structure of noncommutative algebras.

This stimulating period in the history of mathematics—including the crisis in its foundations, the emergence of a major female mathematician, greater abstraction, and the coming of age of American mathematics—was interrupted in 1933 by the dispersal of the Göttingen school of mathematics because of Hitler's decree against Jewish professors. Hitler's rise to power in Germany in 1933 boded ill not only for the mathematical community but for the entire world.

SUGGESTIONS FOR FURTHER READING

Max Black, *The Nature of Mathematics: A Critical Survey.* Totowa, New Jersey: Littlefield, Adams, 1965.

Joseph W. Dauben, "Mathematicians and World War I: The International Diplomacy of G. H. Hardy and Gösta Mittag-Leffler as Reflected in Their Personal Correspondence." *Historia Mathematica* 7 (1980): 261-288.

Thomas Hawkins, *Lebesgue's Theory of Integration: Its origins and Development.* Madison: University of Wisconsin Press, 1970.

Henri Lebesgue, *Measure and the Integral.* Edited by Kenneth O. May. San Francisco: Holden-Day, 1966.

Edith H. Luchins and Abraham S. Luchins, "Logicism." *Scripta Mathematica* 27 (1965): 223-243.

Jerome H. Manheim, *The Genesis of Point Set Topology.* New York: Macmillan, 1964.

Ernest Nagel and James R. Newman, *Gödel's Proof.* New York: New York University Press, 1974.

Constance Reid, *Hilbert.* New York: Springer-Verlag, 1970.

Walter P. Van Stigt, "The Rejected Parts of Brouwer's Dissertation on the Foundations of Mathematics," *Historia Mathematica* 6 (1979): 385-404.

Hermann Weyl, "David Hilbert and His Mathematical Work." *Bulletin of the American Mathematical Society* 50 (1944): 612-654.

Hermann Weyl, "Emmy Noether." *Scripta Mathematica* 3 (1935): 201-220.

Chapter IX
The Early 20th Century to 1932

Section A
Creativity and the Paris Problems

(JULES-)HENRI POINCARÉ (1854–1912)

The dominant figure in the world of mathematics at the turn of the 20th century was the French theoretician Henri Poincaré. He was, like Gauss who exerted similar influence in the early 19th century, a universal mathematician who contributed substantially to mathematical physics, theoretical astronomy, and the philosophy of science. Poincaré came from an upper bourgeous family; his father, Leon, was a physician and university professor. The family had a distinguished record of service to French government. Raymond Poincaré, a first cousin of Henri, was several times president of the Third French Republic.

Henri Poincaré was a precocious child, who was ambidexterous, nearsighted, and suffered for a time from diphtheria. His able mother tutored him before he entered elementary school. In elementary school he excelled in written composition and during adolescence first displayed a deep interest and ability in mathematics. At the lycée (secondary school) in Nancy he won top honors in mathematics, including first prizes in the *concours général,* prize competitions for students from all French lycées. A teacher at Nancy reportedly referred to him as a "monster of mathematics." Before the lycée studies were completed Poincaré's outstanding scholarly abilities were evident. He had displayed an exceptionally retentive memory and a faculty to visualize experiments, to perform in his head complex mathematical computations, and to write papers quickly without needing extensive revisions. He was, however, undistinguished in physical exercise and art. During the Franco-Prussian War (1870–71) he learned German and saw firsthand the suffering caused to soldiers and the populace in the patients his father treated.

In 1873 Poincaré received a first place in the entrance examinations for École Polytechnique and fifth for École Normale Supérieure. He chose to attend the Polytechnique. After graduating in 1875, he enrolled in the École National Supérieure des Mines. He worked briefly as an engineer before receiving his doctorate in mathematics in 1879 for a thesis on differential equations entitled "sur les propriétés des fonctions définies par des équations aux dérivées partielles."

Poincaré pursued a university career. After being appointed an instructor in mathematical analysis at the University of Caen in 1879, he went in 1881 to the University of Paris, remaining there for the rest of his life.

At the University of Paris he taught mathematical analysis from 1881 to 1885, supervised a course on experimental and physical mechanics from 1885 to 1886, was professor of mathematical physics and probability from 1886 to 1896, and was professor of mathematical astronomy and celestial mechanics from 1896 until his death. He also worked at the École Polytechnique teaching analysis from 1883 to 1897 and was professor of general astronomy from 1904 to 1908. He taught electricity at École Supérieure des Postes et Telegraphes from 1902 to 1912.

As his prestige and influence in French science increased, Poincaré received numerous honors. In 1887 the Paris Academy of Sciences elected him a member, and in 1889 King Oscar II of Sweden awarded him a prize for his powerful solution using differential equations to the n-body problem, an extension of the classical three-body problem, (*e.g.*, the gravitational interaction of the Sun, Moon, and Earth). The solution proved only partially correct. Also, in 1889 he was named a knight of the French Legion of Honor. In 1904 he visited the United States to lecture at St. Louis Exposition. In 1906 the Paris Academy of Sciences elected him president, and two years later he was made a member of the French Academy, the highest honor given a French author.

After 1900 Poincaré turned his mastery of French prose more and more to describing the importance of science and mathematics to the general public. With his clarity of mind and vivid writing style he proved an excellent as well as prolific expositor and commentator. (He wrote over 400 books, articles, and notes on scientific subjects.) Among his books were *Science and Hypothesis* (1903), *The Value of Science* (1905), *Science and Method* (1908), and *Final Thoughts* (1913). Thousands of people from all walks of life read his writings. Some were translated into English, German, Hungarian, Japanese, Spanish, and Swedish.

Poincaré made fundamental contributions to many branches of mathematics. Perhaps most importantly he discovered by the age of 30 automorphic functions of one complex variable and later developed a general theory for them that left little for his successors to do. He called automorphic functions Fuchsians, after the German mathematician Immanuel L. Fuchs, a founder of the theory of differential equations, and associated their domain with transformations arising in non-Euclidean geometries. In this research he worked independently and ahead of Felix Klein. After reading papers of Riemann and Weierstrass he turned to another branch of function theory, working to generalize particular cases of Abelian functions and proving the "complete reducibility" theorem.

Throughout his career the theory of differential equations and its application to the dynamics were central to Poincaré's mathematical research. In an extremely creative period from 1880 to 1883 he invented the qualitative theory of differential equations. His summary of the application of such new methods to dynamics in *Les Méthodes nouvelles de la méchanique céleste* (3 vols., 1892, 1893, and 1899) inaugurated the rigorous treatment of celestial mechanics.

Poincaré had many other achievements. He probably first used Cantor's set theory in mathematical analysis. In pursuing "continuity" he prepared *Analysis situs* (1895), a pioneering systematic treatment of algebraic topology. In it he generalized the Euler theorem for polyhedra (now known as the Euler-Poincaré formula). A student of Hermite, he also applied himself to the theory of numbers, recasting Gauss's conception of binary quadratic forms into geometric form (1901–04). In 1905 and 1906 he obtained results in the special theory of relativity independently of Einstein.

In examining the psychology of mathematical discovery and invention, Poincaré stressed the subconscious. He believed that the apparently sudden illumination preceding mathematical creation occurs only after long subconscious incubation. His belief that some mathematical induction is *a priori* and independent of logic anticipated the later doctrines of the modern intuitionist school in the foundations of mathematics. Indeed he often poked fun at the program of formalization by disciples of Peano and Russell and less often at that of Hilbert and his followers. The Bourbaki group subsequently expanded upon his foundations research.

118. From *Science and Method* (1908)*

HENRI POINCARÉ

MATHEMATICAL CREATION

The genesis of mathematical creation is a problem which should intensely interest the psychologist. It is the activity in which the human mind seems to take least from the outside world, in which it acts or seems to act only of itself and on itself, so that in studying the procedure of geometric thought we may hope to reach what is most essential in man's mind.

This has long been appreciated, and some time back the journal called *L'enseignement mathématique,* edited by Laisant and Fehr, began an investigation of the mental habits and methods of work of different mathematicians. I had finished the main outlines of this article when the results of that inquiry were published, so I have hardly been able to utilize them and shall confine myself to saying that the majority of witnesses confirm my conclusions; I do not say all, for when the appeal is to universal suffrage unanimity is not to be hoped.

A first fact should surprise us, or rather would surprise us if we were not so used to it. How does it happen there are people who do not understand mathematics? If mathematics invokes only the rules of logic, such as are accepted by all normal minds; if its evidence is based on principles common to all men, and that none could deny without being mad, how does it come about that so many persons are here refractory?

That not every one can invent is nowise mysterious. That not every one can retain a demonstration once learned may also pass. But that not every one can understand mathematical reasoning when explained appears very surprising when we think of it. And yet those who can follow this reasoning only with difficulty are in the majority: that is undeniable, and will surely not be gainsaid by the experience of secondary-school teachers.

And further: how is error possible in mathematics? A sane mind should not be guilty of a logical fallacy, and yet there are very fine minds who do not trip in brief reasoning such as occurs in the ordinary doings of life, and who are incapable of following or repeating without error the mathematical demonstrations which are longer, but which after all are only an accumulation of

*Source: H. Poincaré, *The Foundations of Science,* trans. by George Bruce Halsted (1913), 383-394. This selection is reprinted by permission of the Science Press.

brief reasonings wholly analogous to those they make so easily. Need we add that mathematicians themselves are not infallible?

The answer seems to me evident. Imagine a long series of syllogisms, and that the conclusions of the first serve as premises of the following: we shall be able to catch each of these syllogisms, and it is not in passing from premises to conclusion that we are in danger of deceiving ourselves. But between the moment in which we first meet a proposition as conclusion of one syllogism, and that in which we reencounter it as premise of another syllogism occasionally some time will elapse, several links of the chain will have unrolled; so it may happen that we have forgotten it, or worse, that we have forgotten its meaning. So it may happen that we replace it by a slightly different proposition, or that, while retaining the same enunciation, we attribute to it a slightly different meaning, and thus it is that we are exposed to error.

Often the mathematician uses a rule. Naturally he begins by demonstrating this rule; and at the time when this proof is fresh in his memory he understands perfectly its meaning and its bearing, and he is in no danger of changing it. But subsequently he trusts his memory and afterward only applies it in a mechanical way; and then if his memory fails him, he may apply it all wrong. Thus it is, to take a simple example, that we sometimes make slips in calculation because we have forgotten our multiplication table.

According to this, the special aptitude for mathematics would be due only to a very sure memory or to a prodigious force of attention. It would be a power like that of the whist-player who remembers the cards played; or, to go up a step, like that of the chess-player who can visualize a great number of combinations and hold them in his memory. Every good mathematician ought to be a good chess-player, and inversely; likewise he should be a good computer. Of course that sometimes happens; thus Gauss was at the same time a geometer of genius and a very precocious and accurate computer.

But there are exceptions; or rather I err; I can not call them exceptions without the exceptions being more than the rule. Gauss it is, on the contrary, who was an exception. As for myself, I must confess, I am absolutely incapable even of adding without mistakes. In the same way I should be but a poor chess-player; I would perceive that by a certain play I should expose myself to a certain danger; I would pass in review several other plays, rejecting them for other reasons, and then finally I should make the move first examined, having meantime forgotten the danger I had foreseen.

In a word, my memory is not bad, but it would be insufficient to make me a good chess-player. Why then does it not fail me in a difficult piece of mathematical reasoning where most chess-players would lose themselves? Evidently because it is guided by the general march of the reasoning. A mathematical demonstration is not a simple juxtaposition of syllogisms, it is syllogisms *placed in a certain order,* and the order in which these elements are placed is much more important than the elements themselves. If I have the feeling, the intuition, so to speak, of this order, so as to perceive at a glance the reasoning as a whole, I need no longer fear lest I forget one of the elements, for each of them will take its allotted place in the array, and that without any effort of memory on my part.

It seems to me then, in repeating a reasoning learned, that I could have invented it. This is often only an illusion; but even then, even if I am not so gifted as to create it by myself, I myself reinvent it in so far as I repeat it.

We know that this feeling, this intuition of mathematical order, that makes us divine hidden harmonies and relations, can not be possessed by every one. Some will not have either this delicate feeling so difficult to define, or a strength of memory and attention beyond the ordinary, and then they will be absolutely incapable of understanding higher mathematics. Such are the majority. Others will have this feeling only in a slight degree, but they will be gifted with an uncommon memory and a great power of attention. They will learn by heart the details one after another; they can understand mathematics and sometimes make applications, but they cannot create. Others, finally, will possess in a less or greater degree the special intuition referred to, and then not only can they understand mathematics even if their memory is nothing extraordinary, but they may become creators and try to invent with more or less success according as this intuition is more or less developed in them.

In fact, what is mathematical creation? It does not consist in making new combinations with mathematical entities already known. Any one could do that, but the combinations so made would be infinite in number and most of them absolutely without interest. To create consists precisely in not making useless combinations and in making those which are useful and which are only a small minority. Invention is discernment, choice.

How to make this choice I have before explained; the mathematical facts worthy of being studied are those which, by their analogy with other facts, are capable of leading us to the knowledge of a mathematical law just as experimental facts lead us to the knowledge of a physical law. They are those which reveal to us unsuspected kinship between other facts, long known, but wrongly believed to be strangers to one another.

Among chosen combinations the most fertile will often be those formed of elements drawn from domains which are far apart. Not that I mean as sufficing for invention the bringing together of objects as disparate as possible; most combinations so formed would be entirely sterile. But certain among them, very rare, are the most fruitful of all.

To invent, I have said, is to choose; but the word is perhaps not wholly exact. It makes one think of a purchaser before whom are displayed a large number of samples, and who examines them, one after the other, to make a choice. Here the samples would be so numerous that a whole lifetime would not suffice to examine them. This is not the actual state of things. The sterile combinations do not even present themselves to the mind of the inventor. Never in the field of his consciousness do combinations appear that are not really useful, except some that he rejects but which have to some extent the characteristics of useful combinations. All goes on as if the inventor were an examiner for the second degree who would only have to question the candidates who had passed a previous examination.

But what I have hitherto said is what may be observed or inferred in reading the writings of the geometers, reading reflectively.

It is time to penetrate deeper and to see what goes on in the very soul of the mathematician. For this, I believe, I can do best by recalling memories of my own. But I shall limit myself to telling how I wrote my first memoir on Fuchsian functions. I beg the reader's pardon; I am about to use some technical expressions, but they need not frighten him, for he is not obliged to understand them. I shall say, for example, that I have found the demonstration of such a theorem under such circumstances. This theorem will have a barbarous name, unfamiliar to many, but that is unimportant; what is of interest for the psychologist is not the theorem but the circumstances.

For fifteen days I strove to prove that there could not be any functions like those I have since called Fuchsian functions. I was then very ignorant; every day I seated myself at my work table, stayed an hour or two, tried a great number of combinations and reached no results. One evening, contrary to my custom, I drank black coffee and could not sleep. Ideas rose in crowds; I felt them collide until pairs interlocked, so to speak, making a stable combination. By the next morning I had established the existence of a class of Fuchsian functions, those which come from the hypergeometric series; I had only to write out the results, which took but a few hours.

Then I wanted to represent these functions by the quotient of two series; this idea was perfectly conscious and deliberate, the analogy with elliptic functions guided me. I asked myself what properties these series must have if they existed, and I succeeded without difficulty in forming the series I have called theta-Fuchsian.

Just at this time I left Caen, where I was then living, to go on a geological excursion under the auspices of the school of mines. The changes of travel made me forget my mathematical work. Having reached Coutances, we entered an omnibus to go some place or other. At the moment when I put my foot on the step the idea came to me, without anything in my former thoughts seeming to have paved the way for it, that the transformations I had used to define the Fuchsian functions were identical with those of non-Euclidean geometry. I did not verify the idea; I should not have had time, as, upon taking my seat in the omnibus, I went on with a conversation already commenced, but I felt a perfect certainty. On my return to Caen, for conscience' sake I verified the result at my leisure.

Then I turned my attention to the study of some arithmetic questions apparently without much success and without a suspicion of any connection with my preceding researches. Disgusted with my failure, I went to spend a few days at the seaside, and thought of something else. One morning, walking on the bluff, the idea came to me, with just the same characteristics of brevity, suddenness and immediate certainty, that the arithmetic transformations of indeterminate ternary quadratic forms were identical with those of non-Euclidean geometry.

Returned to Caen, I meditated on this result and deduced the consequences. The example of quadratic forms showed me that there were Fuchsian groups other than those corresponding to the hypergeometric series; I saw that I could apply to them the theory of theta-Fuchsian series and that consequently there existed Fuchsian functions other than those from the hypergeometric series, the ones I then knew. Naturally I set myself to form all these functions. I made a systematic attack upon them and carried all the outworks, one after another. There was one however that still held out, whose fall would involve that of the whole place. But all my efforts only served at first the better to show me the difficulty, which indeed was something. All this work was perfectly conscious.

Thereupon I left for Mont-Valérien, where I was to go through my military service; so I was very differently occupied. One day, going along the street, the solution of the difficulty which had stopped me suddenly appeared to me. I did not try to go deep into it immediately, and only after my service did I again take up the question. I had all the elements and had only to arrange them and put them together. So I wrote out my final memoir at a single stroke and without difficulty.

I shall limit myself to this single example; it is useless to multiply them. In regard to my other researches I would have to say analogous things, and the observations of other mathematicians given in *L'enseignement mathématique* would only confirm them.

Most striking at first is this appearance of sudden illumination, a manifest sign of long, unconscious prior work. The role of this unconscious work in mathematical invention appears to me incontestable, and traces of it would be found in other cases where it is less evident. Often when one works at a hard question, nothing good is accomplished at the first attack. Then one takes a rest, longer or shorter, and sits down anew to the work. During the first half-hour, as before, nothing is found, and then all of a sudden the decisive idea presents itself to the mind. It might be said that the conscious work has been more fruitful because it has been interrupted and the rest has given back to the mind its force and freshness. But it is more probable that this rest has been filled out with unconscious work and that the result of this work has afterwards revealed itself to the geometer just as in the cases I have cited; only the revelation, instead of coming during a walk or a journey, has happened during a period of conscious work, but independently of this work which plays at most a role of excitant, as if it were the goad stimulating the results already reached during rest, but remaining unconscious, to assume the conscious form.

There is another remark to be made about the conditions of this unconscious work: it is possible, and of a certainty it is only fruitful, if it is on the one hand preceded and on the other hand followed by a period of conscious work. These sudden inspirations (and the examples already cited sufficiently prove this) never happen except after some days of voluntary effort which has appeared absolutely fruitless and whence nothing good seems to have come, where the way taken seems totally astray. These efforts then have not been as sterile as one thinks; they have set agoing the unconscious machine and without them it would not have moved and would have produced nothing.

The need for the second period of conscious work, after the inspiration, is still easier to understand. It is necessary to put in shape the results of this inspiration, to deduce from them the immediate consequences, to arrange them, to word the demonstrations, but above all is verification necessary. I have spoken of the feeling of absolute certitude accompanying the inspiration; in the cases cited this feeling was no deceiver, nor is it usually. But do not think this a rule without exception; often this feeling deceives us without being any the less vivid, and we only find it out when we seek to put on foot the demonstration. I have especially noticed this fact in regard to ideas coming to me in the morning or evening in bed while in a semi-hypnagogic state.

Such are the realities; now for the thoughts they force upon us. The unconscious, or, as we say, the subliminal self plays an important role in mathematical creation; this follows from what we have said. But usually the subliminal self is considered as purely automatic. Now we have seen that mathematical work is not simply mechanical, that it could not be done by a machine, however perfect. It is not merely a question of applying rules, of making the most combinations possible according to certain fixed laws. The combinations so obtained would be exceedingly numerous, useless and cumbersome. The true work of the inventor consists in choosing among these combinations so as to eliminate the useless ones or rather to avoid the trouble of making them, and the rules which must guide this choice are extremely fine and delicate. It is almost impossible to state them precisely; they are felt rather than formulated. Under these conditions, how imagine a sieve capable of applying them mechanically?

A first hypothesis now presents itself: the subliminal self is in no way inferior to the conscious self; it is not purely automatic; it is capable of discernment; it has tact, delicacy; it knows how to choose, to divine. What do I say? It knows better how to divine than the conscious self, since it succeeds where that has failed. In a word, is not the subliminal self superior to the conscious self? You recognize the full importance of this question. Boutroux in a recent lecture has shown how it came up on a very different occasion, and what consequences would follow an affirmative answer. . . .

Is this affirmative answer forced upon us by the facts I have just given? I confess that, for my part, I should hate to accept it. Reexamine the facts then and see if they are not compatible with another explanation.

It is certain that the combinations which present themselves to the mind in a sort of sudden illumination, after an unconscious working somewhat prolonged, are generally useful and fertile combinations, which seem the result of a first impression. Does it follow that the subliminal self, having divined by a delicate intuition that these combinations would be useful, has formed only these, or has it rather formed many others which were lacking in interest and have remained unconscious?

In his second way of looking at it, all the combinations would be formed in consequence of the automatism of the subliminal self, but only the interesting ones would break into the domain of consciousness. And this is still very mysterious. What is the cause that, among the thousand products of our unconscious activity, some are called to pass the threshold, while others remain below? Is it a simple chance which confers this privilege? Evidently not; among all the stimuli of our senses, for example, only the most intense fix our attention, unless it has been drawn to them by other causes. More generally the privileged unconscious phenomena, those susceptible of becoming conscious, are those which, directly or indirectly, affect most profoundly our emotional sensibility.

It may be surprising to see emotional sensibility invoked *à propos* of mathematical demonstrations which, it would seem, can interest only the intellect. This would be to forget the feeling of mathematical beauty, of the harmony of numbers and forms, of geometric elegance. This is a true esthetic feeling that all real mathematicians know, and surely it belongs to emotional sensibility.

Now, what are the mathematic entities to which we attribute this character of beauty and elegance, and which are capable of developing in us a sort of esthetic emotion? They are those whose elements are harmoniously disposed so that the mind without effort can embrace their totality while realizing the details. This harmony is at once a satisfaction of our esthetic needs and an aid to the mind, sustaining and guiding. And at the same time, in putting under our eyes a well-ordered whole, it makes us foresee a mathematical law. Now, as we have said above, the only mathematical facts worthy of fixing our attention and capable of being useful are those which can teach us a mathematical law. So that we reach the following conclusion: The useful combinations are precisely the most beautiful. I mean those best able to charm this special sensibility that all mathematicians know, but of which the profane are so ignorant as often to be tempted to smile at it.

What happens then? Among the great numbers of combinations blindly formed by the subliminal self, almost all are without interest and without utility; but just for that reason they are also without effect upon the esthetic sensibility. Consciousness will never know them; only certain ones are harmonious, and, consequently, at once useful and beautiful. They will be capable of touching this special sensibility of the geometer of which I have just spoken, and which, once aroused, will call our attention to them, and will bring them into our consciousness.

This is only a hypothesis, and yet here is an observation which may confirm it: when a sudden illumination seizes upon the mind of the mathematician, it usually happens that it does not deceive him, but it also sometimes happens, as I have said, that it does not stand the test of verification; well, we almost always notice that this false idea, had it been true, would have gratified our natural feeling for mathematical elegance.

Thus it is this special esthetic sensibility which plays the role of the delicate sieve of which I spoke, and that sufficiently explains why the one lacking it will never be a real creator.

Yet all the difficulties have not disappeared. The conscious self is narrowly limited, and as for the subliminal self we know not its limitations, and this is why we are not too reluctant in supposing that it has been able in a short time to make more different combinations than the whole life of a conscious being could encompass. Yet these limitations exist. Is it likely that it is able to form all the possible combinations, whose number would frighten the imagination? Nevertheless that would seem necessary, because if it produces only a small part of these combinations, and if it makes them at random, there would be small chance that the *good,* the one we should choose, would be found among them.

Perhaps we ought to seek the explanation in that preliminary period of conscious work which always precedes all fruitful unconscious labor. Permit me a rough comparison. Figure the future elements of our combinations as something like the hooked atoms of Epicurus. During the complete repose of the mind, these atoms are motionless, they are, so to speak, hooked to the wall; so this complete rest may be indefinitely prolonged without the atoms meeting, and consequently without any combination between them.

On the other hand, during a period of apparent rest and unconscious work, certain of them are detached from the wall and put in motion. They flash in every direction through the space (I was about to say the room) where they are enclosed, as would, for example, a swarm of gnats or, if you prefer a more learned comparison, like the molecules of gas in the kinematic theory of gases. Then their mutual impacts may produce new combinations.

What is the role of the preliminary conscious work? It is evidently to mobilize certain of these atoms, to unhook them from the wall and put them in swing. We think we have done no good, because we have moved these elements a thousand different ways in seeking to assemble them, and have found no satisfactory aggregate. But, after this shaking up imposed upon them by our will, these atoms do not return to their primitive rest. They freely continue their dance.

Now, our will did not choose them at random; it pursued a perfectly determined aim. The mobilized atoms are therefore not any atoms whatsoever; they are those from which we might reasonably expect the desired solution. Then the mobilized atoms undergo impacts which make them enter into combinations among themselves or with other atoms at rest which they struck against in their course. Again I beg pardon, my comparison is very rough, but I scarcely know how otherwise to make any thought understood.

However it may be, the only combinations that have a chance of forming are those where at least one of the elements is one of those atoms freely chosen by our will. Now, it is evidently among these that is found what I called the *good combination.* Perhaps this is a way of lessening the paradoxical in the original hypothesis.

Another observation. It never happens that the unconscious work gives us the result of a somewhat long calculation *all made,* where we have only to apply fixed rules. We might think the wholly automatic subliminal self particularly apt for this sort of work, which is in a way exclusively mechanical. It seems that thinking in the evening upon the factors of a multiplication we might hope to find the product ready made upon our awakening, or again that an algebraic calculation, for example a verification, would be made unconsciously. Nothing of the sort, as observation proves. All one may hope from these inspirations, fruits of unconscious work, is a point of departure for such calculations. As for the calculations themselves, they must be made in the second period of conscious work, that which follows the inspiration, that in which one verifies the results of this inspiration and deduces their consequences. The rules of these calculations are strict and complicated. They require discipline, attention, will, and therefore consciousness. In the subliminal self, on the contrary, reigns what I should call liberty, if we might give this name to the simple absence of discipline and to the disorder born of chance. Only, this disorder itself permits unexpected combinations.

I shall make a last remark: when above I made certain personal observations, I spoke of a night of excitement when I worked in spite of myself. Such cases are frequent, and it is not necessary that the abnormal cerebral activity be caused by a physical excitant as in that I mentioned. It seems, in such cases, that one is present at his own unconscious work, made partially perceptible to the over-excited consciousness, yet without having changed its nature. Then we vaguely comprehend what distinguishes the two mechanisms or, if you wish, the working methods of the two egos. And the psychologic observations I have been able thus to make seem to me to confirm in their general outlines the views I have given.

Surely they have need of it, for they are and remain in spite of all very hypothetical: the interest of the questions is so great that I do not repent of having submitted them to the reader.

Chapter IX
The Early 20th Century to 1932

Section A
Creativity and the Paris Problems

DAVID HILBERT (1862–1943)

Hilbert is the leading mathematician of the 20th century. He was the eldest child of Otto Hilbert, a country judge, and his wife Maria. The Hilberts were pietist Protestants. It is believed that the young David received his inclination to study mathematics from his mother who was enthusiastic about philosophy, astronomy, and arithmetic. After graduating from the Wilhelms Gymnasium, where he earned exceptional grades, he attended Königsberg University (1880-84), except for a semester at Heidelberg in 1884. Königsberg, where Kant had taught, had recently been turned into a center of mathematical learning by Carl Jacobi. Hilbert studied under Heinrich Weber, Ferdinand Lindemann, and Adolf Hurwitz becoming a close friend of his fellow student Hermann Minkowski. The work of Leopold Kronecker profoundly influenced him. After receiving his Ph.D. in 1885, he visited Leipzig and Paris. He qualified as *Privatdozent* at Königsberg University (1885), became associate professor (1892), and full professor (1893). In 1892 he married Käthe Jerosch. The following year he attended the International Mathematical Congress in Chicago.

From 1895 until his retirement in 1930, Hilbert was a named professor of mathematics at the University of Göttingen. At the Second International Mathematical Congress in Paris in 1900 he presented 23 problems for research in the 20th century. (They have continued to engage the attention of mathematicians to the present time, and in 1980 five were still unsolved.) In 1902 he had a new chair in mathematics created for his friend Minkowski at Göttingen. Among Hilbert's favorite amusements were dancing and mixing socially with young women. He was also fond of bicycle riding for a time.

Hilbert displayed a strong personality and independent mind, albeit that of an East-Prussian, who often favored political conservatism and abhorred nationalist emotions. Because of his international reputation in the sciences, he was asked early in World War I to sign a "Manifesto to the Civilized World." The German invasion of neutral Belgium had provoked outrage in Western Europe and the United States. In response the "Manifesto" attempted to absolve German militarism of blame. It included a list of categorical sentences, each beginning with the words "It is not true that, . . ." as well as denigration of Slavic people. Eighty-three German scientists, including Felix Klein, Max Planck, and Wilhelm Röntgen, signed but not Hilbert. He could not ascertain the truth of its statements. Moreover, he thought the war was stupid and said so.

After 1925, when he fell ill with pernicious anemia, Hilbert's scientific activity lessened. He died in Germany during World War II. He had a characteristic optimism that new discoveries would continually be made, and that these discoveries were necessary for the vitality of mathematics. The motto on the marker placed over his grave in Göttingen reflected his optimistic spirit. It read:

> Wir mussen wissen,
> Wir werden wissen.
> (We must know,
> We will know.)

The scientific activity of Hilbert covered several branches of mathematics. He began his work in algebraic forms. In this area he formulated his basis theorem and irreducibility theorem for the theory of invariants. By adapting a new, direct, nonalgorithmic method to the study of invariants, he prepared the way for modern abstract algebra. He thus situated invariants in a broader context. In the mid-1890s he turned to algebraic number theory, a field whose beauty has irresistibly attracted the elite of mathematicians. He reorganized and reshaped number theory under new unifying viewpoints in his lucid treatise *Der Zahlbericht* ("The Number Report," 1897). His number theoretic work centered on his new proof of the law of biquadratic reciprocity and culminated in class field theory. In 1898–99 he delivered a series of lectures that were later published as *Grundlagen der Geometrie* ("The Foundations of Geometry"). In these lectures Hilbert went beyond Euclid's axiomatic system and developed the refined formal axiomatics of the present. To achieve his goals in axiomatics of consistency and independence, he skillfully used algebraic models and countermodels as tools. He also endeavored to make axiomatic systems more general. There should not be, he believed, just one interpretation of terms; rather it should be possible to talk about anything. He once stated: "It must be possible to replace in all geometric statements the words *point, line, plane* by *table, chair, beer mug.*"

Next, in mathematical analysis, Hilbert salvaged the Dirichlet principle (1904) from the discredit caused by Weierstrass' criticism and also enriched the calculus of variations. His chief contributions to analysis came in integral equations, where he introduced what is now called Hilbert space and coined the term "spectrum." He also proved Waring's hypothesis that every positive integer can be represented as a sum of, at most, m l^{th} powers, with m depending on l alone.

Hilbert's last two major scientific interests were theoretical physics and the foundations of mathematics. He concentrated on theoretical physics from 1909 to 1914 with results that were generally less profound and inventive than his work in pure mathematics. He published on relativity (1905), kinetic gas theory, and axiomatics of radiation. After 1918 he once again explored the foundations of mathematics, continuing research that he had pursued in geometry at the turn of the century. His personal intuition, so fruitful in the past, did not serve him well in foundations. He proposed a formalist program, entailing the reduction of mathematics to a finitary game with an infinite but finitely defined group of formulas. Hilbert and his collaborators, Paul Bernays and Wilhelm Ackermann, attempted to develop a formalist program that included a "proof theory" that would provide a direct test for consistency in elementary number theory and ultimately in all of classical mathematical analysis. Their expectation that a sufficiently rich formalized deductive system would prove the consistency of mathematics was contested by L. E. J. Brouwer, who asserted that truth must take precedence over formal consistency in mathematics. Brouwer's intuitionism (including constructivism) was reinforced in 1931 when Kurt Gödel proved that rigorous formal proof of the consistency of elementary number theory is in principle unattainable.

Although the goal of a formal consistency proof cannot, therefore, be fully met, Hilbert's championing of the axiomatic method within the formalist program admirably achieved another important goal—greater uniformity in modern mathematics. The importance of uniformity as a goal had been stated by Hilbert at the end of his Paris paper on the famous 23 problems.

119. From "Mathematical Problems: Lecture Delivered Before The International Congress of Mathematicians at Paris in 1900"*

(Paris Problems and the Formalist Program)

DAVID HILBERT

Who of us would not be glad to lift the veil behind which the future lies hidden; to cast a glance at the next advances of our science and at the secrets of its development during future centuries? What particular goals will there be toward which the leading mathematical spirits of coming generations will strive? What new methods and new facts in the wide and rich field of mathematical thought will the new centuries disclose?

History teaches the continuity of the development of science? We know that every age has its own problems, which the following age either solves or casts aside as profitless and replaces by new ones. If we would obtain an idea of the probable development of mathematical knowledge in the immediate future, we must let the unsettled questions pass before our minds and look over the problems which the science of today sets and whose solution we expect from the future. To such a review of problems the present day, lying at the meeting of the centuries, seems to me well adapted. For the close of a great epoch not only invites us to look back into the past but also directs our thoughts to the unknown future.

The deep significance of certain problems for the advance of mathematical science in general and the important rôle which they play in the work of the individual investigator are not to be denied. As long as a branch of science offers an abundance of problems, so long is it alive; a lack of problems foreshadows extinction or the cessation of independent development. Just as every human undertaking pursues certain objects, so also mathematical research requires its problems. It is by the solution of problems that the investigator tests the temper of his steel; he finds new

*Source: This translation by Mary Winston Newson is reprinted from the *Bulletin* of the American Mathematical Society, vol. 8 (1902), 437-479.

methods and new outlooks, and gains a wider and freer horizon.

It is difficult and often impossible to judge the value of a problem correctly in advance; for the final award depends upon the gain which science obtains from the problem. Nevertheless we can ask whether there are general criteria which mark a good mathematical problem. An old French mathematician said: "A mathematical theory is not to be considered complete until you have made it so clear that you can explain it to the first man whom you meet on the street." This clearness and ease of comprehension, here insisted on for a mathematical theory, I should still more demand for a mathematical problem if it is to be perfect; for what is clear and easily comprehended attracts, the complicated repels us.

Moreover a mathematical problem should be difficult in order to entice us, yet not completely inaccessible, lest it mock our efforts. It should be to us a guide post on the mazy paths to hidden truths, and ultimately a reminder of our pleasure in the successful solution.

The mathematicians of past centuries were accustomed to devote themselves to the solution of difficult particular problems with passionate zeal. They knew the value of difficult problems. I remind you only of the "problem of the line of quickest descent," proposed by John Bernoulli. Experience teaches, explains Bernoulli in the public announcement of this problem that lofty minds are led to strive for the advance of science by nothing more than by laying before them difficult and at the same time useful problems, and he therefore hopes to earn the thanks of the mathematical world by following the example of men like Mersenne, Pascal, Fermat, Viviani and others and laying before the distinguished analysts of his time a problem by which, as a touchstone, they may test the value of their methods and measure their strength. The calculus of variations owes its origin to this problem of Bernoulli and to similar problems.

Fermat had asserted, as is well known, the diophantine equation $x^n + y^n = z^n$ (x, y and z integers) is unsolvable—except in certain self-evident cases. The attempt to prove this impossibility offers a striking example of the inspiring effect which such a very special and apparently unimportant problem may have upon science. For Kummer, incited by Fermat's problem was led to the introduction of ideal numbers and to the discovery of the law of the unique decomposition of the numbers of a circular field into ideal prime factors—a law which to-day, in its generalization to any algebraic field by Dedekind and Kronecker, stands at the center of the modern theory of numbers and whose significance extends far beyond the boundaries of number theory into the realm of algebra and the theory of functions.

To speak of a very different region of research, I remind you of the problem of three bodies. The fruitful methods and the far-reaching principles which Poincaré has brought into celestial mechanics and which are today recognized and applied in practical astronomy are due to the circumstance that he undertook to treat anew that difficult problem and to approach nearer a solution.

The two last mentioned problems—that of Fermat and the problem of the three bodies—seem to us almost like opposite poles—the former a free invention of pure reason, belonging to the region of abstract number theory, the latter forced upon us by astronomy and necessary to an understanding of the simplest fundamental phenomena of nature.

But it often happens also that the same special problem finds application in the most unlike branches of mathematical knowledge. So, for example, the problem of the shortest line plays a chief and historically important part in the foundations of geometry, in the theory of curved lines and surfaces, in mechanics, and in the calculus of variations. And how convincingly has F. Klein, in his work on the icosahedron, pictured the significance which attaches to the problem of the regular polyhedra in elementary geometry, in group theory, in the theory of equations and in that of linear differential equations.

In order to throw light on the importance of certain problems, I may also refer to Weierstrass, who spoke of it as his happy fortune that he found at the outset of his scientific career a problem so important as Jacobi's problem of inversion on which to work.

Having now recalled to mind the general importance of problems in mathematics, let us turn to the question from what sources this science derives its problems. Surely the first and oldest problems in every branch of mathematics spring from experience and are suggested by the world of external phenomena. Even the rules of calculation with integers must have been discovered in this fashion in a lower stage of human civilization, just as the child of today learns the application of these laws by empirical methods. The same is true of the first problems of geometry, the problems bequeathed us by antiquity, such as the duplication of the cube, the squaring of the circle; also the oldest problems in the theory of the solution of numerical equations, in the theory of curves and the differential and integral calculus, in the calculus of variations, the theory of Fourier series and the theory of potential—to say nothing of the further abundance of problems properly belonging to mechanics, astronomy and physics.

But, in the further development of a branch of mathematics, the human mind, encouraged by the success of its solutions, becomes conscious of its independence. It evolves from itself alone, often without appreciable influence from without, by means of logical combination, generalization, specialization, by separating and collecting ideas in fortunate ways, new and fruitful problems, and appears then itself as the real questioner. Thus arose the problem of prime numbers and the other problems of number theory, Galois's theory of equations, the theory of algebraic invariants, the theory of abelian and automorphic functions; indeed almost all the nicer questions of modern arithmetic and function theory arise in this way.

In the meantime, while the creative power of pure reason is at work, the outer world again comes into play, forces upon us new questions from actual experience, opens up new branches of mathematics, and while we seek to conquer these new fields of knowledge for the realm of pure thought, we often find the answers to old unsolved problems and thus at the same time advance most successfully the old theories. And it seems to me that the numerous and surprising analogies and that apparently prearranged harmony which the mathematician so often perceives in the questions, methods and ideas of the various branches of his science, have their origin in this ever-recurring interplay between thought and experience.

It remains to discuss briefly what general requirements may be justly laid down for the solution of a mathematical problem. I should say first of all, this: that it shall be possible to establish the correctness of the solution by means of a finite number of steps based upon a finite number of hypotheses which are implied in the statement of the problem and which must always be exactly formulated. This requirement of logical deduction by means of a finite number of processes is simply the requirement of rigor in reasoning. Indeed the requirement of rigor, which has become proverbial in mathematics, corresponds to a universal philosophical necessity of our understanding; and, on the other hand, only by satisfying this requirement do the thought content and the suggestiveness of the problem attain their full effect. A new problem, especially when it comes from the world of outer experience, is like a young twig, which thrives and bears fruit only when it is grafted carefully and in accordance with strict horticultural rules upon the old stem, the established achievements of our mathematical science.

Besides it is an error to believe that rigor in the proof is the enemy of simplicity. On the contrary we find it confirmed by numerous examples that the rigorous method is at the same time the simpler and the more easily comprehended. The very effort for rigor forces us to find out simpler methods of proof. It also frequently leads the way to methods which are more capable of development than the old methods of less rigor. Thus the theory of algebraic curves experienced a considerable simplification and attained greater unity by means of the more rigorous function-theoretical methods and the consistent introduction of transcendental devices. Further, the proof that the power series permits the application of the four elementary arithmetical operations as well as the term by term differentiation and integration, and the recognition of the utility of the power series depending upon this proof contributed materially to the simplification of all analysis, particularly of the theory of elimination and the theory of differential equations, and also of the existence proofs demanded in those theories. But the most striking example for my statement is the calculus of variations. The treatment of the first and second variations of definite integrals required in part extremely complicated calculations, and the processes applied by the old mathematicians had not the necessary rigor. Weierstrass showed us the way to a new and sure foundation of the calculus of variations. By the examples of the simple and double integral I will show briefly, at the close of my lecture, how this way leads at once to a surprising simplification of the calculus of variations. For in the demonstration of the necessary and sufficient criteria for the occurrence of a maximum and minimum, the calculation of the second variation and in part, indeed, the wearisome reasoning connected with the first variation may be completely dispensed with—to say nothing of the advance which is involved in the removal of the restriction to variations for which the differential coefficients of the function vary but slightly.

While insisting on rigor in the proof as a requirement for a perfect solution of a problem, I should like, on the other hand, to oppose the opinion that only the concepts of analysis, or even those of arithmetic alone, are susceptible of a fully rigorous treatment. This opinion, occasionally advocated by eminent men, I consider entirely erroneous. Such a one-sided interpretation of the requirement of rigor would soon lead to the ignoring of all concepts arising from geometry, mechanics and physics, to a stoppage of the flow of new material from the outside world, and finally, indeed, as a last consequence, to the rejection of the ideas of the continuum and of the irrational number. But what an important nerve, vital to mathematical science, would be cut by the extirpation of geometry and mathematical physics! On the contrary I think that wherever, from the side of the theory of knowledge or in geometry, or from the theories of natural or physical science, mathematical ideas come up, the problem arises for mathematical science to investigate the principles underlying these ideas and so to establish them upon a simple and complete system of axioms, that the exactness of the new ideas and their applicability to deduction shall be in no respect inferior to those of the old arithmetical concepts.

To new concepts correspond, necessarily, new signs. These we choose in such a way that they remind us of the phenomena which were the occasion for the formation of the new concepts.

So the geometrical figures are signs or mnemonic symbols of space intuition and are used as such by all mathematicians. Who does not always use along with the double inequality $a > b > c$ the picture of three points following one another on a straight line as the geometrical picture of the idea "between"? Who does not make use of drawings of segments and rectangles enclosed in one another, when it is required to prove with perfect rigor a difficult theorem on the continuity of functions or the existence of points of condensation? Who could dispense with the figure of the triangle, the circle with its center, or with the cross of three perpendicular axes? Or who would give up the representation of the vector field, or the picture of a family of curves or surfaces with its envelope which plays so important a part in differential geometry, in the theory of differential equations, in the foundation of the calculus of variations and in other purely mathematical sciences?

The arithmetical symbols are written diagrams and the geometrical figures are graphic formulas; and no mathematician could spare these graphic formulas, any more than in calculation the insertion and removal of parentheses or the use of other analytical signs.

The use of geometrical signs as a means of strict proof presupposes the exact knowledge and complete mastery of the axioms which underlie those figures; and in order that these geometrical figures may be incorporated in the general treasure of mathematical signs, there is necessary a rigorous axiomatic investigation of their conceptual content. Just as in adding two numbers, one must place the digits under each other in the right order, so that only the rules of calculation, *i.e.*, the axioms of arithmetic, determine the correct use of the digits, so the use of geometrical signs is determined by the axioms of geometrical concepts and their combinations.

The agreement between geometrical and arithmetical thought is shown also in that we do not habitually follow the chain of reasoning back to the axioms in arithmetical, any more than in geometrical discussions. On the contrary we apply, especially in first attacking a problem, a rapid, unconscious, not absolutely sure combination, trusting to a certain arithmetical feeling for the behavior of the arithmetical symbols, which we could dispense with as little in arithmetic as with the geometrical imagination in geometry. As an example of an arithmetical theory operating rigorously with geometrical ideas and signs, I may mention Minkowski's work, *Die Geometrie der Zahlen* (1896).

Some remarks upon the difficulties which mathematical problems may offer, and the means of surmounting them, may be in place here.

If we do not succeed in solving a mathematical problem, the reason frequently consists in our failure to recognize the more general standpoint from which the problem before us appears only as a single link in a chain of related problems. After finding this standpoint, not only is this problem frequently more accessible to our investigation, but at the same time we come into possession of a method which is applicable also to related problems. The introduction of complex paths of integration by Cauchy and of the notion of the IDEALS in number theory by Kummer may serve as examples. This way for finding general methods is certainly the most practicable and the most certain; for he who seeks for methods without having a definite problem in mind seeks for the most part in vain.

In dealing with mathematical problems, specialization plays, as I believe, a still more important part than generalization. Perhaps in most cases where we seek in vain the answer to a question, the cause of the failure lies in the fact that problems simpler and easier than the one in hand have been either not at all or incompletely solved. All depends, then, on finding out these easier problems, and on solving them by means of devices as perfect as possible and of concepts capable of generalization. This rule is one of the most important levers for overcoming mathematical difficulties and it seems to me that it is used almost always, though perhaps unconsciously.

Occasionally it happens that we seek the solution under insufficient hypotheses or in an incorrect sense, and for this reason do not succeed. The problem then arises: to show the impossibility of the solution under the given hypotheses, or in the sense contemplated. Such proofs of impossibility were effected by the ancients, for instance when they showed that the ratio of the hypotenuse to the side of an isosceles right triangle is irrational. In later mathematics, the question as to the impossibility of certain solutions plays a preëminent part, and we perceive in this way that old and difficult problems, such as the proof of the axiom of parallels, the squaring of the circle, or the solution of equations of the fifth degree by radicals have finally found fully satisfactory and rigorous solutions, although in another sense than that originally intended. It is probably this important fact along with other philosophical reasons that gives rise to the conviction (which every mathematician shares, but which no one has as yet supported by a proof) that every definite mathematical problem must necessarily be susceptible of an exact settlement, either in the form of an actual answer to the question asked, or by the proof of the impossibility of its solution and therewith the necessary failure of all attempts. Take any definite unsolved problem, such as the question as to the irrationality of the Euler-Mascheroni constant C, or the existence of an infinite number of prime numbers of the form $2^n + 1$. However unapproachable these problems may seem to us and however helpless we stand before them, we have, nevertheless, the firm conviction that their solution must follow by a finite number of purely logical processes.

Is this axiom of the solvability of every problem a peculiarity characteristic of mathematical thought alone, or is it possibly a general law inherent in the nature of the mind, that all questions which it asks must be answerable? For in other sciences also one meets old problems which have been settled in a manner most satisfactory and most useful to science by the proof of their impossibility. Consider the problem of perpetual motion. After seeking in vain for the construction of a perpetual motion machine, the relations were investigated which must subsist between the forces of nature if such a machine is to be impossible;[1] and this inverted question led to the discovery of the law of the conservation of energy, which, again, explained the impossibility of perpetual motion in the sense originally intended.

This conviction of the solvability of every mathematical problem is a powerful incentive to the worker. We hear within us the perpetual call: There is the problem. Seek its solution. You can find it by pure reason, for in mathematics there is no *ignorabimus*.

The supply of problems in mathematics is inexhaustible, and as soon as one problem is solved numerous others come forth in its place. Permit me in the following, tentatively as it were, to mention particular definite problems, drawn from various branches of mathematics, from the discussion of which an advancement of science may be expected.

Let us look at the principles of analysis and geometry. The most suggestive and notable achievements of the last century in this field are, as it seems to me, the arithmetical formulation of the concept of the continuum in the works of Cauchy, Bolzano and Cantor, and the discovery of non-euclidean geometry by Gauss, Bolyai, and Lobachevsky. I therefore first direct your attention to some problems belonging to these fields.

1. CANTOR'S PROBLEM OF THE CARDINAL NUMBER OF THE CONTINUUM

Two systems, *i.e.*, two assemblages of ordinary real numbers or points, are said to be (according to Cantor) equivalent or of equal *cardinal number*, if they can be brought into a relation to one another such that to every number of the one assemblage corresponds one and only one definite number of the other. The investigations of Cantor on such assemblages of points suggest a very plausible theorem, which nevertheless, in spite of the most strenuous efforts, no one has succeeded in proving. This is the theorem:

Every system of infinitely many real numbers, *i.e.*, every assemblage of numbers (or points), is either equivalent to the assemblage of natural integers, 1, 2, 3, . . . or to the assemblage of all real numbers and therefore to the continuum, that is, to the points of a line; *as regards equivalence there are, therefore, only two assemblages of numbers, the countable assemblage and the continuum.*

From this theorem it would follow at once that the continuum has the next cardinal number beyond that of the countable assemblage; the proof of this theorem would, therefore, form a new bridge between the countable assemblage and the continuum.

Let me mention another very remarkable statement of Cantor's which stands in the closest connection with the theorem mentioned and which, perhaps, offers the key to its proof. Any system of real numbers is said to be ordered, if for every two numbers of the system it is determined which one is the earlier and which the later, and if at the same time this determination is of such a kind that, if a is before b and b is before c, then a always comes before c. The natural arrangement of numbers of a system is defined to be that in which the smaller precedes the larger. But there are, as is easily seen, infinitely many other ways in which the numbers of a system may be arranged.

If we think of a definite arrangement of numbers and select from them a particular system of these numbers, a so-called partial system or assemblage, this partial system will also prove to be ordered. Now Cantor considers a particular kind of ordered assemblage which he designates as a well ordered assemblage and which is characterized in this way, that not only in the assemblage itself but also in every partial assemblage there exists a first number. The system of integers 1, 2, 3, . . . in their natural order is evidently a well ordered assemblage. On the other hand the system of real numbers, *i.e.*, the continuum in its natural order, is evidently not well ordered. For, if we think of the points of a segment of a straight line, with its initial point excluded, as our partial assemblage, it will have no first element.

The question now arises whether the totality of all numbers may not be arranged in another manner so that every partial assemblage may have a first element, *i.e.*, whether the continuum cannot be considered as a well ordered assemblage—a question which Cantor thinks must be answered in the affirmative. It appears to me most desirable to obtain a direct proof of this remarkable statement of Cantor's, perhaps by actually giving an arrangement of numbers such that in every partial system a first number can be pointed out.

2. THE COMPATIBILITY OF THE ARITHMETICAL AXIOMS

When we are engaged in investigating the foundations of a science, we must set up a system of axioms which contains an exact and complete description of the relations subsisting between the elementary ideas of that science. The axioms so set up are at the same time the definitions of those elementary ideas; and no statement within the realm of the science whose foundation we are testing is held to be correct unless it can be derived from those axioms by means of a finite number of logical steps. Upon closer consideration the question arises: *Whether, in any way, certain statements of single axioms depend upon one another, and whether the axioms may not therefore contain certain parts in common, which must be isolated if one wishes to arrive at a system of axioms that shall be altogether independent of one another.*

But above all I wish to designate the following as the most important among the numerous questions which can be asked with regard to the axioms: *To prove that they are not contradictory, that is, that a finite number of logical steps based upon them can never lead to contradictory results.*

In geometry, the proof of the compatibility of the axioms can be effected by constructing a suitable field of numbers, such that analogous relations between the numbers of this field correspond to the geometrical axioms. Any contradiction in the deductions from the geometrical axioms must thereupon be recognizable in the arithmetic of this field of numbers. In this way the desired proof for the compatibility of the geometrical axioms is made to depend upon the theorem of the compatibility of the arithmetical axioms.

On the other hand a direct method is needed for the proof of the compatibility of the arithmetical axioms. The axioms of arithmetic are essentially nothing else than the known rules of calculation, with the addition of the axiom of continuity. I recently collected them[2] and in so doing replaced the axiom of continuity by two simpler axioms, namely, the well-known axiom of Archimedes, and a new axiom essentially as follows: that numbers form a system of things which is capable of no further extension, as long as all the other axioms hold (axiom of completeness). I am convinced that it must be possible to find a direct proof for the compatibility of the arithmetical axioms, by means of a careful study and suitable modification of the known methods of reasoning in the theory of irrational numbers.

To show the significance of the problem from another point of view, I add the following observation: If contradictory attributes be assigned to a concept, I say, that *mathematically the concept does not exist*. So, for example, a real number whose square is -1 does not exist mathematically. But if it can be proved that the attributes assigned to the concept can never lead to a contradiction by the application of a finite number of logical processes, I say that the mathematical existence of the concept (for example of a number or a function which satisfies certain conditions) is thereby proved. In the case before us, where we are concerned with the axioms of real numbers in arithmetic, the proof of the compatibility of the axioms is at the same time the proof of the mathematical existence of the complete system of real numbers or of the continuum. Indeed, when the proof for the compatibility of the axioms shall be fully accomplished, the doubts which have been expressed occasionally as to the existence of the complete system of real numbers will become totally groundless. The totality of real numbers, *i.e.*, the continuum according to the point of view just indicated, is not the totality of all possible series in decimal fractions, or of all possible laws according to which the elements of a fundamental sequence may proceed. It is rather a system of things whose mutual relations are governed by the axioms set up and for which all propositions, and only those, are true which can be derived from the axioms by a finite number of logical processes. In my opinion, the concept of the continuum is strictly logically tenable in this sense only. It seems to me, indeed, that this corresponds best also to what experience and intuition tell us. The concept of the continuum or even that of the system of all functions exists, then, in exactly the same sense as the system of integral, rational numbers, for example, or as Cantor's higher classes of numbers and cardinal numbers. For I am convinced that the existence of the latter, just as that of the continuum, can be proved in the sense I have described; unlike the system of *all* cardinal numbers or of *all* Cantor's alephs, for which, as may be shown, a system of axioms, compatible in my sense, cannot be set up. Either of these systems is, therefore, according to my terminology, mathematically non-existent.

From the field of the foundations of geometry I should like to mention the following problem:

3. THE EQUALITY OF THE VOLUMES OF TWO TETRAHEDRA OF EQUAL BASES AND EQUAL ALTITUDES

In two letters to Gerling, Gauss[3] expresses his regret that certain theorems of solid geometry depend upon the method of exhaustion, *i.e.*, in modern phraseology, upon the axiom of continuity (or upon the axiom of Archimedes). Gauss mentions in particular the theorem of Euclid, that triangular pyramids of equal altitudes are to each other as their bases. Now the analogous problem in the plane has been solved.[4] Gerling also succeeded in proving the equality of volume of symmetrical polyhedra by dividing them into congruent parts. Nevertheless, it seems to me probable that a general proof of this kind for the theorem of Euclid just mentioned is impossible, and it should be our task to give a rigorous proof of its impossibility. This would be obtained, as soon as we succeeded in *specifying two tetrahedra of equal bases and equal altitudes which can in no way be split up into congruent tetrahedra, and which cannot be combined with congruent tetrahedra to form two polyhedra which themselves could be split up into congruent tetrahedra.*[5]

4. PROBLEM OF THE STRAIGHT LINE AS THE SHORTEST DISTANCE BETWEEN TWO POINTS

Another problem relating to the foundations of geometry is this: If from among the axioms necessary to establish ordinary euclidean geometry, we exclude the axiom of parallels, or assume it as not satisifed, but retain all other axioms, we obtain, as is well known, the geometry of Lobachevsky (hyperbolic geometry). We may therefore say that this is a geometry standing next to euclidean geometry. If we require further that that axiom be not satisfied whereby, of three points of a straight line, one and only one lies between the other two, we obtain Riemann's (elliptic) geometry, so that this geometry appears to be the next after Lobachevsky's. If we wish to carry out a similar investigation with respect to the axiom of Archimedes, we must look upon this as not satisfied, and we arrive thereby at the non-archimedean geometries which have been investigated by Veronese and myself. The more general question now arises: Whether from other suggestive standpoints geometries may not be devised which, with equal right, stand next to euclidean geometry. Here I should like to direct your attention to a theorem which has, indeed, been employed by many authors as a definition of a straight line, viz., that the straight line is the shortest distance between two points. The essential content of this statement reduces to the theorem of Euclid that in a triangle the sum of two sides is always greater than the third side—a theorem which, as is easily seen, deals solely with elementary concepts, *i.e.*, with such as are derived directly from the axioms, and is therefore more accessible to logical investigation. Euclid proved this theorem, with the help of the theorem of the exterior angle, on the basis of the congruence theorems. Now it is readily shown that this theorem of Euclid cannot be proved solely on the basis of those congruence theorems which relate to the application of segments and angles, but that one of the theorems on the congruence of triangles is necessary. We are asking, then, for a geometry in which all the axioms of ordinary euclidean geometry hold, and in particular all the congruence axioms except the one of the congruence of triangles (or all except the theorem of the equality of the base angles in the isosceles triangle), and in which, besides, the proposition that in every triangle the sum of two sides is greater than the third is assumed as a particular axiom.

One finds that such a geometry really exists and is no other than that which Minkowski constructed in his book, Geometrie der Zahlen,[6] and made the basis of his arithmetical investigations. Minkowski's is therefore also a geometry standing next to the ordinary euclidean geometry; it is essentially characterized by the following stipulations:

1. The points which are at equal distances from a fixed point O lie on a convex closed surface of the ordinary euclidean space with O as a center.
2. Two segments are said to be equal when one can be carried into the other by a translation of the ordinary euclidean space.

In Minkowski's geometry the axiom of parallels also holds. By studying the theorem of the straight line as the shortest distance between two points, I arrived[7] at a geometry in which the parallel axiom does not hold, while all other axioms of Minkowski's geometry are satisfied. The theorem of the straight line as the shortest distance between two points and the essentially equivalent theorem of Euclid about the sides of a triangle, play an important part not only in number theory but also in the theory of surfaces and in the calculus of variations. For this reason, and because I believe that the thorough investigation of the conditions for the validity of this theorem will throw a new light upon the idea of distance, as well as upon other elementary ideas, e.g., upon the idea of the plane, and the possibility of its definition by means of the idea of the straight line, *the construction and systematic treatment of the geometries here possible seem to me desirable.*

5. LIE'S CONCEPT OF A CONTINUOUS GROUP OF TRANSFORMATIONS WITHOUT THE ASSUMPTION OF THE DIFFERENTIABILITY OF THE FUNCTIONS DEFINING THE GROUP

It is well known that Lie, with the aid of the concept of continuous groups of transformations, has set up a system of geometrical axioms and, from the standpoint of his theory of groups, has proved that this system of axioms suffices for geometry. But since Lie assumes, in the very foundation of his theory, that the functions defining his group can be differentiated, it remains undecided in Lie's development, whether the assumption of the differentiability in connection with the question as to the axioms of geometry is actually unavoidable, or whether it may not appear rather as a consequence of the group concept and the other geometrical axioms. This consideration, as well as certain other problems in connection with the arithmetical axioms, brings before us the more general question: *How far Lie's concept of continuous groups of transformations is approachable in our investigations without the assumption of the differentiability of the functions.*

Lie defines a finite continuous group of transformations as a system of transformations

$$x_i' = f_i(x_1, \cdots, x_n; a_1, \cdots, a_r) \qquad (i = 1, \cdots, n)$$

having the property that any two arbitrarily chosen transformations of the system, as

$$x_i' = f_i(x_1, \cdots, x_n; a_1, \cdots, a_r),$$
$$x_i'' = f_i(x_1', \cdots, x_n'; b_1, \cdots, b_r),$$

applied successively result in a transformation which also belongs to the system, and which is therefore expressible in the form

$$x_i'' = f_i\{f_1(x, a), \cdots, f_n(x, a); b_1, \cdots, b_r\}$$
$$= f_i(x_1, \cdots, x_n; c_1, \cdots, c_r),$$

where $c_1, \cdots, c_r$ are certain functions of $a_1, \cdots, a_r$ and $b_1, \cdots, b_r$. The group property thus finds its full expression in a system of functional equations and of itself imposes no additional restrictions upon the functions $f_1, \cdots, f_n; c_1, \cdots, c_r$. Yet Lie's further treatment of these functional equations, viz., the derivation of the well-known fundamental differential equations assumes necessarily the continuity and differentiability of the functions defining the group.

As regards continuity: this postulate will certainly be retained for the present—if only with a view to the geometrical and arithmetical applications, in which the continuity of the functions in question appears as a consequence of the axiom of continuity. On the other hand the differentiability of the functions defining the group contains a postulate which, in the geometrical axioms, can be expressed only in a rather forced and complicated manner. Hence there arises the question whether through the introduction of suitable new variables and parameters, the group can always be transformed into one whose defining functions are differentiable; or whether, at least with the help of certain simple assumptions, a transformation is possible into groups admitting Lie's methods. A reduction to analytic groups is, according to a theorem announced by Lie[8] but first proved by Schur,[9] always possible when the group is transitive and the existence of the first and certain second derivatives of the functions defining the group is assumed.

For infinite groups the investigation of the corresponding question is, I believe, also of interest. Moreover we are thus led to the wide and interesting field of functional equations which have been heretofore investigated usually only under the assumption of the differentiability of the functions involved. In particular the functional equations treated by Abel[10] with so much ingenuity, the difference equations, and other equations occurring in the literature of mathematics, do not directly involve anything which necessitates the requirement of the differentiability of the accompanying functions. In the search for certain existence proofs in the calculus of variations I came directly upon the problem: To prove the differentiability of the function under consideration from the existence of a difference equation. In all these cases, then, the problem arises: *In how far are the assertions which we can make in the case of differentiable functions true under proper modifications without this assumption?*

It may be further remarked that H. Minkowski in his above-mentioned *Geometrie der Zahlen* starts with the functional equation

$$f(x_1 + y_1, \cdots, x_n + y_n) \leqq f(x_1, \cdots, x_n) + f(y_1, \cdots, y_n)$$

and from this actually succeeds in proving the existence of certain differential quotients for the function in question.

On the other hand I wish to emphasize the fact that there certainly exist analytical functional equations whose sole solutions are non-differentiable functions. For example a uniform continuous non-differentiable function $\phi(x)$ can be constructed which represents the only solution of the two functional equations

$$\phi(x + \alpha) - \phi(x) = f(x), \phi(x + \beta) - \phi(x) = 0,$$

where α and β are two real numbers, and $f(x)$ denotes, for all the real values of x, a regular analytic uniform function. Such functions are obtained in the simplest manner by means of trigonometrical series by a process similar to that used by Borel (according to a recent announcement of Picard)[11] for the construction of a doubly periodic, nonanalytic solution of a certain analytic partial differential equation.

6. MATHEMATICAL TREATMENT OF THE AXIOMS OF PHYSICS

The investigations on the foundations of geometry suggest the problem: *To treat in the same manner, by means of axioms, those physical sciences in which mathematics plays an important part; in the first rank are the theory of probabilities and mechanics.*

As to the axioms of the theory of probabilities,[12] it seems to me desirable that their logical investigation should be accompanied by a rigorous and satisfactory development of the method of mean values in mathematical physics, and in particular in the kinetic theory of gases.

Important investigations by physicists on the foundations of mechanics are at hand; I refer to the writings of Mach,[13] Hertz,[14] Boltzmann[15] and Volkmann.[16] It is therefore very desirable that the discussion of the foundations of mechanics be taken up by mathematicians also. Thus Boltzmann's work on the principles of mechanics suggests the problem of developing mathematically the limiting processes, there merely indicated, which lead from the atomistic view to the laws of motion of continua. Conversely one might try to derive the laws of the motion of rigid bodies by a limiting process from a system of axioms depending upon the idea of continuously varying conditions of a material filling all space continuously, these conditions being defined by parameters. For the question as to the equivalence of different systems of axioms is always of great theoretical interest.

If geometry is to serve as a model for the treatment of physical axioms, we shall try first by a small number of axioms to include as large a class as possible of physical phenomena, and then by adjoining new axioms to arrive gradually at the more special theories. At the same time Lie's a principle of subdivision can perhaps be derived from profound theory of infinite transformation groups. The mathematician will have also to take account not only of those theories coming near to reality, but also, as in geometry, of all logically possible theories. He must be always alert to obtain a complete survey of all conclusions derivable from the system of axioms assumed.

Further, the mathematician has the duty to test exactly in each instance whether the new axioms are compatible with the previous ones. The physicist, as his theories develop, often finds himself forced by the results of his experiments to make new hypotheses, while he depends, with respect to the compatibility of the new hypotheses with the old axioms, solely upon these experiments or upon a certain physical intuition, a practice which in the rigorously logical building up of a theory is not admissible. The desired proof of the compatibility of all assumptions seems to me also of importance, because the effort to obtain such proof always forces us most effectually to an exact formulation of the axioms.

So far we have considered only questions concerning the foundations of the mathematical sciences. Indeed, the study of the foundations of a science is always particularly attractive, and the testing of these foundations will always be among the foremost problems of the investigator. Weierstrass once said, "The final object always to be kept in mind is to arrive at a correct understanding of the foundations of the science. . . . But to make any progress in the sciences the study of particular problems is, of course, indispensable." In fact, a thorough understanding of its special theories is necessary to the successful treatment of the foundations of the science. Only that architect is in the position to lay a sure foundation for a structure who knows its purpose thoroughly and in detail. So we turn now to the special problems of the separate branches of mathematics and consider first arithmetic and algebra.

7. IRRATIONALITY AND TRANSCENDENCE OF CERTAIN NUMBERS

Hermite's arithmetical theorems on the exponential function and their extension by Lindemann are certain of the admiration of all generations of mathematicians. Thus the task at once presents itself to penetrate further along the path here entered, as A. Hurwitz has already done in two interesting papers,[17] "Ueber arithmetische Eigenschaften gewisser transzendenter Funktionen." I should like, therefore, to sketch a class of problems which, in my opinion, should be attacked as here next in order. That certain special transcendental functions, important in analysis, take algebraic values for certain algebraic arguments, seems to us particularly remarkable and worthy of thorough investigation. Indeed, we expect transcendental functions to assume, in general, transcendental values for even algebraic arguments; and, although it is well known that there exist integral transcendental functions which even have rational values for all algebraic arguments, we shall still consider it highly probable that the exponential function $e^{i\pi z}$, for example, which evidently has algebraic values for all rational arguments z, will on the other hand always take transcendental values for irrational algebraic values of the argument z. We can also give this statement a geometrical form, as follows:

If, in an isosceles triangle, the ratio of the base angle to the angle at the vertex be algebraic but not rational, the ratio between base and side is always transcendental.

In spite of the simplicity of this statement and of its similarity to the problems solved by Hermite and Lindemann, I consider the proof of this theorem very difficult; as also the proof that

The expression α^{β}, for an algebraic base α and an irrational algebraic exponent β, e.g., the number $2^{\sqrt{2}}$ or $e^{\pi} = i^{-2i}$, always represents a transcendental or at least an irrational number.

It is certain that the solution of these and similar problems must lead us to entirely new methods and to a new insight into the nature of special irrational and transcendental numbers.

8. PROBLEMS OF PRIME NUMBERS

Essential progress in the theory of the distribution of prime numbers has lately been made by Hadamard, de la Vallée-Poussin, Von Mangoldt and others. For the complete solution, however, of the problems set us by Riemann's paper "Ueber die Anzahl der Primzahlen unter einer gegebenen Grösse," it still remains to prove the correctness of an exceedingly important statement of Riemann, viz., *that the zero points of the function $\zeta(s)$ defined by the series*

$$\zeta(s) = 1 + \frac{1}{2^s} + \frac{1}{3^s} + \frac{1}{4^s} + \cdots$$

all have the real part ½, except the well-known negative integral real zeros. As soon as the proof has been successfully established, the next problem would consist in testing more exactly Riemann's infinite series for the number of primes below a given number and, especially, *to decide whether the difference between the number of primes below a number x and the integral logarithm of x does in fact become infinite of an order not greater than ½ in x.*[18] Further, we should determine whether the occasional condensation of prime numbers which has been noticed in counting primes is really due to those terms of Riemann's formula which depend upon the first complex zeros of the function $\zeta(s)$.

After an exhaustive discussion of Riemann's prime number formula, perhaps we may sometime be in a position to attempt the rigorous solution of Goldbach's problem,[19] viz., whether every integer is expressible as the sum of two positive prime numbers; and further to attack the well-known question, whether there are an infinite number of pairs of prime numbers with the difference 2, or even the more general problem, whether the linear diophantine equation

$$ax + by + c = 0$$

(with given integral coefficients each prime to the others) is always solvable in prime numbers x and y.

But the following problem seems to me of no less interest and perhaps of still wider range: *To apply the results obtained for the distribution of rational prime numbers to the theory of the distribution of ideal primes in a given number-field k*—a problem which looks toward the study of the function $\zeta_k(s)$ belonging to the field and defined by the series

$$\zeta_k(s) = \Sigma \frac{1}{n(j)^s},$$

where the sum extends over all ideals j of the given realm k, and $n(j)$ denotes the norm of the ideal j.

I may mention three more special problems in number theory: one on the laws of reciprocity, one on diophantine equations, and a third from the realm of quadratic forms.

9. PROOF OF THE MOST GENERAL LAW OF RECIPROCITY IN ANY NUMBER FIELD

For any field of numbers the law of reciprocity is to be proved for the residues of the lth power, when l denotes an odd prime, and further when l is a power of 2 or a power of an odd prime.

The law, as well as the means essential to its proof, will, I believe, result by suitably generalizing the theory of the field of the lth roots of unity,[20] developed by me, and my theory of relative quadratic fields.[21]

10. DETERMINATION OF THE SOLVABILITY OF A DIOPHANTINE EQUATION

Given a diophantine equation with any number of unknown quantities and with rational integral numerical coefficients: *To devise a process according to which it can be determined by a finite number of operations whether the equation is solvable in rational integers.*

11. QUADRATIC FORMS WITH ANY ALGEBRAIC NUMERICAL COEFFICIENTS

Our present knowledge of the theory of quadratic number fields[22] puts us in a position *to attack successfully the theory of quadratic forms with any number of variables and with any algebraic numerical coefficients*. This leads in particular to the interesting problem: to solve a given quadratic equation with algebraic numerical coefficients in any number of variables by integral or fractional numbers belonging to the algebraic realm of rationality determined by the coefficients.

The following important problem may form a transition to algebra and the theory of functions:

12. EXTENSION OF KRONECKER'S THEOREM ON ABELIAN FIELDS TO ANY ALGEBRAIC REALM OF RATIONALITY

The theorem that every abelian number field arises from the realm of rational numbers by the composition of fields of roots of unity is due to Kronecker. This fundamental theorem in the theory of integral equations contains two statements, namely:

First. It answers the question as to the number and existence of those equations which have a given degree, a given abelian group and a given discriminant with respect to the realm of rational numbers.

Second. It states that the roots of such equations form a realm of algebraic numbers which coincides with the realm obtained by assigning to the argument z in the exponential function $e^{i\pi z}$ all rational numerical values in succession.

The first statement is concerned with the question of the determination of certain algebraic numbers by their groups and their branching. This question corresponds, therefore, to the known problem of the determination of algebraic functions corresponding to given Riemann surfaces. The second statement furnishes the required numbers by transcendental means, namely, by the exponential function $e^{i\pi z}$.

Since the realm of the imaginary quadratic number fields is the simplest after the realm of rational numbers, the problem arises, to extend Kronecker's theorem to this case. Kronecker himself has made the assertion that the abelian equations in the realm of a quadratic field are given by the equations of transformation of elliptic functions with singular moduli, so that the elliptic function assumes here the same role as the exponential function in the former case. The proof of Kronecker's conjecture has not yet been furnished; but I believe that it must be obtainable without very great difficulty on the basis of the theory of complex multiplication developed by H. Weber[23] with the help of the purely arithmetical theorems on class fields which I have established.

Finally, the extension of Kronecker's theorem to the case that, *in place of the realm of rational numbers or of the imaginary quadratic field, any algebraic field whatever is laid down as realm of rationality*, seems to me of the greatest importance. I regard this problem as one of the most profound and far-reaching in the theory of numbers and of functions.

The problem is found to be accessible from many standpoints. I regard as the most important key to the arithmetical part of this problem the general law of reciprocity for residues of the lth powers within any given number field.

As to the function-theoretical part of the problem, the investigator in this attractive region will be guided by the remarkable analogies which are noticeable between the theory of algebraic functions of one variable and the theory of algebraic numbers. Hensel[24] has proposed and investigated the analogue in the theory of algebraic numbers to the development in power series of an algebraic function; and Landsberg[25] has treated the analogue of the Reimann-Roch theorem. The analogy between the deficiency of a Riemann surface and that of the class number of a field of numbers is also evident. Consider a Riemann surface of deficiency $p = 1$ (to touch on the simplest case only) and on the other hand a number field of class $h = 2$. To the proof of the existence of an integral everywhere finite on the Riemann surface, corresponds the proof of the existence of an integer α in the number field such that the number $\sqrt{\alpha}$ represents a quadratic field, relatively unbranched with respect to the fundamental field. In the theory of algebraic functions, the method of boundary values (*Randwerthaufgabe*) serves, as is well known, for the proof of Riemann's existence theorem. In the theory of number fields also, the proof of the existence of just this number α offers the greatest difficulty. This proof succeeds with indispensable assistance from the theorem that in the number field there are always prime ideals corresponding to given residual properties. This latter fact is therefore the analogue in number theory to the problem of boundary values.

The equation of Abel's theorem in the theory of algebraic functions expresses, as is well known, the necessary and sufficient condition that the points in question on the Riemann surface are the zero points of an algebraic function belonging to the surface. The exact analogue of Abel's theorem, in the theory of the number field of class $h = 2$, is the equation of the law of quadratic reciprocity[26]

$$\left(\frac{\alpha}{j}\right) = +1,$$

which declares that the ideal j is then and only then a principal ideal of the number field when the quadratic residue of the number α with respect to the ideal j is positive.

It will be seen that in the problem just sketched the three fundamental branches of mathematics, number theory, algebra and function theory, come into closest touch with one another, and I am certain that the theory of analytical functions of several variables in particular would be notably enriched if one should succeed *in finding and discussing those functions which play the part for any algebraic number field corresponding to that of the exponential function in the field of rational numbers and of the elliptic modular functions in the imaginary quadratic number field.*

Passing to algebra, I shall mention a problem from the theory of equations and one to which the theory of algebraic invariants has led me.

13. IMPOSSIBILITY OF THE SOLUTION OF THE GENERAL EQUATION OF THE 7TH DEGREE BY MEANS OF FUNCTIONS OF ONLY TWO ARGUMENTS

. . .

14. PROOF OF THE FINITENESS OF CERTAIN COMPLETE SYSTEMS OF FUNCTIONS

In the theory of algebraic invariants, questions as to the finiteness of complete systems of forms deserve, as it seems to me, particular interest. L. Maurer[27] has lately succeeded in extending the theorems on finiteness in invariant theory proved by P. Gordan and myself, to the case where, instead of the general projective group, any subgroup is chosen as the basis for the definition of invariants.

An important step in this direction has been taken already by A. Hurwitz,[28] who, by an ingenious process, succeeded in effecting the proof, in its entire generality, of the finiteness of the system of orthogonal invariants of an arbitrary ground form. . . .

From the boundary region between algebra and geometry, I will mention two problems. The one concerns enumerative geometry and the other the topology of algebraic curves and surfaces.

15. RIGOROUS FOUNDATION OF SCHUBERT'S ENUMERATIVE CALCULUS

The problem consists in this: *To establish rigorously and with an exact determination of the limits of their validity those geometrical numbers which Schubert[29] especially has determined on the basis of the so-called principle of special position, or conservation of number, by means of the enumerative calculus developed by him.*

Although the algebra of today guarantees, in principle, the possibility of carrying out the processes of elimination, yet for the proof of the theorems of enumerative geometry decidedly more is requisite, namely, the actual carrying out of the process of elimination in the case of equations of special form in such a way that the degree of the final equations and the multiplicity of their solutions may be foreseen.

16. PROBLEM OF THE TOPOLOGY OF ALGEBRAIC CURVES AND SURFACES

The maximum number of closed and separate branches which a plane algebraic curve of the nth order can have has been determined by Harnack.[30] There arises the further question as to the relative position of the branches in the plane. As to curves of the 6th order, I have satisfied myself—by a complicated process, it is true—that of the eleven branches which they can have according to Harnack, by no means all can lie external to one another, but that one branch must exist in whose interior one branch and in whose exterior nine branches lie, or inversely. *A thorough investigation of the relative position of the separate branches when their number is the maximum seems to me to be of very great interest, and not less so the corresponding investigation as to the number, form, and position of the sheets of an algebraic surface in space.* Till now, indeed, it is not even known what is the maximum number of sheets which a surface of the 4th order in three dimensional space can really have.[31]

17. EXPRESSION OF DEFINITE FORMS BY SQUARES

A rational integral function or form in any number of variables with real coefficients such that it becomes negative for no real values of these variables, is said to be *definite*. The system of all definite forms is invariant with respect to the operations of addition and multiplication, but the quotient of two definite forms—in case it should be an integral function of the variables—is also a definite form. The square of any form is evidently always a definite form. But since, as I have shown,[32] not every definite form can be compounded by addition from squares of forms, the question arises—which I have answered affirmatively for ternary forms[33]—whether every definite form may not be expressed as a quotient of sums of squares of forms. At the same time it is desirable, for certain questions as to the possibility of certain geometrical constructions, to know whether the coefficients of the forms to be used in the expression may always be taken from the realm of rationality given by the coefficients of the form represented.[34]

I mention one more geometrical problem:

18. BUILDING UP OF SPACE FROM CONGRUENT POLYHEDRA

If we enquire for those groups of motions in the plane for which a fundamental region exists, we obtain various answers, according as the plane considered is Riemann's (elliptic), Euclid's, or Lobachevsky's (hyperbolic). In the case of the elliptic plane there is a finite number of essentially different kinds of fundamental regions, and a finite number of congruent regions suffices for a complete covering of the whole plane; the group consists indeed of a finite number of motions only. In the case of the hyperbolic plane there is an infinite number of essentially different kinds of fundamental regions, namely, the well-known Poincaré polygons. For the complete covering of the plane an infinite number of congruent regions is necessary. The case of Euclid's plane stands between these; for in this case there is only a finite number of essentially different kinds of groups of motions with fundamental regions, but for a complete covering of the whole plane an infinite number of congruent regions is necessary.

Exactly the corresponding facts are found in space of three dimensions. The fact of the finiteness of the groups of motions in elliptic space is an immediate consequence of a fundamental theorem of C. Jordan,[35] whereby the number of essentially different kinds of finite groups of linear substitutions in n variables does not surpass a certain finite limit dependent upon n. The groups of motions with fundamental regions in hyperbolic space have been investigated by Fricke and Klein in the lectures on the theory of automorphic functions,[36] and finally Fedorov,[37] Schoenflies[38] and lately Rohn[39] have given the proof that there are, in euclidean space, only a finite number of essentially different kinds of groups of motions with a fundamental region. Now, while the results and methods of proof applicable to elliptic and hyperbolic space hold directly for n-dimensional space also, the generalization of the theorem for euclidean space seems to offer decided difficulties. The investigation of the following question is therefore desirable: *Is there in n-dimensional euclidean space also only a finite number of essentially different kinds of groups of motions with a fundamental region?*

A fundamental region of each group of motions, together with the congruent regions arising from the group, evidently fills up space completely. The question arises: *Whether polyhedra also exist which do not appear as fundamental regions of groups of motions, by means of which nevertheless by a suitable juxtaposition of congruent copies a complete filling up of all space is possible.* I point out the following question, related to the preceding one, and important to number theory and perhaps sometimes useful to physics and chemistry: How can one arrange most densely in space an infinite number of equal solids of given form, e.g., spheres with given radii or regular tetrahedra with given edges (or in prescribed position), that is, how can one so fit them together that the ratio of the filled to the unfilled space may be as great as possible?

If we look over the development of the theory of functions in the last century, we notice above all the fundamental importance of that class of functions which we now designate as analytic functions—a class of functions which will probably stand permanently in the center of mathematical interest.

There are many different standpoints from which we might choose, out of the totality of all conceivable functions, extensive classes worthy of a particularly thorough investigation. Consider, for example, *the class of functions characterized by ordinary or partial algebraic differential equations*. It should be observed that this class does not contain the functions that arise in number theory and whose investigation is of the greatest importance. For example, the before-mentioned function $\zeta(s)$ satisfies no algebraic differential equation, as is easily seen with the help of the well-known relation between $\zeta(s)$ and $\zeta(1 - s)$, if one refers to the theorem proved by Hölder,[40] that the function $\Gamma(x)$ satisfies no algebraic differential equation. Again, the function of the two variables s and x defined by the infinite series

$$\zeta(s, x) = x + \frac{x^2}{2^s} + \frac{x^3}{3^s} + \frac{x^4}{4^s} + \cdots,$$

which stands in close relation with the function $\zeta(s)$, probably satisfies no algebraic partial differential equation. In the investigation of this question the functional equation

$$x \frac{\partial \zeta(s, x)}{\partial x} = \zeta(s - 1, x)$$

will have to be used.

If, on the other hand, we are lead by arithmetical or geometrical reasons to consider the class of all those functions which are continuous and indefinitely differentiable, we should be obliged in its investigation to dispense with that pliant instrument, the power series, and with the circumstance that the function is fully determined by the assignment of values in any region, however small. While, therefore, the former limitation of the field of functions was too narrow, the latter seems to me too wide.

The idea of the analytic function on the other hand includes the whole wealth of functions most important to science, whether they have their origin in number theory, in the theory of differential equations or of algebraic functional equations, whether they arise in geometry or in mathematical physics; and, therefore, in the entire realm of functions, the analytic function justly holds undisputed supremacy.

19. ARE THE SOLUTIONS OF REGULAR PROBLEMS IN THE CALCULUS OF VARIATIONS ALWAYS NECESSARILY ANALYTIC?

One of the most remarkable facts in the elements of the theory of analytic functions appears to me to be this: That there exist partial differential equations whose integrals are all of necessity analytic functions of the independent variables, that is, in short, equations susceptible of none but analytic solutions. The best known partial differential equations of this kind are the potential equation

$$\frac{\partial^2 f}{\partial x^2} + \frac{\partial^2 f}{\partial y^2} = 0$$

and certain linear differential equations investigated by Picard;[41] also the equation

$$\frac{\partial^2 f}{\partial x^2} + \frac{\partial^2 f}{\partial y^2} = e^f,$$

the partial differential equation of minimal surfaces, and others. Most of these partial differential equations have the common characteristic of being the lagrangian differential equations of certain problems of variation, viz., of such problems of variation

$$\int\int F(p, q, z; x, y)\, dx\, dy = \text{minimum}$$

$$\left[p = \frac{\partial z}{\partial x}, q = \frac{\partial z}{\partial y}\right],$$

as satisfy, for all values of the arguments which fall within the range of discussion, the inequality

$$\frac{\partial^2 F}{\partial p^2} \cdot \frac{\partial^2 F}{\partial q^2} - \left(\frac{\partial^2 F}{\partial p \partial q}\right)^2 > 0,$$

F itself being an analytic function. We shall call this sort of problem a *regular* variation problem. It is chiefly the regular variation problems that play a rôle in geometry, in mechanics, and in mathematical physics; and the question naturally arises, whether all solutions of regular variation problems must necessarily be analytic functions. In other words, *does every lagrangian partial differential equation of a regular variation problem have the property of admitting analytic integrals exclusively?* And is this the case even when the function is constrained to assume as, *e.g.*, in Dirichlet's problem on the potential function, boundary values which are continuous, but not analytic?

I may add that there exist surfaces of constant *negative* gaussian curvature which are representable by functions that are continuous and possess indeed all the derivatives, and yet are not analytic; while on the other hand it is probable that every surface whose gaussian curvature is constant and *positive* is necessarily an analytic surface. And we know that the surfaces of positive constant curvature are most closely related to this regular variation problem: To pass through a closed curve in space a surface of minimal area which shall inclose, in connection with a fixed surface through the same closed curve, a volume of given magnitude.

20. THE GENERAL PROBLEM OF BOUNDARY VALUES

An important problem closely connected with the foregoing is the question concerning the existence of solutions of partial differential equations when the values on the boundary of the region are prescribed. This problem is solved in the main by the keen methods of H. A. Schwarz, C. Neumann, and Poincaré for the differential equation of the potential. These methods, however, seem to be generally not capable of direct extension to the case where along the boundary there are prescribed either the differential coefficients or any relations between these and the values of the function. Nor can they be extended immediately to the case where the inquiry is not for potential surfaces but, say, for surfaces of least area, or surfaces of constant positive gaussian curvature, which are to pass through a prescribed twisted curve or to stretch over a given ring surface. It is my conviction that it will be possible to prove these existence theorems by means of a general principle whose nature is indicated by Dirichlet's principle. This general principle will then perhaps enable us to approach the question: *Has not every regular variation problem a solution, provided certain assumptions regarding the given boundary conditions are satisfied* (say that the functions concerned in these boundary conditions are continuous and have in sections one or more derivatives), *and provided also if need be that the notion of a solution shall be suitably extended?*[42]

21. PROOF OF THE EXISTENCE OF LINEAR DIFFERENTIAL EQUATIONS HAVING A PRESCRIBED MONODROMIC GROUP

In the theory of linear differential equations with one independent variable z, I wish to indicate an important problem, one which very likely Riemann himself may have had in mind. This problem is as follows: *To show that there always exists a linear differential equation of the Fuchsian class, with given singular points and monodromic group*. The problem requires the production of n functions of the variable z, regular throughout the complex z plane except at the given singular points; at these points the functions may become infinite of only finite order, and when z describes circuits about these points the functions shall undergo the prescribed linear substitutions. The existence of such differential equations has been shown to be probable by counting the constants, but the rigorous proof has been obtained up to this time only in the particular case where the fundamental equations of the given substitutions have roots all of absolute magnitude unity. L. Schlesinger has given this proof,[43] based upon Poincaré's theory of the Fuchsian ζ-functions. The theory of linear differential equations would evidently have a more finished appearance if the problem here sketched could be disposed of by some perfectly general method.

22. UNIFORMIZATION OF ANALYTIC RELATIONS BY MEANS OF AUTOMORPHIC FUNCTIONS

As Poincaré was the first to prove, it is always possible to reduce any algebraic relation between two variables to uniformity by the use of automorphic functions of one variable. That is, if any algebraic equation in two variables be given, there can always be found for these variables two such single valued automorphic functions of a single variable that their substitution renders the given algebraic equation an identity. The generalization of this fundamental theorem to any analytic non-algebraic relations whatever between two variables has likewise been attempted with success by Poincaré,[44] though by a way entirely different from that which served him in the special problem first mentioned. From Poincaré's proof of the possibility of reducing to uniformity an arbitrary analytic relation between two variables, however, it does not become apparent whether the resolving functions can be determined to meet certain additional conditions. Namely, it is not shown whether the two single valued functions of the one new variable can be so chosen that, while this variable traverses the *regular* domain of those functions, the totality of all regular points of the given analytic field are actually reached and represented. On the contrary it seems to be the case, from Poincaré's investigations, that there are beside the branch points certain others, in general infinitely many other discrete exceptional points of the analytic field, that can be reached only by making the new variable approach certain limiting points of the functions. *In view of the fundamental importance of Poincaré's formulation of the question it seems to me that an elucidation and resolution of this difficulty is extremely desirable.*

In conjunction with this problem comes up the problem of reducing to uniformity an algebraic or any other analytic relation among three or more complex variables—a problem which is known to be solvable in many particular cases. Toward the solution of this the recent investigations of Picard on algebraic functions of two variables are to be regarded as welcome and important preliminary studies.

23. FURTHER DEVELOPMENT OF THE METHODS OF THE CALCULUS OF VARIATIONS

So far, I have generally mentioned problems as definite and special as possible, in the opinion that it is just such definite and special problems that attract us the most and from which the most lasting influence is often exerted upon science. Nevertheless, I should like to close with a general problem, namely with the indication of a branch of mathematics repeatedly mentioned in this lecture—which, in spite of the considerable advancement lately given it by Weierstrass, does not receive the general appreciation which, in my opinion, is its due—I mean the calculus of variations.[45]

The lack of interest in this is perhaps due in part to the need of reliable modern text books. So much the more praiseworthy is it that A. Kneser in a very recently published work has treated the calculus of variations from the modern points of view and with regard to the modern demand for rigor.[46]

The calculus of variations is, in the widest sense, the theory of the variation of functions, and as such appears as a necessary extension of the differential and integral calculus. In this sense, Poincaré's investigations on the problem of three bodies, for example, form a chapter in the calculus of variations, in so far as Poincaré derives from known orbits by the principle of variation new orbits of similar character. . . .

The problems mentioned are merely samples of problems, yet they will suffice to show how rich, how manifold and how extensive the mathematical science of today is, and the question is urged upon us whether mathematics is doomed to the fate of those other sciences that have split up into separate branches, whose representatives scarcely understand one another and whose connection becomes ever more loose. I do not believe this nor wish it. Mathematical science is in my opinion an indivisible whole, an organism whose vitality is conditioned upon the connection of its parts. For with all the variety of mathematical knowledge, we are still clearly conscious of the similarity of the logical devices, the *relationship* of the *ideas* in mathematics as a whole and the numerous analogies in its different departments. We also notice that, the farther a mathematical theory is developed, the more harmoniously and uniformly does its construction proceed, and unsuspected relations are disclosed between hitherto separate branches of the science. So it happens that, with the extension of mathematics, its organic character is not lost but only manifests itself the more clearly.

But, we ask, with the extension of mathematical knowledge will it not finally become impossible for the single investigator to embrace all departments of this knowledge? In answer let me point out how thoroughly it is ingrained in mathematical science that every real advance goes hand in hand with the invention of sharper tools and simpler methods which at the same time assist in understanding earlier theories and cast aside older more complicated developments. It is therefore possible for the individual investigator, when he makes these sharper tools and simpler methods his own, to find his way more easily in the various branches of mathematics than is possible in any other science.

The organic unity of mathematics is inherent in the nature of this science, for mathematics is the foundation of all exact knowledge of natural phenomena. That it may completely fulfil this high mission, may the new century bring it gifted masters and many zealous and enthusiastic disciples.

NOTES

1. See Helmholtz, "Ueber die Wechselwirkung der Naturkräefte und die darauf bezüglichen neuesten Ermittelungen der Physik"; Vortrag, gehalten in Königsberg, 1854.

2. *Jahresbericht der Deutschen Mathematiker-Vereinigung*, vol. 8 (1900), p. 180.

3. Werke, vol. 8, pp. 241 and 244.

4. Cf., beside earlier literature, Hilbert, *Grundlagen der Geometrie,* Leipzig, 1899, ch. 4. [Translation by Townsend, Chicago, 1902.]

5. Since this was written Mr. Dehn has succeeded in proving this impossibility. See his note: "Ueber raumgleiche Polyeder," in *Nachrichten d. K. Gesellsch. d. Wiss. zu Göttingen*, 1900, and a paper soon to appear in the *Math. Annalen* [vol. 55, pp. 465-478].

6. Leipzig, 1896.

7. *Math. Annalen*, vol. 46, p. 91.

8. Lie-Engel, *Theorie der Transformationsgruppen,* vol. 3, Leipzig, 1893, §§82, 144.

9. "Ueber den analytischen Charakter der eine endliche Kontinuierliche Transformationsgruppen darstellenden Funktionen," *Math. Annalen*, vol. 41.

10. Werke, vol. 1, pp. 1, 61, 389.

11. "Quelques théories fondamentales dans l'analyse mathématique," Conférences faites à Clark University, *Revue générale des Sciences*, 1900, p. 22.

12. Cf. Bohlmann, "Ueber Versicherungsmathematik", from the collection: Klein and Riecke, *Ueber angewandte Mathematik und Physik,* Leipzig, 1900.

13. *Die Mechanik in ihrer Entwickelung,* Leipzig, 4th edition, 1901.

14. *Die Prinzipien der Mechanik,* Leipzig, 1894.

15. *Vorlesungen* über die Principe der Mechanik, Leipzig, 1897.

16. *Einführung in das Studium der theoretischen Physik,* Leipzig, 1900.

17. *Math. Annalen*, vols. 22, 32 (1883, 1888).

18. Cf. an article by H. von Koch, which is soon to appear in the *Math. Annalen* [Vol. 55, p. 441].

19. Cf. P. Stäckel: "Über Goldbach's empirisches Theorem," *Nachrichten d. K. Ges. d. Wiss. zu Göttingen*, 1896, and Landau, *ibid.*, 1900.

20. *Jahresber. d. Deutschen Math.-Vereinigung*, "Ueber die Theorie der algebraischen Zahlkörper," vol. 4 (1897), Part V.

21. *Math Annalen*, vol. 51 and *Nachrichten d. K. Ges. d. Wiss. zu Göttingen*, 1898.

22. Hilbert, "Ueber den Dirichlet'schen biquadratischen Zahlenkörper," *Math. Annalen*, vol. 45; "Ueber die Theorie der relativquadratischen Zahlkörper," *Jahresber. d. Deutschen Mathematiker-Vereinigung*, 1897, and *Math. Annalen*, vol. 51; "Ueber die Theorie der relative-Abelschen Körper," *Nachrichten d. K. Ges. d. Wiss. zu Göttingen*, 1898; *Grundlagen der Geometrie,* Leipzig, 1899, Chap. VIII, §83. Cf. also the dissertation of G. Rückle, Göttingen, 1901.

23. *Elliptische Functionen und algebraische Zahlen.* Braunschweig, 1891.

24. *Jahresber. d. Deutschen Math.-Vereinigung*, vol. 6, and an article soon to appear in the *Math. Annalen* [Vol. 55, p. 301]: "Ueber die Entwickelung der algebraischen Zahlen in Potenzreihen."

25. *Math. Annalen*, vol. 50 (1898).

26. Cf. Hilbert, "Ueber die Theorie der relativ-Abelschen Zahlkörper," *Gött. Nachrichten*, 1898.

27. Cf. *Sitzungsber. d. K. Acad. d. Wiss. zu München*, 1899, and an article about to appear in the *Math. Annalen*.

28. "Ueber die Erzeugung der Invarianten durch Integration," *Nachrichten d. K. Gesellschaft d. Wiss. zu Göttingen*, 1897.

29. *Kalkül der abzählenden Geometrie,* Leipzig, 1879.

30. *Math. Annalen*, vol. 10.

31. Cf. Rohn, "Flacheu vierter Ordnung," Preisschriften der Fürstlich Jablonowskischen Gesellschaft, Leipzig, 1886.

32. *Math. Annalen*, vol. 32.

33. *Acta Mathematica*, vol. 17.

34. Cf. Hilbert: *Grundlagen der Geometrie,* Leipzig, 1899, Chap. 7 and in particular § 38.

35. *Crelle's Journal*, vol. 84 (1878), and *Atti d. Reale Acad. di Napoli* 1880.

36. Leipzig, 1897. Cf. especially Abschnitt I, Chapters 2 and 3.

37. *Symmetrie der regelmässigen Systeme von Figuren,* 1890.

38. *Krystallsysteme und Krystallstruktur,* Leipzig, 1891.

39. *Math. Annalen*, vol. 53.

40. *Math. Annalen*, vol. 28.

41. *Jour. de l' Ecole Polytech.*, 1890.

42. Cf. my lecture on Dirichlet's principle in the *Jahresber. d. Deutschen Math.-Vereinigung*, vol. 8 (1900), p. 184.

43. *Handbuch der Theorie der linearen Differentialgleichungen,* vol. 2, part 2, No. 366.

44. *Bull. de la Soc. Math. de France*, vol. 11 (1883).

45. Text-books: Moigno-Lindelöf, *Leçons du calcul des variations,* Paris, 1861, and A. Kneser, *Lehrbuch der Variations rechnung,* Braunschweig, 1900.

46. As an indication of the contents of this work, it may here be noted that for the simplest problems Kneser derives sufficient conditions of the extreme even for the case that one limit of integration is variable, and employs the envelope of a family of curves satisfying the differential equations of the problem to prove the necessity of Jacobi's conditions of the extreme. Moreover, it should be noticed that Kneser applies Weierstrass's theory also to the inquiry for the extreme of such quantities as are defined by differential equations.

Chapter IX
The Early 20th Century to 1932

Section A
Creativity and the Paris Problems

ERNST (FRIEDRICH FERDINAND) ZERMELO (1871–1953)

The German mathematician Ernst Zermelo was the son of Ferdinand Zermelo, a college professor, and Maria Zieger. He grew up in Berlin, receiving his secondary education at the Luisenstädtisches Gymnasium. After graduating in 1889, he studied mathematics, physics, and philosophy at universities in Berlin, Halle, and Freiburg. Among his teachers were Max Planck, Herman A. Schwarz, and Edmund Husserl. The University of Berlin awarded him the doctorate in 1894 for his dissertation *Untersuchungen zur Variationsrechnung* ("Investigations in the Calculus of Variations"), which extended the method of Karl Weierstrass for the extrema of integrals and which carefully defined the concept of neighborhood in the space of curves. He retained a lifelong interest in the calculus of variations, often lecturing on it and contributing to its progress.

Upon receiving his doctorate Zermelo went to the University of Göttingen, which appointed him *Privatdozent* in 1899 after he submitted his *Habilitationsschrift* on hydrodynamics. During the winter semester of 1900–01 he lectured on set theory. His thorough study of Cantor's work and discussion of it with one of his former teachers, Erhard Schmidt, led in 1904 to his sensational proof of the well-ordering theorem. A year later Göttingen named him a titular professor.

In 1910 Zermelo accepted a professorship at Zurich, but he was forced to resign in 1916 because of poor health. Fortunately, David Hilbert in Göttingen had earlier come to his assistance. Hilbert had helped to arrange that interest from the Wolfskehl Fund, a sum of 5000 marks be awarded to Zermelo in recognition of his results in set theory. With this income Zermelo was able to leave Zurich to reside in the Black Forest, where his health gradually improved. In 1926 the University of Freiburg in Breisgau appointed him an honorary professor. He accepted and taught at Freiburg until 1935, when he renounced any connection with the university because he disapproved of the Hitler regime. After World War II he asked to be reinstated; this was granted in 1946.

Zermelo contributed decisively to the development of set theory. He first proved the well-ordering theorem, which holds that every set can be well-ordered. This is a valuable tool for establishing correspondence between sets and thus comparing their sizes. Cantor, who in 1883 had called the theorem a "fundamental law of thought of great consequence," had unsuccessfully sought a proof. In his Paris Problems of 1900, Hilbert listed its proof as one of the most important tasks facing mathematicians. In 1904 Zermelo completed a relatively simple proof that utilized the powerful new axiom of choice following a suggestion of E. Schmidt. His proof provoked a great deal of criticism, most of it unjustified. A number of mathematicians, including Borel, Hadamard, Lebesgue, and Peano, found the use of the axiom of choice objectionable. In 1908 Zermelo provided a second proof based on this same axiom with fewer remaining set theoretic assumptions. After this proof appeared the debate subsided, and it was generally accepted that the axiom of choice implies the well-ordering theorem.

Zermelo also set forth the first axiomatic or formal set theory. At the turn of the 20th century a number of contradictions or paradoxes had appeared in Cantor's naive set theory because it failed to restrict the concept of a set. In 1897 C. Burali-Forti raised a paradox when he observed that if one assumes that the set of ordinals is well-ordered and hence has an ordinal, this ordinal turns out to be both an element of the set of ordinals and yet greater than any ordinal in the set. Russell, Whitehead, and Poincaré called unpredicative such definitions with objects defined in terms of a class of objects that contain the objects. To avoid the paradoxes Russell proposed in 1908 his ramified theory of types, a far-reaching logical theory, while Zermelo offered seven clear and explicit axioms to clarify what is meant by a set and define the properties that a set has. His axioms admitted only those safe classes that seemed to preclude contradictions in the system. Zermelo's axioms constituted an effort in predicative mathematics to meet the needs of working mathematicians.

Although nonaxiomatic Cantorean set theory flourished in the decade after 1908, especially the branch leading to point-set topology, there was no progress in formal set theory until 1921 when Abraham Adolph Fraenkel improved upon Zermelo's pioneering work. The resulting Zermelo–Fraenkel system placed some restrictions on the formation of sets that were necessary to avoid paradoxes but at the same time admitted enough sets to serve as a foundation for practically all of classical analysis. Subsequently, Thoralf Skolem and John von Neumann have modified Zermelo's formal set theory.

120. From a letter to David Hilbert (September 24, 1904)*

ERNST ZERMELO

PROOF THAT EVERY SET CAN BE WELL-ORDERED

. . . The proof in question grew out of conversations that I had last week with Mr. Erhard Schmidt, and it is as follows.

(1) Let M be an arbitrary set of cardinality m, let m denote an arbitrary ele-

*Source: This translation by Stefan Bauer-Mengelberg is taken from Jean van Heijenoort, ed., *From Frege to Gödel: A Source Book in Mathematical Logic, 1879-1931* (1967), 139-141. It is reprinted by permission of Springer Verlag.

ment of it, let M', of cardinality m′, be a subset of M that contains at least one element m and may even contain all elements of M, and let $M - M'$ be the subset "complementary" to M'. Two subsets are regarded as distinct if one of them contains some element that does not occur in the other. Let the set of all subsets M' be denoted by M.

(2) *Imagine that with every subset M' there is associated an arbitrary element m'_1 that occurs in M' itself; let m'_1 be called the "distinguished" element of M'.* This yields a "covering" γ of the set M by certain elements of the set M. The numbers of these coverings γ is equal to the product $\Pi m'$ taken over all subsets M' and is therefore certainly different from 0. In what follows we take an arbitrary covering γ and derive from it a definite well-ordering of the elements of M.

(3) *Definition.* Let us apply the term "γ-set" to any well-ordered set M_γ that consists entirely of elements of M and has the following property: whenever a is an arbitrary element of M_γ and A is the "associated" segment, which consists of the elements x of M such that $x < a$, a is the distinguished element of $M - A$.

(4) *There are γ-sets included in M.* Thus, for example, [the set containing just] m_1, the distinguished element of M' when $M' = M$, is itself a γ-set; so is the (ordered) set $M_2 = (m_1, m_2)$, where m_2 is the distinguished element of $M - m_1$.

(5) *Whenever M'_γ and M''_γ are any two distinct γ-sets* (associated, however, with the same covering γ chosen once for all!), *one of the two is identical with a segment of the other.*

For, of the two well-ordered sets, let M'_γ be the one for which there exists a similar mapping onto the other, M''_γ, or onto one of its segments. Then any two elements corresponding to each other under this mapping must be identical. For the first element of *every* γ-set is m_1, since the associated segment A contains no element and therefore $M - A = M$. If now m' were the *first* element of M'_γ that differs from the corresponding element m'', the associated segments A' and A'' would still have to be identical, consequently also the complements $M - A'$ and $M - A''$, and thus their distinguished elements m' and m'' themselves, contrary to assumption.

(6) *Consequences.* If two γ-sets have an element a in common, they also have the segment A of the preceding elements in common. If they have *two* elements a and b in common, then either in *both* sets $a < b$ or in *both* sets $b < a$.

(7) If we call any element of M that occurs in some γ-set a "γ-element," the following theorem holds: *The totality L_γ of all γ-elements can be so ordered that it will itself be a γ-set, and it contains all elements of the original set M. M* itself is thereby well-ordered.

(I) If a and b are two arbitrary γ-elements and if M'_γ and M''_γ are any two γ-sets to which they respectively belong, then according to (5) the larger of the two γ-sets contains both elements and determines whether the order relation is $a < b$ or $b < a$. According to (6) this order relation is independent of the γ-sets selected.

(II) If a, b, and c are three arbitrary γ-elements and if $a < b$ and $b < c$, then always $a < c$. For according to (6) every γ-set containing c also contains b, hence also a, and then, since it is simply ordered, within the set, $a < c$ indeed follows from $a < b$ and $b < c$. The set L_γ is therefore *simply ordered.*

(III) If L'_γ is an arbitrary subset of L and a is one of its elements, belonging, say, to the γ-set M_γ, then according to (6) M_γ contains all elements preceding a, hence includes the subset L''_γ that is obtained from L'_γ when all elements following a are removed; L''_γ, being a subset of the well-ordered set M_γ, possesses a *first* element, which is also the first element of L'_γ. L_γ is therefore also *well-ordered.*

(IV) If a is an arbitrary γ-element and A the totality of *all* preceding elements $x < a$, then according to (6), in every set

M_γ containing a, A is the segment associated with a; according to (3), consequently, a is the distinguished element of $M - A$. *Therefore* L_γ *is itself a* γ*-set*.

(V) If there existed an element of M that belonged to *no* γ-set, that consequently was an element of $M - L_\gamma$, there would also exist a distinguished element m'_1 of $M - L_\gamma$, and the ordered set (L_γ, m'_1), in which every γ-element precedes the element m'_1, would itself according to (3) be a γ-set. Then m'_1 too would be a γ-element, contrary to assumption; so really $L_\gamma = M$, and thus M is itself a *well-ordered set*.

Accordingly, to every covering γ there corresponds a definite well-ordering of the set M, even if the well-orderings that correspond to two distinct coverings are not always themselves distinct. There must at any rate exist *at least one* such well-ordering, and every set for which the totality of subsets, and so on, is meaningful may be regarded as well-ordered and its cardinality as an "aleph." It therefore follows that, for every transfinite cardinality,

$$\mathfrak{m} = 2\mathfrak{m} = \aleph_0\mathfrak{m} = \mathfrak{m}^2, \text{ and so forth;}$$

and any two sets are "comparable"; that is, one of them can always be mapped one-to-one onto the other or one of its parts.

The present proof rests upon the assumption that coverings γ actually do exist, hence upon the principle that even for an infinite totality of sets there are always mappings that associate with every set one of its elements, or, expressed formally, that the product of an infinite totality of sets, each containing at least one element, itself differs from zero. This logical principle cannot, to be sure, be reduced to a still simpler one, but it is applied without hesitation everywhere in mathematical deduction. For example, the validity of the proposition that the number of parts into which a set decomposes is less than or equal to the number of all of its elements cannot be proved except by associating with each of the parts in question one of its elements.

I owe to Mr. Erhard Schmidt the idea that, by invoking this principle, we can take an *arbitrary* covering γ as a basis for the well-ordering; the proof, as I carried it through, then rests upon the fusion of the various possible "γ-sets," that is, of the well-ordered segments resulting from the ordering principle.

Chapter IX
The Early 20th Century to 1932

Section B
Foundational Crisis and Undecidability

121. From Correspondence Between Bertrand Russell and Gottlob Frege (June 16 and 22, 1902)*

(Russell Paradox and the Logicist School)

RUSSELL TO FREGE

For a year and a half I have been acquainted with your *Grundgesetze der Arithmetik*, but it is only now that I have been able to find the time for the thorough study I intended to make of your work. I find myself in complete agreement with you in all essentials, particularly when you reject any psychological element [Moment] in logic and when you place a high value upon an ideography [Begriffsschrift] for the foundations of mathematics and of formal logic, which, incidentally, can hardly be distinguished. With regard to many particular questions, I find in your work discussions, distinctions, and definitions that one seeks in vain in the works of other logicians. Especially so far as function is concerned (§ 9 of your *Begriffsschrift*), I have been led on my own to views that are the same even in the details. There is just one point where I have encountered a difficulty. You state (p. 17 . . .) that a function, too, can act as the indeterminate element. This I formerly believed, but now this view seems doubtful to me because of the following contradiction. Let *w* be the predicate: to be a predicate that cannot be predicated of itself. Can *w* be predicated of itself? From each answer its opposite follows. Therefore we must conclude that *w* is not a predicate. Likewise there is no class (as a totality) of those classes which, each taken as a totality, do not belong to themselves. From this I conclude that under certain circumstances a definable collection [Menge] does not form a totality.

I am on the point of finishing a book on the principles of mathematics and in it I should like to discuss your work very thoroughly.[1] I already have your books or shall buy them soon, but I would be very grateful to you if you could send me reprints of your articles in various periodicals. In case this should be impossible, however, I will obtain them from a library.

The exact treatment of logic in fun-

*Source: These translations from the German originals by Beverly Woodward are taken from Jean van Heijenoort (ed.), *From Frege to Gödel: A Source Book in Mathematical Logic, 1879-1931*, 124–125 and 127–128. They are reprinted by permission of Harvard University Press. A biography of Bertrand Russell is given before selection 117, and a biography of Frege is given before selection 114.

damental questions, where symbols fail, has remained very much behind; in your works I find the best I know of our time, and therefore I have permitted myself to express my deep respect to you. It is very regrettable that you have not come to publish the second volume of your *Grundgesetze;* I hope that this will still be done.

PS. The above contradiction, when expressed in Peano's ideography, reads as follows:

$$w = \mathrm{cls} \quad x \ni (x \sim\varepsilon\, x) \quad : w\, \varepsilon\, w \,.=.\, w \sim\varepsilon\, w.$$

I have written to Peano about this, but he still owes me an answer.

FREGE TO RUSSELL

Many thanks for your interesting letter of 16 June. I am pleased that you agree with me on many points and that you intend to discuss my work thoroughly. In response to your request I am sending you the following publications:

1. "Kritische Beleuchtung" [*1895*],
2. "Ueber die Begriffsschrift des Herrn Peano" [*1896*],
3. "Ueber Begriff und Gegenstand" [*1892*],
4. "Über Sinn und Bedeutung" [*1892a*],
5. "Ueber formale Theorien der Arithmetik" [*1885*].

I received an empty envelope that seems to be addressed by your hand. I surmise that you meant to send me something that has been lost by accident. If this is the case, I thank you for your kind intention. I am enclosing the front of the envelope.

When I now read my *Begriffsschrift* again, I find that I have changed my views on many points, as you will see if you compare it with my *Grundgesetze der Arithmetik*. I ask you to delete the paragraph beginning "Nicht minder erkennt man" on page 7 of my *Begriffsschrift* ["It is no less easy to see" . . .], since it is incorrect; incidentally, this had no detrimental effects on the rest of the booklet's contents.

Your discovery of the contradiction caused me the greatest surprise and, I would almost say, consternation, since it has shaken the basis on which I intended to build arithmetic. It seems, then, that transforming the generalization of an equality into an equality of courses-of-values [die Umwandlung der Allgemeinheit einer Gleichheit in eine Werthverlaufsgleichheit] (§ 9 of my *Grundgesetze*) is not always permitted, that my Rule V (§ 20, p. 36) is false, and that my explanations in § 31 are not sufficient to ensure that my combinations of signs have a meaning in all cases. I must reflect further on the matter. It is all the more serious since, with the loss of my Rule V, not only the foundations of my arithmetic, but also the sole possible foundations of arithmetic, seem to vanish. Yet, I should think, it must be possible to set up conditions for the transformation of the generalization of an equality into an equality of courses-of-values such that the essentials of my proofs remain intact. In any case your discovery is very remarkable and will perhaps result in a great advance in logic, unwelcome as it may seem at first glance.

Incidentally, it seems to me that the expression "a predicate is predicated of itself" is not exact. A predicate is as a rule a first-level function, and this function requires an object as argument and cannot have itself as argument (subject). Therefore I would prefer to say "a notion is predicated of its own extension." If the function $\Phi(\xi)$ is a concept, I denote its extension (or the corresponding class) by "$\grave{\varepsilon}\Phi(\varepsilon)$" (to be sure, the justification for this has now become questionable to me). In "$\Phi(\grave{\varepsilon}\Phi(\varepsilon))$" or "$\grave{\varepsilon}\Phi(\varepsilon) \cap \grave{\varepsilon}\Phi(\varepsilon)$"[1] we then have a case in which the concept $\Phi(\xi)$ is predicated of its own extension.

The second volume of my *Grundgesetze* is to appear shortly. I shall no doubt have to add an appendix in which your discovery is taken into account. If only I already had the right point of view for that!

NOTES

1. This was done in *Russell The Principles of Mathematics* (1903), Appendix A, "The logical and arithmetical doctrines of Frege."

2. " " is a sign used by Frege for reducing second-level functions to first-level functions. See *Frege 1893*, § 34.

122. From a Letter to Jean van Heijenoort
Concerning the Publication of the Russell–Frege Correspondence*
(November 23, 1962)

I should be most pleased if you would publish the correspondence between Frege and myself, and I am grateful to you for suggesting this. As I think about acts of integrity and grace, I realise that there is nothing in my knowledge to compare with Frege's dedication to truth. His entire life's work was on the verge of completion, much of his work had been ignored to the benefit of men infinitely less capable, his second volume was about to be published, and upon finding that his fundamental assumption was in error, he responded with intellectual pleasure clearly submerging any feelings of personal disappointment. It was almost superhuman and a telling indication of that of which men are capable if their dedication is to creative work and knowledge instead of cruder efforts to dominate and be known.

Yours sincerely,
Bertrand Russell

**Source:* This letter is taken from Jean van Heijenoort (ed.), *From Frege to Gödel*, 127. It is reprinted by permission of Professor Heijenoort and Harvard University Press.

123. From *Principia Mathematica* (1910)*

(Russell's Solution to the Burali-Forti Paradox and to His Own)

ALFRED NORTH WHITEHEAD
BERTRAND RUSSELL

CHAPTER II
THE THEORY OF LOGICAL TYPES

The theory of logical types, to be explained in the present Chapter, recommended itself to us in the first instance by its ability to solve certain contradictions, of which the one best known to mathematicians is Burali-Forti's concerning the greatest ordinal. But the theory in question is not wholly dependent upon this indirect recommendation: it has also a certain consonance with common sense which makes it in-

**Source:* Alfred North Whitehead and Bertrand Russell, *Principia Mathematica to * 56* (1910, reprinted edition 1970), 37-39 and 66-71. This selection is reprinted by permission of the Cambridge University Press.

herently credible. In what follows, we shall therefore first set forth the theory on its own account, and then apply it to the solution of the contradictions.

I. THE VICIOUS-CIRCLE PRINCIPLE

An analysis of the paradoxes to be avoided shows that they all result from a certain kind of vicious circle. The vicious circles in question arise from supposing that a collection of objects may contain members which can only be defined by means of the collection as a whole. Thus, for example, the collection of *propositions* will be supposed to contain a proposition stating that "all propositions are either true or false." It would seem, however, that such a statement could not be legitimate unless "all propositions" referred to some already definite collection, which it cannot do if new propositions are created by statements about "all propositions." We shall, therefore, have to say that statements about "all propositions" are meaningless. More generally, given any set of objects such that, if we suppose the set to have a total, it will contain members which presuppose this total, then such a set cannot have a total. By saying that a set has "no total," we mean, primarily, that no significant statement can be made about "all its members." Proposition, as the above illustration shows, must be a set having no total. The same is true, as we shall shortly see, of propositional functions, even when these are restricted to such as can significantly have as argument a given object *a*. In such cases, it is necessary to break up our set into smaller sets, each of which is capable of a total. This is what the theory of types aims at effecting.

The principle which enables us to avoid illegitimate totalities may be stated as follows: "Whatever involves *all* of a collection must not be one of the collection"; or, conversely: "If, provided a certain collection had a total, it would have members only definable in terms of that total, then the said collection has no total." We shall call this the "vicious-circle principle," because it enables us to avoid the vicious circles involved in the assumption of illegitimate totalities. Arguments which are condemned by the vicious-circle principle will be called "vicious-circle fallacies." Such arguments, in certain circumstances, may lead to contradictions, but it often happens that the conclusions to which they lead are in fact true, though the arguments are fallacious. Take, for example, the law of excluded middle, in the form "all propositions are true or false." If from this law we argue that, because the law of excluded middle is a proposition, therefore the law of excluded middle is true or false, we incur a vicious-circle fallacy. "All propositions" must be in some way limited before it becomes a legitimate totality, and any limitation which makes it legitimate must make any statement about the totality fall outside the totality. Similarly, the imaginary sceptic, who asserts that he knows nothing, and is refuted by being asked if he knows that he knows nothing, has asserted nonsense, and has been fallaciously refuted by an argument which involves a vicious-circle fallacy. In order that the sceptic's assertion may become significant, it is necessary to place some limitation upon the things of which he is asserting his ignorance, because the things of which it is possible to be ignorant form an illegitimate totality. But as soon as a suitable limitation has been placed by him upon the collection of propositions of which he is asserting his ignorance, the proposition that he is ignorant of every member of this collection must not itself be one of the collection. Hence any significant scepticism is not open to the above form of refutation.

The paradoxes of symbolic logic concern various sorts of objects: propositions, classes, cardinal and ordinal numbers, etc. All these sorts of objects, as we shall show, represent illegitimate totalities, and are therefore capable of giving rise to vicious-circle fallacies. But by means of the theory (to be explained in Chapter III) which reduces statements that are verbally concerned with classes and relations to statements that are concerned with propositional functions, the paradoxes are reduced to such as are concerned with propositions and propositional functions. The paradoxes that concern propositions are only indirectly relevant to mathematics, while those that more nearly concern the mathematician are all concerned with *propositional functions*. We shall therefore proceed at once to the consideration of propositional functions.

II. THE NATURE OF PROPOSITIONAL FUNCTIONS

By a "propositional function" we mean something which contains a variable x, and expresses a *proposition* as soon as a value is assigned to x. That is to say, it differs from a proposition solely by the fact that it is ambiguous: it contains a variable of which the value is unassigned. It agrees with the ordinary functions of mathematics in the fact of containing an unassigned variable; where it differs is in the fact that the values of the function are propositions. Thus *e.g.*, "x is a man" or "$\sin x = 1$" is a propositional function. We shall find that it is possible to incur a vicious-circle fallacy at the very outset, by admitting as possible arguments to a propositional function terms which presuppose the function. This form of the fallacy is very instructive, and its avoidance leads, as we shall see, to the hierarchy of types.

CHAPTER III INCOMPLETE SYMBOLS: DESCRIPTIONS

By an "incomplete" symbol we mean a symbol which is not supposed to have any meaning in isolation, but is only defined in certain contexts. In ordinary mathematics, for example, d/dx and $\int_a^b$ are incomplete symbols: something has to be supplied before we have anything signficant. Such symbols have what may be called a "definition in use." Thus, if we put

$$\nabla^2 = \frac{\partial^2}{\partial x^2} + \frac{\partial^2}{\partial y^2} + \frac{\partial^2}{\partial z^2} \quad \text{Df,}$$

we define the *use* of ∇^2, but ∇^2 by itself remains without meaning. This distinguishes such symbols from what (in a generalized sense) we may call *proper names*: "Socrates," for example, stands for a certain man and therefore has a meaning by itself, without the need of any context. If we supply a context, as in "Socrates is mortal," these words express a fact of which Socrates himself is a constituent: there is a certain object, namely Socrates, which does have the property of mortality, and this object is a constituent of the complex fact which we assert when we say "Socrates is mortal." But in other cases this simple analysis fails us. Suppose we say: "The round square does not exist." It seems plain that this is a true proposition, yet we cannot regard it as denying the existence of a certain object called "the round square." For if there were such an object, it would exist: we cannot first assume that there is a certain object and then proceed to deny that there is such an object. Whenever the grammatical subject of a proposition can be supposed not to exist, without rendering the proposition meaningless, it is plain that the grammatical subject is not a proper name, that is, not a name directly representing some object. Thus, in all such cases, the proposition must be capable of being so analyzed that what was the grammatical subject shall have disappeared. Thus, when we say "the round square does not exist," we may, as a first attempt at such analysis, substitute "it is false that there is an object x which is both round and square." Generally, when "the so-and-so" is said not to exist, we have a proposition of the form[1]

$$\text{“}\sim E!(\imath x)(\phi x),\text{”}$$

that is,

$$\sim\{(\exists c) : \phi x . \equiv_x . x = c\},$$

or some equivalent. Here the apparent grammatical subject $(\imath x)(\phi x)$ has completely disappeared; thus, in "$\sim E!(\imath x)(\phi x)$," $(\imath x)(\phi x)$ is an *incomplete* symbol.

By an extension of the above argument, it can easily be shown that $(\imath x)(\phi x)$ is *always* an incomplete symbol. Take, for example, the following proposition: "Scott is the author of Waverley." (Here "the author of Waverley" is $(\imath x)(x$ wrote Waverly).) This proposition expresses an identity; thus, if "the author of Waverley" could be taken as a proper name and supposed to stand for some object c, the proposition would be "Scott is c." But if c is any one except Scott, this proposition is false; while if c *is* Scott, the proposition is "Scott is Scott," which is trivial and plainly different from "Scott is the author of Waverley." Generalizing, we see that the proposition $a = (\imath x)(\phi x)$ is one which may be true or may be false, but is never merely trivial, like $a = a$; whereas, if $(\imath x)(\phi x)$ were a proper name, $a = (\imath x)(\phi x)$ would necessarily be either false or the same as the trivial proposition $a = a$. We may express this by saying that $a = (\imath x)(\phi x)$ is not a value of the propositional function $a = y$, from which it follows that $(\imath x)(\phi x)$ is not a value of y. But, since y may be anything, it follows that $(\imath x)(\phi x)$ is nothing. Hence, since in use it has meaning, it must be an incomplete symbol.

It might be suggested that "Scott is the author of Waverley" asserts that "Scott" and "the author of Waverley" are two names for the same object. But a little reflection will show that this would be a mistake. For if that were the meaning of "Scott is the author of Waverley," what would be required for its truth would be that Scott should have been *called* the author of Waverley: if he had been so called, the proposition would be true, even if some one else had written Waverley; while, if no one called him so, the proposition would be false, even if he had written Waverley. But in fact he was the author of Waverley at a time when no one called him so, and he would not have been the author if every one had called him so but some one else had written Waverley. Thus the proposition "Scott is the author of Waverley" is not a proposition about names, like "Napoleon is Bonaparte"; and this illustrates the sense in which "the author of Waverley" differs from a true proper name.

Thus all phrases (other than propositions) containing the word *the* (in the singular) are incomplete symbols: they have a meaning in use, but not in isolation. For "the author of Waverley" cannot mean the same as "Scott," or "Scott is the author of Waverley" would mean the same as "Scott is Scott," which it plainly does not; nor can "the author of Waverley" mean anything other than "Scott," or "Scott is the author of Waverley" would be false. Hence "the author of Waverley" means nothing.

It follows from the above that we must not attempt to define "$(\imath x)(\phi x)$," but must define the *uses* of this symbol, that is, the propositions in whose symbolic expression it occurs. Now in seeking to define the uses of this symbol, it is important to observe the import of propositions in which it occurs. Take as an illustration: "The author of Waverley was a poet." This implies (1) that Waverley was written, (2) that it was written by one man, and not in collaboration, (3) that the one man who wrote it was a poet. If any one of these fails, the proposition is false. Thus "the author of "Slawkenburgius on Noses' was a poet" is false, because no such book was ever written; "the author of 'The Maid's Tragedy' was a poet" is false, because this play was written by Beaumont and Fletcher jointly. These two possibilities of falsehood do not arise if we say "Scott was a poet." Thus our interpretation of the uses of $(\imath x)(\phi x)$ must be such as to allow for them. Now taking ϕx to replace "x wrote Waverley," it is plain that any statement apparently about $(\imath x)(\phi x)$ requires (1) $(\exists x).(\phi x)$ and (2) $\phi x . \phi y . \supset_{x,y} . x = y$; here (1) states that *at least* one object satisfies ϕx, while (2) states that *at most* one object satisfies ϕx. The two together are equivalent to

$$(\exists c): \phi x . \equiv_x . x = c,$$

which we defined as $E!(\imath x)(\phi x)$. Thus "$E!(\imath x)(\phi x)$" must be part of what is affirmed by any proposition about $(\imath x)(\phi x)$. If our proposition is $f\{(\imath x)(\phi x)\}$, what is further affirmed is fc, if $\phi x . \equiv_x . x = c$. Thus we have

$$f\{(\imath x)(\phi x)\} . = : (\exists c): \phi x . \equiv_x . x = c : fc \text{ Df},$$

that is, "the x satisfying ϕx satisfies fx" is to mean: "There is an object c such that ϕx is true when, and only when, x is c, and fc is true," or, more exactly: "There is a c such that 'ϕx' is always equivalent to 'x is c' and fc." In this, "$(\imath x)(\phi x)$" has completely disappeared; thus "$(\imath x)(\phi x)$" is merely symbolic, and does not directly represent an object, as single small Latin letters are assumed to do.[2]

The proposition "$a = (\imath x)(\phi x)$" is easily shown to be equivalent to "$\phi x . \equiv_x . x = a$." For, by the definition, it is

$$(\exists c): \phi x . \equiv_x . x = c : a = c,$$

that is, "there is a c for which $\phi x . \equiv_x . x = c$, and this c is a," which is equivalent to "$\phi x . \equiv_x . x = a$." Thus "Scott is the author of Waverley" is equivalent to "'x wrote Waverley' is always equivalent to 'x is Scott,'" that is, "x wrote Waverley" is true when x is Scott and false when x is not Scott.

Thus, although "$(\imath x)(\phi x)$" has no meaning by itself, it may be substituted for y in any propositional function fy, and we get a significant proposition, though not a value of fy.

When $f\{(\imath x)(\phi x)\}$, as above defined, forms part of some other proposition, we shall say that $(\imath x)(\phi x)$ has a *secondary* occurrence. When $(\imath x)(\phi x)$ has a secondary occurrence, a proposition in which it occurs may be true even when $(\imath x)(\phi x)$ does not exist. This applies, for example, to the proposition: "There is no such person as the King of France." We may interpret this as

$$\sim\{E!(\imath x)(\phi x)\},$$

or as

$$\sim\{(\exists c). c = (\imath x)(\phi x)\},$$

if "ϕx" stands for "x is King of France." In either case, what is asserted is that a proposition p in which $(\imath x)(\phi x)$ occurs is false, and this proposition p is thus part of a larger proposition. The same applies to such a proposition as the following: "If France were a monarchy, the King of France would be of the House of Orleans."

It should be observed that such a proposition as

$$\sim f\{(\imath x)(\phi x)\}$$

is ambiguous; it may deny $f\{(\imath x)(\phi x)\}$, in which case it will be true if $(\imath x)(\phi x)$ does not exist, or it may mean

$$(\exists c): \phi x . \equiv_x . x = c : \sim fc,$$

in which case it can be true only if $(\imath x)(\phi x)$ exists. In ordinary language, the latter interpretation would usually be adopted. For example, the proposition "the King of France is not bald" would usually be rejected as false, being held to mean "the King of France exists and is not bald," rather than "it is false that the King of France exists and is bald." When $(\imath x)(\phi x)$ exists, the two interpretations of the ambiguity give equivalent results; but, when $(\imath x)(\phi x)$ does not exist, one interpretation is true and one is false. It is necessary to be able to distinguish these in our notation; and generally, if we have such propositions as

$$\psi(\imath x)(\phi x) . \supset . p,$$

$$p . \supset . \psi(\imath x)(\phi x),$$

$$\psi(\imath x)(\phi x) . \supset . \chi(\imath x)(\phi x),$$

and so on, we must be able by our notation to distinguish whether the whole or only part of the proposition concerned is to be treated as the "$f(\imath x)(\phi x)$" of our definition. For this purpose, we will put "$[(\imath x)(\phi x)]$" followed by dots at the beginning of the part (or whole) which is to be taken as $f(\imath x)(\phi x)$, the dots being sufficiently numerous to bracket off the $f(\imath x)(\phi x)$; that is, $f(\imath x)(\phi x)$ is to be everything following the dots until we reach an equal number of dots not signifying a logical product, or a greater number signifying a logical product, or the end of the sentence, or the end of a bracket enclosing "$[(\imath x)(\phi x)]$". Thus

$$[(\imath x)(\phi x)] . \psi(\imath x)(\phi x) . \supset . p$$

will mean

$$(\exists c):\phi x .\equiv_x. x = c:\psi c: \supset . p,$$

but

$$[(\imath x)(\phi x)]:\psi(\imath x)(\phi x) . \supset . p$$

will mean

$$(\exists c):\phi x .\equiv_x. x = c:\psi c . \supset . p.$$

It is important to distinguish these two, for, if $(\imath x)(\phi x)$ does not exist, the first is true and the second false. Again

$$[(\imath x)\phi x)] . \sim\psi(\imath x)(\phi x)$$

will mean

$$(\exists c):\phi x .\equiv_x. x = c:\sim\psi c,$$

while

$$\sim\{[(\imath x)(\phi x)].\psi(\imath x)(\phi x)\}$$

will mean

$$\sim\{(\exists c):\phi x .\equiv_x. x = c:\psi c\}.$$

Here again, when $(\imath x)(\phi x)$ does not exist, the first is false and the second true.

In order to avoid this ambiguity in propositions containing $(\imath x)(\phi x)$, we amend our definition, or rather our notation, putting

$$[(\imath x)(\phi x)]\ f(\imath x)(\phi x) .=: (\exists c):\phi x .\equiv_x. x = c:fc \quad \text{Df.}$$

By means of this definition, we avoid any doubt as to the portion of our whole asserted proposition which is to be treated as the "$f(\imath x)(\phi x)$" of the definition. This portion will be called the *scope* of $(\imath x)(\phi x)$. Thus in

$$[(\imath x)(\phi x)] . f(\imath x)(\phi x) . \supset . p$$

the scope of $(\imath x)(\phi x)$ is $f(\imath x)(\phi x)$; but in

$$[(\imath x)(\phi x)] : f(\imath x)(\phi x) . \supset . p$$

the scope is

$$f(\imath x)(\phi x) . \supset . p;$$

in

$$\sim\{[(\imath x)(\phi x)] . f(\imath x)(\phi x)\}$$

the scope is $f(\imath x)(\phi x)$; but in

$$[(\imath x)(\phi x)] . \sim f(\imath x)(\phi x)$$

the scope is

$$\sim f(\imath x)(\phi x).$$

It will be seen that, when $(\imath x)(\phi x)$ has the whole of the proposition concerned for its scope, the proposition concerned cannot be true unless $E!(\imath x)(\phi x)$; but, when $(\imath x)(\phi x)$ has only part of the proposition concerned for its scope, it may often be true even when $(\imath x)(\phi x)$ does not exist. It will be seen further that, when $E!(\imath x)(\phi x)$, we may enlarge or diminish the scope of $(\imath x)(\phi x)$ as much as we please without altering the truth value of any proposition in which it occurs.

If a proposition contains two descriptions, say $(\imath x)(\phi x)$ and $(\imath x)(\psi x)$, we have to distinguish which of them has the larger scope, that is, we have to distinguish

(1) $[(\imath x)(\phi x)] : [(\imath x)(\psi x)] . f\{(\imath x)(\phi x), (\imath x)(\psi x)\}$,

(2) $[(\imath x)(\psi x)] : [(\imath x)(\phi x)] . f\{(\imath x)(\phi x), (\imath x)(\psi x)\}$.

The first of these, eliminating $(\imath x)(\phi x)$, becomes

(3)
$$(\exists c):\phi x .\equiv_x. x = c:[(\imath x)(\psi x)] . f\{c, (\imath x)(\psi x)\},$$

which, eliminating $(\imath x)(\psi x)$, becomes

(4)
$$(\exists c):.\phi x .\equiv_x. x = c:.(\exists d):\psi x .\equiv_x. x = c:f(c, d),$$

and the same proposition results if, in (1) we eliminate first $(\iota x)(\psi x)$ and then $(\iota x)(\phi x)$. Similarly (2) becomes, when $(\iota x)(\phi x)$ and $(\iota x)(\psi x)$ are eliminated,

(5)
$$(Ed):.\psi x \,.\!\equiv_x.\, x = d:.(Ec):\phi x \,.\!=_x.\, x = c:f(c,d).$$

(4) and (5) are equivalent, so that the truth value of a proposition containing two descriptions is independent of the question which has the larger scope.

It will be found that, in most cases in which descriptions occur, their scope is, in practice, the smallest proposition enclosed in dots or other brackets in which they are contained. Thus, for example,

$$[(\iota x)(\phi x)].\psi(\iota x)(\phi x)\ .\supset.\ [(\iota x)(\phi x)].\chi(\iota x)(\phi x)$$

will occur much more frequently than

$$[(\iota x)(\phi x)]:\psi(\iota x)(\phi x)\ .\supset.\ \chi(\iota x)(\phi x).$$

For this reason it is convenient to decide that, when the scope of an occurrence of $(\iota x)(\phi x)$ is the smallest proposition, enclosed in dots or other brackets, in which the occurrence in question is contained, the scope need not be indicated by "$[(\iota x)(\phi x)]$." Thus, for example,

$$p\ .\supset.\ a = (\iota x)(\phi x)$$

will mean

$$p\ .\supset.\ [(\iota x)(\phi x)].a = (\iota x)(\phi x);$$

and

$$p\ .\supset.\ (\exists a).a = (\iota x)(\phi x)$$

will mean

$$p\ .\supset.\ (\exists a).[(\iota x)(\phi x)].a = (\iota x)(\phi x);$$

and

$$p\ .\supset.\ a \neq (\iota x)(\phi x)$$

will mean

$$p\ .\supset.\ [(\iota x)(\phi x)].\sim\{a = (\iota x)(\phi x)\};$$

but

$$p\ .\supset.\ \sim\{a = (\iota x)(\phi x)\}$$

will mean

$$p\ .\supset.\ \sim\{[(\iota x)(\phi x)].a = (\iota x)(\phi x)\}.$$

This convention enables us, in the vast majority of cases that actually occur, to dispense with the explicit indication of the scope of a descriptive symbol; and it will be found that the convention agrees very closely with the tacit conventions of ordinary language on this subject. Thus, for example, if "$(\iota x)(\phi x)$" is "the so-and-so," "$a \neq (\iota x)(\phi x)$" is to be read "a is not the so-and-so," which would ordinarily be regarded as implying that "the so-and-so" exists; but "$\sim\{a = (\iota x)(\phi x)\}$" is to be read "it is not true that *a* is the so-and-so," which would generally be allowed to hold if "the so-and-so" does not exist. Ordinary language is, of course, rather loose and fluctuating in its implications on this matter; but, subject to the requirement of definiteness, our convention seems to keep as near to ordinary language as possible.

In the case when the smallest proposition enclosed in dots or other brackets contains two or more descriptions, we shall assume, in the absence of any indication to the contrary, that one which typographically occurs earlier has a larger scope than one which typographically occurs later. Thus

$$(\iota x)(\phi x) = (\iota x)(\psi x)$$

will mean

$$(\exists c):\phi x \,.\!\equiv_x.\, x = c:[(\iota x)(\psi x)].c = (\iota x)(\psi x),$$

while

$$(\iota x)(\psi x) = (\iota x)(\phi x)$$

will mean

$$(\exists d):\psi x \,.\!\equiv_x.\, x = d:[(\iota x)(\phi x)].(\iota x)(\phi x) = d.$$

These two propositions are easily shown to be equivalent.

NOTES

1. See *Whitehead and Russell 1910*, pp. 31-32. . . .
2. We shall generally write "$f(\iota x)(\phi x)$" rather than "$f\{(\iota x)(\phi x)\}$" in future.

Chapter IX
The Early 20th Century to 1932

Section B
Foundational Crisis and Undecidability

L(UITZEN) E(GBERTUS) J(AN) BROUWER (1881–1966)

L. E. J. Brouwer, the systematic founder of the modern intuitionist school of mathematical thought, displayed exceptional intellectual ability from his youth. After graduating from the high school in Hoorn, Netherlands, at age 14, Brouwer studied Greek and Latin for the next two years to qualify for entrance into a university. After his family moved to Haarlem he took and passed the entrance examination for its municipal Gymnasium in 1897 but entered the University of Amsterdam instead that same year. For the next seven years Brouwer studied mathematics. D. J. Korteweg taught him current mathematics, and G. Mannoury introduced him to topology and the foundations of mathematics. From 1904 to 1907 he also closely examined philosophy, especially mysticism. These studies led him to write *Leven, Kunst, en Mystiek* ("Life, Art, and Mysticism," 1905). He graduated from the University of Amsterdam in 1907 with the degree of Doctor of Mathematics and Physics for his dissertation *Over de Grondslagen der Wiskunde* ("On the Foundations of Mathematics"). In the first chapter of his dissertation, Brouwer began to unfold his intuitionist program, asserting that he showed therein "how the fundamental parts of mathematics can be built from units of perception." His third and final chapter severely criticized the two other major approaches to mathematical foundations, namely Peano and Russell's logistic thesis and Hilbert's evolving formalist program that still lacked a definitive *Beweistheorie* ("proof theory"). He also rejected Cantor's theory of transfinite numbers.

Brouwer spent his entire career in research and teaching at the University of Amsterdam, as an unsalaried lecturer from 1909 to 1912 and a professor of mathematics from 1912 until his retirement in 1951. Among the subjects he taught were set theory, function theory, and axiomatics. In mathematical circles, he quickly gained international recognition for his ideas, even though they were not widely accepted. In 1912 the Royal Netherlands Academy of Science elected him a member. Subsequently the Prussian Academy of Science in Berlin, the American Philosophical Society, and the Royal Society of London also elected him a member. In addition, he received honorary doctorates from various universities including Oslo and Cambridge.

Brouwer made fundamental contributions to topology and the foundations of mathematics. Inspired by Hilbert's Paris Problems of 1900 and Schoenflies' report on the development of set theory, he immersed himself in mathematical research from 1907 to 1913. His investigation of Hilbert's Fifth Problem—*i.e.*, treating the theory of continuous groups independently of assumptions of differentiability—yielded fragmentary results. That problem is a natural consequence of Klein's Erlanger Program. Brouwer's examination of the Erlanger Program, function theory, and the mapping studies of Cantor and Peano, led to the discovery of the plane translation theorem with its homotopic characterization of topological mappings of the Cartesian plane in 1910. The following year he discovered the first fixed point theorem about one-to-one continuous (topological) mappings. Fixed points are singular or invariant points of vector fields on a manifold. He also found indecomposable continua in the plane to be common boundary of denumerably many, simply connected domains and defined the dimension of topological spaces.

At the turn of the 20th century, there was a crisis in the foundations of mathematics. Brouwer's intuitionistic mathematics, whose origins may be traced to Immanuel Kant and Leopold Kronecker, represents a distinctive response. Mathematics for Brouwer was a free, creative process of the mind that constructs mathematical objects independently of the universe of experience and is restricted only in that it starts from self-evident, primitive intuitive notions. In his treatise *Over de Onbetrouwbaarheid der logische Principes* ("On the Untrustworthiness of Logical Principles," 1908) he extended the program he had begun in his doctoral dissertation by rejecting as invalid mathematical proofs that used the principle of the excluded middle. The principle of the excluded middle, A or not A, in other words A is true or A is false, so important for *reductio ad absurdum* proofs could only be applied for finite sets. Hilbert and Poincaré emphatically disagreed. If Brouwer were correct, many basic arguments and definitions in mathematical analysis would become unsupported and would have to be reexamined. Since mathematical analysis in turn is essential for solving equations in other sciences (such as differential equations of mathematical physics), many saw Brouwer's attack on classical logic as undermining a vast body of scientific knowledge. This did not deter Brouwer whose continuing criticism was directed chiefly against Hilbert's formalist program. He agreed with its finitist methods for proofs but rejected the belief that freedom from contradiction (mere consistency) justifies formal systems. Mathematical existence, he said, was based not on consistency but intuitive constructibility. Debates with logicists and formalists were sometimes heated. In an article published in 1912, Brouwer despaired of there being any meaningful communication between the three groups because they were not speaking the same language.

Brouwer's intuitionistic work may be divided into at least two distinct phases. The first, before World War I, was primarily negative with vigorous criticisms; the second, after World War I, was positive. In the second phase he endeavored to reconstruct mathematics on intuitionistic tenets. Brouwer developed a set theory independent of the principle of the excluded middle, a constructive theory of measure, and a constructive theory of real numbers and real functions. He also derived his fundamental theorem on finitary spreads (the fan theorem). Unlike axiomatic set theory, his constructive theory did not accept the notion of set as primitive. Instead, it depended on a law specifying sequences of choice—the specifying law is called a spread. The spread in essence gives the intuitionistic continuum. Although Brouwer's research in the 1920s laid the groundwork for redeveloping major portions of classical algebra, analysis, geometry, and topology, he won few converts to the intuitionist viewpoint. The character of his exposition may partly explain this response. Initially it was extremely polemical, and in the 1920s it involved obscure and unfamiliar notions as well as familiar terminology with new meanings.

In 1931 Kurt Gödel's famous work on undecidability showed the inherent limitations in Hilbert's formalist program. In a sense Brouwer's criticisms were borne out. Gödel's findings and the rise of the theory of recursive functions led subsequently to a revival of the intuitionistic foundations of mathematics, principally in the pioneering studies of Stephen Cole Kleene.

124. From "Intuitionism and Formalism" (1912)*

L. E. J. BROUWER

The subject for which I am asking your attention deals with the foundations of mathematics. To understand the development of the opposing theories existing in this field one must first gain a clear understanding of the concept "science"; for it is as a part of science that mathematics originally took its place in human thought.

By science we mean the systematic cataloguing by means of laws of nature of causal sequences of phenomena, *i.e.* sequences of phenomena which for individual or social purposes it is convenient to consider as repeating themselves identically,—and more particularly of such causal sequences as are of importance in social relations.

That science lends such great power to man in his action upon nature is due to the fact that the steadily improving cataloguing of ever more causal sequences of phenomena gives greater and greater possibility of bringing about desired phenomena, difficult or impossible to evoke directly, by evoking other phenomena connected with the first by causal sequences. And that man always and everywhere creates order in nature is due to the fact that he not only isolates the causal sequences of phenomena (*i.e.*, he strives to keep them free from disturbing secondary phenomena) but also supplements them with phenomena caused by his own activity, thus making them of wider applicability. Among the latter phenomena the results of counting and measuring take so important a place, that a large number of the natural laws introduced by science treat only of the mutual relations between the results of counting and measuring. It is well to notice in this connection that a natural law in the statement of which measurable magnitudes occur can only be understood to hold in nature with a certain degree of approximation; indeed natural laws as a rule are not proof against sufficient refinement of the measuring tools.

The exceptions to this rule have from ancient times been practical arithmetic and geometry on the one hand, and the dynamics of rigid bodies and celestial

*Source: From the Inaugural Address given by L. E. J. Brouwer at the University of Amsterdam on October 14, 1912. This translation by Arnold Dresden is reprinted from the *Bulletin* of the American Mathematical Society, vol. 20 (1913-1914), 81-96.

mechanics on the other hand. Both these groups have so far resisted all improvements in the tools of observation. But while this has usually been looked upon as something accidental and temporal for the latter group, and while one has always been prepared to see these sciences descend to the rank of approximate theories, until comparatively recent times there has been absolute confidence that no experiment could ever disturb the exactness of the laws of arithmetic and geometry; this confidence is expressed in the statement that mathematics is "the" exact science.

On what grounds the conviction of the unassailable exactness of mathematical laws is based has for centuries been an object of philosophical research, and two points of view may here be distinguished, *intuitionism* (largely French) and *formalism* (largely German). In many respects these two viewpoints have become more and more definitely opposed to each other; but during recent years they have reached agreement as to this, that the exact validity of mathematical laws as laws of nature is out of the question. The question where mathematical exactness does exist, is answered differently by the two sides; the intuitionist says: in the human intellect, the formalist says: on paper.

In Kant we find an old form of intuitionism, now almost completely abandoned, in which time and space are taken to be forms of conception inherent in human reason. For Kant the axioms of arithmetic and geometry were synthetic a priori judgments, *i.e.,* judgments independent of experience and not capable of analytical demonstration; and this explained their apodictic exactness in the world of experience as well as in abstracto. For Kant, therefore, the possibility of disproving arithmetical and geometrical laws experimentally was not only excluded by a firm belief, but it was entirely unthinkable.

Diametrically opposed to this is the view of formalism, which maintains that human reason does not have at its disposal exact images either of straight lines or of numbers larger than ten, for example, and that therefore these mathematical entities do not have existence in our conception of nature any more than in nature itself. It is true that from certain relations among mathematical entities, which we assume as axioms, we deduce other relations according to fixed laws, in the conviction that in this way we derive truths from truths by logical reasoning, but this non-mathematical conviction of truth or legitimacy has no exactness whatever and is nothing but a vague sensation of delight arising from the knowledge of the efficacy of the projection into nature of these relations and laws of reasoning. For the formalist therefore mathematical exactness consists merely in the method of developing the series of relations, and is independent of the significance one might want to give to the relations or the entities which they relate. And for the consistent formalist these meaningless series of relations to which mathematics are reduced have mathematical existence only when they have been represented in spoken or written language together with the mathematical-logical laws upon which their development depends, thus forming what is called symbolic logic.

Because the usual spoken or written languages do not in the least satisfy the requirements of consistency demanded of this symbolic logic, formalists try to avoid the use of ordinary language in mathematics. How far this may be carried is shown by the modern Italian school of formalists, whose leader, Peano, published one of his most important discoveries concerning the existence of integrals of real differential equations in the *Mathematische Annalen* in the language of symbolic logic; the result was that it could only be read by a few of the initiated and that it did not become generally available until one of these had translated the article into German.

The viewpoint of the formalist must

lead to the conviction that if other symbolic formulas should be substituted for the ones that now represent the fundamental mathematical relations and the mathematical-logical laws, the absence of the sensation of delight, called "consciousness of legitimacy," which might be the result of such substitution would not in the least invalidate its mathematical exactness. To the philosopher or to the anthropologist, but not to the mathmatician, belongs the task of investigating why certain systems of symbolic logic rather than others may be effectively projected upon nature. Not to the mathematician, but to the psychologist, belongs the task of explaining why we believe in certain systems of symbolic logic and not in others, in particular why we are averse to the so-called contradictory systems in which the negative as well as the positive of certain propositions are valid.[1]

As long as the intuitionists adhered to the theory of Kant it seemed that the development of mathematics in the nineteenth century put them in an ever weaker position with regard to the formalists. For in the first place this development showed repeatedly how complete theories could be carried over from one domain of mathematics to another; projective geometry, for example, remained unchanged under the interchange of the rôles of point and straight line, an important part of the arithmetic of real numbers remained valid for various complex number fields and nearly all the theorems of elementary geometry remained true for non-archimedian geometry, in which there exists for every straight line segment another such segment, infinitesimal with respect to the first. These discoveries seemed to indicate indeed that of a mathematical theory only the logical form was of importance and that one need no more be concerned with the material than it is necessary to think of the significance of the digit groups with which one operates, for the correct solution of a problem in arithmetic.

But the most serious blow for the Kantian theory was the discovery of non-euclidean geometry, a consistent theory developed from a set of axioms differing from that of elementary geometry only in this respect that the parallel axiom was replaced by its negative. For this showed that the phenomena usually described in the language of elementary geometry may be described with equal exactness, though frequently less compactly in the language of non-euclidean geometry; hence, it is not only impossible to hold that the space of our experience has the properties of elementary geometry but it has no significance to ask for *the* geometry which would be true for the space of our experience. It is true that elementary geometry is better suited than any other to the description of the laws of kinematics of rigid bodies and hence of a large number of natural phenomena, but with some patience it would be possible to make objects for which the kinematics would be more easily interpretable in terms of non-euclidean than in terms of euclidean geometry.[2]

However weak the position of intuitionism seemed to be after this period of mathematical development, it has recovered by abandoning Kant's apriority of space but adhering the more resolutely to the apriority of time. This neo-intuitionism considers the falling apart of moments of life into qualitatively different parts, to be reunited only while remaining separated by time as the fundamental phenomenon of the human intellect, passing by abstracting from its emotional content into the fundamental phenomenon of mathematical thinking, the intuition of the bare two-oneness. This intuition of two-oneness, the basal intuition of mathematics, creates not only the numbers one and two, but also all finite ordinal numbers, inasmuch as one of the elements of the two-oneness may be thought of as a new two-oneness, which process may be finite numbers of the intuitionist on account of their construction, shows at the same time that the former will never be able to justify his choice of axioms by replacing the unsatisfactory appeal to inexact practice or to intuition equally inexact for him by a proof of the non-contradictoriness of his theory. For in order to prove that a contradiction can never arise among the infinitude of conclusions that can be drawn from the axioms he is using, he would first have to show that if no contradiction has as yet arisen with the nth conclusion then none could arise with the $(n + 1)$th conclusion, and secondly he would have to apply the principle of complete induction intuitively. But it is this last step which the formalist may not take, even though he should have proved the principle of complete induction; for this would require mathematical certainty that the set of properties obtained after the nth conclusion had been reached, would satisfy for an arbitrary n his definition for finite sets,[3] and in order to secure this certainty he would have to have recourse not only to the unpermissible application of a symbolic criterion to a concrete example but also to another intuitive application of the principle of complete induction; this would lead him to a vicious circle reasoning.

In the domain of finite sets in which the formalistic axioms have an interpretation perfectly clear to the intuitionists, unreservedly agreed to by them, the two tendencies differ solely in their method, not in their results; this becomes quite different however in the domain of infinite or transfinite sets, where, mainly by the application of the axiom of inclusion, quoted above, the formalist introduces various concepts, entirely meaningless to the intuitionist, such as for instance *"the set whose elements are the points of space," "the set whose elements are the continuous functions of a variable," "the set whose elements are the discontinuous functions of a variable,"* and so forth. In the course of these formalistic developments it turns out that the consistent application of the axiom of inclusion leads inevitably to contradictions. A clear illustration of this fact is furnished by the so-called paradox of Burali-Forti.[4] To exhibit it we have to lay down a few definitions.

A set is called ordered if there exists between any two of its elements a relation of "higher than" or "lower than" with this understanding that if the element a is higher than the element b, then the element b is lower than the element a, and if the element b is higher than a and c is higher than b, then c is higher than a.

A well-ordered set (in the formalistic sense) is an ordered set, such that every subset contains an element lower than all others.

Two well-ordered sets that may be brought into one-to-one correspondence under invariance of the relations of "higher than" and "lower than" are said to have the same ordinal number.

If two ordinal numbers A and B are not equal, then one of them is greater than the other one, let us say A is greater than B; this means that B may be brought into one-to-one correspondence with an initial segment of A under invariance of the relations of "higher than" and "lower than." We have introduced above, from the intuitionist viewpoint, the smallest infinite ordinal number ω, *i.e.*, the ordinal number of the set of all finite ordinal numbers arranged in order of magnitude.[5] Well-ordered sets having the ordinal number ω are called elementary series.

It is proved without difficulty by the formalist that an arbitrary subset of a well-ordered set is also a well-ordered set, whose ordinal number is less than or equal to that of the original set; also, that if to a well-ordered set that does not contain all mathematical objects a new element be added that is defined to be higher than all elements of the original set, a new well-ordered set arises whose ordinal number is greater than that of the first set.

We construct now on the basis of the axiom of inclusion the *set s which contains as elements all the ordinal num-*

bers arranged in order of magnitude; then we can prove without difficulty, on the one hand that s is a well-ordered set whose ordinal number can not be exceeded by any other ordinal number in magnitude, and on the other hand that it is possible, since not all mathematical objects are ordinal numbers, to create an ordinal number greater than that of s by adding a new element to s,—a contradiction.[6]

Although the formalists must admit contradictory results as mathematical if they want to be consistent, there is something disagreeable for them in a paradox like that of Burali-Forti because at the same time the progress of their arguments is guided by the principium contradictionis, *i.e.*, by the rejection of the simultaneous validity of two contradictory properties. For this reason the axiom of inclusion has been modified to read as follows: *"If for all elements of a set it is decided whether a certain property is valid for them or not, then the set contains a subset containing nothing but those elements for which the property does hold."*[7]

In this form the axiom permits only the introduction of such sets as are subsets of sets previously introduced; if one wishes to operate with other sets, their existence must be explicitly postulated. Since however in order to accomplish anything at all the existence of a certain collection of sets will have to be postulated at the outset, the only valid argument that can be brought against the introduction of a new set is that it leads to contradictions; indeed the only modifications that the discovery of paradoxes has brought about in the practice of formalism has been the abolition of those sets that had given rise to these paradoxes. One continues to operate without hesitation with other sets introduced on the basis of the old axiom of inclusion; the result of this is that extended fields of research, which are without significance for the intuitionist are still of considerable interest to the formalist. An example of this is found in the theory of potencies, of which I shall sketch the principal features here, because it illustrates so clearly the impassable chasm which separates the two sides.

Two sets are said to possess the same potency, or power, if their elements can be brought into one-to-one correspondence. The power of set A is said to be greater than that of B, and the power of B less than that of A, if it is possible to establish a one-to-one correspondence between B and a part of A, but impossible to establish such a correspondence between A and a part of B. The power of a set which has the same power as one of its subsets, is called infinite, other powers are called finite. Sets that have the same power as the ordinal number ω are called denumerably infinite and the power of such sets is called aleph-null: it proves to be the smallest infinite power. According to the statements previously made, this power aleph-null is the only infinite power of which the intuitionists recognize the existence.

Let us now consider the concept: "denumerably infinite ordinal number." From the fact that this concept has a clear and well-defined meaning for both formalist and intuitionist, the former infers the right to create the "set of all denumerably infinite ordinal numbers," the power of which he calls aleph-one, a right not recognized by the intuitionist. Because it is possible to argue to the satisfaction of both formalist and intuitionist, first, that denumerably infinite sets of denumerably infinite ordinal numbers can be built up in various ways, and second, that for every such set it is possible to assign a denumerably infinite ordinal number, *not* belonging to this set, the formalist concludes: "aleph-one is greater than aleph-null," a proposition, that has no meaning for the intuitionist. Because it is possible to argue to the satisfaction of both formalist and intuitionist that it is impossible to construct[8] a set of denumerably infinite ordinal numbers, which could be proved to have a power less than that of aleph-one, but greater than that of aleph-null, the formalist concludes: "aleph-one is the second smallest infinite ordinal number," a proposition that has no meaning for the intuitionist.

Let us consider the concept: "real number between 0 and 1." For the formalist this concept is equivalent to "elementary series of digits after the decimal point,"[9] for the intuitionist it means "law for the construction of an elementary series of digits after the decimal point, built up by means of a finite number of operations." And when the formalist creates the "set of all real numbers between 0 and 1," these words are without meaning for the intuitionist, even whether one thinks of the real numbers of the formalist, determined by elementary series of freely selected digits, or of the real numbers of the intuitionist, determined by finite laws of construction. Because it is possible to prove to the satisfaction of both formalist and intuitionist, first, that denumerably infinite sets of real numbers between 0 and 1 can be constructed in various ways, and second that for every such set it is possible to assign a real number between 0 and 1, not belonging to the set, the formalist concludes: "the power of the continuum, *i.e.*, the power of the set of real numbers between 0 and 1, is greater than aleph-null," a proposition which is without meaning for the intuitionist; the formalist further raises the question, whether there exist sets of real numbers between 0 and 1, whose power is less than that of the continuum, but greater than aleph-null, in other words, "whether the power of the continuum is the second smallest infinite power," and this question, which is still waiting for an answer, he considers to be one of the most difficult and most fundamental of mathematical problems.

For the intuitionist, however, the question as stated is without meaning; and as soon as it has been so interpreted as to get a meaning, it can easily be answered.

If we restate the question in this form: "Is it impossible to construct[10] infinite sets of real numbers between 0 and 1, whose power is less than that of the continuum, but greater than aleph-null?," then the answer must be in the affirmative; for the intuitionist can only construct denumerable sets of mathematical objects and if, on the basis of the intuition of the linear continuum, he admits elementary series of free selections as elements of construction, then each non-denumerable set constructed by means of it contains a subset of the power of the continuum.

If we restate the question in the form: "Is it possible to establish a one-to-one correspondence between the elements of a set of denumerably infinite ordinal numbers on the one hand, and a set of real numbers between 0 and 1 on the other hand, both sets being indefinitely extended by the construction of new elements, of such a character that the correspondence shall not be disturbed by any continuation of the construction of both sets?," then the answer must also be in the affirmative, for the extension of both sets can be divided into phases in such a way as to add a denumerably infinite number of elements during each phase.[11]

If however we put the question in the following form: "Is it possible to construct a law which will assign a denumerably infinite ordinal number to every elementary series of digits and which will give certainty a priori that two different elementary series will never have the same denumerably infinite ordinal number corresponding to them?," then the answer must be in the negative; for this law of correspondence must prescribe in some way a construction of certain denumerably infinite ordinal numbers at each of the successive places of the elementary series; hence there is for each place c_ν a well-defined largest denumerably infinite number α_ν, the construction of which is suggested by that particular place; there is then also a well-defined denumerably infinite ordinal number α_ω, greater than all α_ν's and that can not therefore be exceeded by any of the ordinal numbers involved by the law of correspondence; hence the power of that set of ordinal numbers can not exceed aleph-null.

As a means for obtaining ever greater powers, the formalists define with every power μ a "set of all the different ways in which a number of selections of power μ may be made," and they prove that the power of this set is greater than μ. In particular, when it has been proved to the satisfaction of both formalist and intuitionist that it is possible in various ways to construct laws according to which functions of a real variable different from each other are made to correspond to all elementary series of digits, but that it is impossible to construct a law according to which an elementary series of digits is made to correspond to every function of a real variable and in which there is certainty a priori that two different functions will never have the same elementary series corresponding to them, the formalist concludes: "the power c' of the set of all functions of a real variable is greater than the power c of the continuum," a proposition without meaning to the intuitionist; and in the same way in which he was led from c to c', he comes from c' to a still greater power c''.

A second method used by the formalists for obtaining ever greater powers is to define for every power μ, which can serve as a power of ordinal numbers, "the set of all ordinal numbers of power μ," and then to prove that the power of this set is greater than μ. In particular they denote by aleph-two the power of the set of all ordinal numbers of power aleph-one and they prove that aleph-two is greater than aleph-one and that it follows in magnitude immediately after aleph-one. If it should be possible to interpret this result in a way in which it would have meaning for the intuitionist, such interpretation would not be as simple in this case as it was in the preceding cases. . . .

So far my exposition of the fundamental issue, which divides the mathematical world. There are eminent scholars on both sides and the chance of reaching an agreement within a finite period is practically excluded. To speak with Poincaré:

> Les hommes ne s'entendent pas, parce qu'ils ne parlent pas la même langue et qu'il y a des langues qui ne s'apprennent pas.

NOTES

1. See Mannoury, "Methodologisches und Philosophisches zur Elementarmathematik," pp. 149-154.

2. See Poincaré, *La Science et l'Hypothèse*, p. 104.

3. Compare Poincaré, *Revue de Métaphysique et de Morale*, 1905, p. 834.

4. Compare *Rendiconti del Circolo Matematico di Palermo*, 1897.

5. The more general ordinal numbers of the intuitionist are the numbers constructed by means of Cantor's two principles of generation (compare *Math. Annalen*, vol. 49, p. 226).

6. It is without justice that the paradox of Burali-Forti is sometimes classed with that of Richard, which in a somewhat simplified form reads as follows: "Does there exist a *least integer, that can not be defined by a sentence of at most twenty words?* On the one hand *yes*, for the number of sentences of at most twenty words is of course finite; on the other hand *no*, for if it should exist, it would be defined by the sentence of fifteen words formed by the words italicized above."

The origin of this paradox does not lie in the axiom of inclusion but in the variable meaning of the word *"defined"* in the italicized sentence, which makes it possible to define by means of this sentence an infinite number of integers in succession.

7. Compare Zermelo, *Math. Annalen*, vol. 65, p. 263.

8. If "construct" were here replaced by "define" (in the formalistic sense), the proof would *not* be satisfactory to the intuitionist.

For, in Cantor's argument in *Math. Annalen*, vol. 49, it is not allowed to replace the words "können wir bestimmen" (p. 214, line 17 from top) by the words "muss es geben."

9. Here as everywhere else in this paper, the assumption is tacitly made that there are an infinite number of digits different from 9.

10. If "construct" were here replaced by "define" (in the formalistic sense), and if we suppose that the problem concerning the pairs of digits in the decimal fraction development of π, discussed [later in this paper] *can not be solved*, then the question of the test must be answered negatively. For, let us denote by Z the set of those infinite binary fractions, whose nth digit is 1, if the nth pair of digits in the decimal fraction development of π consists of unequal digits; let us further denote by X the set of all finite binary fractions. Then the power of $Z + X$ is greater than aleph-null, but less than that of the continuum.

11. Calling *denumerably unfinished* all sets of which the elements can be individually realized, and in which for every denumerably infinite subset there exists an element not belonging to this subset, we can say in general, in accordance with the definitions of the text: *"All denumerably unfinished sets have the same power."*

125. From "Intuitionistic Reflections on Formalism" (1927)*

L. E. J. BROUWER

The disagreement over which is correct, the formalistic way of founding mathematics anew or the intuitionistic way of reconstructing it, will vanish, and the choice between the two activities be reduced to a matter of taste, as soon as the following insights, which pertain primarily to formalism but were first formulated in the intuitionistic literature, are generally accepted. The acceptance of these insights is only a question of time, since they are the results of pure reflection and hence contain no disputable element, so that anyone who has once understood them must accept them. Two of the four insights have so far been understood and accepted in the formalistic literature. When the same state of affairs has been reached with respect to the other two, it will mean the end of the controversy concerning the foundations of mathematics.

First Insight. *The differentiation, among the formalistic endeavors, between a construction of the "inventory of mathematical formulas" (formalistic view of mathematics) and an intuitive (contentual) theory of the laws of this construction, as well as the recognition of the fact that for the latter theory the intuitionistic mathematics of the set of natural numbers is indispensable.*

Second Insight. *The rejection of the thoughtless use of the logical principle of excluded middle, as well as the recognition, first, of the fact that the investigation of the question why the principle mentioned is justified and to what extent it is valid constitutes an essential object of research in the foundations of mathematics, and, second, of the fact*

Source: This translation of Brouwer's "Intuitionische Betrachtungen über den Formalismus" by Stefan Bauer-Mengelberg is taken from Jean van Heijenoort (ed.), *From Frege to Gödel: A Source Book in Mathematical Logic, 1879-1931* (1967), 490-492. It is reprinted by permission of Walter de Gruyter and Co. The writings cited by year in this selection are listed in the section on references in the Heijenoort book.

that in intuitive (contentual) mathematics this principle is valid only for finite systems.

Third Insight. *The identification of the principle of excluded middle with the principle of the solvability of every mathematical problem.*

Fourth Insight. *The recognition of the fact that the (contentual) justification of formalistic mathematics by means of the proof of its consistency contains a vicious circle, since this justification rests upon the (contentual) correctness of the proposition that from the consistency of a proposition the correctness of the proposition follows, that is, upon the (contentual) correctness of the principle of excluded middle.*

1. The first insight is still lacking in *Hilbert 1904*, see in particular Section V, pp. 184–185, which is in contradiction with it. After having been strongly prepared by Poincaré, it first appears in the literature in *Brouwer 1907*, where on pp. 173-174 the terms *mathematical language* and *mathematics of the second order* are used to distinguish between the parts of formalistic mathematics mentioned above and where the intuitive character of the latter part is emphasized.[1] This insight penetrated into the formalistic literature with *Hilbert 1922* (see in particular p. 165 and p. 174), where mathematics of the second order was given the name *metamathematics*. The claim of the formalistic school to have reduced intuitionism to absurdity by means of this insight, borrowed from intuitionism, is presumably not to be taken seriously.

2. The thoughtless use of the logical principle of excluded middle is still to be found in *Hilbert 1904* and *1917* (see, for example, *1917*, p. 413, ll. 11u-4u, and in particular *1904:* p. 182, ll. 16–19; p. 182, l. 2u, to p. 183, l. 2; p. 184, ll. 21u-13u; in each of these places the principle of excluded middle is regarded as essentially equivalent to the principle of contradiction). The second insight is found in the literature for the first time in *Brouwer 1908* and then at greater or lesser length in *Brouwer 1912, 1914, 1917, 1919b, 1923b,* and *1923d*. Except for the recognition, most intimately connected with it, of the intuitionistic consistency of the principle of excluded middle, it penetrates the formalistic literature with *Hilbert 1922a*,[2] where, on the one hand, the limited contentual validity of the principle of excluded middle is acknowledged (see in particular pp. 155–156) and, on the other, the task is posed of consistently combining a logical formulation of the principle of excluded middle with other axioms in the framework of formalistic mathematics. The limited contentual validity of the principle of excluded middle is pointed out with particular eloquence in *Hilbert 1925* (pp. 173–174), where, however, the goal is overshot when the area called into question is extended to include the remaining Aristotelian laws.

3. During the period of the thoughtless use of the principle of excluded middle in the formalistic literature, the principle of the solvability of every mathematical problem is first advanced in *Hilbert 1900b*, p. 52, as an axiom or a conviction and then in *Hilbert 1917*, pp. 412-413, in two different forms (in which, instead of "solvability," "solvability in principle" and, after that, "decidability by means of a finite number of operations" are mentioned) as the object of problems still to be settled. But even after the discussion of the third insight in *Brouwer 1908*, p. 156, *1914*, p. 80, *1919b*, pp. 203–204, and the penetration of the second insight into the formalistic literature, we find that in *Hilbert 1925*, p. 180—where the problem of the consistency of the axiom of the solvability of any mathematical problem is offered as an example of a "problem of a fundamental character that falls within the domain of mathematics but formerly could not even be approached"—this question is presented as still open, irrespective of whether the foundations of the science of mathematics (which also comprise

the consistency of the principle of excluded middle) be secured or not.

4. The fourth insight is expressed in *Brouwer 1927*, p. 64. No trace of it is to be found thus far in the formalistic literature but many an utterance contradicting it, for example in *Hilbert 1900b*, pp. 55–56, and above all in *Hilbert 1925*, where on pp. 162–163 we still find the exclamation: "No, if justifying a procedure means anything more than proving its consistency, it can only mean determining whether the procedure is successful in fulfilling its purpose."

According to what precedes, formalism has received nothing but benefactions from intuitionism and may expect further benefactions. The formalistic school should therefore accord some recognition to intuitionism, instead of polemicizing against it in sneering tones while not even observing proper mention of authorship. Moreover, the formalistic school should ponder the fact that in the framework of formalism *nothing* of mathematics proper has been secured up to now (since, after all, the meta-mathematical proof of the consistency of the axiom system is lacking, now as before), whereas intuitionism, on the basis of its constructive definition of set[3] and the fundamental property it has exhibited for finitary sets,[4] has already erected anew several of the theories of mathematics proper in unshakable certainty. If, therefore, the formalistic school, according to its utterance in *Hilbert 1925*, p. 180 has detected modesty on the part of intuitionism, it should seize the occasion not to lag behind intuitionism with respect to this virtue.

NOTES

1. An oral discussion of the first insight took place in several conversations I had with Hilbert in the autumn of 1909.

2. After attention had already been paid to the principle of excluded middle in *Hilbert 1922*, p. 160.

3. [Later Brouwer uses the word "spread" for this notion; here the word "Menge," translated as "set," suggests that Brouwer considers spreads to be constructive substitutes for classical sets.]

4. See *Brouwer* 1927, p. 66, Theorem 2.

Chapter IX
The Early 20th Century to 1932

Section B
Foundational Crisis and Undecidability

KURT GÖDEL (1906–78)

Kurt Gödel, the leading mathematical logician of the 20th century, was born in Bruenn (in what is today Czechoslovakia); his parents were Rudolf and Marianne Handschuh Gödel. After receiving his doctorate at the University of Vienna in 1930, he began teaching there. He held the position of *Privatdozent* from 1933 until 1938, when the Austrian Republic fell, as Austria was fused with Nazi Germany in a political union *(Anschluss)*. During that year Gödel married Adele Porkert and emigrated to the U.S.

Gödel was to spend the rest of his life in the U.S. at the Institute for Advanced Study at Princeton. He was no stranger to the Institute, having served as a member at its founding in 1933 and again in 1935. After 1938 he continued to be a member until 1952 when he was appointed professor. He retired as professor emeritus in 1976. During these years Gödel pursued research in philosophy, mathematics, and physics. He wrote *Rotating Universes in General Relativity Theory* (1950) and was widely honored by the scientific community. A friend and colleague of Albert Einstein, Gödel was the co-recipient of the first Einstein Award for Achievement in the Natural Sciences in 1951. He was also a member of the National Academy of Sciences, the Association for Symbolic Logic, and the Royal Society.

In mathematics Gödel made three landmark contributions. In his doctoral thesis, "Über die Vollständigkeit des Logikkalküls" ("The Completeness of the Axioms of the Functional Calculus of Logic," 1930), he definitively treated the completeness of first order logic. That is, proceeding from the axioms of the *Principia Mathematica* of Whitehead and Russell, he proved that every valid formula of the propositional calculus is either refutable or $\aleph_0-$ satisfiable. His proof thus yielded not only completeness, which requires that any valid formula be refutable or satisfiable, but also the Löwenheim-Skolem theorem, wherein a satisfiable formula is $\aleph_0-$ satisfiable. Using the arithmetization procedure introduced by Gödel in 1931, David Hilbert and Paul Bernays subsequently extended Gödel's completeness theorem to every consistent system S that remains consistent when number theory axioms and any verifiable formulas of the theory are added to the original axioms of S.

In his monumental paper, "Über formal unentscheidbare Sätze der *Principia mathematica* und verwandter Systeme I" ("On Formally Undecidable Propositions of *Principia Mathematica* and Related Systems I," 1931), Gödel demonstrated that a formal system combining Peano's axioms for natural numbers and the logic of *Principia Mathematica* (1910–13) lacked the desired completeness. His finding profoundly revised Hilbert's formalist foundations of mathematics. The formalist system required simple consistency; such a system could not contain formula A and its negation. With his metamathematics, or proof theory, Gödel showed that there are formally undecidable propositions—there is no *a priori* reason why every true statement has to be provable. His results were general, they applied to the Zermelo-Fraenkel axiom system, to that of von Neumann, to set theory, and to the Peano axioms for natural numbers. Alonzo Church, Stephen Kleene, and Alan Turing generalized Gödel's incompleteness theorem.

In his third major contribution, Gödel established between 1938 and 1940 the consistency of the axiom of choice and of the general continuum hypothesis with the other axioms of set theory. Georg Cantor has posed the continuum hypothesis. In one form it holds that each subset of the continuum has the cardinal $\aleph_0$ of the natural numbers or the cardinal 2 $_0$ of the entire continuum; in another form $2^{\aleph_0} = \aleph_1$. At first it was possible that pairs of cardinals incomparable to each other might exist. But after Hilbert placed the continuum problem first in his famous Paris problems, Zermelo had proved that if one assumes the axiom of choice, every set can be well ordered. This meant that every cardinal is an aleph and every two cardinals are comparable. Working in the context of axiomatic set theory, Gödel saw three possibilities to the axiomatization of the continuum hypothesis, $2^{\aleph_0} = \aleph_1$:

1. It can be proved from the axioms.
2. It can be refuted from the axioms.
3. It is undecidable from the axioms.

After Gödel eliminated the second possibility, Paul J. Cohen eliminated the first in 1963–64. Taken together their work opened a new era in the theory of sets. Mathematicians are now constructing models to investigate a host of problems relating to the consistency or independence of different conjectures in set theory relative to this or that set of axioms.

126. From "Einige metamathematische Resultate über Entscheidungsdefinitheit und Widerspruchsfreiheit" (1930)*

KURT GÖDEL

If to the Peano axioms we add the logic of *Principia mathematica*[1] (with the natural numbers as the individuals) together with the axiom of choice (for all types), we obtain a formal system *S*, for which the following theorems hold:

I. The system *S* is *not* complete [entscheidungsdefinit]; that is, it con-

**Source:* This translation by Stefan Bauer-Mengelberg is reprinted by permission of the publisher from *From Frege to Gödel: A Source Book in Mathematical Logic, 1879-1931,* edited by Jean van Heijenoort, Cambridge, Mass.: Harvard University Press, Copyright © 1967 by the President and Fellows of Harvard College.

tains propositions A (and we can in fact exhibit such propositions) for which neither A nor $\bar{A}$ is provable and, in particular, it contains (even for decidable properties F of natural numbers) undecidable problems of the simple structure $(Ex)F(x)$, where x ranges over the natural numbers.[2]

II. Even if we admit all the logical devices of *Principia mathematica* (hence in particular the extended functional calculus[1] and the axiom of choice) in metamathematics, there does *not* exist a *consistency proof* for the system S (still less so if we restrict the means of proof in any way). Hence a consistency proof for the system S can be carried out only by means of modes of inference that are not formalized in the system S itself, and analogous results hold for other formal systems as well, such as the Zermelo-Fraenkel axiom system of set theory.[3]

III. Theorem I can be sharpened to the effect that, even if we add finitely many axioms to the system S (or infinitely many that result from a finite number of them by "type elevation"), we do *not* obtain a complete system, provided the extended system is ω-consistent. Here a system is said to be ω-consistent if, for no property $F(x)$ of natural numbers, $F(1)$, $F(2)$, . . ., $F(n)$, . . . ad infinitum as well as $(Ex)\overline{F(x)}$ are provable. (There are extensions of the system S that, while consistent, are not ω - consistent.)

IV. Theorem I still holds for all ω-consistent extensions of the system S that are obtained by the addition of *infinitely many* axioms, provided the added class of axioms is decidable [entscheidungsdefinit], that is, provided it is metamathematically decidable [entscheidbar] for every formula whether it is an axiom or not (here again we suppose that the logic used in metamathematics is that of *Principia mathematica*).

Theorems I, III, and IV can be extended also to other formal systems for example, to the Zermelo-Fraenkel axiom system of set theory, provided the systems in question are ω - consistent.

The proofs of these theorems will appear in *Monatschefte für Mathematik und Physik*.

NOTES

1. With the axiom of reducibility or without ramified theory of types.

2. Furthermore, S contains formulas of the restricted functional calculus such that neither universal validity nor existence of a counterexample is provable for any of them.

3. This result, in particular, holds also for the axiom system of classical mathematics, as it has been constructed. For example, by von Neumann (1927).

127. From "Über formal unentscheidbare Sätze der *Principia mathematica* und verwandter Systeme I" (1931)*[1]

(Incompleteness Theorem: Undecidability)

KURT GÖDEL

The development of mathematics toward greater precision has led, as is well known, to the formalization of large tracts of it, so that one can prove any theorem using nothing but a few mechanical rules. The most comprehensive formal systems that have been set up hitherto are the system of *Principia mathematica (PM)*[2] on the one hand and the Zermelo-Fraenkel axiom system of set theory (further developed by J. von Neumann)[3] on the other. These two systems are so comprehensive that in them all methods of proof today used in mathematics are formalized, that is, reduced to a few axioms and rules of inference. One might therefore conjecture that these axioms and rules of inference are sufficient to decide *any* mathematical question that can at all be formally expressed in these systems. It will be shown below that this is not the case, that on the contrary there are in the two systems mentioned relatively simple problems in the theory of integers[4] that cannot be decided on the basis of the axioms. This situation is not in any way due to the special nature of the systems that have been set up but holds for a wide class of formal systems; among these, in particular, are all systems that result from the two just mentioned through the addition of a finite number of axioms,[5] provided no false propositions of the kind specified in footnote 4 become provable owing to the added axioms.

Before going into details, we shall first sketch the main idea of the proof, of course without any claim to complete precision. The formulas of a formal system (we restrict ourselves here to the system *PM*) in outward appearance are finite sequences of primitive signs (variables, logical constants, and parentheses or punctuation dots), and it is easy to state with complete precision *which* sequences of primitive signs are meaningful formulas and which are not.[6] Similarly, proofs, from a formal point of view, are nothing but finite sequences of formulas (with certain specifiable properties.) Of course, for metamathematical considerations it does not matter what objects are chosen as primitive signs, and we shall assign natural numbers to this use.[7] Consequently, a formula will be a finite sequence of natural numbers,[8] and a proof array a finite sequence of finite sequences of natural numbers. The metamathematical notions (propositions) thus become notions (propositions) about natural numbers or sequences of them;[9] therefore they can (at least in part) be expressed by the symbols of the system *PM* itself. In particular, it can be shown

**Source:* This translation by Stefan Bauer-Mengelberg is reprinted by permission of the publisher from *From Frege to Gödel: A Source Book in Mathematical Logic, 1879–1931,* edited by Jean van Heijenoort, Cambridge, Mass.: Harvard University Press,

that the notions "formula," "proof array," and "provable formula" can be defined in the system *PM;* that is, we can, for example, find a formula $F(v)$ of *PM* with one free variable v (of the type of a number sequence)[10] such that $F(v)$, interpreted according to the meaning of the terms of *PM*, says: v is a provable formula. We now construct an undecidable proposition of the system *PM*, that is, a proposition A for which neither A nor *not-A* is provable, in the following manner.

A formula of *PM* with exactly one free variable, that variable being of the type of the natural numbers (class of classes), will be called a *class sign*. We assume that the class signs have been arranged in a sequence in some way,[11] we denote the nth one by $R(n)$, and we observe that the notion "class sign," as well as the ordering relation R, can be defined in the system *PM*. Let α be any class sign; by $[\alpha; n]$ we denote the formula that results from the class sign α when the free variable is replaced by the sign denoting the natural number n. The ternary relation $x = [y; z]$, too, is seen to be definable in *PM*. We now define a class K of natural numbers in the following way:

$$n \in K \equiv \overline{Bew}\ [R(n); n] \qquad (1)$$

(where $Bew\ x$ means: x is a provable formula).[11a] Since the notions that occur in the definiens can all be defined in *PM*, so can the notion K formed from them; that is, there is a class sign S such that the formula $[S; n]$, interpreted according to the meaning of the terms of *PM*, states that the natural number n belongs to K.[12] Since S is a class sign, it is identical with some $R(q)$; that is, we have

$$S = R(q)$$

for a certain natural number q. We now show that the proposition $[R(q); q]$ is undecidable in *PM*.[13] For let us suppose that the proposition $[R(q); q]$ were provable; then it would also be true. But in that case, according to the definitions given above, q would belong to K, that is, by (1), $\overline{Bew}\ [R(q); q]$ would hold, which contradicts the assumption. If, on the other hand, the negation of $[R(q); q]$ were provable, then $\overline{q \in K}$,[13a] that is, $Bew\ [R(q); q]$, would hold. But then $[R(q); q]$, as well as its negation, would be provable, which again is impossible.

The analogy of this argument with the Richard antinomy leaps to the eye. It is closely related to the "Liar" too;[14] for the undecidable proposition $[R(q); q]$ states that q belongs to K, that is, by (1), that $[R(q); q]$ is not provable. We therefore have before us a proposition that says about itself that it is not provable [in *PM*].[15] The method of proof just explained can clearly be applied to any formal system that, first, when interpreted as representing a system of notions and propositions, has at its disposal sufficient means of expression to define the notions occurring in the argument above (in particular, the notion "provable formula") and in which, second, every provable formula is true in the interpretation considered. The purpose of carrying out the above proof with full precision in what follows is, among other things, to replace the second of the assumptions just mentioned by a purely formal and much weaker one.

From the remark that $[R(q); q]$ says about itself that it is not provable it follows at once that $[R(q); q]$ is true, for $[R(q); q]$ *is* indeed unprovable (being undecidable). Thus, the proposition that is undecidable *in the system PM* still was decided by metamathematical considerations. The precise analysis of this curious situation leads to surprising results concerning consistency proofs for formal systems, results that will be discussed in more detail in Section 4 (Theorem XI). . . .

The general result about the existence of undecidable propositions reads as follows:

Theorem VI. *For every ω - consistent recursive class κ of* FORMULAS *there are recursive* CLASS SIGNS *r such that neither* v Gen *r nor* Neg(v Gen *r*) *belongs to* Flg(κ) (where v is the FREE VARIABLE of *r*).

Proof. Let κ be any recursive ω - consistent class of FORMULAS. We define

$$Bw_\kappa(x) \equiv (n)[n \leqq l(x) \rightarrow Ax(n\ Gl\ x) \vee (n\ Gl\ x) \epsilon \kappa \vee$$
$$(Ep, q)\{0 < p,q, < n \ \&\ Fl(n\ Gl\ x,\ p\ Gl\ x,\ q\ Gl\ x)\}] \ \&\ l(x) > 0 \tag{5}$$

(see the analogous notion 44),

$$x\ B_\kappa y \equiv Bw_\kappa(x) \ \&\ [l(x)]\ Gl\ x = \tag{6}$$
$$\text{Bew}_\kappa(x) \equiv (Ey) y\ B_\kappa\ x \tag{6.1}$$

(see the analogous notions 45 and 46).

We obviously have

$$(x)[\text{Bew}_\kappa(x) \sim x \ \epsilon\ \text{Flg}(\kappa)] \tag{7}$$

and

$$(x)[\text{Bew}(x) \rightarrow \text{Bew}_\kappa(x)]. \tag{8}$$

We now define the relation

$$Q(x, y) \equiv \overline{x\ B_\kappa [Sb(y^{19}_{Z(y)})]}. \tag{8.1}$$

Since $x\ B_\kappa y$ (by (6) and (5)) and $Sb(y^{19}_{Z(y)})$ (by Definitions 17 and 31) are recursive, so is $Q(x, y)$. Therefore, by Theorem V and (8) there is a RELATION SIGN q (with the FREE VARIABLES 17 and 19) such that

$$\overline{x\ B_\kappa [Sb(y^{19}_{Z(y)})]} \rightarrow \text{Bew}_\kappa [Sb(q^{17\ \ 19}_{Z(x)\ Z(y)})], \tag{9}$$

and

$$x\ B_\kappa [Sb(y^{19}_{Z(y)})] \rightarrow \text{Bew}_\kappa [\text{Neg}(Sb(q^{17\ \ 19}_{Z(x)\ Z(y)}))]. \tag{10}$$

We put

$$p = 17 \text{ Gen } q \tag{11}$$

(p is a class sign with the free variable 19) and

$$r = Sb(q^{19}_{Z(p)}) \tag{12}$$

(r is a recursive CLASS SIGN[43] with the FREE VARIABLE 17).
Then we have

$$Sb(p^{19}_{Z(p)}) = Sb([17 \text{ Gen } q]^{19}_{Z(p)}) = 17 \text{ Gen } Sb(q^{19}_{Z(p)}) = 17 \text{ Gen } r \tag{13}$$

(by (11) and (12));[44] furthermore

$$Sb(q^{17\ \ 19}_{Z(x)\ Z(p)}) = Sb(r^{17}_{Z(x)}) \tag{14}$$

(by (12)). If we now substitute p for y in (9) and (10) and take (13) and (14) into account, we obtain

$$\overline{x\ B_\kappa (17 \text{ Gen } r)} \rightarrow \text{Bew}_\kappa [Sb(r^{17}_{Z(x)})], \tag{15}$$
$$x\ B_\kappa (17 \text{ Gen } r) \rightarrow \text{Bew}_\kappa [Neg(Sb(r^{17}_{Z(x)}))]. \tag{16}$$

This yields:

1. 17 Gen r is not κ-PROVABLE.[45] For, if it were, there would (by (6.1)) be an n

such that $n\ B_\kappa(17 \text{ Gen } r)$. Hence by (16) we would have $\text{Bew}_\kappa[\text{Neg}(Sb(r^{17}_{Z(n)}))]$, while, on the other hand, from the κ-PROVABILITY of 17 Gen r that of $Sb(r^{17}_{Z(n)})$ follows. Hence, κ would be inconsistent (and a fortiori ω - inconsistent).

2. Neg(17 Gen r) is not κ-PROVABLE. Proof: As has just been proved, 17 Gen r is not κ-PROVABLE; that is (by (6.1)), $(n)\overline{n\ B_\kappa(17 \text{ Gen } r)}$ holds. From this, $(n)\ \text{Bew}_\kappa\ [Sb(r^{17}_{Z(n)})]$ follows by (15), and that, in conjunction with $\text{Bew}_\kappa[\text{Neg}(17 \text{ Gen } r)]$, is incompatible with the ω - consistency of κ.

17 Gen r is therefore undecidable on the basis of κ, which proves Theorem VI. . . .

The results of Section 2 have a surprising consequence concerning a consistency proof for the system P (and its extensions), which can be stated as follows:

Theorem XI. *Let κ be any recursive consistent*[63] *class of* FORMULAS; *then the* SENTENIAL FORMULA *stating that κ is consistent is not* κ-PROVABLE; in particular, the consistency of P is not provable in P,[64] provided P is consistent (in the opposite case, of course, every proposition is provable [in P]).

The proof (briefly outlined) is as follows. Let κ be some recursive class of FORMULAS chosen once and for all for the following discussion (in the simplest case it is the empty class). As appears from 1, page 608, only the consistency of κ was used in proving that 17 Gen r is not κ-PROVABLE;[65] that is, we have

$$\text{Wid}(\kappa) \rightarrow \overline{\text{Bew}_\kappa}(17 \text{ Gen } r), \tag{23}$$

that is, by (6.1),

$$\text{Wid}(\kappa) \rightarrow (x)\ \overline{x\ B_\kappa(17 \text{ Gen } r)}.$$

By (13), we have

$$17 \text{ Gen } r = Sb(p^{19}_{Z(p)}),$$

hence

$$\text{Wid}(\kappa) \rightarrow (x)\ \overline{x\ B_\kappa Sb(p^{19}_{Z(p)})},$$

that is, by (8.1),

$$\text{Wid}(\kappa) \rightarrow (x) Q(x),\ p). \tag{24}$$

We now observe the following: all notions defined (or statements proved) in Section 2,[66] and in Section 4 up to this point, are also expressible (or provable) in P. For throughout we have used only the methods of definition and proof that are customary in classical mathematics, as they are formalized in the system P. In particular, κ (like every recursive class) is definable in P. Let w be the SENTENTIAL formula by which Wid(κ) is expressed in P. According to (8.1), (9), and (10), the relation $Q(x, y)$ is expressed by the RELATION SIGN q, hence $Q(x, p)$ by r (since, by (12), $r = Sb(q^{19}_{Z(p)})$), and the proposition $(x)Q(x\ p)$ by 17 Gen r.

Therefore, by (24), w Imp (17 Gen r) is provable in P[67] (and a fortiori κ-PROVABLE). If now w were κ-PROVABLE, then 17 Gen r would also be κ-PROVABLE, and from this it would follow, by (23), that κ is not consistent.

Let us observe that this proof, too, is constructive; that is, it allows us to actually derive a contradiction from κ, once a PROOF of w from κ is given. The entire proof of Theorem XI carries over word for word to the axiom system of set theory, M, and to that of classical mathematics,[68] A, and here, too, it yields the result: There is no consistency proof for M, or for A, that could be formalized in M, or A, respectively, provided M, or A, is consistent. I wish to note expressly that Theorem XI (and the corresponding results for M and A) do not contradict Hilbert's formalistic viewpoint. For this viewpoint presupposes only the existence of a consistency proof in which nothing but finitary means of proof is used, and it is conceivable that there exist finitary proofs that *cannot* be expressed in the formalism of *P* (or of *M* or *A*).

Since, for any consistent class κ, *w* is not κ-PROVABLE, there always are propositions (namely *w*) that are undecidable (on the basis of κ) as soon as Neg(*w*) is not κ-PROVABLE; in other words, we can, in Theorem VI, replace the assumption of ω-consistency by the following: The proposition "κ is inconsistent" is not κ-PROVABLE. (Note that there are consistent κ for which this proposition is κ-PROVABLE.)

In the present paper we have on the whole restricted ourselves to the system *P*, and we have only indicated the applications to other systems. The results will be stated and proved in full generality in a sequel to be published soon[68a] In that paper, also, the proof of Theorem XI, only sketched here, will be given in detail.

Note added 28 August 1963. In consequence of later advances, in particular of the fact that due to A. M. Turing's work[69] a precise and unquestionably adequate definition of the general notion of formal system[70] can now be given, a completely general version of Theorems VI and XI is now possible. That is, it can be proved rigorously that in *every* consistent formal system that contains a certain amount of finitary number theory there exist undecidable arithmetic propositions and that, moreover, the consistency of any such system cannot be proved in the system.

NOTES

1. A summary of the results in this paper appear in the preceding paper by Gödel.

2. *Whitehead and Russell 1925*. Among the axioms of the system *PM* we include also the axiom of infinity (in this version: there are exactly denumerably many individuals), the axiom of reducibility, and the axiom of choice (for all types).

3. See *Fraenkel 1927* and *von Neumann 1925, 1928,* and *1929*. We note that in order to complete the formalization we must add the axioms and rules of inference of the calculus of logic to the set-theoretic axioms given in the literature cited. The considerations that follow apply also to the formal systems (so far as they are available at present) constructed in recent years by Hilbert and his collaborators. See *Hilbert 1922, 1922a, 1927, Bernays 1923, von Neumann 1927,* and *Ackermann 1924*.

4. That is, more precisely, there are undecidable propositions in which, besides the logical constants $^{-}$ (not), V (or), *(x)* (for all), and = (identical with), no other notions occur but + (addition) and · (multiplication), both for natural numbers, and in which the prefixes *(x)*, too, apply to natural numbers only.

5. In *PM* only axioms that do not result from one another by mere change of type are counted as distinct.

6. Here and in what follows we always understand by "formula of *PM*" a formula written without abbreviations (that is, without the use of definitions). It is well known that [in *PM*] definitions serve only to abbreviate notations and therefore are dispensable in principle.

7. That is, we map the primitive signs one-to-one onto some natural numbers. . . .

8. That is, a number-theoretic function defined on a initial segment of the natural numbers. (Numbers, of course, cannot be arranged in a spatial order.)

9. In other words, the procedure described above yields an isomorphic image of the system *PM* in the domain of arithmetic, and all metamathematical arguments can just as well be carried out in this isomorphic image. This is what we do below when we sketch the proof; that is, by "formula," "proposition," "variable," and so on, *we must always understand the corresponding objects of the isomorphic image*.

10. It would be very easy (although somewhat cumbersome) to actually write down this formula.

11. For example, by increasing sum of the finite sequence of integers that is the "class sign," and lexicographically for equal sums.

11a. The bar denotes negation.

12. Again, there is not the slightest difficulty in actually writing down the formula *S*.

13. Note that "[*R*(*q*); *q*]" (or, which means the same, "[*S*; *q*]") is merely a *metamathematical description* of the undecidable proposition. But, as soon as the formula *S* has been obtained, we can, of course, also determine the number *q* and, therewith, actually write down the undecidable proposition itself. [This makes no difficulty in principle. However, in order not to run into formulas of entirely unmanageable lengths and to avoid practical difficulties in the computation of the number *q*, the construction of the undecidable proposition would have to be slightly modified, unless the technique of abbreviation by definition used throughout in *PM* is adopted.]

13a. [The German text reads $\overline{n \epsilon K}$, which is a misprint.]

14. Any epistemological antinomy could be used for a similar proof of the existence of undecidable propositions.

15. Contrary to appearances, such a proposition involves no faulty circularity, for initially it [only] asserts that a certain well-defined formula (namely, the one obtained from the *q*th formula in the lexicographic order by a certain substitution) is unprovable. Only subsequently (and so to speak by chance) does it turn out that this formula is precisely the one by which the proposition itself was expressed.

[*Footnotes 16–42 omitted.*]

43. Since *r* is obtained from the recursive relation sign *q* through the replacement of a variable by a definite number, *p*. [Precisely stated the final part of this footnote (which refers to a side remark unnecessary for the proof) would read thus: "replacement of a variable by the numeral for *p*."]

44. The operations Gen and *Sb*, of course, can always be interchanged in case they refer to different VARIABLES.

45. By "*x* is κ-PROVABLE" we mean $x \epsilon \mathrm{Flg}(\kappa)$, which, by (7), means the same thing as $\mathrm{Bew}_{\kappa}(x)$.

$\overline{63}$. "κ is consistent" (abbreviated by "Wid(κ)") is defined thus: $\mathrm{Wid}(\kappa) \equiv (Ex)(\mathrm{Form}(x)\ \&\ \overline{\mathrm{Bew}_{\kappa}(x)})$.

64. This follows if we substitute the empty class of FORMULAS for κ.

65. Of course, *r* (like *p*) depends on κ.

66. From the definition of "recursive" . . . to the proof of Theorem VI inclusive.

67. That the truth of *w* Imp (17 Gen *r*) can be inferred from (23) is simply due to the fact that the undecidable proposition 17 Gen *r* asserts its own unprovability, as was noted at the very beginning.

68. See *von Neumann 1927*.

68a. [This explains the "I" in the title of the paper. The author's intention was to publish this sequel in the next volume of the *Monatshefte*. The prompt acceptance of his results was one of the reasons that made him change his plan.]

69. See *Turing 1937*, p. 249.

70. In my opinion the term "formal system" or "formalism" should never be used for anything but this notion. In a lecture at Princeton (mentioned in *Princeton University 1946*, p. 11 [see *Davis 1965*, pp. 84–88]) I suggested certain transfinite generalizations of formalisms, but these are something radically different from formal systems in the proper sense of the term, whose characteristic property is that reasoning in them, in principle, can be completely replaced by mechanical devices.

128. From "Über Vollständikeit und Widerspruchsfreiheit" ("On Completeness and Consistency," 1931)*

KURT GÖDEL

Let Z be the formal system that we obtain by supplementing the Peano axioms with the schema of definition by recursion (on one variable) and the logical rules of the *restricted* functional calculus. Hence Z is to contain no variables other than variables for individuals (that is, natural numbers), and the principle of mathematical induction must therefore be formulated as a rule of inference. Then the following hold:

1. Given any formal system S in which there are finitely many axioms and in which the sole principles of inference are the rule of substitution and the rule of implication, if S contains[1] Z, S is incomplete, that is, there are in S propositions (in particular, propositions of Z) that are undecidable on the basis of the axioms of S, provided that S is ω-consistent. Here a system is said to be ω-consistent if, for no property F of natural numbers, $(Ex)\overline{Fx}$ as well as all the formulas $F(i)$, $i = 1, 2, \ldots,$ are provable.

2. In particular, in every system S of the kind just mentioned the proposition that S is consistent (more precisely, the equivalent arithmetic proposition that we obtain by mapping the formulas one-to-one on natural numbers) is unprovable.

Theorems 1 and 2 hold also for systems in which there are infinitely many axioms and in which there are other principles of inference than those mentioned above, provided that when we enumerate the formulas (in order of increasing length and, for equal length, in lexicographical order) the class of numbers assigned to the axioms is definable and decidable [entscheidungsdefinit] in the system Z, and that the same holds of the following relation $R(x_1, x_2, \ldots, x_n)$ between natural numbers: "the formula with number x_1 follows from the formulas with numbers $x_2, \ldots, x_n$ by a single application of one of the rules of inference." Here a relation (class) $R(x_1, x_2, \ldots, x_n)$ *is said to be decidable in* Z if for every n-tuple $(k_1, k_2, \ldots, k_n)$ *of natural numbers either* $R(k_1, k_2, \ldots, k_n)$ or $\overline{R}(k_1, k_2, \ldots, k_n)$ is provable in Z. (At present no decidable number-theoretic relation is known that is not definable and decidable already in Z.)

If we imagine that the system Z is successively enlarged by the introduction of variables for classes of numbers, classes of classes of numbers, and so forth, together with the corresponding comprehension axioms, we obtain a sequence (continuable into the transfinite) of formal systems that satisfy the assumptions mentioned above, and it turns out that the consistency (ω-consistency) of any of those systems is provable in all subsequent systems.

Also, the undecidable propositions constructed for the proof of Theorem 1 become decidable by the adjunction of higher types and the corresponding axioms; however, in the higher systems we can construct other undecidable propositions by the same procedure, and so forth. To be sure, all the propositions thus constructed are expressible in *Z* (hence are number-theoretic propositions); they are, however, not decidable in *Z*, but only in higher systems, for example, in that of analysis. In case we adopt a type-free construction of mathematics, as is done in the axiom system of set theory, axioms of cardinality (that is, axioms postulating the existence of sets of ever higher cardinality) take the place of the type extensions, and it follows that certain arithmetic propositions that are undecidable in *Z* become decidable by axioms of cardinality, for example, by the axiom that there exist sets whose cardinality is greater than every α_n, where $\alpha_0 = \aleph_0$, $\alpha_{n+1} = 2^{\alpha_n}$.

NOTE

1. That a formal system *S* contains another formal system *T* means that every proposition expressible (provable) in *T* is expressible (provable) also in *S*.

[Remark by the author. 18 May 1966:]

[This definition is not precise, and, if made precise in the straightforward manner, it does not yield a sufficient condition for the non-demonstrability in *S* of the consistency of *S*. A sufficient condition is obtained if one uses the following definition: "*S* contains *T* if and only if every meaningful formula (or axiom or rule (of inference, of definition, or of construction of axioms)) of *T* is a meaningful formula (or axiom, and so forth) of *S*, that is, if *S* is an extension of *T*."

Under the weaker hypothesis that *Z* is recursively one-to-one translatable into *S*, with demonstrability preserved in this direction, the consistency, even of very strong systems *S*, *may* be provable in *S* and even in primitive recursive number theory. However, what can be shown to be unprovable in *S* is the fact that the rules of the equational calculus applied to equations, between primitive recursive terms, demonstrable in *S* yield only correct numerical equations (provided that *S* possesses the property that is asserted to be unprovable). Note that it is necessary to prove this "outer" consistency of *S* (which for the usual systems is trivially equivalent with consistency) in order to "justify," in the sense of Hilbert's program, the transfinite axioms of a system *S*. ("Rules of the equational calculus" in the foregoing means the two rules of substituting primitive recursive terms for variables and substituting one such term for another to which it has been proved equal.)

The last-mentioned theorem and Theorem 1 of the paper remain valid for much weaker systems than *Z*, in particular for primitive recursive number theory, that is, what remains of *Z* if quantifiers are omitted. With insignificant changes in the wording of the conclusions of the two theorems they even hold for any recursive translation into *S* of the equations between primitive recursive terms, under the sole hypothesis of ω-consistency (or outer consistency) of *S* in this translation.]

Chapter IX
The Early 20th Century to 1932

Section C
Selected Topics: The Development of General Abstract Theories

129. From the preface to "Analysis Situs" (1895)*

(Algebraic Topology)

HENRI POINCARÉ

". . . It has been said many times that Geometry is the art of correct reasoning supported by incorrect figures, but in order not to be misleading, these figures must satisfy certain conditions. The proportions may be altered but the relative positions of its various parts must not be changed. The use of figures is there mainly to clarify the relations between the objects that we study and these relations belong to a branch of Geometry called Analysis situs which describes the relative positions of lines and surfaces without regard to their size. There are relations of the same kind in hyperspace; as shown by Riemann and Betti, there is also an Analysis situs in spaces of more than three dimensions. Through this branch of science we know these kinds of relations although our knowledge cannot be intuitive; our senses fail us. In certain cases it gives us the service that we demand from geometric figures. . . ."

**Source:* Lars Garding, *Encounters with Mathematics* (1977), 123. This extract is reprinted by permission of Springer-Verlag. The biography of Poincaré is given before selection 118.

130. From *An Introduction to the Foundations and Fundamental Concepts of Mathematics* (1966)*

HOWARD EVES
CARROLL V. NEWSOM

4.2 HILBERT'S *GRUNDLAGEN DER GEOMETRIE*

The eminent German mathematician, Professor David Hilbert (1862-1943), gave a course of lectures on the foundations of Euclidean geometry at the University of Göttingen during the 1898–99 winter term. These lectures, which concerned themselves with a postulational discussion of Euclidean geometry, were rearranged and published in a slender volume in June, 1899, under the title *Grundlagen der Geometrie* ("Foundations of Geometry"). This work, in its various improved revisions, is today a classic in its field; it has done far more than any other single work since the discovery of non-Euclidean geometry to promote the modern axiomatic method and to shape the character of a good deal of present-day mathematics. The influence of the book was immediate. A French edition appeared soon after the German publication, and an English version, translated by E. J. Townsend, appeared in 1902.[1] The work went through seven German editions during the author's lifetime, the seventh edition appearing in 1930. An eighth German edition, a revision and enlargement by Paul Bernays, appeared in 1956, and a ninth edition, also by Bernays, appeared in 1962.

By developing a postulate set for plane and solid Euclidean geometry that does not depart too greatly in spirit from Euclid's own, and by employing a minimum of symbolism, Hilbert succeeded in convincing mathematicians, to a far greater extent than had Pasch and Peano, of the purely hypothetico-deductive nature of geometry. But the influence of Hilbert's work went far beyond this, for, backed by the author's great mathematical authority, it firmly implanted the postulational method, not only in the field of geometry, but also in essentially every other branch of mathematics of the twentieth century. The stimulus to the development of the foundations of mathematics provided by Hilbert's little book is difficult to overestimate. Lacking the strange symbolism of the works of Pasch and Peano, Hilbert's work can be read, in great part, by any intelligent student of high school geometry.

Whereas Euclid made a distinction between "axioms" and "postulates," modern mathematicians consider these two terms synonymous, and designate all the assumed propositions of a logical discourse by either term. From this point of view, Hilbert's treatment of plane and solid Euclidean geometry rests upon 21 axioms or postulates, and these involve

*Source: Howard Eves and Carroll V. Newsom, *An Introduction to the Foundations and Fundamental Concepts of Mathematics* (Revised Edition, 1966), 94-97. This selection is reprinted by permission of Holt, Rinehart and Winston. The biography of Hilbert is given before selection 119.

six primitive, or undefined, terms. For simplicity we shall here consider only those postulates of Hilbert's set which apply to *plane* geometry. Under this limitation there are only 15 postulates and five primitive terms.

The primitive terms in Hilbert's treatment of plane Euclidean geometry are *point, line* (meaning *straight line*), *on* (a relation between a point and a line), *between* (a relation between a point and a pair of points), and *congruent* (a relation between pairs of points and between configurations called *angles*, which are explicitly defined in the treatment). For convenience of language, the phrase "point *A* is on line *m*" will frequently be stated alternatively by the equivalent phrases, "line *m* passes through point *A*" or "line *m* contains point *A*."

We shall now list Hilbert's 15 postulates for plane geometry, interspersing these postulates with occasional definitions when needed. The statements of the postulates are taken, with some slight modifications for the sake of clarity, from the seventh (1930) edition of Hilbert's *Grundlagen der Geometrie*. The postulates will be presented in certain related groups.

GROUP I: POSTULATES OF CONNECTION

I-1. *There is one and only one line passing through any two given distinct points.*

I-2. *Every line contains at least two distinct points, and for any given line there is at least one point not on the line.*

GROUP II: POSTULATES OF ORDER

II-1. *If point* C *is between points* A *and* B, *then* A, B, C *are all on the same line, and* C *is between* B *and* A, *and* B *is not between* C *and* A, *and* A *is not between* C *and* B.

II-2. *For any two distinct points* A *and* B *there is always a point* C *which is between* A *and* B, *and a point* D *which is such that* B *is between* A *and* D.

II-3. *If* A, B, C *are three distinct points on the same line, then one of the points is between the other two.*

DEFINITION. By the *segment AB* is meant the points *A* and *B* and all points which are between *A* and *B*. Points *A* and *B* are called the *end points* of the segment. A point *C* is said to be *on* the segment *AB* if it is *A* or *B* or some point between *A* and *B*.

DEFINITION. Two lines, a line and a segment, or two segments, are said to *intersect* if there is a point which is on both of them.

DEFINITION. Let *A, B, C* be three points not on the same line. Then by the *triangle ABC* is meant the three segments *AB, BC, CA*. The segments *AB, BC, CA* are called the *sides* of the triangle, and the points *A, B, C* are called the *vertices* of the triangle.

II-4. (Pasch's postulate) *A line which intersects one side of a triangle but does not pass through any of the vertices of the triangle must also intersect another side of the triangle.*

GROUP III: POSTULATES OF CONGRUENCE

III-1. *If* A *and* B *are distinct points and if* A′ *is a point on a line* m, *then there are two and only two distinct points* B′, B″ *on* m *such that the pair of points* A′, B′ *is congruent to the pair* A, B *and the pair of points* A′, B″ *is congruent to the pair* A, B; *moreover,* A′ *is between* B′ *and* B″.

III-2. *If two pairs of points are congruent to the same pair of points, then they are congruent to each other.*

III-3. *If point* C *is between points* A *and* B *and point* C′ *is between points* A′ *and* B′, *and if the pair of points* A, C *is congruent to the pair* A′, C′, *and the pair of points* C, B *is congruent to the pair* C′, B′, *then the pair of points* A, B *is congruent to the pair* A′, B′.

DEFINITION. Two segments are said to be *congruent* if the end points of the segments are congruent pairs of points.

DEFINITION. By the *ray AB* is meant the set of all points consisting of those which are between *A* and *B*, the point *B* itself, and all points *C* such that *B* is between *A* and *C*. The ray *AB* is said to *emanate from* point *A*.

THEOREM. *If* B′ *is any point on the ray* AB, *then the rays* AB′ *and* AB *are identical.*

DEFINITION. By an *angle* is meant a point (called the *vertex* of the angle) and two rays (called the *sides* of the angle) emanating from the point. By virtue of the above theorem, if the vertex of the angle is point *A* and if *B* and *C* are any two points other than *A* on the two sides of the angle, we may unambiguously speak of the angle *BAC* (or *CAB*).

DEFINITION. If *ABC* is a triangle, then the three angles *BAC, CBA, ACB* are called the *angles* of the triangle. Angle *BAC* is said to be *included* by the sides *AB* and *AC* of the triangle.

III-4. *If* BAC *is an angle whose sides do not lie in the same line, and if* A′ *and* B′ *are two distinct points, then there are two and only two distinct rays,* A′C′ *and* A′C″, *such that angle* B′A′C′ *is congruent to angle* BAC *and angle* B′A′C″ *is congruent to angle* BAC; *moreover, if* D′ *is any point on the ray* A′C′ *and* D″ *is any point on the ray* A′C″, *then the segment* D′D″ *intersects the line determined by* A′ *and* B′.

III-5. *Every angle is congruent to itself.*

III-6. *If two sides and the included angle of one triangle are congruent, respectively, to two sides and the included angle of another triangle, then each of the remaining angles of the first triangle is congruent to the corresponding angle of the second triangle.*

GROUP IV: POSTULATE OF PARALLELS

IV-1. (Playfair's postulate) *Through a given point* A *not on a given line* m *there passes at most one line which does not intersect* m.

GROUP V: POSTULATES OF CONTINUITY

V-1. (Postulate of Archimedes) *If* A, B, C, D *are four distinct points, then there is, on the ray* AB, *a finite set of distinct points* A_1 A_2 · · · , A_n *such that* (1) *each of the pairs* A, A_1; A_1, A_2; A_2, A_3; · · · ; A_{n-1}, A_n is congruent to the pair C, D, *and* (2) B *is between* A *and* A_n.

V-2. (Postulate of completeness) *The points of a line constitute a system of points such that no new points can be assigned to the line without causing the line to violate at least one of the nine postulates* I-1, I-2, II-1, II-2, II-3, II-4, III-1, III-2, V-1.

Upon these fifteen postulates rests the entire extensive subject of plane Euclidean geometry! To develop the geometry appreciably from these postulates is too long a task for us to undertake here, but we shall add a few words concerning the significance of some of the postulates.

The postulates of the first group define implicitly the idea expressed by the primitive term "on," and they establish a connection between the two primitive entities, "points" and "lines."

The postulates of the second group were first studied by Pasch, and they define implicitly the idea expressed by the primitive term "between." In particular, they assure us of the existence of an infinite number of points on a line and that a line is not terminated at any point, and they guarantee that the order of points on a line is serial rather than cyclical. Postulate II-4 (Pasch's postulate) differs from the other postulates of the group, for, since it involves points not all on the same line, it gives information about the plane as a whole. The postulates of order are of historical interest inasmuch as Euclid completely failed to recognize any of them. It is this serious omission on Euclid's part that permits one, using only Euclid's list of assumptions, to derive paradoxes by applying apparently sound reasoning to misconceived figures.

The postulates of the third group define implicitly the idea expressed by the primitive term "congruence" as applied to pairs of points and to angles. These postulates are included in order to circumvent the necessity of dealing with the concept of motion. For example, it is interesting to note how, in Postulate III-6, Hilbert introduces the congruency of triangles without employing Euclid's method of superposition, still common in many high school textbooks.

The Playfair parallel postulate appears as the only postulate in Group IV; it is, of course, equivalent to Euclid's parallel postulate. Using the postulates of the first three groups one can prove that there is at least one line through the given point A and not intersecting the given line m.

The first postulate of the last group (the postulate of Archimedes) corresponds to the familiar process of estimating the distance from one point of a line to another by the use of a measuring stick; it guarantees that if we start at the one point and lay off toward the second point a succession of equal distances (equal to the length of the measuring stick) we will ultimately pass the second point. Upon this postulate can be made to depend the entire theory of measurement and, in particular, Euclid's theory of proportion. The final postulate (the postulate of completeness) is not required for the derivation of the theorems of Euclidean geometry, but it makes possible the establishment of a one-to-one correspondence between the points on any line and the set of all real numbers, and is necessary for the free use of the real number system in analytic, or coordinate, geometry. . . .

NOTE:

1. David Hilbert, *The Foundations of Geometry,* trans. by E. J. Townsend (Chicago: The Open Court Publishing Company, 3rd edition 1938).

Chapter IX
The Early 20th Century to 1932

Section C
Selected Topics: The Development of General Abstract Theories

HENRI (LÉON) LEBESGUE (1875–1941)

One of the leading mathematicians of the early 20th century, Henri Lebesgue was born into a middle class family near Paris, France. His father worked in typography and his mother taught at an elementary school. The parents both had intellectual interests that led them to have a substantial library in their home, and the intellectual ability of young Lebesgue quickly became apparent to his teachers. Following the early death of his father, local philanthropy made it possible for him to continue his education.

Lebesgue began higher education at the École Normale Supérieure where he studied from 1894 to 1897. Paul Langevin taught him the physical sciences, and the student displayed an independent, critical mind that led him to question the statements of professors. From 1897 to 1899 he worked in the library of the École and published four papers. One of these, "Sur une generalisation de l'intégrale définie" ("On a Generalization of the Definite Integral," 1900), contained the famous definition of what has come to be known as the Lebesgue integral. He also provided a simpler proof of Weierstrass' approximation theorem and studied functions without derivatives. Hermite and others, who considered such functions scandalous, criticized his innovative studies of them. Lebesgue taught and continued his research at the Lycée Centrale in Nancy from 1899 to 1902. The results of this research on the theory of integration appeared in his doctoral dissertation, "Intégrale, Longeur, Aire" ("Integral, Length, Surface"), which was presented in 1902 at the Sorbonne.

The first two university appointments of Lebesgue were at Rennes (1902–06) and at Poitiers (1906–10). While at Rennes he twice gave the Cours Peccot at the Collège de France, the first on his new integral and the second on trigonometric series. The two resulting monographs—*Leçons sur l'integration ...* (1904) and *Leçons sur les series trigonometrique* (1906)—made his ideas better known just prior to their widespread acceptance and development. In 1910 he moved to the Sorbonne. He was *maître de conférences* (lecture master) in mathematical analysis until 1919, when he became *professeur d'application de la géométrie à l'analyse*. In 1921 he was appointed professor at the Collège de France and the next year was elected to the Paris Académie des Sciences. In 1930 he was elected a foreign member of the Royal Society of London.

Lebesgue's election to the Paris Académie marks a dividing point in his career. By then he had published nearly 90 books and articles mainly on the theory of integration, the structure of sets and functions, calculus of variations, theory of surface areas, and dimension theory. Based on this work, his reputation had steadily grown and he had received numerous honors, including Prix Poncelet (1914), Prix Saintour (1917), and Prix Petit d'Ormoy (1919). He remained active after 1922, but his writings reflected broadening interests in pedagogical and historical questions as well as in basic geometry. He advocated teaching mathematics in a simple, genetic style that drew upon the history of ideas. He stated that the role of the teacher in class was "to think in front of his students" and that students "gain nothing from a solution that is satisfying from the logical, but not from the human point of view." He upheld the cumulative, historic character of mathematical research, writing in the preface to his *Lessons on Integration* (1928): "In order to do useful work it is necessary to march along paths opened by previous workers; acting otherwise, one runs too great a risk of creating a science without links with the rest of mathematics."

The outstanding contribution of Lebesgue to mathematics was the theory of integration that now bears his name. At the close of the 19th century, mathematical analysis was effectively limited to continuous functions because the Riemann method of integration applied only to continuous and a few discontinuous functions. While still a university student, Lebesgue drew upon René Baire's study of discontinuous functions, Camille Jordan's measure-theoretical treatment of Riemann's integral, and Émile Borel's definitions of measure and measurability to formulate a new theory of measure and to frame a new definition of the definite integral which generalized the Riemann integral (1854). The basis for its generality is that a Lebesgue integrable function does not have to be continuous nearly everywhere (except on a set of measure 0). It needs only to have pointwise convergence. The Lebesgue integral is a major achievement in modern real analysis. It is the standard integral of mathematical analysis today and will surely hold this position well into the future.

Importantly Lebesgue recognized that his new integral was an analytical tool capable of overcoming theoretical difficulties that had beset Riemann's theory of integration, most notably those concerning Fourier analysis and the fundamental theorem of the calculus, $\int_a^b f'(x)dx = f(b) - f(a)$.

In Fourier analysis Lebesgue proved that for any bounded series of Lebesgue-integrable functions term-by-term integration is permissible (this is not always valid for Riemann-integrable functions) and that if a trigonometric series can represent a bounded function, that series is the Fourier series of the function. In the late 19th century, mathematicians had with increasing frequency identified functions with bounded derivatives that are not integrable in Riemann's sense. For these functions the fundamental theorem was meaningless. Lebesgue proved that these difficulties with bounded derivatives disappeared when his new integral was applied. He thereby provided the "almost everywhere" version of the fundamental theorem of the calculus.

In other mathematical research, Lebesgue coined the term "summable," made progress in the theory of multiple integrals, and derived his pavement theorem, which is significant in topology. With others he also endeavored to generalize and extend his theory of integration. Johann

Radon achieved a notable generalization when he defined the Lebesgue–Stieltjes integral in 1913. The development of abstract theories of measure and integration has come to dominate recent mathematical research. Such theories have applications in harmonic analysis (generalized Fourier analysis), ergodic theory, spectral theory, and the theory of probability.

131. From "The Development of the Integral Concept" (1926)*

HENRI LEBESGUE

Leaving aside all technicalities, we are going to examine the successive modifications and enrichments of the concept of the integral and the appearance of other notions used in recent research on functions of a real variable.

Before Cauchy there was no definition of the integral in the modern meaning of the word "definition." One merely said which areas had to be added or subtracted in order to obtain the integral $\int_a^b f(x)\,dx$.

For Cauchy a definition was necessary, because with him there appeared the concern for rigor which is characteristic of modern mathematics. Cauchy defined continuous functions and their integrals in about the same way as we do today. In order to arrive at the integral of $f(x)$ it suffices to form the sums (Fig. 1)

$$S = \Sigma f(\xi_i)(x_{i+1} - x_i), \tag{1}$$

which surveyors and mathematicians have always used to approximate area, and then deduce the integral $\int_a^b f(x)\,dx$ by passage to the limit.

Although the legitimacy of such a passage to the limit was evident for one who thought in terms of area, Cauchy

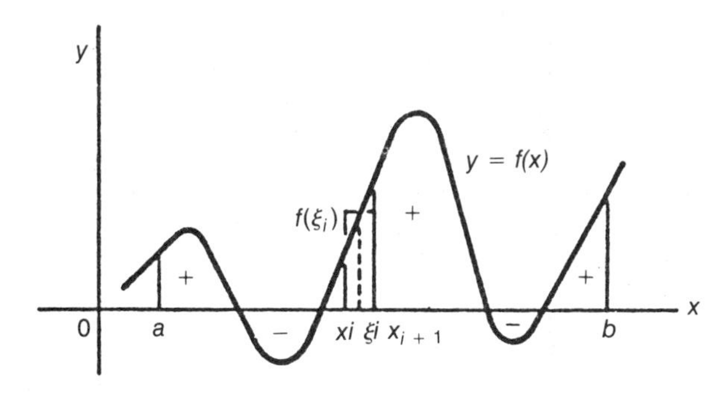

FIGURE 1

*Source: This translation of "Sur le developpement de la notion d'intégrale" is taken from part II of Henri Lebesgue, *Measure and the Integral*, edited with a biographical essay by Kenneth O. May (1966), 177-183. It is reprinted by permission of Holden-Day, Inc.

had to demonstrate that S actually tended to a limit in the conditions he considered. A similar necessity appears every time one replaces an experimental notion by a purely logical defintion. One should add that the interest of the defined object is no longer obvious, it can be developed only from a study of the properties following from the definition. This is the price of logical progress.

What Cauchy did is so substantial that it has a kind of philosophic sweep. It is often said that Descartes reduced geometry to algebra. I would say more willingly that by the use of coordinates he reduced all geometries to that of the straight line, and that the straight line, in giving us the notions of continuity and irrational number, has permitted algebra to attain its present scope.

In order to achieve the reduction of all geometries to that of the straight line, it was necessary to eliminate a certain number of concepts related to geometries of several dimensions such as the length of a curve, the area of a surface, and the volume of a body. The progress realized by Cauchy lies precisely here. After him, in order to complete the arithmetization of mathematics it was sufficient for the arithmeticians to construct the linear continuum from the natural numbers.

And now, should we limit ourselves to doing analysis? No. Certainly, everything that we do can be translated into arithmetical language, but if we renounce direct, geometrical, and intuitive views, if we are reduced to pure logic which does not permit a choice among things that are correct, then we would hardly think of many questions, and certain concepts, for example, most of the ideas that we are going to examine here today, would escape us completely.

For a long time certain discontinuous functions have been integrated. Cauchy's definition still applies to these integrals, but it is natural to examine, as did Riemann, the exact capacity of this definition.

If $\underline{f_i}$ and $\overline{f_i}$ represent the lower and upper bounds of $f(x)$ in (x_i, x_{i+1}), then S lies between

$$\underline{S} = \Sigma \underline{f_i}(x_{i+1} - x_i) \text{ and } \overline{S} = \Sigma \overline{f}_i(x_{i+1} - x_i).$$

Riemann showed that for the definition of Cauchy to apply it is sufficient that

$$\overline{S} - \underline{S} = \Sigma(\overline{f_i} - \underline{f_i})(x_{i+1} - x_i)$$

tends toward zero for a particular sequence of partitions of the interval from a to b into smaller and smaller subdivisions (x_i, x_{i+1}). Darboux added that under the usual operation of passage to the limit $\underline{S}$ and $\overline{S}$ always give two definite numbers

$$\underline{\int_a^b} f(x)\,dx \text{ and } \overline{\int_a^b} f(x)\,dx.$$

These numbers are generally different and are equal only when the Cauchy-Riemann integral exists.

From a logical point of view, these are very natural definitions aren't they? However, one can say that from a practical point of view they have been useless. In particular, Riemann's definition has the drawback of applying only rarely and in a sense by chance.

It is evident that breaking up the interval (a, b) into smaller and smaller subintervals (x_i, x_{i+1}) makes the differences $\overline{f_i} - \underline{f_i}$ smaller and smaller if $f(x)$ is continuous, and that the continued refinement of the subdivision will make $\overline{S} - \underline{S}$ tend toward zero if there are only a few points of discontinuity. But we have no reason to hope that the same thing will happen for a function that is discontinuous everywhere. To take smaller intervals (x_i, x_{i+1}), that is to say values of $f(x)$ corresponding to values of x closer together, does not in any way guarantee that one takes values of $f(x)$ whose differences become smaller.

Let us be guided by the goal to be attained—to collect approximately equal values of $f(x)$. It is clear then that we must break up not (a, b), but the interval $(\underline{f}, \overline{f})$ bounded by the lower and upper bounds of $f(x)$ in (a, b). Let us do this with the aid of number y_i differing among themselves by less than ϵ. We are led to consider the values of $f(x)$ defined by

$$y_i \leq (x) \leq y_{i+1}.$$

The corresponding values of x form a set E_i. In Figure 2 this set E_i consists of four intervals. With some continuous functions it might consist of an infinity of intervals. For an arbitrary function it might be very complicated. But this matters little. It is this set E_i which plays the role analogous to the interval (x_i, x_{i+1}) in the usual definition of the integral of continuous functions, since it tells us the values of x which give to $f(x)$ approximately equal values.

If η_i is any number whatever taken between y_i and y_{i+1}, $y_i \leq \eta_i \leq y_{i+1}$, the values of $f(x)$ for points of E_i differ from η_i by less than ϵ. The number η_i is going to play the role which $f(\xi_i)$ played in formula (1). As to the role of the length or measure $x_{i+1} - x_i$ of the interval (x_i, x_{i+1}), it will be played by a measure $m(E_i)$ which we shall assign to the set E_i in a moment. In this way we form the sum

$$S = \Sigma \eta_i m(E_i). \qquad (2)$$

Let us look closely at what we have just done and, in order to understand it better, repeat it in other terms.

The geometers of the seventeenth century considered the integral of $f(x)$—the word "integral" had not been invented, but that does not matter—as the sum of an infinity of indivisibles, each of which was the ordinate, positive or negative, of $f(x)$. Very well! We have simply grouped together the indivisibles of comparable size. We have, as one says in algebra, collected similar terms. One could say that, according to Riemann's procedure, one tried to add the indivisibles by taking them in the order in which they were furnished by the variation in x, like an unsystematic merchant who counts coins and bills at random in the order in which they came to hand, while we operate like a methodical merchant who says:

I have $m(E_1)$ pennies which are worth $1 \cdot m(E_1)$,
I have $m(E_2)$ nickels worth $5 \cdot m(E_2)$,
I have $m(E_3)$ dimes worth $10 \cdot m(E_3)$, etc.

Altogether then I have

$$S = 1 \cdot m(E_1) + 2 \cdot m(E_2) + 5 \cdot m(E_3) + \cdots.$$

The two procedures will certainly lead the merchant to the same result because no matter how much money he has there is only a finite number of coins or bills to count. But for us who must add an infinite number of indivisibles the difference between the two methods is of capital importance.

We now consider the definition of the number $m(E_i)$ attached to E_i. The analogy of this measure to length, or even to a number of coins, leads us naturally to say that, in the example of Fig. 2, $m(E_i)$

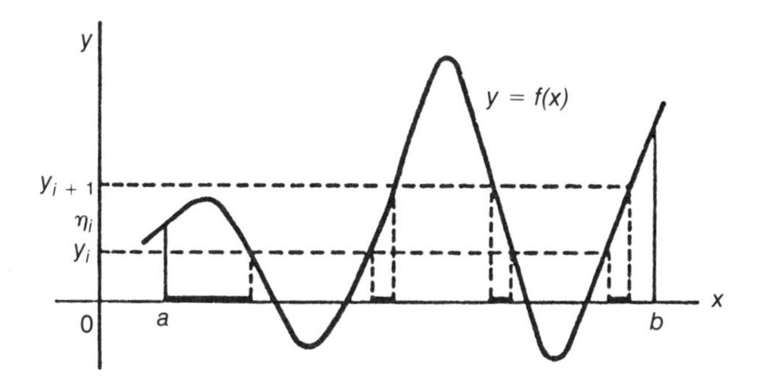

FIGURE 2

will be the sum of the lengths of the four intervals that make up E_i, and that, in an example where E_i is formed from an infinity of intervals, $m(E_i)$ will be the sum of the length of all these intervals. In the general case it leads us to proceed as follows. Enclose E_i, in a finite or denumerably infinite number of intervals, and let $l_1, l_2, \ldots$ be the length of these intervals. We obviously wish to have

$$m(E_i) \leq l_1 + l_2 + \cdots.$$

If we look for the greatest lower bound of the second member for all possible systems of intervals that cover E_i, this bound will be an upper bound of $m(E_i)$. For this reason we represent it by $\overline{m(E_i)}$, and we have

$$m(E_i) \leq \overline{m(E_i)}. \qquad (3)$$

If C is the set of points of the interval (a, b) that do not belong to E_i, we have similarly

$$m(C) \leq \overline{m(C)}.$$

Now we certainly wish to have

$$m(E_i) + m(C) = m[(a, b)] = b - a;$$

and hence we must have

$$m(E_i) \geq b - a - \overline{m(C)}. \qquad (4)$$

The inequalities (3) and (4) give us upper and lower bounds for $m(E_i)$. One can easily see that these two inequalities are never contradictory. When the lower and upper bounds for E_i are equal, $m(E_i)$ is defined, and we say then that E_i is measurable.[1]

A function $f(x)$ for which the sets E_i are measurable for all choices of y_i is called measurable. For such a function formula (2) defines a sum S. It is easy to prove that when the y_i vary so that ϵ tends toward zero, the S tend toward a definite limit which is, by definition,[2] $\int_a^b f(x)\,dx$.

This first extension of the notion of the definite integral led to many others. Let us suppose that it is a question of integrating a function $f(x, y)$ of two variables. Proceeding exactly as before, we construct sets E_i which are now sets of points in the plane and no longer on a line. To these sets we must now attribute a plane measure, and this measure is deduced from the area of rectangles

$$\alpha \leq x \leq \beta; \quad \gamma \leq y \leq \delta$$

in exactly the same way as the linear measure was derived from the length of intervals. Once measure is defined, formula (2) gives the sums S from which the integral is obtained by passage to the limit. Hence the definition that we have considered extends immediately to functions of several variables.

Here is another extension which applies equally well regardless of the number of variables, but which I explain only in the case where it is a question of integrating $f(x)$ in the interval (a, b). I have said that it is a question of summing indivisibles represented by the various ordinates at points x, $y = f(x)$. A moment ago, we collected these indivisibles according to their sizes. Now let us merely group them according to their signs. We will have to consider then the set E_p of points in the plane whose ordinates are positive, and the set E_n of points whose ordinates are negative. As I recalled at the beginning of my lecture, for the simple case where $f(x)$ is continuous, even before Cauchy's time one wrote

$$\int_a^b f(x)\,dx = \text{area}\,(E_p) - \text{area}\,(E_n).$$

This leads us to assert

$$\int_a^b f(x)\,dx = m_s(E_p) - m_s(E_n),$$

where m_s stands for a plane measure. This new definition is equivalent to the preceding one. It brings us back to the intuitive method before Cauchy, but the definition of measure puts it on a solid logical foundation.

NOTES

1. The definition of measure of sets used here is that of C. Jordan, *Cours d'analyse de*

l'École Polytechnique, Vol. I, but with this modification, essential for our purpose, that we enclose the set E_i to be measured in intervals whose number may be infinite, while Jordan employed only a finite number. This use of a denumerable infinity in place of a finite number of intervals was suggested by the work of Borel, who himself had utilized this idea in order to get a definition of measure (*Leçons sur la théorie des fonctions*).

2. *C. R. Acad. Sci. Paris,* **132**, 1900, pp. 1025–1028. Definitions equivalent to that given here have been proposed by various authors. The most interesting are due to W. H. Young, *Phil. Trans. Roy. Soc. London,* **204**, 1905, pp. 221–252, and *Proc. London Math. Soc.,* **9,** 1911, pp. 15–50. See also, for example, the notes of Borel and M. F. Riesz, *C. R. Acad. Sci. Paris,* **154**, 1912, pp. 413–415, 641–643.

Chapter IX
The Early 20th Century to 1932

Section C
Selected Topics: The Development of General Abstract Theories

(AMALIE) EMMY NOETHER (1882–1935)

Emmy Noether, who is probably the most creative woman mathematician in history, grew up in the south German town of Erlangen amid a family drawn to the study of the sciences. She was the eldest child of Max Noether, a guiding spirit of 19th-century algebraic geometry, and Ida Amalia Kaufmann. Two of her three brothers were also to pursue scientific careers: Alfred as a chemist and Fritz as a physicist. As a child Emmy was acutely nearsighted and did not exhibit exceptional intellectual ability.

In 1900, three years after completing studies at a girls' high school, Emmy passed a test to qualify as a teacher of French and English. However, she had already begun to move away from languages and desired to take up university studies, even though most universities did not admit women students at the turn of the 20th century. In the winter semester of 1900 she was one of two women allowed to audit courses in foreign languages, in history—but mainly in mathematics—at the University of Erlangen, where her father taught. In 1903, still an auditor, she moved to the University of Göttingen to listen to the lectures of Hermann Minkowski, Felix Klein, and David Hilbert. She left after only one semester because the University of Erlangen permitted her to register officially as a student in 1904. The only doctoral student of the garrulous and eccentric algorithmist Paul Gordan, she passed the doctoral oral *summa cum laude* in 1907 with a dissertation on algebraic invariant theory, "On Complete Systems of Invariants for Ternary Biquadratic Forms." Until 1915 she remained with her family at Erlangen, where Gordan's successor, Ernst Fischer, also strongly influenced her.

In 1915 David Hilbert invited Emmy Noether to the University of Göttingen. When she arrived the next year, he asked her to teach courses given under his name and to apply her invariant-theoretic knowledge to solve problems concerning the nascent general theory of relativity which he and Klein were investigating. As a result she devised elegant mathematical formulations for some of the concepts of general relativity. Under the influence of Hilbert and his circle her primary interests shifted from invariant theory to modern algebra. In theoretical physics she formulated Noether's theorem that connects conservation laws with the symmetry properties of a physical system. Despite repeated attempts by Hilbert to make her a *Privatdozent*, the prejudice against women at Göttingen prevented her "habilitation" during World War I. She received no regular financial remuneration and had to live quite frugally. She generally took no part in political life, except in 1918 when the Weimar Republic was established. She supported one of its founding groups, the Social Democrats, a nominally Marxist working class party with an international outlook, and expressed pacifist ideas.

Emmy Noether finally became a *Privatdozent* at Göttingen in 1919. Three years later she was made an unofficial associate professor (honorary) and was given a modest salary. She continued to teach at Göttingen until 1933 except for leaves as a visiting professor at the University of Moscow for the winter semester of 1928–29 and at Frankfurt for the summer of 1930. In 1924 she was fortunate to have as a student B. L. van der Waerden, who later promoted her ideas widely. From 1927 her influence on contemporary mathematics grew steadily. This influence must be measured by more than her teaching, articles, position at the center of mathematical research in Göttingen by 1930, and her help in editing the *Mathematische Annalen*. Through suggestions and casual comments she also freely shared insights that appeared in the writings of her students and colleagues. In 1932 the International Mathematical Congress in Zurich recognized the importance of her ideas. It was the high point of her international scientific reputation.

The rise to power of Adolf Hitler altered the life of Emmy Noether, just as it did the lives of many other Jewish scholars. The Nazi Party attacked both Jews and independent minded mathematicians. In April 1933 Göttingen summarily dismissed Noether. On the recommendation of Hermann Weyl, Bryn Mawr College offered her a visiting professorship, and she departed for the U.S. in October 1933. She spent the last 18 months of her life lecturing and doing research at Bryn Mawr and at the Institute for Advanced Study at Princeton.

The chief contributions of Emmy Noether were in the area of modern abstract algebra. Building upon Dedekind's approach—in which ideals are not numbers but sets of numbers—she developed from 1920 to 1927 the abstract axiomatic "theory of ideals." With this theory she reinstated into algebra Euclid's theorem on the unique decomposition into prime factors, which had broken down in algebraic number fields. In addition, she related the ideals to the generalized structures of "group" and "ring" and formulated the concept of primary ideals, thereby generalizing Dedekind's prime ideals.

Noether's later work was principally inspired by the writings of Hamilton and Grassmann. With a novel, unified, and purely conceptual approach she examined generalized systems that arise if one or more formal properties of algebra (*e.g.*, the associative, commutative, or distributive) is not assumed. In this way she was to construct a theory of noncommutative algebras. Her development of the theory of representations from 1925 to 1927 attempted to attain noncommutative rings (or algebras) by using matrices or linear transformations. This had to be done while preserving all relations involving the ring addition and/or multiplications. Put another way, she investigated the homomorphisms of a given ring into a ring of matrices. Two fundamental articles from her last years are "Hypercomplexe Grössen und Darstellungstheorie" (1929), which treats the general arithmetic of hypercomplex numbers, and "Nichtcommutative Algebra" (1933), in which Noether skillfully probes the structure of noncommutative algebras with her concept of the "cross" product (verschränktes). Both articles were published in *Mathematische Zeitschrift*. In 1932 she proved with Richard Brauer and Helmut Hasse that "every 'simple' algebra over an ordinary algebraic number field is cyclic"—a theorem that Hermann Weyl called "a high watermark in the history of algebra."

132. Proof of a Fundamental Theorem in the Theory of Algebras (1932)*

R. BRAUER (Königsberg)
H. HASSE (Marburg)
E. NOETHER (Göttingen)[1]

Finally our united efforts have succeeded in proving the correctness of the following theorem which is of fundamental importance for the structure theory of algebras over algebraic number fields, as well as beyond:

Fundamental Theorem. *Every central* [[*normale*]] *division algebra over an algebraic number field is cyclic (or, as one also says, of Dickson type).*

It is a particular joy for us to present this result as a success to be credited essentially to the p-adic method, to Mr. Kurt Hensel, the originator of this method, on the occasion of his 70th birthday.

Our proof consists of three reductions, of which each of us has contributed one.[2]

1. The first reduction was given by H. Hasse on the basis of the theory of cyclic algebras over algebraic number fields, which was recently developed by him.[3]

Reduction 1. *The fundamental theorem is proved if the following is shown:*

I. *Every everywhere-splitting algebra over* Ω *is* $\sim \Omega$.

To shorten the presentation we stipulate that here as in what follows "Algebra A over Ω" shall always mean "central simple algebra A over the algebraic number field Ω" ("simple hypercomplex system with the center Ω"). Further, $A \sim \Omega$ indicates that A is a complete matrix algebra in Ω, thus the membership of A in the particular class determined by Ω itself as a division algebra over Ω in the sense of similarity (equality of the corresponding division algebras over Ω) [H, 5]. Finally, one understands by an "everywhere-splitting" algebra over Ω one for which for every prime place $\mathfrak{p}$ of Ω the $\mathfrak{p}$-adic extension (*i.e.*, the algebra obtained by extension of the coefficient field Ω to the corresponding $\mathfrak{p}$-adic field $\Omega_{\mathfrak{p}}$) $A_{\mathfrak{p}}$ is $\sim \Omega_{\mathfrak{p}}$.

Proof. Let D be a division algebra over Ω. If then Z is a cyclic field over Ω such that for every prime place $\mathfrak{p}$ of Ω the $\mathfrak{p}$-degree of Z is a multiple of the $\mathfrak{p}$-index of D [cf. the note in H, 17 Bb], then for the corresponding prime divisors $\mathfrak{P}$ in Z the $\mathfrak{P}$-adic field $Z_{\mathfrak{P}}$ is in each case a splitting field of $D_{\mathfrak{p}}$ [H, 18.1]. It results from this that the algebra

*Source: From *Journal für die reine und angewandte Mathematik,* vol. 167 (1932), pp. 399–404. This selection was translated by Dr. Kurt Bing, Professor Emeritus of Mathematical Sciences at Rensselaer Polytechnic Institute, with the assistance of the editor. Dr. Bing also wrote the translator's note at the end of this selection.

D_Z over Z which is obtained by extension of the coefficient field Ω of D to Z is everywhere splitting. If now **I** is already established (for every Ω), then $D_Z \sim$ Z follows. This amounts to saying that D possesses the cyclic splitting field Z, *i.e.*, is cyclically representable [H, 5]. But then D also possesses such a cyclic splitting field whose degree coincides with the degree of D itself, *i.e.*, D is cyclic [H, 6, Theorem 6]. Under hypothesis **I** the fundamental theorem is therefore proved.

2. R. Brauer gave the decisive impetus [[*Anstoss*]] leading to the second reduction by informing H. Hasse by letter about the way in which the related problem of the precise value of the exponent of a division algebra (which, moreover, is solved at the same time by the fundamental theorem, see below) can be reduced to the case of a solvable splitting field by means of the Sylow group theorem. This idea could be used as well for a corresponding further reduction of the cyclicity problem.

Reduction 2. *Theorem* **I** *is proved if the following is shown:*

II. *Every everywhere-splitting algebra with a solvable splitting field over Ω is $\sim \Omega$.*

Proof. Let A be an everywhere-splitting algebra over Ω. If then K is a normal [[*galoisscher*]] splitting field for A, and further p is any prime number and Σ one of the Sylow fields of K over Ω related to p (*i.e.*, a field of invariants of a Sylow group belonging to p of the Galois group), then A_Σ is a likewise everywhere-splitting algebra over Σ, which possesses the solvable splitting field K. If now **II** is already established (for every Ω), then $A_\Sigma \sim \Sigma$ follows. This amounts to saying that A possesses the splitting field Σ of a degree prime to p. The index of A, being a divisor of this degree, is therefore prime to p. Since this is true for every prime number p, the index is therefore equal to 1, *i.e.*, $A \sim \Omega$. Under hypothesis **II, I** is therefore proved.[4]

3. The third reduction, which as H. Hasse—in possession of the first two reductions—recognized, leads to the final proof. That reduction was given by E. Noether. She was motivated by a communication from H. Hasse, in which the fundamental theorem was proved for the special case of an Abelian splitting field, namely by means of the general reduction **1** and reduction of the corresponding factor system according to formula. As the core of this E. Noether recognized, in fact, the third reduction. Independently of E. Noether, R. Brauer had also already thought out this third reduction.

Reduction 3. *Theorem* **II** *is proved if the following is shown:*

III. *Every everywhere-splitting algebra with a cyclic splitting field of prime degree over Ω is $\sim \Omega$.*

Proof. Let A be an everywhere-splitting algebra with a solvable splitting field K over Ω. Then let $K = \Lambda_0 > \Lambda_1 > \cdots > \Lambda_r = \Omega$ be a chain of fields between K and Ω such that always Λ_i is cyclic of prime degree over Λ_{i+1}. Then, first, A_{Λ_1}, is a likewise everywhere-splitting algebra over Λ_1, which possesses the cyclic splitting field of prime degree $K = \Lambda_0$. If now **III** is already proved (for every Ω), then $A_{\Lambda_1} \sim \Lambda_1$ follows. This amounts to saying that Λ_1 is a splitting field for A. Now one proves, starting with Λ_1 instead of Λ_0, exactly in the same way that $A_{\Lambda_2} \sim \Lambda_2$, etc., until finally $A_{\Lambda_r} \sim \Lambda_r$, *i.e.*, $A \sim \Omega$. Under hypothesis **III, II** is therefore proved.

4. However, **III** is true according to the results of H. Hasse [H, 24]. Therefore the correctness of the fundamental theorem follows by the reductions **3, 2, 1** in the reverse order.

In this connection, E. Noether points out the following in addition:

The correctness of **III**—more generally for cyclic splitting fields of *arbitrary degree*—comes back to Hasse's Norm Theorem.[5] But for Reduction **3** only the special case of the prime degree, thus the original Hilbert-Furtwängler Norm Theorem[6] is used. Hence Reduction **III** provides a new simple proof of the

Hasse Norm Theorem. This proof would run as follows:

Let Z be a cyclic field over Ω and α a number from Ω, for which the norm rest symbol $\left(\frac{\alpha, Z}{\mathfrak{p}}\right) = 1$ for every prime place $\mathfrak{p}$ of Ω. Then every cyclic algebra $A = (\alpha, Z)$[7] is everywhere splitting [H, 17.7]. By Reduction **3** there follows by the sole use of the Hilbert-Furtwängler Norm Theorem, $A \sim \Omega$. But then α is the norm of an element from Z [H, 15.4].

In connection with the Norm Theorem we remark further that Theorem **I** is to be considered as its correct generalization to higher (also non-Abelian) cases, while indeed the verbatim generalization, as H. Hasse showed,[5] is not generally correct.

CONSEQUENCES (H. HASSE)

5. As the most obvious consequence of the Fundamental Theorem one may remark that now the theory of cyclic algebras over algebraic number fields which was developed by H. Hasse has acquired the significance of a general theory of the structure and invariants of the central simple algebras (in particular of the central division algebras) over algebraic number fields. In particular, the following is now proved in general:

Theorem 1. *The exponent of a central simple division algebra over an algebraic number field is equal to its index.*

Further, there results from Theorem **I** the following remarkable fact:

Theorem 2. *The basic ideal* [[*Grundideal*]] *of a proper central division algebra over Ω is divisible at least by one prime place of Ω (and even at least by two).*

Here the basic ideal is defined as, say, the (reduced) [[(*red.*)]] norm of the (reduced) different.[8] Besides, an infinite prime place is admitted into the basic ideal if and only if for it the algebra reduces to the quaternion algebra (and not the field of real or complex numbers).

Proof. According to the results of H. Hasse[9] a prime place $\mathfrak{p}$ of Ω divides the basic ideal if and only if the $\mathfrak{p}$-index is different from 1. Now if all $\mathfrak{p}$-indices were equal to 1, then the algebra would be everywhere splitting, and according to Theorem **I**, one would not have a proper division algebra. (That one $\mathfrak{p}$-index alone cannot be different from 1 either results from the fact that for the norm remainder symbols, whose orders are the $\mathfrak{p}$-indices [H, 17.7], the product theorem (Law of Reciprocity) holds [H, 3.8].)

6. Further, Theorem **I** makes possible the determination (arithmetical characterization) of all splitting fields K belonging to an algebra A over Ω. There results in exact generalization of the state of affairs already established by H. Hasse in the special case of cyclic A, K [H, 6, Satz 2]:

Theorem 3. *For an algebra A over Ω an algebraic number field K over Ω is a splitting field if and only if for all prime divisors $\mathfrak{P}_i$ in K of all prime places $\mathfrak{p}$ of Ω the corresponding $\mathfrak{P}_i$-degree $n_{\mathfrak{P}_i}$ of K is a multiple of the $\mathfrak{p}$-index $m_\mathfrak{p}$ of A.*

Here the $\mathfrak{P}_i$-degree of K is the degree of the $\mathfrak{P}_i$-adic field $\mathrm{K}_{\mathfrak{P}_i}$ belonging to $\mathfrak{P}_i$ over the $\mathfrak{p}$-adic field $\Omega_\mathfrak{p}$, *i.e.*, if

$$\mathfrak{p} = \prod_i \mathfrak{P}_i^{e_{\mathfrak{P}_i}}, \quad \mathrm{N}_{\mathrm{K}\Omega}(\mathfrak{P}_i) = \mathfrak{p}^{f_{\mathfrak{P}_i}}$$

is the decomposition of $\mathfrak{p}$ in K, the product of the degree $f_{\mathfrak{P}_i}$ and the ramification order $e_{\mathfrak{P}_i}$ of $\mathfrak{P}_i$ relative to $\mathfrak{p}$, $n_{\mathfrak{P}_i} = f_{\mathfrak{P}_i} e_{\mathfrak{P}_i}$ [H, 4]. And the $\mathfrak{p}$-index of A is the index $m_\mathfrak{p}$ of the $\mathfrak{p}$-adic extension $A_\mathfrak{p}$ [H, 6].

Proof. That K is a splitting field for A, *i.e.*, that $A_\mathrm{K} \sim \mathrm{K}$ holds, is, according to Theorem **I**, equivalent to $A_{\mathrm{K}_{\mathfrak{P}_i}} \sim \mathrm{K}_{\mathfrak{P}_i}$ holding for every $\mathfrak{P}_i$.

Now suppose that in the case of a *finite* prime place $\mathfrak{p}$, $A_\mathfrak{p} \sim (\pi, W_\mathfrak{p})$[7] is the arithmetically distinguished cyclic representation of $A_\mathfrak{p}$ [H, 16; besides, 1.c.[9], Satz 38], hence $W_\mathfrak{p}$ is the unramified field of degree $m_\mathfrak{p}$ over $\Omega_\mathfrak{p}$ and π is a number from $\Omega_\mathfrak{p}$ which is divisible exactly by $\mathfrak{p}^1$. Furthermore, let $K_{\mathfrak{P}_i}$ be the composite [[Kompositum]] and $\Delta_{\mathfrak{P}_i}$ the intersection of $W_{\mathfrak{p}}$ and $K_{\mathfrak{P}_i}$. Since the greatest unramified subfield contained in $K_{\mathfrak{P}_i}$ is of degree $f_{\mathfrak{P}_i}$ over $\Omega_{\mathfrak{p}}$, and since for each degree there exists only *one* unramified field over $\Omega_{\mathfrak{p}}$, $\Delta_{\mathfrak{P}_i}$ is of degree $d_{\mathfrak{P}_i} = (m_{\mathfrak{p}}, f_{\mathfrak{P}_i})$ over $\Omega_{\mathfrak{p}}$. Hence $K_{\mathfrak{P}_i}$ is of degree $\frac{m_{\mathfrak{p}}}{d_{\mathfrak{P}_i}}$ over $K_{\mathfrak{P}_i}$, and at the same time unramified.

Now $A_{K_{\mathfrak{P}_i}} = (A_{\mathfrak{p}})_{K_{\mathfrak{P}_i}} \sim (\pi, K_{\mathfrak{P}_i})$ [H, 15.5]. It follows that $A_{K_{\mathfrak{P}_i}} \sim K_{\mathfrak{P}_i}$ is equivalent to π being a norm from $K_{\mathfrak{P}_i}$ relative to $K_{\mathfrak{P}_i}$. But, because $K_{\mathfrak{P}_i}$ is unramified over $K_{\mathfrak{P}_i}$, this is the case if and only if the order number $e_{\mathfrak{P}_i}$ of π in $\mathfrak{P}_i$ is a multiple of the degree $\frac{m_{\mathfrak{p}}}{d_{\mathfrak{P}_i}}$ of $K_{\mathfrak{P}_i}$ over $K_{\mathfrak{P}_i}$, or therefore, what amounts to the same thing because of $\left(\frac{m_{\mathfrak{p}}}{d_{\mathfrak{P}_i}}, \frac{f_{\mathfrak{P}_i}}{d_{\mathfrak{P}_i}}\right) = 1$, if $\frac{f_{\mathfrak{P}_i}}{d_{\mathfrak{P}_i}} e_{\mathfrak{P}_i} = \frac{n_{\mathfrak{P}_i}}{d_{\mathfrak{P}_i}}$ is a multiple of $\frac{m_{\mathfrak{p}}}{d_{\mathfrak{P}_i}}$, *i.e.*, if $n_{\mathfrak{P}_i}$ is a multiple of $m_{\mathfrak{p}}$, as asserted.

For the case that $\mathfrak{p}$ is an *infinite* prime place and that one does not have the trivial case $m_{\mathfrak{p}} = 1$, one has $m_{\mathfrak{p}} = 2$ and $A_{\mathfrak{p}}$ is the quaternion algebra over the real number field $\Omega_{\mathfrak{p}}$. That $A_{K_{\mathfrak{P}_i}} = (A_{\mathfrak{p}})_{K_{\mathfrak{P}_i}} \sim K_{\mathfrak{P}_i}$ holds is then equivalent to $K_{\mathfrak{P}_i}$ being the complex number field, *i.e.*, with $n_{\mathfrak{P}_i} = f_{\mathfrak{P}_i} = 2 = m_{\mathfrak{p}}$ ($e_{\mathfrak{P}_i}$ does not appear here), which here, too, again yields the assertion ($n_{\mathfrak{P}_i}$, like $m_{\mathfrak{p}}$, may assume only the values 1 or 2).

7. One comes to significant results if one reverses the inquiry after all splitting fields K of a fixed algebra A, answered in Theorem 3, namely, if one considers, for a fixed algebraic field K over Ω, the totality of the algebras A over Ω split by K. These algebras A form a subgroup $\mathfrak{K}$, determined by K, of the R. Brauer group $\mathfrak{A}$ of all algebras (more precisely, all classes of similar algebras) over Ω.[9a] If one assumes that K is normal over Ω, one thus comes to theorems which are to be considered as a **generalization of fundamental theorems of class field theory** (theory of relatively Abelian number fields) **to general relatively normal number fields:**

Theorem 4. (Decomposition theorem). *The relative degree f of the prime divisors in* K *of a prime ideal* $\mathfrak{p}$ *of* Ω *which does not divide the relative discriminant of* K *is equal to the earliest exponent for which one has* $A_{\mathfrak{p}}^f \sim \Omega_{\mathfrak{p}}$ *for all algebras A of the group* $\mathfrak{K}$ *associated with* K.

Proof. Since f is the $\mathfrak{p}$-degree of K (which is the same for all prime divisors of $\mathfrak{p}$ in K), while on the other hand the $\mathfrak{p}$-index of A is equal to the index, therefore to the exponent of $A_{\mathfrak{p}}$, it follows by Theorem 3 that f is at any rate a multiple of that earliest exponent, and for the proof it is sufficient to show in addition that in the group $\mathfrak{K}$ there exist algebras A with the exact $\mathfrak{p}$-index f.

According to the density theorem of Frobenius there now exists certainly yet a further prime ideal $\mathfrak{p}'$ in Ω, not a divisor of the relative discriminant of K, whose prime divisors have in K the relative degree f. Then let Z be a cyclic field over Ω whose $\mathfrak{p}$-degree and $\mathfrak{p}'$-degree has the value f. Further, let α be a number in Ω for which the norm rest symbols $\left(\frac{\alpha, Z}{\mathfrak{p}}\right)$ and $\left(\frac{\alpha, Z}{\mathfrak{p}'}\right)$ have reciprocal values of the order f, while for all remaining divisors $\mathfrak{q}$ of the conductor of Z one has $\left(\frac{\alpha, Z}{\mathfrak{q}}\right) = 1$. According to the generalized theorem of the arithmetic progression α can, in addition, be chosen so that besides, possibly, $\mathfrak{p}$, $\mathfrak{p}'$, and the $\mathfrak{q}$, it contains only a unique prime ideal $\mathfrak{r}$ of Ω exactly in the first power. For this one then has, by the product theorem for the norm rest symbol (reciprocity law), likewise $\left(\frac{\alpha, Z}{\mathfrak{r}}\right) = 1$ [H, 3.8–10]. Every algebra $(\alpha, Z) = A$ then has the $\mathfrak{p}$-index and $\mathfrak{p}'$-index f, but for all other prime places of Ω it has the index 1 [H, 17.7]. By Theorem 3 it follows from this, because of the assumption about $\mathfrak{p}$ and the choice of $\mathfrak{p}'$, that

K is a splitting field for A. Thus one has indeed proved the existence of algebras A with the $\mathfrak{p}$-index f in the group $\mathfrak{K}$.

Theorem 5. (Uniqueness and ordering theorem.) *If* $\mathrm{K} \leq \mathrm{K}'$, *then* $\mathfrak{K} \leq \mathfrak{K}'$ *and conversely.*

In particular, the correspondence of the groups $\mathfrak{K}$ *of algebras to the normal fields* K *is one-to-one.*

Proof. a) From $\mathrm{K} \leq \mathrm{K}'$ there follows trivially $\mathfrak{K} \leq \mathfrak{K}'$; for every algebra A split by K is, a fortiori, an algebra A' split by K'.

b) Suppose that, conversely, $\mathfrak{K} \leq \mathfrak{K}'$. Then, by the decomposition theorem, one has $f_{\mathfrak{p}} \mid f'_{\mathfrak{p}}$, for every $\mathfrak{p}$ which does not divide the relative discriminant. In particular, one then has $f_{\mathfrak{p}} = 1$ for almost all $\mathfrak{p}$ with $f'_{\mathfrak{p}} = 1$. But from this there follows by the well-known analytic proof procedure[10] $\mathrm{K} \leq \mathrm{K}'$.

8. Finally, we wish to show that the fundamental theorem provides an essential advance for the problem, treated by I. Schur,[11] of the number fields in which the absolutely irreducible representations of a finite group are possible:

Theorem 6. *All absolutely irreducible representations of a finite group* $\mathfrak{G}$ *are possible in cyclotomic fields, e.g., at any rate always in the field of the* n^h*-th roots of unity if* n *is the order of* $\mathfrak{G}$, *and* h *is sufficiently large.*

Proof. If one passes to the group ring G of $\mathfrak{G}$ with rational numbers, the absolutely irreducible representations Γ_i of $\mathfrak{G}$ become the absolutely irreducible representations of the simple parts G_i of the semisimple algebra G, and their centers are in each case the fields Ω_i of the corresponding characters.[12] Thus, because of the finiteness of $\mathfrak{G}$, the centers are in any case cyclotomic fields, and, in fact, certainly subfields of the field of the n-th roots of unity.

Now, according to Theorem 3, a cyclic field Z_i over Ω_i is a splitting field for G_i if for every prime place $\mathfrak{p}$ of Ω_i its $\mathfrak{p}$-degree $n_{i\mathfrak{p}}$ is a multiple of the $\mathfrak{p}$-index $m_{i\mathfrak{p}}$ of G_i. The $\mathfrak{p}$-index $m_{i\mathfrak{p}}$ is different from 1 only for the prime divisors of the basic ideal of G_i relative to Ω_i, thus certainly only for the prime divisors of the absolute discriminant of G. Since this discriminant divides n^n—n^n is the discriminant of a (non-maximal) order in G[13]—it follows that $m_{i\mathfrak{p}}$ is different from 1 at most for the prime divisors $\mathfrak{p}$ of n. In order to make $n_{i\mathfrak{p}}$ a multiple of $m_{i\mathfrak{p}}$ for these $\mathfrak{p}$, it is sufficient to prescribe that the $\mathfrak{p}$ should be ramified in Z_i of an order divisible in each case by $m_{i\mathfrak{p}}$. But this is achieved, as one easily sees for oneself, by the field of the n^h-th roots of unity for sufficiently large h.

In the last theorem of the paper cited, I. Schur states that in all known cases one already finds the field of the n-th roots of unity sufficient. The problem whether this is *always* true, and whether the methods developed here suffice to decide this question, remains reserved for further investigations.

NOTES

1. The writing of this note was done by H. Hasse.

2. These are rendered in the order in which they arose, which is opposite to the systematic order.

3. H. Hasse, "Theorie der zyklischen Algebren über einem algebraischen Zahlkörper," *Gött. Nachr.* 1931. A detailed presentation of the proofs will appear shortly in *Trans. Amer. Math. Society* ("Theory of cyclic algebras over an algebraic number field"). In that presentation, the theory of splitting fields and crossed products, which was developed by E. Noether in a course of lectures and which is fundamental there as here, will in particular also be developed. The Transactions' paper will be cited by H in what follows. H, 1-6 constitute—apart from the difference in language—the first-mentioned note.

4. The idea of the reduction to a solvable splitting field by means of the Sylow group theorem was applied already earlier by R. Brauer, namely, in order to show that every prime divisor of the index also occurs in the exponent ("Über den Zusammenhang von arithmetischen und invariantentheoretischen Eigenschaften von Gruppen linearer Substitutionen," *Berl. Akad-Ber.* 1926). Recently

A. A. Albert has developed simple proofs, which are independent of representation theory, for this idea as well as on the whole for a number of general theorems of the theory due to R. Brauer and E. Noether (1. "On direct products, cyclic division algebras, and pure Riemann matrices;" 2. "On direct products;" both in *Trans. Amer. Math. Soc.* **33** (1931); for the reduction referred to here see in particular Theorem 23 in 2.).

Added in press. Further, A. A. Albert, in possession of the communication by letter from H. Hasse that the fundamental theorem had been proved by him for Abelian algebras (see immediate sequel in the text), inferred from this immediately, independently of us, the following facts:

a) the fundamental theorem for degrees of the form 2^e,

b) Theorem 1 (exponent = index), which follows below,

c) besides the basic idea of Reduction 2 also in addition that of the subsequent Reduction 3, naturally without reference to Reduction 1, and correspondingly with the result: For division algebras D of prime power degree p^e over Ω there exists an extention field Ω' of degree prime to p over Ω such that $D_{\Omega'}$ is cyclic.

Naturally, all three results are now outdated because of our proof of the fundamental theorem which was carried out in the meantime. They show, however, that A. A. Albert, too, has an independent share in the proof of the fundamental theorem.

Finally, A. A. Albert (knowing our proof of the fundamental theorem) has remarked, in addition, that our central Theorem **I** follows in a few lines from Theorems 13, 10, 9 of a paper by him which is in press (*Bull. Amer. Math. Soc.* **37** (1931)). The proof of these theorems is based essentially on the same inferences as our reductions 2 and 3.

5. Cited in H, 3.11; proved in: H. Hasse, "Beweis eines Satzes und Widerlegung einer Vermutung über das allgemeine Normenrestsymbol," *Gött. Nachr.* 1931.

6. See H. Hasse, "Bericht über neuere Untersuchungen und Probleme in der Theorie der algebraischen Zahlkörper II," *Jahresber. der D.M.-V.*, Erg.-Bd. **6** (1930), § 8, and also the literature cited there.

7. In the notation of H, 1.—The specification of the automorphism S of Z connected with α was omitted here as unimportant.

8. In the special case of the rational quaternion algebras this amounts to the concept of "basic number" [[*Grundzahl*]] introduced by H. Brandt ("Idealtheorie in Quaternionenalgebren," *Math. Ann.* **99** (1928)). Since here the subject of discussion is not a *number*, but an *ideal* we had to say "basic ideal" [[*Grundideal*]]; one should not confuse this with Dedekind's term of basic ideal and basic number for different and discriminant which, exactly because of the reason cited, proves unserviceable for relative fields.—Concerning the definition of the (reduced) different see the following footnote. The (reduced) norm of an ideal is likewise defined by E. Noether "prime place by prime place" [[*"primstellenweise"*]], and in fact—corresponding to the fact that for the individual prime places every ideal is a principal ideal—simply by formation of the (reduced) number norms of the principal ideal basis numbers for the individual prime places.

9. H. Hasse, "Über p-adische Schiefkörper und ihre Bedeutung für die Arithmetik hyperkomplexer Zahlsysteme," *Math. Ann.* **104** (1931); see Satz 42, 59 there.

9a. R. Brauer, "Über Systeme hyperkomplexer Grössen," *Jahresber. d. D.M.-V.* **38** (1929), pp. 47/48.—Also see H, 13.1.

10. Concerning this, see, say, H. Hasse [loc. cit. Note 6], § 25, III.

11. I. Schur, "Arithmetische Untersuchungen über endliche Gruppen linearer Substitutionen," *Berl. Akad.-Ber.* 1906.

12. Concerning this, see: a) R. Brauer und E. Noether, "Über minimale Zerfällungskörper irreduzibler Darstellungen," *Berl. Akad.-Ber.* 1927; § 1. b) R. Brauer, "Über Systeme hyperkomplexer Zahlen," *Math. Zeitschr.* **30** (1929); Satz 3. c) E. Noether, "Hyperkomplexe Grössen und Darstellungstheorie," *Math. Zeitschr.* **30** (1929); §§ 21, 24, 26.

13. See E. Noether [loc. cit. Note 7c] [[it seems this should read "12 c"]], § 26.

[Received November 11, 1931.]

TRANSLATOR'S NOTES

1. This paper should be of considerable interest to the student of mathematics, and in fact, to anyone interested in mathematics as a human activity, and thus in the nature of mathematics, for several reasons.

Most obviously, it illustrates the nature of mathematics as a cumulative, and, in modern times, a cooperative enterprise. The fundamental theorem presented in this paper is the result of three "reductions," each one contributed by one of the three authors, it uses a method created earlier by a fourth, to whom the paper is dedicated, and it mentions related results obtained by a fifth, who had contributed to the subject matter and who independently found related results when he was informed by one of the authors that he had proved a special case of the fundamental theorem. Finally, it mentions consequences of the fundamental theorem for related theories developed earlier; all this in addition to the usual references to earlier work.

The method of solving a problem by "reducing" it to another problem is one frequently used in mathematics. (Such methods of problem-solving are considered in G. Polya's "How to solve it," [Polya] in the references appended to these notes.) Here the method consists in showing that a theorem which one wishes to prove, say Theorem A, can be proved if another theorem, say Theorem B, is assumed to be proved, and then proving Theorem B. The method is iterated three times; iteration, "doing the same thing" a number of times for the purpose of simplification (decomposition, analysis) or complication (composition, synthesis) is another frequently used method; reduction is a kind of simplification. The principal theorem is shown to be true if Theorem I is; then the same is shown for Theorem I and Theorem II, and again for Theorem II and Theorem III; and finally, a reference establishing Theorem III is given. Each of Theorems II and III has a stronger hypothesis than the preceding theorem, and the last reduction uses itself an iteration with an unspecified number r of steps, essentially the principle of mathematical induction (see [Kleene, 19-22], [Hilbert and Bernays, both editions, 20-23], and, for a different treatment, VI, 67-70 of [Dedekind], a selection from which is contained in this anthology.)

(An important, more recent example of the solution of a major problem by cooperation and reduction is the negative solution of Hilbert's tenth problem (see [Hilbert], a selection from which is contained in this anthology). Research on this problem was begun around 1950 on the basis of various ideas and results which go back to [Gödel], a selection from which is contained in this anthology. By 1961, the work of the three mathematicians who had carried out this research had shown that Hilbert's tenth problem is unsolvable if a relation $P(u, v)$ of non-negative integers with certain properties exists, and in 1970, Matijasevič], a fourth mathematician, proved the unsolvability of Hilbert's problem by showing that such a relation does exist, see [Matijasevič]. More details of this history are in [Davis], in [Robinson 1] and [Robinson 2], and positive aspects of the negative solution of the problem are discussed in [Davis - Matijasevič - Robinson]).

The paper translated above is also remarkable by the way in which it is written. It is scrupulous in giving credit for ideas and results. As stated in the footnotes, it was written by one of the three authors, and the three reductions which led to the proof of the fundamental theorem are presented in the order in which they were found. This order is the reverse of the logical order, in which Theorem III would be proved or referred to first and the desired result be reached at the end (see Footnote 2 and paragraph 4 of the paper). The student is likely to overlook or to forget the fact that a mathematical result or theory does not come ready-made in the neat orderly form in which it may be presented, but must first be found and verified, and that this is a creative human activity (for a study of it, see [Hadamard]); he should not be surprised that in the present case the desired result came first, rather than last (see the work of Polya mentioned above).

The presentation of an important proof or theory in the order in which it was found may make it easier to understand the motivation of the parts and to see the work as a whole, and it may even help the reader in his own work; in the present case it, together with details such as those in Footnote 4, conveys some of the excitement of discovery, and it lends to this paper a freshness and zest which, as well as details such as those mentioned at the beginning of this note, may be lost in the kind of treatment which is necessary in a comprehensive text.

2. The translator has been helped by consulting [van der Waerden], together with its translation, and [Hecke].

REFERENCES

1. [Davis] Martin Davis, "Hilbert's tenth problem is unsolvable," *American Mathe-*

matical Monthly, vol. 80 (1973), 233-269. (See 264-267).

2. [Davis - Matijasevič - Robinson] Martin Davis, Yuri Matijasevič, and Julia Robinson, "Hilbert's tenth problem. Diophantine equations: Positive aspects of a negative solution," *Proceedings of Symposia in Pure Mathematics*, vol. 28 (1976), 323-378.

3. [Dedekind] Richard Dedekind, *Was sind und was sollen die Zahlen?* (1888), trans. Beman (1901), 29-115.

4. [Gödel] Kurt Gödel, "Über formal unentscheidbare Sätze der *Principia Mathematica* und verwandter Systeme I," (1931), trans. and ed. van Heijenoort, *From Frege to Gödel: A Source Book in Mathematical Logic, 1879-1931* (1967), 596-616.

5. [Hadamard] Jacques Hadamard, *The Psychology of Invention in the Mathematical Field* (1945, 1954).

6. [Hecke] Erich Hecke, *Theorie der algebraischen Zahlen* (1923).

7. [Hilbert] David Hilbert, "Mathematical problems: Lecture delivered before the International Congress of Mathematicians at Paris in 1900," trans. Newson, *Bulletin of American Mathematical Society*, vol. 8 (1902), 437-479. Reprinted in *Proceedings of Symposia in Pure Mathematics*, vol. 28 (1976), 1-34.

8. [Hilbert and Bernays] D. Hilbert and P. Bernays, *Grundlagen der Mathematik*, vol. 1, 1934; second edition, 1968.

9. [Kleene] Stephen Cole Kleene, *Introduction to metamathematics* (1952).

10. [Matijasevič] Ju. V. Matijasevič, "Enumerable sets are Diophantine," trans. from the Russian with improvements, A. Doohovskoy, *Soviet Math. Doklady*, vol. 11 (1970), 354-358.

11. [Polya] G. Polya, *How to solve it. A new Aspect of Mathematical Method* (1945).

12. [Robinson 1] Julia Robinson, "Diophantine decision problems," *MAA Studies in Mathematics*, vol. 6: *Studies in Number Theory*, 76-116 (1969).

13. [Robinson 2] ———, "Hilbert's tenth problem," *Proceedings of Symposia in Pure Mathematics*, vol. 20 (1971), 191-194.

14. [van der Waerden] B. L. van der Waerden, *Algebra*. vol. 1 (1966); vol. 2 (1967). Trans. vol. 1, Fred Blum and John R. Schulenberger (1970); vol. 2, John R. Schulenberger (1970).

Chapter IX
The Early 20th Century to 1932

Section C
Selected Topics: The Development of General Abstract Theories

GEORGE DAVID BIRKHOFF (1884–1944)

The U.S. mathematician George David Birkhoff was the eldest of six children of David Birkhoff, a physician, and Jane (or Jennie) Gertrude Droppers. Both parents were of Netherlands extraction and were members of the Dutch Reformed Church. When their son was two they moved from Michigan to Chicago, where young George grew up. After studying at the Lewis Institute (now Illinois Institute of Technology) from 1896 to 1902, he entered the University of Chicago. After a year he transferred to Harvard University, where Professor Maxime Bôcher strongly influenced him. Bôcher introduced him to classical analysis and algebra. Birkhoff earned a bachelor's degree at Harvard in 1905 and a master's in 1906. He thereupon returned to the University of Chicago to earn his doctorate under the guidance of E. H. Moore. He received the Ph.D. summa cum laude in 1907 for a dissertation treating the boundary value and asymptotic expansion problem for ordinary differential equations of arbitrary value (Sturm–Liouville theory).

Birkhoff began his career in research and teaching as a mathematics instructor at the University of Wisconsin from 1907 to 1909. At Wisconsin Professor E. B. Van Vleck deepened his interest in linear differential equations. In 1908 he married Margaret Elizabeth Grafius, known to their later friends as Marjorie. They were to have three children. In 1909 he moved to Princeton as a preceptor and rose in two years to the rank of professor. His initial research at Princeton dealt with difference equations but his interest soon turned to celestial mechanics and differential equations. To prepare himself in these fields, he read C. F. Gauss, B. Riemann, and above all the French theoretician Henri Poincaré, whose work in these two fields he hoped to extend. In 1912 he came across the article "Sur un théorème de géométrie," in which Poincaré stated but could not prove his "last geometric theorem." This topological theorem bears on the restricted problem of three mutually gravitating bodies in dynamics. Birkhoff set out to prove it and quickly succeeded. His proof was a striking mathematical achievement; its publication in 1913 brought him international acclaim.

In 1912 Birkhoff had returned to Harvard, where he was to spend the rest of his career, and was an assistant professor of mathematics from 1912 to 1919. His creative powers now reached full development and his brilliance was gradually recognized more widely in scientific circles in Europe and the United States. His paper entitled "The Restricted Problem of Three Bodies" (1915) won the prize of the Royal Venice Institute of Science, and his memoir entitled "Dynamical Systems with Two Degrees of Freedom" (1917) garnered the Bôcher Prize of the American Mathematical Society. In 1918 he was elected a member of the National Academy of Sciences. The next year he rose to the rank of full professor at Harvard.

By the end of World War I Birkhoff had become one of the most influential men in mathematical circles in the United States together with Oswald Veblen, Luther Eisenhart, Griffith Evans, and Roland Richardson. Because of his eminence in research he was considered first among equals and was often asked to represent the U.S. mathematical community in larger circles. Thus, after the war he was instrumental in channeling American support to the new mathematical institutes created at Göttingen and Paris. In 1924 he was elected to a two-year term as president of the American Mathematical Society.

Birkhoff continued to teach at Harvard, where he was named Perkins Professor in 1932. He was also Dean of the Faculty of Arts and Sciences from 1935 to 1939. Although his lectures were often not polished and sometimes unprepared, he was a stimulating teacher who showed students a powerful mind at work in solving problems in class. He was an extraordinary director of research who brought out the creative talents of his most able students. Six of his doctoral students subsequently became members of the National Academy of Sciences—an unusually large number for one teacher. His students and colleagues knew the conservative Birkhoff for his natural charm and frankness in expressing his views. They also knew that he held his opinions strongly but not rigidly and that these opinions were tempered by a kind, judicious approach.

Birkhoff received many honors during his career. The American Philosophical Society and the American Academy of Arts and Sciences elected him a member. The *Accademei dei Lincei*, the Paris Academy of Sciences, and the Pontifical Academy named him a foreign member. Among the many U.S. and foreign universities conferring honorary doctorates upon him were Athens, Brown, Buenos Aires, Chicago, Harvard, Lima, Paris, Pennsylvania, St. Andrews, and Sophia. In 1937 the American Association for the Advancement of Science elected him president.

As a result of his outstanding research and professional activities, Birkhoff is considered the leading U.S. mathematician of the early 20th century. He chiefly contributed to mathematical analysis and analysis applied to dynamics. Paul Smith's and his definition of metric transitivity (1928), which implies the existence of transitive motion, proved an important concept in modern dynamics and its branches of symbolic and topological dynamics. His most celebrated result was the formulation in 1931 of the famous pointwise ergodic theorem (the so-called strong form), which has had fruitful consequences in dynamics, functional analysis, group theory, and probability theory. For example, his theorem proved that systems in the kinetic theory of gases are ergodic. (Boltzmann had named ergodic "those mechanical systems which had the property that each particular motion, when continued indefinitely, passes through every configuration and state of motion of the system which is compatible with the value of the total energy.") Ludwig Boltzmann and Clerk Maxwell had first made the conjecture that systems in the kinetic theory of gases are ergodic, but its proof had baffled theoretical physicists for over 50 years. Suggestions in Poincaré's writings and John von Neumann's establishment of the mean ergodic theorem (the so-called weak form) stimulated Birkhoff's resolution. In other research he contributed to the theory of map coloring and the calculus of variations.

Birkhoff also wrote on newly emerging branches of modern physics, elementary geometry, and aesthetics. During the mid-1920s he critically examined the foundations of the theory of relativity, posing illuminating critiques and controversial physical models that avoided employing Riemannian geometry. He also studied quantum theory. Two of his books on physics are *Relativity and Modern Physics* (1923) and *Dynamical Systems* (1927). In 1929 he commenced work with Ralph Beatley on a beginning textbook in geometry that was published under the title *Basic Geometry* (1940). During the 1920s he also began studying the quantitative bases suggested since ancient times for canons of beauty in music, art, and poetry. He expressed his conclusions in the book *Aesthetic Measure* (1933).

133. From "Proof of the Ergodic Theorem" (1931)*

GEORGE DAVID BIRKHOFF

Let

$$\frac{dx_i}{dt} = X_i(x_1, \ldots x_n) \qquad (i = 1, \ldots n)$$

be a system of n differential equations valid on a closed analytic manifold M, possessing an invariant volume integral, and otherwise subject to the same restrictions as in the preceding note, except that the hypothesis of strong transitivity is no longer made.

We propose to establish first that, without this hypothesis, we have

$$\lim_{n=\infty} \frac{t_n(P)}{n} = \tau(P) \tag{1}$$

for all points P of the surface σ save for points of a set of measure 0. In other words, there is a "mean time $\tau(P)$, of crossing" of σ for the general trajectory.

The proof of the "ergodic theorem," that there is a time-probability p that a point P of a general trajectory lies in a given volume v of M, parallels that of the above recurrence theorem, as will be seen.

The important recent work of von Neumann (not yet published) shows only that there is convergence *in the mean*, so that (1) is not proved by him to hold for any

*Source: George D. Birkhoff, "Proof of the Ergodic Theorem," in *Proceedings of the National Academy of Sciences*, vol. 17 (1931), 656-660. This article is reprinted by permission of Dover Publications, Inc.

point P, and the time-probability is not established in the usual sense for any trajectory. A *direct* proof of von Neumann's results (not yet published) has been obtained by E. Hopf.

Our treatment will be based upon the following *lemma:* If S_λ $[S'_\lambda]$ is a measurable set on σ, which is invariant under T, except possibly for a set of measure 0, and if for any point P of this set

$$\lim_{n=\infty} \sup \frac{t_n(P)}{n} \geqq \lambda > 0 \left[\lim_{n=\infty} \inf \frac{t_n(P)}{n} \leqq \lambda > 0 \right] \tag{2}$$

then

$$\int_{S_\lambda} t_n(P) dP \geqq \lambda \int_{S_\lambda} dP \; [\int_{S'_\lambda} t(P) dP \leqq \lambda \int_{S_\lambda} dP. \tag{3}$$

We consider only the first case, for the proof of the second case is entirely similar. In analogy with the preceding note, define the distinct measurable sets U_1, U_2, . . . on S_λ so that for P in U_n

$$t_n(P) > n(\lambda - \epsilon) \; (P \text{ not in } U_1, U_2, \ldots, U_{n-1}) \tag{4}$$

The quantity $\epsilon > 0$ is taken arbitrarily. It is, of course, clear that for every point P of S_λ

$$t_n(P) > n(\lambda - \epsilon)$$

for infintely many values of n, so that all such points belong to at least one of the sets U_1, U_2, ... Now, by the argument of the earlier note, we infer

$$\int_{S^k_\lambda} t(P) dP > (\lambda - \epsilon) \int_{S^k_\lambda} dP$$

where $S^k_\lambda = U_1 + U_2 + \ldots + U_k$. But S^k_λ is, for every value of k, a measurable part of the invariant set S_λ and increases toward a limit $U_1 + U_2 + \ldots$ which contains every point of S_λ. Consequently we obtain by a limiting process

$$\int_{S_\lambda} t(P) dP \geqq (\lambda - \epsilon) \int_{S_\lambda} dP$$

for any $\epsilon > 0$, whence the inequality of the lemma.

The recurrence theorem stated results directly from this lemma.

Consider the measurable invariant set of points P on σ for which

$$t_n(P) \geqq n\lambda \tag{5}$$

for infinitely many values of n (see the preceding note). This is a set S_λ to which the lemma applies. Similarly the set of points P on σ for which

$$t_n(P) < n\lambda \tag{6}$$

for infinitely many values of n is a set S'_λ of the kind specified in the lemma.

The set S_λ diminishes and the set S'_λ increases with σ, and both sets taken together exhaust σ. The measure of the set S_λ must tend toward 0 as λ increases. Otherwise it would tend toward an invariant measurable set of positive measure, S^*, for which the inequality of the lemma holds for $\lambda = \Lambda$, an arbitrarily large positive quantity, and we should infer

$$\int_{S^*} t(P) dP \geqq \Lambda \int_{S^*} dP$$

for any Λ, which is absurd. Moreover, when λ tends toward 0, S_λ becomes vacuous, since there is a least time of crossing, λ_0. In a similar way, S'_λ increases with λ from a set of zero measure for $\lambda < \lambda_0$ toward the set σ.

If then S_λ and S'_λ are not essentially complementary parts of σ, one decreasing, the other increasing, they must, for certain values of λ, have a common measurable component S^*_λ of positive measure, also invariant under T.

Consider the set of points belonging to S^*_λ such that

$$t_n(P) > n\mu \qquad (\mu > \lambda)$$

for infinitely many values of n. These form an invariant measurable subset $S^*_{\lambda\mu}$ of S^*_λ, which must be of measure 0 for any such μ. Otherwise the inequalities of the lemma would give us simultaneously

$$\int_{S^*_{\lambda\mu}} t(P)dP \geqq \mu \int_{S^*_{\lambda\mu}} dP, \qquad \int_{S^*_{\lambda\mu}} t(P)dP \leqq \lambda \int_{S^*_{\lambda\mu}} dP,$$

which are mutually contradictory.

Hence we infer that all of the points P of S^*_λ save for a set of measure 0, satisfy the inequality

$$t_n(P) \leqq n\mu$$

for any $\mu > \lambda$ and for $n = n_P$ sufficiently large, that is,

$$\lim_{n=\infty} \sup \frac{t_n(P)}{n} \leqq \lambda .$$

Likewise we infer that for all of the points of S^*_λ, save for a set of measure 0, we have

$$\liminf_{n=\infty} \frac{t_n(P)}{n} \geqq \lambda .$$

It follows then that for points P of S^*_λ, with the usual exception,

$$\lim_{n=\infty} \frac{t_n(P)}{n} = \lambda . \tag{7}$$

Two such sets S^*_λ belonging to different λ's are evidently distinct except for a set of measure 0. Hence there can exist only a numerable set $S^*_{\lambda_i} (i = 1, 2, \ldots)$ of such sets since each has a positive measure. Except for these values λ_i of λ, S'_λ and S_λ are complementary parts of σ aside from a set of measure 0.

Choose now any two values of λ, say λ,μ with $\lambda < \mu$, not belonging to this numerable set, and consider the points of S_λ which do not belong to S_μ. These form an invariant measurable set $S_{\lambda,\mu}$, such that for any point P of this set

$$\lambda \leqq \lim_{n=\infty} \sup \frac{t_n(P)}{n} \quad \mu \tag{8}$$

and also

$$\lambda \leqq \lim_{n=\infty} \inf \frac{t_n(P)}{n} \leqq \mu , \tag{9}$$

since $S_{\lambda,\mu}$ is essentially identical with the part of S'_μ not in S_λ. We infer then that $t_n(P)/n$ oscillates between λ and μ as n tends toward ∞, for all points P of $S_{\lambda,\sigma}$ except a set of measure 0.

By choosing a set of values such as λ,μ sufficiently near together we infer then that for all of the points of σ except a set of measure 0, the oscillation of $t_n(P)/n$, as n becomes infinite, is less than an arbitrary $\delta > 0$.

Obviously then the stated recurrence theorem is true.

It should also be noted that if t_n/P denotes the time to the nth crossing as time decreases, the same result holds if n tends toward $\pm\infty$, *with the same limit* except for a set of points P of measure 0. This follows at once from the fact that (8) may be written

$$\lambda \leqq \lim_{n=-\infty} \sup \frac{t_n(P)}{n} \cdot \leqq \mu ,$$

where P of $S_{\lambda,\mu}$ is replaced by $T^n(P)$; and (9) may be given a corresponding form.

This theorem of recurrence admits of certain evident extensions. In the first place there is no need to restrict attention to the analytic case. Moreover, instead of a single surface σ, any measurable set σ^*, imbedded in a numerable set of distinct ordinary surface elements with $v \cos\theta > d > 0$, throughout, will serve, in which case $t^*(P)$ denotes the time from P on σ^* to the first later crossing of σ^*.

In order to prove the "ergodic theorem" we observe first that a set σ^* can be found which cuts every trajectory except those corresponding to equilibrium and others of total measure 0. This is possible; for a numerable set of distinct ordinary surface elements $\sigma_1, \sigma_2, \ldots$ with $v \cos\theta > d > 0$ can be found which cut every trajectory not corresponding to equilibrium. If we define σ_k as the limit of

$$\sigma_1 + \sigma_{12} + \sigma_{123} + \ldots + \sigma_{1\ldots k}$$

where σ_{12} denotes the set of points P of σ_2 not on a trajectory cutting σ_1, σ_{123} denotes the set of points of σ_3 not on a trajectory cutting σ_1 or σ_2, etc., it will have the desired properties.

Now let v denote any "measurable" volume in the manifold M, and let $\bar{t}(P)$ denote the interval of time during which the point on the trajectory which issues from P on such a set σ^* lies in v before the point $T(P)$ of σ^* is reached. Thus $\bar{t}(P) \leqq t(P)$ in all cases. In addition, $\bar{t}_n(P)$ satisfies the same functional equation as $t(P)$

$$\bar{t}_n(P) = \bar{t}(T^{n-1}(P)) + \bar{t}_{n-1}(P).$$

Hence the same reasoning as before is applicable to show that, except for a set of points P of measure,

$$\lim_{n=\pm\infty} \frac{\bar{t}_n(P)}{n} = \bar{t}(P),$$

where $\bar{\tau}(P) \leqq \tau(P)$; while at the same time, of course,

$$\lim_{n\pm\infty} \frac{t_n(P)}{n} = \tau(P) > 0.$$

We conclude that the following "ergodic theorem" holds:

For any dynamical system of type (1) *there is a definite "time probability" p that any moving point, excepting those of a set of measure 0, will lie in a region v; that is,*

$$\lim_{n=\pm\infty} \frac{\bar{t}}{t} = p \leqq 1$$

will exist, where t denotes total elapsed time measured from a fixed point and $\bar{t}$ the elapsed time in v.

For a strongly transitive system p is, of course, the ratio of the volume of v to V.

Evidently the germ of the above argument is contained in the lemma. The abstract character of this lemma is to be observed, for it shows that the theorem above will extend at once to function space under suitable restrictions.

It is obvious that $\tau(P)$ and $\bar{\tau}(P)$ as defined above satisfy functional relations of the following type:

$$\int_0^\lambda \lambda\, dm(S_\lambda) = \int_{S_\lambda} t(P)dP$$

where the integral on the left is a Stieltjes integral, $m(S_\lambda)$ being the measure of S_λ.